Advances in Experimental Medicine and Biology

Volume 1310

Advances in Experimental Medicine and Biology provides a platform for scientific contributions in the main disciplines of the biomedicine and the life sciences. This series publishes thematic volumes on contemporary research in the areas of microbiology, immunology, neurosciences, biochemistry, biomedical engineering, genetics, physiology, and cancer research. Covering emerging topics and techniques in basic and clinical science, it brings together clinicians and researchers from various fields.

Advances in Experimental Medicine and Biology has been publishing exceptional works in the field for over 40 years, and is indexed in SCOPUS, Medline (PubMed), Journal Citation Reports/Science Edition, Science Citation Index Expanded (SciSearch, Web of Science), EMBASE, BIOSIS, Reaxys, EMBiology, the Chemical Abstracts Service (CAS), and Pathway Studio.

2019 Impact Factor: 2.450 5 Year Impact Factor: 2.324

More information about this series at http://www.springer.com/series/5584

Jun Ki Kim • Jeong Kon Kim
Chan-Gi Pack
Editors

Advanced Imaging and Bio Techniques for Convergence Science

Springer

Editors
Jun Ki Kim
Department of Convergence Medicine
University of Ulsan College of Medicine
Asan Institute for Life Sciences
Asan Medical Center
Seoul
Korea (Republic of)

Jeong Kon Kim
Department of Radiology
University of Ulsan College of Medicine
Asan Medical Center
Seoul
Korea (Republic of)

Chan-Gi Pack
Department of Convergence Medicine
University of Ulsan College of Medicine
Asan Institute for Life Sciences
Asan Medical Center
Seoul
Korea (Republic of)

ISSN 0065-2598 ISSN 2214-8019 (electronic)
Advances in Experimental Medicine and Biology
ISBN 978-981-33-6063-1 ISBN 978-981-33-6064-8 (eBook)
https://doi.org/10.1007/978-981-33-6064-8

This Springer imprint is published by the registered company Springer Nature Singapore Pte Ltd.
The registered company address is: 152 Beach Road, #21-01/04 Gateway East, Singapore 189721, Singapore

Contents

Confocal Laser Scanning Microscopy and Fluorescence Correlation Methods for the Evaluation of Molecular Interactions

Chan-Gi Pack

Abbreviations

CLSM	confocal laser scanning microscopy
CPM	count rate per molecule
CPS	count rate per second
DsRed	discosoma red fluorescence protein
ER	estrogen receptor
FCCS	fluorescence cross-correlation spectroscopy
FCS	fluorescence correlation spectroscopy
FITC	fluorescein-5-isothiocyanate
HER2	human epithelial receptor 2
HS	heat shock
mAb	monoclonal antibody
mGFP	monomeric green fluorescence protein
mRFP	monomeric red fluorescence protein
NP	nanoparticle
PR	progesterone receptor
QD	quantum dot
RCA	relative cross-correlation amplitude
Rh123	rhodamine 123
Rh6G	rhodamine 6G
RITC	Rhodamine B isothiocyanate

C.-G. Pack (✉)
Convergence Medicine Research Center (CREDIT), Asan Institute for Life Sciences, Asan Medical Center, Seoul, South Korea

Department of Convergence Medicine University of Ulsan College of Medicine, Asan Institute for Life Sciences, Asan Medical Center, Seoul, Korea (Republic of)
e-mail: changipack@amc.seoul.kr

J. K. Kim et al. (eds.), *Advanced Imaging and Bio Techniques for Convergence Science*, Advances in Experimental Medicine and Biology,
https://doi.org/10.1007/978-981-33-6064-8_1

Si-FNP silica-based fluorescent nanoparticle
Si-NP silica-based nanoparticle
TRITC tetramethylrhodamine

Introduction

Fluorescence correlation spectroscopy (FCS) and dual-color fluorescence cross-correlation spectroscopy (FCCS) are based on the fluctuation analysis of fluorescence intensity to characterize fluorescent probe at the single molecular level in live cells as well as in aqueous solutions. They allow the measurement of important physical parameters for biological and biophysical science; the average number of molecules in confocal detection volume (i.e., absolute concentration), the translational diffusion time (i.e., diffusion coefficient), and the fractional ratio of molecular interaction (Fig. 1) (Aragon and Pecora 1976; Bacia et al. 2006; Elson and Magde 1974; Pack et al. 2014a, b; Park et al. 2008). Details regarding FCS and FCCS theory can be found elsewhere (Aragon and Pecora 1976; Bacia et al. 2006; Capoulade et al. 2011; Elson and Magde 1974; Kinjo and Rigler 1995). Since diffusional mobility reflects the molecular size (i.e., hydrodynamic radius) and shape of probe molecules, FCS is useful for investigating the dynamic properties of various probe molecules in solution and in live cells.

In solution, FCS can indirectly analyze the binding interactions of biomolecules that lead to the formation of large complexes between small fluorescent molecules and much larger molecules by detecting changes in diffusional mobility. However, FCS application for protein–protein binding interaction is difficult and only a few applications have been reported (Klingler and Friedrich 1997; Pack et al. 1999). In most cases, protein–protein interactions or dimerization are not accompanied by any detectable change in diffusional mobility because of the similar molecular weight of the proteins. On the contrary, aggregation or oligomerization of proteins, which is accompanied by large changes in diffusional mobility, can easily be applicable for FCS (Kawai-Noma et al. 2006, 2010).

For live cells, biophysical studies using inert probes such as green fluorescent protein (GFP), fluorescently labeled dextran, and Ficoll have demonstrated that cellular microenvironments, including the local viscosity of the cytosolic and nuclear compartment, can be characterized by the diffusion coefficient (D) of the probe molecule (Luby-Phelps et al. 1987; Pack et al. 2006; Verkman 2002), even though the aqueous viscosities of subcellular compartments are considered to be greater than those of water. Diffusion analyses of the macromolecules in living cells have demonstrated that the apparent viscosities of cytosol and nucleoplasm are three to four times higher than the viscosity of pure water (~1 cP) (Pack et al. 2006; Seksek et al. 1997; Wachsmuth et al. 2008). Interestingly, the viscosity of the nucleolus is much higher than that of the cytosol and nucleoplasm depending on probe size and ranges from 10 to 30 cP (Park et al. 2015). The quantitative comparison of the D values of inert probes and functional proteins supports the hypothesis that the slow

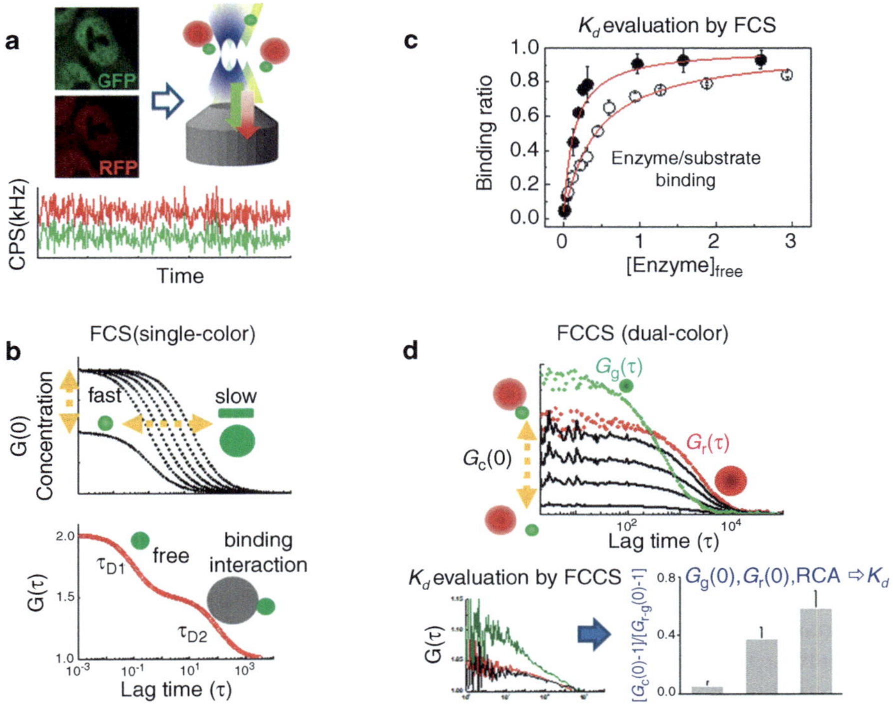

Fig. 1 Schematic diagram for FCS and FCCS measurements and parameters. (**a**) Single-color (green or red) fluorescence autocorrelation spectroscopy (FCS) and dual-color (red and green) cross-correlation spectroscopy (FCCS) are based on the fluctuation analysis of averaged fluorescence intensity (count per second; CPS in kHz) detected from a tiny open volume (i.e., confocal volume in each channel). (**b**) FCS provides a fluorescence autocorrelation function (G_g for green channel or G_r for red channel), allowing the measurement of two important biophysical parameters; the average number of fluorescent molecules ($G\,(0) \sim 1/N$) in a detection volume, and the translational diffusion time (τ_D) of the molecules through the open volume. The diffusion time (i.e., diffusion coefficient, D) of molecules reflects the molecular size (molecular weight or R_h), the shape of fluorescent species, and the apparent viscosity in a solution or in a live cell. Moreover, the molecular interaction of small probes (τ_{D1}) with large targets (τ_{D2}) can be detected by two-component model analysis. (**c**) An FCS titration experiment for evaluating the dissociation constant (K_d) between fluorescently labeled substrate protein and large enzyme protein is shown. (**d**) Dual-color FCCS provides a cross-correlation function (Gcross, represented by G_c in the main text) and two autocorrelation functions (G_g and G_r), respectively. G_c allows the measurement of direct interaction or colocalization of green and red fluorescent molecules transiting through the open volumes of each channel, in addition to the parameters for single-color FCS, as described in b. Subscripts g and r represent green and red fluorescent probes, respectively. Since FCCS provides the absolute concentration and binding ratio (i.e., RCA value represented by $[G_c(0)–1]/[G_{r,\,g}(0)–1]$) of two probes, the K_d value can be directly evaluated from the concentration of free and bound probes

diffusion of functional proteins originates from their specific interactions, such as complex formation and transient binding interactions with immobile target molecules (Pack and Ahn 2015; Pack et al. 2017; Phair and Misteli 2000; Takada et al. 2018). This means that FCS provides useful information regarding the functional status of proteins, whether they interact with target complexes or not.

FCCS is an advanced FCS method that uses two differently colored probes to directly detect an interaction between them, in addition to the acquisition of each of the FCS parameters of the two probe molecules (Capoulade et al. 2011; Pack et al. 2014a, b). As various dye and fluorescent protein colors are now available, binding interaction analyses between ligands and cells or proteins, in addition to the quantification of mobility, are informative for understanding various extra- and intercellular processes of target biomolecules. This chapter provides an introduction to the experimental practices of FCS (FCCS) and gives an overview of live cell applications of CLSM and FCS (FCCS). Special attention will be paid to cellular phenomena related to the activation of heat shock transcription factors (HSFs) and reactions between drug and single live cells. HSFs are the important target molecules for drug development (Dong et al. 2019; Neef et al. 2011; Whitesell and Lindquist 2009; Vydra et al. 2014), even though the molecular mechanisms of their complexes and multiple functions are not currently fully understood (Anckar and Sistonen 2011; Pirkkala et al. 2001). Analyzing the reaction processes of drugs using living cells is important for understanding the efficacy of drugs such as antibodies and nanocarriers. In these molecular situations, FCS and FCCS analyses provide new insights for understanding the cellular phenomena at a single molecular level, and for quantifying the interaction and reaction systems. For a detailed description of the theoretical FCS and FCCS principles, other cited literature may be useful (Elson and Magde 1974; Magde et al. 1972).

Live Cell Imaging by CLSM

Fluorescence imaging of live cells expressing fluorescent proteins was performed using a confocal microscope (LSM 510 or 780, Carl Zeiss). A water-immersion objective (C-Apochromat, 40×, 1.2NA; Carl Zeiss) was used for the two LSM systems. Each confocal dish (i.e., a glass-bottom dish or a glass-bottom chamber) was imaged in the incubation chamber unit of the microscope, which was maintained in the presence of 5% CO_2 at 37 °C (Live-cell instrument, LCI) for CLSM observation without FCS. For the FCS measurement after CLSM observation, each dish was maintained in air at 25 °C. Green fluorescent dyes and green fluorescent protein (GFP) were excited with a wavelength at 488 nm and detected in the range of 500–530 nm for LSM510 or 500–550 nm for LSM780. Red fluorescent dyes and red fluorescent protein (RFP) such as monomeric RFP (mRFP), DsRed, and mCherry, were imaged with an excitation wavelength of 543 nm for LSM510 and detected at a wavelength higher than 560 nm. For LSM780, red fluorescent probes were imaged with an excitation of 561 nm then detected at a wavelength higher than 580 nm. The pinhole diameters for confocal imaging on LSM 510 were adjusted to

70 and 78 μm for green fluorescent probes and red fluorescent probes, respectively. For LSM780, the pinhole diameters were adjusted to 37 μm for both GFP and RFP. To avoid bleed-through effects in the double-scanning experiments, GFP and RFP expressed in co-transfected cells and other dual-colored cells were scanned independently in the multitracking mode.

FCS and FCCS Measurements on CLSM and Data Analysis

All CLSM cell imaging and corresponding FCS and FCCS measurements were taken at 25 °C with an LSM 510 combined with a ConfoCor2 system or with an LSM 780 with Confocor3 (Carl Zeiss) (Kang et al. 2018; Pack et al. 2006, 2014a, b). For the LSM 510 system, green fluorescent probes such as Rhodamine 6G (Rh6G) and GFP were excited at the 488 nm laser line, and red fluorescent probes, such as tetramethyl rhodamine (TRITC) and mCherry were excited at the 543-nm laser line. The confocal pinhole diameter for CLSM and FCS/FCCS was adjusted to 70 mm for 488 nm and 78 nm for 543 nm. The emitted light was collimated then split using an NFT570 dichroic mirror and detected at 505–530 nm by the green channel for the green fluorescent probes, and 600–650 nm by the red channel for the red fluorescent probes. For the LSM 780 system, green fluorescent probes, such as Rh6G, fluorescein isothiocyanate (FITC), and GFP, were excited at the 488-nm laser line, and red fluorescent probes such as rhodamine B isothiocyanate (RITC) and mCherry, were excited at the 561-nm laser line. Minimal and optimal excitation power were selected to obtain sufficiently high signal-to-noise ratios for the FCS and FCCS analyses while preventing artifacts such as saturation, blinking, and photobleaching that might affect the analysis of diffusion time and interaction. The confocal pinhole diameter was adjusted to 37 μm for the 488- and 543-nm lasers. Emission signals were split using a dichroic mirror (488/561-nm beam splitter) and detected at 500–550 nm in the green channel for green fluorescent probes and at 600–660 nm in the red channel for red fluorescent probes. FCS and FCCS data were analyzed with ConfoCor2 software installed on the LSM510 system and with analytical software installed in the ZEN 2012 acquisition software for LSM780 (Kang et al. 2018; Pack et al. 2006). Briefly, the fluorescence autocorrelation functions (FAFs) of the red and green channels, $G_r(\tau)$ and $G_g(\tau)$, and the corresponding fluorescence cross-correlation function (FCF), $G_c(\tau)$, were calculated from Eq. 1:

$$G_x(\tau) = 1 + \frac{\delta I_i(t) \cdot \delta I_j(t+\tau)}{I_i(t) I_j(t)} \tag{1}$$

where τ denotes the time delay (lag time) in μsec, I_i is the detected fluorescence intensity of the red channel (i = r) or green channel (i = g), and $G_r(\tau)$, $G_g(\tau)$, and $G_c(\tau)$ denote the FAFs of the red ($i = j = x$ = r), green ($i = j = x$ = g), and cross (i = r, j = g, x = c), respectively. The acquired $G_x(\tau)$s were fitted by a model equation represented as per Eq. 2:

$$G_x(\tau) = 1 + \frac{1}{N}\sum_i F_i\left(1+\frac{\tau}{\tau_i}\right)^{-1}\left(1+\frac{\tau}{s^2\tau_i}\right)^{-1/2} \tag{2}$$

where F_i and τ_i are the fraction and diffusion time of component i, respectively. N is the average number of fluorescent probes in the excitation–detection volume (i.e., confocal detection volume) defined by radius w_0 and length $2z_0$, and s is the structure parameter representing the ratio $s = z_0/w_0$. The detection volume defined by the beam radius w_0 and axial radius $2z_0$ was approximated as a cylinder (volume, $V_0 = 2\pi w_0^2 z_0$). Moreover, N is related to absolute concentration by a simple equation:

$$c = \frac{N}{N_A\left(2\pi w_0^2 z_0\right)} \tag{3}$$

where c and N_A are the absolute concentration in molar concentration and the Avogadro number (6.022×10^{23} mol^{-1}), respectively. For the FCS control experiment, solution samples of Rh6G or Rhodamine 123 (Rh123) with a concentration of 10^{-7} M were excited with a 488-nm laser. For the FCCS control experiment, the solution sample of Rh6G was simultaneously excited at the 488- and 543- or 488- and 561-nm laser lines. The pinhole adjustment of the FCS and FCCS setups, the structure parameters, and the detection volumes were calibrated each day through FCS and FCCS measurements of an Rh6G solution at 25 °C (Pack et al. 2006, 2000, 1999). When the structure parameter value was stable and ranged from 4 to 7, the average value was fixed for the FCS and FCCS analyses of all data obtained within the same day under the same conditions. In this study, $G_x(\tau)$ values were fitted using a one-($i = 1$) or two-($i = 2$) component model. If necessary, $G_x(\tau)$s were fitted using a three-component model ($i = 3$) for a complex molecular system in live cells. All the FAFs and FCFs in aqueous solutions were measured for 10 s, 10 times. Since FCS and FCCS are usually combined with confocal imaging, the users can define the detection volume at any position of interest inside cells, such as the plasma membrane, cytosol, and nucleus. For the extra- or intracellular measurements, all $G_x(\tau)$ functions were sequentially measured six times in live cells for 10 s at one position. Photobleaching of the immobile probe in the cellular compartments and the very slow drift of fluorescence intensity originating from cell mobility or the cover glass must be carefully checked during cell measurements. Data obtained during photobleaching and drift were discarded from the analysis. The photochemical term of the fluorescent probe molecule represented by the triplet state was omitted in Eq. 2 for mathematical simplicity, even though the triplet term was analyzed for all FCS and FCCS measurements in the software. The diffusion time of component i, τ_i, is related to the translational diffusion coefficient, D, of component i by

$$\tau_i = \frac{w^2}{4D_i} \tag{4}$$

The diffusion of a spheroidal molecule is related to various physical parameters by the Stokes–Einstein equation:

$$D_i = \frac{k_B T}{6\pi\eta R_{hi}} \tag{5}$$

where T is the absolute temperature (i.e., 298.15 K in this study), R_{hi} is the hydrodynamic radius of the spheroidal molecule, η is the fluid-phase viscosity of the solvent, and k_B is the Boltzmann constant ($1.381 \times 10^{-23} m^2 kg s^{-2} K^{-1}$). The known viscosity of pure water at 25 °C (0.9 cP) was used as a reference.

$$\tau_{\text{probe}} = \tau_{Rh6G} \left(\frac{MW_{\text{probe}}}{MW_{Rh6G}} \right)^{1/3} \tag{6}$$

The diffusion coefficients of fluorescent probes were easily calculated from the diffusion coefficient of Rh6G (D = 280 μm²/s) and measured diffusion times of Rh6G and fluorescent probes as follows:

$$\frac{D_{\text{probe}}}{D_{Rh6G}} = \frac{\tau_{Rh6G}}{\tau_{\text{probe}}} \tag{7}$$

where D_{probe} and τ_{probe} denote the diffusion coefficients and the diffusion times of probe molecules, respectively. The single diffusion coefficient obtained from one-component analyses ($i = 1$ in Eq. 3) was described by D, and the two diffusion coefficients from two-component analyses ($i = 2$ in Eq. 3) were described by D_{fast} and D_{slow}. The subscripts of fast and slow signify the fast-free and slow-restricted diffusion components, respectively. To quantitatively evaluate the binding interaction and dissociation constant from FCCS measurements, the amplitude of the FCF was normalized by the amplitude of the FAF of the red or green fluorescent molecules to calculate the relative cross-correlation amplitude (RCA = $[G_c (0)-1]/[G_{r,g} (0)-1]$) (Pack et al. 2014a, b; Tiwari et al. 2014). To obtain the stationary fluctuating signals with minimal photobleaching, we selected the cells that expressed each fluorescent protein at a low level and the media that contained the fluorescent probe under a low concentration of 1 μM. In addition, the excitation of fluorescent probes in the cell compartment was reduced as much as possible to minimize the effect of photobleaching on the FCS and FCCS analyses (Pack and Ahn 2015; Pack et al. 2014a, b).

Experimental Result and Discussion

Practice of FCS with Solution Samples

FCS analysis provides physical parameters of diffusion coefficient, hydrodynamic size, viscosity, and absolute concentration of fluorescent probe as described in Eqs. 1 to 6, if a stable structure parameter value is experimentally determined using a standard probe with known diffusion coefficient and tuning LSM and FCS setup (Fig. 2) (Magde et al. 1974; Pack et al. 2006, 1999; Rigler et al. 1993). The absolute concentration of fluorescent probes can be determined by evaluating N from FCS

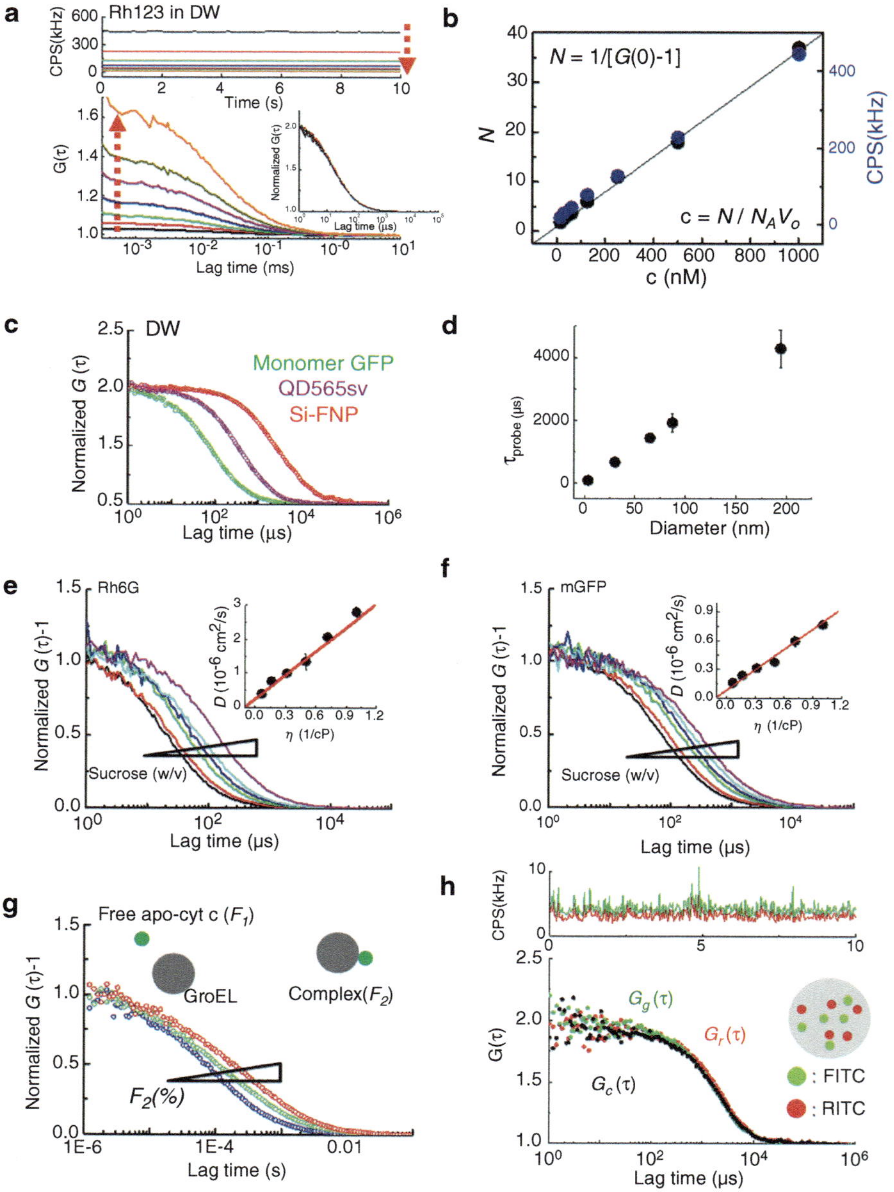
a
Rh123 in DW
CPS(kHz)
Time (s)
G(τ)
Lag time (ms)
Normalized G(τ)
Lag time (μs)
b
N = 1/[G(0)-1]
c = N / N_A V_o
N
CPS(kHz)
c (nM)
c
DW
Monomer GFP
QD565sv
Si-FNP
Normalized G (τ)
Lag time (μs)
d
τ_probe (μs)
Diameter (nm)
e
Rh6G
Normalized G (τ)-1
D (10^-6 cm^2/s)
η (1/cP)
Sucrose (w/v)
Lag time (μs)
f
mGFP
Normalized G (τ)-1
D (10^-6 cm^2/s)
η (1/cP)
Sucrose (w/v)
Lag time (μs)
g
Free apo-cyt c (F_1)
GroEL
Complex(F_2)
F_2(%)
Normalized G (τ)-1
Lag time (s)
h
CPS(kHz)
G(τ)
G_g(τ)
G_r(τ)
G_c(τ)
: FITC
: RITC
Lag time (μs)

Fig. 2 Practice of FCS evaluation with standard fluorescent probes in solution. (**a**) Representative fluorescence intensity traces, FAF curves, and normalized FAF curves of Rh123 in a distilled water solution are shown. An Rh123 solution sample with different concentrations ranging from 15 nM to 1 μM was prepared and measured by FCS. Arrows show an inverse relationship between *G* (0) and averaged fluorescent intensity (CPS). (**b**) The linear proportional relationship between the known sample concentration of Rh123 (**c**) and the number of particles (*N*) obtained from the FCS measurement is shown. (**c**) FAF curves of monomeric GFP, quantum dot (QD), and silica-based fluorescent nanoparticles (Si-FNPs) in distilled water solutions are shown. For the comparison of mobility (i.e., diffusion time), all functions were normalized to the same amplitude, $G(0) = 2$. (**d**) The diffusion time (Eq. 6) of the GFP molecule and nanoparticles with known diameters in a water solution was evaluated by FCS. To demonstrate the inverse linear relationship between the diffusion coefficient and viscosity of the medium, (**e**) Rh6G (479 Da) and (**f**) monomeric GFP (28 kDa) in sucrose–water solutions with different concentrations ranging from 0% to 50%, w/v were measured by FCS. Representative normalized FAF curves of Rh6G (**e**) and mGFP (**f**) in sucrose–water solutions are shown, respectively. The inset shows the diffusion coefficient plot measured from FCS versus the known viscosity of the medium. To compare the mobility change according to viscosity, all functions were normalized to the same amplitude, $G(0)–1 = 1$. (**g**) Typical FAFs in the titration of diluted chaperonin GroEL (1127 kDa) solutions to sample solution of apo-cytochrome c (11 kDa) labeled with tetramethylrhodamine-5-isothiocyanate (TRITC). To compare the mobility change according to the interaction between GroEL and apo-cytochrome c, all functions were normalized to the same amplitude, $G(0)-1 = 1$. (**h**) Synthesized silica-based fluorescent nanoparticles (Si-FNPs) containing both FITC and RICT dyes were measured by FCCS. Because the two dyes physically trapped in the NPs cannot be released from Si-FNP (i.e., ideal dual-colored complex), a maximum RCA value of 1.0 is maintained. The *G* (0) values of the FAF and FCF are not normalized

measurements (Eqs. 2 and 3). The absolute concentration of Rh123 in water was experimentally evaluated by FCS measurements, varying the known concentration ranging from 10 to 1000 nM (Fig. 2a). As the known concentration decreased, the average fluorescence intensity (count per second, CPS in kHz) of the sample solution decreased and the *G* (0) value of the FAF increased, even though the diffusion time did not change. This result indicates that values of CPS and N from FCS measurements were proportional to the absolute concentration, according to Eq. 3 (Fig. 2b). Organic fluorescent dyes such as Rh6G are generally adsorptive to the surface of cover glass. Rh123 was selected as a standard sample because of its minimized adsorptive property. Moreover, the sample chamber was treated with a blocker prior to FCS measurement to prevent the nonspecific adsorption of dyes (Pack et al. 2000). To evaluate the hydrodynamic size of the fluorescent probe by FCS, well-characterized probes, such as monomeric GFP (mGFP), quantum dot, and silica-based fluorescent nanoparticles (Si-FNPs) with different sizes were examined in water or PBS buffer solution (Fig. 2c, d) (Pack et al. 2006, 2012, 2014a, b). According to the Stokes–Einstein relationship (Eqs. 4 and 5), the measured diffusion times of the probes were linearly proportional to the hydrodynamic radius. Equation 5 indicates that the fluidic viscosity of the media solution can be experimentally determined by FCS measurements if the well-characterized probe of a fixed hydrodynamic size is exploited. To demonstrate that the viscosity of the media solution is inversely proportional to the diffusion coefficient (D) of the probe, various sucrose–water solution samples of Rh6G or mGFP ranging from 0% to 50% w/v sucrose concentration were prepared and evaluated by FCS analysis. FAFs shifted to the right according to increases in sucrose concentration (i.e., viscosity of sample solution) and the calculated D values of probes were inversely proportional to the viscosities (Fig. 2e, f). The result demonstrates that the in vitro FCS assay is a reliable method for the evaluation of fluidic viscosity of the media solution, as well as absolute concentration and hydrodynamic size of various molecules and particles. Interestingly, a recent study demonstrated that the viscosity of a glycerol–water solution is proportional to the refractive index of the solution, and the physical parameters can be represented with the D value of the fluorescent probe in the solution (Kim et al. 2019).

The shape of the FAF obtained from the FCS measurement changes or shifts when the physical parameters (i.e., molecular status) of the target probe are changed. Nevertheless, FCS is not sensitive to changes in molecular weight and temperature but is sensitive to changes in viscosity. In practice, an eightfold increase of molecular weight accompanies a twofold increase of diffusion time (Eq. 6) or a twofold decrease of D (Eq. 7). For example, the homer- or heterodimer formation of proteins are barely detected by FCS analysis because of insignificant changes in the molecular weight of the probe protein. On the contrary, large complex and aggregate formations by proteins, which accompany large changes in molecular weights, can be easily detected by the shape change of the FAF. Then, FCS analysis provides D_i values (Eq. 4) of diffusional components of the complex and probe molecules in addition to the fractional ratios (F_i in Eq. 2) of the component in the sample solution (Fig. 2g) (Kawai-Noma et al. 2006; Pack et al. 2000, 2017, 1999). Although the shape of FAFs originate from a one-component model (Fig. 2e, f, h) depending on diffusional mobility, the shape of FAFs originating from a multicomponent model (green and red curve in Fig. 2g) depends not only on the diffusional mobility but also on the fractional ratio of the mobile species. To overcome the size limitation of FCS for interaction analyses, FCCS analysis using a dual-color system is useful for specific interaction between proteins, all of which do not significantly differ in molecular weight. Principally, the RCA value of $[G_c(0)\text{-}1]/[G_{r,g}(0)\text{-}1]$ ranges from 0 to 1 depending on the strength or affinity of the molecular interaction. In an in vitro assay, model FCCS measurements for demonstrating various RCA values of dual-colored probes can be accomplished, if such ideal probes with well-defined affinities exist. For example, Si-FNPs containing FITC and RITC can be exploited to detect an RCA value of 1.0 with the same D values for each color channel (Fig. 2h). Such ideal dual-colored Si-FNPs are useful for tuning and optimizing the FCCS setup (Bacia et al. 2006). In contrast, it is easy to demonstrate FCCS measurements for producing an RCA value of 0 by using an aqueous solution containing two different organic dyes.

For the live-cell assay, the dual-color FCCS measurements of proteins are accomplished by genetically tagging target proteins with GFP and RFP. However, GFP and RFP expressed in live cells are vulnerable to laser excitation even at minimal power and are easily photobleached or tend to blink compared to organic dye probes (Fig. 3). Irreversible photobleaching and reversible blinking of the fluorescent

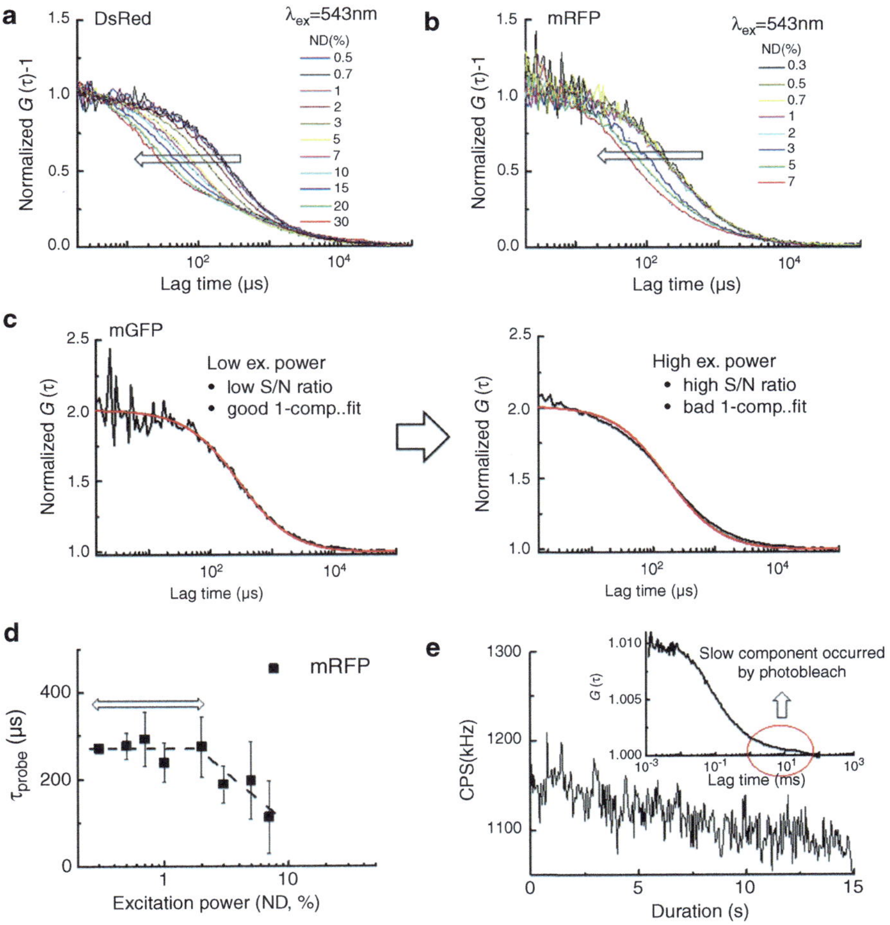

Fig. 3 The influence of excitation laser power on FAF curves and diffusion analyses. (**a, b**) Changes of the FAFs of DsRed and mRFP expressed in HeLa cells under varying excitation power are shown, respectively. Arrows indicate the FAF shift direction according to high power. (**c**) The calculated FAFs (black) with a fit-curve (red) of a one-component model for mGFP detected in the nucleus under varying excitation power are shown, respectively. Note that the FAF curve obtained with a high excitation power is not well fitted to the curve of the one-component model. (**d**) The diffusion time change of mRFP expressed in HeLa cells depending on excitation power is summarized. ND; neutral-density filter. (**e**) Averaged fluorescence intensity during FCS measurements on mGFP expressed in HeLa cells. The inset shows the corresponding FAF obtained from the intensity fluctuation

molecules means that the bleached molecules are permanently extinct, and the blinking molecules are intermittently extinct (Dickson et al. 1997; Hodges et al. 2018). The extinction of fluorescence can induce decreases in the RCA value. In such cases, specific interactions of proteins are underestimated according to the strength of the extinctions. A previous study using FCCS and genetic engineering suggested that two subunits of a proteasome particle tagged by GFP and RFP strongly interact with each other, showing RCA values larger than those of tandemly linked GFP and RFP (i.e., the positive control has a maximal RCA value) (Pack et al. 2014a, b). Due to the strong photobleaching of GFP and RFP tagged in proteasome subunits compared to the positive control, it is likely that the evaluated RCA values for the interaction between subunits are underestimated. Figure 3 summarizes practical examples of blinking and photobleaching when various fluorescent proteins respectively expressed in HeLa cells are measured by FCS with varying excitation power. The shape of FAFs of the fluorescent proteins is largely changed and shifted to the left by strong increases in excitation power due to blinking of the molecules (Fig. 3a, b, c). Any changes of the FAF by overloaded laser power induces a faulty FCS analysis, including a mismatched fit result, decrease in diffusion time, and the occurrence of very slow diffusional components (Fig. 3c, d, e). As shown in Fig. 3, FCS and FCCS measurements on live cells must be carried out at an optimized excitation power for each fluorescent probe used because all fluorescent probes are sensitive to excitation power in different media or cellular conditions. In summary, once optimized conditions of FCS (FCCS) measurements are obtained, FCS and FCCS analyses will provide reliable and quantitative parameter values for characterizing target probes in live cells as well as in aqueous solutions as described in previous studies.

Interactions between Heat Shock Transcriptional Factors (HSFs)

The molecular response of cells to various stresses begins with the activation of HSFs, transcriptional activators of heat shock (HS) genes, which exist as three mammalian isoforms (i.e., HSF1, HSF2, and HSF4) (Akerfelt et al. 2010; Anckar and Sistonen 2011; Pirkkala et al. 2001). HSFs activated by cellular stress bind to heat-shock sequence elements (HSEs) throughout the genome in the nucleus to increase the synthesis of ubiquitous heat shock proteins (HSPs) (Ahn and Thiele 2003; Lindquist 1986). Among the three mammalian HSFs, HSF1 is activated by classical stresses, such as HS, and is closely associated with the stress-induced expression of HSPs. The activation of HSF1 is a molecular process that involves trimerization, the acquisition of HSEs-binding activity, and phosphorylation (Anckar and Sistonen 2011). Unlike HSF1, HSF2 lacks intrinsic classical stress responsiveness and has been proposed to be related to the specific HSP expression observed during developmental processes (Pirkkala et al. 2001). HSF4, which lacks the carboxyl-terminal hydrophobic repeat domain necessary for the suppression of HSF trimer formation, shows constitutive DNA binding activity (Akerfelt et al. 2007; Nakai et al. 1997; Pirkkala et al. 2001). Previous studies have suggested that the interactions between HSFs can regulate their function and play an important role in activating their target genes (Kim et al. 2012; Ostling et al. 2007; Sandqvist et al. 2009). These studies reflect that HSF family members can commonly regulate various cellular stresses via their combinational interaction, even though the molecular interaction and related regulation mechanisms are not fully understood. In this section, FCCS analysis for detecting the molecular interactions among the three HSFs in the nuclei of live HeLa cells before and after HS is described (Pack and Ahn 2015).

To evaluate the interaction properties among HSFs in live cells, plasmid constructs of HSF1, HSF2, and HSF4 tagged with GFP or mCherry were prepared, and HeLa cells co-expressing two pairs of HSFs were used for FCCS measurements. To detect and compare the interaction between HSFs before and after HS, and during recovery, a procedure consisting of CLSM and FCCS measurements with three steps was carried out as described in Fig. 4a (upper). HeLa cells co-expressing HSF1-GFP and mCherry-HSF1 were observed by CLSM and measured by FCCS before and after HS treatment, respectively (Fig. 4a bottom). Regardless of the type of fluorescent protein, HSF1 was diffusely localized in the nuclei of HeLa cells, and the cells did not show nuclear stress granules (nSGs) of HSF1 in the nucleus prior to HS. After HS, both HSF1-GFP and mCherry-HSF1 formed stress granules with strong fluorescent intensity and showed high colocalization in the nucleus (i.e., nSG) indicating that the HSF1s were activated by HS stress. The result of nuclear granule formation is consistent with that of previous studies (Cotto et al. 1997; Jolly et al. 1999). The nuclear granules were maintained for 20 min, even after resetting the heat-shocked cells on the microscope at 25 °C, and gradually diminished with time (Fig. 4b) (Cotto et al. 1997; Jolly et al. 1999). The result of CLSM observation demonstrated that FCCS, which required several seconds for measurement, could be used to detect changes in the interactions between HSFs during the delayed recovery time after HS. In practice, previous FCCS studies have demonstrated that several seconds of FCCS measurement combined with CLSM observation is sufficient to detect the mobility and interaction of target molecules (Bacia et al. 2006; Park et al. 2008). However, the procedure for the detection of HSF activation did not detect changes in the molecular state of HSF molecules upon and soon after HS. Consequently, FCCS measurements were carried out and compared before and after HS by considering the nuclear granule of HSF1 as a marker of HS activation.

To identify characteristic interactions between HSF1s in live cells, HeLa cells co-expressing HSF1-GFP and mCherry-HSF1 were analyzed according to the procedure shown in Fig. 4. For simplicity, only procedures from (2) to (3) in Fig. 4a observing and measuring the same cell will be described and discussed. The heat-shocked cells that showed stress granules, a marker of HS activation, were selected, and FCCS measurements were carried out on the nucleoplasmic positions, excluding nuclear stress granules (Fig. 5a). Interestingly, G_c (0) of the nuclear HSF1s and the corresponding RCA values were high in cells containing stress granules. This result indicates that nuclear HSF1 molecules strongly bond with each other and form stable homo-complexes after HS. Moreover, the autocorrelation curves of G_g (τ) and G_r (τ) were largely shifted to the right after HS. Interestingly, two-component analyses of FAFs showed that a slightly slower mobility (D_{fast} = 8.3 μm^2/s) with a

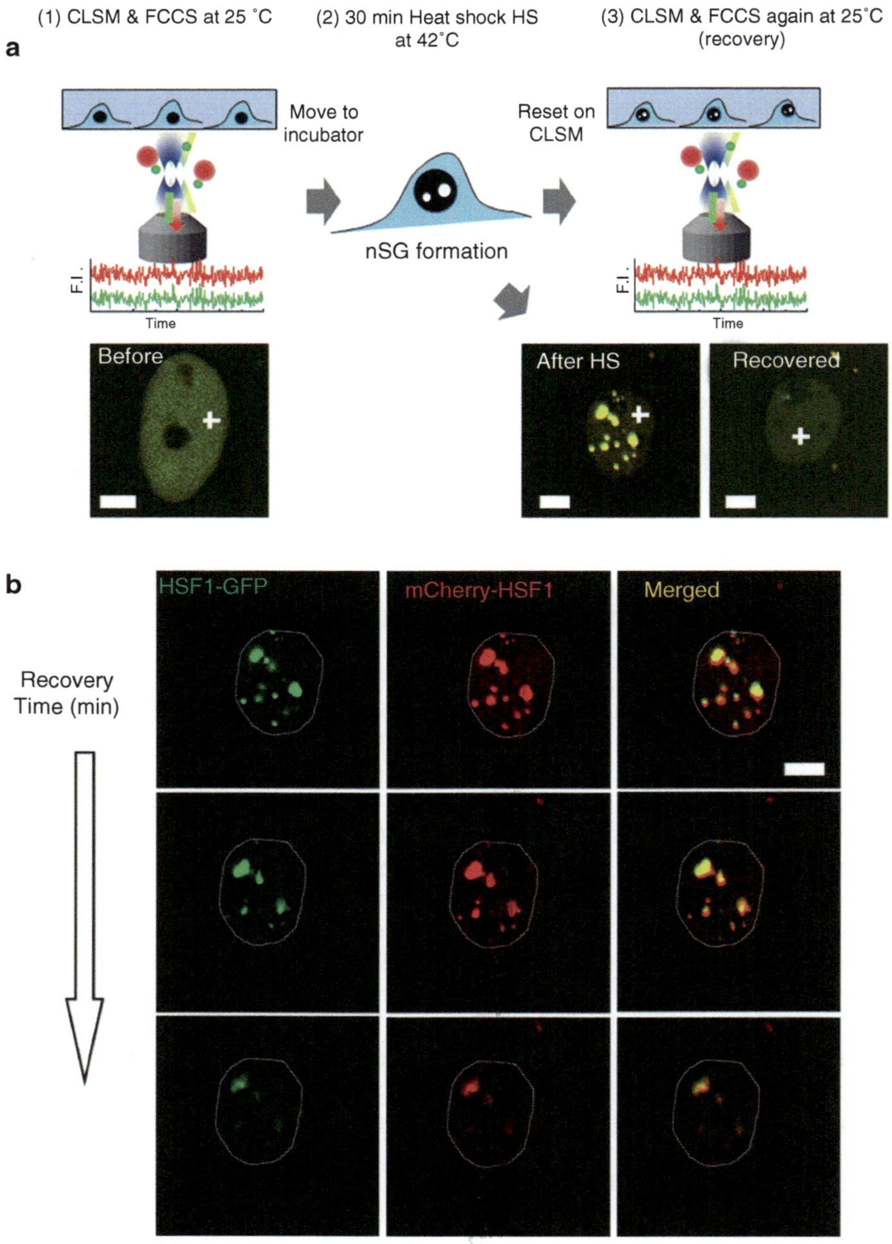
a
(1) CLSM & FCCS at 25 °C
(2) 30 min Heat shock HS at 42°C
(3) CLSM & FCCS again at 25°C (recovery)
Move to incubator
Reset on CLSM
nSG formation
F.I.
Time
Before
After HS
Recovered
b
HSF1-GFP
mCherry-HSF1
Merged
Recovery Time (min)

Fig. 4 Confocal image-based FCS and FCCS measurements of HSFs in live cells. (**a**) The procedures for the CLSM, FCS, and FCCS measurements on nuclear HSF molecules in live HeLa cells before and after heat shock are shown schematically (upper). (1) First, HeLa cells transiently expressing GFP-tagged HSF1 or co-expressing two types of HSFs tagged with GFP or mCherry were observed by CLSM. Then, the positions in the cellular compartments were measured by FCS or FCCS. (2) The glassed chamber containing cells was then moved to the incubator to receive heat shock at 42 °C for 30 min, which induces stress granules of HSF1 in the nucleus (white balls). nSG; nuclear stress granule formed by HSFs. (3) The chamber containing heat-shocked cells was again set up for CLSM and maintained at 25 °C, and single cells containing stress granules were then selected and measured continuously by FCS or FCCS until the nSGs completely vanished. A representative merged CLSM image of a HeLa cell co-expressing HSF1-GFP and mCherry-HSF1 at 25 °C and images of another cell after heat shock and recovery are shown (bottom). Selected positions in the CLSM images were measured by FCCS (crosshair). Scale bar: 5 μm. (**b**) HeLa cells expressing HSF1-GFP and mCherry-HSF1 were heat-shocked at 42 °C for 30 min. The recovery process of heat-shocked HeLa cells maintained at 25 °C was traced by CLSM and FCCS during nSG diminishing. Scale bar: 5 μm

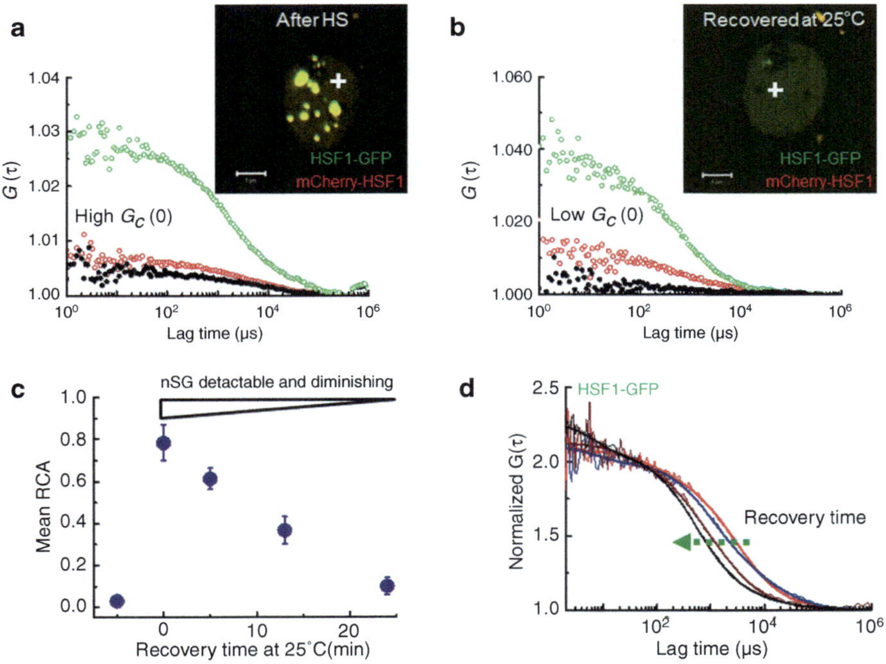

Fig. 5 Direct detection of HSF1 homo-complex by FCCS. (**a, b**) Representative FCFs measured from the heat-shocked and fully recovered cells, respectively (left). Confocal images of a HeLa cell expressing HSF1-GFP and mCherry-HSF1 after heat shock at 42 °C for 30 min and after recovery at 25 °C (right). For heat-shocked cells, positions in the nuclear section excluding stress granules were selected and measured by FCCS (crosshair). Scale bar: 5 μm. (**c**) Changes in the mean RCA value, a degree of interaction amplitude, during the recovery process are summarized. The data point at −5 min represents the mean RCA value obtained from HeLa cells under normal culture conditions. (**d**) Changes of the normalized FAFs of HSF1-GFP (arrow) during the recovery process are shown

Table 1 Summary of D values of GFP-tagged HSF1 in live HeLa cells

	Before heat shock		After heat shock	
HSF1-GFP	D_{fast}, μm²/s (F, %)	D_{slow}, μm²/s (F, %)	D_{fast}, μm²/s (F, %)	D_{slow}, μm²/s (F, %)
Cytosol	13.7 ± 0.6 (99)	0.3 ± 0.2 (1)	15.7 ± 0.9 (99)	0.3 ± 0.1 (1)
Nucleoplasm	12.0 ± 0.4 (98)	0.5 ± 0.2 (2)	8.3 ± 0.7 (56)	0.4 ± 0.1 (44)

fractional ratio of 56%, and a much slower mobility (D_{slow} = 0.4 μm²/s) with a fractional ratio of 44%, were newly induced. The value of D_{slow} after HS was 20-fold slower than the D_{fast} value of HSF1-GFP (Table 1). Because the change in molecular weight via the trimerization of HSF1 induces only a 1.4-fold decrease in the diffusion coefficient (Anckar and Sistonen 2011), it is expected that trimerization induces a small shift in the autocorrelation function to the right. However, the two-component analysis of FAFs suggests that a homo-complex of fluorescently tagged HSF1s after HS also interacts with other molecules, such as immobile target DNAs, which considerably slows the diffusion of the HSF1 homo-complex, and therefore, induces a large shift of the corresponding FAF to the right. After observing the strong interaction between fluorescently tagged HSF1s due to HS stress, the binding kinetics between nuclear HSF1s were traced during recovery after HS (Fig. 5a–d). Interestingly, mean RCA values linearly decreased thereafter for a recovery time of 25 min (Fig. 5c), during which the fluorescence intensity of granules also gradually decreased and then vanished (Fig. 4b). Moreover, the significantly slowed diffusional mobility of HSF1-GFP or mCherry-HSF1 gradually changed to fast mobility according to the interaction decrease (Fig. 5c, d). This result demonstrates that the binding interaction between HSF1s and the slowed diffusion of the HSF1 homo-complex originated from the strong interactions with target HSEs that were coupled to each other.

Recently, the interaction between HSF2 and HSF1 on the stress granules formed after HS has been reported (Sandqvist et al. 2009). In addition, molecular interactions between HSF2 and HSF4a have been suggested by using conventional in vitro assays (Kim et al. 2012). These studies demonstrate that the functional interplay between HSF family members can be controlled via the hetero-complex formation between them, followed by the interaction with target HSEs. However, these binding interactions in the nuclei of live cells and the mechanistic relationships of the interplay during HS response remain unclear (Anckar and Sistonen 2011; Pirkkala et al. 2001). In addition to detecting the interaction between nuclear HSF1s, FCCS indicated the strong hetero-complex formation between HSF1 and others, such as nuclear HSF2 and HSF4 after HS (Pack and Ahn 2015). To investigate protein–protein interaction between HSF2 and HSF4, GFP-tagged HSF2 and mCherry-tagged HSF4 were expressed in HeLa cells and measured by CLSM and FCCS (Fig. 6). Interestingly, CLSM observation showed colocalization of HSF2 and HSF4 on granular structures even in the absence of HS (Fig. 6a), even though it is unclear whether the intranuclear granule structure is the same as the nuclear stress granules produced by homo-complexes of HSF1s after HS. Moreover, FCCS showed that

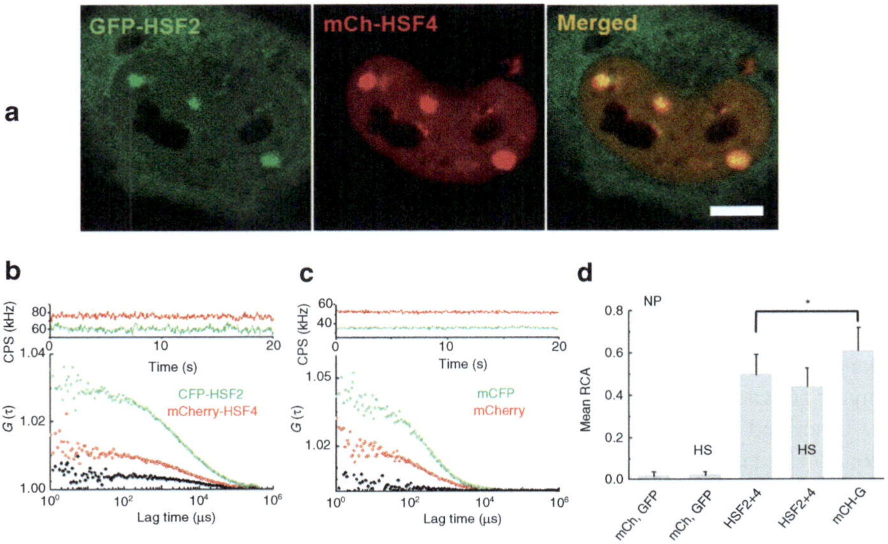

Fig. 6 FCCS analyses of interactions between HSF2 and HSF4. (**a**) Representative CLSM image of a HeLa cell expressing GFP-HSF2 and mCherry-HSF4 at 37 °C is shown. (**b**) Representative FCF measured from a HeLa cell co-expressing GFP-HSF2 and mCherry-HSF4 without heat shock is shown. Positions in the nuclear section excluding bright fluorescent granules were selected and measured by FCCS. (**b**) Representative FCF measured from a HeLa cell co-expressing mGFP and mCherry after heat shock at 42 °C for 30 s is shown. (**c**) Mean RCA values representing interaction amplitudes between GFP-HSF2 and mCherry-HSF4 in the nucleus are summarized. *NP* nucleoplasm, *HS* heat shock

GFP-tagged HSF2 was tightly bound to mCherry-tagged HSF4 in the nucleoplasm excluding the granular structure even without HS, and diffusional mobility of HSF2 and HSF4 was similar to that of heat-shocked HSF1 (Fig. 6b). Based on the D_{slow} of the slow mobile fractions of nuclear HSFs after HS, the estimate of nuclear viscosity, and the assumption that these fractions are spherical in shape, their estimated molecular weights were found to correspond to several GDa (~250 nm in diameter) (Pack et al. 2006; Shimi et al. 2008). These estimates of molecular size of the slow-moving components cannot be attributable to the free diffusion of large protein complexes, as it has been shown to be the cutoff for diffusion in the nucleus (Gorisch et al. 2005). Therefore, it is reasonable to assume that the hetero-complex of HSF2 and HSF4 before HS is stable and intrinsically interacts with DNA, even though it is not known whether the interaction with DNA is specific. On the contrary, HeLa cells co-expressing mGFP and mCherry used as negative control showed no interaction and fast diffusional mobility with one-component analysis (Fig. 6c). The RCA values for the interaction between HSF2 and HSF4 and the negative control were not significantly changed before or after HS (Fig. 6d). The average strength of the hetero-complex formation between HSF2 and HSF4 (mean RCA = 0.43) was much lower than that of the homo-complex between HSF1s (Fig. 5c, mean RCA = 0.80).

These results shown in Figs. 5 and 6 demonstrate that HSF family members of HSF1, HSF2, and HSF4 specifically interact with each other, suggesting that the formation of an HSF1 homo-complex is more favorable and dynamically changed in the HS response.

In recent years, in the development of cancer drugs, those that directly inhibit transcription factor protein–protein interactions are drawing attention compared to the drugs target membrane receptor proteins (Fontaine et al. 2015). At present, HSF is attracting attention as an anticancer drug and drug target for various diseases, such as neurodegenerative diseases (Neef et al. 2011; Vydra et al. 2014; Whitesell and Lindquist 2009). Although the procedure of CLSM and FCCS measurement described in this section is primitive in technical aspects and the results are preliminary, they will be useful for quantitatively investigating the efficacy of drugs inhibiting or activating the interactions among HSFs and between HSFs and target DNA as described in Fig. 7. It is noteworthy that a sophisticated and fast temperature-controlled incubator for CLSM and FCCS would allow more accurate FCCS analysis during HS response (Wachsmuth et al. 2015). Various microscope accessories for rapid measurements or small-scale drug screening are currently in development based on CLSM and FCCS technology.

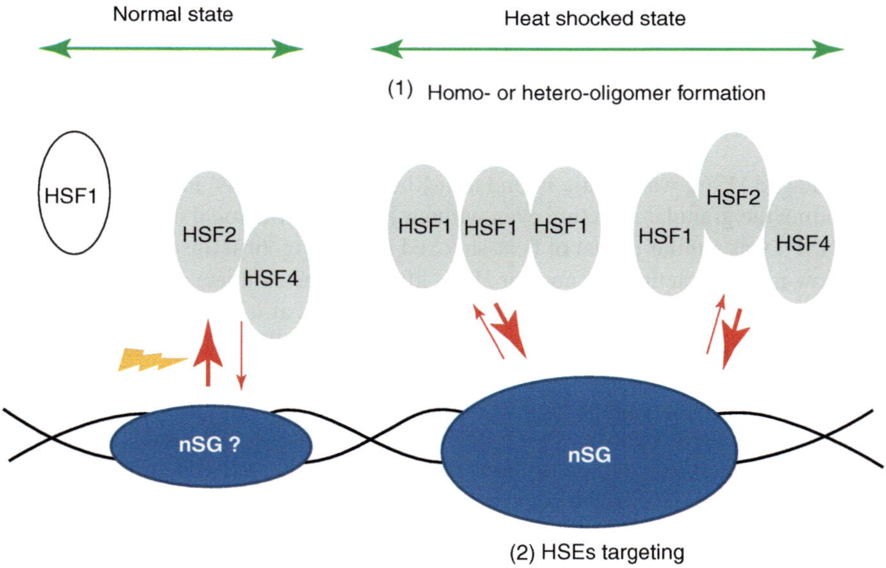

Fig. 7 Schematic diagram of HSF family activation. HSF1 molecules are freely mobile in the cell nucleus and interaction with DNA is minimized under normal culture conditions. On the contrary, HSF2 and HSF4 form a hetero-complex with each other and interact with DNA, even without heat shock. Heat shock induces strong interactions among HSF1, HSF2, and HSF4. From FCS and FCCS analysis, it is reasonable to assume that homo- and hetero-oligomer complexes (1) of the HSF family strongly interact with target HSEs in the nucleus and form nSGs (2)

Drug Single-Cell Reactions Evaluated by FCS and FCCS

Recently, there have been substantial developments in the field of drug delivery systems to allow therapeutic compounds to reach their target location (Lee et al. 2019; Miele et al. 2012; Pack et al. 2005; Patra et al. 2018). Furthermore, various NPs, including silica-based nanoparticles (Si-NPs), quantum dots (QDs), and gold NPs have been developed for biomedical applications, such as imaging agents or markers at a cellular and tissue level (Chithrani et al. 2006; Derfus et al. 2004; Kim et al. 2008; Lu et al. 2009; Xiao et al. 2010). In such applications, NPs are directly or indirectly introduced to cells either as pure NPs or specific NP carriers containing target drugs, and information is required regarding how NPs effectively enter the cells and the quantity of NPs or drugs that reach the target compartments inside the cells (Fig. 8a). Therefore, quantitative parameters directly obtained from the uptake process in living cells will be important for further development of these NPs for desired biomedical or other cellular applications. On the other hand, quantitative analyses based on fluorescence imaging and spectroscopy to monitor dynamics and interactions of fluorescent probes, fluorescently tagged biomolecules and NPs in living cells are notable challenges in the fields of nanomedicine as well as cell biology and biophysics (Braeckmans et al. 2010; Pack et al. 2014a, b). At present, there are various technologies of fluorescence detection and analysis available, including CLSM, super-resolution microscopy, single-molecule detection, and FCS. Among them, FCS is considered one of the most useful techniques due to its high sensitivity and quantitative ability. FCS was recently applied for spatiotemporal evaluation of the cellular uptake process using highly soluble and monodisperse Si-FNPs of 50 nm in diameter. The previous study demonstrated that the Si-FNP probes were freely mobile in live cells, and the dynamic range of the D value of the NPs was broad, ranging from 0.01 to 2 μm^2/sec (Pack et al. 2012). On the contrary, the dynamic range of QDs was narrow, ranging from 0.01 to 0.15 μm^2/sec. Moreover, it was suggested that the mobility of Si-FNP can be categorized into fast mobility corresponding to single Si-FNPs, and slow mobility corresponding to Si-FNPs trapped in the vesicle. However, the spatiotemporal change in the intracellular dynamics of NPs and interactions with the cellular constituents after cellular uptake are still unclear at the single-particle level. Parameters, such as spatial distribution, concentration, temporal dynamics (i.e., mobility), and interactions with targets are important for designing and evaluating the efficiency and cellular toxicity of drugs and delivery nanocarriers. For the quantitative and spatiotemporal analysis of cellular uptake, intracellular dynamics, and interactions of various delivery agents, analytical methods that integrate the complementary techniques such as CLSM, super-resolution microscopy, TEM, and FCS (FCCS) can be useful (Pack et al. 2014a, b). For detail regarding the NP cell uptake and interaction analysis, refer to the cited literature.

In this section, analytical methods of FCS and FCCS to evaluate interaction between antibody-based cancer drugs and live cells are proposed and discussed (Fig. 8b). Figure 8b illustrates potential strategies of FCS and FCCS for quantifying drug–cell reactions. Monoclonal antibodies (mAbs) targeted against membrane receptors are now one of the most important classes of therapeutic agents, which are mainly used to treat cancer diseases (Awwad and Angkawinitwong 2018; Nahta and

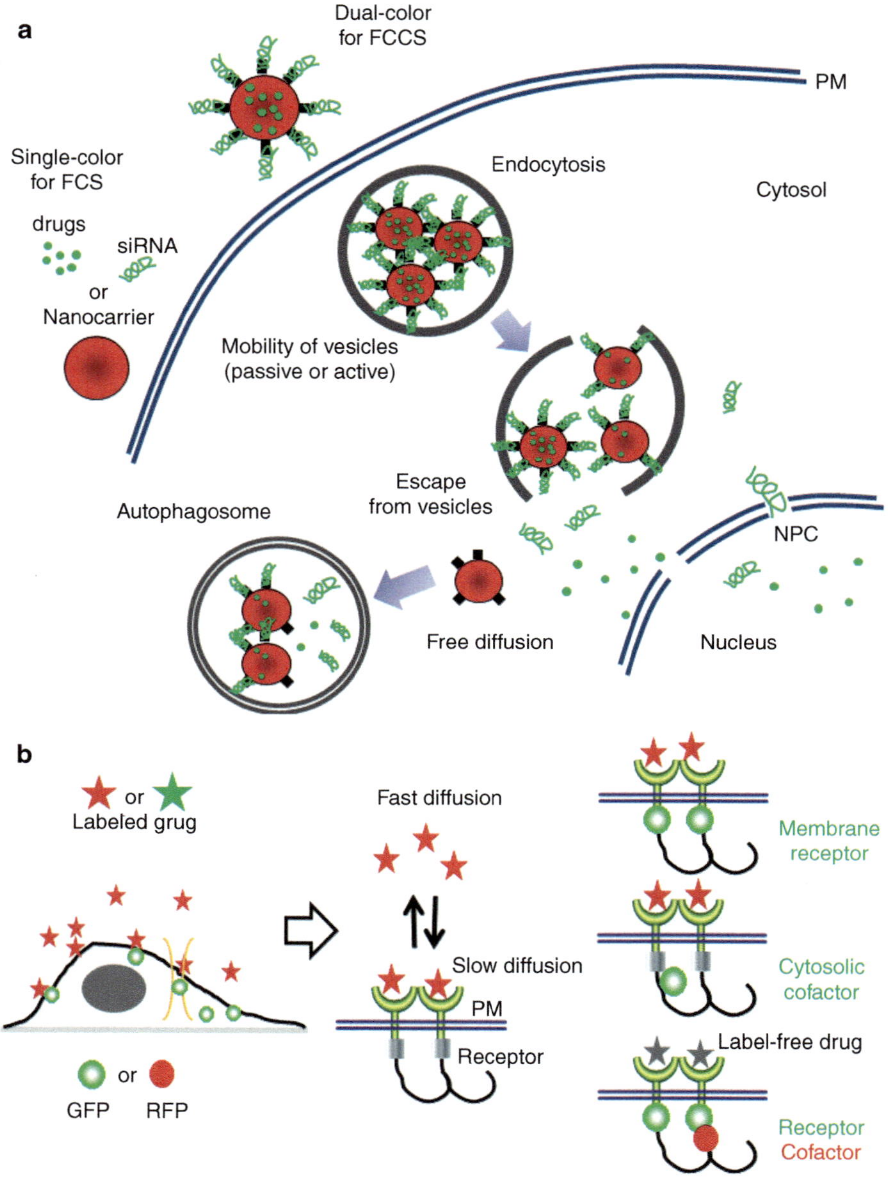

Fig. 8 Schematic illustrations describing the delivery process of drugs and nanocarriers into cells and drug molecule-single cell interaction, which can be probed by FCS and FCCS. (**a**) Fluorescently labeled probes, such as drugs, antibodies, siRNA molecules, and nanocarriers can be traced during cellular uptake and degradation by FCS in addition to 3D and time-lapse confocal imaging. In addition, fluorescent nanocarrier complexed with labeled drugs with different colors can be quantitatively traced by using FCCS analysis if the drug molecules are released into the cytosol or nucleus. NPC; nuclear pore complex, No; the nucleolus. (**b**) Fluorescent drugs, such as antibodies can be treated into culture media and interact with cells through a target of membrane receptor protein. Diffusional dynamics and interactions of drugs outside and inside cells can be detected by FCS or FCCS as shown in (**a**). Colored and gray stars represent fluorescently labeled and label-free drugs, respectively. Green and red balls represent GFP and RFP expressed in cells, respectively. *PM* plasma membrane

Esteva 2006). Because of the ease of fluorescent labeling and high affinity with target proteins, the specific interactions between mAbs and target molecules are one of the positive applications for in vitro FCS or FCCS assay (Fujii and Kinjo 2007; Pack et al. 1999). As described in Fig. 8b, FCS detects the slowed diffusion of labeled mAbs induced by interaction with target receptors, which can be differentiated from the free and fast diffusion in the extracellular region of culture media, due to much slower diffusion of membrane receptors on the plasma membrane of the cell (Sako et al. 2012). In this case, if the mAbs form oligomer complexes with the receptor, the brightness of the labeled drug molecule will also change and the difference in the brightness in the medium may be detected by FCS. On the other hand, FCCS can be useful when the target proteins are known and are tagged by fluorescent proteins for cellular expression since the specific interaction between mAb and the target molecule can be easily detected and evaluated.

Trastuzumab, known as Herceptin, is a HER2 receptor targeting mAb used to treat breast cancer (Boekhout et al. 2011; Nahta and Esteva 2006). Therefore, it is specifically used for targeting breast cancer cells that are HER2 receptor-positive. Since Trastuzumab will strongly interact with HER2 expressed on the plasma membrane of living cells, it is expected that FCS and CLSM analyses for Trastuzumab–cell reactions are easily established by labeling the mAb (Fig. 9a). As demonstrated in Fig. 2, various parameters, such as hydrodynamic property, labeling efficiency,

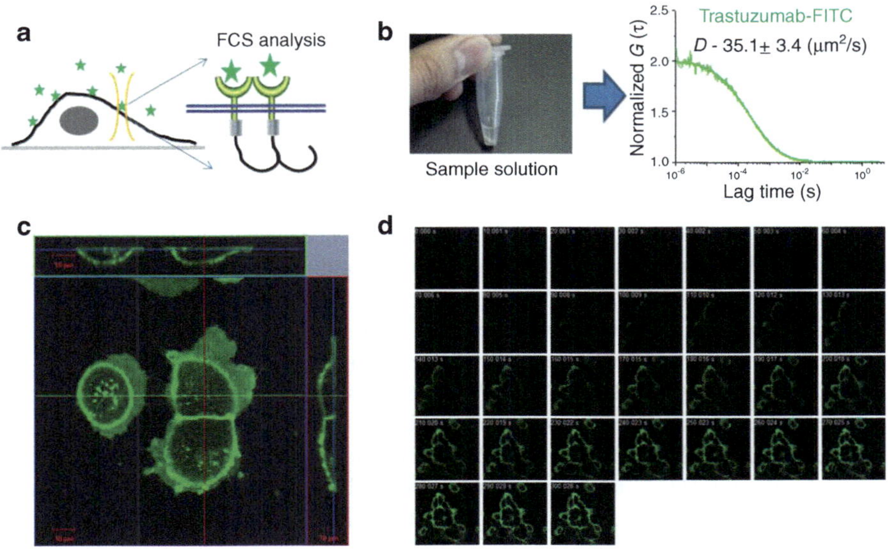

Fig. 9 Quality check of labeled drug by FCS evaluation and 3D and time-lapse confocal imaging for drug–cell interaction evaluation. (**a**) A schematic diagram of the labeled drug (green) interaction with label-free target receptors on a cell is described. (**b**) The solution sample of FITC-labeled trastuzumab and the measured FAF curve of the sample are shown. From FCS evaluation, FITC is fully labeled into trastuzumab without free FITC molecules in the sample solution. (**c**) The 3D confocal image of the MCF7 cells after treatment of FITC-labeled trastuzumab into culture media is shown. Scale bar: 10 μm. (**d**) Serial confocal images of cells after treatment of labeled trastuzumab during time-lapse observations are shown

Table 2 Types of breast cancer cell lines used for FCS analysis of trastuzumab–cell interaction

Cell lines	ER	PR	Her2	Subtype
SKBR3	–	–	+	HER2-enriched
BT474	+	+	+	Luminal B
MDAMB231	–	–	–	Triple-negative B
MDAMB361	+	–	+	Luminal B

and sample concentration, were evaluated with prepared solution samples of labeled Trastuzumab (Fig. 9b). Moreover, cellular localization of the labeled mAbs and temporal change of the localization were traced after the treatment of the labeled mAb with known concentration (Fig. 9c, d). The result of the labeled Trastuzumab being localized into the cell membrane suggests that the mAb is strongly bound to the cell membrane of MCF7 cells, a well-recognized breast cancer cell lines (Lattrich et al. 2008; Merlin et al. 2002), even though MCF7 is categorized into luminal A cell line, a subtype of estrogen receptor (ER), and/or progesterone receptor (PR) positive and human epithelial receptor 2 (HER2) negative (Dai et al. 2017).

Gene expression profiling has been broadly applied to categorize breast cancer cell lines. Genetic characterization of the breast cancer cell lines is now based on the expression status of three important receptors: ER, PR, and HER2. There are five main molecular subtypes of breast cancer cell lines that are based on the genes that cancer cells express (Dai et al. 2017; Holliday and Speirs 2011). Among them, we selected four cell lines with different subtypes for FCS analyses of labeled trastuzumab–cell interaction as shown in Table 2. CLSM measurements demonstrated that the FITC-labeled mAb bound to the cell membrane of each cell subtype, even MDAMB231 cells, a triple-negative cell type (Fig. 10). On the contrary, the mAb is weakly bound to the cell membrane of MDAMB361 cells which are HER2 positive. Intriguingly, SKBR3 cells showed inhomogeneous localization of labeled trastuzumab on the cell membrane (Fig.10a), while other cell lines were evenly localized on the cell membrane (Fig. 10b–d). An immunofluorescence assay for HER2 would assist in the understanding of the correlation between trastuzumab and HER2 molecules in each cell type.

In addition to CLSM, FCS measurements were carried out on the cell membrane of each cell type treated with labeled trastuzumab (Fig. 11). As shown in Fig. 11a, FCS detected the slowed diffusion of labeled trastuzumab induced by interaction with target receptors, because of the much slower diffusion of membrane receptors on the plasma membrane of each cell type. FCS analyses showed that the values of slowed diffusion (D_{slow}) on the membranes of each cell type were not significantly different (Fig. 11b). Instead, the fractional ratio of D_{slow} for SKBR3 and BT474 cells was much larger than those for MDAMB231 and MDAMB361 suggesting that more trastuzumab molecules were bound to target molecules distributed in the cell membrane. This result is consistent with that obtained from CLSM measurements as shown in Fig. 10. Average fluorescence intensity represented by count per second (CPS in kHz) and brightness of the molecule represented by count per molecule (CPM in kHz) are summarized in Fig. 11c, d. The results show that fluorescence

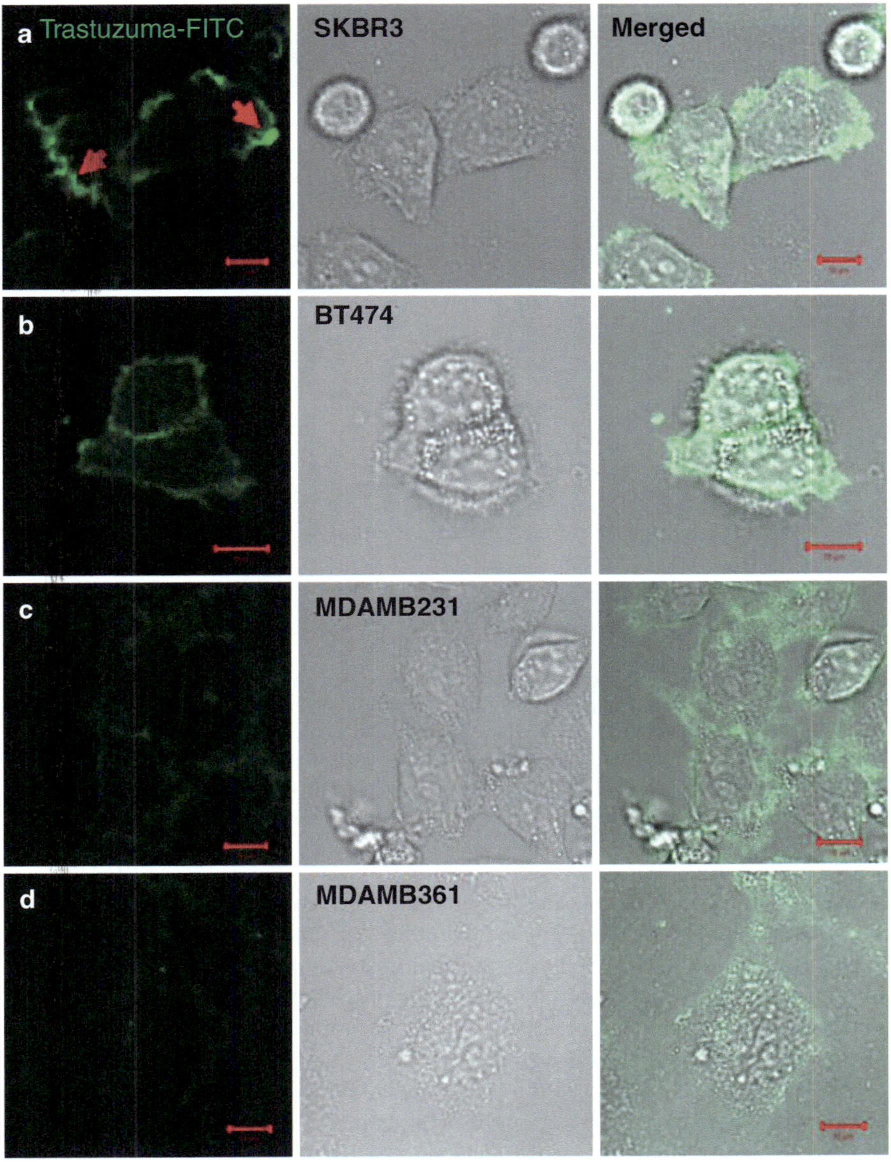

Fig. 10 Comparison of labeled trastuzumab distribution on various breast cancer cells. Confocal images of four different breast cancer cell lines, (**a**) SKBR3, (**b**) BT474, (**c**) MDAMB231, (**d**) MDAMB361, after treatment of FITC-labeled trastuzumab (75 ng/ml) in the culture media are shown. Arrows indicate positions of high fluorescence intensity on the plasma membrane. Scale bar: 10 μm

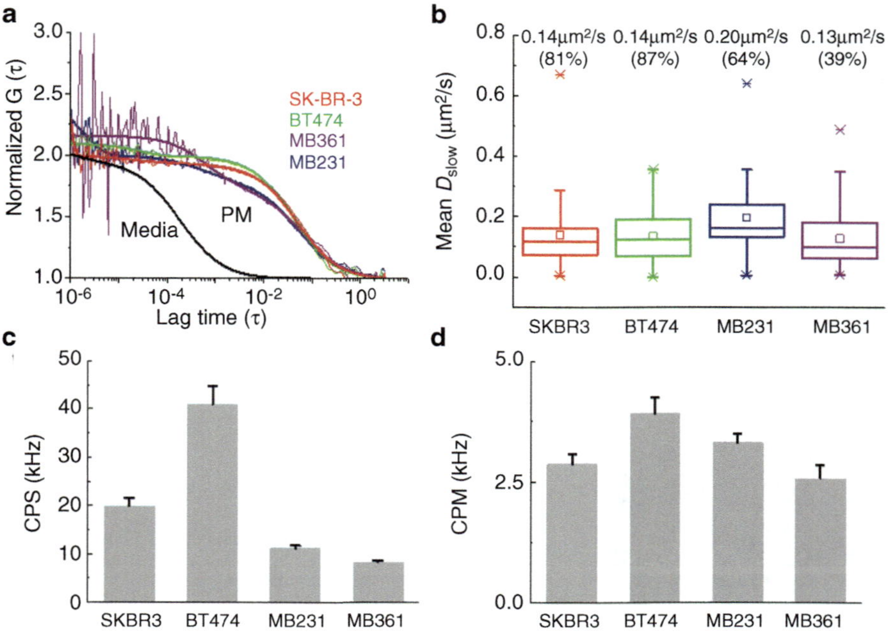

Fig. 11 Summary of the results obtained from FCS measurement on the cells. (**a**) Representative FAFs of FITC-labeled trastuzumab on the four cancer cell lines and in culture media are shown. (**b**) The mean D_{slow} value of FITC-labeled trastuzumab and mean fractional ratio of the D_{slow} component detected on the plasma membrane are shown. Fast diffusion (D_{fast}) in the media solution was fixed in the model fit. (**c, d**) Averaged fluorescence intensity and averaged fluorescence intensity per molecule on the plasma membrane of the cells are shown, respectively. MB231: MDAMB231; MB361: MDAMB361 cell

intensity (i.e., concentration) and brightness of the molecule on the membrane of BT474 cells are significantly greater than those of other cell types. High average values of CPM on BT474 cells suggest that trastuzumab molecules are multimerized or oligomerized compared to other cell types. The region of high fluorescence intensity on the membrane of SKBR-3 cells was rapidly photobleached, and FCS analysis on such positions was excluded. It is likely that HER2 molecules expressed on the membrane of SKBR3 cells are immobilized at the region, and labeled trastuzumab strongly bound to the immobilized target molecules experience rapid photobleaching. Since single-molecule analysis is complementary to FCS analysis, combinational use of FCS and single-molecule detection will assist in evaluating molecular kinetics and dynamics on cell membranes (Sako et al. 2012).

Figure 12 shows an FCCS experiment to detect the specific binding interaction between TRITC-labeled trastuzumab and the GFP-tagged HER2 receptor protein on the membrane of MDAMB231. CLSM observation showed colocalization of the drug and HER2-GFP on the membrane. Moreover, FCCS detected a strong binding interaction between them on the membrane and the evaluated value of dissociation constant (i.e., K_d) was 6 nM. The concentration of endogenous HER2 in the cell was

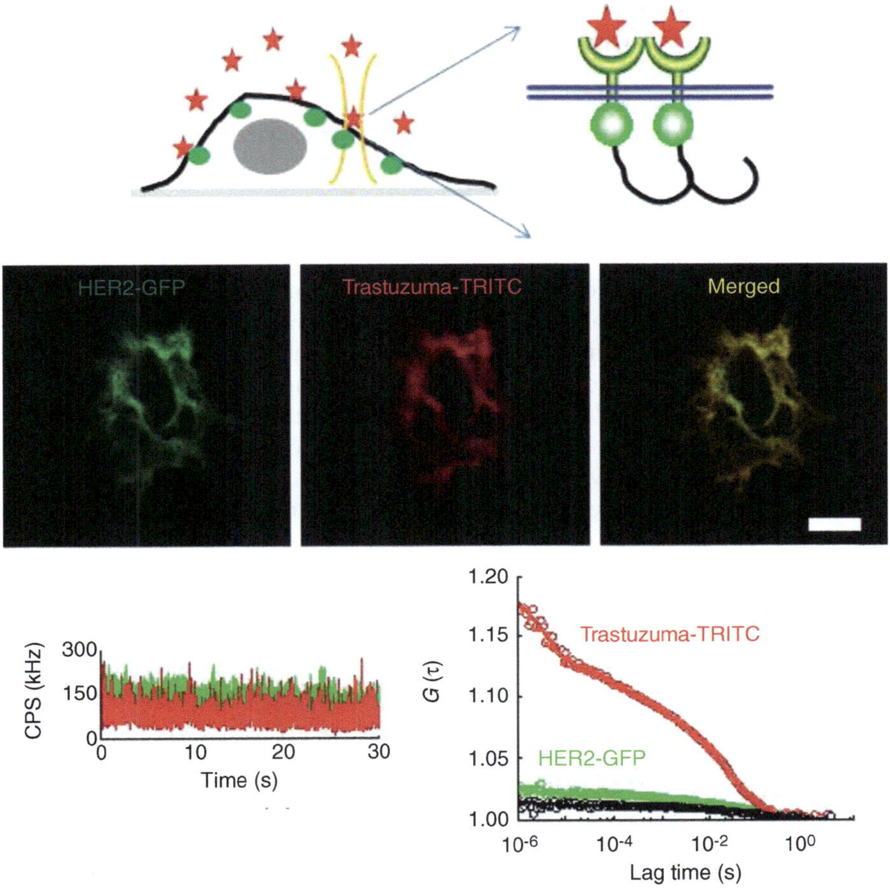

Fig. 12 Dual-color FCCS experiment for the direct detection of a drug–receptor interaction in live cells. A schematic diagram of FCCS measurements for detecting the interaction between labeled trastuzumab and GFP-tagged receptor expressed on plasma membranes is shown (upper). Confocal images of a MDAMB231 cell expressing HER2-GFP after treatment of TRITC (tetramethylrhodamine) labeled trastuzumab into culture media (center) are shown (center). Scale bar: 10 μm. Representative time trace of averaged fluorescent intensity and FCF measured on the cell are shown, respectively (lower)

ignored for simplicity. In practice, there are various limitations for FCCS analyses in live cells. Although fluorescence technologies have advanced dramatically, it is not easy to transfect cells with exogenous genes coding for GFP/RFP-tagged proteins. This is because the transfection efficiency varies widely depending on the cancer cell type. In addition, since the expressed GFP-tagged protein competes with the endogenous protein, correction is necessary to calculate a true binding constant, but this is also not easy. If stable cell lines expressing fluorescently tagged proteins endogenously can be developed using genome editing technology, such as CRISPR (refer Chapter "The Genetic Basis of Reporter Mouse Strains" in this book), the FCCS assay described above can be applied accurately and efficiently in many cell types. Moreover, as discussed in Section "Interactions Between Heat Shock Transcriptional Factors (HSFs)", efforts are being challenged to develop confocal microscope accessories for the evaluation of drug efficacy at a single-cell level.

Conclusion

FCS and FCCS combined on CLSM are versatile techniques that have become the standard for detecting molecular dynamics and interactions in cells as well as in vitro, which allows potential applications for convergence science and preclinical studies, such as nanomedicine, drug delivery, and drug efficacy assessment. Two experimental approaches described in this study detecting protein–protein interactions are now applicable for investigating the efficacy of newly developed drugs at a cellular and molecular level, even though there are challenges in various technical aspects and efficiency, from measurements to the analysis of FCCS for high-throughput/content-screening level analysis. Due to the dynamic and strong interaction kinetics of HSFs, the system of protein–protein interaction can be used as a barometer to identify effective small compounds of inhibitors or activators. In addition to gene therapy, which exploits siRNA and DNA, antibody/peptide-based therapy for cancer and other diseases is now one of the most successful strategies for treating patients. Therefore, various antibodies or peptide drugs as well as siRNA and DNA used in gene therapy will be developed and must be evaluated at various basic and preclinical stages. Due to easy labeling of synthetic peptides, antibodies, RNA, and DNA, CLSM-based FCS and FCCS will be useful for the direct evaluation of the cellular delivering and targeting efficacy of the drugs. In addition, various types of synthetic nanomaterials (i.e., nanocarriers) for drug delivery can also be quantitatively characterized and evaluated for delivery efficacy in detail. The analytical FCS and FCCS methods can be combined with other imaging techniques described in other chapters such as correlative light and electron microscopy (CLEM), and label-free imaging methods, such as Raman and optical diffraction tomography (ODT). It is expected that the conceptual and experimental examples provided in this chapter will aid in the design of application strategies that are required in the field of convergence studies.

Acknowledgments I thank the Confocal Microscope and Optical Imaging core facilities at the ConveRgence mEDIcine research cenTer (CREDIT), Asan Medical Center for support and instrumentation. This chapter was supported by the Basic Science Research Program (NRF-2018R1D1A1B07048696) through the National Research Foundation of Korea (NRF) funded by the Ministry of Education. This study was also supported by grants (2020IP0028) from the Asan Institute for Life Sciences, Asan Medical Center.

References

Ahn SG, Thiele DJ (2003) Redox regulation of mammalian heat shock factor 1 is essential for Hsp gene activation and protection from stress. Genes Dev 17(4):516–528. https://doi.org/10.1101/gad.1044503

Akerfelt M, Trouillet D, Mezger V, Sistonen L (2007) Heat shock factors at a crossroad between stress and development. Ann N Y Acad Sci 1113:15–27. https://doi.org/10.1196/annals.1391.005

Akerfelt M, Morimoto RI, Sistonen L (2010) Heat shock factors: integrators of cell stress, development and lifespan. Nat Rev Mol Cell Biol 11(8):545–555. https://doi.org/10.1038/nrm2938

Anckar J, Sistonen L (2011) Regulation of HSF1 function in the heat stress response: implications in aging and disease. Annu Rev Biochem 80:1089–1115. https://doi.org/10.1146/annurev-biochem-060809-095203

Aragon SR, Pecora R (1976) Fluorescence correlation spectroscopy as a probe of molecular dynamics. J Chem Phys 64:1791–1803

Awwad S, Angkawinitwong U (2018) Overview of antibody drug delivery. Pharmaceutics 10(3):83. https://doi.org/10.3390/pharmaceutics10030083

Bacia K, Kim SA, Schwille P (2006) Fluorescence cross-correlation spectroscopy in living cells. Nat Methods 3(2):83–89. https://doi.org/10.1038/nmeth822

Boekhout AH, Beijnen JH, Schellens JH (2011) Trastuzumab. Oncologist 16(6):800–810. https://doi.org/10.1634/theoncologist.2010-0035

Braeckmans K, Buyens K, Naeye B, Vercauteren D, Deschout H, Raemdonck K, Remaut K, Sanders NN, Demeester J, De Smedt SC (2010) Advanced fluorescence microscopy methods illuminate the transfection pathway of nucleic acid nanoparticles. J Control Release 148(1):69–74. https://doi.org/10.1016/j.jconrel.2010.08.029

Capoulade J, Wachsmuth M, Hufnagel L, Knop M (2011) Quantitative fluorescence imaging of protein diffusion and interaction in living cells. Nat Biotechnol 29(9):835–839. https://doi.org/10.1038/nbt.1928

Chithrani BD, Ghazani AA, Chan WC (2006) Determining the size and shape dependence of gold nanoparticle uptake into mammalian cells. Nano Lett 6(4):662–668. https://doi.org/10.1021/nl052396o

Cotto J, Fox S, Morimoto R (1997) HSF1 granules: a novel stress-induced nuclear compartment of human cells. J Cell Sci 110(Pt 23):2925–2934

Dai X, Cheng H, Bai Z, Li J (2017) Breast cancer cell line classification and its relevance with breast tumor subtyping. J Cancer 8(16):3131–3141. https://doi.org/10.7150/jca.18457

Derfus AM, Chan WCW, Bhatia SN (2004) Probing the cytotoxicity of semiconductor quantum dots. Nano Lett 4(1):11–18. https://doi.org/10.1021/nl0347334

Dickson RM, Cubitt AB, Tsien RY, Moerner WE (1997) On/off blinking and switching behaviour of single molecules of green fluorescent protein. Nature 388(6640):355–358. https://doi.org/10.1038/41048

Dong B, Jaeger AM, Thiele DJ (2019) Inhibiting heat shock factor 1 in cancer: a unique therapeutic opportunity. Trends Pharmacol Sci 40(12):986–1005. https://doi.org/10.1016/j.tips.2019.10.008

Elson EL, Magde D (1974) Fluorescence correlation spectroscopy. I. Conceptual basis and theory. Biopolymers 13:1–17

Fontaine F, Overman J, Francois M (2015) Pharmacological manipulation of transcription factor protein-protein interactions: opportunities and obstacles. Cell Regen (Lond) 4(1):2. https://doi.org/10.1186/s13619-015-0015-x

Fujii F, Kinjo M (2007) Detection of antigen protein by using fluorescence cross-correlation spectroscopy and quantum-dot-labeled antibodies. Chembiochem 8(18):2199–2203. https://doi.org/10.1002/cbic.200700399

Gorisch SM, Wachsmuth M, Toth KF, Lichter P, Rippe K (2005) Histone acetylation increases chromatin accessibility. J Cell Sci 118(Pt 24):5825–5834. https://doi.org/10.1242/jcs.02689
Hodges C, Kafle RP, Hoff JD, Meiners JC (2018) Fluorescence correlation spectroscopy with photobleaching correction in slowly diffusing systems. J Fluoresc 28(2):505–511. https://doi.org/10.1007/s10895-018-2210-y
Holliday DL, Speirs V (2011) Choosing the right cell line for breast cancer research. Breast Cancer Res 13(4):215. https://doi.org/10.1186/bcr2889
Jolly C, Usson Y, Morimoto RI (1999) Rapid and reversible relocalization of heat shock factor 1 within seconds to nuclear stress granules. Proc Natl Acad Sci U S A 96(12):6769–6774. https://doi.org/10.1073/pnas.96.12.6769
Kang MH, Choi H, Oshima M, Cheong JH, Kim S, Lee JH, Park YS, Choi HS, Kweon MN, Pack CG, Lee JS, Mills GB, Myung SJ, Park YY (2018) Estrogen-related receptor gamma functions as a tumor suppressor in gastric cancer. Nat Commun 9(1):1920. https://doi.org/10.1038/s41467-018-04244-2
Kawai-Noma S, Ayano S, Pack CG, Kinjo M, Yoshida M, Yasuda K, Taguchi H (2006) Dynamics of yeast prion aggregates in single living cells. Genes Cells 11(9):1085–1096. https://doi.org/10.1111/j.1365-2443.2006.01004.x
Kawai-Noma S, Pack CG, Kojidani T, Asakawa H, Hiraoka Y, Kinjo M, Haraguchi T, Taguchi H, Hirata A (2010) In vivo evidence for the fibrillar structures of Sup35 prions in yeast cells. J Cell Biol 190(2):223–231. https://doi.org/10.1083/jcb.201002149
Kim J, Kim HS, Lee N, Kim T, Kim H, Yu T, Song IC, Moon WK, Hyeon T (2008) Multifunctional uniform nanoparticles composed of a magnetite nanocrystal core and a mesoporous silica shell for magnetic resonance and fluorescence imaging and for drug delivery. Angew Chem Int Ed Eng 47(44):8438–8441. https://doi.org/10.1002/anie.200802469
Kim SA, Yoon JH, Ahn SG (2012) Heat shock factor 4a (HSF4a) represses HSF2 expression and HSF2-mediated transcriptional activity. J Cell Physiol 227(1):1–6. https://doi.org/10.1002/jcp.22948
Kim TK, Lee BW, Fujii F, Lee KH, Lee S, Park Y, Kim JK, Lee SW, Pack CG (2019) Mitotic chromosomes in live cells characterized using high-speed and label-free optical diffraction tomography. Cell 8(11):1368. https://doi.org/10.3390/cells8111368
Kinjo M, Rigler R (1995) Ultrasensitive hybridization analysis using fluorescence correlation spectroscopy. Nucleic Acids Res 23(10):1795–1799. https://doi.org/10.1093/nar/23.10.1795
Klingler J, Friedrich T (1997) Site-specific interaction of thrombin and inhibitors observed by fluorescence correlation spectroscopy. Biophys J 73(4):2195–2200. https://doi.org/10.1016/S0006-3495(97)78251-1
Lattrich C, Juhasz-Boess I, Ortmann O, Treeck O (2008) Detection of an elevated HER2 expression in MCF-7 breast cancer cells overexpressing estrogen receptor beta1. Oncol Rep 19(3):811–817
Lee SH, Park OK, Kim J, Shin K, Pack CG, Kim K, Ko G, Lee N, Kwon SH, Hyeon T (2019) Deep tumor penetration of drug-loaded nanoparticles by click reaction-assisted immune cell targeting strategy. J Am Chem Soc 141(35):13829–13840. https://doi.org/10.1021/jacs.9b04621
Lindquist S (1986) The heat-shock response. Annu Rev Biochem 55:1151–1191. https://doi.org/10.1146/annurev.bi.55.070186.005443
Lu F, Wu SH, Hung Y, Mou CY (2009) Size effect on cell uptake in well-suspended, uniform mesoporous silica nanoparticles. Small 5(12):1408–1413. https://doi.org/10.1002/smll.200900005
Luby-Phelps K, Castle PE, Taylor DL, Lanni F (1987) Hindered diffusion of inert tracer particles in the cytoplasm of mouse 3T3 cells. Proc Natl Acad Sci U S A 84(14):4910–4913. https://doi.org/10.1073/pnas.84.14.4910
Magde D, Elson E, Webb WW (1972) Thermodynamic fluctuations in a reacting system-measurement by fluorescence correlation spectroscopy. Phys Rev Lett 29:705–708
Magde D, Elson EL, Webb WW (1974) Fluorescence correlation spectroscopy. II. An experimental realization. Biopolymers 13(1):29–61. https://doi.org/10.1002/bip.1974.360130103
Merlin JL, Barberi-Heyob M, Bachmann N (2002) In vitro comparative evaluation of trastuzumab (Herceptin) combined with paclitaxel (Taxol) or docetaxel (Taxotere) in HER2-expressing

human breast cancer cell lines. Ann Oncol 13(11):1743–1748. https://doi.org/10.1093/annonc/mdf263

Miele E, Spinelli GP, Miele E, Di Fabrizio E, Ferretti E, Tomao S, Gulino A (2012) Nanoparticle-based delivery of small interfering RNA: challenges for cancer therapy. Int J Nanomedicine 7:3637–3657. https://doi.org/10.2147/IJN.S23696

Nahta R, Esteva FJ (2006) HER2 therapy: molecular mechanisms of trastuzumab resistance. Breast Cancer Res 8(6):215. https://doi.org/10.1186/bcr1612

Nakai A, Tanabe M, Kawazoe Y, Inazawa J, Morimoto RI, Nagata K (1997) HSF4, a new member of the human heat shock factor family which lacks properties of a transcriptional activator. Mol Cell Biol 17(1):469–481. https://doi.org/10.1128/mcb.17.1.469

Neef DW, Jaeger AM, Thiele DJ (2011) Heat shock transcription factor 1 as a therapeutic target in neurodegenerative diseases. Nat Rev Drug Discov 10(12):930–944. https://doi.org/10.1038/nrd3453

Ostling P, Bjork JK, Roos-Mattjus P, Mezger V, Sistonen L (2007) Heat shock factor 2 (HSF2) contributes to inducible expression of hsp genes through interplay with HSF1. J Biol Chem 282(10):7077–7086. https://doi.org/10.1074/jbc.M607556200

Pack CG, Ahn SG (2015) Heat shock-induced interactions among nuclear HSFs detected by fluorescence cross-correlation spectroscopy. Biochem Biophys Res Commun 463(3):303–308. https://doi.org/10.1016/j.bbrc.2015.05.056

Pack CG, Nishimura G, Tamura M, Aoki K, Taguchi H, Yoshida M, Kinjo M (1999) Analysis of interaction between chaperonin GroEL and its substrate using fluorescence correlation spectroscopy. Cytometry 36(3):247–253. https://doi.org/10.1002/(sici)1097-0320(19990701)36:3<247::aid-cyto15>3.3.co;2-r

Pack CG, Aoki K, Taguchi H, Yoshida M, Kinjo M, Tamura M (2000) Effect of electrostatic interactions on the binding of charged substrate to GroEL studied by highly sensitive fluorescence correlation spectroscopy. Biochem Biophys Res Commun 267(1):300–304. https://doi.org/10.1006/bbrc.1999.1864

Pack DW, Hoffman AS, Pun S, Stayton PS (2005) Design and development of polymers for gene delivery. Nat Rev Drug Discov 4(7):581–593. https://doi.org/10.1038/nrd1775

Pack C, Saito K, Tamura M, Kinjo M (2006) Microenvironment and effect of energy depletion in the nucleus analyzed by mobility of multiple oligomeric EGFPs. Biophys J 91(10):3921–3936. https://doi.org/10.1529/biophysj.105.079467

Pack CG, Song MR, Tae EL, Hiroshima M, Byun KH, Kim JS, Sako Y (2012) Microenvironments and different nanoparticle dynamics in living cells revealed by a standard nanoparticle. J Control Release 163(3):315–321. https://doi.org/10.1016/j.jconrel.2012.07.036

Pack CG, Jung MK, Song MR, Kim JS, Sako Y (2014a) Use of engineered nanoparticle based on fluorescence methods for live cell phenomena. Fluorescence microscopy: super-resolution and other novel techniques, 1st edn. Elsevier, London

Pack CG, Yukii H, Toh-e A, Kudo T, Tsuchiya H, Kaiho A, Sakata E, Murata S, Yokosawa H, Sako Y, Baumeister W, Tanaka K, Saeki Y (2014b) Quantitative live-cell imaging reveals spatio-temporal dynamics and cytoplasmic assembly of the 26S proteasome. Nat Commun 5:3396. https://doi.org/10.1038/ncomms4396

Pack CG, Inoue Y, Higurashi T, Kawai-Noma S, Hayashi D, Craig E, Taguchi H (2017) Heterogeneous interaction network of yeast prions and remodeling factors detected in live cells. BMB Rep 50(9):478–483. https://doi.org/10.5483/bmbrep.2017.50.9.084

Park H, Pack C, Kinjo M, Kaang BK (2008) In vivo quantitative analysis of PKA subunit interaction and cAMP level by dual color fluorescence cross correlation spectroscopy. Mol Cell 26(1):87–92

Park H, Han SS, Sako Y, Pack CG (2015) Dynamic and unique nucleolar microenvironment revealed by fluorescence correlation spectroscopy. FASEB J 29(3):837–848. https://doi.org/10.1096/fj.14-254110

Patra JK, Das G, Fraceto LF, Campos EVR, Rodriguez-Torres MDP, Acosta-Torres LS, Diaz-Torres LA, Grillo R, Swamy MK, Sharma S, Habtemariam S, Shin HS (2018) Nano based drug delivery systems: recent developments and future prospects. J Nanobiotechnol 16(1):71. https://doi.org/10.1186/s12951-018-0392-8

Phair RD, Misteli T (2000) High mobility of proteins in the mammalian cell nucleus. Nature 404(6778):604–609. https://doi.org/10.1038/35007077

Pirkkala L, Nykanen P, Sistonen L (2001) Roles of the heat shock transcription factors in regulation of the heat shock response and beyond. FASEB J 15(7):1118–1131. https://doi.org/10.1096/fj00-0294rev

Rigler R, Mets U, Widengren J, Kask P (1993) Fluorescence correlaton spectroscopy with high count rate and low background: analysis of translational diffusion. Eur Biophys J 22:169–175. https://doi.org/10.1007/BF00185777

Sako Y, Hiroshima M, Pack CG, Okamoto K, Hibino K, Yamamoto A (2012) Live cell single-molecule detection in systems biology. Wiley Interdiscip Rev Syst Biol Med 4(2):183–192. https://doi.org/10.1002/wsbm.161

Sandqvist A, Bjork JK, Akerfelt M, Chitikova Z, Grichine A, Vourc'h C, Jolly C, Salminen TA, Nymalm Y, Sistonen L (2009) Heterotrimerization of heat-shock factors 1 and 2 provides a transcriptional switch in response to distinct stimuli. Mol Biol Cell 20(5):1340–1347. https://doi.org/10.1091/mbc.E08-08-0864

Seksek O, Biwersi J, Verkman AS (1997) Translational diffusion of macromolecule-sized solutes in cytoplasm and nucleus. J Cell Biol 138(1):131–142. https://doi.org/10.1083/jcb.138.1.131

Shimi T, Pfleghaar K, Kojima S, Pack CG, Solovei I, Goldman AE, Adam SA, Shumaker DK, Kinjo M, Cremer T, Goldman RD (2008) The A- and B-type nuclear Lamin networks: microdomains involved in chromatin organization and transcription. Genes Dev 22(24):3409–3421. https://doi.org/10.1101/gad.1735208

Takada R, Mii Y, Krayukhina E, Maruyama Y, Mio K, Sasaki Y, Shinkawa T, Pack CG, Sako Y, Sato C, Uchiyama S, Takada S (2018) Assembly of protein complexes restricts diffusion of Wnt3a proteins. Commun Biol 1:165. https://doi.org/10.1038/s42003-018-0172-x

Tiwari M, Mikuni S, Muto H, Kinjo M (2014) Determination of dissociation constant of the NFkB p50/p65 heterodimer using fluorescence cross-correlation spectroscopy in the living cell. Biochem Biophys Res Commun 436:430–435

Verkman AS (2002) Solute and macromolecule diffusion in cellular aqueous compartments. Trends Biochem Sci 27(1):27–33

Vydra N, Toma A, Widlak W (2014) Pleiotropic role of HSF1 in neoplastic transformation. Curr Cancer Drug Targets 14(2):144–155. https://doi.org/10.2174/1568009614666140122155942

Wachsmuth M, Caudron-Herger M, Rippe K (2008) Genome organization: balancing stability and plasticity. Biochim Biophys Acta 1783(11):2061–2079. https://doi.org/10.1016/j.bbamcr.2008.07.022

Wachsmuth M, Conrad C, Bulkescher J, Koch B, Mahen R, Isokane M, Pepperkok R, Ellenberg J (2015) High-throughput fluorescence correlation spectroscopy enables analysis of proteome dynamics in living cells. Nat Biotechnol 33(4):384–389. https://doi.org/10.1038/nbt.3146

Whitesell L, Lindquist S (2009) Inhibiting the transcription factor HSF1 as an anticancer strategy. Expert Opin Ther Targets 13(4):469–478. https://doi.org/10.1517/14728220902832697

Xiao Y, Dane KY, Uzawa T, Csordas A, Qian J, Soh HT, Daugherty PS, Lagally ET, Heeger AJ, Plaxco KW (2010) Detection of telomerase activity in high concentration of cell lysates using primer-modified gold nanoparticles. J Am Chem Soc 132(43):15299–15307. https://doi.org/10.1021/ja106513f

Number and Brightness Analysis: Visualization of Protein Oligomeric State in Living Cells

Ryosuke Fukushima, Johtaro Yamamoto, and Masataka Kinjo

Introduction

Protein Oligomer

Brightness analysis (Macdonald et al. 2013) is used to monitor the protein oligomers in a living cell. The protein oligomers can be found in signal transduction and neurodegenerative diseases. In signal transduction, some receptor proteins assemble as oligomers by specific ligands stimulation and trigger the downstream cellular functions. Several researchers have reported that protein oligomerization and aggregation are related to neurodegenerative diseases (Kitamura and Kinjo 2018; Fang et al. 2014). An effective characterization of the oligomeric state can enhance the understanding of protein functions in living cells. Oligomerization is a dynamic and transient state change, and it demonstrates spatial heterogeneity in a living cell. A technique focused on identifying this spatiotemporal heterogeneity of the oligomeric state is required for accurate understanding. In this chapter, we introduce the number and brightness (N&B) analysis method based on fluorescence fluctuation spectroscopy (FFS). N&B method can provide a way to characterize the oligomeric state in a living cell.

R. Fukushima
Laboratory of Molecular Cell Dynamics, Graduate School of Life Science, Hokkaido University, Sapporo, Japan
e-mail: rfukushima@eis.hokudai.ac.jp

J. Yamamoto
Bioimaging Research Group, Health and Medical Research Institute, The National Institute of Advanced Industrial Science and Technology (AIST), Tsukuba, Japan
e-mail: yamamoto-jtr@aist.go.jp

M. Kinjo (✉)
Laboratory of Molecular Cell Dynamics, Faculty of Advanced Life Science, Hokkaido University, Sapporo, Japan
e-mail: kinjo@sci.hokudai.ac.jp

J. K. Kim et al. (eds.), *Advanced Imaging and Bio Techniques for Convergence Science*, Advances in Experimental Medicine and Biology,
https://doi.org/10.1007/978-981-33-6064-8_2

Fluorescence Fluctuation Spectroscopy (FFS)

Brightness analysis can be performed by fluorescence microscopy coupled with fluorescence fluctuation spectroscopy (FFS) (Youker and Teng 2014). Fluorescence microscopy is utilized to monitor the spatial localization and temporal changes in a fluorescently labeled molecule. The objective lens of the microscope creates an open measurement volume (Fig. 1a) through which the labeled molecules diffuses (Fig. 1b). When these molecules are passing through the measurement volume, the number of photons is counted within a sampling time (Fig. 1c). The intensity of the recorded photon count signal fluctuates in time and space. FFS is used to analyze these fluctuations, which can help in determining the parameters describing concentration, oligomeric state, or diffusion coefficient, etc. Elson and Magde demonstrated FFS for the first time in 1974; they referred it to as fluorescence correlation spectroscopy (FCS) (Elson and Magde 1974). FCS involves the analysis of fluorescence signal by using an autocorrelation function, which provides information about the brightness and the diffusion time. Here, brightness stands for the emitted

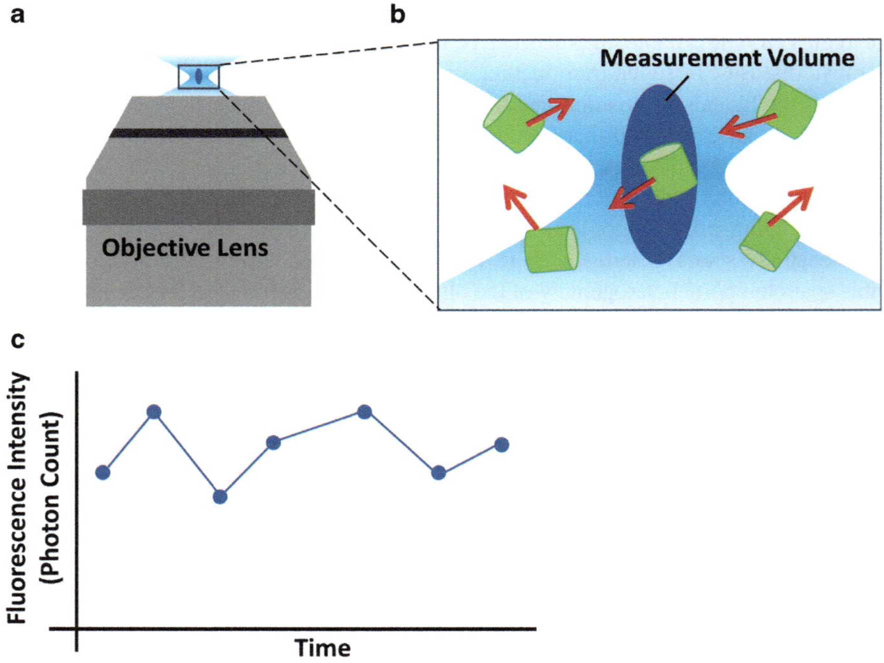

Fig. 1 Fluorescence fluctuation measurement using a confocal fluorescence microscope. (**a**) The objective lens creates an open measurement volume by excitation light (light blue). The dark blue ellipse is the measurement volume (confocal region). (**b**) Diffusion of fluorescent proteins (green cylinder) through the measurement volume. The red arrows indicate the direction of the diffusion. (**c**) Fluorescence intensity, i.e., the intensity of the photons emitted by the diffused proteins

photon count from a single particle (counts per particle (CPP) or counts per molecule (CPM)), and the diffusion time implies characteristic residence time in the measurement volume. Both these parameters can be used to detect oligomerization. However, FCS is a single-spot measurement, and it is difficult to monitor changes as a two-dimensional image even though multiple point measurement are performed (Krmpot et al. 2015; Yamamoto et al. 2018). In addition, diffusion time analysis of FCS is not sensitive enough to distinguish a small change in the oligomerization, e.g., monomer to dimer conversion. Here, we assume that monomer and dimer are spheres with molecular weights of M_w and $2M_w$, respectively. Considering the diffusion time of the monomer to be τ_D, Stokes–Einstein relation implies that the diffusion time of dimer is $\sqrt[3]{2}\tau_D$. It is difficult to detect this small ratio in FCS to distinguish monomer and dimers (Meseth et al. 1999). In contrast, brightness analysis can help to distinguish a small change in the oligomerization compared to diffusion time.

Brightness Analysis

Brightness analysis is based on a general assumption that the recorded photon count is a random variable following the product of two Poisson distribution (Qian and Elson 1990a). The first Poisson distribution originates from the number fluctuation of diffusing particles in a measurement volume. The second distribution originates from photon emission fluctuation from a single particle and photoelectron cascade fluctuation in the detector. For a one-component system, brightness (details are provided in Section "Number and Brightness (N&B) Analysis") analysis can help to characterize the oligomeric state. We consider the brightness analysis for the one-component system containing fluorescent protein, fluorescently labeled monomer, and dimer, respectively, as shown in Fig. 2. If the fluorescence intensity for monomers and a dimer is the same ($2\overline{X}$ and $2\overline{X}$, respectively), the particle number for the dimer is a half of monomers. The dimer carries two fluorescent proteins; therefore, dimeric brightness is two times higher than monomeric brightness (B for monomer, $2B$ for dimer). The brightness of monomer normalized by brightness of fluorescent protein can give normalized brightness of 1, and that of dimer gives normalized brightness of 2. The ratio of the normalized brightness is larger than that of the diffusion time for monomer and dimer.

Introduction for Number and Brightness Analysis

Number and brightness (N&B) analysis (Digman et al. 2008; Nolan et al. 2017b) can help to investigate the temporal fluctuations in fluorescence image series. It determines the apparent brightness across the whole image and generates a brightness map with pixel resolution. In contrast to FCS, it does not provide information about the characteristic residence time in the measurement volume. N&B analysis

	Fluorescent Protein	Monomer	Dimer
		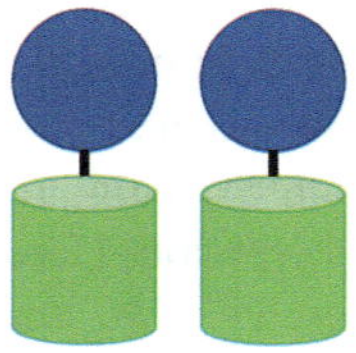	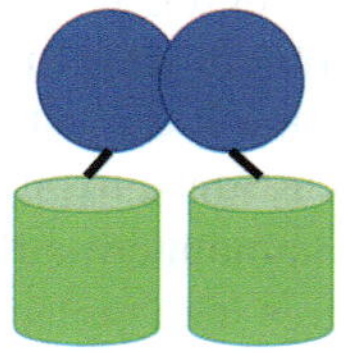
Fluorescence Intensity	$\overline{X}$	$2\overline{X}$	$2\overline{X}$
Particle Number	1	2	1
Brightness	B	B	$2B$

Fig. 2 Reflection of oligomeric state by brightness. Brightness analysis of a fluorescent protein (green cylinder), fluorescently labeled two monomers (blue circle), and dimerized protein. This brightness is the fluorescence intensity from a single particle. If a fluorescent protein has an intensity of $\overline{X}$, the measured fluorescence intensities for the monomer and the dimer are $2\overline{X}$ and $2\overline{X}$, respectively. The particle number becomes half by the dimerization. The dimer carries two fluorescent proteins. Therefore, the brightness of the dimer is two times higher

can help to resolve spatially heterogeneous oligomer distribution. It cannot resolve the mixture of different brightness in the same pixel (details are provided in Section "Number and Brightness (N&B) Analysis"). An advantage of this technique as compared to other analysis methods is that N&B analysis can be applied to fast-diffusing molecules, such as soluble proteins in a living cell. In Imaging FCS, the applicability is limited to slow-diffusing molecules, such as membrane binding protein (details are provided in Section "Comparison with Other Techniques").

N&B analysis has been used in various research areas including oligomeric receptors and aggregation forming proteins. Zamai et al. revealed that fibroblast growth factor-2 (FGF2) and neural cell adhesion molecule (NCAM) form dynamic and transient oligomer at cell membrane (Zamai et al. 2019). Ojosnegros et al. applied N&B analysis to demonstrate clustering models of Eph receptors (Ojosnegros et al. 2017; Cutrale et al. 2019) and proposed a polymerization-condensation model. Ossato et al. monitored inclusion formation in huntingtin peptide (Ossato et al. 2010). Monitoring of protein transition into an oligomer is of paramount importance. As mentioned before, FCS fails to monitor the oligomerization because it is a single-spot measurement. Therefore, N&B analysis could serve as a better way to monitor the oligomerization and its localization over time in a living cell.

Organization of the Chapter

Section "Number and Brightness (N&B) Analysis" of this chapter describes the basic theory of N&B analysis along with an investigation of various parameters. Section "Notation Used in N&B Analysis" describes the notations used in N&B analysis implemented with confocal laser scanning microscopy. Apparent brightness is prone to be a biased parameter. Various factors, such as sampling time, additional intensity and fluctuations, frame number, and nonfluorescent state of labeled protein can lead to this bias. We have also derived equations to evaluate the influence on apparent brightness under a low frame number. Section "Comparison with Other Techniques" provides the comparison of N&B analysis and other FFS techniques.

Number and Brightness (N&B) Analysis

The basic theory of N&B analysis was first proposed in 1990 (Qian and Elson 1990a, b) and it was demonstrated in a living cell in 2008 (Digman et al. 2008). N&B analysis estimates the apparent brightness and apparent number from the fluctuations in photon count. The analysis is conducted on every pixel of a fluorescence image and it produces an apparent brightness map with a single-pixel resolution. Digman et al. performed N&B analysis on fluorescence image series using a confocal laser scanning microscope (CLSM) equipped with two-photon excitation measurement and a photomultiplier tube (Digman et al. 2008). It was also demonstrated that it is possible to use images obtained from a detector operating in analog mode (Dalal et al. 2008) and total internal reflection fluorescence microscopy equipped with an electron-multiplied charge-coupled device (EMCCD) (Unruh and Gratton 2008).

We assume that the measurement sample is a freely diffusing fluorescent particle and measure the photon counts with a sampling time (pixel time) equal to T_{sampling} [s], and collect m images. Let X_i ($i = 1, 2, 3, \ldots, m$) be the number of photon counts in the ith image at a pixel. We define estimators as the ratio of factorial cumulants, and it can be written by sample raw moments as following (Müller 2004):

$$B := \frac{\kappa_{[2]}}{\kappa_{[1]}} = \frac{\overline{X^2} - \overline{X}^2 - \overline{X}}{\overline{X}}, \tag{1}$$

$$N := \frac{{\kappa_{[1]}}^2}{\kappa_{[2]}} = \frac{\overline{X}^2}{\overline{X^2} - \overline{X}^2 - \overline{X}}, \tag{2}$$

where, B and N are the apparent brightness and apparent number, respectively. $\kappa_{[1]}$ and $\kappa_{[2]}$ are first and second-order factorial cumulant, respectively. $\overline{X}$ is sample mean (Eq. 16), and $\overline{X^2} - \overline{X}^2$ is sample variance (Eq. 17). The estimation of apparent brightness has been reported in various articles; however, information about the apparent number and quantitative comparison is scarce. To our knowledge, the apparent number has low precision and exhibits outliers in their map compared to apparent brightness.

The estimators can be expressed as a function of particle brightness ε and particle number n:

$$B = \frac{\gamma_2 \sum_{l=1}^{c} \varepsilon_l^{\,2} n_l}{\sum_{l=1}^{c} \varepsilon_l n_l}, \tag{3}$$

$$N = \frac{\left(\sum_{l=1}^{c} \varepsilon_l n_l\right)^2}{\gamma_2 \sum_{l=1}^{c} \varepsilon_l^{\,2} n_l}, \tag{4}$$

where, particle brightness (or molecular brightness) is the average photon count emitted from a single particle in sampling time, $\varepsilon_l/T_{\text{sampling}}$ is particle brightness in Hz [Hz/particle], particle number n_l is the average number of particle in measurement volume, the subscript l represents l^{th} component, and γ_2 is the γ-factor, which is a constant that depends on the shape of the point spread function (Thompson 1991; Chen et al. 1999) (Eq. 42).

If we assume that the measurement system is composed of a single component ($c = 1$), estimators B and N can be expressed as follows:

$$B = \gamma_2 \varepsilon_1, \tag{5}$$

$$N = \frac{1}{\gamma_2} n_1. \tag{6}$$

N&B analysis users should note that apparent brightness is just a weighted average of multiple particle brightness without any constraint on the number of components. N&B analysis cannot resolve different components. The users need to consider the appropriate constraint. If the system is composed of two components, the particle number ratio can be determined by independent reference measurements (Fukushima et al. 2018). For further investigation in multiple components systems, higher-order moment estimation is required.

Studies estimating higher-order moments have been reported in the literature. With respect to FCS, Palmer and Thompson derived a higher-order correlation function based on higher moments (Palmer and Thompson 1987, 1989a, b). Melnykov and Hall improved the formulation based on cumulants (Melnykov and Hall 2009). This made the formulation simple and facilitated its application toward the characterization of chemical reactions. Abdollah-Nia et al. proposed methods to reduce the bias caused by detector dead time and after pulse by improving data collection and analysis (Abdollah-Nia et al. 2017a, b). The disadvantage of higher-order moments and cumulants is that a larger number of sampling (frame number) is required to obtain a reliable estimated value. Furthermore, error analysis of the estimator (variance of estimator) is necessary to evaluate the reliability (uncertainty).

Notation Used in N&B Analysis

In this section, we describe the necessary conditions for quantitative measurements. Apparent brightness is easy to be biased (see Sections "Optimal Sampling Time and Lag Time, Additional Photon Counts and Signal Fluctuations, Frame Number, and Non-fluorescent State of Fluorescent Proteins"), and normalized brightness does not always correspond to an oligomeric state. We strongly recommend brightness calibration using monomer fluorescent label and tandem-linked oligomers (dimer, trimer, tetramer, pentamer, etc.) (Pack et al. 2006) for every experiment (Macdonald et al. 2013; Chen et al. 2010). Apparent brightness depends on the laser power and optical system, which is prone to day-to-day variations.

For later discussions, we assume that N&B analysis is implemented with confocal laser scanning microscopy (CLSM) and photon counting images unless otherwise noted. For the analog detector system, readers are referred to the following article describing additional factors (Dalal et al. 2008).

Optimal Sampling Time and Lag Time

In image acquisition, a pixel value is obtained within a sampling time, and a pixel size is the x and y length of a pixel in a CLSM image. The x length is a scanning distance in the sampling time, and the y length depends on the x length. An image is obtained by scanning measurement volume across a sample (Fig. 3a), and image series is obtained by repeating the image acquisition multiple times (Fig. 3b). Let us define the notations as follows. Sampling time (T_{sampling}) is defined as the time required to measure photon count that corresponds to a pixel value (pixel dwell time). T_{frame} is described as the time duration required to obtain an image, and T_{wait} is the waiting time between two frame acquisitions. T_{lag} is the lag time, i.e., the time between two continuous pixel acquisitions (Fig. 3c). T_{sampling} and T_{lag} are needed to be set appropriately for quantitative measurements. τ_D is the characteristic residence time in measurement volume for diffusing particles. T_{sampling} should be smaller than τ_D ($T_{\text{sampling}} \ll \tau_D$) (Digman et al. 2008). Furthermore, T_{lag} should to be larger than τ_D ($\tau_D \ll T_{\text{lag}}$) (Nolan et al. 2017b).

The photon count fluctuations are required to obtain the particle number fluctuations as hidden parameters. For pixel acquisition, if the sampling time is not small enough, the number of particles during this time would change by diffusion. In addition, N&B analysis does not assume data-dependent sampling at each photon count. If the lag time is shorter than the characteristic residence time, the fluorescence particle would appear to be immobile (see Section "Background Intensity"). Therefore, a longer waiting time is required to avoid the data-dependent sampling.

Optimization of sampling time is required for a precise estimation. Measurements with short sampling time lead to small photon counts, and reduced signal-to-noise ratio. However, as mentioned before, the sampling time is often restricted by a characteristic residence time. In FFS, Müller derived a theory to extend the sampling time in fluorescence cumulant analysis (FCA) (Müller 2004) and time-integrated

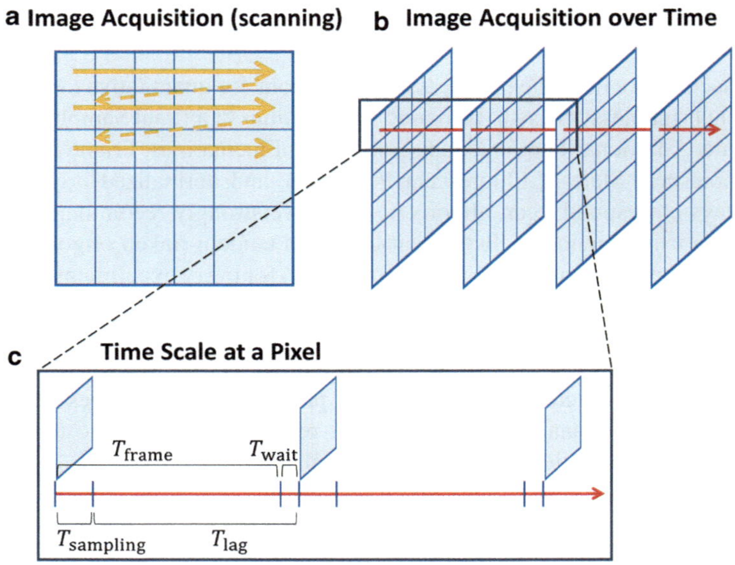

Fig. 3 Image acquisition in laser scanning microscopy. (**a**) A pixel value is obtained within a sampling time, and an image is obtained by scanning the measurement volume. The solid and dotted arrows indicate the scanning and moving directions, respectively. (**b**) Image series is obtained by repeating the image acquisition. The red arrow indicates the passage of time. (**c**) Sampling time (T_{sampling}) is the duration required to obtain a pixel value. The frame time (T_{frame}) is the time required to obtain an image. The waiting time (T_{wait}) is the duration between two frame acquisitions. The lag time (T_{lag}) is the time between a pixel acquisition and the next acquisition at the same pixel

fluorescence cumulant analysis (TIFCA) (Wu and Müller 2005). They evaluated the influence of sampling time on cumulants and corrected the diffusion effect during the sampling time. A similar theory is applied in mean-segmented Q (MSQ) analysis (Hur and Mueller 2015) and time-integrated multipoint moment analysis (TIMMA) (Oh et al. 2011) (see Section “Comparison with Other Techniques”).

Additional Photon Counts and Signal Fluctuations

If the recorded photon counts and the signal fluctuations originate from diffusing fluorescent particles only under a stationary signal, the theory described previously works well. However, there are several additional sources of photon counts and signal fluctuation. In this section, we have described the influence of some of these factors, such as (1) background, (2) nonstationary signal (intensity drift and photobleaching), (3) detector dead time and afterpulse, and (4) additional factors. These factors may introduce bias if they are not corrected or removed.

Background Intensity

In brightness analysis, the background signal affects the estimated value. For example, dark count, stray light, autofluorescence, and immobile fraction contribute to the background signal. Some of these sources can be removed by modifying the condition of measurement. If the background signal is quite low as compared to the target fluorescence, the influence of the background would be negligible. Otherwise, appropriate correction is recommended.

Dark count is defined as the inherent photon count of detectors in the absence of incident light. Stray light originates from the optical system. The dark count and stray light can be corrected by subtracting the average photon count obtained by a reference measurement (Digman et al. 2008).

$$B_{\text{bgCorr}} := \frac{\overline{X^2} - \overline{X}^2 - \overline{X}}{\overline{X} - \overline{X_{\text{bg}}}}, \tag{7}$$

$$N_{\text{bgCorr}} := \frac{\left(\overline{X} - \overline{X_{\text{bg}}}\right)^2}{\overline{X^2} - \overline{X}^2 - \overline{X}}, \tag{8}$$

where, B_{bgCorr} and N_{bgCorr} are the background-corrected estimators for apparent brightness and apparent number, respectively. $\overline{X_{\text{bg}}}$ is the average photon count of the background.

Autofluorescence correction is essentially difficult, as the autofluorescence intensity is different depending on cell and cellular components. It is recommended to confirm that the autofluorescence is quite low compared to target fluorescence intensity in all measurements.

N&B analysis determines the brightness of only diffusing particles. The basic theory described in Section "Number and Brightness (N&B) Analysis" cannot determine the brightness of immobile particles, such as particle binding to cellular structure. Qian and Elson evaluated the immobile fraction by introducing a third moment for the system composed of a mobile and immobile fraction (Qian and Elson 1990a). In this evaluation, it was assumed that photobleaching is negligible for both fractions. However, immobile particles are continuously excited and easy to be photobleached. Skinner et al. evaluated the influence of immobile fraction on brightness in FFS under similar conditions (Skinner et al. 2008).

Nonstationary Signal (Intensity Drift and Photobleaching)

The brightness estimator gives apparent brightness accurately under stationary signal. However, various factors distort the signal. For example, in a measurement of fluorescently labeled soluble proteins in a living cell, the presence of large organelles or vesicles coming in and out of measurement volume can cause a transient intensity drift. In addition, the cellular movements and focus drift can also contribute to the intensity drift. Another factor is photobleaching which can decrease the measured intensity over time. To obtain a robust estimate, the bias caused by the intensity drift can be reduced by a segmentation, and the bias caused by the photobleaching can be reduced by the segmentation along with detrending with a line or a curve.

Segmentation divides a signal sequence into multiple segments in the time direction, and apparent brightness values are estimated at each segment (Fig. 4a). If the intensity drifts are rare, only a few segments would be biased. Therefore, the average of multiple-segmented brightness values would be less biased as compared to their unsegmented counterparts (Chen et al. 2002; Macdonald et al. 2013). This approach could also reduce the bias caused by mild photobleaching as described later. In FFS, an effective application of the segmentation can be found in mean-segmented Q (MSQ) analysis (Hur and Mueller 2015). Hur and Mueller used different lengths of the segmentation to obtain a robust estimate of brightness.

Boxcar filtering is a similar way to reduce the bias caused by the intensity drifts and mild photobleaching (Trullo et al. 2013). The difference is the presence or absence of the overlap in each segment. A boxcar filter divides a signal sequence into multiple segments by moving sliding window frame by frame (Fig. 4b). For example, if we consider a signal sequence consisting of 10 photon counts, segmentation with a length of 5 generates 2 segments. In contrast, boxcar filtering with a length of 5 generates 6 segments.

Photobleaching (photodepletion) is an irreversible process in which fluorescence molecules turn into nonfluorescent molecules. In contrast, the decrease could be alleviated by diffusing fluorescent molecules coming out of the measurement volume (Petrášek and Schwille 2008). The estimated brightness is overestimated under photobleaching (Hur et al. 2014). Generally, it is recommended to reduce excitation laser power. When photobleaching cannot be avoided, bias correction is recommended.

Caccia et al. removed the decreasing trend with the sum of two exponential decay (Caccia et al. 2005). Hur et al. evaluated the estimation bias caused by

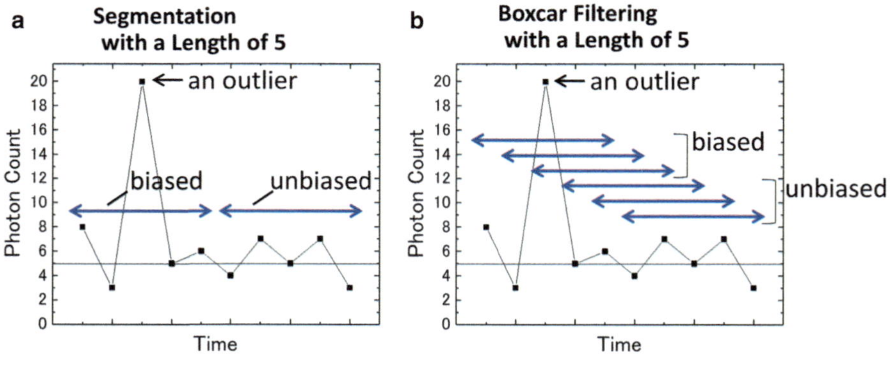

Fig. 4 Segmentation and boxcar filtering. The signal sequence has 10 photon counts which contain 1 outlier. Estimation with a length of 10 leads to biased brightness by the outlier. (**a**) Segmentation divides the signal sequence into multiple segments without overlap. Segmentation with a length of 5 divides the signal sequence into 2 segments. (**b**) Boxcar filtering divides the signal sequence with overlap. Boxcar filtering with a length of 5 divides the signal sequence into 6 segments

photobleaching by assuming single exponential decay (Hur et al. 2014), and demonstrated that the bias depends on the initial number of molecules in measurement volume and "photodepletion fraction." The photodepletion fraction is a decreasing function of the initial fluorescence intensity during the measurement. This indicates that segmentation is also effective to reduce the bias. Nolan et al. discussed the application of exponential filtering to remove the decreasing trend and automatic parameter selection for filtering (Nolan et al. 2017a).

In addition to brightness overestimation, another problem of photobleaching is the decrease in brightness of the oligomer. For example, in a dimer complex carrying 2 EGFPs, photobleaching on the half side of EGFP will produce another bright species, that is, a dimer complex carrying a single EGFP. Hur et al. derived an equation to obtain initial brightness under single exponential decay (Hur et al. 2014).

Detector Dead Time and Afterpulsing

The presence of detector dead time and afterpulse can lead to concentration-dependent brightness (Hillesheim and Müller 2003; Hendrix et al. 2013). Dead time is the period after the detection of a photon in which detectors cannot detect another photon, leading to a loss of photon count (Fig. 5a). Afterpulse is a spurious pulse following a genuine pulse, which results in the addition of photon counts sometimes (Fig. 5b). O'Donnell and Campbell evaluated photon count distribution affected by dead time (O'Donnell 1986) and afterpulsing (Campbell 1992). Hellesheim and Müller evaluated the relative error of the brightness estimator as Mandel's Q factor (Hillesheim and Müller 2003). They claimed that afterpulsing can be neglected in typical experiments equipped with avalanche photodiode (APD). Sanchez-Andres et al. reported that time-shifted covariance estimation is useful to reduce this effect (Sanchez-Andres et al. 2005). An application of time-shifted covariance is also found in N&B analysis with image series obtained by EMCCD (Unruh and Gratton 2008). However, the time-shifted covariance estimation cannot be applied to image series obtained by CLSM because the time-series data is independent. We

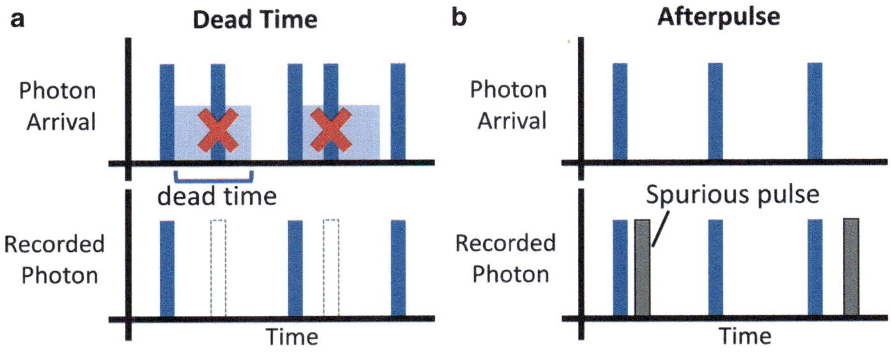

Fig. 5 Dead Time and afterpulsing. The blue bars show the arrival time of the photon. (**a**) Dead time is the time period after photon detection in which detectors cannot detect any photon. The gray dashed bars are photon which are not recorded. (**b**) Afterpulse is a spurious pulse (gray bar) following a genuine pulse, which occasionally results in the increase of photon count

previously reported that a useful way to reduce the effect of dead time and afterpulsing in N&B analysis is covariance estimation between two detectors (Fukushima et al. 2018). In the two-detector system, a half mirror splits the fluorescence signal into two channels. Consequently, covariance estimation reduced the effect of dead time and afterpulsing.

Additional Factors

The geometry of excitation light can also lead to biased brightness. It has been reported that the use of oil-immersion objective can cause focal depth-dependent bias (Macdonald et al. 2013). Photon count collection at deep positions into the sample can cause spherical aberration, leading to biased brightness. The use of a water-immersion objective with a well-adjusted correction collar allows focal depth independent estimation.

In addition, the geometry of the sample can also lead to biased brightness. Basic FFS theory assumes that the measurement volume completely overlaps with the sample. However, this is not true for measurements at the peripheral regions of the living cell. The measurement volume overlaps with the peripheral region as a thin layer. Macdonald et al. reported z-scan FFS to remove this bias (Macdonald et al. 2010). They suggested that the thin layer can be approximated by a rectangular slab, and derived an equation to estimate the thickness of the slab by scanning along z-direction. Consequently, they recovered unbiased brightness by correcting γ-factor.

Only the large-scale variations due to intensity drifts and photobleaching can be identified through signal intensity changes. However, a small signal fluctuation process leading to biased brightness was recently reported. Hennen and coworkers reported that nuclear membrane undulation leads to biased brightness (Hennen et al. 2017). The nuclear membrane consisted of an inner nuclear membrane and an outer nuclear membrane. The membranes were separated by lumen. They found an additional exponential correlation term in signal fluctuation due to lumen localized protein.

Frame Number

Frame number is the number of frames in fluorescence image series for estimation. The time-dependent change of parameters is not considered in N&B analysis. A low frame number increases the time resolution. Ojosnegros et al. applied boxcar filtering to increase the time resolution (Ojosnegros et al. 2017). However, a low frame decreases precision as well as accuracy of measurement. To our knowledge, very few reports are available on the effect of frame numbers. The determination of an appropriate frame number depends on the judgment of users. We derived the following equation to evaluate the influence of frame numbers on accuracy. This is the approximate solution for expected apparent brightness and its relative error for one-component system (see Appendix 1 for derivation).

$$E[B]=\gamma_2\varepsilon_1-\frac{1}{m}(\gamma_2\varepsilon_1+1)-\frac{1}{m}\frac{1}{\varepsilon_1 n_1}\left(\gamma_3\varepsilon_1^{\ 2}-\gamma_2^{\ 2}\varepsilon_1^{\ 2}+\gamma_2\varepsilon_1\right)+\mathrm{O}\left(\left(\frac{1}{m}\right)^2\right), \quad (9)$$

$$\frac{E[B]-\gamma_2\varepsilon_1}{\gamma_2\varepsilon_1}=-\frac{1}{m}\frac{1}{\gamma_2\varepsilon_1}\left\{(\gamma_2\varepsilon_1+1)+\frac{1}{\varepsilon_1 n_1}\left(\gamma_3\varepsilon_1^{\ 2}-\gamma_2^{\ 2}\varepsilon_1^{\ 2}+\gamma_2\varepsilon_1\right)\right\}+\mathrm{O}\left(\left(\frac{1}{m}\right)^2\right), \quad (10)$$

where, $E[B]$ is the expected value of B, m is the frame number, and $\varepsilon_1 n_1$ is average photon count in a sampling time. A similar approximation was previously reported in the ratio estimation (van Kempen and van Vliet 2000), autocorrelation function (Schätzel et al. 1988; Saffarian and Elson 2003), and mean-segmented Q (MSQ) analysis (Hur and Mueller 2015). Eq. (9) indicates that the estimator is asymptotically equal to the true value with enough frame number and average photon count, but the bias is not negligible when the frame number or average photon count is less. Eq. (10) indicates that the relative error becomes smaller for high particle brightness. Figure 6a shows the dependence of frame number on the relative error with different particle brightness. Here, symbols correspond to simulated results, and the dashed lines represent the theoretical value expected by Eq. (10). A relative error of 0 implies that there is no deviation between the true and estimated values. The relative error approaches zero with increasing frame number. The estimated brightness was critically underestimated, especially for lower particle brightness. This suggests that reducing the excitation laser power to avoid photobleaching or decreasing the frame number to improve time resolution in a careless manner could cause bias. Fig. 6b shows the dependence of the average photon count ($\overline{X}$) on relative error. The relative error decreased for average photon count equal to 1, and it saturated when the average photon count exceeded 1. A simple solution to reduce the bias is correction. A less biased estimator B_{corr} can be defined as follows:

$$B_{\mathrm{corr}}:=B+\frac{1}{m}(B+1), \quad (11)$$

$$E[B_{\mathrm{corr}}]=\gamma_2\varepsilon_1-\frac{1}{m}\frac{1}{\varepsilon_1 n_1}\left(\gamma_3\varepsilon_1^{\ 2}-\gamma_2^{\ 2}\varepsilon_1^{\ 2}+\gamma_2\varepsilon_1\right)+\mathrm{O}\left(\left(\frac{1}{m}\right)^2\right), \quad (12)$$

$$\frac{E[B_{\mathrm{corr}}]-\gamma_2\varepsilon_1}{\gamma_2\varepsilon_1}=-\frac{1}{m}\frac{1}{\gamma_2\varepsilon_1}\frac{1}{\varepsilon_1 n_1}\left(\gamma_3\varepsilon_1^{\ 2}-\gamma_2^{\ 2}\varepsilon_1^{\ 2}+\gamma_2\varepsilon_1\right)+\mathrm{O}\left(\left(\frac{1}{m}\right)^2\right). \quad (13)$$

B_{corr} is defined as the transposed and substituted equation of $E[B]$. The second term in Eq. (9) at the right-hand side is transposed to the left-hand side, and estimator B is substituted into the transposed true value. In fact, B_{corr} is still a biased estimator, but it is less biased compared to B. Comparing Eqs. (9) and (12), the influence

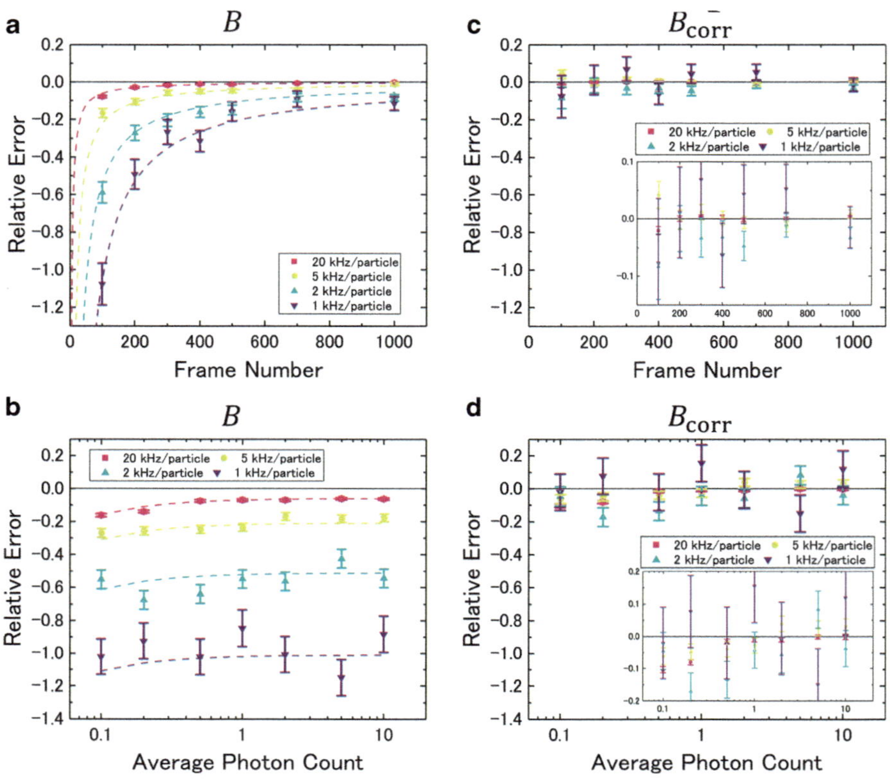

Fig. 6 Dependence of frame number and average photon count on accuracy. Average and standard error estimated from the simulations. The dotted line is the expected value. Dependence of frame number on the relative error with different particle brightness analyzed by (**a**) conventional estimator and (**c**) corrected estimator. Dependence of average photon count on the relative error with different particle brightness analyzed by (**b**) conventional estimator and (**d**) corrected estimator. Insets in (**c**) and (**d**) show the magnified plots

from the second term in Eq. (9) is vanished, but that from the third term still remains. Figure 6c, d show the improvement by the correction. The bias in the expected value is clearly reduced here. Further correction can be implemented to reduce the influence from the third term in Eq. (9). This influence is remarkable when the average photon count is less than 1. The bias in low average photon count has been reported even though the background intensity is low (Youker and Teng 2014; Gambin et al. 2016; Fukushima et al. 2018). However, the bias correction needs to estimate the third sample raw moment, which makes the precision low. N&B analysis users are suggested to confirm the optimal range of average photon count and frame number for quantitative measurements.

Nonfluorescent State of Fluorescent Proteins

Variation in Normalized Brightness of Tandem Dimer

Theoretically, the normalized brightness of the tandem dimer should be 2, but a variation in this value is also reported (1.5–2.0). A value below 2 suggests that some fluorescent proteins are in a nonfluorescent state. Let us define fluorescent probability as a probability for a fluorescent protein to be fluorescent. We have listed some articles that reported the normalized or monomeric brightness and dimeric brightness in Tables 1 and 2. Here, the brightness unit and the sample vary according to the source article.

Possible Explanations for Nonfluorescent State of Proteins

There are several possible explanations for the nonfluorescent state of proteins: (1) Photophysical and photochemical processes, such as long-lived dark states (Widengren et al. 1999), multiple brightness state (Wu et al. 2009), blinking or flickering (Hendrix et al. 2008), and photobleaching (Hur et al. 2014); (2) Chromophore formation (protein maturation, incorrect folding of fluorescent protein) (Balleza et al. 2018); and (3) Unlabeled endogenous proteins.

It has been reported that photophysical processes resulting from one-photon or and two-photon excitation are different. Hendrix et al. reported the presence of blinking in mCherry due to one-photon excitation (Hendrix et al. 2008). However, Wu et al. reported the absence of blinking due to two-photon excitation (Wu et al. 2009). Regarding chromophore formation, Dunsing et al. reported

Table 1 Normalized brightness close to 2.0 (100% fluorescent state)

Sample	Analysis methods	Tandem monomer	Tandem dimer	Unit	Ratio	References
EGFP	PCH [1] and moment analysis	4000 ± 300	7800 ± 508	Cpsm (counts per second per molecule)	–	Chen et al. (2003)
EYFP	TIFCA	680 ± 50	1380 ± 80	Cps	–	Wu et al. (2009)
GPI-mEGFP	N&B analysis	0.103 ± 0.004	0.205 ± 0.006	Counts/molecule×dwell time	–	Zamai et al. (2019)
GFP	PCH [1]	–	–	–	~2	Slaughter et al. (2007)
mEGFP	PCH [1]	–	–	–	~2	Huet et al. (2010)
EGFP	FCS	–	–	–	2	Foo et al. (2012)

[1] Photon counting histogram (Chen et al. 1999)

Table 2 Normalized brightness lower than 2.0 (presence of a non-fluorescence state)

Sample	Analysis methods	Tandem monomer	Tandem dimer	Unit	Ratio	References
mEGFP	N&B analysis or FCS	–	–	–	1.69 ± 0.05	Dunsing et al. (2018)
mEYFP	Same as above	–	–	–	1.63 ± 0.05	Same as above
mCherry	Same as above	–	–	–	1.41 ± 0.04	Same as above
EGFP	FCS	–	–	–	[1]	Vámosi et al. (2016)
EGFP	FCS and N&B analysis	–	–	–	[2]	Fukushima et al. (2018)

[1] Estimated fluorescent probabilities were lower than 1.0. Probability decreased with increasing laser power
[2] Normalized brightness was lower than the number of subunits in EGFP oligomers in vivo

that there are negligible differences in the brightness between HEK, A549, CHO, and HeLa cells, and between 23 °C and 37 °C (Dunsing et al. 2018). Foo et al. used FCS and reported that cycloheximide treatment increased the fluorescent probability in mCherry (Foo et al. 2012). This indicates that a fraction of mCherry had not matured. These reports suggest that the normalized brightness for tandem dimer is not always 2.

Subunit counting (Ulbrich and Isacoff 2007; Das et al. 2007) is a method to estimate the number of subunits in a complex. It observes stepwise photobleaching and counts the number of the steps. The photobleaching in each fluorescent protein occurs independently. Subunit counting indicates the following fluorescent probability: 79.5% for EGFP-tagged CNG channel (Ulbrich and Isacoff 2007), 80% for Kv4.3-EGFP (Zhou et al. 2015), 78% for GFP-ASIC1a, and 77% for GFP-ASIC2a (Bartoi et al. 2014).

Correction for Nonfluorescent State of Proteins

If the fluorescent probability for a protein is assumed to be constant, a simple correction method can be used (Vámosi et al. 2016; Dunsing et al. 2018). This method assumes two states of the protein: fluorescent state (on) and nonfluorescent state (off). It also assumes that the number of fluorescent subunits in an oligomer follows a binomial distribution. Let p be the fluorescent probability, and B_s be the apparent brightness for one-component system that is comprised of oligomers with s subunits (for example, B_1 is apparent brightness for monomer and B_2 is that for tandem dimer). The apparent brightness with a nonfluorescent state can be written as following (Vámosi et al. 2016):

$$B_s = \{1 + (s-1)p\}\gamma_2\varepsilon_1, \tag{14}$$

$$\frac{B_s}{B_1} = 1 + (s-1)p, \tag{15}$$

where, ε_1 is particle brightness for a monomer subunit. $B_1 = \gamma_2\varepsilon_1$ and normalized brightness $B_s/B_1 = s$ for $p = 1.0$. Figure 7 shows the dependence of a number of subunits on the normalized brightness with different fluorescent probabilities. The normalized brightness increases linearly with the number of subunits, and the slope of the line is smaller in low fluorescent probability. The fluorescent probability can be determined from independent reference measurements for one-component system, and the estimated number of subunits can be corrected (Dunsing et al. 2018). Vamosi et al. reported that fluorescent probability depends on pH and laser power (Vámosi et al. 2016). Dunsing et al. confirmed that the corrected number of subunits provides better estimation by using dimeric and trimeric Influenza A virus HA protein variants and 12-meric E. Coli GlnA, respectively (Dunsing et al. 2018).

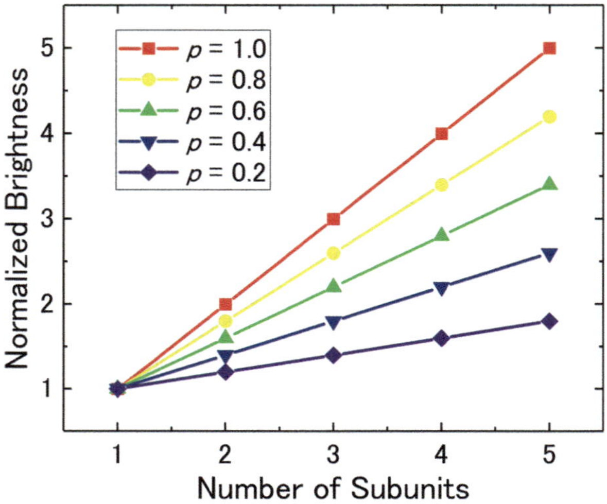

Fig. 7 Non-correspondence of normalized brightness and the number of subunits in the presence of nonfluorescent state. *p* indicates the probability for the fluorescent state. Dependence of the number of subunits on the normalized brightness with different fluorescence probabilities. This plot is a visual representation of Eq. (15)

Comparison with Other Techniques

FCS coupled with a fast frame rate camera, such as an electron-multiplied charge-coupled device (EMCCD), is called Imaging FCS (Kannan et al. 2006; Bag and Wohland 2014). This technique enables a spatially simultaneous measurement of fluorescence intensity over time. Imaging FCS determines not only concentration and brightness distribution but also the diffusion time by implementing autocorrelation analysis at each pixel. However, its applicability is limited to slow-diffusing molecules, such as membrane binding protein. A necessary condition for autocorrelation analysis is that frame time must be sufficiently smaller than the diffusion time. In EMCCD, frame time is restricted to milliseconds order. In contrast, the diffusion time of EGFP is several hundred microseconds. Therefore, it is difficult to apply this method to EGFP labeled soluble protein unless the diffusion time of the labeled protein is sufficiently long. An advantage of Imaging FCS technique as compared to N&B analysis is that the former facilitates the visualization of diffusion time distribution, and helps to distinguish diffusion modes on cellular membrane (Wawrezinieck et al. 2005). The FCS diffusion law calculates the dependence of spatial scale on diffusion coefficient and estimates the contribution from the diffusion mode, free diffusion, raft partitioning, and meshwork.

Time-integrated multipoint moment analysis (TIMMA) (Oh et al. 2011) is an application of N&B analysis and TIFCA. TIMMA determines the distribution of concentration, brightness, and diffusion time. Oh et al. used a stationary spinning disk confocal microscope equipped with an EMCCD, an acousto-optic tunable filter (AOTF), and a 50/50 beam splitter. They used the beam splitter to divide the signal into two parts. In EMCCD, the frame time primarily consists of sampling time (exposure time) and readout time. They measured the fluorescence signal with different sampling times to estimate the average and covariance. TIFCA facilitates the measurement of the fluorescence signal with a longer sampling time and determination of diffusion time. Oh et al. demonstrated that TIMMA enables the measurement of soluble EGFP. It is difficult to measure soluble EGFP in Imaging FCS implemented with an EMCCD. The long readout time of EMCCD hampers the application. On the contrary, a necessary condition to apply the basic N&B analysis theory is to reduce the sampling time as compared to the diffusion time (see Section "Optimal Sampling Time and Lag Time"). Consequently, it can be applied even if the readout time is long.

Software

Nolan et al. described software for N&B analysis in their article (Nolan et al. 2017b).

Simulations

Photon count images were numerically simulated by using Microsoft Visual Studio Community 2017 version 15.9.10 (Microsoft, USA). All programs were written in Visual C++. Here, we assumed that photon count is a random variable following the product of two Poisson distribution along with a uniform illumination ($\gamma_r = 1$). The photon counts were simulated with a sampling time of 10 μs. The frame size of simulated images was 128 × 128, and the brightness map was obtained in the same size. An average and a standard error of relative error were estimated from the brightness map. For Fig. 6a, photon count images were simulated with a fixed true particle number ($n = 20$) and different true particle brightness (ε=20, 5, 2, and 1 kHz/particle), and they were analyzed with different frame numbers (m=100, 200, 300, 400, 500, 700, and 1000). For Fig. 6b, photon count images were simulated with different average photon count ($\varepsilon_1 n_1$=0.1, 0.2, 0.4, 1, 2, 4, and 10 counts/sampling time) and different true particle brightness (ε=20, 5, 2, and 1 kHz/particle), and they were analyzed with fixed frame number ($m = 100$). In Fig. 6a, b, each plot is obtained from independent simulations. In Fig. 6c, d, photon count images for each plot in Fig. 6a, b are analyzed with Eq. (11).

Conclusion

N&B analysis facilitates the determination of apparent brightness per pixel using fluorescence image series. This brightness is defined as the emitted photon count from a single particle, and it reflects the dynamic oligomer formation of protein. Therefore, it helps in visualizing the spatially heterogeneous oligomeric state of diffusing particles over time in a living cell. However, N&B analysis does not resolve multiple brightness components. In addition, apparent brightness is easy to be biased by several factors, such as improper sampling and lag time, background intensity, nonstationary intensity, detector dead time and afterpulse, and insufficient frame number. Even if all these biases are removed, normalized brightness does not reflect the oligomeric state (the number of subunits in complex) directly in the presence of a nonfluorescent state. These facts suggest that brightness calibration must be performed for each experiment. Otherwise, the experimental results may not be reliable.

N&B analysis has the potential to enhance the understanding of the controlling mechanism in the protein function, especially in signal transduction and neurodegenerative diseases. Moreover, N&B analysis coupled with time-integrated fluorescence cumulant analysis (TIFCA) can help in determining the diffusion time by measurements at different sampling times. Future advancement may help in analyzing other kinetics that are slower than diffusion, such as quenching affected by circumstance.

Appendix 1

We evaluate the dependence of relative error on frame number. Suppose that we obtain m images with a sampling time T_{sampling} by fluorescence microscopy. Let X_i ($i = 1, 2, 3, \ldots, m$) be independent and identically distributed random variables with expected photon count value equal to μ for in i^{th} image at one pixel. We assume that the number of fluorescent particles in a measurement volume does not change during the sampling time. We define first sample raw moment $\overline{X}$ (sample mean) and second sample raw moment $\overline{X^2}$ as follows:

$$\overline{X} := \frac{1}{m}\sum_{i=1}^{m} X_i, \tag{16}$$

$$\overline{X^2} := \frac{1}{m}\sum_{i=1}^{m} X_i^2. \tag{17}$$

Sample variance can be expressed as follows:

$$\frac{1}{m}\sum_{i=1}^{m}\left(X_i - \overline{X}\right)^2 = \overline{X^2} - \overline{X}^2. \tag{18}$$

The photon count X_i can be written as the sum of the first raw moment μ (population mean) and instantaneous fluctuation around the first raw moment ΔX_i, i.e.:

$$X_i = \mu + \Delta X_i. \tag{19}$$

Since the expectation value of the photon count is similar to the first raw moment ($E[X] = \mu$), the expectation value of the instantaneous fluctuation becomes 0 ($E[\Delta X] = 0$). Substituting Eq. (19) into Eqs. (16) and (17), we get

$$\overline{X} = \mu + \overline{\Delta X}, \tag{20}$$

$$\overline{X^2} = \mu^2 + 2\mu\overline{\Delta X} + \overline{\Delta X^2}, \tag{21}$$

where,

$$\overline{\Delta X} := \frac{1}{m}\sum_{i=1}^{m}\Delta X_i, \tag{22}$$

$$\overline{\Delta X^2} := \frac{1}{m}\sum_{i=1}^{m}\Delta X_i^{\,2}. \tag{23}$$

The average instantaneous fluctuation $\left(\overline{\Delta X}\right)$ can be removed by considering large samples (m). In other words, the expectation value of $\overline{\Delta X}$ is 0 $\left(E\left[\overline{\Delta X}\right] = 0\right)$. Next, we consider the brightness estimator, which can be defined from Eqs. (20) and (21) as follows:

$$B := \frac{\overline{X^2} - \overline{X}^2 - \overline{X}}{\overline{X}} = \frac{\overline{\Delta X^2} - \overline{\Delta X}^2}{\mu + \overline{\Delta X}} - 1 = \frac{1}{\mu}\frac{\overline{\Delta X^2} - \overline{\Delta X}^2}{1 + \dfrac{\overline{\Delta X}}{\mu}} - 1. \tag{24}$$

Taylor series expansion of the denominator up to second order around 0 gives

$$\frac{1}{1 + \dfrac{\overline{\Delta X}}{\mu}} = 1 - \frac{\overline{\Delta X}}{\mu} + \left(\frac{\overline{\Delta X}}{\mu}\right)^2 - O\left(\left(\frac{\overline{\Delta X}}{\mu}\right)^3\right). \tag{25}$$

Therefore,

$$B = \frac{1}{\mu}\left(\overline{\Delta X^2} - \overline{\Delta X}^2\right)\left(1 - \frac{\overline{\Delta X}}{\mu} + \left(\frac{\overline{\Delta X}}{\mu}\right)^2 - O\left(\left(\frac{\overline{\Delta X}}{\mu}\right)^3\right)\right) - 1. \tag{26}$$

The expectation value of the brightness estimator is

$$E[B] = \frac{1}{\mu}\left\{ \begin{array}{l} \left(E\left[\overline{\Delta X^2}\right] - E\left[\overline{\Delta X}^2\right]\right) - \frac{1}{\mu}\left(E\left[\overline{\Delta X}\,\overline{\Delta X^2}\right] - E\left[\overline{\Delta X}^3\right]\right) \\ + \frac{1}{\mu^2}\left(E\left[\overline{\Delta X}^2\,\overline{\Delta X^2}\right] - E\left[\overline{\Delta X}^4\right]\right) + \dots \end{array} \right\}$$

$$= \frac{1}{\mu}\left\{ \begin{array}{l} \left(1 - \frac{1}{m}\right)E\left[\Delta X^2\right] - \frac{1}{\mu}\frac{1}{m}\left(1 - \frac{1}{m}\right)E\left[\Delta X^3\right] + \\ \frac{1}{\mu^2}\frac{1}{m^2}\left(1 - \frac{1}{m}\right)E\left[\Delta X^4\right] + \\ \frac{1}{\mu^2}\frac{1}{m}\left(1 - \frac{1}{m}\right)\left(E\left[\Delta X^2\right]\right)^2 + \dots \end{array} \right\} - 1$$

$$= \frac{E\left[\Delta X^2\right]}{\mu} - 1 - \frac{1}{m}\left\{ \begin{array}{l} \frac{E\left[\Delta X^2\right]}{\mu} + \\ \frac{1}{\mu}\left(\frac{E\left[\Delta X^3\right]}{\mu} - \left(\frac{E\left[\Delta X^2\right]}{\mu}\right)^2\right) \end{array} \right\}$$

$$+ \mathrm{O}\left(\left(\frac{1}{m}\right)^2\right)$$

$$= \frac{\kappa_{[2]}}{\kappa_{[1]}} - \frac{1}{m}\left(\begin{array}{l} \frac{\kappa_{[2]}}{\kappa_{[1]}} + 1 + \\ \frac{1}{\kappa_{[1]}}\left(\frac{\kappa_{[3]}}{\kappa_{[1]}} - \left(\frac{\kappa_{[2]}}{\kappa_{[1]}}\right)^2 + \frac{\kappa_{[2]}}{\kappa_{[1]}}\right) \end{array} \right) + \mathrm{O}\left(\left(\frac{1}{m}\right)^2\right). \tag{27}$$

where we used the solution obtained in Appendix 2 shown below. For one-component system, the expectation value of the brightness estimator is

$$E[B] = \gamma_2\varepsilon_1 - \frac{1}{m}\left(\gamma_2\varepsilon_1 + 1\right) - \frac{1}{m}\frac{1}{\varepsilon_1 n_1}\left(\gamma_3\varepsilon_1^{\ 2} - \gamma_2^{\ 2}\varepsilon_1^{\ 2} + \gamma_2\varepsilon_1\right) + \mathrm{O}\left(\left(\frac{1}{m}\right)^2\right) \tag{28}$$

Appendix 2

We consider the expected values in the first row of Eq. (27). The expected values are written in terms of central moments ($E[\Delta X]$, $E[\Delta X^2]$, $E[\Delta X^3]$, $E[\Delta X^4]$) as follows:

$$E\left[\overline{\Delta X^2}\right] = \frac{1}{m}E\left[\sum_{i=1}^{m}\Delta X_i^{\ 2}\right] = \frac{1}{m}\sum_{i=1}^{m}E\left[\Delta X_i^{\ 2}\right] = E\left[\Delta X^2\right], \tag{29}$$

$$
\begin{aligned}
E\left[\overline{\Delta X}^2\right] &= \frac{1}{m^2} E\left[\left(\sum_{i=1}^{m} \Delta X_i\right)^2\right] = \frac{1}{m^2} E\left[\left(\sum_{i=1}^{m} \Delta X_i\right)\left(\sum_{j=1}^{m} \Delta X_j\right)\right] \\
&= \frac{1}{m^2} E\left[\sum_{i=1}^{m}\sum_{j=1}^{m} \Delta X_i \Delta X_j\right] \\
&= \frac{1}{m^2} E\left[\sum_{i=1}^{m} \Delta X_i^2 + 2\sum_{i=1}^{m}\sum_{j=1<i}^{m} \Delta X_i \Delta X_j\right] \\
&= \frac{1}{m^2}\sum_{i=1}^{m} E\left[\Delta X_i^2\right] + \frac{2}{m^2}\sum_{i=1}^{m}\sum_{j=1<i}^{m} E\left[\Delta X_i \Delta X_j\right] \\
&= \frac{1}{m} E\left[\Delta X^2\right] + \frac{2}{m^2}\frac{m^2-m}{2} E\left[\Delta X\right]^2 \\
&= \frac{1}{m} E\left[\Delta X^2\right],
\end{aligned} \tag{30}
$$

where we used the following facts: (1) We considered that A_iB_j is the element of $m \times m$ matrix, and $\sum_{i=1}^{m}\sum_{j=1}^{m} A_i B_j$ is the summation of all matrix elements. The summation of matrix elements is separated into the summation of diagonal and non-diagonal elements. (2) We assume that C and D are independent random variables, $E[CD] = E[C]E[D]$. (3) $E[\Delta X] = 0$.

Similarly, we have

$$E\left[\overline{\Delta X}\,\overline{\Delta X^2}\right] = \frac{1}{m} E\left[\Delta X^3\right], \tag{31}$$

$$E\left[\overline{\Delta X}^3\right] = \frac{1}{m^2} E\left[\Delta X^3\right], \tag{32}$$

$$E\left[\overline{\Delta X}^2\,\overline{\Delta X^2}\right] = \frac{1}{m^2} E\left[\Delta X^4\right] + \frac{1}{m}\left(1 - \frac{1}{m}\right)\left(E\left[\Delta X^2\right]\right)^2, \tag{33}$$

$$E\left[\overline{\Delta X}^4\right] = \frac{1}{m^3} E\left[\Delta X^4\right] + \frac{1}{m^2}\left(1 - \frac{1}{m}\right)\left(E\left[\Delta X^2\right]\right)^2. \tag{34}$$

The expectation value of X and central moment are written in terms of cumulants as follows (Abdollah-Nia 2016):

$$E\left[X\right] = \mu = \kappa_1, \tag{35}$$

$$E\left[\Delta X^2\right] = \kappa_2, \tag{36}$$

$$E\left[\Delta X^3\right] = \kappa_3. \tag{37}$$

In addition, cumulants are written in terms of factorial cumulants as follows:

$$\kappa_1 = \kappa_{[1]}, \tag{38}$$

$$\kappa_2 = \kappa_{[2]} + \kappa_{[1]}, \tag{39}$$

$$\kappa_3 = \kappa_{[3]} + 3\kappa_{[2]} + \kappa_{[1]}. \tag{40}$$

k^{th} factorial cumulants of photon count is expressed in terms of particle brightness ε, particle number n, and γ-factor as follows (Müller 2004):

$$\kappa_{[k]} = \gamma_k \sum_{l=1}^{c} {\varepsilon_l}^k n_l, \tag{41}$$

where,

$$\gamma_k := \frac{\int_V \left(\overline{\text{PSF}}(\vec{r})\right)^k d\vec{r}}{\int_V \overline{\text{PSF}}(\vec{r}) d\vec{r}}, \tag{42}$$

$\overline{\text{PSF}}(\vec{r})$ is normalized point spread function at position $\vec{r}$ with $\overline{\text{PSF}}(0) = 1$, V is measurement volume.

References

Abdollah-Nia F (2016) Cumulant-based formulation of higher-order fluorescence correlation spectroscopy. arXiv.org. e-Print Arch

Abdollah-Nia F, Gelfand MP, Van Orden A (2017a) Artifact-free and detection-profile-independent higher-order fluorescence correlation spectroscopy for microsecond-resolved kinetics. 1. Multidetector and sub-binning approach. J Phys Chem B 121:2373–2387. https://doi.org/10.1021/acs.jpcb.7b00407

Abdollah-Nia F, Gelfand MP, Van Orden A (2017b) Artifact-free and detection-profile-independent higher-order fluorescence correlation spectroscopy for microsecond-resolved kinetics. 2. Mixtures and reactions. J Phys Chem B 121:2388–2399. https://doi.org/10.1021/acs.jpcb.7b00408

Bag N, Wohland T (2014) Imaging fluorescence fluctuation spectroscopy: new tools for quantitative bioimaging. Annu Rev Phys Chem 65:225–248. https://doi.org/10.1146/annurev-physchem-040513-103641

Balleza E, Kim JM, Cluzel P (2018) Systematic characterization of maturation time of fluorescent proteins in living cells. Nat Methods 15:47–51. https://doi.org/10.1038/nmeth.4509

Bartoi T, Augustinowski K, Polleichtner G, Grunder S, Ulbrich MH (2014) Acid-sensing ion channel (ASIC) 1a/2a heteromers have a flexible 2:1/1:2 stoichiometry. Proc Natl Acad Sci 111:8281–8286. https://doi.org/10.1073/pnas.1324060111

Caccia M, Camozzi E, Collini M, Zaccolo M, Chirico G (2005) Photon moment analysis in cells in the presence of photo-bleaching. Appl Spectrosc 59:227–236. https://doi.org/10.1366/0003702053084981

Campbell L (1992) Afterpulse measurement and correction. Rev Sci Instrum 63:5794–5798. https://doi.org/10.1063/1.1143365

Chen Y, Müller JD, So PTC, Gratton E (1999) The photon counting histogram in fluorescence fluctuation spectroscopy. Biophys J 77:553–567. https://doi.org/10.1016/S0006-3495(99)76912-2

Chen Y, Müller JD, Ruan Q, Gratton E (2002) Molecular brightness characterization of EGFP in vivo by fluorescence fluctuation spectroscopy. Biophys J 82:133–144. https://doi.org/10.1016/S0006-3495(02)75380-0

Chen Y, Wei L-N, Muller JD (2003) Probing protein oligomerization in living cells with fluorescence fluctuation spectroscopy. Proc Natl Acad Sci 100:15492–15497. https://doi.org/10.1073/pnas.2533045100

Chen Y, Johnson J, Macdonald P, Wu B, Mueller JD (2010) Observing protein interactions and their stoichiometry in living cells by brightness analysis of fluorescence fluctuation experiments. In: Single molecule tools: fluorescence based approaches, Part A. Elsevier, San Diego, CA, pp 345–363

Cutrale F, Rodriguez D, Hortigüela V, Chiu C-L, Otterstrom J, Mieruszynski S, Seriola A, Larrañaga E, Raya A, Lakadamyali M, Fraser SE, Martinez E, Ojosnegros S (2019) Using enhanced number and brightness to measure protein oligomerization dynamics in live cells. Nat Protoc 14:616–638. https://doi.org/10.1038/s41596-018-0111-9

Dalal RB, Digman MA, Horwitz AF, Vetri V, Gratton E (2008) Determination of particle number and brightness using a laser scanning confocal microscope operating in the analog mode. Microsc Res Tech 71:69–81. https://doi.org/10.1002/jemt.20526

Das SK, Darshi M, Cheley S, Wallace MI, Bayley H (2007) Membrane protein stoichiometry determined from the step-wise photobleaching of dye-labelled subunits. Chembiochem 8:994–999. https://doi.org/10.1002/cbic.200600474

Digman MA, Dalal R, Horwitz AF, Gratton E (2008) Mapping the number of molecules and brightness in the laser scanning microscope. Biophys J 94:2320–2332. https://doi.org/10.1529/biophysj.107.114645

Dunsing V, Luckner M, Zühlke B, Petazzi RA, Herrmann A, Chiantia S (2018) Optimal fluorescent protein tags for quantifying protein oligomerization in living cells. Sci Rep 8:10634. https://doi.org/10.1038/s41598-018-28858-0

Elson EL, Magde D (1974) Fluorescence correlation spectroscopy. I. Conceptual basis and theory. Biopolymers 13:1–27. https://doi.org/10.1002/bip.1974.360130102

Fang Y-S, Tsai K-J, Chang Y-J, Kao P, Woods R, Kuo P-H, Wu C-C, Liao J-Y, Chou S-C, Lin V, Jin L-W, Yuan HS, Cheng IH, Tu P-H, Chen Y-R (2014) Full-length TDP-43 forms toxic amyloid oligomers that are present in frontotemporal lobar dementia-TDP patients. Nat Commun 5:4824. https://doi.org/10.1038/ncomms5824

Foo YH, Naredi-Rainer N, Lamb DC, Ahmed S, Wohland T (2012) Factors affecting the quantification of biomolecular interactions by fluorescence cross-correlation spectroscopy. Biophys J 102:1174–1183. https://doi.org/10.1016/j.bpj.2012.01.040

Fukushima R, Yamamoto J, Ishikawa H, Kinjo M (2018) Two-detector number and brightness analysis reveals spatio-temporal oligomerization of proteins in living cells. Methods 140–141:161–171. https://doi.org/10.1016/j.ymeth.2018.03.007

Gambin Y, Polinkovsky M, Francois B, Giles N, Bhumkar A, Sierecki E (2016) Confocal spectroscopy to study dimerization, oligomerization and aggregation of proteins: a practical guide. Int J Mol Sci 17:655. https://doi.org/10.3390/ijms17050655

Hendrix J, Flors C, Dedecker P, Hofkens J, Engelborghs Y (2008) Dark states in monomeric red fluorescent proteins studied by fluorescence correlation and single molecule spectroscopy. Biophys J 94:4103–4113. https://doi.org/10.1529/biophysj.107.123596

Hendrix J, Schrimpf W, Höller M, Lamb DC (2013) Pulsed interleaved excitation fluctuation imaging. Biophys J 105:848–861. https://doi.org/10.1016/j.bpj.2013.05.059

Hennen J, Hur K-H, Saunders CA, Luxton GWG, Mueller JD (2017) Quantitative brightness analysis of protein oligomerization in the nuclear envelope. Biophys J 113:138–147. https://doi.org/10.1016/j.bpj.2017.05.044

Hillesheim LN, Müller JD (2003) The photon counting histogram in fluorescence fluctuation spectroscopy with non-ideal photodetectors. Biophys J 85:1948–1958. https://doi.org/10.1016/S0006-3495(03)74622-0

Huet S, Avilov SV, Ferbitz L, Daigle N, Cusack S, Ellenberg J (2010) Nuclear import and assembly of influenza a virus RNA polymerase studied in live cells by fluorescence cross-correlation spectroscopy. J Virol 84:1254–1264. https://doi.org/10.1128/JVI.01533-09

Hur K-H, Mueller JD (2015) Quantitative brightness analysis of fluorescence intensity fluctuations in E. Coli PLoS One 10:e0130063. https://doi.org/10.1371/journal.pone.0130063

Hur K-H, Macdonald PJ, Berk S, Angert CI, Chen Y, Mueller JD (2014) Quantitative measurement of brightness from living cells in the presence of photodepletion. PLoS One 9:e97440. https://doi.org/10.1371/journal.pone.0097440

Kannan B, Har JY, Liu P, Maruyama I, Ding JL, Wohland T (2006) Electron multiplying charge-coupled device camera based fluorescence correlation spectroscopy. Anal Chem 78:3444–3451. https://doi.org/10.1021/ac0600959

Kitamura A, Kinjo M (2018) State-of-the-art fluorescence fluctuation-based spectroscopic techniques for the study of protein aggregation. Int J Mol Sci 19:964. https://doi.org/10.3390/ijms19040964

Krmpot AJ, Nikolić SN, Vitali M, Papadopoulos DK, Oasa S, Thyberg P, Tisa S, Kinjo M, Nilsson L, Gehring WJ, Terenius L, Rigler R, Vukojević V (2015) Quantitative confocal fluorescence microscopy of dynamic processes by multifocal fluorescence correlation spectroscopy. In: Advanced microscopy techniques IV; and Neurophotonics II. OSA, Washington, DC, p 95360O

Macdonald PJ, Chen Y, Wang X, Chen Y, Mueller JD (2010) Brightness analysis by Z-scan fluorescence fluctuation spectroscopy for the study of protein interactions within living cells. Biophys J 99:979–988. https://doi.org/10.1016/j.bpj.2010.05.017

Macdonald P, Johnson J, Smith E, Chen Y, Mueller JD (2013) Brightness analysis. In: Methods in enzymology. Elsevier, Amsterdam, pp 71–98

Melnykov AV, Hall KB (2009) Revival of high-order fluorescence correlation analysis: generalized theory and biochemical applications. J Phys Chem B 113:15629–15638. https://doi.org/10.1021/jp906539k

Meseth U, Wohland T, Rigler R, Vogel H (1999) Resolution of fluorescence correlation measurements. Biophys J 76:1619–1631. https://doi.org/10.1016/S0006-3495(99)77321-2

Müller JD (2004) Cumulant analysis in fluorescence fluctuation spectroscopy. Biophys J 86:3981–3992. https://doi.org/10.1529/biophysj.103.037887

Nolan R, Alvarez LAJ, Elegheert J, Iliopoulou M, Jakobsdottir GM, Rodriguez-Muñoz M, Aricescu AR, Padilla-Parra S (2017a) Nandb—number and brightness in R with a novel automatic detrending algorithm. Bioinformatics 33:3508–3510. https://doi.org/10.1093/bioinformatics/btx434

Nolan R, Iliopoulou M, Alvarez L, Padilla-Parra S (2017b) Detecting protein aggregation and interaction in live cells: a guide to number and brightness. Methods 140:1–6. https://doi.org/10.1016/j.ymeth.2017.12.001

O'Donnell KA (1986) Correction of dead-time effects in photoelectric-counting distributions. J Opt Soc Am A 3:113. https://doi.org/10.1364/JOSAA.3.000113

Oh D, Zidovska A, Xu Y, Needleman DJ (2011) Development of time-integrated multipoint moment analysis for spatially resolved fluctuation spectroscopy with high time resolution. Biophys J 101:1546–1554. https://doi.org/10.1016/j.bpj.2011.08.013

Ojosnegros S, Cutrale F, Rodríguez D, Otterstrom JJ, Chiu CL, Hortigüela V, Tarantino C, Seriola A, Mieruszynski S, Martínez E, Lakadamyali M, Raya A, Fraser SE (2017) Eph-ephrin signaling modulated by polymerization and condensation of receptors. Proc Natl Acad Sci 114:13188–13193. https://doi.org/10.1073/pnas.1713564114

Ossato G, Digman MA, Aiken C, Lukacsovich T, Marsh JL, Gratton E (2010) A two-step path to inclusion formation of huntingtin peptides revealed by number and brightness analysis. Biophys J 98:3078–3085. https://doi.org/10.1016/j.bpj.2010.02.058

Pack C, Saito K, Tamura M, Kinjo M (2006) Microenvironment and effect of energy depletion in the nucleus analyzed by mobility of multiple oligomeric EGFPs. Biophys J 91:3921–3936. https://doi.org/10.1529/biophysj.105.079467

Palmer AG, Thompson NL (1987) Molecular aggregation characterized by high order autocorrelation in fluorescence correlation spectroscopy. Biophys J 52:257–270. https://doi.org/10.1016/S0006-3495(87)83213-7

Palmer AG, Thompson NL (1989a) High-order fluorescence fluctuation analysis of model protein clusters. Proc Natl Acad Sci 86:6148–6152. https://doi.org/10.1073/pnas.86.16.6148

Palmer AG, Thompson NL (1989b) Intensity dependence of high-order autocorrelation functions in fluorescence correlation spectroscopy. Rev Sci Instrum 60:624–633. https://doi.org/10.1063/1.1140374

Petrášek Z, Schwille P (2008) Photobleaching in two-photon scanning fluorescence correlation spectroscopy. ChemPhysChem 9:147–158. https://doi.org/10.1002/cphc.200700579

Qian H, Elson EL (1990a) On the analysis of high order moments of fluorescence fluctuations. Biophys J 57:375–380. https://doi.org/10.1016/S0006-3495(90)82539-X

Qian H, Elson EL (1990b) Distribution of molecular aggregation by analysis of fluctuation moments. Proc Natl Acad Sci 87:5479–5483. https://doi.org/10.1073/pnas.87.14.5479

Saffarian S, Elson EL (2003) Statistical analysis of fluorescence correlation spectroscopy: the standard deviation and Bias. Biophys J 84:2030–2042. https://doi.org/10.1016/S0006-3495(03)75011-5

Sanchez-Andres A, Chen Y, Müller JD (2005) Molecular brightness determined from a generalized form of Mandel's Q-parameter. Biophys J 89:3531–3547. https://doi.org/10.1529/biophysj.105.067082

Schätzel K, Drewel M, Stimac S (1988) Photon correlation measurements at large lag times: improving statistical accuracy. J Mod Opt 35:711–718. https://doi.org/10.1080/09500348814550731

Skinner JP, Chen Y, Müller JD (2008) Fluorescence fluctuation spectroscopy in the presence of immobile fluorophores. Biophys J 94:2349–2360. https://doi.org/10.1529/biophysj.107.115642

Slaughter BD, Schwartz JW, Li R (2007) Mapping dynamic protein interactions in MAP kinase signaling using live-cell fluorescence fluctuation spectroscopy and imaging. Proc Natl Acad Sci 104:20320–20325. https://doi.org/10.1073/pnas.0710336105

Thompson NL (1991) Fluorescence correlation spectroscopy. In: Lakowics JR (ed) Topics in fluorescence spectroscopy, Volume 1 Techniques. Plenum Press, New York, pp 337–378

Trullo A, Corti V, Arza E, Caiolfa VR, Zamai M (2013) Application limits and data correction in number of molecules and brightness analysis. Microsc Res Tech 76:1135–1146. https://doi.org/10.1002/jemt.22277

Ulbrich MH, Isacoff EY (2007) Subunit counting in membrane-bound proteins. Nat Methods 4:319–321. https://doi.org/10.1038/nmeth1024

Unruh JR, Gratton E (2008) Analysis of molecular concentration and brightness from fluorescence fluctuation data with an electron multiplied CCD camera. Biophys J 95:5385–5398. https://doi.org/10.1529/biophysj.108.130310

Vámosi G, Mücke N, Müller G, Krieger JW, Curth U, Langowski J, Tóth K (2016) EGFP oligomers as natural fluorescence and hydrodynamic standards. Sci Rep 6:33022. https://doi.org/10.1038/srep33022

van Kempen GMP, van Vliet LJ (2000) Mean and variance of ratio estimators used in fluorescence ratio imaging. Cytometry 39:300–305. https://doi.org/10.1002/(SICI)1097-0320(20000401)39:4<300::AID-CYTO8>3.0.CO;2-O

Wawrezinieck L, Rigneault H, Marguet D, Lenne P-F (2005) Fluorescence correlation spectroscopy diffusion Laws to probe the submicron cell membrane organization. Biophys J 89:4029–4042. https://doi.org/10.1529/biophysj.105.067959

Widengren J, Mets Ü, Rigler R (1999) Photodynamic properties of green fluorescent proteins investigated by fluorescence correlation spectroscopy. Chem Phys 250:171–186. https://doi.org/10.1016/S0301-0104(99)00255-4

Wu B, Müller JD (2005) Time-integrated fluorescence cumulant analysis in fluorescence fluctuation spectroscopy. Biophys J 89:2721–2735. https://doi.org/10.1529/biophysj.105.063685

Wu B, Chen Y, Müller JD (2009) Fluorescence fluctuation spectroscopy of mCherry in living cells. Biophys J 96:2391–2404. https://doi.org/10.1016/j.bpj.2008.12.3902

Yamamoto J, Mikuni S, Kinjo M (2018) Multipoint fluorescence correlation spectroscopy using spatial light modulator. Biomed Opt Express 9:5881. https://doi.org/10.1364/BOE.9.005881

Youker RT, Teng H (2014) Measuring protein dynamics in live cells: protocols and practical considerations for fluorescence fluctuation microscopy. J Biomed Opt 19:090801. https://doi.org/10.1117/1.JBO.19.9.090801

Zamai M, Trullo A, Giordano M, Corti V, Arza Cuesta E, Francavilla C, Cavallaro U, Caiolfa VR (2019) Number and brightness analysis reveals that NCAM and FGF2 elicit different assembly and dynamics of FGFR1 in live cells. J Cell Sci 132:jcs220624. https://doi.org/10.1242/jcs.220624

Zhou J, Tang Y, Zheng Q, Li M, Yuan T, Chen L, Huang Z, Wang K (2015) Different KChIPs compete for Heteromultimeric assembly with pore-forming Kv4 subunits. Biophys J 108:2658–2669. https://doi.org/10.1016/j.bpj.2015.04.024

In-Cell Single-Molecule Analysis of Molecular State and Reaction Kinetics Coupling

Michio Hiroshima and Yasushi Sako

Single-Molecule Analysis in Cells

The single-molecule analysis of biological processes was first reported by Funatsu et al. in 1995, in which an enzymatic reaction of myosin was visualized in vitro as the association with fluorescently labeled ATP (substrates) in vitro. To detect the fluorescence signal from a single fluorophore at the video-rate (33 fps), a fluorescence spot, corresponding to several hundred photons, must be recognized from the dim surrounding space. To achieve the high signal-to-noise ratio necessary for single-molecule detection, one solution is to spatially limit the excitation illumination around the observation plane to suppress the fluorescence outside. Evanescent light satisfies this condition, because of its shallow penetration through a medium (<200 nm) and up to fourfold amplified fluorescence intensity. Total internal reflection (TIR) illumination, which is easily incorporated into a normal fluorescence microscope, is ideal for this condition. In early studies, TIR illumination was produced by setting a prism on two coverslips sandwiching a solution with a narrow gap to refract the incident light beyond the critical angle of the glass–water interface (Fig. 1a). This geometry requires the specimen to be placed upside down on the upper coverslip and observed by an inverted objective lens under the lower coverslip, a setup that makes it difficult to image thick samples, exchange solutions, or manipulate the samples. Stout and Axelrod demonstrated another method for TIR illumination by using an objective lens with a high numerical aperture (Axelrod

M. Hiroshima (✉)
RIKEN Center for Biological Dynamics Research, Suita, Osaka, Japan

Cellular Informatics Laboratory, RIKEN, Wako, Saitama, Japan
e-mail: m_hiroshima@riken.jp

Y. Sako
Cellular Informatics Laboratory, RIKEN, Wako, Saitama, Japan
e-mail: sako@riken.jp

J. K. Kim et al. (eds.), *Advanced Imaging and Bio Techniques for Convergence Science*, Advances in Experimental Medicine and Biology,
https://doi.org/10.1007/978-981-33-6064-8_3

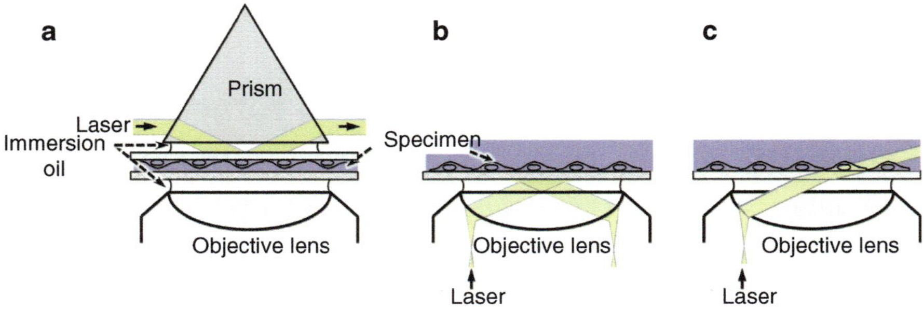

Fig. 1 Configurations of the specimen and illumination optics for single-molecule imaging. (**a**) Prism type and (**b**) objective type TIR illuminations. (**c**) Oblique illumination based on (**b**)

1989). Here, the incident light is reflected at the TIR angle by simply shifting the light to pass through the peripheral of the objective pupil (Fig. 1b). The resulting "objective type" TIR makes a working space above the sample that overcomes the problems of "prism type" TIR and is now widely used in single-molecule cell imaging at the basal (ventral) plasma membrane. When the incident angle is slightly less than the TIR angle for epi-illumination, a thin oblique illumination with a low angle can be produced (Fig. 1c), which enables single-molecule imaging on the apical (dorsal) plasma membrane (Teramura et al. 2006; Hiroshima et al. 2012) or at the surface of cytoplasmic organelles (Tokunaga et al. 2008).

For single-molecule analysis in cells, a target molecule should be labeled with a fluorescent probe. When the purified molecule is added to the extracellular medium, chemical modifications with a fluorescent dye is possible. For the labeling of molecules expressed in cells, antibody-conjugated fluorescent particles (e.g., Qdot) (Chung et al. 2010; Low-Nam et al. 2011) or chemical fluorophores bound to a protein tag (e.g., HaloTag) are typically used. This method can partially label the target protein to detect individual molecules at a suitable area density (2–3 μm^{-2}) by adjusting the concentration of the fluorescent probe. Otherwise, genetically fusing the target protein and a fluorescent protein (e.g., GFP) is an option (Lommerse et al. 2006; Xiao et al. 2008; Zhang et al. 2009; Schibany et al. 2018). Because the target proteins are intrinsically labeled one-to-one without any additional labeling process and are labeled if they are not endogenously expressed (i.e., target protein null cells or gene knock-out), thus permitting quantitative single-molecule analysis. In all labeling methods, the function and localization of the labeled molecules must be confirmed in the cells.

The first single-molecule analysis in living cells (Sako et al. 2000a, b) showed association and dissociation events between fluorescently labeled ligands and their receptor, and intermolecular FRET (fluorescence resonance energy transfer) between dimerized receptor molecules. This achievement suggested that single-molecule analysis can obtain quantitative insights of the dynamic, kinetic, and structural aspects of cell signaling. Single-molecule imaging of GFP-fused signaling proteins was first achieved for the Ras and Rho family of small GTPases (Sako et al. 2000a, b; Hibino et al. 2003). Studies of the cellular phenomenon at the single-molecule level have the following advantages in comparison with conventional ensemble measurements. 1) The spatiotemporal information of an individual molecule is directly determined as spatial changes in the cell (Fig. 2a). The spatiotemporal distribution of a molecule can be linked to characteristic parameters of its behavior, structure, and other properties. 2) Molecular reactions in cells are observed from start to end for every molecule. Different from conventional methods (e.g., stopped-flow), in which the reaction is synchronized by precisely mixing purified chemical species, single-molecule analysis does not need to align the start point experimentally and statistically handles the data obtained separately from each molecule in a natural cellular condition (Fig. 2b). 3) A parameter distribution corresponding to the molecular behavior can be obtained from individual molecules, enabling statistical interpretation of the distribution (Fig. 2c). Typical ensemble measurements mostly acquire only average values of a parameter; therefore, when a number of distribution subsets coexist, the mechanism driving the behavior may be misinterpreted.

Using single-molecule analysis, membrane receptors have been widely studied in living cells. The primary behaviors of receptors are lateral diffusive motion, ligand binding with subsequent activation, and interactions with other signaling molecules, all of which occur on the plasma membrane, thus making the analysis complementary with TIR or oblique illumination. Epidermal growth factor receptor (EGFR) is a 180 KDa receptor tyrosine kinase (RTK) and a popular model in single-molecule studies. Its association with its ligand, EGF, triggers a signaling cascade and induces various cell responses such as proliferation, migration, apoptosis, and so forth. EGFR is also known as ErbB1, a member of the ErbB family of four RTKs (ErbB1-B4). Crystallographic structures indicate that ErbBs consist of three regions: an extracellular (ectodomain) region containing four subdomains (I–IV) concerning ligand binding, an α-helical transmembrane (TM) region, and a cytosolic region

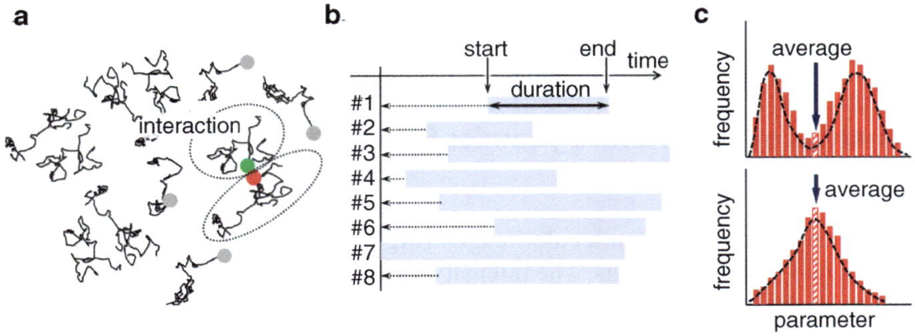

Fig. 2 Advantages of single-molecule analysis. (**a**) Direct observation of individual molecules provides spatiotemporal information of molecular behavior and interactions. (**b**) Start points of a reaction for individual molecules in a cell can be synchronized after the experiments. (**c**) The distribution of a parameter from individual molecules enables correct interpretation of the observed data, e.g., the same averages but from different distributions

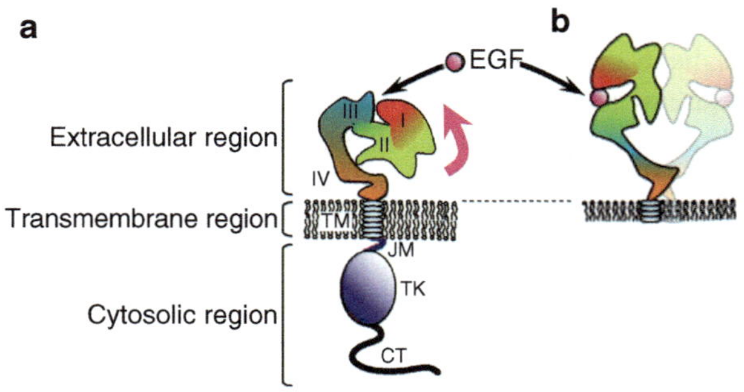

Fig. 3 Structure of the EGFR molecule. (**a**) I–IV denote the subdomains of the extracellular region. The cytosolic region consists of juxtamembrane (JM), tyrosine kinase (TK), and C-terminal phosphorylation (CT) domains. EGF (shown as circle) binds at a site between domains I and III (arrow). The domains I and II of the EGF-bound EGFR rise up along the direction denoted by a thick arrow. (**b**) The extended extracellular regions are bridged in the ligand-induced back-to-back dimer

containing juxtamembrane (JM), tyrosine kinase (TK), and C-terminal (CT) domains (Garrett et al. 2002; Ogiso et al. 2002) (Fig. 3a). When no ligand is bound, subdomains II and IV in the extracellular region are tethered, keeping the ligand-binding site including the subdomains I and III open for ligand entry. Specific ligands are known for each ErbB, except ErbB2, and have different affinities. After ligand association, receptor conformation exposes a dimerization loop in subdomain II that links to the same or different ErbB family member, generating a homo- or heterodimer, respectively, in a "back-to-back" configuration (Fig. 3b). Subsequently, tyrosine residues in the CT domain are phosphorylated by the kinase activity of the dimeric counterpart except ErbB3. Although ErbBs have been studied in detail, the relationship between their spatiotemporal behaviors and the regulation of cell signaling remains poorly understood, especially the coupling of ligand binding, lateral motility, dimer/oligomer formation (clustering), phosphorylation, and membrane structure. Single-molecule analysis offers new information on how these different phenomena couple together.

Analysis of Behavioral Dynamics from Single-Molecule Tracking Data

The analysis of single-molecule images (Fig. 4a) starts with the recognition of individual fluorescent spots. The intensity profile of each fluorescent spot spreads over several pixels and obeys a point spread function (PSF), which is usually fitted to a two-dimensional Gaussian function, in which the centroid corresponds to a molecular position within a localization error. The trajectory of a molecule (Fig. 4b) is obtained by connecting frame-to-frame the centroids and provides spatiotemporal characteristics of the lateral mobility. The total pixel intensity of

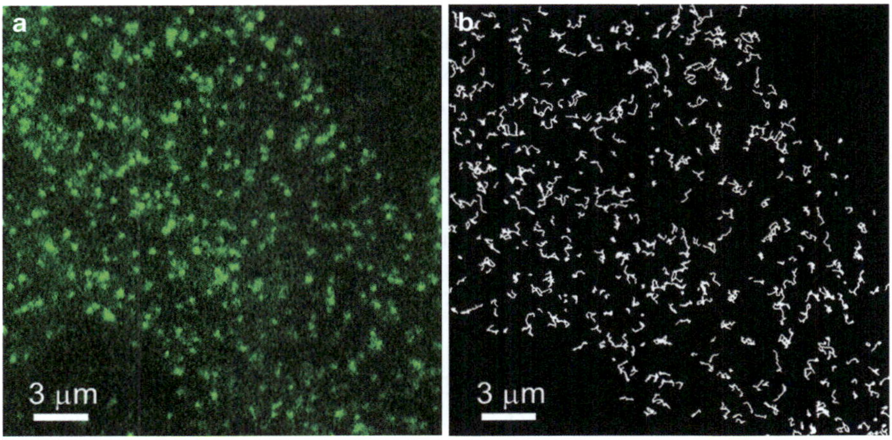

Fig. 4 (**a**) Single-molecule fluorescence image of EGFR–EGFP in a CHO cell. (**b**) Single-molecule trajectories

a spot defines the brightness, which is equivalent to a value calculated by the Gaussian integral with the fitted parameters. The brightness from a single fluorescent dye or protein is represented as a single normal distribution. Therefore, when more than one fluorescent molecule exists within a region of comparable size to the optical resolution, such as dimers and oligomers (clusters), it is observed as a single fluorescent spot with a brightness proportional to the number of molecules. The molecular behavior changes with mobility and clustering along the molecular trajectory through a variable cellular microenvironment. This variability makes investigation of the spatiotemporal behavioral dynamics difficult. Variability in the molecular displacement and brightness, which are intrinsically caused by the stochasticity of the lateral diffusion and fluorescence emission of the observed molecule, should be considered when calculating the motional and clustering states at each time point. Therefore, conventional methods collect displacement and fluorescence intensity from all trajectories independent of time and location, and the distributions are fitted with stochastic functions to determine the number and fractions of states. However, even though this approach confirms the existence of several molecular states, it is impossible to associate spatiotemporal information to the states.

Recent progress in single-molecule analysis techniques has overcome this information loss by utilizing machine learning theory to infer the molecular state at each step along the trajectory. Therefore, researchers can know where, when, and how a molecule changes its behavior in a cell based on changes in the molecular and/or membrane properties. The hidden Markov model (HMM) (Rabiner 1989) has been applied to EGFR mobility (Chung et al. 2010; Low-Nam et al. 2011). By defining the obtained time series of single-molecule data of the molecular position or fluorescence intensity as $\boldsymbol{X} = \{x_1, \ldots, x_N\}$, where N is the total number of frames, the corresponding series of molecular states is given as $\boldsymbol{Z} = \{z_1, \ldots, z_N\}$ and zn = $\{z_{n1}, \ldots, z_{nK}\}$, where the total number of states is K. Our aim is to estimate the state series $\boldsymbol{Z}$ from the observed data $\boldsymbol{X}$, in which $\boldsymbol{Z}$ is assumed to obey the Markov process with transition matrix $\boldsymbol{A}$. Probabilities of the state at the initial and nth frame are

$$p\left(z_1\middle|\pi\right)=\prod_{i=1}^{K}\pi_i^{z_{1i}} \quad and \quad p\left(z_n\middle|z_{n-1},A\right)=\prod_{i=1}^{K}\prod_{j=1}^{K}A_{ij}^{z_{(n-1)i}\,z_{nj}}, \tag{1}$$

respectively. Here, $z_{nk} = 1$ and otherwise 0 for a molecule at the nth frame and kth state. A_{ij} is an element of A from the ith to jth state and satisfies $0 \leq A_{ij} \leq 1$. $\pi_k \equiv p(z_{1i} = 1)$ satisfies $0 \leq i \leq 1$ and $\sum_i \pi_i = 1$. The probability which links an observed phenomenon and the molecular state is described as:

$$p\left(x_n\middle|z_n,\varphi\right)=\prod_{k=1}^{K}p\left(x_n\middle|z_n,\varphi\right)^{z_{nk}}. \tag{2}$$

ϕ denotes parameters of the probability function. In the case of motional states, the probability is a two-dimensional diffusion equation,

$$p\left(x_n\middle|z_n,D_k\right)=\frac{x_n}{2D_k\delta t}\exp\left(-\frac{x_n^{\;2}}{4D_k\delta t}\right), \tag{3}$$

where, x_n and D_k are the step displacement between two adjacent frames for a duration of δt (30 ms) and the diffusion coefficient of state k, respectively. For the clustering states, the probability is described with a Gaussian function,

$$p\left(x_n\middle|z_n,\mu,\sigma\right)=\frac{1}{\sqrt{2\pi k\sigma}}\exp\left[-\frac{\left(x_n-k\mu\right)^2}{2k\sigma^2}\right], \tag{4}$$

where, μ and σ are the average and standard deviation of the fluorescence intensity of a single fluorescent molecule, respectively. k is the cluster size, which means the number of molecules contained in a fluorescent spot. The joint probability distribution, $p(\boldsymbol{X}, \boldsymbol{Z}\,|\theta)$, is described as

$$p\left(X,Z\middle|\theta\right)=p\left(z_1\middle|\pi\right)\times\prod_{n=2}^{K}p\left(z_n\middle|z_{n-1},A\right)\times\prod_{m=1}^{K}p\left(x_m\middle|z_m,\phi\right). \tag{5}$$

where, $\theta = \{\pi, A, \phi\}$. The variational Bayesian (VB) method (Beal 2003; Okamoto and Sako 2012) is used to estimate the molecular state Z and parameter θ, which satisfy the maximum value of the logarithmic likelihood function, $p(X)$. The VB method is able to prevent the parameters from falling into local minima more effectively than the expectation-maximization (EM)-HMM method (X). When the distribution of θ is specified by model M,

$$\ln p\left(X\right)=\ln\sum_{Z}\int p\left(X,Z,\theta\middle|M\right)d\theta. \tag{6}$$

Eq. 6 can be rewritten as:

$$\begin{aligned}\ln p(X) &= \ln \sum_{Z} \int q(Z,\theta) \frac{p(X,Z,\theta|M)}{q(Z,\theta)} \mathrm{d}\theta \\ &= \sum_{Z} \int q(Z,\theta) \ln \frac{q(Z,\theta)}{p(Z,|,\theta,|,X,|,M)} \mathrm{d}\theta + \sum_{Z} \int q(Z,\theta) \ln \frac{p(X,Z,\theta|M)}{q(Z,\theta)} \mathrm{d}\theta \\ &= KL(q \quad p) + L_q. \end{aligned} \tag{7}$$

Here, $KL(q \| p)$ is the Kullback–Leibler divergence between prior (p) and posterior (q). Because $KL(q \| p)$ has been fixed for model M and observation X, L_q corresponds to the lower bound of $\ln p(X)$. When $q(\mathbf{Z},\theta)$ is assumed to be factorized,

$$\begin{aligned} q(Z,\theta) &= q(Z) q(\theta) \\ &= q(Z) \prod_{i=1}^{I} q(\theta_i), \end{aligned} \tag{8}$$

where, I is the number of parameters. Therefore, L_q can be described as:

$$L_q = \sum_{Z} \int q(Z) q(\theta) \ln \frac{p(X,Z,\theta|M)}{q(Z)} \mathrm{d}\theta - \sum_{i=1}^{I} \int q(\theta_i) \ln q(\theta_i) \ \mathrm{d}\theta_i. \tag{9}$$

For optimization of the distribution functions $q(Z)$ and $q(\theta)$, the variational Bayesian expectation-maximization (VB-EM) algorithm is applied (Beal 2003). The VB-E and VB-M steps maximize L_q against $q(Z)$ and $q(\theta)$, respectively. The VB-E step corresponds to the calculation of

$$\begin{aligned} \ln q(Z) &= E_\theta \left[\ln p(X,Z,\theta|M) \right] + const \\ &= \int q(\theta) \ln p(X,Z,\theta|M) \ \mathrm{d}\theta + const \end{aligned} \tag{10}$$

Here, $E_\theta[...]$ is the expectation with respect to θ. Thus,

$$q(Z) \propto \exp\left\{ \int q(\theta) \ln p(X,Z,\theta|M) \ \mathrm{d}\theta \right\}. \tag{11}$$

By taking Eq. 5 into Eq. 11 and incorporating M,

$$q(Z) \propto \tilde{p}(z_1|\pi,M) \times \prod_{n=2}^{K} \tilde{p}(z_n|z_{n-1},A,M) \times \prod_{m=1}^{K} \tilde{p}(x_m|z_m,\varphi,M). \tag{12}$$

Based on Eqs. 1 and 2, each term in Eq. 12 becomes

$$\begin{aligned} \tilde{p}(z_1|\pi,M) &= \prod_{i=1}^{K} \exp\left(\overline{\ln \pi_i}\right)^{z_{1i}}, \\ \tilde{p}(z_n|z_{n-1},A,M) &= \prod_{i=1}^{K} \prod_{j=1}^{K} \exp\left(\overline{\ln A_{ij}}\right)^{z_{n-1,i}\, z_{nj}}, \\ \tilde{p}(x_m|z_m,\phi,M) &= \prod_{k=1}^{K} \exp\left(\overline{\ln p(x_m|z_m,\phi,M)}\right), \end{aligned} \tag{13}$$

The overhead lines denote averages. $q(Z)$ is optimized by the forward–backward algorithm using the equations in Eq. 13.

The VB-M step corresponds to the calculation of

$$\begin{aligned}\ln q\left(\theta_j\right) &= E_Z\left[\ln p\left(X,Z,\theta \middle| M\right)\right] + const \\ &= \sum_Z q(Z) \int \prod_{i \neq j} q(\theta_i) \ln p\left(X,Z,\theta \middle| M\right)\, \mathrm{d}\theta_i + const.\end{aligned} \tag{14}$$

Thus,

$$q\left(\theta_j\right) \propto \exp\left\{\sum_Z q(Z) \int \prod_{i \neq j} q(\theta_i) \ln p\left(X,Z,\theta \middle| M\right)\, \mathrm{d}\theta_i\right\}, \tag{15}$$

in which $q(\theta_j)$ is factorized to separate terms for each parameter. By optimizing $q(\theta_j)$, the expectations of the parameters are obtained and used as updated values in the next VB-E step.

The Dirichlet distribution is used for the given prior functions of the initial state and transition probability, of which the calculated posterior is also the Dirichlet distribution. Therefore, the log of the expectation of the initial state is obtained as:

$$\left\langle \ln \pi_i \right\rangle = \psi\left(u_i^\pi + \overline{z_{1,i}}\right) - \psi\left(\sum_i^K u_i^\pi + 1\right). \tag{16}$$

where, $\Psi(x)$ is the digamma function

$$\psi(x) = \frac{d}{dx} \ln \Gamma(x) = \frac{\Gamma'(x)}{\Gamma(x)}, \tag{17}$$

and u_i^π is a hyperparameter and takes 1, giving a flat probability distribution. The log of the expectation of the transition probability is

$$\left\langle \ln A_{ij} \right\rangle = \psi\left(u^A{}_{ij} + \sum_{n=2}^N \overline{z_{n-1,i} z_{n,j}}\right) - \psi\left(\sum_j^K \left(u^A{}_{ij} + \sum_{n=2}^N \overline{z_{n-1,i} z_{n,j}}\right)\right). \tag{18}$$

Here, $u^A{}_{ij}$ is a hyperparameter and also takes 1. For the emission probability of a two-dimensional diffusion equation (Eq. 3), the prior function is given by a gamma distribution. When representing the diffusion coefficient as D_k, the log-expectation of parameter $\tau^D{}_k$ ($= 1/2D_k\delta t$) is

$$\left\langle \ln \tau^D{}_k \right\rangle = \psi\left(a_D + \sum_{n=1}^N \overline{z_{n,i}}\right) - \ln\left(b_D + \tfrac{1}{2} \sum_{n=1}^N \overline{z_{n,i}} x_n{}^2\right). \tag{19}$$

Here, a_D and b_D are hyperparameters assigned values to maximize the lower bound, L_p. In the case of the clustering size being derived from a Gaussian distribution of the fluorescence intensities (Eq. 4), the prior function is described by the average (μ_k) and standard deviation (σ_k). The distributions of μ_k and $\tau^G{}_k$ ($= 1/\sigma_k^2$) are provided as normal and gamma distributions, respectively. The log-expectations of the parameters are calculated as:

$$\langle \ln \mu_k \rangle = \frac{\lambda_0 \mu_0 + \sum_{n=1}^{N} \overline{z_{n,i}} x_n}{\lambda_0 + \sum_{n=1}^{N} \overline{z_{n,i}}}, \tag{20}$$

$$\langle \ln \tau^G{}_K \rangle = \psi\left(\tfrac{1}{2}\left(a_G + \sum_{n=1}^{N} \overline{z_{n,i}} \right)\right) - \ln\left(\tfrac{1}{2}\left(b_G + \sum_{n=1}^{N} \overline{z_{n,i}} x_n{}^2 + \lambda_0 \mu_0{}^2 - \left(\lambda_0 + \sum_{n=1}^{N} \overline{z_{n,i}} \right) \mu_i{}^2 \right)\right). \tag{21}$$

where a_G, b_G, μ_0, and λ_0 are hyperparameters selected to maximize L_p, which can be rewritten as:

$$\begin{aligned} L_q &= E\left[\ln p(\pi)\right] + E\left[\ln p(A)\right] + E\left[\ln p(\varphi)\right] \\ &\quad - E\left[\ln q(\pi)\right] - E\left[\ln q(A)\right] - E\left[\ln q(\varphi)\right] + \sum_{n=1}^{N} \ln c_n. \end{aligned} \tag{22}$$

Here, c_n is the scaling factor calculated in the VB-E step.

The first process of the state estimation is setting initial values of the parameters for every state. When the number of states is set to N, the obtained single-molecule data is divided into N groups and the initial values are calculated for the distribution of every group. In this process, a clustering method (e.g., K-means) can be used. Then, the VB-E step and VB-M step are sequentially executed to optimize $q(Z)$ using the forward–backward algorithm and to update the parameters by Eqs. 16–21, respectively. The iteration is continued until the lower bound, L_q, converges. The suitable number of states is determined by comparing L_q for different numbers of N. At last, the state sequence is optimized, for example, by the Viterbi algorithm, which chooses the most likely path from the first to last frames.

In previous single-molecule studies of EGFR, antibody-modified Qdots were often used for the fluorescent labeling to enable long tracking suitable for the mobility analysis. We adopted EGFR tagged with GFP at the cytoplasmic tail (EGFR–GFP, Carter and Sorkin 1998) expressed in CHO-K1 (Chinese hamster ovary) cells without endogenous EGFR. This experimental system is useful for the quantitative analysis of clustering, which requires visualizing all EGFR molecules in a cell. Although the observable time period is limited by the bleaching rate of GFP (> 0.1 sec^{-1}), our state estimation of EGFR mobility and clustering showed a sufficient number of transitions. The cluster size as well as the motional state could be assigned to the same steps in the trajectories, providing new information about the molecule by single-molecule analysis.

State transitions were seen between three states in mobility more than four states in clustering (Fig. 5a). The transition rate constants of the motional and clustering states were calculated from the transition probability.

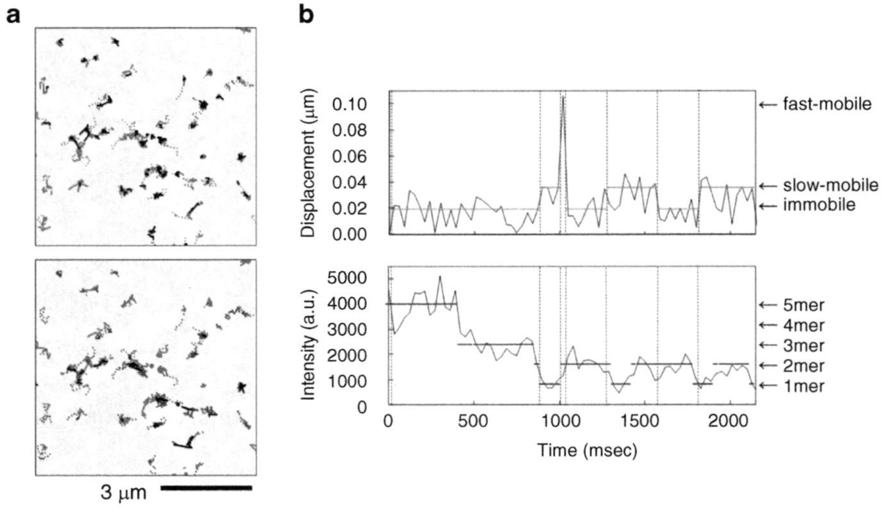

Fig. 5 Transitions between the molecular states of EGFR in a cell. (**a**) Spatial distributions of the states. Upper, three motional states are shown as dotted gray, gray, and black lines for immobile, slow-mobile, and fast-mobile states, respectively. Lower, clustering states indicated as dotted gray, gray, dark gray, and black lines for monomer, dimer, trimer, and larger oligomers, respectively. (**b**) Temporal changes of the states. Upper, transitions between motional states. Lower, transitions between clustering states. Dotted line means transition time points of motional states

$$k_{A\to B} = {-\ln\left(1-P_{A\to B}\right)}\Big/{\Delta t}, \tag{23}$$

where, Δt is the frame time (30 ms). When a cluster splits into two, tracking is continued on one molecule and often stopped for the other. Thus, changes in the cluster size as $(N-n)$ *mer* + n *mer* → N *mer* were observed as $(N-n)$ *mer* → N *mer* or n *mer* → N *mer*. In these cases, the association rate was the sum of the two observed rates. The dissociation equilibrium constant between $(N-n)$ mer and n mer, which is calculated using the association and dissociation rate constants, is

$$\begin{aligned} K_d &= \frac{k_2^n + k_2^{N-n}}{k_1^n + k_1^{N-n}} \\ &= \frac{k_2^n + k_2^{N-n}}{\dfrac{\left[(N-n)\ mer\right]\cdot k_1^n\left[n\ mer\right]+\left[n\ mer\right]k_1^{N-n}\left[(N-n)\ mer\right]}{\left[n\ mer\right]\left[(N-n)\ mer\right]}}. \end{aligned} \tag{24}$$

Here, K_d is the dissociation equilibrium constant. $k^{N-n}{}_1$ and $k^n{}_1$ are the rate constants from $(N-n)$ *mer* to N *mer* and from n *mer* to N *mer*, respectively. $k^{N-n}{}_2$ and $k^n{}_2$ are the dissociation rate constants from N *mer* to $(N-n)$ *mer* and from N *mer* to n *mer*, respectively. $[(N-n)\ mer]$ and $[n\ mer]$ denote their respective fractions.

The motional states were named after their diffusion coefficients: immobile, slow mobile, and fast mobile. Localizations of these states were typically observed as fast-mobile movements linked between slow-mobile regions surrounding immobile regions. Upon comparison with the spatial distribution of the clustering states, the immobile region tended to be overlapped with larger clusters; however, the time points of the state transitions were not completely synchronized between the motional states and cluster sizes in individual trajectories (Fig. 5b). Our results showed the spatiotemporal correlation between the mobility and clustering of a receptor for the first time (Hiroshima et al. 2018). To elucidate the characteristic behavior of each state, segments with specific mobility and clustering states were extracted from the trajectories and calculated as a temporal development of the mean square displacement (MSD) for the identical state (Fig. 6a). We obtained the MSD by the following equation for every combination of the three motional states and four cluster sizes:

$$\mathrm{MSD}_{\mathrm{p,\,q}}\left(\mathrm{n}\cdot\delta t\right)=\frac{1}{\mathrm{N}-1-\mathrm{n}}\sum_{\mathrm{j}=1}^{\mathrm{N}-1-\mathrm{n}}\left\{\begin{matrix}\left[\mathrm{x}\left(\mathrm{j}\cdot\delta t+\mathrm{n}\cdot\delta t\right)-\mathrm{x}\left(\mathrm{j}\cdot\delta t\right)\right]^2+\\ \left[\mathrm{y}\left(\mathrm{j}\cdot\delta t+\mathrm{n}\cdot\delta t\right)-\mathrm{y}\left(\mathrm{j}\cdot\delta t\right)\right]^2\end{matrix}\right\} \quad (25)$$

Here, N, n, δt, j, and (x, y) are total frames, present frame, time duration, initial frame of a state, and molecule position, respectively. p is a motional state and q is a clustering state. As shown in Fig. 6b for the resting condition, all clusters of the immobile states were confined with a confinement length of 70 ± 5 nm. The slow-mobile receptors also showed confined diffusion, and the confinement length was shorter for monomers (170 nm) than for higher-order clusters (300–420 nm). That mobility was dependent on cluster size suggested that interactions between the receptor and some membrane factor, which might constrain the diffusion, were more frequent for monomers than for clusters. The fast-mobile state was, on the other hand, attributable to free diffusion. The confined region composed of immobile and slow-mobile states was almost consistent in size with the membrane subdomain (Jacobson et al. 2007; Lingwood and Simons 2010). When EGF was added to cells to trigger cell signaling, the fraction of immobile and large clusters increased, and the confinement size was slightly reduced (Fig. 6c). This method revealed the coupling of protein motility and clustering for signal transduction; however, the significance of the receptor dynamics was not understood. Therefore, we applied a two-color single-molecule analysis on the interactions between EGFR and a downstream signaling protein.

Kinetic Analysis in Signal Transduction at the Single-Molecule Level

EGFR associates cytoplasmic proteins via its phosphorylated tyrosines in the C-terminal region to transmit signals. Growth factor receptor-bound protein 2 (Grb2) has no enzymatic activity but is an adaptor protein that bridges EGFR and

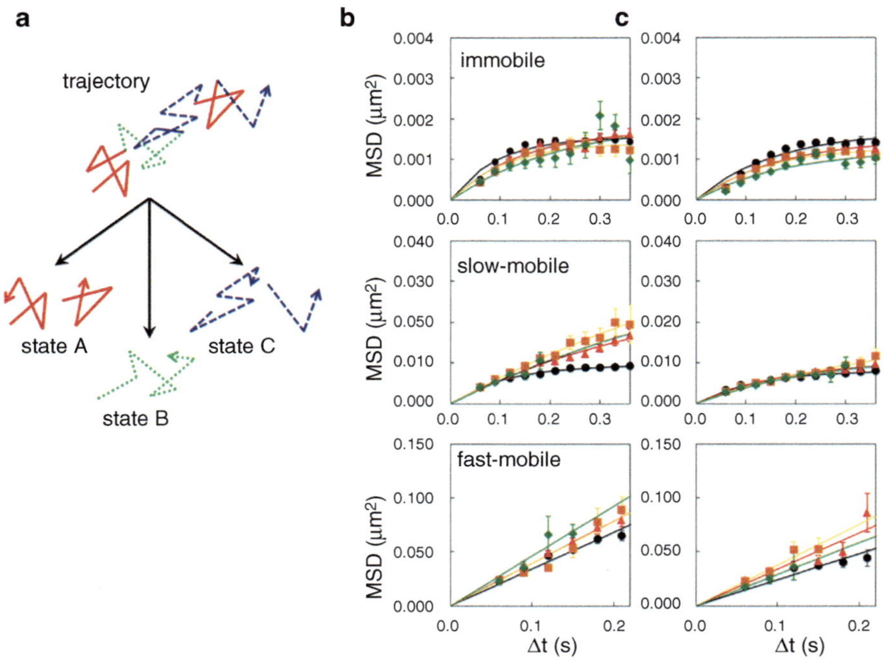

Fig. 6 Motional state-dependent diffusion modes. (**a**) MSD was calculated for each state. (**b**) and (**c**) MSD for resting and EGF-stimulated cells, respectively. Monomers, dimers, trimers, and tetramers are indicated as circles, triangles, squares, and rhomboids, respectively. Lines show the fitted results using the following equation for the confined diffusions, $\mathrm{MSD} = L^{3/2}\left(1-\exp\left(-12\frac{D_1\Delta t}{L^2}\right)\right)+4D_2\Delta t$, D_1 and D_2 are the diffusion coefficients of EGFR in the confined area and of the confined area itself, respectively. L is the confinement length, and Δt is the time duration. The obtained D_2 indicated that the confined area did not diffuse regardless of L. The equation for free diffusion is $\mathrm{MSD} = 4D_3\Delta t$, where, D_3 is the diffusion coefficient

another signaling protein, SOS, a RAS GTP exchange factor, via its Src homology 2 (SH2) domain and one of its two SH3 domains (Lowenstein et al. 1992). The primary Grb2-binding site of EGFR is the phosphorylated residue of pY1068 and the secondary site is pY1086. We observed EGFR–Grb2 interactions by two-color single-molecule tracking of EGFR–EGFP and Grb2-HaloTag labeled with tetramethylrhodamine (TMR), which can directly detect their interactions (Fig. 7a). When the EGFR and Grb2 distance was less than the sum of the localization accuracies of the fluorescence probes and the merging error of the two fluorescence channels, an association was determined to occur. The merging error was defined as the standard deviation of the position differences between images from the same particle in the EGFP and TMR channels.

To analyze the kinetics of the EGFR–Grb2 interaction, two kinds of time durations were measured; duration from the association to dissociation of a Grb2 molecule, and duration from dissociation to the next association of different Grb2 molecules. The duration histogram was fitted with an exponential function with a decay constant corresponding to the association or dissociation rate. Because not all Grb2-HaloTag molecules were labeled with TMR and because endogenous Grb2 molecules exist in cells, we observed only interactions of EGFR–EGFP with fluorescent Grb2-HaloTag::TMR. Therefore, the association rate constant obtained from the visualized events had to be corrected for nonfluorescent Grb2. We categorized the observed EGFR–EGFP fluorescent spots as no Grb2 bound (x), Grb2-HaloTag::TMR bound (y), and unobservable Grb2 bound (w) fractions. When we consider the time from the dissociation of the observable Grb2 to the next association, the reaction equations are described below (Fig. 7b).

$$x \xrightarrow{k_1} y$$

$$x \xrightarrow{k_2} w$$

$$w \xrightarrow{d} x.$$

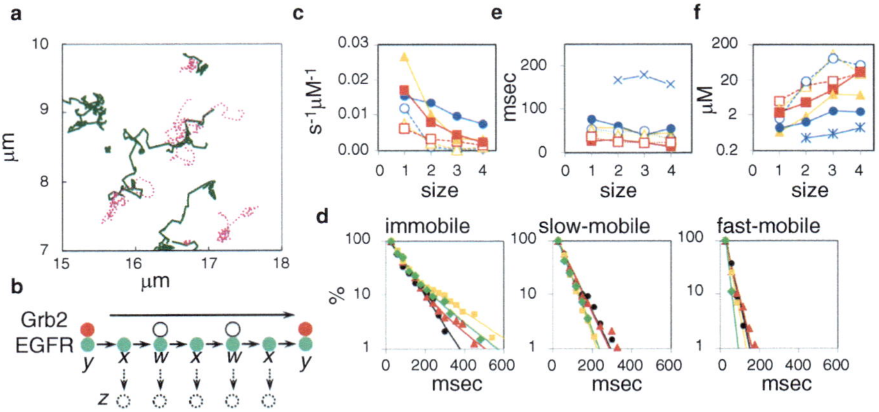

Fig. 7 Single-molecule analysis of interactions between EGFR and Grb2. (**a**) Trajectories of EGFR (solid line) and Grb2 (dotted line) molecules. (**b**) To calculate the association rate constants, the associations of nonfluorescent Grb2 (hollowed circles) between the association of fluorescent Grb2 (upper solid circles) should be considered. Sometimes, EGFR (lower solid circles) take forms that cannot be analyzed (dotted circles). (**c**) Association rate constants. (**d**) Histograms of Grb2 resident time under EGF stimulation for the motional states. Monomers, dimers, trimers, and tetramers are indicated by circles, triangles, squares, and rhomboids, respectively. (**e**) Dissociation rate constants. (**f**) Dissociation equilibrium constants. In (**c**, **e**, and **f**), plots for the immobile, slow-mobile, and fast-mobile states are shown as circles, triangles, and squares, respectively. Data from resting cells are plotted as open symbols

Here, k_1 and k_2 are the first-order association rate constants of Grb2-HaloTag::TMR and unobservable Grb2, respectively. The relationships of k_1 and k_3 with the true association rate, k, are $k_1 = k\,[Grb2_{TMR}]$ and $k_2 = k\,[Grb_{white}]$, in which $[Grb2_{TMR}]$ and $[Grb_{white}]$ are the concentrations of Grb2-HaloTag::TMR and unobservable Grb2, respectively. d is the dissociation rate constant. The fraction of EGFR–EGFP that could not be analyzed, such as molecules after bleaching, after changing to a different state, or without association event, is denoted as z and the transition rate constant from the observable EGFR–EGFP is k_L. Then,

$$x \xrightarrow{k_L} z$$
$$w \xrightarrow{k_L} z.$$

Differential equations between the fractions are described as follows:

$$\begin{aligned} dx/dt &= -(k_1 + k_2 + k_L)\ x + dw \\ dy/dt &= k_1 x \\ dz/dt &= k_L (x + w) \\ dw/dt &= k_2 x - (k_L + d) w \end{aligned} \tag{26}$$

Here, $x + y + z + w = 1$ and $x(0) = y(0) = z(0) = 0,\ w(0) = 1$..
$x(t)$ can be solved as

$$x(t) = \exp\left(-\gamma/2\, t\right) \frac{\left(-q+\sqrt{p}\right)\exp\left(\sqrt{p}/2\ t\right) + \left(q+\sqrt{p}\right)\exp\left(-\sqrt{p}/2\ t\right)}{2\sqrt{p}} \exp\left(-k_L t\right). \tag{27}$$

Here, $p = (k_1 + k_2)^2 - 2d(k_1 - k_2) + d^2$ *and* $q = k_1 + k_2 - d,\ \ \gamma = k_1 + k_2 + d$..
The distribution of the duration time between Grb2-HaloTag::TMR associations is

$$\begin{aligned} dy/dt &= k_1 x(t) \\ &= k_1 \frac{\left(-q+\sqrt{p}\right)\exp\left(-\frac{\gamma-\sqrt{p}}{2}\mathrm{t}\right) + \left(q+\sqrt{p}\right)\exp\left(-\frac{\gamma+\sqrt{p}}{2}\mathrm{t}\right)}{2\sqrt{p}} \exp(-k_L t). \end{aligned} \tag{28}$$

The apparent rate constants, k_{obs}, obtained experimentally by using Eq. 23 correspond to the inverse of the expectation of Eq. 28. Thus, the expectation, $\tau_{obs,}$ is

$$\tau_{obs} = \frac{\left\{\begin{array}{l}\int_0^\infty \left(-q+\sqrt{p}\right)\exp\left[-\left(\frac{\gamma-\sqrt{p}}{2}+\mathrm{k_L}\right)\mathrm{t}\right]dt\cdot\left(\frac{\gamma-\sqrt{p}}{2}+\mathrm{k_L}\right)^{-1} \\ \qquad +\int_0^\infty \left(q+\sqrt{p}\right)\exp\left[-\left(\frac{\gamma+\sqrt{p}}{2}+\mathrm{k_L}\right)\mathrm{t}\right] \\ \qquad dt\cdot\left(\frac{\gamma+\sqrt{p}}{2}+\mathrm{k_L}\right)^{-1}\end{array}\right\}}{\begin{array}{l}\int_0^\infty \left(-q+\sqrt{p}\right)\exp\left[-\left(\frac{\gamma-\sqrt{p}}{2}+\mathrm{k_L}\right)\mathrm{t}\right]dt+ \\ \int_0^\infty \left(q+\sqrt{p}\right)\exp\left[-\left(\frac{\gamma+\sqrt{p}}{2}+\mathrm{k_L}\right)\mathrm{t}\right]dt\end{array}} \tag{29}$$

When the distribution of the duration between the associations of fluorescent Grb2 is an exponential function with a rate constant of k but varies by a stochastic observation limiting factor (e.g., photobleaching) with a rate constant of k_L, the observed fraction, a, which is the ratio of observed Grb2 associating with EGFR to all fluorescent Grb2, corresponds to

$$a = \int_\Psi^\infty k\exp\left\{-\left(k+k_L\right)\, t\right\}dt.$$

Thus,

$$k+k_L = {}^{k}\!/_{a}. \tag{30}$$

According to Eq. 30, terms in the parentheses of the exponential function can be substituted as

$$\frac{\gamma-\sqrt{p}}{2}+\mathrm{k_L} = \frac{1}{a}\left(\frac{\gamma-\sqrt{p}}{2}\right) \quad and \quad \frac{\gamma+\sqrt{p}}{2}+\mathrm{k_L} = \frac{1}{a}\left(\frac{\gamma+\sqrt{p}}{2}\right). \tag{31}$$

τ_{obs} is calculated using Eqs. 30 and 31 as

$$\tau_{obs} = a\frac{k_2+d}{k_1 d}.$$

Thus,

$$k_{obs} = \frac{d}{a\left(k_2+d\right)}k_1. \tag{32}$$

Noting $k_2 = \frac{\left[Grb2_{white}\right]}{\left[Grb2_{TMR}\right]}k_1$, Eq. 32 is rewritten as follows:

$$k_{obs} = \frac{d}{a\left(d + \frac{[Grb2_{white}]}{[Grb2_{TMR}]}k_1\right)}k_1. \quad (33)$$

From Eq. 33, k_1 is

$$k_1 = \frac{adk_{obs}}{d - ak_{obs}\frac{[Grb2_{white}]}{[Grb2_{TMR}]}} = \frac{1}{k_{obs}^{-1} - ad^{-1}\frac{[Grb2_{white}]}{[Grb2_{TMR}]}}a. \quad (34)$$

In Eq. 34, k_{obs}^{-1} and d^{-1} are the expectations of the duration time between associations and the bindings of Grb2-HaloTag::TMR, respectively, and obtained experimentally. a in Eq. 33 is substituted in Eq. 34. The calculated k_1 is used to acquire the true association rate constant, k, as follows:

$$k = {k_1}/{[Grb2_{TMR}]} = \frac{1}{k_{obs}^{-1} - ad^{-1}\frac{[Grb2_{white}]}{[Grb2_{TMR}]}}\frac{a}{[Grb2_{TMR}]}. \quad (35)$$

This equation can be also derived such that the duration between fluorescent Grb2 associations, k_{obs}^{-1}, is considered to contain the associations of nonfluorescent Grb2 at a frequency determined by the concentration ratio $[Grb2_{white}]/[Grb2_{TMR}]$, the rate-limiting factor, k_L, and the dissociation rate constant, d. A detailed description of the derivation was reported (Hiroshima et al. 2018).

To calculate the association rate constant, the concentrations of endogenous Grb2 and Grb2-HaloTag in the cells should be determined. Western blotting with anti-Grb2 antibody was carried out on cell extract. The band intensities were converted to amounts of proteins by referring to a standard curve between the band intensities and known amounts of purified Grb2 and GST-Grb2. The concentrations were determined by taking the cell volume and transfection efficiency into consideration. Because not all Grb2-HaloTag molecules were labeled with TMR, the concentration of Grb2–HaloTag::TMR was measured in the cytoplasm by fluorescence correlation spectroscopy (FCS). The obtained concentrations of TMR-labeled Grb2 ($[Grb2_{TMR}]$) and non-labeled Grb2 ($[Grb2_{white}]$) were applied to Eq. 33 with the apparent rates of association (k_{obs}) and dissociation (d) observed by single-molecule imaging and the observed fraction (a).

By combining our approach with VB-HMM analysis, we correlated the EGFR mobility and clustering states with the kinetics of Grb2 interactions. For the association rate, the obtained histogram was well fitted with one component exponential function for every combination of three motional and four clustering states regardless of the EGF stimulation. The association rate constants were finally determined by subtracting those of an EGFR mutant lacking the cytoplasmic region as values for nonspecific interactions. The obtained rate constants suggested that EGF facilitated the associations of Grb2 with EGFR, but the receptors in larger clusters showed less frequent associations probably due to steric hindrance (Fig. 7c). The dissociation rates showed two components for EGF-bound immobile clusters and one component otherwise (Fig. 7d). One of the two components corresponding to the shorter association did not change before and after EGF stimulation, denoting an interaction unrelated to the signaling. The other component of a longer residence time (170 ms), which was close to the previously reported duration (400–500 ms) (Morimatsu et al. 2007), appeared after the EGF stimulation. Thus, phosphorylated immobile clusters with high molecular density cause the effective retention of Grb2. The two components described above could not be detected in the association rates by our measurement, therefore, both fractions of the immobile clusters were assumed to have the same association rate constant. The dissociation equilibrium constants were calculated from the association and dissociation rate constants for every EGFR state (Fig. 7e). These values were similar to those reported previously (100–700 nM) (Cussac et al. 1994; Chook 1996; Morimatsu et al. 2007). Furthermore, we distinguished two different equilibrium constants in the immobile EGFR clusters after ligand stimulation, which were yielded from the two dissociation rate constants. One of the affinities was especially high due to not only the lower dissociation rates but also the relatively high association rates of the immobile state, implying that the immobile clusters in particular are concerned with the signal transduction.

The single-molecule tracking and kinetics study revealed that characteristic EGFR mobility and clustering are dynamically coupled, producing a significant platform for the downstream signaling after EGF introduction. EGFR mobility reflects the plasma membrane structures and varies with EGFR phosphorylation, probably through changes in the molecular conformation and electrostatic properties (Maeda et al. 2018). That the multiple affinities between EGFR and Grb2 are dependent on mobility and clustering suggest that the phosphorylation domain of EGFR adopts multiple conformations due to lateral interactions, as shown with single-molecule imaging of fluorescent Grb2 in vitro (Morimatsu et al. 2007). The EGFR behavioral dynamics fulfills a flexible regulation of cell signaling, which is required for preventing the breakdown of homeostasis seen in serious diseases such as carcinogenesis.

Extended Applications of Single-Molecule Analysis

Because single-molecule analysis is an effective method for studying signaling molecule dynamics and kinetics in cells, a certain number of studies (Morimatsu et al. 2007; Hibino et al. 2011; Nakamura et al. 2017; Yoshizawa et al. 2017) employed this analysis to investigate well-known molecules. However, it has been difficult to apply single-molecule analysis at a large scale and comprehensively due to the requisite expertise and low measurement efficiency. To overcome these critical problems, we have developed a novel system that automates all the imaging processes, from focusing on samples to the image acquisition, and utilized artificial intelligence (AI) to search suitable cells/regions for the single-molecule analysis (Yasui et al. 2018). AI greatly simplified the process of the image filter configuration, as researchers have only to present original data to the computer with teaching images, in which suitable regions are binarily painted, instead of determining the number of filter parameters. This apparatus, called "Auto in-cell Single-molecule Imaging System (AiSIS)," achieved single-molecule analysis of 1600 cells in one day, which is several tenfolds higher efficiency than manual operation. The performance of AiSIS enables a large-scale analysis of comprehensive signal transduction pathways. Time courses of the fraction of EGFR mobility and clustering states and the interaction duration of EGFR-Grb2 were measured at many time points (Fig. 8a), suggesting plausible schemes of regulated signaling (Fig. 8b).

Beyond cell biology, AiSIS is capable of a large-scale analysis to the applied sciences. Based on our single-molecule studies introduced in the previous sections, the activation (phosphorylation) of receptors could be detected by molecular behavioral changes becoming slower, confined, or more clustered (Hiroshima et al. 2018; Yanagawa et al. 2018). The obtained parameters (e.g., diffusion coefficient, MSD, and/or cluster size) were suggested to be indices for whether a ligand/agonist or inhibitor/antagonist functions after assessing hundreds of cells in a well plate. We confirmed that ligand-stimulated cells could be completely distinguished from resting cells by referring to these parameters (Fig. 9). Furthermore, the MSD at a time duration of Δt was indicated to change corresponding to the efficacies and concentrations of the ligand and inhibitor, as shown in Fig. 10a for EGF and AG1478, a tyrphostin that blocks the ATP-binding pocket. This property could be applied to obtain the half-maximal effective and inhibitory concentrations (EC_{50} and IC_{50}) by

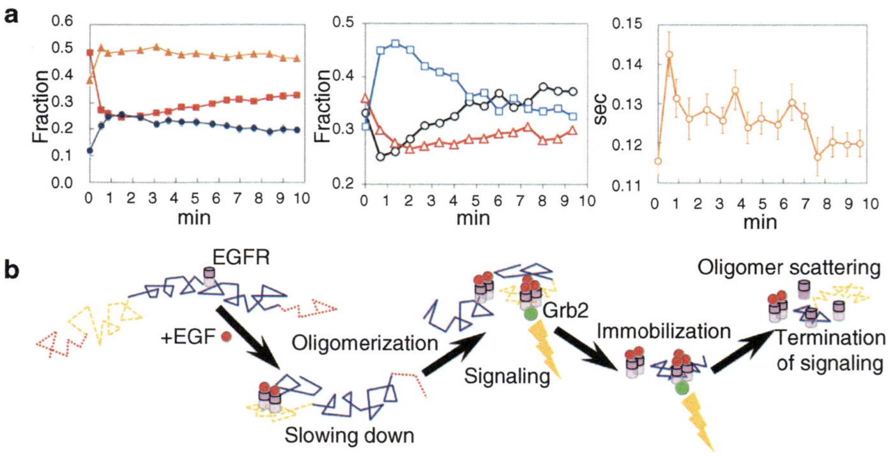

Fig. 8 EGFR signaling analyzed by the automated system. (**a**) Left, changes in the fractions of immobile (circle), slow-mobile (triangle), and fast-mobile (square) states. Middle, time course of the monomer (circle), dimer (triangle), and larger oligomers (square) fractions. Right, resident time of Grb2 on the plasma membrane. (**b**) Proposed scheme of EGF signaling. Mobile EGFR transitions between three motional states (solid, dashed, and dotted lines) to slow down and cluster with EGF, leading to GRb2 translocation and signaling. The signaling is terminated when the immobilized signaling clusters scatter

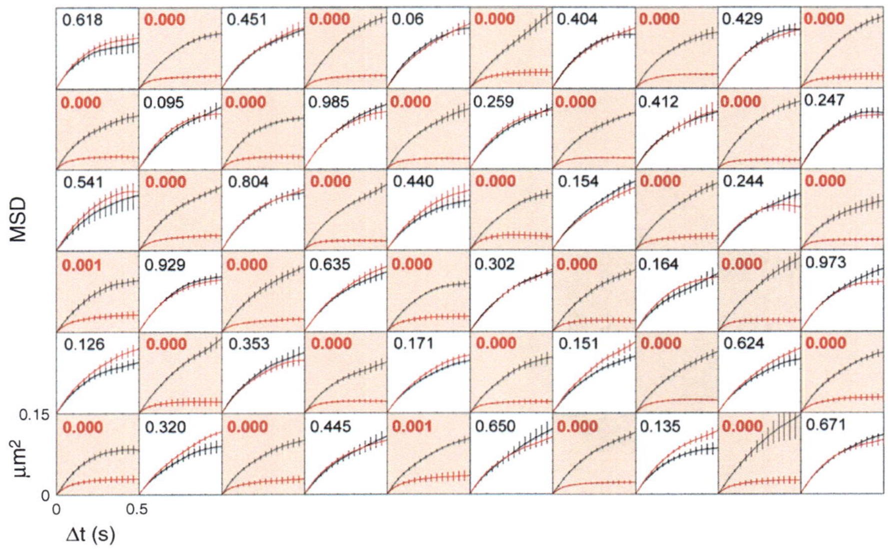

Fig. 9 Recognition of EGF-stimulated cells in a well plate by the behavioral parameters of EGFR. The shaded and white squares indicate wells with EGF and mock solutions, respectively. MSD values became smaller when cells are stimulated with EGF. Significant differences in MSDs ($\Delta t = 66$ ms) between before and after the addition of EGF or mock solution were assessed, and p-values are indicated

fitting the MSD distribution with a reasonable model. Because the binding sites of EGFR for EGF and AG1478 are on opposite sides of the plasma membrane, the noncompetitive inhibition model, described as follows, was employed (Fig. 10b).

$$\mathrm{MSD} = MSD_{\max} - \frac{\left(MSD_{\max} - MSD_{\min}\right)\left[EGF\right]}{\left[EGF\right] + EC_{50} + \dfrac{EC_{50}\left[AG1478\right]}{IC_{50}}} \tag{36}$$

Here, MSD_{max} and MSD_{min} denote the upper and lower converged values of the MSD. The obtained EC_{50} and IC_{50} values are similar to those reported in previous studies, validating the applicability of AiSIS to pharmacological applications such as drug screening.

As shown in this chapter, the single-molecule analysis provides more informative outputs than conventional methods and suggests extensive applications beyond the field of basic science. Our fully automated apparatus, AiSIS, which achieved highly efficient imaging, enabled a large-scale and comprehensive single-molecule analysis, and showed feasibility for novel drug/molecule screening using behavioral changes of the target proteins. However, there still remain significant problems to be solved for generalizing the method, such as unknown mechanisms bridging single-molecule behavior and signal activation. Only partial information contained in the data has been utilized due to the incompleteness of the present analysis method.

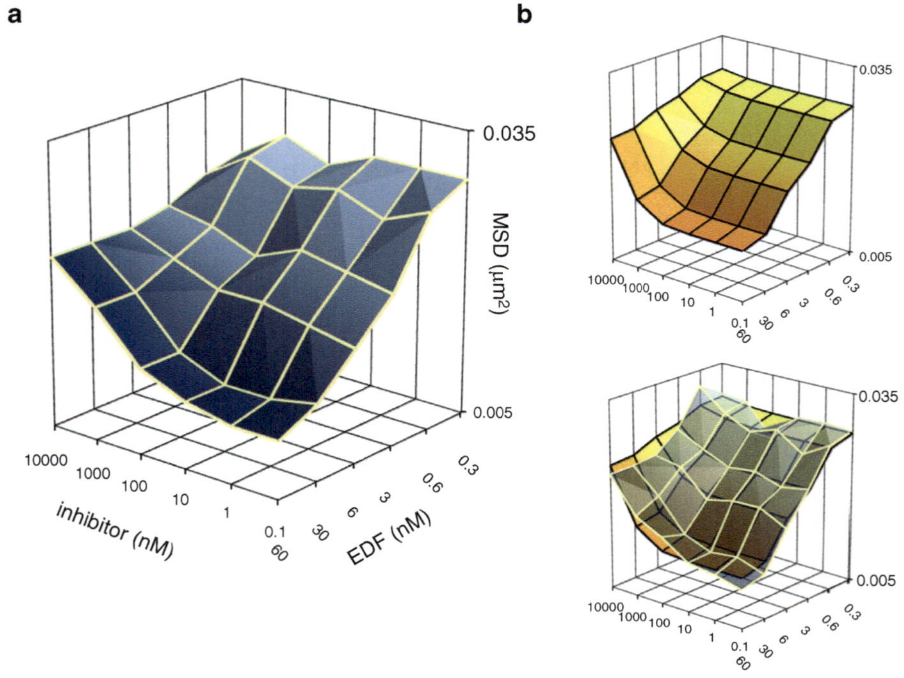

Fig. 10 Pharmacological parameters obtained from single-molecule behavior. (**a**) MSD at $\Delta t = 66$ ms depends on the EGF and inhibitor concentrations. (**b**) Upper, fitted surface to the data in **a** by a noncompetitive inhibition model. Lower, merged graph of the data and the fitted surface

Combination with other microscopies, mathematical modeling, or molecular dynamics simulations, would enhance the potential of single-molecule analysis to obtain deeper insights on molecular mechanisms corresponding to various cellular phenomena.

References

Axelrod D (1989) Total internal reflection fluorescence mimoscopy. Methods Cell Biol 30:245–270

Beal M (2003) Variational algorithms for approximate Bayesian inference. (Doctoral dissertation, UCL (University College London))

Carter and Sorkin (1998) Endocytosis of functional epidermal growth factor receptor-green fluorescent protein chimera. J Biol Chem 273:35000–7

Chook YM (1996) The Grb2-mSos1 complex binds phosphopeptides with higher affinity than Grb2. J Biol Chem 271:30472–30478. https://doi.org/10.1074/jbc.271.48.30472

Chung I, Akita R, Vandlen R et al (2010) Spatial control of EGF receptor activation by reversible dimerization on living cells. Nature 464:783–787. https://doi.org/10.1038/nature08827

Cussac D, Frech M, Chardin P (1994) Binding of the Grb2 SH2 domain to phosphotyrosine motifs does not change the affinity of its SH3 domains for Sos proline-rich motifs. EMBO J 13:4011–4021

Funatsu T, Harada Y, Tokunaga M et al (1995) Imaging of single fluorescent molecules and individual ATP turnovers by single myosin molecules in aqueous solution. Nature 374:555–559

Garrett TPJ, McKern NM, Lou M et al (2002) Crystal structure of a truncated epidermal growth factor receptor extracellular domain bound to transforming growth factor α. Cell 110:763–773. https://doi.org/10.1016/S0092-8674(02)00940-6

Hibino K, Watanabe TM, Kozuka J et al (2003) Single- and multiple-molecule dynamics of the signaling from H-Ras to cRaf-1 visualized on the plasma membrane of living cells. ChemPhysChem 4:748–753. https://doi.org/10.1002/cphc.200300731

Hibino K, Shibata T, Yanagida T, Sako Y (2011) Activation kinetics of RAF protein in the ternary complex of RAF, RAS-GTP, and kinase on the plasma membrane of living cells: single-molecule imaging analysis. J Biol Chem 286:36460–36468. https://doi.org/10.1074/jbc.M111.262675

Hiroshima M, Saeki Y, Okada-Hatakeyama M, Sako Y (2012) Dynamically varying interactions between heregulin and ErbB proteins detected by single-molecule analysis in living cells. Proc Natl Acad Sci U S A 109:13984–13989. https://doi.org/10.1073/pnas.1200464109

Hiroshima M, Pack C, Kaizu K et al (2018) Transient acceleration of epidermal growth factor receptor dynamics produces higher-order signaling clusters. J Mol Biol 430:1386–1401. https://doi.org/10.1016/j.jmb.2018.02.018

Jacobson et al (2007) Lipid rafts: at a crossroad between cell biology and physics. Nat Cell Biol 9:7–14

Lingwood and Simons (2010) Lipid rafts as a membrane-organizing principle. Science 327:46–50

Lommerse PHM, Vastenhoud K, Pirinen NJ et al (2006) Single-molecule diffusion reveals similar mobility for the Lck, H-ras, and K-ras membrane anchors. Biophys J 91:1090–1097. https://doi.org/10.1529/biophysj.105.079053

Low-Nam ST, Lidke KA, Cutler PJ et al (2011) ErbB1 dimerization is promoted by domain co-confinement and stabilized by ligand binding. Nat Struct Mol Biol 18:1244–1249. https://doi.org/10.1038/nsmb.2135

Lowenstein et al (1992) The SH2 and SH3 domain-containing protein GRB2 links receptor tyrosine kinases to ras signaling. Cell 70:431–42

Maeda R, Sato T, Okamoto K et al (2018) Lipid-protein interplay in dimerization of the juxtamembrane domains of epidermal growth factor recepto. Biophys J 114:893–903

Morimatsu M, Takagi H, Ota KG et al (2007) Multiple-state reactions between the epidermal growth factor receptor and Grb2 as observed by using single-molecule analysis. Proc Natl Acad Sci U S A 104:18013–18018. https://doi.org/10.1073/pnas.0701330104

Nakamura Y, Umeki N, Abe M, Sako Y (2017) Mutation-specific mechanisms of Hyperactivation of Noonan syndrome SOS molecules detected with single-molecule imaging in living cells. Sci Rep 7:1–10. https://doi.org/10.1038/s41598-017-14190-6

Ogiso H, Ishitani R, Nureki O et al (2002) Crystal structure of the complex of human epidermal growth factor and receptor extracellular domains. Cell 110:775–787. https://doi.org/10.1016/S0092-8674(02)00963-7

Okamoto K, Sako Y (2012) Variational Bayes analysis of a photon-based hidden Markov model for single-molecule FRET trajectories. Biophys J 103:1315–1324. https://doi.org/10.1016/j.bpj.2012.07.047

Rabiner LR (1989) A tutorial on hidden Markov models and selected applications in speech recognition. Proc IEEE 77:257–286. https://doi.org/10.1109/5.18626

Sako Y, Hibino K, Miyauchi T et al (2000a) Single-molecule imaging of signaling molecules in living cells. Single Mol 1:159–163. https://doi.org/10.1002/1438-5171(200006)1:2<159::AID-SIMO159>3.0.CO;2-4

Sako Y, Minoghchi S, Yanagida T (2000b) Single-molecule imaging of EGFR signalling on the surface of living cells. Nat Cell Biol 2:168–172. https://doi.org/10.1038/35004044

Schibany S, Kleine Borgmann LAK, Rösch TC et al (2018) Single molecule tracking reveals that the bacterial SMC complex moves slowly relative to the diffusion of the chromosome. Nucleic Acids Res 46:7805–7819. https://doi.org/10.1093/nar/gky581

Teramura Y, Ichinose J, Takagi H et al (2006) Single-molecule analysis of epidermal growth factor binding on the surface of living cells. EMBO J 25:4215–4222. https://doi.org/10.1038/sj.emboj.7601308

Tokunaga M, Imamoto N, Sakata-Sogawa K (2008) Highly inclined thin illumination enables clear single-molecule imaging in cells. Nat Methods 5:159–161. https://doi.org/10.1038/nmeth1171

Xiao Z, Ma X, Jiang Y et al (2008) Single-molecule study of lateral mobility of epidermal growth factor receptor 2/HER2 on activation. J Phys Chem B 112:4140–4145. https://doi.org/10.1021/jp710302j

Yanagawa M, Hiroshima M, Togashi Y et al (2018) Single-molecule diffusion-based estimation of ligand effects on G protein – coupled receptors. Sci Signal 11:548. https://doi.org/10.1126/scisignal.aao1917

Yasui M, Hiroshima M, Kozuka J et al (2018) Automated single-molecule imaging in living cells. Nat Commun 9:3061. https://doi.org/10.1038/s41467-018-05524-7

Yoshizawa R, Umeki N, Yanagawa M et al (2017) Single-molecule fluorescence imaging of RalGDS on cell surfaces during signal transduction from Ras to Ral. Biophys Physicobiology 14:75–84. https://doi.org/10.2142/biophysico.14.0_75

Zhang W, Jiang Y, Wang Q et al (2009) Single-molecule imaging reveals transforming growth factor-beta-induced type II receptor dimerization. Proc Natl Acad Sci U S A 106:15679–15683. https://doi.org/10.1073/pnas.0908279106

Imaging Sphingomyelin- and Cholesterol-Enriched Domains in the Plasma Membrane Using a Novel Probe and Super-Resolution Microscopy

Mitsuhiro Abe and Toshihide Kobayashi

Abbreviations

Chol	cholesterol
ELISA	enzyme-linked immunosorbent assay
HA	hemagglutinin
HILO	highly inclined laminated optical sheet
MDCK	*Madin–Darby canine kidney*
NA	neuraminidase
PALM	photoactivated localization microscopy
PH domain	pleckstrin homology domain
PIP_2	phosphatidylinositol 4,5-bisphosphate
PIP_3	phosphatidylinositol 3,4,5-trisphosphate
PLC	phospholipase C
PS	phosphatidylserine
SIM	structured illumination microscopy
SM	sphingomyelin

M. Abe (✉)
Cellular Informatics Laboratory, RIKEN, Wako, Saitama, Japan
e-mail: abemitsu@riken.jp

T. Kobayashi
Cellular Informatics Laboratory, RIKEN, Wako, Saitama, Japan

UMR 7021 CNRS, Faculté de Pharmacie, Université de Strasbourg, Illkirch, France
e-mail: toshihide.kobayashi@unistra.fr

J. K. Kim et al. (eds.), *Advanced Imaging and Bio Techniques for Convergence Science*, Advances in Experimental Medicine and Biology,
https://doi.org/10.1007/978-981-33-6064-8_4

Introduction

Lipid rafts are defined as sphingolipid- and sterol-enriched domains (Pike 2006) that form rigid structures in the plasma membrane. The proposed size of the raft is on the order of 10–200 nm (Jacobson et al. 2007; Pike 2006). Lipid rafts are thought to play pivotal roles in membrane trafficking, signal transduction, and the entry of infectious pathogens into cells (Lingwood and Simons 2010; Simons and Toomre 2000). However, the size, function, molecular organization, and dynamics of lipid rafts are not precisely understood for two main reasons.

First, the proposed size of the raft is below the diffraction limit of light (Mizuno et al. 2011). Therefore, conventional fluorescence microscopy is unable to clarify their structure in detail. However, super-resolution optical microscopy systems have recently been developed and provide a resolution beyond the diffraction limit (Toomre and Bewersdorf 2010). These include photoactivated localization microscopy (PALM) (Betzig et al. 2006; Hess et al. 2006), stochastic optical reconstruction microscopy (STORM) (Rust et al. 2006), direct STORM (*d*STORM) (Heilemann et al. 2008; Van De Linde et al. 2011), stimulated emission depletion microscopy (STED) (Willig et al. 2006), structured illumination microscopy (SIM) (Heilemann et al. 2009, 2008), ground-state depletion microscopy followed by individual molecule return (GSDIM) (Folling et al. 2008), blink microscopy (Steinhauer et al. 2008; Vogelsang et al. 2009), and points accumulation for imaging in nanoscale topography (PAINT) (Sharonov and Hochstrasser 2006). Various protein clusters have been revealed with super-resolution microscopy (Hess et al. 2007; Lillemeier et al. 2010; Owen et al. 2010; Schermelleh et al. 2019), and this technique should provide an effective tool for studying lipid rafts in the plasma membrane.

Second, there are no specific probes for lipid rafts (Kraft and Klitzing 2014; Munro 2003). Considerable effort has been made to visualize the lipid components of the rafts in mammalian cells using lipid probes, and both specific lipid-binding proteins (Table 1) and lipid analogs have been widely used as these probes. Cholesterol (Chol) is virtually the exclusive sterol of mammalian cells, which is typically present at 30–40 mol % of plasma membrane lipids. Using the C-terminal domain of perfringolysin O (also referred to as "θ toxin"), which binds the Chol-rich domains in membranes without cytotoxicity (Heuck et al. 2007; Shimada et al. 2002; Verherstraeten et al. 2015), the Chol-rich domains were localized in both the outer and inner leaflets of the mammalian plasma membrane (Abe et al. 2012; Wilhelm et al. 2019). Boron-dipyrromethene (BODIPY)- or polyethylene glycol (PEG)-labeled Chol was developed to monitor the molecular dynamics of sterols in living cells (Hullin-Matsuda et al. 2009; Ikonen and Blom 2016). Sphingomyelin (SM) is one of the major sphingolipids in mammalian cells, which accounts for 10–20 mol% of plasma membrane lipids. To visualize SM in the outer leaflet of the plasma membrane, proteins that specifically bind SM, lysenin, and equinatoxin II were used for the biological analysis of cells (Abe et al. 2012; Bakrac et al. 2010; Kiyokawa et al. 2005). Because lysenin and equinatoxin II preferentially interact with clustered SM-rich domains and dispersed SM-rich domains, respectively, different types of SM domains could be distinguished in the plasma membrane (Makino et al. 2015). The term "clustered SM," used here, indicates small

Table 1 Nontoxic and protein-based probes for observing lipids at the plasma membrane

Target lipid	Binding protein
Chol	C-terminal domain of perfringolysin O (θ toxin)[b]
Clustered SM[a]	Fragment 161–297 amino acid of lysenin[b]
Dispersed SM	V8C K69C mutant of equinatoxin II[b]
GM1	B subunit of cholera toxin
PS	C2 domain of lactadherin, PH domain of evectin-2
PIP_2	PH domain of PLC-δ[b]
PIP_3	PH domain of GRP1, PH domain of Akt
SM/Chol complex	Nakanori[b]

aSmall aggregates of fewer than 10 SM molecules
bPlasmid can be obtained from RIKEN BRC (http://dna.brc.riken.jp/)

aggregates of fewer than 10 SM molecules. ATTO647N-labeled SM was used for the analysis of single molecules in the plasma membrane of living cells (Eggeling et al. 2009). To analyze the ultrastructure of the ganglioside GM1, another sphingolipid in mammalian cells, cholera toxin B, was used for immunoelectron microscopy with sodium dodecyl sulfate (SDS)-treated freeze–fracture replica labeling (SDS-FRL) (Fujita et al. 2007). The distribution of BODIPY–GM1 in the membrane monolayers was examined with a combination of atomic force microscopy and near-field scanning optical microscopy (Coban et al. 2007). The precursors of sphingolipids were also labeled isotopically, and the distribution of sphingolipids in the plasma membrane was mapped with a secondary ion mass spectrometer (Frisz et al. 2013). However, none of these methods detected the complexes of sphingolipid and sterol. Because lipid rafts are defined as complexes of sphingolipid- and sterol-enriched domains (Pike 2006), a protein that specifically binds to complexes of sphingolipid and sterol would be a valuable marker of lipid rafts.

Recently, we identified novel proteins that bind to mixtures of SM and Chol (Bhat et al. 2013; Makino et al. 2017). In extracts from edible mushrooms, we screened for proteins that specifically bind to SM/Chol liposomes. Among them, nakanori was isolated from *Grifola frondosa* and showed strong affinity and specificity for SM/Chol liposomes. Moreover, because nakanori is not cytotoxic to mammalian cells, this protein should be a reliable marker of SM/Chol domains in living cells. Here, we describe a method of visualizing SM- and Chol-rich membrane domains with a combination of nakanori and super-resolution microscopy.

A Novel Protein, Nakanori Strongly Binds SM/Chol Mixtures

In the proteins extracted from the edible mushroom *G. frondosa* (Fig. 1a), we screened for those that specifically bound to liposomes containing equimolar SM

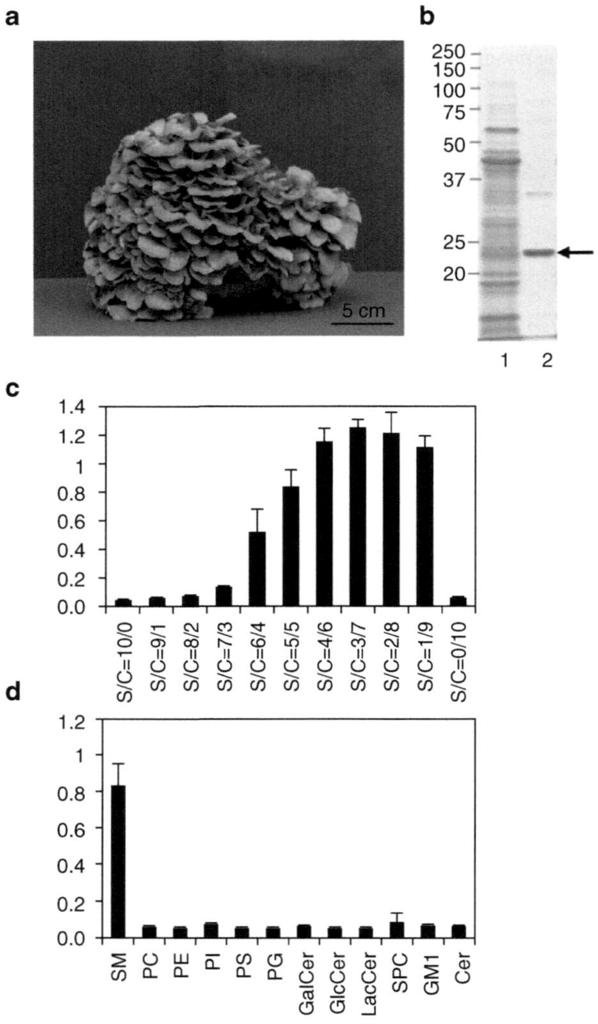

Fig. 1 Isolation and characterization of SM/Chol-binding protein, nakanori, from mushroom *Grifola frondosa*. (**a**) Mushroom *Grifola frondosa*. Scale bar = 5 cm. (**b**) Screening of SM/Chol-binding protein from *G. frondosa*. The supernatant (lane 1) and pellet (lane 2) fractions after incubation of 11,000 rpm supernatant of *G. frondosa* extract with SM/Chol liposomes. Arrow indicates the position of nakanori. (**c**) Binding of purified nakanori to liposomes composed of different SM(S)/Chol(C) ratios by ELISA. (**d**) Binding of nakanori to indicated lipids in the presence of equimolar Chol by ELISA. Data are mean of triplicate experiments ± SD. *SM* porcine brain sphingomyelin, *PC* egg phosphatidylcholine, *PE* 1-palmitoyl-2-oleoyl-*sn*-glycero-3-phosphoethanolamine, *PS* porcine brain phosphatidylserine, *PI* bovine liver phosphatidylinositol, *PG* 1,2-dioleoyl-*sn*-glycero-3-phospho-(1′-*rac*-glycerol), *GalCer* porcine brain cerebroside, *GlcCer* human glucosylceramide, *LacCer* lactosylceramide, *SPC* sphingosylphosphorylcholine, *GM1* bovine brain ganglioside GM1, *Cer* porcine brain ceramide. Adapted from reference (Makino et al. 2017)

and Chol. After incubation with SM/Chol liposomes, several proteins were identified as SM/Chol-binding proteins. A mass spectrometric analysis and molecular cloning determined the sequence of the major protein (arrow in Fig. 1b), showing that it contains 202 amino acid residues, with no obvious consensus motifs. We designated this protein "nakanori," derived from the Japanese for "mid-raft rider." After the recombinant nakanori protein was purified from *Escherichia coli*, its binding specificity for several lipids was examined with enzyme-linked immunosorbent assays (ELISAs) (Fig. 1c and d). Nakanori bound neither SM alone nor Chol alone but significantly bound a mixture of SM and Chol (Fig. 1c). The ELISA also revealed that nakanori interacts with the SM/Chol complex but not with the other lipid/Chol mixtures (Fig. 1d), indicating its high specificity for SM/Chol mixtures. The affinity (K_d) of nakanori for equimolar SM/Chol membranes was determined. Measurement with a quartz crystal microbalance revealed a K_d value of 141 ± 4 nM (Makino et al. 2017).

SIM Analysis Suggesting That Influenza Virus Buds from the Rim of the SM/Chol Domain

It has been proposed that the influenza virus buds from cells at sites around the lipid rafts, where the viral proteins, hemagglutinin (HA) and neuraminidase (NA) assemble (Gerl et al. 2012; Scheiffele et al. 1999). To observe the budding of the influenza virus from the plasma membranes of mammalian cells, *Madin–Darby canine kidney* (MDCK) cells were infected with the virus for 24 h in the presence of a NA inhibitor. Nakanori was fused with the fluorescent protein mKate, and the fusion protein (mKate–nakanori) was expressed in *E. coli* and purified (Makino et al. 2017). The specificity of mKate–nakanori for SM was confirmed with a biochemical assay. After mKate–nakanori was bound to the SM- and Chol-rich complexes in living MDCK cells, the protein interacting with SM/Chol was chemically fixed. The cells were then stained with an anti-influenza A H1N1 antibody, and SM/Chol and the influenza virus were observed at the plasma membrane with confocal microscopy (Fig. 2a). Although distributions of SM/Chol and the influenza virus were detected in the plasma membrane, their detailed structures and co-localization were far from obvious because of the diffraction limit of light. To clarify them, we observed the patterns with super-resolution microscopy. Structured illumination microscopy (SIM) images indicated that the virus budded from the rim of the SM/Chol domain (Fig. 2b). Here, we use the term "SM/Chol domain" to indicate the specific area of the membrane where the labeling density of nakanori is high. Several lines of evidence suggest that the influenza viral proteins, HA and NA, associate with lipid rafts, whereas the viral M2 protein does not (Blom et al. 2001; Gerl et al. 2012; Zhang et al. 2000). Taken together, these data imply that associations of raft- and non-raft proteins form at the rim of the SM/Chol domain, followed by the budding and fission of the influenza virus.

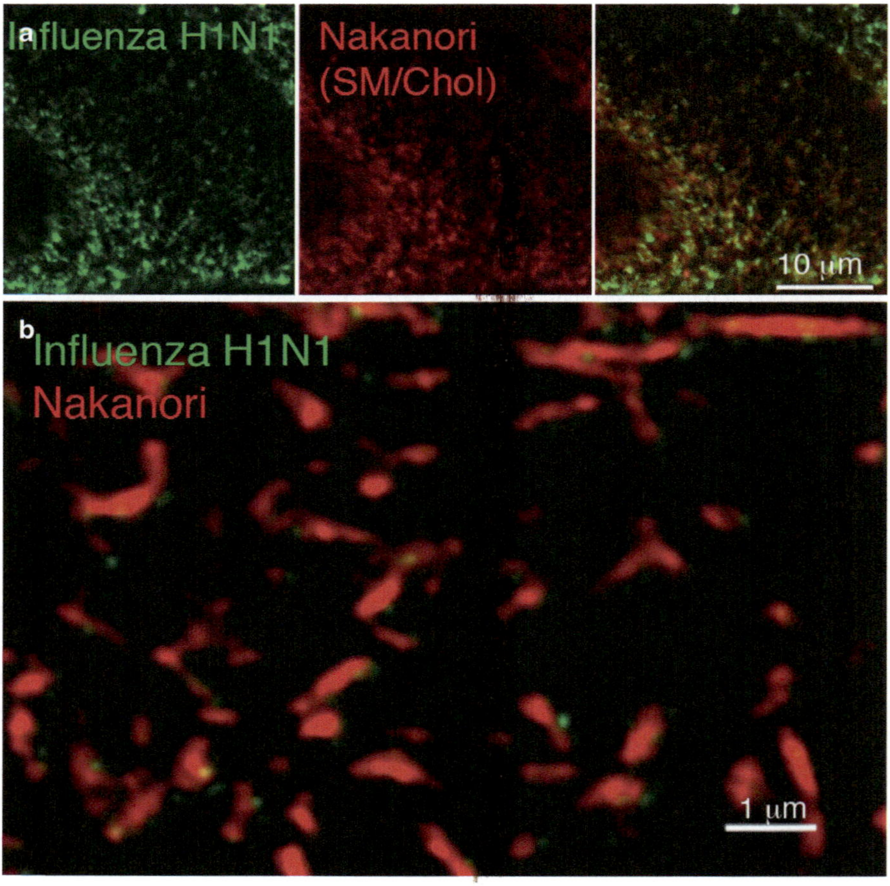

Fig. 2 Images of SM/Chol domains and influenza virus in the apical plasma membrane in MDCK cells. MDCK cells were incubated for 24 h in the presence of NA inhibitor after infection with influenza virus. Cells were then labeled with mKate–nakanori (red) and anti-influenza A H1N1 antibody (MAB8256; EMD–Millipore) (green). (**a**) Confocal images. Scale bar = 10 μm. (**b**) SIM images. Images were obtained at the apical face of the plasma membrane. Scale bar = 1 μm. Adapted from reference (Makino et al. 2017)

PALM Analysis Indicating the Transbilayer Co-localization Between SM/Chol Domains and PIP_2 Domains

We tried to observe the SM/Chol domains in the apical plasma membrane using another super-resolution microscopy, PALM. For this purpose, nakanori was fused with the photoswitchable fluorescent protein PAmCherry1 (Subach et al. 2009), and the recombinant protein (PAmCherry–nakanori) was purified from *E. coli* (Makino et al. 2017). Because total internal reflection fluorescence (TIRF) illumination restricts the excitation region to a few hundred nanometers above the glass, it is unsuitable for observing structures in the apical membrane. Therefore, we used instead highly inclined laminated optical sheet (HILO) illumination (Tokunaga et al. 2008), which excites a higher region above the glass with low background. The SM- and Chol-rich domains stained with PAmCherry–nakanori were observed in the apical membrane with a PALM analysis with HILO illumination (Fig. 3). The SM/Chol in the outer leaflet was visualized with PALM after it was labeled with exogenous PAmCherry–nakanori (Fig. 3, red). Small domains of SM/Chol were observed at the apical plasma membrane.

Several analyses have suggested that phosphatidylinositol 4,5-bisphosphate (PIP_2) associates with the lipid raft fraction of membranes (Brown and London 1998; Hope and Pike 1996; Laux et al. 2000). PIP_2 occurs in the inner leaflet in mammalian cells, whereas the SM/Chol complex is enriched in the outer leaflet of the plasma membrane (Fujita et al. 2009; Murate et al. 2015). To address the possibility that PIP_2 forms domains in the inner leaflet just below the SM/Chol domains, we examined the localization of SM/Chol and PIP_2 simultaneously (Fig. 3) (Makino et al. 2017). To observe PIP_2 in the inner leaflet with PALM, pleckstrin homology (PH) domain of phospholipase C δ (PLC-δ) was fused with the photoswitchable fluorescent protein Dronpa (Ando et al. 2004), and the fusion protein (Dronpa–PLC-δPH) was expressed in LLC-PK1 cells (Fig. 3, green). Small colocalized domains of PAmCherry–nakanori and Dronpa–PLC-δPH were observed at the apical plasma membrane (Fig. 3). These results indicate that PIP_2 forms domains in the inner leaflet, just below the SM/Chol domains in the outer leaflet.

There are several possible explanations of why the PIP_2 domains form just beneath the SM/Chol domains. First, SM/Chol and PIP_2 may interact, directly or

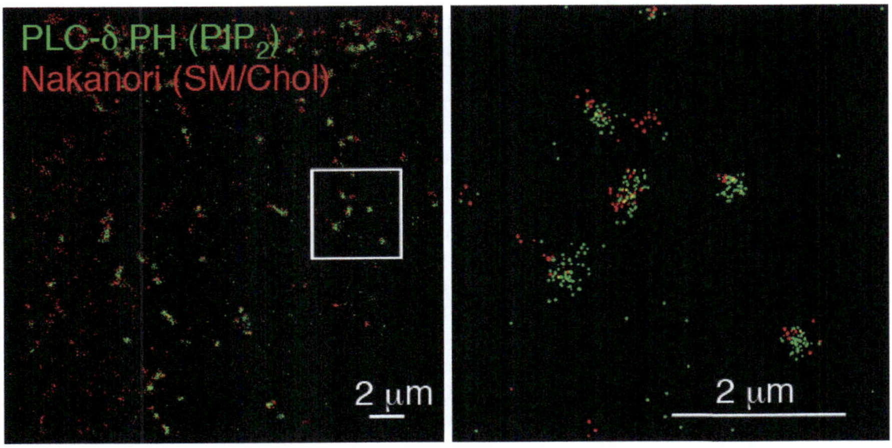

Fig. 3 Transbilayer co-localization between the SM-rich domains on the outer leaflet and PIP_2-rich domains on the inner leaflet of the plasma membrane. LLC-PK1 cells expressing Dronpa–PLC-δPH (green) were stained with PAmCherry–nakanori (red). Bars, 2 μm. Adapted from reference (Makino et al. 2017)

indirectly. Indeed, it has been shown that long-acyl-chain lipids in both leaflets stabilize cholesterol-dependent transbilayer interactions (Raghupathy et al. 2015). Second, the production of PIP_2 may be restricted to just below the SM/Chol domains. We have shown that the kinase required for the production of PIP_2 is enriched in the inner leaflet, on the opposite side to the clustered SM domains (Abe et al. 2012). Therefore, if this is the case, why the mobility of PIP_2 is restricted to around the SM/Chol domains must be explained. Further experiments are required to understand the molecular mechanisms underlying the transbilayer co-localization between SM/Chol domains and PIP_2 domains.

Acknowledgments I thank Toshiyuki Watanabe and Carl Zeiss Microscopy Co. Ltd. Japan for SIM images, and Hideaki Mizuno for PALM images. We are grateful to Asami Makino and Takehiko Inaba for biochemical analysis of nakanori. The original work was supported by the Lipid Dynamics Program of RIKEN, and the Grant-in-Aid from the Ministry of Education, Culture, Sports, Science, and Technology of Japan (22390018, 25293015) and Naito Foundation.

References

Abe M, Makino A, Hullin-Matsuda F et al (2012) A role for sphingomyelin-rich lipid domains in the accumulation of phosphatidylinositol-4,5-bisphosphate to the cleavage furrow during cytokinesis. Mol Cell Biol 32:1396–1407

Ando R, Mizuno H, Miyawaki A (2004) Regulated fast nucleocytoplasmic shuttling observed by reversible protein highlighting. Science 306:1370–1373

Bakrac B, Kladnik A, Macek P et al (2010) A toxin-based probe reveals cytoplasmic exposure of Golgi sphingomyelin. J Biol Chem 285:22186–22195

Betzig E, Patterson GH, Sougrat R et al (2006) Imaging intracellular fluorescent proteins at nanometer resolution. Science 313:1642–1645

Bhat HB, Kishimoto T, Abe M et al (2013) Binding of a pleurotolysin ortholog from Pleurotus eryngii to sphingomyelin and cholesterol-rich membrane domains. J Lipid Res 54:2933–2943

Blom TS, Koivusalo M, Kuismanen E et al (2001) Mass spectrometric analysis reveals an increase in plasma membrane polyunsaturated phospholipid species upon cellular cholesterol loading. Biochemistry 40:14635–14644

Brown DA, London E (1998) Functions of lipid rafts in biological membranes. Annu Rev Cell Dev Biol 14:111–136

Coban O, Burger M, Laliberte M et al (2007) Ganglioside partitioning and aggregation in phase-separated monolayers characterized by bodipy GM1 monomer/dimer emission. Langmuir 23:6704–6711

Eggeling C, Ringemann C, Medda R et al (2009) Direct observation of the nanoscale dynamics of membrane lipids in a living cell. Nature 457:1159–1162

Folling J, Bossi M, Bock H et al (2008) Fluorescence nanoscopy by ground-state depletion and single-molecule return. Nat Methods 5:943–945

Frisz JF, Lou K, Klitzing HA et al (2013) Direct chemical evidence for sphingolipid domains in the plasma membranes of fibroblasts. Proc Natl Acad Sci USA 110:E613–E622

Fujita A, Cheng J, Hirakawa M et al (2007) Gangliosides GM1 and GM3 in the living cell membrane form clusters susceptible to cholesterol depletion and chilling. Mol Biol Cell 18:2112–2122

Fujita A, Cheng J, Tauchi-Sato K et al (2009) A distinct pool of phosphatidylinositol 4,5-bisphosphate in caveolae revealed by a nanoscale labeling technique. Proc Natl Acad Sci USA 106:9256–9261

Gerl MJ, Sampaio JL, Urban S et al (2012) Quantitative analysis of the lipidomes of the influenza virus envelope and MDCK cell apical membrane. J Cell Biol 196:213–221
Heilemann M, Van De Linde S, Schuttpelz M et al (2008) Subdiffraction-resolution fluorescence imaging with conventional fluorescent probes. Angew Chem Int Ed Engl 47:6172–6176
Heilemann M, Van De Linde S, Mukherjee A et al (2009) Super-resolution imaging with small organic fluorophores. Angew Chem Int Ed Engl 48:6903–6908
Hess ST, Girirajan TP, Mason MD (2006) Ultra-high resolution imaging by fluorescence photoactivation localization microscopy. Biophys J 91:4258–4272
Hess ST, Gould TJ, Gudheti MV et al (2007) Dynamic clustered distribution of hemagglutinin resolved at 40 nm in living cell membranes discriminates between raft theories. Proc Natl Acad Sci USA 104:17370–17375
Heuck AP, Savva CG, Holzenburg A et al (2007) Conformational changes that effect oligomerization and initiate pore formation are triggered throughout perfringolysin O upon binding to cholesterol. J Biol Chem 282:22629–22637
Hope HR, Pike LJ (1996) Phosphoinositides and phosphoinositide-utilizing enzymes in detergent-insoluble lipid domains. Mol Biol Cell 7:843–851
Hullin-Matsuda F, Ishitsuka R, Takahashi M et al (2009) Imaging lipid membrane domains with lipid-specific probes. Methods Mol Biol 580:203–220
Ikonen E, Blom T (2016) Lipoprotein-mediated delivery of BODIPY-labeled sterol and sphingolipid analogs reveals lipid transport mechanisms in mammalian cells. Chem Phys Lipids 194:29–36
Jacobson K, Mouritsen OG, Anderson RG (2007) Lipid rafts: at a crossroad between cell biology and physics. Nat Cell Biol 9:7–14
Kiyokawa E, Baba T, Otsuka N et al (2005) Spatial and functional heterogeneity of sphingolipid-rich membrane domains. J Biol Chem 280:24072–24084
Kraft ML, Klitzing HA (2014) Imaging lipids with secondary ion mass spectrometry. Biochim Biophys Acta 1841:1108–1119
Laux T, Fukami K, Thelen M et al (2000) GAP43, MARCKS, and CAP23 modulate PI(4,5)P(2) at plasmalemmal rafts, and regulate cell cortex actin dynamics through a common mechanism. J Cell Biol 149:1455–1472
Lillemeier BF, Mortelmaier MA, Forstner MB et al (2010) TCR and Lat are expressed on separate protein islands on T cell membranes and concatenate during activation. Nat Immunol 11:90–96
Lingwood D, Simons K (2010) Lipid rafts as a membrane-organizing principle. Science 327:46–50
Makino A, Abe M, Murate M et al (2015) Visualization of the heterogeneous membrane distribution of sphingomyelin associated with cytokinesis, cell polarity, and sphingolipidosis. FASEB J 29:477–493
Makino A, Abe M, Ishitsuka R et al (2017) A novel sphingomyelin/cholesterol domain-specific probe reveals the dynamics of the membrane domains during virus release and in Niemann-Pick type C. FASEB J 31:1301–1322
Mizuno H, Abe M, Dedecker P et al (2011) Fluorescent probes for superresolution imaging of lipid domains on the plasma membrane. Chem Sci 2:1548–1553
Munro S (2003) Lipid rafts: elusive or illusive? Cell 115:377–388
Murate M, Abe M, Kasahara K et al (2015) Transbilayer distribution of lipids at nano scale. J Cell Sci 128:1627–1638
Owen DM, Rentero C, Rossy J et al (2010) PALM imaging and cluster analysis of protein heterogeneity at the cell surface. J Biophotonics 3:446–454
Pike LJ (2006) Rafts defined: a report on the keystone symposium on lipid rafts and cell function. J Lipid Res 47:1597–1598
Raghupathy R, Anilkumar AA, Polley A et al (2015) Transbilayer lipid interactions mediate nanoclustering of lipid-anchored proteins. Cell 161:581–594
Rust MJ, Bates M, Zhuang X (2006) Sub-diffraction-limit imaging by stochastic optical reconstruction microscopy (STORM). Nat Methods 3:793–795
Scheiffele P, Rietveld A, Wilk T et al (1999) Influenza viruses select ordered lipid domains during budding from the plasma membrane. J Biol Chem 274:2038–2044

Schermelleh L, Ferrand A, Huser T et al (2019) Super-resolution microscopy demystified. Nat Cell Biol 21:72–84

Sharonov A, Hochstrasser RM (2006) Wide-field subdiffraction imaging by accumulated binding of diffusing probes. Proc Natl Acad Sci USA 103:18911–18916

Shimada Y, Maruya M, Iwashita S et al (2002) The C-terminal domain of perfringolysin O is an essential cholesterol-binding unit targeting to cholesterol-rich microdomains. Eur J Biochem 269:6195–6203

Simons K, Toomre D (2000) Lipid rafts and signal transduction. Nat Rev Mol Cell Biol 1:31–39

Steinhauer C, Forthmann C, Vogelsang J et al (2008) Superresolution microscopy on the basis of engineered dark states. J Am Chem Soc 130:16840–16841

Subach FV, Patterson GH, Manley S et al (2009) Photoactivatable mCherry for high-resolution two-color fluorescence microscopy. Nat Methods 6:153–159

Tokunaga M, Imamoto N, Sakata-Sogawa K (2008) Highly inclined thin illumination enables clear single-molecule imaging in cells. Nat Methods 5:159–161

Toomre D, Bewersdorf J (2010) A new wave of cellular imaging. Annu Rev Cell Dev Biol 26:285–314

Van De Linde S, Loschberger A, Klein T et al (2011) Direct stochastic optical reconstruction microscopy with standard fluorescent probes. Nat Protoc 6:991–1009

Verherstraeten S, Goossens E, Valgaeren B et al (2015) Perfringolysin O: the underrated *Clostridium perfringens* toxin? Toxins (Basel) 7:1702–1721

Vogelsang J, Cordes T, Forthmann C et al (2009) Controlling the fluorescence of ordinary oxazine dyes for single-molecule switching and superresolution microscopy. Proc Natl Acad Sci USA 106:8107–8112

Wilhelm LP, Voilquin L, Kobayashi T et al (2019) Intracellular and plasma membrane cholesterol labeling and quantification using filipin and GFP-D4. Methods Mol Biol 1949:137–152

Willig KI, Kellner RR, Medda R et al (2006) Nanoscale resolution in GFP-based microscopy. Nat Methods 3:721–723

Zhang J, Pekosz A, Lamb RA (2000) Influenza virus assembly and lipid raft microdomains: a role for the cytoplasmic tails of the spike glycoproteins. J Virol 74:4634–4644

Clinically Compatible Fluorescence Microscopy Based on Moxifloxacin Antibiotic

Seunghoon Lee and Ki Hean Kim

Introduction

High-resolution optical microscopy is an indispensable tool in biology. Optical microscopy has been around since 1600s and has helped the advance of biological science. Recently, optical microscopy has made revolutionary steps with the advance of other technologies such as light sources and molecular probes and with the need of live-cell imaging. On the one hand, live-cell imaging at super resolutions and deep inside of intact tissues expanded our knowledge about the universe of life. On the other hand, optical microscopy techniques for medical applications have made relatively slow progress compared to the ones for biological studies due to safety restriction and inherent limitation of microscopy such as a small imaging field of view (FOV). Recently, high-speed, high-contrast optical microscopy techniques have been developed for the rapid and precise diagnosis and the real-time guidance of precision surgery. Optical microscopy techniques for medical applications are typically label-free methods for noninvasive imaging by using intrinsic optical contrasts such as light reflection, autofluorescence (AF), Raman scattering, and others. Optical imaging techniques based on light reflection are reflectance confocal microscopy (RCM) and optical coherence tomography (OCT). RCM is a

S. Lee
Department of Mechanical Engineering, Pohang University of Science and Technology, Pohang, Gyeongbuk, Republic of Korea
e-mail: laserman@postech.ac.kr

K. H. Kim (✉)
Department of Mechanical Engineering, Pohang University of Science and Technology, Pohang, Gyeongbuk, Republic of Korea

Division of Integrative Biosciences and Biotechnology, Pohang University of Science and Technology, Pohang, Gyeongbuk, Republic of Korea
e-mail: kiheankim@postech.ac.kr

J. K. Kim et al. (eds.), *Advanced Imaging and Bio Techniques for Convergence Science*, Advances in Experimental Medicine and Biology,
https://doi.org/10.1007/978-981-33-6064-8_5

three-dimensional (3D) imaging technique based on laser scanning and confocal detection of reflected light through a pinhole (Wilson and Sheppard 1984) and it can visualize tissue microstructures including cell distribution and vasculature. Cells are visible via light reflection at the interface of cell compartments such as the cell nucleus and membrane. Extracellular matrix (ECM) is visualized via relatively strong reflection compared to cells. RCM has limitation in the imaging depth due to light scattering and absorption in turbid tissues so that it is good for visualizing cellular structures in the superficial region of turbid tissues. After its long development and clinical studies, RCM has been adapted in ophthalmology and dermatology (Batta et al. 2015). In ophthalmology, RCM is used to examine the anterior surface of eye including the cornea and conjunctiva, and the posterior surface including the retina in the cellular level. In dermatology, RCM is used to detect nonmelanoma skin cancers and their boundary for noninvasive diagnosis and surgery guidance. Microscopic screening of clinically relevant large regions based on high-speed imaging and mosaicking showed the pretty good detection sensitivity of skin cancers such as basal cell carcinoma (BCC). OCT is another 3D imaging technique based on light reflection and interferometric detection (Huang et al. 1991), and OCT has been used for visualizing cross-sectional microstructure of tissues including the retina in the eye and several internal organs such as the coronary artery, gastrointestinal tract, etc. OCT visualizes layered neuronal structure and vasculature in the retina with the depth resolution of less than 10 μm (Nassif et al. 2004). It runs at high imaging speeds owing to the abundance of reflected light and high detection sensitivity, and it is currently used as a standard method for the retina examination.

Fluorescence microscopies visualize tissue microstructure and composition based on fluorescence, which could be either intrinsic AF or extrinsic fluorescence from fluorescent labeling agents. Fluorescence microscopies provide more specific information of tissues than reflection-based methods. Label-free fluorescence microscopies use various properties of AF including the intensity, spectrum, and fluorescence lifetime to assess cells and tissues. Fluorescence microscopies can use FDA-approved fluorescent dyes such as fluorescein and indocyanine green (ICG) to visualize cellular structure and vasculature in high contrasts. Fluorescence can be generated by both linear single-photon (1P) excitation and nonlinear multiphoton excitation including two-photon (2P) and three-photon (3P) excitations. Two-photon microscopy (2PM) is a 3D microscopy that utilizes 2P excitation of fluorophores (Zipfel et al. 2003) and second harmonic generation (SHG) of collagen (Freund and Deutsch 1986). 2PM can image deeper in turbid tissues than fluorescence confocal microscopy (FCM) by using the longer excitation wavelength. Label-free 2PM can visualize cells and ECM, and assess the detailed metabolic status of cells including cancer cells by both AF spectroscopic redox imaging and fluorescence lifetime imaging (Provenzano et al. 2009). However, label-free fluorescence microscopy cannot image fast and can screen limited areas only due to weak AF signal levels and phototoxicity of the excitation light sources. Confocal laser endomicroscopy uses fluorescein to examine cellular structure and vasculature in the internal organs in high contrasts (Sanai et al. 2011). However, fluorescein-based cellular imaging provides limited contrasts, because fluorescein does not label cells but fill the extracellular space. Fluorescence microscopies with exogenous cell-labeling agents have been developed for the rapid pathology examination of excised tissues during surgery (Progatzky et al. 2013). Cells on the tissue surface are quickly labeled with fluorescent labeling agents of the cell nucleus and cytoplasm, and then the tissue surfaces are screened by rapid microscopic imaging. This fluorescence examination method is to compensate the limitations of current pathology examination practice, which requires various tissue processing procedures. The current procedures during surgery include sectioning excised tissues into thin slices after freezing, labeling tissue slices with hematoxylin and eosin (H&E) dyes, and then microscopic examination. This procedure is not only time consuming but also cannot do the comprehensive examination of the excised tissue due to time limitation and tissue loss during preparation. 3D fluorescence microscopy with fluorescent cell labeling can rapidly screen the surfaces of excised tissues in the cell level. Quick cell-labeling methods of fresh tissues including labeling dyes and high-speed 3D fluorescence microscopies are developed and tested. Fluorescence microscopy–based pathology examination methods are getting adapted in some applications which take advantage of the comprehensive screening of tissue surfaces, although its sensitivity and specificity need to be improved to the level of conventional pathology.

Various optical microscopy techniques have been developed to address the needs of noninvasive and cell-level imaging for early and high-precision diagnosis and surgery guiding. Among the microscopy techniques, only reflection-based optical imaging techniques are currently adapted in clinic with the advantages of noninvasiveness and high imaging speeds. Fluorescence imaging techniques have not been used much due to the safety issue of molecular agents and the specificity limitation of the FDA-approved labeling agents, although fluorescence imaging provides better contrasts than reflection-based imaging techniques. Some breakthrough such as clinically compatible cell labeling agents will be useful for high-contrast and high-speed imaging. Recently, moxifloxacin was introduced as a potential cell-labeling agent with clinical compatibility for fluorescence microscopy. Moxifloxacin is a fourth-generation fluoroquinolone antibiotic to both treat and prevent bacterial infections, including conjunctivitis, pneumonia, etc. Moxifloxacin showed good pharmacokinetic properties such as high tissue penetration (Robertson et al. 2005) and high intracellular distribution (Pascual et al. 1999), and intrinsic fluorescence under ultraviolet(UV) excitation (Ocana et al. 2000). Fluorescence imaging with topical administration of moxifloxacin ophthalmic solution showed good tissue penetration and enhanced intracellular fluorescence, indicating the possibility as the cell-labeling agent. Excitation of moxifloxacin was demonstrated not only by linear 1P excitation but also by nonlinear 2P and 3P excitations. In this chapter, the development and applications of moxifloxacin-based fluorescence microscopy will be discussed. Properties of moxifloxacin such as fluorescence characteristics and tissue penetration will be described. Linear and nonlinear fluorescence microscopy techniques based on moxifloxacin will be demonstrated in various tissues including the cornea, skin, brain, small intestine, and colon. A major advantage of moxifloxacin-based fluorescence microscopy is clinical compatibility, and specific clinical applications of moxifloxacin-based fluorescence microscopy will be explained with preliminary preclinical study results.

Characteristics of Moxifloxacin as the Cell-Labeling Agent

Pharmacokinetic Properties

Moxifloxacin hydrochloride, hereafter mentioned to as moxifloxacin, is a fourth-generation fluoroquinolone antibiotic, which suppress DNA replication of both Gram-positive and Gram-negative bacteria by restraining its gyrase and topoisomerase IV activity (Drlica and Zhao 1997). Moxifloxacin is currently used in clinic to both treat and prevent bacterial infections by oral, intravenous, and topical administration. Pharmacokinetic studies of moxifloxacin have shown good tissue penetration and high cellular-to-extracellular and tissue-to-serum concentration ratios compared to other fluoroquinolone antibiotics, attributing to its higher lipophilicity and aqueous solubility. Pharmacokinetic study of moxifloxacin has been studied in both the cells and tissues. Analysis of moxifloxacin penetration into cells (phagocytic and nonphagocytic cells; human neutrophils and epithelial cells) showed much higher intracellular concentrations than extracellular ones in the cellular-to-extracellular concentration ratios of 9–10 (Pascual et al. 1999). The ocular penetration and pharmacokinetics of moxifloxacin was analyzed in comparison with other fluoroquinolones. Moxifloxacin showed the faster penetration and higher maximum concentration than any other fluoroquinolones in ocular tissues (Robertson et al. 2005; Solomon et al. 2005). Analysis of moxifloxacin penetration into gastrointestinal (GI) mucosal tissues in patients showed high tissue concentration with the tissue-to-serum ratio up to 9.7 (Wirtz et al. 2004). The pharmacokinetics studies showed that moxifloxacin penetrated tissues well and was distributed intracellularly.

Fluorescence Properties of Moxifloxacin Ophthalmic Solution

Fluorescence properties of moxifloxacin were measured by using moxifloxacin ophthalmic solution (Vigamox, Alcon) and various fluorescence microscopies. Vigamox has 0.5% moxifloxacin hydrochloride (5 mg/mL) as an active ingredient. Fluorescence properties were measured in both the 1P excitation and nonlinear 2P excitation, and 3P excitation conditions (Fig. 1) (Lee et al. 2016a, b, 2018a, b, 2019). In the case of single-photon excitation, a fluorophotometer was used to measure excitation/emission spectra. Maximum fluorescence intensities were found at two excitation wavelengths of 290 nm and 340 nm as the first and second peaks, respectively. Emission spectrum had its peak intensity at 465 nm with the excitation wavelength of 290 nm. Efficiency of 1P excitation of moxifloxacin at over 400 nm excitation wavelengths was smaller than 1% of the first peak. Emission spectrum at 405 nm excitation wavelength had its emission peak at 500 nm and its full-width

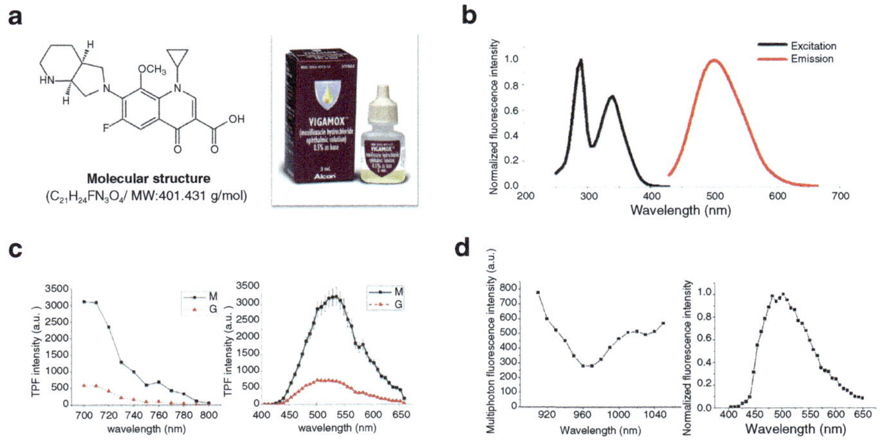

Fig. 1 Fluorescence characteristics of moxifloxacin ophthalmic solution. (**a**) Moxifloxacin molecular structure and moxifloxacin ophthalmic solution, which was used for moxifloxacin-based fluorescence microscopy. (**b–d**) 1P, 2P and 3P fluorescence spectra of moxifloxacin ophthalmic solution. Reproduced with permission from Ref. (Lee et al. 2016a, b, 2018a, b, 2019). The figures from Ref. (Lee et al. 2016a, b, 2018a, b) are licensed under a Creative Common Attribution 4.0 International License (http://creativecommons.org/licenses/by/4.0/). They are attributed to Lee, S. et al. and Springer Nature

half maximum of 90 nm approximately. Although excitation efficiency at over 400 nm was quite low, 405 nm light was used for the imaging because it is the shortest wavelengths available in commercial confocal microscopes. Moxifloxacin might have sufficiently high fluorescence in the short visible wavelengths just as commercial cell nuclear dyes such as DAPI and Hoechst, which have excitation peaks in UV and still have strong fluorescence at 405 nm excitation. 1P fluorescence intensities of moxifloxacin and Hoechst were compared at 405 nm excitation wavelength. Moxifloxacin showed approximately six times higher fluorescence intensity than Hoechst at the same concentration of 1 mM. Because Hoechst has its fluorescence enhancement upon DNA binding by approximately 30 times, moxifloxacin would be dimmer than Hoechst in tissue imaging. However, moxifloxacin still showed strong fluorescence compatible with that of the commercial dye. Fluorescence properties of moxifloxacin ophthalmic solution in 2P and 3P excitation conditions were measured by using a femto-second Ti–Sapphire pulse laser, which was used for 2P and 3P imaging later on. 2P excitation spectrum of moxifloxacin ophthalmic solution had the spectral range from 700 nm to 800 nm and had maximum intensities at 700 nm. Excitation spectra shorter than 700 nm could not be measured, because of leakage of the excitation source in the current microscope setup and the limitation of laser tuning range. Emission spectrum, measured at 780 nm excitation wavelength, had its spectral range from 450 nm to 660 nm and have maximum intensities at approximately 510 nm. In order to verify that fluorescence of these ophthalmic solutions is from the active pharmaceutical ingredient (API), the aqueous solution of moxifloxacin hydrochloride was prepared, and 2P spectra were measured. The aqueous solution showed similar excitation and emission spectra as those of the corresponding ophthalmic solution. Fluorescence intensity as the function of excitation power was measured and results showed that fluorescence intensity was proportional to 1.89 power of excitation power, indicating that the measured fluorescence was 2P process. Fluorescence intensity of moxifloxacin under 2P excitation was approximately eight times smaller than that of fluorescein (pH 7.4) at 780 nm. In the case of 3P excitation, excitation spectrum was measured from 960 nm to 1050 nm, and excitation spectrum of moxifloxacin showed a slow increase with wavelength range over 1000 nm. The increasing trend of moxifloxacin excitation spectrum indicated the new 3P excitation process. 3P excitation was verified by measuring the relationship of fluorescence intensity with excitation power. Wavelength of 1030 nm was chosen as the excitation wavelength in this measurement, because this wavelength is currently available in commercial light sources such as ytterbium (Yb) fiber femtosecond lasers. Moxifloxacin fluorescence intensity increased nonlinearly with excitation power, and its slope in log–log plot was measured to be 2.96. Emission spectrum of moxifloxacin was measured at 1030 nm excitation wavelength. Emission spectrum of moxifloxacin had broad width with its peak at 510 nm. 3P active cross-section of moxifloxacin was measured in comparison with that of DAPI at the excitation wavelength of 1000 nm (Xu et al. 1996). The 3P active cross-section of moxifloxacin was 0.2×10^{-83} $cm^6/(s/photon)^2$ which corresponds to 0.8 of the 3P active cross-section of DAPI.

Moxifloxacin-Based Fluorescence Microscopy

After the fluorescence properties were characterized, moxifloxacin ophthalmic solution was used for the fluorescence microscopy imaging of various tissues (Fig. 2) (Lee et al. 2018a, b, 2019; Wang et al. 2016). Pharmacokinetic studies of moxifloxacin showed that moxifloxacin could be used as a nonspecific cell-labeling agent with good tissue penetration and high intracellular concentration. Both linear and nonlinear fluorescence microscopies were tested. As linear fluorescence microscopy, fluorescence confocal microscopy (FCM) with 405 nm excitation wavelength was used (Lee et al. 2019). Although 405 nm was off-the-peak excitation wavelength whose fluorescence intensity was only 1/14 of the one with the second peak excitation wavelength, the wavelength was used because this wavelength is accessible in commercial confocal microscopy systems. High-speed imaging up to the video rate was possible with the excitation energy much below the damage threshold (Ramakrishnan et al. 2016) owing to the high efficiency of 1P excitation. High-speed tissue imaging of moxifloxacin-based confocal microscopy was demonstrated in ex vivo mouse brain sections. Fresh mouse brains were sliced in the coronal plane, and topically treated by moxifloxacin ophthalmic solution for 5 min, and then FCM was conducted. Video-rate imaging speed (29.4 frames per second) was possible at the excitation power of 2.3 mW. The confocal images visualized the 3D cell structure in the brain via moxifloxacin labeling. The confocal images showed a sparsely distributed cells in the cortex, and a densely and band-like distributed cells in the hippocampus. Individual cells exhibited round shapes, and some cells appeared to have thin branches. The round shape could be either cell bodies or cell

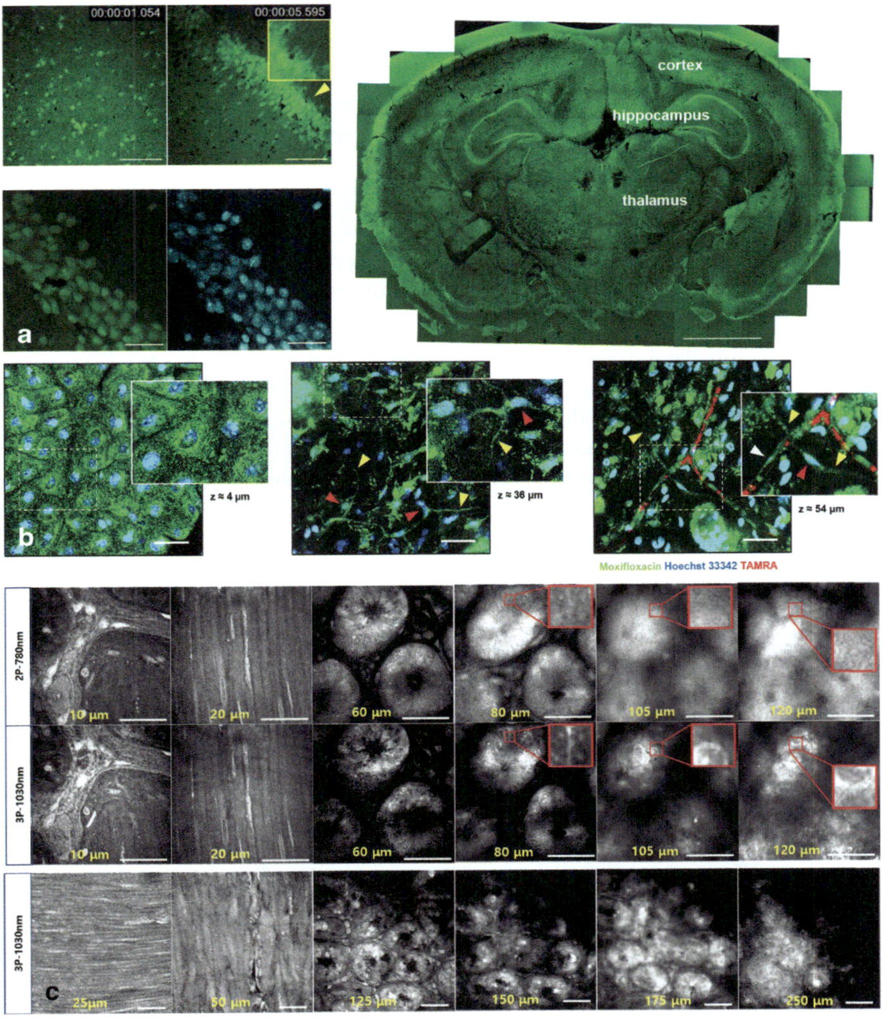

Fig. 2 Example tissue images acquired by moxifloxacin-based fluorescence microscopies. (**a**) Example moxifloxacin-based 1P fluorescence images: confocal images of freshly excised mouse brain section. Confocal images in the cortex, hippocampus, and the entire mouse brain section are shown with high-speed imaging capability. Moxifloxacin (green color) and acridine orange (cyan color, commercial cell nuclear dye) based images of the mouse brain were compared for the analysis of moxifloxacin labeling characteristics of brain cells. (**b**) Example moxifloxacin-based 2P fluorescence images: 2P images of the mouse skin in both the epidermis and dermis. Moxifloxacin, Hoechst (cell nucleus), and TAMRA (blood vessel) were co-labeled. (**c**) Example moxifloxacin-based 3P fluorescence images: 2P and 3P images of the ex vivo mouse small intestine in the top 2 rows, and 3P images of in vivo mouse small intestine in the bottom row. Reproduced with permission from Ref. (Lee et al. 2018a, b, 2019; Wang et al. 2016). The figure from Ref. (Lee et al. 2018a, b) is licensed under a Creative Common Attribution 4.0 International License (http://creativecommons.org/licenses/by/4.0/). It is attributed to Lee, S. et al. and Springer Nature. The figure from Ref. (Wang et al. 2016) is licensed under a Creative Commons Attribution-NonCommercial-ShareAlike 4.0 International License (http://creativecommons.org/licenses/by-nc-sa/4.0/). It is attributed to Wang, T. et al. and Springer Nature

nuclei. The images showed some background intensities in extracellular space, and the background came from intercellularly distributed moxifloxacin. Although moxifloxacin-based confocal microscopy could do high-speed imaging, its imaging depth was limited to approximately 50 μm from the surface in the brain due to excitation light scattering. Moxifloxacin-based fluorescence microscopy would be useful for high-speed cellular imaging on the tissue surface.

As nonlinear microscopy with moxifloxacin, both 2PM and 3PM were demonstrated. Both moxifloxacin-based 2PM and 3PM used a wavelength-tunable Ti–Sapphire femtosecond oscillator as the light source. In the case of 3PM deep tissue imaging, the oscillator could not provide enough power to compensate the scattering effect of excitation light with the imaging depth, and a femtosecond amplifier at 1030 nm was used instead. Moxifloxacin-based 2PM used the excitation wavelength from 700 nm to 820 nm. Moxifloxacin-based 2PM was used to test the cell-labeling properties of moxifloxacin (Wang et al. 2016). A variety of in vitro cell lines of the mouse and human were imaged before and after the treatment of 10 μM moxifloxacin. Moxifloxacin-treated cell lines were detectable by approximately 7–14 times greater intracellular signal than that of AF at the same excitation laser power of 2.52 mW. Because all the tested cells were imaged with the enhanced fluorescence intensities after moxifloxacin application, it was concluded that moxifloxacin labeled the cells. Moxifloxacin-based 2PM was tested in various tissues. A freshly extracted, moxifloxacin-treated rat cornea was imaged by both 2PM and RCM, a standard method to examine the cornea in the cell level. In both cases, individual cells in the epithelium, stroma, and endothelium appeared in good spatial co-registration. Moxifloxacin-based 2PM was applied to the in vivo skin imaging for comparison with label-free AF-based 2PM. AF-based 2PM provides cellular information of the skin based on AF of cells. The skin of mouse hindlimb was topically treated with moxifloxacin ophthalmic solution (12.4 mM) after mild tape stripping to remove the stratum corneum, then imaged with 2PM, and compared with the result of AF-based 2P imaging. While the layers of the epidermis were clearly visible by AF, the application of moxifloxacin significantly enhanced the visualization of cellular structures in the dermis, permitting the reduction of excitation laser power by more than three times (from 19 mW to 5 mW). 2PM images of moxifloxacin-labeled dermis showed approximately 10 times higher fluorescence intensities in cells than those of the unlabeled skin dermis. To confirm whether the signals shown in the dermis comes from intracellular moxifloxacin fluorescence, the skin specimen was counterstained with Hoechst 33342 (b2261, Sigma-Aldrich, Saint Louis, US) for cell nucleus and tetramethylrhodamine (TAMRA, D-7139, ThermoFisher Scientific, Waltham, US) for vasculature. Indeed, within each structure labeled by moxifloxacin, the cell nucleus was present. Counterstained nuclei visualized cellular morphology and distribution of cells in the stratum spinosum of the epidermis. Below the epidermis, cells with thin fibrous shape branch were shown. Considering their morphology and location, these fibrous structures might be nerves. Cells running parallel with the TAMRA signal was identified as capillary endothelial cells owing to the narrow and elongated morphology.

Moxifloxacin-based 3PM was characterized in comparison with moxifloxacin-based 2PM in biological tissues (Lee et al. 2018a, b). Both the moxifloxacin-based 2PM and 3PM imaging of the mouse small intestine were conducted by changing the excitation wavelength of the Ti–Sapphire oscillator between 780 nm and 1030 nm, respectively. Several depths of the small intestine were imaged from the serosa by both the moxifloxacin-based 2PM and 3PM. Cellular structures such as the myenteric plexus, muscle, and intestinal glands were visualized at various depths. Cells in the myenteric plexus were clearly visualized, and spindle-shaped cells in the muscle layer were also visible in both moxifloxacin-based 2PM and 3PM images. Intestinal glands in small and large intestine are depressed tubular glands in the lining of epithelium which covers the lumen of intestine. Both moxifloxacin-based 2PM and 3PM images showed epithelial cells well at the depth of 60 μm. However, moxifloxacin-based 2PM image at 80 μm depth from the surface barely resolved epithelial cells due to the reduced contrast associated with excitation light scattering, and moxifloxacin-based 2PM images at the higher depth could not exhibit individual epithelial cells. Only several cells in the laminar propria (white arrowhead) were shown at 120 μm deep from the surface. However, moxifloxacin-based 3PM images maintained the image contrast and showed individual cells in the glands at deep regions. In vivo moxifloxacin-based 3P imaging of the mouse small intestine were conducted by using the femtosecond Yb fiber amplifier laser as the light source. This Yb fiber laser had enough energy (10 μJ at 1034 nm) for 3P excitation, and its repetition rate was 102.9 kHz. This laser allowed to carry out the moxifloxacin-based 3PM imaging of deep tissues by compensating the energy loss with the increase of imaging depth. In vivo moxifloxacin-based 3PM images at various depths of the mouse small intestine from the serosa were demonstrated. The in vivo images showed similar cellular structures to the image acquired in the ex vivo condition. Cells emitting bright fluorescence were visible at 150 μm depth from the surface, and the membranes of epithelial cells in intestinal glands were also shown at 175 μm. It was confirmed that cells were resolved even down to 250 μm depth from the surface.

Applications of Moxifloxacin-Based Fluorescence Microscopy

Moxifloxacin-based fluorescence microscopy could visualize cellular structures within tissues at high imaging speeds and be clinically compatible by using the FDA-approved moxifloxacin. Therefore, moxifloxacin-based fluorescence microscopy was applied to several preclinical studies with animal models and ex vivo patient tissues. Moxifloxacin-based fluorescence microscopy was tested in the detection of cancer and cancer margin in the brain tumor (Le et al. 2017; Lee et al. 2019) and skin cancer (Chang et al. 2019) based on the difference of cytoarchitecture. Although moxifloxacin labeling does not provide any cellular specificity by labeling all the cells, morphological information could be used to identify cells with unique morphology such as stretched neurons, round leukocytes, and filamentous fungal cells. Moxifloxacin-based fluorescence microscopy was applied to the detection of fungal keratitis in animal models (Lee et al. 2018a, b). Although moxifloxacin labeled all the cells without specificity, some specific cells were found to be strongly labeled by moxifloxacin and these cells were Paneth cells in the small intestine (Jang et al. 2018) and goblet cells in the colon and conjunctiva. Moxifloxacin-based fluorescence microscopy was used to image these cells within tissues and detect cellular changes in abnormal conditions.

Moxifloxacin-Based Fluorescence Microscopy for the Detection of Brain Tumor Margin

Moxifloxacin-based fluorescence microscopy was used to detect brain tumor and its boundary by visualizing cytoarchitecture for the potential application of surgery guiding in situ (Le et al. 2017; Lee et al. 2019). Malignant brain tumor is a deadly cancer that results in finite patient survival due to its recurrence. The primary treatment for the malignant brain tumor is surgical removal, and the more part of tumor resection gives the longer survival to patient (Lacroix et al. 2001). However, excessive resection including normal brain tissue could induce neurological deficits (McGirt et al. 2009), and it lower the quality of life and results in poor prognosis. Therefore, complete removal of tumor tissue without removing normal brain tissue is important.

Current medical imaging methods such as magnetic resonance imaging, computerized tomography, and interaoperative ultrasound could localize bulky brain tumor tissue, but they are not available for precise tumor margin delineation due to low resolution. For real-time and high-sensitivity delineation of brain tumors several optical imaging methods have been developed. Confocal laser endomicroscopy, which can visualize cells, has been tested in the clinic. However, its image contrast has been limited partly due to the limitation of FDA-approved fluorescent dyes and their delivery to tumors (Milej et al. 2017; Valdés et al. 2012). OCT detected tumors by label-free imaging of brain tissue, which showed the reduced light scattering and microstructure change in brain tissues (Kut et al. 2015). However, its contrast and resolution were not enough for the precise delineation of tumor margins. Coherent Raman scattering (CRS) microscopy is another label-free optical imaging technuqie, which uses Raman scattering of molecules such as lipids and proteins. Since CRS could show different lipid and protein composition between normal and tumor tissue, it detected brain tumors (Ji et al. 2013; Uckermann et al. 2014). Although CRS microscopy had advantages in good chemical specificity and imaging speeds, the methods are expensive owing to the requirement of costly pico-second pulse lasers, and the development of a small imaging device for CRS is challenging. In the study, moxifloxacin-based confocal microscopy (CM) was tested for brain tumor detection and tumor margin delineation. Moxifloxacin-based CM was tested on normal mouse brain, and then used in a brain tumor mouse model and human brain tumor tissues for brain tumor delienation.

Moxifloxacin-based CM images of normal mouse brain visualized the 3D cell structure in the brain. Moxifloxacin-based CM images exhibited sparsely distributed cells in the cortex, and dense and band-like distributed cells in the hippocampus. Brain tumor delineation by moxifloxacin-based CM was demonstrated in brain tumor mouse models (Fig. 3) (Lee et al. 2019). For the brain tumor mouse model, DsRed fluorescent protein expressing cancer cells were implanted in the mouse brain so that cancer cells could be traced via DsRed fluorescence imaging. After 5 min administration of moxifloxacin to fresh brain slices of brain tumor

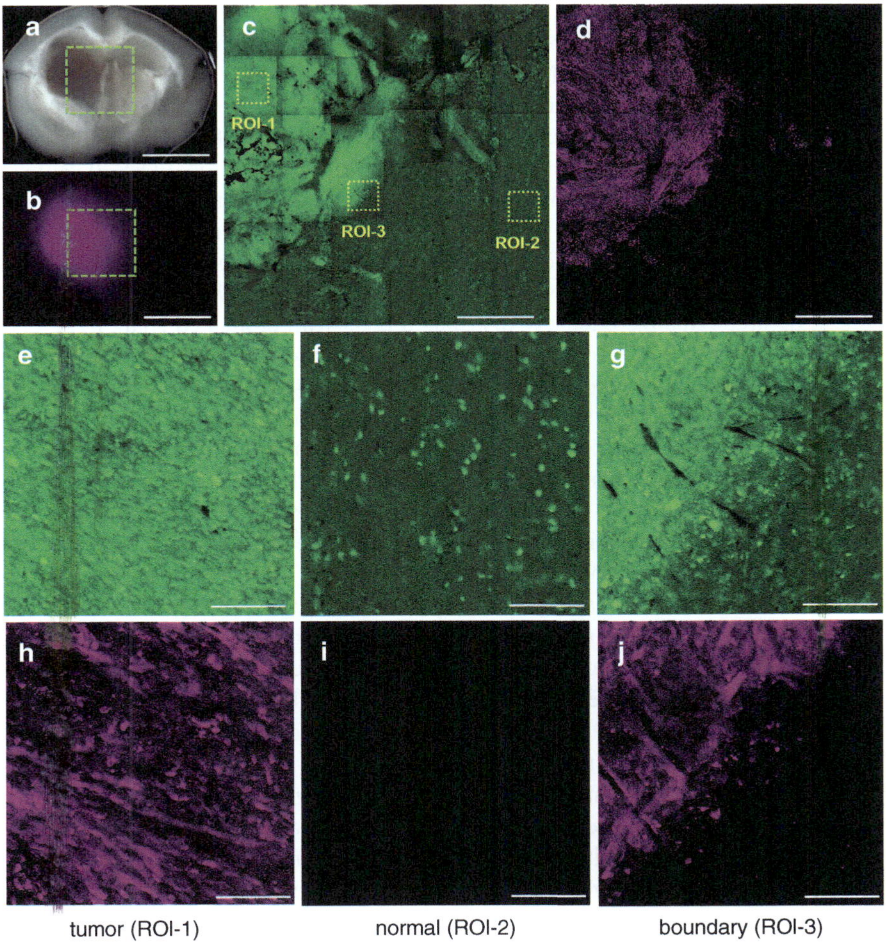

Fig. 3 Example moxifloxacin-based CM images of brain tumor. Bright-field, wide-field fluorescence, and moxifloxacin/DsRed-based confocal images of a brain tumor mouse model, ex vivo. (**a**, **b**) A bright-field image of a mouse brain slice containing DsRed protein expressing cancer cells and a corresponding wide-field DsRed fluorescence image of the mouse brain slice. Scale bar indicates 3 mm. (**c**, **d**) Large-sectional mosaic moxifloxacin-based and DsRed-based CM images just below the surface. Scale bar indicates 1 mm. (**e**–**g** and **h**–**j**) Magnified moxifloxacin-based and DsRed-based CM images in the tumor, normal, and tumor-boundary region, respectively. Scale bar indicates 100 μm. Reproduced with permission from Ref. (Lee et al. 2019)

mouse models, moxifloxacin-based CM imaging was conducted. A bright-field microscopic image of a fresh brain slice showed a dim region on the left side, and the dim region corresponded to DsRed fluorescence expressing region in wide-field DsRed fluorescence image, which confirms that the dim region was tumor. Moxifloxacin-based CM was conducted in a tissue containing both the tumor and tumor-neighboring areas to test brain tumor delineation. Moxifloxacin and DsRed fluorescence were concurrently imaged by spectrally separating their fluorescences into two channels. Confocal images of moxifloxacin and DsRed channels were mosaicked by translating sample stages, and large-sectional confocal images were generated. The large-sectional images visualized the tumor distribution in both moxifloxacin and DsRed channels. The left side of the large-sectional moxifloxacin image showed a strong fluorescent structure. The strong moxifloxacin fluorescence resulted from dense moxifloxacin-labeled cells. The large-sectional DsRed channel image confirmed that the strong fluorescent structure in the moxifloxacin channel was a tumor based on DsRed fluorescence expression of cancer cells. Magnified confocal images showed detail cytoarchitecture in several regions of interest (ROIs). Magnified moxifloxacin image at ROI-1, which was located in the strong fluorescent structure, visualized densely distributed cells with elongated shape. The corresponding magnified DsRed image exhibited densely distributed cancer cells with elongated shape. The cancer cells in the magnified DsRed image were less dense than the cells in the magnified moxifloxacin image. This was partly due to the non-uniform DsRed expression levels of cancer cells, and the visualization of additional cells including inflammatory cells beside cancer cells. Magnified images at ROI-2, which was located in the low moxifloxacin fluorescence intensity region of the large-sectional moxifloxacin image, showed sparse cell distribution in the moxifloxacin channel and the absence of cancer cells in the DsRed channel, which indicated that ROI-2 was normal region. Furthermore, a magnified moxifloxacin image at ROI-3 exhibited two distinctive regions: one with densely distributed cells on the upper left side and the other with sparsely distrubuted cells on the lower right side. The two regions with different cell distribution and fluorescence intensity in the moxifloxacin image were confirmed to be tumor and normal regions based on DsRed image. Moxifloxacin-based CM easily detected the implanted tumor and its boundary in the brain tumor mouse model by using moxifloxacin fluorescence intensity and cytoarchitecture. After demonstrating brain tumor demarcation with moxifloxacin-based CM in the mouse model, brain tumor demarcation with moxifloxacin-based CM was demonstrated in two types of ex vivo human brain tumor specimens, glioblastoma and metastatic cancer. Moxifloxacin-based CM detected tumors in both the human brain tumor specimens similarly to the case of brain tumor mouse model. Moxifloxacin-based CM imaging of the normal brain tissue showed regularly distributed cells at various cell densities. Most area of the normal brain had sparse cell distribution, and only hippocampus had dense cell distribution. However, the cell distribution in both hippocampus and other areas was regular. Brain tumor was detected by visualizing the different cytoarchitecture compared to that of the normal brain. Densely and irregularly distributed cells were the characteristic of tumor in all the cases. Owing to dense cell distribution, tumor regions had stronger moxifloxacin fluorescence intensity than normal or tumor-infiltrating neighboring regions. 3D confocal imaging provided cellular-level information such as cell orientation and distribution. The demonstration of tumor delineation on brain tumor specimens via moxifloxacin-based CM clearly distinguished brain tumors from the tumor-surrounding brain regions. This study showed the potential of moxifloxacin-based CM for being applied to brain tumor demarcation in clinic.

Moxifloxacin-Based Fluorescence Microscopy in the Characterization of Human Skin Cancer Specimens

Skin cancer specimens were characterized similarly to brain tumor specimens by moxifloxacin-based fluorescence microscopy (Chang et al. 2019). Microscopic examination of cancer cells is indispensable for both the skin cancer diagnosis and the successful surgical treatment like Mohs micrographic surgery (MMS) by exhibiting tumor-free margins. The standard biopsy procedure is time consuming so that various optical imaging technologies including RCM have been developed for rapid examination at the cellular level either in vivo or fresh ex vivo condition. RCM can scan the large section of skin via fast 3D imaging and mosaicking. Skin examination methods based on RCM have been applied to skin cancer surgery guide (González and Tannous 2002; Guitera et al. 2012; Rajadhyaksha et al. 2017). However, RCM typically had limited contrasts of cancer cells, because light reflection from cells is relatively low compared to that of surrounding ECM in the skin lesions where cancer cells are found in the depth of skin dermis. To overcome the low cell contrast of RCM, it gets help from the aceto-whitening method and FCM utilizing fluorescent cell-labeling agents. FCM employed exogenous cell-labeling agents such as acridine orange (Arndt-Jovin and Jovin 1989; Ragazzi et al. 2014), 4′,6-diamidino-2-phenylindole (DAPI) (Chazotte 2011a; Suzuki et al. 1997) and Hoechst (Chazotte 2011b) for the high-contrast cellular imaging. FCM studies showed the superior contrast than RCM and the sensitive detection of small skin cancer lesions (Gareau et al. 2012; Jain et al. 2017; Karen et al. 2009). However, FCM is usually applied to the excised skin tissues only, since the exogenous cell-labeling agents are not FDA approved due to safety issue. Label-free 2PM based on endogenous fluorophores such as NADH and FAD could visualize cellular structures in biological tissues without cell labeling agents, and it has been demonstrated in various skin disorders including psoriatic lesion (Kapsokalyvas et al. 2014), squamous cell carcinoma (SCC) (Paoli et al. 2008), dermatofibrosarcoma protuberans (DFSP) (Wu et al. 2018), BCC (Balu et al. 2015; Lin et al. 2006), and melanoma (Balu et al. 2014). However, label-free 2PM had a limitation in the imaging speed due to the weak fluorescence of the endogenous fluorophores. In this study, moxifloxacin-based 2PM was tested in ex vivo human skin specimens in comparison with both label-free 2PM and FCM. Then, moxifloxacin 2PM was applied to several types of skin cancer specimens and the moxifloxacin 2PM images were compared with histopathological features.

Moxifloxacin-based 2PM of the ex vivo human skin was compared with label-free 2PM. Label-free 2PM visualized cells and surrounding ECM structures such as collagen of the skin with AF and SHG contrasts, respectively. Moxifloxacin-based 2PM of the ex vivo normal human skin visualized cells and collagen similarly to label-free 2PM, but the imaging speed of moxifloxacin-based 2PM was more than 10 times higher than that of label-free 2PM by using moxifloxacin fluorescence. Moxifloxacin-based 2PM and acridine orange labeling–based FCM (acridine orange–based FCM) was compared in human basal cell carcinoma (BCC) specimens. After labeling the human BCC specimens with both acridine orange and moxifloxacin, they were sequentially imaged with FCM and 2PM. Both acridine orange–based FCM and moxifloxacin-based 2PM visualized BCC net structures similarly with enhanced cellular contrasts. The reason for the similar imaging results was that acridine orange mainly labeled the cell nucleus and moxifloxacin labeled both the cell nucleus and the cell cytoplasm. The comparison confirmed that moxifloxacin-based 2PM could provide almost similar cellular information as the acridine orange–based FCM.

Moxifloxacin-based 2PM was applied to several types of ex vivo human skin cancer specimens including BCC, squamous cell carcinoma (SCC) and dermato fibrosarcoma protuberans (DFSP). Imaging results of moxifloxacin-based 2PM were compared with histopathological findings. A moxifloxacin-based 2PM result of BCC is presented here as an example (Fig. 4) (Chang et al. 2019). In the moxifloxacin-based 2PM image of BCC, small and round BCC cell clusters were

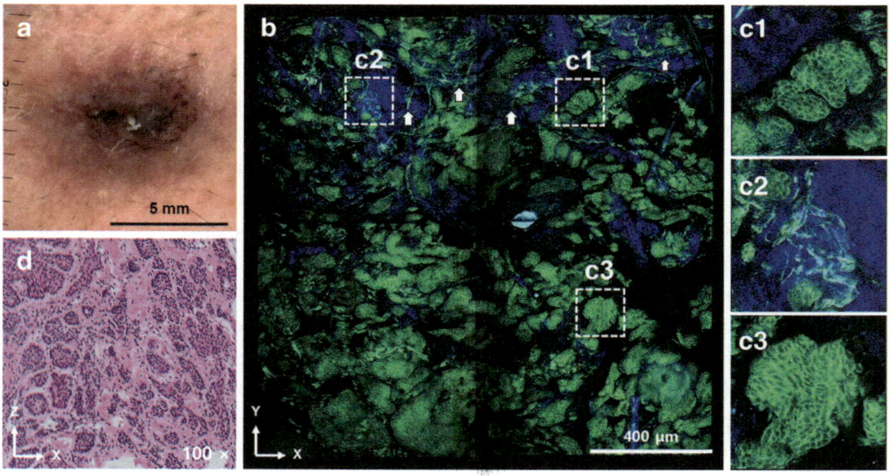

Fig. 4 Example moxifloxacin-based 2PM images of skin cancer: moxifloxacin-based 2PM images of ex vivo human micronodular BCC. (**a**) A bright-field clinical image showing a pigmented dark brown–colored ovoid nodule on the nose (**b**) a large-sectional mosaic moxifloxacin-based 2PM image, (**c**1–3) zoomed images of white-dashed-box regions in (**b**), (**d**) an H&E histological image showing micronodular structures embedded in the fibrovascular dermis. Green color depicts moxifloxacin fluorescence, and blue color does SHG. White arrows in the image mark fibrous structures. Reproduced with permission from Ref. (Chang et al. 2019)

found embedded in the skin dermis. In the magnified moxifloxacin-based 2PM images, individual cells with negative contrast in the cell nucleus were visualized in the clusters. The surrounding ECM with some fibrous structure and collagen enclosed the BCC cell clusters. The H&E histological image of BCC had showed similar features: (1) micronodular structures consisting of small, round aggregates of basal cell carninomas; (2) micronodular structures embedded in the surrouding fibrovascular dermis; (3) the fibrovascular dermis consisting of fibroblasts in a meshwork of fine collagen fibers and abundant ground substance. Moxifloxacin-based 2PM of the several types of skin cancer specimens visualized the characteristic cellular structures of individual cancer types. Because moxifloxacin-based 2PM can provide the high-speed and high-contrast images of cells and surrounding ECM structure in the skin, it might be useful for the cellular examination of the skin. In this study, moxifloxacin 2PM was applied to ex vivo human skin samples only. However, it could be applicable to the in vivo human skin just like label-free 2PM and RCM.

Moxifloxacin-Based Two-Photon Microscopy of Fungal Keratitis Rabbit Models Ex Vivo

Moxifloxacin-based 2PM was applied to the detection of fungal keratitis in rabbit models, ex vivo. Microbial keratitis is corneal inflammation due to fungal, viral, bacterial, and protozoal infection. For the diagnosis, conventional microbiological methods including specimen sampling, culturing, and examination are conducted (Whitcher et al. 2001). However, the microbiological methods take long time to identify pathogens and may delay the diagnosis and proper treatment (Jurkunas et al. 2009; Thomas and Geraldine 2007). The optical microscopy–aided keratitis diagnosis may be an alternative to the time-consuming microbiological methods by the fast and noninvasive visualizion of cellular information within the infected cornea. RCM has been used to detect Acanthamoeba (De Craene et al. 2018) and fungal infections (Brasnu et al. 2007) with noninvasive high-resolution imaging. However, its detection specificity is low due to reflection-based cellular morphology (Steven et al. 2009). RCM also has a shallow imaging depth in opaque tissues such as infected turbid corneas owing to the rapid decrease of image contrast with depth (Kumar et al. 2010; Lee et al. 2015, 2016a, b). Label-free 2PM was tested as an alternative imaging-based detection method by providing information of corneal cells and collagen based on intrinsic AF and SHG, respectively. 2PM can have the higher image contrast in turbid tissues than RCM by using the longer excitation wavelengths (Gibson et al. 2011; König 2000; Lee et al. 2015), and it demonstrated the detection of Acanthamoeba and fungal infection (Tan et al. 2007). However, label-free 2PM is slow due to the weak AF of fungal cells. Therefore, a method to enhance the fluorescence signal in the 2PM imaging of fungal cells will be helpful. Moxifloxacin-based 2PM was applied to the detection of infecting fungal pathogens in rabbit models, ex vivo (Lee et al. 2018a, b). Fungal cell suspensions and ex vivo fungal keratitis rabbit corneas were prepared by using two types of fungal pathogens, *Aspergillus fumigatus* and *Candida albicans*, and 2PM imaging was conducted both with and without moxifloxacin treatment for comparison.

The study results of *Aspergillus fumigatus* are presented here as an example (Fig. 5) (Lee et al. 2018a, b). Moxifloxacin-based 2PM was tested in both fungal cell suspension specimens and fungal keratitis rabbit models. In the fungal cell suspensions, 2PM without moxifloxacin treatment visualized the distinctive filamentous structures of *Aspergillus fumigatus* cells based on AF, and 2PM with moxifloxacin treatment also visualized the fungal cells similarly but with 10 times enhanced fluorescence. There was a slight difference between 2PM with and without moxifloxacin in the fungal cell visualization. The 2PM images of moxifloxacin-treated fungal cells exhibited relatively strong fluorescence in the cell wall with some intensity variation, while the label-free 2PM images showed relatively uniform fluorescence in the fungal cells. Both 2PM with and without moxifloxacin visualized the distinctive morphologies of fungal cells in the suspension specimens. In case of the fungal keratitis rabbit models, 2PM images without moxifloxacin visualized both fungal cells and other cell types in the cornea based on AF. Based on distinctive morphological features, *Aspergillus fumigatus* cells were easily identified from the other cells. 2PM imaging with moxifloxacin also visualized the filamentous structures of the fungal cells with 10 times stronger fluorescence than AF. The morphologcal features of fungal cells were not found in the 2PM images of other cornea specimens such as normal and chemically burned corneas. In the normal and chemically burned rabbit corneal stormas, keratocytes and activated keratocytes were visualized. They had the different morphologies of stellate-shaped in the normal cornea and spindle-shaped cells in the chemically burned cornea. The 2PM results of the fungal keratitis rabbit corneas agreed with histology, in which mixtures of filamentous fungi and inflammatory cells were visualized in the infected region of fungal keratitis rabbit cornea.

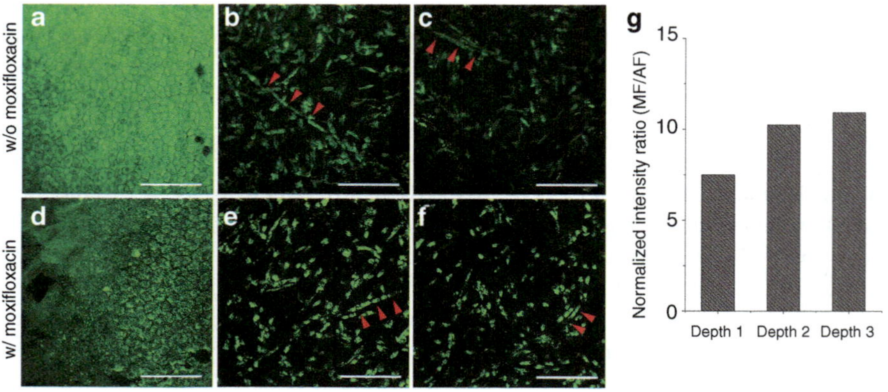

Fig. 5 Example moxifloxacin-based 2P images of *Aspergillus fumigatus*–infected rabbit corneas without and with moxifloxacin application. (**a**, **d**) the surface of the cornea. (**b**, **e**) 40 μm from the surface. (**c**, **f**) 80 μm from the surface. (**g**) Normalized intensity ratio of moxilfoxacin-based 2P to AF 2P. Reproduced with permission from Ref. (Lee et al. 2018a, b)

Moxifloxacin-Based Fluorescence Microscopy of Paneth Cells in the Mouse Small Intestine

Moxifloxacin cell labeling is nonspecific in general by labeling all the cells within tissues. However, it was found recently that some cells expressed much stronger fluorescence than the other cells. The cells strongly labeled by moxifloxacin were Paneth cells in the small intestine, goblet cells in the colon, and conjunctiva. Moxifloxacin-based fluorescence imaging of Paneth cells is explained below (Jang et al. 2018).

Paneth cells are located in the intestinal crypt bases and are one of the principal epithelial cell types in the small intestine. The small intestine is in charge of most digestion and absorption of food. The intestinal epithelium works as a barrier to protect intestine from microbiota and is compsed of various cell types, including absorptive enterocytes, intestinal stem cells, enteroendocrine cells, goblet cells, and Paneth cells (Abreu 2010; Peterson and Artis 2014). Paneth cells are terminally differentiated cells dwelling at the base of intestinal crypts, interspersed with intestinal stem cells. Paneth cells play important roles in maintaining homeostasis of the small intestine against commensal and pathogenic microorganisms by secreting densely packed secretory granules which contains antimicrobial factors such as lysozyme (Lyz) and antimicrobial peptides (AMPs) (Abreu 2010; Clevers and Bevins 2013; Wilson et al. 1999). Paneth cells also contribute to maintain and control intestinal stem cells by secreting regulatory molecules and indispensable growth factors (Sato et al. 2011). Therefore, deficiency or dysfunction of Paneth cells is related with tissue regeneration failure, microbial dysbiosis, and various inflammatory diseases in the intenstine (Cadwell et al. 2009; Wehkamp et al. 2005, 2007). Functional alteration or apoptosis of Paneth cells can be induced by severe inflammation and microbial imbalance which result from environmental stressors or vulnerable genetic predisposition. In spite of physiological importance of Paneth cells, in vivo study of Paneth cells has been limited due to lack of labeling methods. We developed an in vivo Paneth cell imaging method based on moxifloxacin and 2PM (Fig. 6) (Jang et al. 2018). 2PM imaging of the mouse small intestines showed that Paneth cell granules expressed much stronger fluorescence than the other cells. Moxifloxacin labeling of Paneth cell granules was confirmed by using the rhodamine-conjugated *Ulex Europaeus* agglutinin 1 (UEA-1) labeling of small intestine. UEA-1 can stain Paneth cell granules and its membrane by specific binding to glycoproteins containing α-linked fucose that are highly abundant in Paneth cell granules and also exist in Paneth cell membrane. In vivo 2PM imaging of the intact small intestine confirmed the moxiflaxcin labeling of granules by visualizing the granules within the UEA-1-positive Paneth cell membrane.

The in vivo moxifloxacin-based 2PM imaging of Paneth cell granules was conducted on the small intestine of wild-type specific-pathogen-free (SPF) mice. Since Paneth cells play the important function of maintaining the host–microbial homeostasis, intestinal environments change might alter Paneth cell granules. Moxifloxacin-based 2PM was tested in live mice small intestines under different metabolic condtions to demonstrate its ability to detect the alterations of Paneth cell granules.

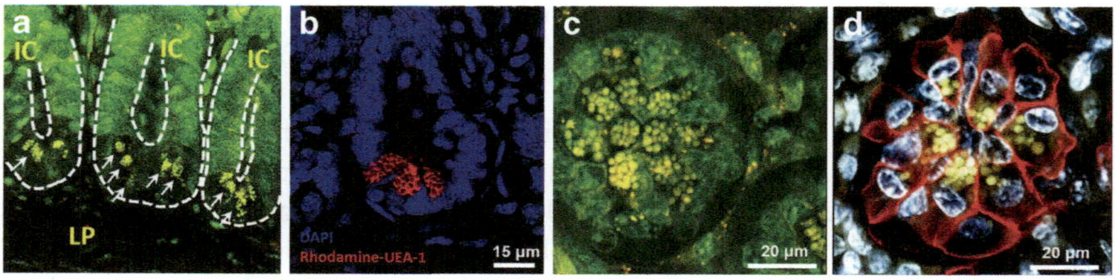

Fig. 6 Example moxifloxacin-based 2P images of Paneth cells in the mouse small intestine. (**a**) A cross-sectional 2P image of the mouse small intestine. Bright granular structures (white arrows) were visible at the base of intestinal crypts owing to the stronger expression of moxifloxacin fluorescence compared with other subcellular structures. (**b**) Corresponding fluorescence image of the sectioned small intestinal tissue showed the location and the morphologies of the Paneth cell granules at the epithelial lining of the small intestine. DAPI (blue) and rohdamine-UEA-1 (red) stained nuclei and Paneth cell granules, respectively, in the fixed and permeabilized small intestinal section. (**c**, **d**) En-face 2P images of the mouse small intestine with moxifloxacin only and the counterstaining of cell membrane with rhodamine-UEA-1 (red) and cell nuclei with Hoechst 33342 (blue). Reproduced with permission from Ref. (Jang et al. 2018). The figure from Ref. (Jang et al. 2018) is licensed under a Creative Common Attribution 4.0 International License (http://creativecommons.org/licenses/by/4.0/). It is attributed to Jang, W. H. et al. and Springer Nature

Paneth cells of two types mice, genetically obese (ob/ob) mice and lean wild-type C57BL/6 mice, raised in SPF condition were compared. 2PM imaging of these mice was conducted with and without moxifloxacin labeling. In the lean SPF mice, moxifloxacin-based 2PM image showed dense Paneth cell granules distribution at the base of intestinal crypts. Compared to lean SPF mice, obese (ob/ob) mice showed almost lack of granular structures. Instead, relatively high moxifloxacin fluorescence was shown in some cells. These results were correlated well with a previous study that lysozyme-containing Paneth cell granules are considerbly decreased in obese individuals (Hodin et al. 2011). Collectively, in vivo moxifloxacin-based 2PM showed the quite different states of Paneth cells between obese (ob/ob) mice and lean SPF mice. Our study demonstrated that 2PM with moxifloxacin labeling have a potential to be used for studying Paneth cells and related diseases. Moxifloxacin-based fluorescence imaging of Paneth cells was the first case of specific cell imaging.

Current Status and Future Directions of Moxifloxacin-Based Fluorescence Microscopy

Moxifloxacin-based fluorescence microscopy is a high-speed cellular imaging method with moxifloxacin labeling. Both linear and nonlinear fluorescence microscopies were demonstrated. This method is unique by using the FDA-approved moxifloxacin for cell labeling and it can be clinically compatible in principle. Moxifloxacin-based fluorescence microscopy was tested in various studies which require the examination of cellular structure for the purpose of diagnosis or surgery guiding. It was tested for detecting brain tumors and boundaries in both the mouse model and human brain tumor specimens. It was used to characterize cellular features of various human skin cancer tissues. It was tested in the detection of fungal keratitis in rabbit models for early and noninvasive diagnosis. Some of the applications where moxifloxacin-based fluorescence microscopy was tested, currently use RCM for the same purpose in clinic. However, RCM provides the limited information of cellular structure and vasculature by using the reflection contrast. The addition of moxifloxacin-based fluorescence microscopy to RCM would improve the detection sensitivity by providing complementary cellular information to the morphological information of RCM. Recently the combined RCM and moxifloxacin-based 2PM was developed for the high-speed multimodal imaging. Moxifloxacin-based fluorescence microscopy worked well in all the applications presented here and have potential as the diagnostic and surgery guiding method. However, all the experimental results are preliminary, and the further development and verification are required for the translation to clinic.

Although moxifloxacin is an FDA-approved drug, the usage of moxifloaxin as the cell-labeling agent for fluorescence imaging has not been approved. Safety needs to be tested and application protocols need to be established. Fluorescence microscopy uses the relatively short wavelength of either short visible or ultraviolet in case of the 1P excitation of moxifloxacin. For nonlinear 2P and 3P excitation, femtosecond pulse lasers are used as the light source. Therefore, the phototoxicity effects of excitation light on biological tissues needs to be examined. Most of the preclinical experiments were conducted by using general microscopy systems. High-performance specialized imaging systems need to be developed in order to proceed to the next step. These systems include microscopic endoscope systems, portable combined RCM and 2PM system, etc. The performance of moxifloxacin-based fluorescence microscopy needs to be verified in terms of sensitivity and specificity, and more specimens need to be imaged for statistical analysis. In vivo clinical studies need to be conducted after the performance verification and the safety clearance.

Conclusions

The developments and applications of moxifloxacin-based fluorescence microscopy were described. Moxifloxacin-based fluorescence microscopy is a high-speed cellular imaging method by using moxifloxacin and can be clinically compatible. Various properties of moxifloxacin for cell labeling and fluorescence imaging were analyzed. Both the linear and nonlinear fluorescence microscopies were demonstrated. Moxifloxacin-based fluorescence microscopy were tested in various preclinical studies including the detection of brain tumors and skin cancers and the early diagnosis of fungal keratitis. Although all the study results were still in the preliminary status, the results were promising for the advance to the next stage. More works are needed in both the technical development and preclinical and clinical studies along with safety check. Moxifloxacin-based fluorescence microscopy has good potential to be used in clinic as either the diagnostic or surgery-guiding method in the future.

Funding

This chapter was supported in part by the National Research Foundation of Korea (NRF) grant (No. 2020R1A2C3009309), the Brain Research Program (No. 2017M3C7A1044964), Korea–Sweden Research Cooperation Program (No. 2017R1A2A1A18070960), and Institute of Information & communications Technology Planning & Evaluation (IITP) grant (No.2020-0-00989) funded by the Korea government (MSIT).

References

Abreu MT (2010) Toll-like receptor signalling in the intestinal epithelium: how bacterial recognition shapes intestinal function. Nat Rev Immunol 10(2):131–144. https://doi.org/10.1038/nri2707

Arndt-Jovin DJ, Jovin TM (1989) Fluorescence labeling and microscopy of DNA. Methods Cell Biol 30:417–448. https://doi.org/10.1016/s0091-679x(08)60989-9

Balu M, Kelly KM, Zachary CB, Harris RM, Krasieva TB, König K et al (2014) Distinguishing between benign and malignant melanocytic nevi by in vivo multiphoton microscopy. Cancer Res 74(10):2688–2697. https://doi.org/10.1158/0008-5472.Can-13-2582

Balu M, Zachary CB, Harris RM, Krasieva TB, König K, Tromberg BJ, Kelly KM (2015) In vivo multiphoton microscopy of basal cell carcinoma. JAMA Dermatol 151(10):1068–1074. https://doi.org/10.1001/jamadermatol.2015.0453

Batta MM, Kessler SE, White PF, Zhu W, Fox CA (2015) Reflectance confocal microscopy: an overview of technology and advances in telepathology. Cutis 95(5):E39–E46

Brasnu E, Bourcier T, Dupas B, Degorge S, Rodallec T, Laroche L et al (2007) In vivo confocal microscopy in fungal keratitis. Br J Ophthalmol 91(5):588–591. https://doi.org/10.1136/bjo.2006.107243

Cadwell K, Patel KK, Komatsu M, Virgin IVHW, Stappenbeck TS (2009) A common role for Atg16L1, Atg5, and Atg7 in small intestinal Paneth cells and Crohn disease. Autophagy 5(2):250–252. https://doi.org/10.4161/auto.5.2.7560

Chang H, Jang WH, Lee S, Kim B, Kim MJ, Kim WO et al (2019) Moxifloxacin labeling-based multiphoton microscopy of skin cancers in Asians. Lasers Surg Med 52(4):373–382. https://doi.org/10.1002/lsm.23138

Chazotte B (2011a) Labeling nuclear DNA using DAPI. Cold Spring Harb Protoc *2011*(1):pdb.prot5556. https://doi.org/10.1101/pdb.prot5556

Chazotte B (2011b) Labeling nuclear DNA with hoechst 33342. Cold Spring Harb Protoc *2011*(1):pdb.prot5557. https://doi.org/10.1101/pdb.prot5557

Clevers HC, Bevins CL (2013) Paneth cells: maestros of the small intestinal crypts. Annu Rev Physiol 75(1):289–311. https://doi.org/10.1146/annurev-physiol-030212-183744

De Craene S, Knoeri J, Georgeon C, Kestelyn P, Borderie VM (2018) Assessment of confocal microscopy for the diagnosis of polymerase chain reaction–positive acanthamoeba keratitis: a case-control study. Ophthalmology 125(2):161–168. https://doi.org/10.1016/j.ophtha.2017.08.037

Drlica K, Zhao X (1997) DNA gyrase, topoisomerase IV, and the 4-quinolones. Microbiol Mol Biol Rev 61(3):377–392

Freund I, Deutsch M (1986) Second-harmonic microscopy of biological tissue. Opt Lett 11(2):94–96. https://doi.org/10.1364/OL.11.000094

Gareau DS, Jeon H, Nehal KS, Rajadhyaksha M (2012) Rapid screening of cancer margins in tissue with multimodal confocal microscopy. J Surg Res 178(2):533–538. https://doi.org/10.1016/j.jss.2012.05.059

Gibson EA, Masihzadeh O, Lei TC, Ammar DA, Kahook MY (2011) Multiphoton microscopy for ophthalmic imaging. J Ophthalmol 2011:11. https://doi.org/10.1155/2011/870879

González S, Tannous Z (2002) Real-time, in vivo confocal reflectance microscopy of basal cell carcinoma. J Am Acad Dermatol 47(6):869–874. https://doi.org/10.1067/mjd.2002.124690

Guitera P, Menzies SW, Longo C, Cesinaro AM, Scolyer RA, Pellacani G (2012) In vivo confocal microscopy for diagnosis of melanoma and basal cell carcinoma using a two-step method: analysis of 710 consecutive clinically equivocal cases. J Investig Dermatol 132(10):2386–2394. https://doi.org/10.1038/jid.2012.172

Hodin CM, Verdam FJ, Grootjans J, Rensen SS, Verheyen FK, Dejong CH et al (2011) Reduced Paneth cell antimicrobial protein levels correlate with activation of the unfolded protein response in the gut of obese individuals. J Pathol 225(2):276–284. https://doi.org/10.1002/path.2917

Huang D, Swanson EA, Lin CP, Schuman JS, Stinson WG, Chang W et al (1991) Optical coherence tomography. Science 254(5035):1178. https://doi.org/10.1126/science.1957169

Jain M, Rajadhyaksha M, Nehal K (2017) Implementation of fluorescence confocal mosaicking microscopy by "early adopter" Mohs surgeons and dermatologists: recent progress. J Biomed Opt 22(2):1–16, 16. https://doi.org/10.1117/1.JBO.22.2.024002

Jang WH, Park A, Wang T, Kim CJ, Chang H, Yang B-G et al (2018) Two-photon microscopy of Paneth cells in the small intestine of live mice. Sci Rep 8(1):14174. https://doi.org/10.1038/s41598-018-32640-7

Ji M, Orringer DA, Freudiger CW, Ramkissoon S, Liu X, Lau D et al (2013) Rapid, label-free detection of brain tumors with stimulated Raman scattering microscopy. Sci Transl Med 5(201):201ra119. https://doi.org/10.1126/scitranslmed.3005954

Jurkunas U, Behlau I, Colby K (2009) Fungal keratitis: changing pathogens and risk factors. Cornea 28(6):638–643. https://doi.org/10.1097/ICO.0b013e318191695b

Kapsokalyvas D, Cicchi R, Bruscino N, Alfieri D, Prignano F, Massi D et al (2014) In-vivo imaging of psoriatic lesions with polarization multispectral dermoscopy and multiphoton microscopy. Biomed Opt Express 5(7):2405–2419. https://doi.org/10.1364/BOE.5.002405

Karen JK, Gareau DS, Dusza SW, Tudisco M, Rajadhyaksha M, Nehal KS (2009) Detection of basal cell carcinomas in Mohs excisions with fluorescence confocal mosaicing microscopy. Br J Dermatol 160(6):1242–1250. https://doi.org/10.1111/j.1365-2133.2009.09141.x

König K (2000) Multiphoton microscopy in life sciences. J Microsc 200(2):83–104. https://doi.org/10.1046/j.1365-2818.2000.00738.x

Kumar RL, Cruzat A, Hamrah P (2010) Current state of in vivo confocal microscopy in management of microbial keratitis. Semin Ophthalmol 25(5–6):166–170. https://doi.org/10.3109/08820538.2010.518516

Kut C, Chaichana KL, Xi J, Raza SM, Ye X, McVeigh ER et al (2015) Detection of human brain cancer infiltration ex vivo and in vivo using quantitative optical coherence tomography. Sci Transl Med 7(292):292ra100. https://doi.org/10.1126/scitranslmed.3010611

Lacroix M, Abi-Said D, Fourney DR, Gokaslan ZL, Shi W, DeMonte F et al (2001) A multivariate analysis of 416 patients with glioblastoma multiforme: prognosis, extent of resection, and survival. J Neurosurg 95(2):190. https://doi.org/10.3171/jns.2001.95.2.0190

Le V-H, Yoo SW, Yoon Y, Wang T, Kim B, Lee S, … Chung E (2017) Brain tumor delineation enhanced by moxifloxacin-based two-photon/CARS combined microscopy. Biomed Opt Express 8(4):2148–2161. https://doi.org/10.1364/BOE.8.002148. (Accession No. 28736661)

Lee JH, Lee S, Gho YS, Song IS, Tchah H, Kim MJ, Kim KH (2015) Comparison of confocal microscopy and two-photon microscopy in mouse cornea in vivo. Exp Eye Res 132:101–108. https://doi.org/10.1016/j.exer.2015.01.013

Lee JH, Lee S, Yoon CJ, Park JH, Tchah H, Kim MJ, Kim KH (2016a) Comparison of reflectance confocal microscopy and two-photon second harmonic generation microscopy in fungal keratitis rabbit model ex vivo. Biomed Opt Express 7(2):677–687. https://doi.org/10.1364/BOE.7.000677. (Accession No. 26977371)

Lee JH, Le V-H, Lee S, Park JH, Lee JA, Tchah H et al (2018a) Two-photon microscopy of fungal keratitis-affected rabbit cornea ex vivo using moxifloxacin as a labeling agent. Exp Eye Res 174:51–58. https://doi.org/10.1016/j.exer.2018.05.018

Lee S, Lee JH, Park JH, Yoon Y, Chung WK, Tchah H et al (2016b) In vivo 3D measurement of moxifloxacin and gatifloxacin distributions in the mouse cornea using multiphoton microscopy. Sci Rep 6(1):25339. https://doi.org/10.1038/srep25339

Lee S, Lee JH, Wang T, Jang WH, Yoon Y, Kim B et al (2018b) Three-photon tissue imaging using moxifloxacin. Sci Rep 8(1):9415. https://doi.org/10.1038/s41598-018-27371-8

Lee S, Park WY, Chang H, Kim B, Jang WH, Kim S et al (2019) Fast and sensitive delineation of brain tumor with clinically compatible moxifloxacin labeling and confocal microscopy. J Biophotonics *13*(1):e201900197. https://doi.org/10.1002/jbio.201900197

Lin S-J, Jee S-H, Kuo C-J, Wu R Jr, Lin W-C, Chen J-S et al (2006) Discrimination of basal cell carcinoma from normal dermal stroma by quantitative multiphoton imaging. Opt Lett 31(18):2756–2758. https://doi.org/10.1364/OL.31.002756

McGirt MJ, Mukherjee D, Chaichana KL, Than KD, Weingart JD, Quinones-Hinojosa A (2009) Association of surgically acquired motor and language deficits on overall survival after resection of glioblastoma multiforme. Neurosurgery 65(3):463–470. https://doi.org/10.1227/01.Neu.0000349763.42238.E9

Milej D, Abdalmalak A, Desjardins L, Ahmed H, Lee T-Y, Diop M, Lawrence KS (2017) Quantification of blood-brain barrier permeability by dynamic contrast-enhanced NIRS. Sci Rep 7(1):1702. https://doi.org/10.1038/s41598-017-01922-x

Nassif NA, Cense B, Park BH, Pierce MC, Yun SH, Bouma BE et al (2004) In vivo high-resolution video-rate spectral-domain optical coherence tomography of the human retina and optic nerve. Opt Express 12(3):367–376. https://doi.org/10.1364/OPEX.12.000367

Ocana JA, Barragan FJ, Callejon M (2000) Spectrofluorimetric determination of moxifloxacin in tablets, human urine and serum. Analyst 125(12):2322–2325. https://doi.org/10.1039/b005991i

Paoli J, Smedh M, Wennberg A-M, Ericson MB (2008) Multiphoton laser scanning microscopy on non-melanoma skin cancer: morphologic features for future non-invasive diagnostics. J Investig Dermatol 128(5):1248–1255. https://doi.org/10.1038/sj.jid.5701139

Pascual A, García I, Ballesta S, Perea EJ (1999) Uptake and intracellular activity of moxifloxacin in human neutrophils and tissue-cultured epithelial cells. Antimicrob Agents Chemother 43(1):12–15. Retrieved from https://www.ncbi.nlm.nih.gov/pubmed/9869557 https://www.ncbi.nlm.nih.gov/pmc/articles/PMC89012/

Peterson LW, Artis D (2014) Intestinal epithelial cells: regulators of barrier function and immune homeostasis. Nat Rev Immunol 14:141. https://doi.org/10.1038/nri3608. https://www.nature.com/articles/nri3608#supplementary-information

Progatzky F, Dallman MJ, Lo Celso C (2013) From seeing to believing: labelling strategies for in vivo cell-tracking experiments. Interface Focus 3(3):20130001. https://doi.org/10.1098/rsfs.2013.0001

Provenzano PP, Eliceiri KW, Keely PJ (2009) Multiphoton microscopy and fluorescence lifetime imaging microscopy (FLIM) to monitor metastasis and the tumor microenvironment. Clin Exp Metastasis 26(4):357–370. https://doi.org/10.1007/s10585-008-9204-0

Ragazzi M, Piana S, Longo C, Castagnetti F, Foroni M, Ferrari G et al (2014) Fluorescence confocal microscopy for pathologists. Mod Pathol 27(3):460–471. https://doi.org/10.1038/modpathol.2013.158

Rajadhyaksha M, Marghoob A, Rossi A, Halpern AC, Nehal KS (2017) Reflectance confocal microscopy of skin in vivo: from bench to bedside. Lasers Surg Med 49(1):7–19. https://doi.org/10.1002/lsm.22600

Ramakrishnan P, Maclean M, MacGregor SJ, Anderson JG, Grant MH (2016) Cytotoxic responses to 405nm light exposure in mammalian and bacterial cells: involvement of reactive oxygen species. Toxicol In Vitro 33:54–62. https://doi.org/10.1016/j.tiv.2016.02.011

Robertson SM, Curtis MA, Schlech BA, Rusinko A, Owen GR, Dembinska O et al (2005) Ocular pharmacokinetics of moxifloxacin after topical treatment of animals and humans. Surv Ophthalmol 50(Suppl 1):S32–S45. https://doi.org/10.1016/j.survophthal.2005.07.001

Sanai N, Eschbacher J, Hattendorf G, Coons SW, Preul MC, Smith KA et al (2011) Intraoperative confocal microscopy for brain tumors: a feasibility analysis in humans. Operat Neurosurg 68(suppl_2):ons282–ons290. https://doi.org/10.1227/NEU.0b013e318212464e

Sato T, van Es JH, Snippert HJ, Stange DE, Vries RG, van den Born M et al (2011) Paneth cells constitute the niche for Lgr5 stem cells in intestinal crypts. Nature 469(7330):415–418. https://doi.org/10.1038/nature09637

Solomon R, Donnenfeld ED, Perry HD, Snyder RW, Nedrud C, Stein J, Bloom A (2005) Penetration of topically applied gatifloxacin 0.3%, moxifloxacin 0.5%, and ciprofloxacin 0.3% into the aqueous humor. Ophthalmology 112(3):466–469. https://doi.org/10.1016/j.ophtha.2004.09.029

Steven, P., Müller, M., Koop, N., Rose, C. & Hüttmann, G. (2009). Comparison of Cornea Module and DermaInspect for noninvasive imaging of ocular surface pathologies. J Biomed Opt, 14(6), 1–8, 8. Retrieved from doi:https://doi.org/10.1117/1.3275475

Suzuki T, Fujikura K, Higashiyama T, Takata K (1997) DNA staining for fluorescence and laser confocal microscopy. J Histochem Cytochem 45(1):49–53. https://doi.org/10.1177/002215549704500107

Tan H-Y, Sun Y, Lo W, Teng S-W, Wu R-J, Jee S-H, … Dong C-Y (2007) Multiphoton fluorescence and second harmonic generation microscopy for imaging infectious keratitis. J Biomed Opt 12(2):1–8, 8. Retrieved from https://doi.org/10.1117/1.2717133

Thomas PA, Geraldine P (2007) Infectious keratitis. Curr Opin Infect Dis 20(2):129–141. https://doi.org/10.1097/QCO.0b013e328017f878

Uckermann O, Galli R, Tamosaityte S, Leipnitz E, Geiger KD, Schackert G et al (2014) Label-free delineation of brain tumors by coherent anti-stokes Raman scattering microscopy in an orthotopic mouse model and human glioblastoma. PLoS One 9(9):e107115. https://doi.org/10.1371/journal.pone.0107115

Valdés PA, Moses ZB, Kim A, Belden CJ, Wilson BC, Paulsen KD et al (2012) Gadolinium- and 5-aminolevulinic acid-induced protoporphyrin IX levels in human gliomas: an ex vivo quantitative study to correlate protoporphyrin IX levels and blood-brain barrier breakdown. J Neuropathol Exp Neurol 71(9):806–813. https://doi.org/10.1097/NEN.0b013e31826775a1

Wang T, Jang WH, Lee S, Yoon CJ, Lee JH, Kim B et al (2016) Moxifloxacin: clinically compatible contrast agent for multiphoton imaging. Sci Rep 6(1):27142. https://doi.org/10.1038/srep27142

Wehkamp J, Salzman NH, Porter E, Nuding S, Weichenthal M, Petras RE et al (2005) Reduced Paneth cell α-defensins in ileal Crohn's disease. Proc Natl Acad Sci USA 102(50):18129–18134. https://doi.org/10.1073/pnas.0505256102

Wehkamp J, Wang G, Kübler I, Nuding S, Gregorieff A, Schnabel A et al (2007) The paneth cell α-defensin deficiency of ileal Crohn's disease is linked to Wnt/Tcf-4. J Immunol 179(5):3109–3118. https://doi.org/10.4049/jimmunol.179.5.3109

Whitcher JP, Srinivasan M, Upadhyay MP (2001) Corneal blindness: a global perspective. Bull World Health Organ 79(3):214–221. Retrieved from https://www.scopus.com/inward/record.uri?eid=2-s2.0-0035084290&partnerID=40&md5=5a154b0d0b870856b7edacf59b8394d0

Wilson CL, Ouellette AJ, Satchell DP, Ayabe T, López-Boado YS, Stratman JL et al (1999) Regulation of intestinal α-defensin activation by the metalloproteinase matrilysin in innate host defense. Science 286(5437):113–117. https://doi.org/10.1126/science.286.5437.113

Wilson T, Sheppard C (1984) Theory and practice of scanning optical microscopy (Vol. 180)

Wirtz M, Kleeff J, Swoboda S, Halaceli I, Geiss HK, Hoppe-Tichy T et al (2004) Moxifloxacin penetration into human gastrointestinal tissues. J Antimicrob Chemother 53(5):875–877. https://doi.org/10.1093/jac/dkh173

Wu S, Huang Y, Li Z, Wu H, Li H (2018) Collagen features of dermatofibrosarcoma protuberans skin base on multiphoton microscopy. Technol Cancer Res Treat 17:1533033818796775. https://doi.org/10.1177/1533033818796775

Xu C, Williams RM, Zipfel W, Webb WW (1996) Multiphoton excitation cross-sections of molecular fluorophores. Bioimaging 4(3):198–207. https://doi.org/10.1002/1361-6374(199609)4:3<198::Aid-bio10>3.0.Co;2-x

Zipfel WR, Williams RM, Webb WW (2003) Nonlinear magic: multiphoton microscopy in the biosciences. Nat Biotechnol 21(11):1369–1377. https://doi.org/10.1038/nbt899

Correlative Light and Electron Microscopy for Nanoparticle–Cell Interaction and Protein Localization

Minkyo Jung, Tae Keun Kim, Ha-Na Woo, Ji Young Mun, Heuiran Lee, and Chan-Gi Pack

Introduction

Correlative light and electron microscopy (CLEM) is a method where a specimen (i.e., cell) is observed using light (or fluorescent) microscopy and electron microscopy (EM), after which the correlation between the light microscopy and electron microscopy images is determined (Fig. 1) (Abandowitz and Geissinger 1975; Rieder and Bowser 1985; Webster et al. 1978). Light microscopy is one of the tools used to observe dynamic biological phenomena, including molecules within a cell. Various fluorescence dyes and fluorescence proteins (FPs) have been used to examine subcellular localization and co-localization of target proteins in living cells. Light microscopy has a key limitation in the form of spatial resolution, mainly due to

M. Jung · J. Y. Mun
Neural Circuits Research Group, Korea Brain Research Institute, Daegu, South Korea

T. K. Kim
Convergence Medicine Research Center (CREDIT), Asan Institute for Life Sciences, Asan Medical Center, Seoul, South Korea

H.-N. Woo · H. Lee
Department of Microbiology, University of Ulsan College of Medicine, Seoul, South Korea

Bio-Medical Institute of Technology, University of Ulsan College of Medicine, Seoul, South Korea

C.-G. Pack (✉)
Convergence Medicine Research Center (CREDIT), Asan Institute for Life Sciences, Asan Medical Center, Seoul, South Korea

Department of Convergence Medicine University of Ulsan College of Medicine, Asan Institute for Life Sciences, Asan Medical Center, Seoul, Korea (Republic of)
e-mail: changipack@amc.seoul.kr

J. K. Kim et al. (eds.), *Advanced Imaging and Bio Techniques for Convergence Science*, Advances in Experimental Medicine and Biology,
https://doi.org/10.1007/978-981-33-6064-8_6

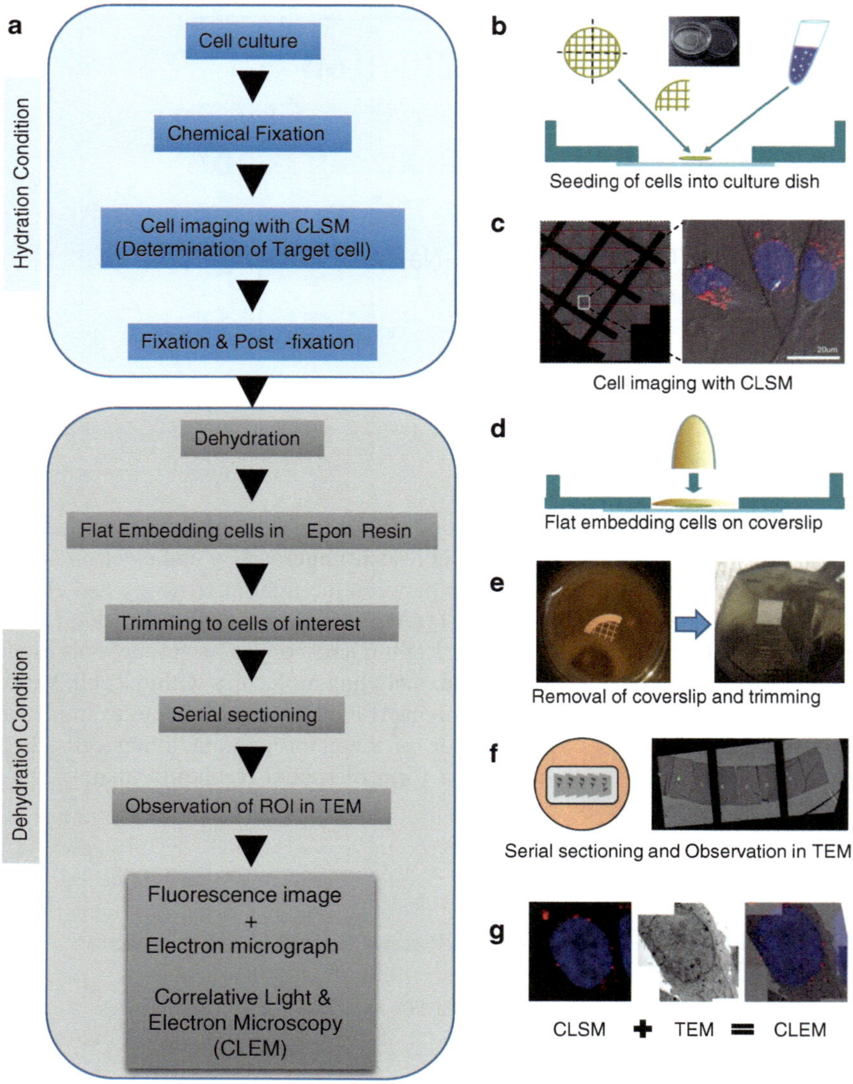

Fig. 1 Workflow of correlative light and transmission electron microscopy. (**a**) Flowchart of the major steps in correlative light and transmission electron microscopy imaging. (**b**) Plastic culture dish with a coverslip attached to the bottom of the dish. The gold grid is cut into pieces and placed at the bottom of the glass. Mammalian cells were spread on the cut gold grid. (**c**) Observation of a HeLa cell. To identify the target cells, a whole grid mesh was scanned using the tile scan function. Three-dimensional fluorescence images of the cell obtained through confocal microscopy using a high-performance lens at intervals of 250 nm in the optical Z-section. (**d**) Flat embedding in Spurr resin. (**e**) Using the transferred gold grid mesh, the block was trimmed so that cells in the area of interest and then serial sections were cut using an ultramicrotome. (**f**) Serial sections were mounted on a formvar carbon-coated single-slot grid or three-hole grid. (**g**) Correlate fluorescent images with electron microscopy (EM) images

optical diffraction, so that fluorescence microscopy (i.e., confocal laser scanning microscopy) can only provide images with relatively low resolution, over approximately 200 nm.

Newly developed super-resolution light microscopy techniques can obtain high spatial resolution in the 20–100 nm range (Sahl et al. 2017). Electron microscopy provides sub-nanometer scale resolution images using resin-embedded biological samples. However, it cannot reveal highly dynamic phenomena within living cells. Therefore, the correlation between light microscopy and electron microscopy in a sample (i.e., single cells) presents dual advantages in the form of dynamic fluorescent imaging of live cells and the high resolutions associated with electron microscopy imaging. The advanced technique has been employed in various biological studies, such as virus–cell interactions, combining live-cell imaging, and immunolabeling of ultrathin cryosections, and cellular localization of quantum dot nanoparticles (Bykov et al. 2016; Han et al. 2019; van Rijnsoever et al. 2008).

The use of tagged FPs and various fluorescent nanoparticle probes has allowed not only the dynamic movement of intracellular organelles but also specific diffusional and interactional properties of the fluorescent proteins to be tracked using confocal microscopy and confocal-related analytical techniques, such as fluorescence recovery after photobleaching (FRAP), fluorescence resonance energy transfer (FRET), fluorescence correlation spectroscopy (FCS), and fluorescence cross-correlation spectroscopy (Ishikawa-Ankerhold et al. 2012; Pack et al. 2012, 2014a, b). Notably, a CLEM method that uses ascorbate peroxidase (APEX) tags to visualize GFP-tagged proteins has been developed over the past few years (Martell et al. 2012, 2017), and it can be combined with the fluorescent techniques described above. Since cells traced using such fluorescent methods could be fixed at the moment of interest and analyzed using CLEM, a lot of information on both the cellular behavior of GFP-tagged target molecules and detailed structural status of cellular microenvironments surrounding the target molecules is available, which is beneficial particularly in the fields of nanomedicine and drug delivery (Jin et al. 2016; Patra et al. 2018). In addition, various clinical anticancer drugs and nanomaterials targeting cells can be fluorescently labeled and then exploited using CLEM and the various fluorescent methods (Darcy et al. 2006). For example, label-free or fluorescently labeled Herceptin, one of the famous anticancer drugs targeting Her2 receptors on cancer cells, strongly interacts with the cell surface through GFP-tagged Her2 receptors (refer to Chap. 1 in this book). Consequently, fluorescently labeled drugs, drug carriers such as nanoparticles, and GFP-tagged proteins are all good candidates for CLEM applications.

Recently, fluorescent NPs have emerged as promising and biocompatible imaging probes in confocal microscopy and have been used as drug carriers for biomedicine and nanomedicine applications. In practice, numerous studies have demonstrated the application of nanomaterials as drug carriers (Kulkarni et al. 2010; Radu et al. 2004; Regier et al. 2012). Among various nanoparticles, silica nanoparticles (Si-NPs) are the most commonly applied nanomaterials in such applications due to their easy surface modification, low cell toxicity, and high levels of uptake into eukaryotic cells (Radu et al. 2004; Sailor and Park 2012). In addition, silica-based fluorescent Si-NPs have been employed as analytical probes for the study of cellular distribution and diffusional dynamics due to their strong fluorescent signals, biological inertness (i.e., low cytotoxicity), and high solubility (Pack et al. 2012). A previous study based on confocal imaging and FCS analysis demonstrated the existence of various types of Si-FNP cellular diffusional movements in peptide-coated and non-coated Si-FNPs. However, the reason for the slowed diffusion of the coated Si-FNPs and the potential interactions between NPs and various cellular organelles were not investigated.

In this chapter, we introduce a straightforward imaging approach using CLEM for the identification of the exact localizations of Si-FMNPs or APEX2 conjugated specific proteins in the cellular compartments, such as the cytosol, nucleus, and outer surface of the plasma membrane at the single-particle level.

Human epithelial cervical carcinoma (HeLa) cells electroporated with Si-FMNP have been imaged using three-dimensional (3D) confocal microscopy and examined under transmission electron microscopy (TEM) to observe the intracellular distribution of NPs. Conversely, cells were incubated after treatment with culture media with NPs, and the distribution of Si-FMNPs on the cell surface was imaged under confocal microscopy and scanning electron microscopy (SEM) (Fig. 2).

Material and Methods

Materials

1. Cell culture dish: Gridded glass-bottom cell culture dishes (35-mm petri dish, No. 1.5 cover glass, P35G-1.5-14-CGRD, MatTek).
2. Silica nanoparticles containing magnetic core (NEO-STEM™ Si-FMNP containing RITC dye, Biterials Co., Ltd., Korea).
3. Glutaraldehyde, 25% in water, electron microscopy grade (Electron Microscopy Sciences, 16200) (see Note 1).
4. 16% Paraformaldehyde (Electron Microscopy Sciences, 15700) (see Note 1).
5. Osmium tetroxide, 4% water solution (Electron Microscopy Sciences, 16320) (see Note 1).
6. Uranyl acetate (Electron Microscopy Sciences, 22400).
7. A graded series of ethanol solutions: 50%, 60%, 70%, 80%, 90%, 95%, 100% (v/v), ethanol (Merck, 100983) in distilled water.
8. Embedding resin (Electron Microscopy Sciences, 14120).
9. Sodium cacodylate buffer, 0.1 M, pH 7.4 (sodium cacodylate trihydrate, Electron Microscopy Sciences, 12300): For 0.1 M sodium cacodylate, dissolve 4.28 g cacodylic acid in 160 ml of distilled water. Adjust pH to 7.4 with 0.1 M HCl then make up to 200 mL with distilled water.
10. Lead citrate (Electron Microscopy Sciences, 17800).
11. Single-slot copper grids (Ted Pella, 01805-F).
12. For a 10× DAB solution, dissolve 50 mg of DAB powder (3,3′-diaminobenzamidine, Sigma-Aldrich, D8001) in 10 mL of 0.1 M HCl solution at room temperature with strong vortexing for 10–20 min. Undissolved material is removed by centrifuging at 13,523×g at room temperature for 10 min.

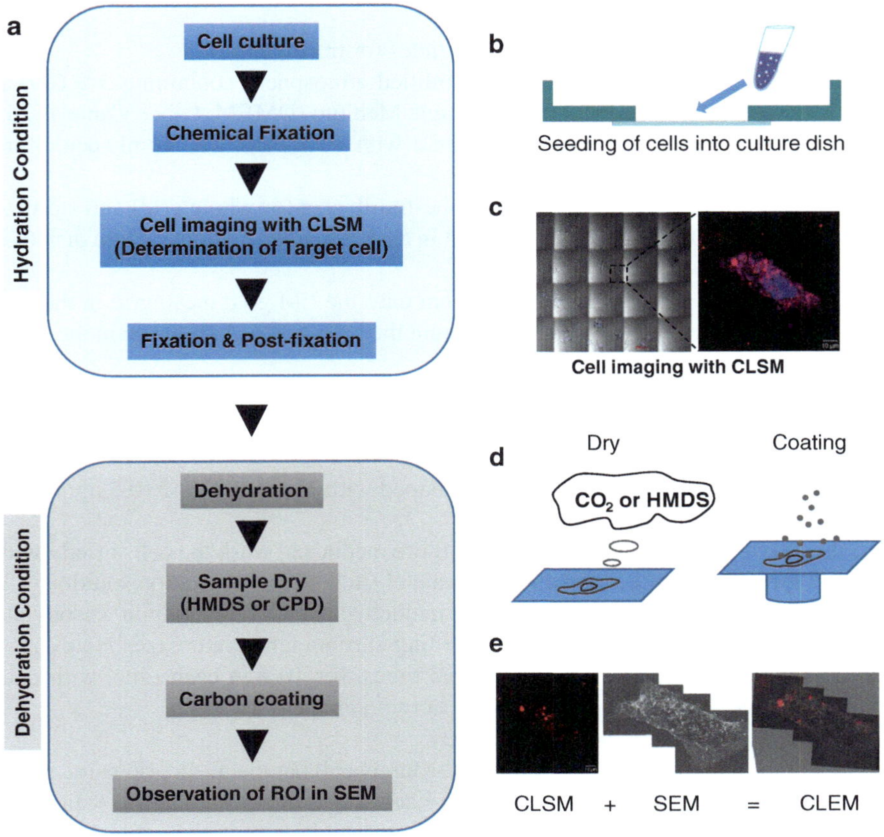

Fig. 2 Workflow of correlative light and scanning electron microscopy (SEM). (**a**) Flowchart of the major steps in the correlative light and scanning electron microscopy imaging procedure. (**b**) Plastic culture dish with a gridded coverslip attached to the bottom of the dish. Mammalian cells were spread on the bottom of the gridded glass coverslip. (**c**) To identify the target cells, a large area was scanned using the tile scan function. High-resolution fluorescence images of the cell were obtained through confocal microscopy using a high-performance lens. (**d**) Specimens were dried using a critical point drier or hexamethyldisilazane (HMDS) and then coated with carbon or metal. (**e**) Correlated fluorescent and SEM images

13. For a 1× DAB solution, 0.3 M cacodylate solution 3.33 mL + 30% H_2O_2 10 μL + 5.67 mL of cold water +1 mL 10× DAB solution.
14. 30% Hydrogen peroxide solution (Merck, 107210).
15. Thiocarbonyldihydrazide (TCH, Sigma, 223220) (see Note 1).
16. Hoechst 33342 (NucBlue Live Ready Probes, Thermo Fisher, R37605).
17. Hexamethyldisilazane (HMDS, Sigma 440191).
18. CellMask Deep Red (Thermo Fisher, C10046).

TEM-Based CLEM Methods

1. Cell culture and treatment of nanoparticles (or transfection).
   1.1. HeLa cells are grown in a humidified atmosphere containing 5% CO_2 at 37 °C in Dulbecco's Modified Eagle Medium (DMEM; Gibco Canada Ltd., Burlington, Canada) supplemented with 10% FBS, 100 U/mL penicillin, and 100 U/mL streptomycin.
   1.2. Cut the EM gold mesh grid to one-fourth size (or whole grid) and sterilize it with 70% alcohol, then place it in a 35-mm glass-bottom culture dish (see Note 2).
   1.3. Seed 1 × 10^5 cells by placing them onto the EM gold mesh grid in the culture dish. It is important to maintain the cells at low confluence in the mesh grid (see Note 3).
   1.4. The day after seeding the cells, to introduce the cellular uptake of Si-FMNP through endocytosis, treat with a Si-FMNP solution (0.2 mg/mL) for 24 h on a culture dish.
   1.5. After 24 h, the cells are counterstained with Hoechst 33342 (0.5 μg/ml).
2. Fixation.
   2.1. To fix the cells, remove all the culture media and wash the cells briefly with pre-warmed PBS and then immediately add 1.5 ml of fixation solution (1% glutaraldehyde and 1% paraformaldehyde in 0.1 M sodium cacodylate solution [pH 7.0]) by gentle pipetting at room temperature (see Note 1).
   2.2. After fixing, wash the cells three times, for 10 min each time, with cold 0.1 M sodium cacodylate buffer at room temperature.
3. Fluorescence (or confocal) microscopy.
   3.1. Observe the overall location of the target cells in the mesh grid using a 40× lens on the confocal microscope. Each single image shot in tile scan mode using the motorizing stage produces a single large image (Fig. 4a). This overview image serves as a map for the detection of specific target cells in the plastic-embedded samples.
   3.2. The target cells that incorporate the nanoparticles are imaged using high-magnification images (Fig. 4b) and z-stack images (Fig. 4c).
4. TEM sample preparation and flat embedding in epoxy resin.
   4.1. Post-fix the cells in 2% (w/v) osmium tetroxide (O_sO_4) and 1.5% (w/v) potassium ferrocyanide in 0.1 M sodium cacodylate buffer for 1 h at room temperature.
   4.2. Remove the fixative and wash the cells three times, for 10 min each time, with distilled water at room temperature.
   4.3. Add 1 ml of 1% uranyl acetate (aqueous), and leave the cells in cold conditions (4 °C).
   4.4. On the following day, wash the cells three times, for 10 min each time, with distilled water at room temperature.
   4.5. Dehydrate the samples in a graded series of 2 mL ethanol solutions, 50%, 60%, 70%, 80%, 90%, 95%, and 100% (w/v), for 20 min each at room temperature.
   4.6. Remove the ethanol and incubate for 1 h in a 25% (v/v) epoxy resin ethanol mixture at room temperature.

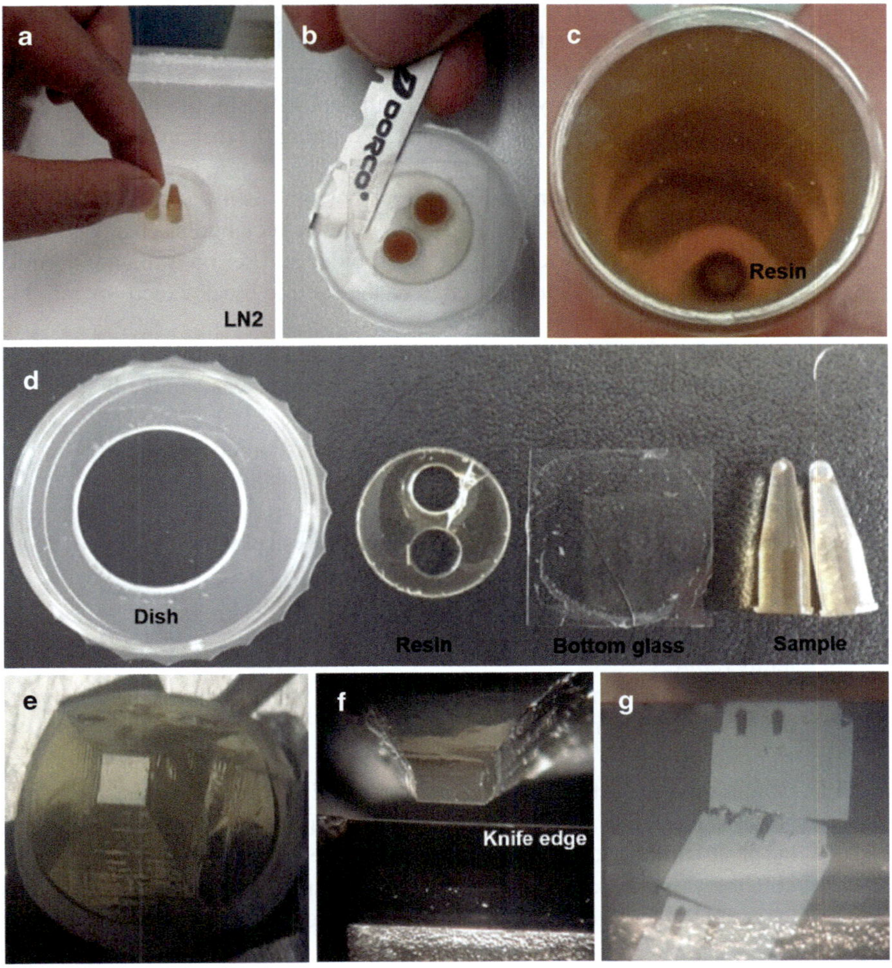

Fig. 3 Removal of the bottom glass and dish from the epoxy resin and serial sectioning. (**a**) Embedded capsules are removed from the glass bottom coverslip by dipping them into liquid nitrogen. (**b**) Using a razor to separate the bottom glass from the plastic dish. (**c**) A resin block well separated from the bottom glass. (**d**) Well-separated bottom glass from the dish and samples are shown. (**e**) Trimming of the four sides of the target position using a razor blade. (**f**) Alignment of the block-face parallel to the knife edge. (**g**) Serial sectioned ribbons

4.7. Remove the medium and incubate for 1 h in a 50% (v/v) epoxy resin ethanol mixture at room temperature.
4.8. Remove the medium and incubate for 1 h in a 75% (v/v) epoxy resin ethanol mixture at room temperature.
4.9. Remove the medium and incubate in a 100% low-viscosity embedding medium mixture overnight at room temperature.
4.10. On the following day, finally, incubate in a 100% low-viscosity embedding medium mixture for 3 h at room temperature.

4.11. Fill a PCR tube or a BEEM capsule with 100% fresh epoxy resin, and place the tube or capsule upside down on the gold grid and incubate for 24 h at 60 °C.

5. Trimming of the sample block and serial sectioning.
   5.1. After the complete polymerization of the epoxy resin, the bottom glass is removed from the epoxy resin by briefly dipping the dish in liquid nitrogen (Fig. 3a–d).
   5.2. Trim the resin block around the target position using a razor blade. The shape of the block-face is made appropriate for rectangular cuts (Fig. 3e).
   5.3. Insert the sample block in the sample holder and set on a diamond knife. Align the leading and trailing edge of the sample block exactly parallel to the diamond knife edge (Fig. 3f).
   5.4. Cut about 50 nm of a thin section using an ultramicrotome. If the perfectly shaped block-face is not cut, the target cell already exists in the first section (Fig. 3g).
   5.5. Carefully collect the sections on the formvar carbon-coated slot grid and let them dry out.
6. Heavy metal staining.
   6.1. For electron microscopy observations, stain the grid with 2% (v/v) aqueous uranyl acetate for 20 min at room temperature.
   6.2. Wash the grid with drops (2 mL) of distilled water.
   6.3. Stain the grid with lead (II) nitrate solution for 10 min at room temperature.
   6.4. Wash the grid with drops (2 mL) of distilled water, and dry the remaining water using filter paper.
7. Electron microscopy.
   7.1. Obtain images of the serial sections under TEM.
   7.2. Observe the cell of interest and the ultrastructure region of interest of a similar position under confocal microscopy image (Fig. 4a, b) in the EM. First, obtain low magnification of a single cell or a small group (3–4 cells) of cells in the EM micrograph and compare with fluorescent images (Fig. 5).

SEM-Based CLEM Methods

1. Cell culture and treatment of nanoparticles.
   1.1. Culture HeLa cells in a humidified atmosphere containing 5% CO_2 at 37 °C in DMEM supplemented with 10% FBS, 100 U/mL penicillin, and 100 U/mL streptomycin.
   1.2. Seed 1×10^5 cells by placing them onto a gridded glass-bottom dish. It is important to maintain the cells at low confluence in the gridded glass-bottom dishes.
   1.3. The day after seeding the cells, to trigger the cellular uptake of Si-FMNP through endocytosis, treat using a solution (0.2 mg/mL) of Si-FMNP for 12 h on a culture dish.

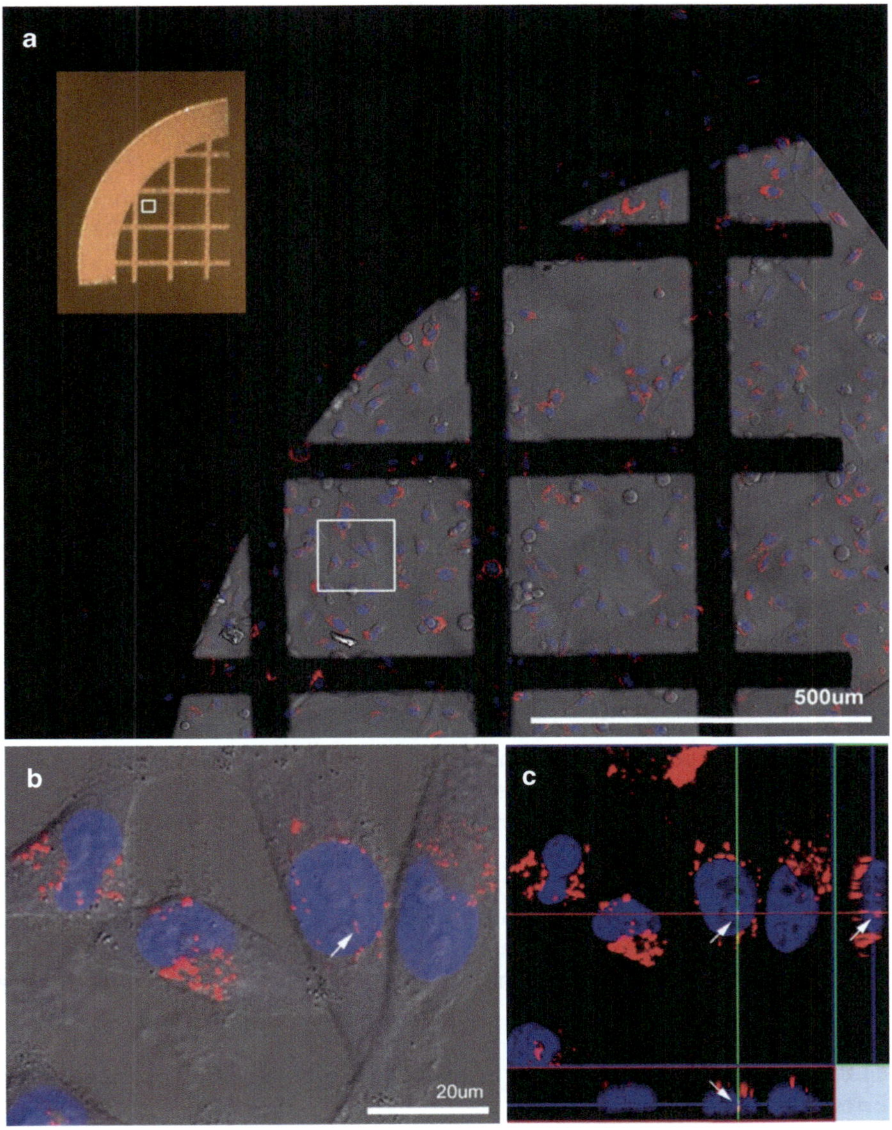

Fig. 4 Identification and selection of cells with silica NPs in the nuclei. Incorporation of Si-FMNPs into the cytosol and nuclei of living HeLa cells after electroporation of 0.2 mg/mL Si-FMNPs. Cells containing Si-FMNPs in the cytosol were imaged 24 h after electroporation. (**a**) A tiled image (9 × 11 stitched images with the total size of the grid) of the entire field of cells on the grid. The white box is a selected area with target cells. (**b**) A cell containing Si-FMNPs in the nucleus (white arrow) as well as in the cytosol was imaged 24 h after electroporation. (**c**) Orthogonal 3D image of the cell shown in B. The orthogonal Z-stack profile along the indicated lines in the x or y direction of the central xy image is shown in the rectangular windows below (x–z) and to the right (y–z) of the center (x–y) image. White arrows indicate aggregates of Si-FMNPs fixed in the nucleus. Red and blue fluorescence are originated from Si-FNMP and DAPI staining, respectively

2. Fixation and confocal light microscopy observation.
   2.1. To stain the plasma membrane and the nucleus, add 1 μL of CellMask Deep Red and 300 nM DAPI stain solution to the culture media. Incubate for 10 min at 37 °C. Remove the staining solution and rinse the cells three times with pre-warmed PBS.
   2.2. To fix the cells, immediately add 1.5 ml of fixation solution (1% glutaraldehyde and 1% paraformaldehyde in 0.1 M sodium cacodylate solution [pH 7.0]) by gentle pipetting for 20 min under dark conditions at 4 °C (see Note 2).
   2.3. After fixing, wash the cells three times, for 10 min each time, with cold 0.1 M sodium cacodylate buffer at room temperature.
   2.4. Observe the overall location of the target cells in the gridded glass-bottom dish by using a 40× lens on the confocal microscope. Shoot each single image in the tile scan mode using the motorizing stage to produce a single large DIC image (Fig. 8a, b). The overview image serves as a map for the identification of specific target cells on the gridded bottom glass dish.
   2.5. Obtain high-magnification fluorescence images of the target cells that incorporate nanoparticles (Fig. 8c).
3. Sample preparation for SEM.
   3.1. Incubate cells for 20 min using a graded ethanol series (50%, 60%, 70%, 80%, 90%, 95%, and 100%) at room temperature.
   3.2. Remove the ethanol and incubate for 1 h in a 25% (v/v) HMDS ethanol mixture at room temperature (see Note 4).
   3.3. Remove the medium and incubate for 1 h in a 50% (v/v) HMDS ethanol mixture at room temperature.
   3.4. Remove the medium and incubate for 1 h in a 75% (v/v) HMDS ethanol mixture at room temperature.
   3.5. Remove the medium and incubate for 1 h in a 100% HMDS solution at room temperature.
   3.6. Air-dry the samples in dust-free conditions.
   3.7. Attach the bottom glass on aluminum stubs and then surface coat them using a 10-nm-thick platinum layer.
   3.8. Observe the glass surface under field emission SEM.

CLEM for the Identification of Specific Proteins Using Genetically Encoded Tag APEX2

Because the genetic probes do not require the permeability of the cell membrane to access the internal structure, they have the advantage of preserving the compact structure of the sample. In many previous studies, several gene probes, such as GFP (Horstmann et al. 2013), miniSOG (Shu et al. 2011), and APEX2 (engineered ascorbate peroxidase), have been reported using CLEM (Lam et al. 2015; Martell et al. 2017). GFP or miniSOG requires special light sources for DAB precipitation; however, APEX2 does not require special equipment. This section describes the procedures for preparing CLEM samples from cell monolayers.

1. Cell culture with patterned grid culture dish and cell transfection with plasmid vector.
   1.1. HeLa cells were cultured in a humidified atmosphere containing 5% CO_2 at 37 °C in DMEM supplemented with 10% FBS, 100 U/mL penicillin, and 100 U/mL streptomycin.
   1.2. Seed 1 × 10^5 cells by placing them onto gridded glass-bottom dishes. It is important to maintain the cells at low confluence in the gridded glass-bottom dishes.
   1.3. On the day after seeding of the cells, introduce the Mito-V5-APEX2 plasmid to the cells with transfection reagent when the cells have grown to 50–60% confluence.
2. Light microscope observation of DAB staining for APEX2.
   2.1. 16–24 h after transfection, to fix the cells, remove all the culture media and wash the cells briefly with pre-warmed PBS. Subsequently, immediately add 1.5 ml of fixation solution (1% glutaraldehyde and 1% paraformaldehyde in 0.15 M sodium cacodylate solution [pH 7.0]) by gentle pipetting at room temperature (see Note 1).
   2.2. Wash fixed cells with cold 0.15 M sodium cacodylate buffer three times on ice.
   2.3. Add 1 ml of cold 50 mM glycine solution and incubate for 10 min on ice, and then wash three times, for 10 min each time, with cold 0.15 M sodium cacodylate buffer.
3. Prepare a fresh 1× DAB solution. Add 500 μL of 1× DAB solution and incubate on ice until a light brown stain is visible under a light microscope for 20 min.
4. Remove the DAB solution and wash three times, for 10 min each time, with cold 0.15 M sodium cacodylate buffer.
5. Observe using a phase-contrast microscope to record the location of the brown stained cells.
6. EM sample preparation and ultra-sections are similar to procedures described in 4.1-7.1.

Results

In the present study, the CLEM method was used to detect Si-FMNP locations in the cell. This protocol presents a method for delivering Si-FMNPs (or specific proteins) into eukaryotic cells (or out of cells) and monitoring their localization and effects on the structures of cellular organelles based on light microscopy and electron microscopy resolutions. Flowcharts of the procedures are presented in Figs. 1 and 2. Confocal microscopy Z-stack images were obtained using a Si-FMNP fluorescence dye, which incorporated Si-FMNPs into the cytosol and nuclei of the HeLa cell (Fig. 4). The cells of interest identified under fluorescence microscopy were well correlated with the low-magnification TEM images (Fig. 5). In addition, the nanoparticles identified in the z-slice fluorescent image were located in the cytosol and nucleus in the high-magnification TEM images (Fig. 6). Based on a 3D

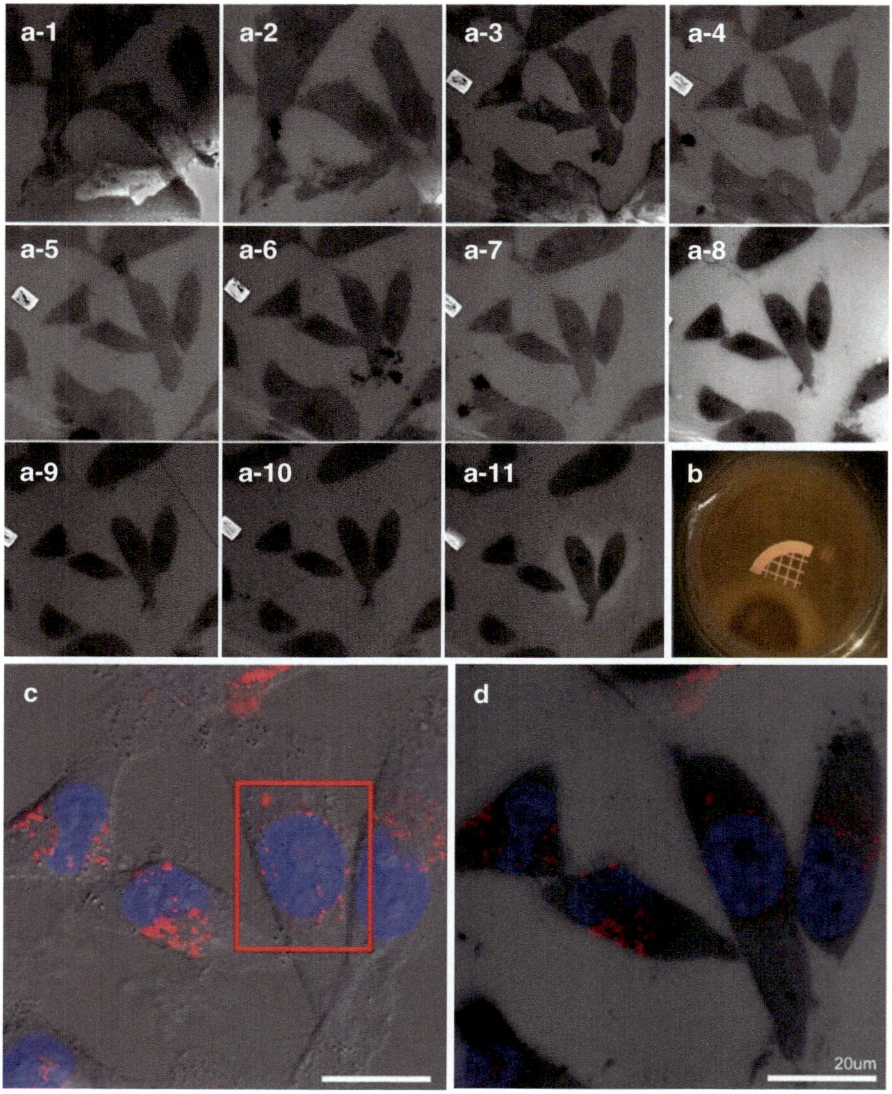

Fig. 5 Low-magnification, low-voltage TEM serial images. (**a**1–**a**11) Images of serial sections of target cells acquired using low-voltage TEM. Section thickness is 250 nm. (**b**) The gold grid mesh contains target of interest. (**c**) Overlaid image of bright field and the fluorescence signatures of RITC-loaded silica NPs is shown. The red box indicates the cell containing NPs in the nucleus as well as in the cytosol that will be visualized using CLEM in the subsequent step. (**d**) Correlated images of a single section shown in panel A8 and fluorescence. Red and blue fluorescence are originated from Si-FNMP and DAPI staining, respectively

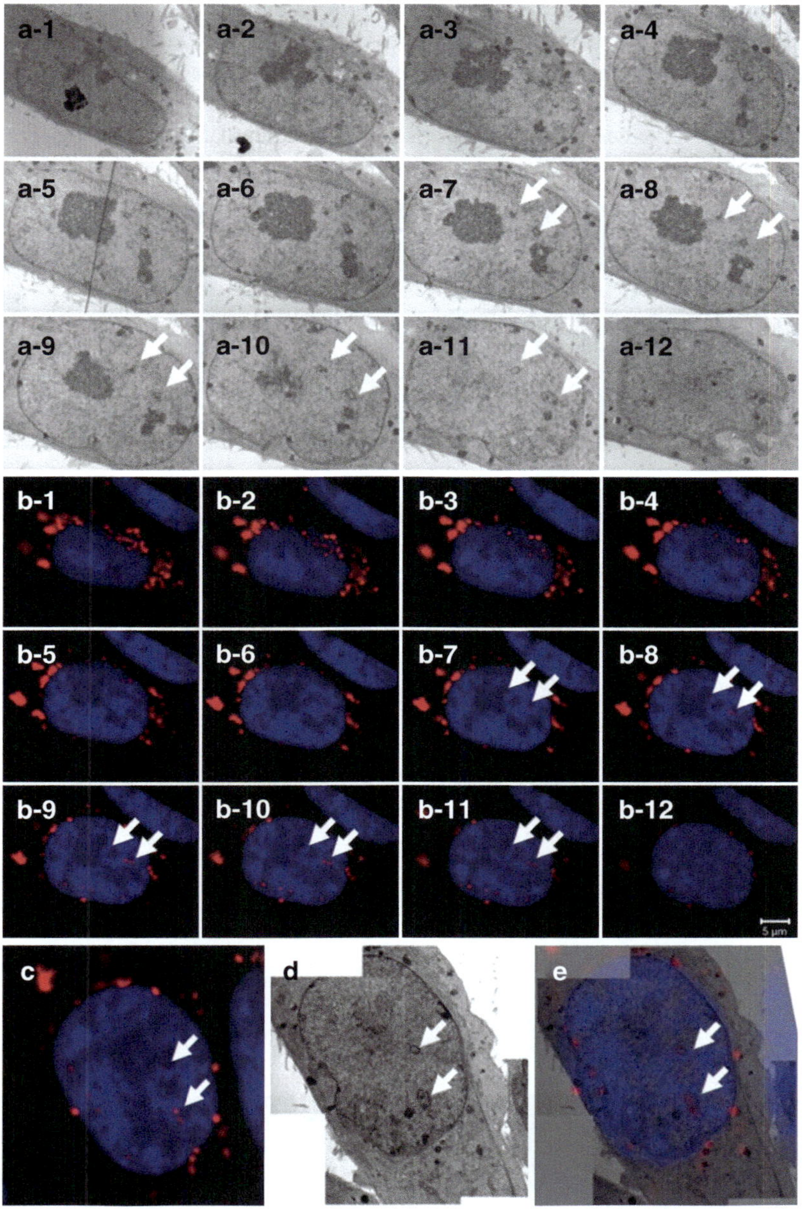

Fig. 6 High-magnification serial images acquired using TEM and z-slice confocal images. (**a**1–**a**12) High-magnification serial images acquired using TEM. (**b**1–**b**12) Z-slice fluorescent image of the cell nucleus. Optical-slice resolution is 0.5 μm. (**c**) White arrows indicate Si-FMNPs localized on the nucleus. (**d**) Electron microscopic image of the target cell corresponding to the fluorescence image shown in C. White arrows indicate an unknown nuclear structure surrounded by membrane. (**e**) Correlated fluorescence and EM images of C and D are shown. Fluorescence of NPs localization in the nucleus is consistent between the fluorescence microscopy and EM images, although some Si-FMNPs located in the cytosol were not

reconstruction of the serial TEM image, we could identify the exact locations of the Si-FMNP particles in a tunnel-like structure with the nuclear envelope in the nucleus (Fig. 7). To confirm the interaction between the NPs and the cell surface, the position of the cell of interest in the gridded glass-bottom dish was identified (Fig. 8), and the NPs on the cell surface were observed under an SEM microscope (Fig. 9).

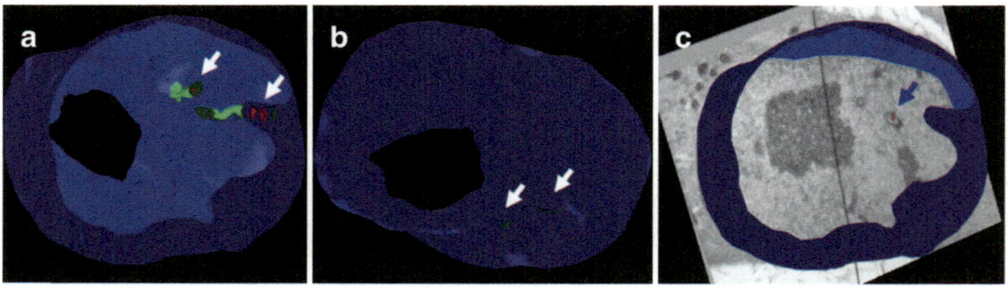

Fig. 7 3D reconstruction image obtained from serial sections of Si-FMNPs incorporated in the nucleus. Surface model rendering of the nuclear membrane (blue), inner tunnel-like structure (green), and nanoparticles (red) was carried out. (**a**) Views from bottom, (**b**) from top, and (**c**) from middle of the sliced images are shown, respectively

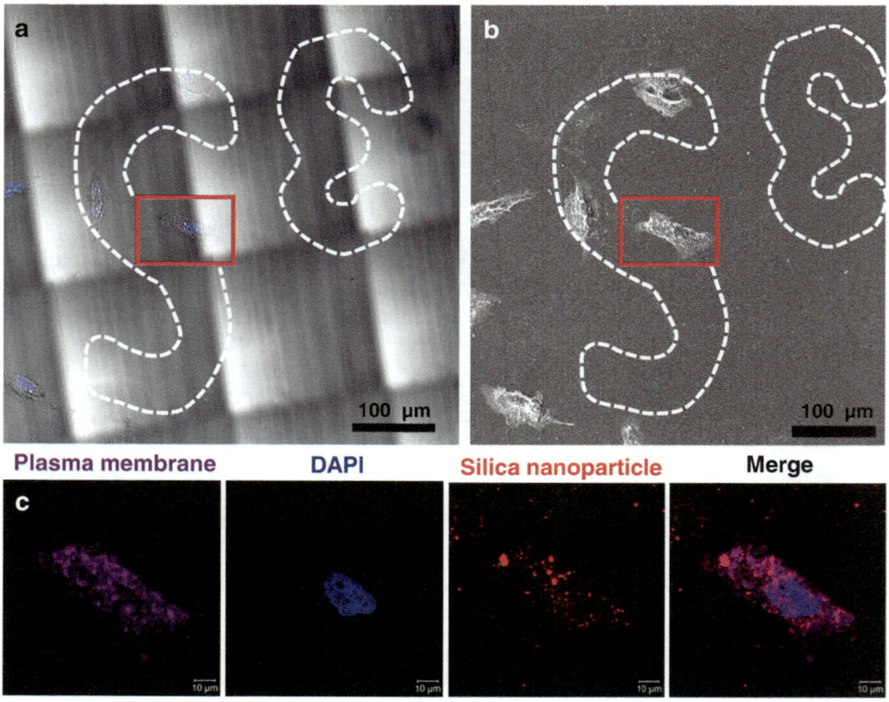

Fig. 8 Identification and selection of cells with silica NPs. (**a**) A tiled image of Si-FNMP treated HeLa cells on the gridded bottom dish based on confocal microscopy. The red box presents a selected target cell for electron microscopy observation. (**b**) Low-magnification SEM image. (**c**) High-resolution fluorescence images of the identical cell shown A and B was obtained through confocal microscopy

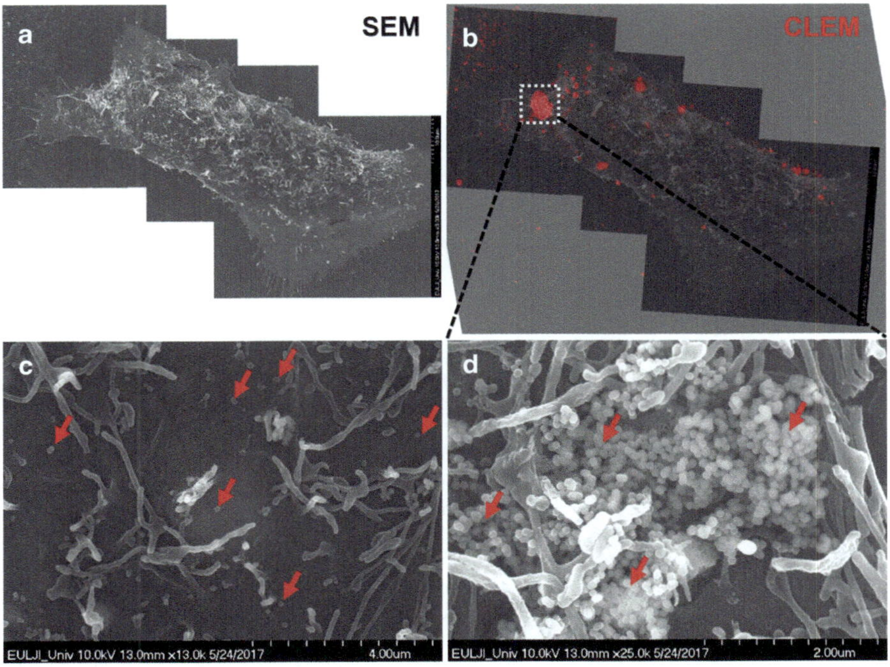

Fig. 9 Correlative micrograph of fluorescence and scanning electron microscopy. (**a**) High-resolution SEM images of the target cell shown in Fig. 8. (**b**) Correlative image from fluorescence microscopy and SEM. (**c**) High-magnification image of a part of the cell surface shown in B, in which single NPs are broadly distributed. (**d**) High-magnification images of the target cell surfaces (dotted line box in **b**). Red arrowheads indicate surface-bound Si-FMNPs

Recently, a genetically modified tag (APEX2) has been developed as a method for identifying the location of a specific protein in a cell at the nanometer scale using EM. We cultured HeLa cells and confirmed the expression of mito-linked APEX2 (mitochondrial matrix protein) in transfected cells (Fig. 10a). A low-magnification EM image showed the APEX2 staining of the mitochondria (Fig. 10c, e) or not (Fig. 10b, d). In high-magnification EM images, the mitochondrial matrix was stained by APEX2, not the intermembrane space (Fig. 10e).

Summary

CLEM is a unique method for the visualization of cellular structures of interest or specific events, which are traced using confocal microscopy observation of a live cell, thus facilitating a detailed EM analysis. Therefore, users can evaluate the specific structures and status of fluorescent probes under EM by providing details of their previous images obtained through confocal microscopy. The procedures and

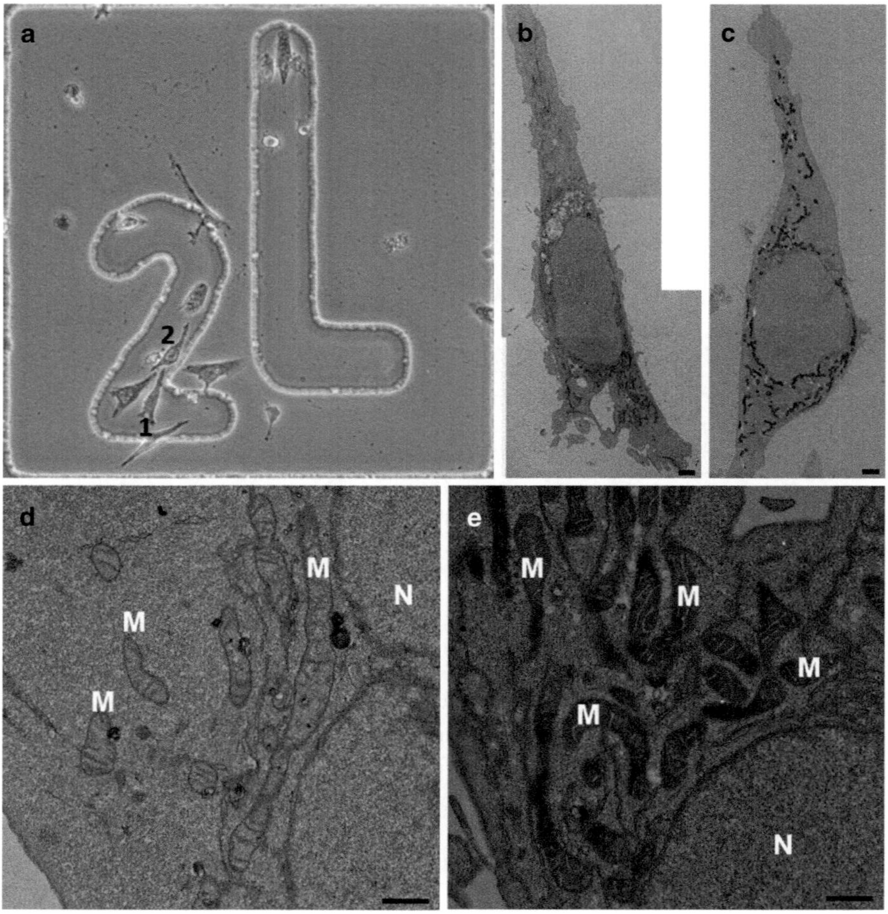

Fig. 10 Genetically encoded reporter gene (APEX2) for correlative light and electron microscopy. (**a**) The phase-contrast images show HeLa cells after DAB staining. The numbers 1 and 2 indicate the two cells to be analyzed. (**b**, **c**) TEM images of the two cells with low magnification. Dark contrast in the cells indicates that DAB is significantly precipitated mitochondria. Scale bar, 2 μm. (**d**) High-magnification TEM image of untransfected HeLa cells is shown. The mitochondria with non-DAB precipitated formation is presented. Scale bar, 500 nm. (**e**) TEM image of a transfected HeLa cell is shown. Dark contrast is apparent in the mitochondria matrix and not cristae. Scale bar, 500 nm. M and N stand for the mitochondria and nucleus, respectively

protocol described here provide a detailed example of 3D confocal imaging, fixation, and labeling, in addition to TEM and SEM processing of CLEM specimens. We believe that our method could be applied in the field of convergence science and in the cell biology research community. We also expect it to be a useful platform for further CLEM development and improvement for the evaluation of the efficacy of drug-loaded nanocarriers in various cells and tissues.

Notes

1. Glutaraldehyde and other chemical fixative solutions are highly toxic. Therefore, one must wear protective clothing and work in a fume hood.
2. The mesh grid facilitates the identification of a region of interest. However, the use of a glass dish with a gridded bottom glass dish is convenient.
3. Cell confluence is vital since a full plate makes it difficult to identify the cells of interest under electron microscopy. To find the proper orientation in the EM images, both the shapes of the cells and empty spaces are used as landmarks.
4. Because the cells in the culture dish are dissolved using HMDS, the bottom glass is separated and placed in a glass petri dish (small diameter). The cell surfaces should face upwards.

Acknowledgments We thank the Confocal Microscope and Electron Microscopy core facilities at the ConveRgence mEDIcine research cenTer (CREDIT), Asan Medical Center and Brain Research Core Facilities in Korea Brain Research Institute for support and instrumentation. This chapter was supported by Basic Science Research Program (NRF-2014R1A1A2058183) through the National Research Foundation of Korea (NRF) funded by the Ministry of Education and KBRI basic research program (21-BR-01-11) through Korea Brain Research Institute funded by the Ministry of Science and ICT. This study was also supported by grants (2020IP0028) from the Asan Institute for Life Sciences, Asan Medical Center.

References

Abandowitz HM, Geissinger HD (1975) Preparation of cells from suspensions for correlative scanning electron and interference microscopy. Histochemistry 45(2):89–94

Bykov YS, Cortese M, Briggs JA, Bartenschlager R (2016) Correlative light and electron microscopy methods for the study of virus-cell interactions. FEBS Lett 590(13):1877–1895

Darcy KJ, Staras K, Collinson LM, Goda Y (2006) An ultrastructural readout of fluorescence recovery after photobleaching using correlative light and electron microscopy. Nat Protoc 1(2):988–994

Han S, Raabe M, Hodgson L, Mantell J, Verkade P, Lasser T, Landfester K, Weil T, Lieberwirth I (2019) High-contrast imaging of nanodiamonds in cells by energy filtered and correlative light-electron microscopy: toward a quantitative nanoparticle-cell analysis. Nano Lett 19(3):2178–2185

Horstmann H, Vasileva M, Kuner T (2013) Photooxidation-guided ultrastructural identification and analysis of cells in neuronal tissue labeled with green fluorescent protein. PLoS One 8(5):e64764

Ishikawa-Ankerhold HC, Ankerhold R, Drummen GP (2012) Advanced fluorescence microscopy techniques – FRAP, FLIP, FLAP, FRET and FLIM. Molecules 17(4):4047–4132

Jin R, Zeng C, Zhou M, Chen Y (2016) Atomically precise colloidal metal nanoclusters and nanoparticles: fundamentals and opportunities. Chem Rev 116(18):10346–10413

Kulkarni RV, Setty CM, Sa B (2010) Polyacrylamide-g-alginate-based electrically responsive hydrogel for drug delivery application: synthesis, characterization, and formulation development. J Appl Polym Sci 115(2):1180–1188

Lam SS, Martell JD, Kamer KJ, Deerinck TJ, Ellisman MH, Mootha VK, Ting AY (2015) Directed evolution of APEX2 for electron microscopy and proximity labeling. Nat Methods 12(1):51–54

Martell JD, Deerinck TJ, Sancak Y, Poulos TL, Mootha VK, Sosinsky GE, Ellisman MH, Ting AY (2012) Engineered ascorbate peroxidase as a genetically encoded reporter for electron microscopy. Nat Biotechnol 30(11):1143–1148

Martell JD, Deerinck TJ, Lam SS, Ellisman MH, Ting AY (2017) Electron microscopy using the genetically encoded APEX2 tag in cultured mammalian cells. Nat Protoc 12(9):1792–1816

Pack CG, Song MR, Tae EL, Hiroshima M, Byun KH, Kim JS, Sako Y (2012) Microenvironments and different nanoparticle dynamics in living cells revealed by a standard nanoparticle. J Control Release 163(3):315–321

Pack CG, Yukii H, Toh-e A, Kudo T, Tsuchiya H, Kaiho A, Sakata E, Murata S, Yokosawa H, Sako Y, Baumeister W, Tanaka K, Saeki Y (2014b) Quantitative live-cell imaging reveals spatio-temporal dynamics and cytoplasmic assembly of the 26S proteasome. Nat Commun 5:3396

Pack C-G, Jung M-K, Song M-R, Kim J-S, Han S-S, Sako Y (2014a) Chapter 11 – Use of engineered nanoparticle-based fluorescence methods for live-cell phenomena. In: Cornea A, Conn PM (eds) Fluorescence microscopy. Academic, Boston, pp 153–169

Patra JK, Das G, Fraceto LF, Campos EVR, Rodriguez-Torres MDP, Acosta-Torres LS, Diaz-Torres LA, Grillo R, Swamy MK, Sharma S, Habtemariam S, Shin HS (2018) Nano based drug delivery systems: recent developments and future prospects. J Nanobiotechnol 16(1):71

Radu DR, Lai CY, Jeftinija K, Rowe EW, Jeftinija S, Lin VS (2004) A polyamidoamine dendrimer-capped mesoporous silica nanosphere-based gene transfection reagent. J Am Chem Soc 126(41):13216–13217

Regier MC, Taylor JD, Borcyk T, Yang Y, Pannier AK (2012) Fabrication and characterization of DNA-loaded zein nanospheres. J Nanobiotechnol 10:44

Rieder CL, Bowser SS (1985) Correlative immunofluorescence and electron microscopy on the same section of epon-embedded material. J Histochem Cytochem 33(2):165–171

Sahl SJ, Hell SW, Jakobs S (2017) Fluorescence nanoscopy in cell biology. Nat Rev Mol Cell Biol 18(11):685–701

Sailor MJ, Park JH (2012) Hybrid nanoparticles for detection and treatment of cancer. Adv Mater 24(28):3779–3802

Shu X, Lev-Ram V, Deerinck TJ, Qi Y, Ramko EB, Davidson MW, Jin Y, Ellisman MH, Tsien RY (2011) A genetically encoded tag for correlated light and electron microscopy of intact cells, tissues, and organisms. PLoS Biol 9(4):e1001041

van Rijnsoever C, Oorschot V, Klumperman J (2008) Correlative light-electron microscopy (CLEM) combining live-cell imaging and immunolabeling of ultrathin cryosections. Nat Methods 5(11):973–980

Webster RE, Osborn M, Weber K (1978) Visualization of the same PtK2 cytoskeletons by both immunofluorescence and low power electron microscopy. Exp Cell Res 117(1):47–61

Immuno-gold Techniques in Biomedical Sciences

Minkyo Jung, Tae Keun Kim, Chan-Gi Pack, and Ji Young Mun

Principle of Immuno-gold Staining

Antigen–antibody reactions are used widely in light and electron microscopy analyses. Among applications of light microscopy, immunocytochemistry is used in studies of cellular contents as a diagnostic tool, whereas electron microscopy is predominantly used to study locations of antigens in cellular organelles with high-resolution data. In these techniques, antibodies recognize antigens, such as proteins, carbohydrates, and lipids, that have several antibody binding sites comprising specific amino acids or polysaccharides in localized regions. For successful immunolabeling, researchers require very specific antibodies, high-performance electron-dense probes, and specific preparation techniques that preserve the antigenicity and morphology of specimens. As antibody, both polyclonal and monoclonal antibodies can be used. Polyclonal antibodies allow strong aldehyde fixation, but

M. Jung · J. Y. Mun (✉)
Neural Circuit Research Group, Korea Brain Research Institute, Daegu, South Korea
e-mail: jymun@kbri.re.kr

T. K. Kim
Convergence Medicine Research Center (CREDIT), Asan Institute for Life Sciences, Asan Medical Center, Seoul, South Korea

C.-G. Pack
Convergence Medicine Research Center (CREDIT), Asan Institute for Life Sciences, Asan Medical Center, Seoul, South Korea

Department of Convergence Medicine University of Ulsan College of Medicine, Asan Institute for Life Sciences, Asan Medical Center, Seoul, Korea (Republic of)
e-mail: changipack@amc.seoul.kr

J. K. Kim et al. (eds.), *Advanced Imaging and Bio Techniques for Convergence Science*, Advances in Experimental Medicine and Biology,
https://doi.org/10.1007/978-981-33-6064-8_7

monoclonal antibodies require lighter fixation because they are highly specific to few amino acids of the antigen. Colloidal gold is the most used probe (Ferguson et al. 1998; Singer 1959), because it provides electron density that is clear (Bendayan et al. 1987) under electron microscopy. Colloidal gold particles are 1–20 nm in diameter (10–20 nm is most common) and allow multiple labeling. Whereas large probes can be used for low magnification images, smaller probes are superior in high magnification applications. Because smaller colloidal gold particles have reduced hindrance and increased diffusion rates, particle sizes are selected on the basis of labeling intensities, efficiency, and magnification of imaging. Depending on the probe, several methods have been used for several decades (Fig. 1). In direct methods, primary antibodies are linked to ferritin or gold particles and directly bind to the antigen and can be observed immediately. These techniques allow produce the highest resolution, with a maximum of 1 gold particle per antigen. Due to their specificity, these techniques are also suited for surface labeling of sections but require well-characterized primary reagents. The most commonly used technique is a two-step method in which immuno-gold conjugates for protein A/G and secondary antibodies are used. Gold particles of various sizes are conjugated to protein A (Bendayan and Zollinger 1983), protein G, and secondary antibodies. When protein A/G is used, one gold probe is bound per primary antibody, producing the highest resolution of two-step labeling, and providing a universally applicable technique for many antibody subclasses. In contrast, when secondary antibodies are used, more gold probes can be bound per primary antibody, facilitating reactions with primary antibody fragments and

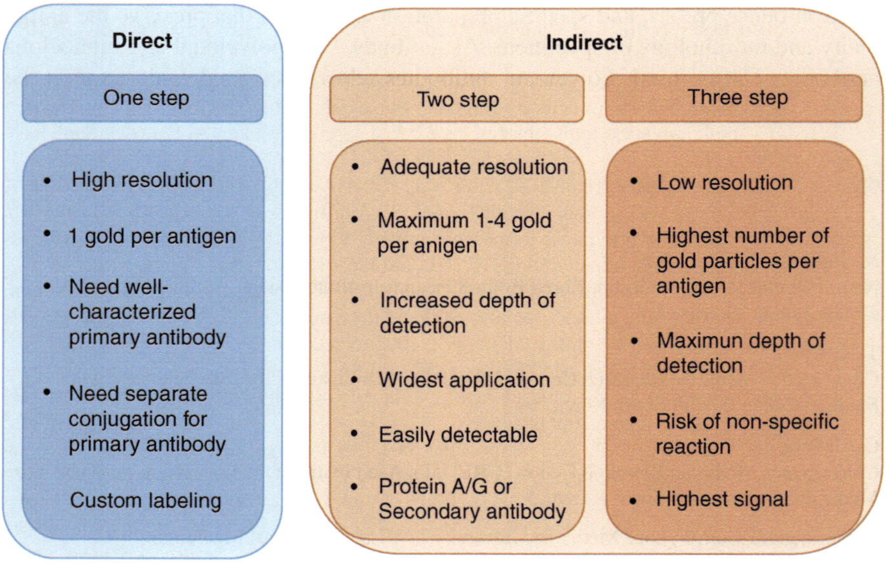

Fig. 1 Detection methods for immuno-gold techniques. For high resolugion, direct method is best. And for easy detect, indirect methods is better techniques

detection at low magnification. Secondary antibodies bind more specifically than protein A/G. In these manipulations, primary antibodies are reacted with antigen, and gold particle-conjugated secondary antibodies are used to detect primary antibodies. This is the most commonly used immuno-gold staining technique. In principle, 1–4 gold particles can be bound per antigen, resulting in advantages of adequate resolution and wide application. Two-step methods are compared briefly in Fig. 1. Other indirect immuno-electron microscopy methods involve three-step processing using horseradish peroxidase (HRP) tagged antibodies, which react with 3,3′-Diaminobenzidine (DAB) signals to produce electron-dense signals. Gold can then be used to produce dot signals in immuno-electron microscopy analyses. The resolution of these methods is better suited for light microscopy, and their advantages include the highest numbers of gold particles per antigen, maximum depths of detection, and the highest signal intensities. But the increased risk of unintended reactions is a clear disadvantage. Commonly, high specificity of reactions, and hence low background signals, are the most important parameters for successful immuno-gold labeling. Because preservation of antigen and ultrastructure is different on basis of sample preparation process such as fixation, embedding media, the order of antigen–antibody reaction, and so on, researchers have to decide which technique is best for their purpose. In principle, antigens are stabilized by fixation during preparation for exposure to the antigen, antigen–antibody reactions are performed before (pre-embedding) or after plastic embedding (post-embedding), and probes are visualized for specific antigen–antibody reactions (Fig. 2). In addition, nonspecific binding sites on samples can be blocked by 1% gelatin, free aldehydes can be blocked by 0.1 M glycine, and nonspecific sites are generally blocked by 1% bovine serum albumin. Reagents and their concentrations vary depending on specimens and are generally identified using trial and error. Procedures for reducing background will be discussed in the troubleshooting section.

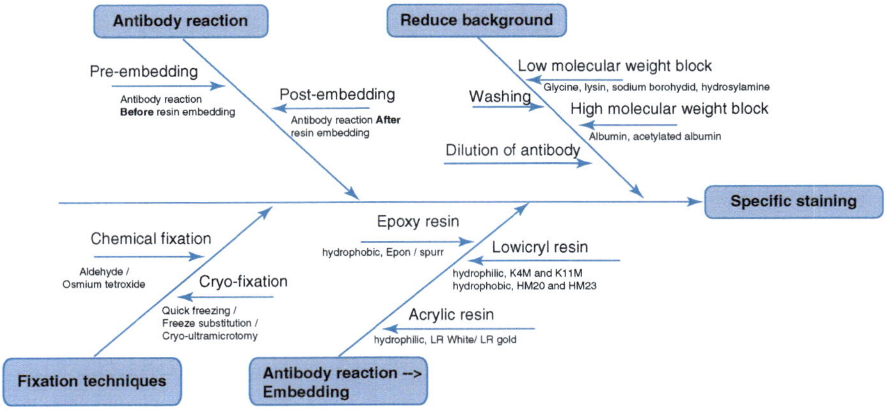

Fig. 2 Considerations for immuno-gold techniques. Techniques for fixation, reducement of background, antibody reaction, and embedding should be selected for specific purpose

Sample Preparation for Antigen Preservation

When preparing samples for antibody labeling, it is important to consider whether ultrastructure or antigen preservation is more important and whether the target is located in intracellular, extracellular, or membrane compartments. Preservation of antigen and ultrastructure is achieved using sample preparation techniques including various fixatives, antigen retrieval followed by antibody binding, and selection of appropriate embedding media. Preparation methods can be divided into pre- or post-embedding techniques depending on when the antibody reacts with antigen, as shown in Fig. 3. In pre-embedding techniques, antigen–antibody reactions are performed before resin embedding of samples, and ultrathin sections are then cut and observed under electron microscopy. In post-embedding methods, antigen–antibody reactions are performed after resin embedding and ultrathin sectioning.

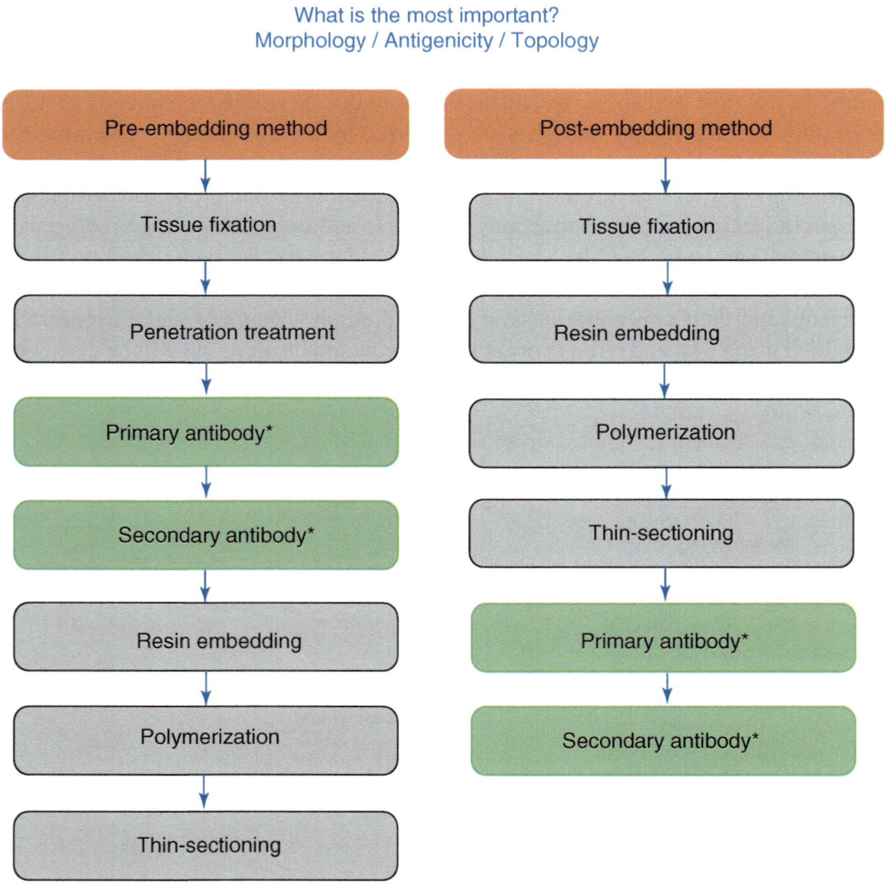

Fig. 3 Schematic workflow of pre- and post-embedding labeling procedures. Pre-embedding favors antigen preservation and post-embedding favors sample morphology

Therefore, pre-embedding methods usually have better sensitivity for antigen–antibody reactions, although weak penetration of antibodies can be a disadvantage. In contrast, post-embedding methods (Roth 1986) give better penetration of antibodies but poor preservation of antigen. These advantages and disadvantages are explained in detail in the following sections.

Pre-embedding Method

Pre-embedding immuno-gold methods have been used for a long time, but their applications are limited due to invasiveness. In general, only surface antigens of samples can be labeled using pre-embedding methods, but low concentrations of detergent or freezing/thawing can help to expose cytoplasmic antigens. Therefore, it is not easy to optimize the balance between morphology and antigen labeling. Pre-embedding has the disadvantage of weak antibody penetration but has the following advantages: First, antigens that may be destroyed during dehydration and embedding by post-embedding methods can be preserved. Therefore, pre-embedding methods are desirable when antigens are very sensitive or are present at low concentrations. Second, the use of osmium tetroxide (OsO_4) after antigen–antibody reactions can be advantageous for membrane fixation in pre-embedding method. Because most antigens are very sensitive to OsO_4, the use of OsO_4 must be avoided or minimized in post-embedding methods. Third, pre-embedding can be applied to correlative light and electron microscopy studies (Polishchuk and Polishchuk 2019) using DAB or green fluorescent protein (GFP). Because advanced techniques show increased resolution from combinations of cryo-techniques, these will be developed with other advanced techniques for the preparation of samples and specific tags (Nisman et al. 2004).

Pre-embedding Method with DAB Reactions and Plastic Embedding

Avidin-biotin-peroxidase reactions (Hsu et al. 1981), DAB reactions, post-fixation, dehydration, plastic embedding, and ultrathin sectioning are performed after antigen–antibody reactions and prior to observations using electron microscopy (Polishchuk and Polishchuk 2019).

1. Perfuse samples of animal tissue with 4% paraformaldehyde and 0.1% glutaraldehyde.
   Caution: Aldehyde fumes are toxic. All work should be conducted in a ventilated fume hood.
   All fixatives must be used on the day they are made up.
   - If the sample is not animal tissue, omit perfusion.
2. Make 100 μm thick sections using a vibratome (Piekut and Casey 1983).
3. Fix samples with 4% PFA and 0.1% glutaraldehyde.

4. Wash samples with phosphate buffer saline (PBS).
5. Use 2.3M sucrose for cryo-protection.
6. For antibody penetration, freeze and thaw by incubating for 1 min in liquid nitrogen.
   - If needed, repeat this step 3–4 times.
7. Wash samples with PBS.
8. Quench free aldehyde by incubating samples in PBS containing 1% $NaBH_4$ for 30 min.
9. Wash samples with PBS
10. To increase cell permeability, incubate samples in 3% H_2O_2.
11. Wash samples with PBS.
12. Block nonspecific antibody binding using 10% normal serum.
13. Incubate samples with primary antibody.
    - Aliquots of antibody must be stored at −20 °C or lower, and diluted solutions must be stored at 4 °C. Because repeated freezing and thawing destroys the antibody, it should be avoided as much as possible.
14. Wash samples with PBS.
15. Block specific antibody binding sites with 2% normal serum for 10 min.
16. Incubate samples with biotinylated secondary antibody.
17. Wash samples with PBS.
18. Incubate samples with avidin for 1 h.
19. Wash samples with PBS.
20. React samples with Ni-DAB.
    - Prepare Ni-DAB reaction solution by mixing 4.6 ml of PB, 50 μl of 0.4% NH_4Cl, 50 μl of 20% D-glucose, 200 μl of 1% Ni ammonium sulfate, 100 μl of DAB (2.5 mg) and glucose oxidase at 1-μl/ml.
21. Wash sample with PBS
22. Post-fix samples with 2.5% glutaraldehyde and 1% OsO_4 in 0.1-M PB for 30 min.
23. Wash sample with PBS
24. Incubate samples in a solution of 3:1 acetone:epoxy embedding mixture for 30 min.
25. Remove the medium and add 1:1 acetone:embedding mixture medium and then incubate for 30 min.
26. Remove the medium and add 1:3 acetone:embedding mixture medium and then incubate for 30 min.
27. Remove the medium and add 100% embedding mixture and incubate overnight at room temperature.
28. Embed samples in pure epoxy embedding mixture and incubate for 24 h at 65 °C.
29. Cut sections of 50–70-nm thickness using an ultramicrotome, and stain with 2% uranyl acetate for observation using tandem electron microscopy (TEM).
    - If cells have been grown on glass coverslips, coverslips can be removed by dropping into liquid nitrogen after the resin is polymerized.

Pre-embedding Method with GFP, Immuno-gold, and Plastic Embedding

In this procedure, sections are imaged under a confocal microscope and locations of GFP fluorescent signals are recorded by labeling with immuno-gold conjugated GFP antibody and analyzing under electron microscopy.

1. Fix sample with 2% paraformaldehyde and 0.1–2% glutaraldehyde in PB for 1 h at room temperature.
2. Wash sample three times in PB buffer for 20 min each.
3. Identify GFP fluorescent signals in sections using confocal microscopy.
4. Wash sample with PB buffer for 20 min.
5. Quench free aldehyde with 50 mM glycine in PB for 20 min.
6. Wash sample with PB buffer for 20 min.
7. Immerse samples in 15% sucrose in 0.1 M PB for 1 h.
8. Immerse samples in 30% sucrose in 0.1 M PB for 2 h.
9. Freeze and thaw samples by incubating for 1 min in liquid nitrogen. Repeat the freeze and thaw if necessary.
10. Wash samples with PB.
11. Wash samples with Tris-buffered saline (TBS).
12. Block nonspecific binding in TBS containing 10% normal serum and 1% Fish skin gelatin (FSG) for 1 h.
13. Incubate samples with primary anti-GFP antibody in TBS containing 1% normal serum, 0.1% FSG, 0.05%, and NaN_3 at 4 °C.
14. Wash samples with TBS.
15. Incubate samples with secondary antibody-conjugated nanogold in TBS containing 1% bovine serum albumin (BSA), 0.1% FSG, and 0.05% NaN_3 at 4 °C.
16. Wash samples with PBS.
17. Fix samples with 1% glutaraldehyde in PBS.
18. Wash samples with PBS.
19. Enhance nanogold signals using a nanoprobe enhancing kit.

Pre-embedding Method with GFP, DAB, and Cryo-techniques

Recently, pre-embedding methods with cryo-techniques, including cryo-fixation, have been applied to genetically engineered samples. In these studies, samples are typically frozen under high pressure and are later processed in the vitreous state directly using a focused ion beam (FIB) system or cryo-ultramicrotomy. After observation of GFP signals under cryo-light microscopy, cryo-tomography can then be applied using a Cryo-TEM. This workflow includes vitreous freezing, cryo-fluorescence microscopy, targeted sample sectioning using FIB or cryo-ultramicrotomy, and high-resolution cryo-electron tomography (Cryo-ET). The ensuing technique allows nanometer-scale 3D modeling of biological structures, such as native proteins, intact cells, lipid rafts, and cellular spaces, in their native cellular contexts. These techniques are, however, limited by section thickness and radiation damage and can require the use of a set of equipments, including cryo-light microscopy, Cryo-TEM, cryo-FIB, and cryo-ultramicrotome instruments.

In studies of cultured cells, specific processes are employed for the preparation of girds and the selection of suitable cells. The best grids are made of gold, because other metals, such as copper and nickel, are cytotoxic and are hence unsuitable for cell cultures. Grid mesh sizes of 200 and 300 mesh are commonly used, and the mesh number should not be >300 Mesh. Although thinner bat mesh maintains a sufficient transparent space between the grid bars when specimens are tilted to high angles, mesh numbers lower than 100 tend to break more frequently during handling and vitrification of specimens. As support films, formvar and carbon are most commonly used. Carbon better facilitates evaporation from both sides, leading to optimal strength, but is disadvantaged by deteriorating resolution with greater thickness. To address this disadvantage, holey carbon films have been developed. The use of these films was supported by reduced background noise during imaging of cells through the holes. Under these conditions, increased hydrophilicity of the grid also improves surface contact with the medium during cell culture. In cryo-electron tomography of whole cells, cell types such as fibroblasts and neurons with thin shapes have been used most frequently. Limits of thickness that can be imaged by TEM depend on the voltages used. In most cases, about 400 nm is considered a maximum thickness for imaging of frozen-hydrated specimens using TEM at 200 kV. However, due to limited penetration of the beam, cubic and columnar shape cells and structures near cell nuclei cannot be imaged using these techniques without sectioning. In other words, bigger cells require sectioning using cryo-ultramicrotomy or FIB.

1. Prepare formvar-coated grids using 1% solutions of chloroform or dichloromethane. Dip glass slides into the solution, dry, and then float formvar layers on water. Subsequently, place grids onto the film layer and retrieve them with a glass slide. Evaporate carbon on both gird sides. Finally, glow discharge the formvar carbon-coated grids after placing on glass slides with the carbon side facing up for 1 min.
2. Sterilize EM grids before cell growth by UV radiation for 15 min. Place cells on EM grids on the bottom of the culture dish overnight using standard cell culture protocol.
3. Use light microscopy to check for proper cell attachment to the grids, confluency, and spreading. When few cells are attached to the grid, increase confluence over further cell culture times.
4. Take images of cells with GFP signals.
5. Treat samples with low concentrations of saponin to enhance penetration of antibody.
   - This procedure can be replaced by freeze–thaw methods (See protocol 1.2)
6. Use anti-GFP antibody for pre-embedding labeling using the Fab fragments of IgG conjugated with ultra-small 1.4 nm gold particles (Hainfeld and Furuya 1992) (See protocol 1.1).

7. Vitrify samples using a plunge freezer, such as a FEI vitrobot or a homemade plunge freezer. Control temperature and humidity before and during blotting at 37 °C and 100% humidity (blot time, 1–2 s; wait time, 0 s; drain time, 0 s; blot force, 0–25; and blot total, 1s).
8. Immerse girds in liquid ethane on liquid nitrogen.
9. Use cryo-ultramicrotomy or cryo-FIB for sectioning if necessary.
10. Transfer vitrified grid from the storage box to a Gatan cryo-electron tomography holder, that has been baked and pumped to low pressure. Subsequently, transfer samples from the cryo station into the cryo-electron microscope chamber.
11. Acquire tilt series under low-dose conditions (<100 E/A^2 cumulative dose). Record images at 1° or 2° tilt increasements between ±90° or ±60°.

Post-embedding Methods

In post-embedding methods, immuno-staining of ultrathin sections is performed on grids after fixation, dehydration, and plastic embedding of samples. Various fixation and embedding techniques (Fig. 4) are used to preserve antigen and ultrastructure. Cryo-fixation, low-temperature embedding using Lowicryl (Acetarin et al. 1986; Roth et al. 1981), room temperature embedding using LR white (Mandry et al. 1993), and cryo-ultramicrotomy (Bos et al. 2011) preserve antigens well, but membrane contrast is better achieved after chemical fixation and epoxy embedding.

Post-Embedding Method with Chemical Fixative, Epon Resin or Lowicryl Embedding

Post-embedding with chemical fixation is typically an excellent starting point for most projects. As embedding media, Lowicryl reportedly enhances immunolabeling (Valentino et al. 1985) and is compatible with post-embedding techniques.

Chemical Fixation Epoxy Embedding

1. Fix samples with 1% glutaraldehyde and 1% PFA in 0.1 M sodium cacodylate solution (pH 7.0) for 1 h.
2. Wash the sample with 0.1 M sodium cacodylate buffer for 10 min three times.
3. Post-fix the sample with 2% Osmium tetroxide for 1 h at 4 °C.
4. Wash the sample three times with water for 10 min each.
5. Incubate samples with a graded acetone series (50%, 60%, 70%, 80%, 90%, 95%, 100%, respectively) for 20 min on the shaker.
6. Incubate samples with 3:1 acetone: epoxy embedding mixture for 30 min.
7. Remove the mixture, add 1:1 acetone: the embedding mixture, then incubate for 30 min.
8. Remove the mixture, add 1:3 acetone: the embedding mixture, then incubate for 30 min.

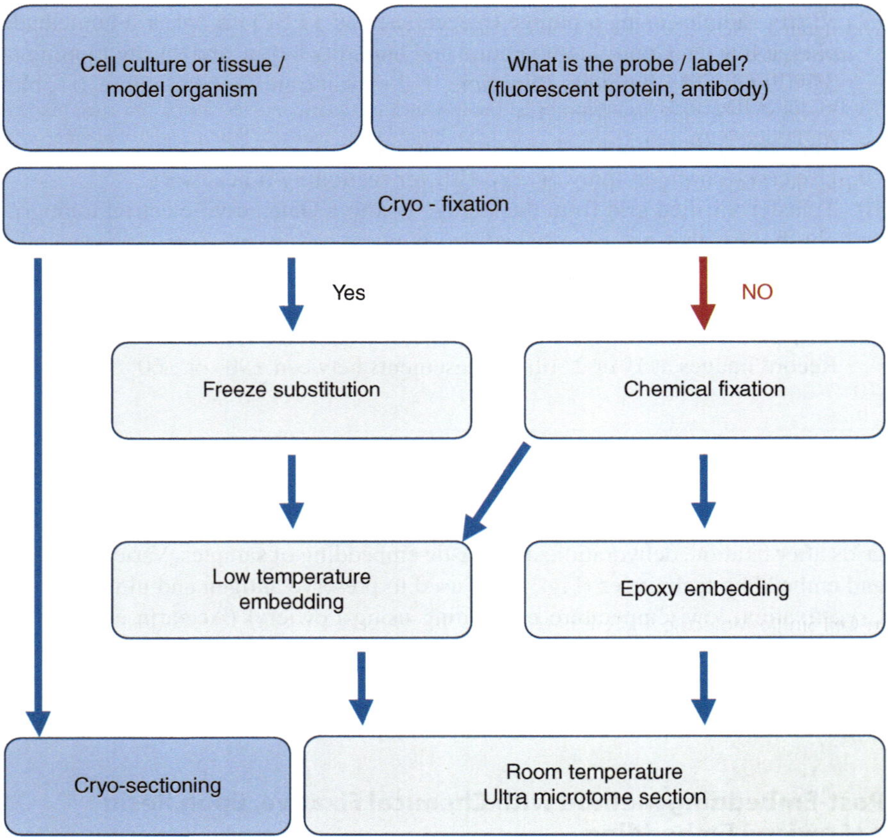

Fig. 4 Different techniques for preservation of antigen and ultrastructures. Depending on more importance among labeling efficiency and ultrastructure, adjust method should be chosen. Contrast is better in chemical fixation, and antigen preservation is better in cryo-fixation

9. Remove the mixture, add 100% embedding mixture, and incubate overnight at room temperature.
10. Embed samples in pure epoxy embedding mixture and bake for 24 h at 65 °C.

Chemical Fixation with Lowicryl HM20 Embedding

1. Fix sample with 1% glutaraldehyde and 1% PFA in 0.1 M sodium cacodylate buffer (pH 7.4) at 4 °C.
2. Wash samples three times with 0.1 M sodium cacodylate buffer (pH 7.4) for 30 min at 4 °C.
3. Post-fix samples with 2% OsO_4 in 0.1 M sodium cacodylate buffer (pH 7.4) for 2 h at 4 °C.

4. Wash samples three times with 0.1 M sodium cacodylate buffer (pH 7.4) for 30 min at 4 °C.
5. Dehydrate samples using ethanol series (70%, 80%, 90%, 95%, and 100% I at 4 °C) and 100% II at room temperature, respectively for 30 min.
6. Incubate samples in 2:1 mixture of ethanol: Lowicryl for 1 h at room temperature.
7. Replace the mixture with 1:1 ethanol: Lowicryl and incubate samples for 1 h at room temperature.
8. Replace the mixture with 1:2 ethanol: Lowicryl and incubate samples for 1 h at room temperature.
9. Remove the mixture and incubate samples in pure Lowicryl mixture I for 1 h at room temperature.
10. Remove the mixture, add pure Lowicryl mixture II, and incubate overnight at room temperature.
11. Embed samples using gelatin capsules and polymerize for 24 h at 60 °C.
12. Fix samples with 1% glutaraldehyde and 1% PFA in 0.1 M sodium cacodylate buffer (pH 7.2) at 4 °C.
13. Wash samples three times in 0.1 M sodium cacodylate buffer for 10 min each.
14. Fix samples in 2% OsO_4 for 1 h at 4 °C.
15. Dehydrate samples using an ethanol series (50%, 60%, 70%, 80%, 90%, 100%) for 20 min, respectively.
16. Transfer samples to Lowicryl resin.
17. After impregnation with the pure Lowicryl, embed into the Lowicryl.

Low Concentration Glutaraldehyde Fixation with LR White Embedding

1. Fix samples in cacodylate buffer (pH 7.4) containing 4% PFA, 0.5% glutaraldehyde, and 0.3% sucrose for 2 h at room temperature.
2. Wash samples three times in cacodylate buffer for 5 min each.
3. Incubate samples for 20 min in a graded ethanol series (50%, 60%, 70%, 80%, 90%, 95%, and 100%, respectively) on a shaker.
4. Incubate samples twice with LR white hard grade for 1 h at 4 °C.
5. Embedded samples using gelatin capsules.
6. Polymerize for 48 h at 25–43 °C.

Sections, Immuno-staining, and Imaging

To prevent background staining, there are general recommendations. Nonspecific binding can be observed in not well-preserved cellular structures due to imperfections of sections such as scratches or chattering. The perfect sections should never become dry during the entire immuno-staining procedure, and grids with the sections (Fig. 5c) should be always floated on the droplets of reagents (Fig. 5d). Immersing them can result in nonspecific sticking of all labeling reagents to the supporting film, resulting in superimposition with the specific label. If buffers containing detergent are used, the washing should be done very gently to prevent detachment of sections from grids.

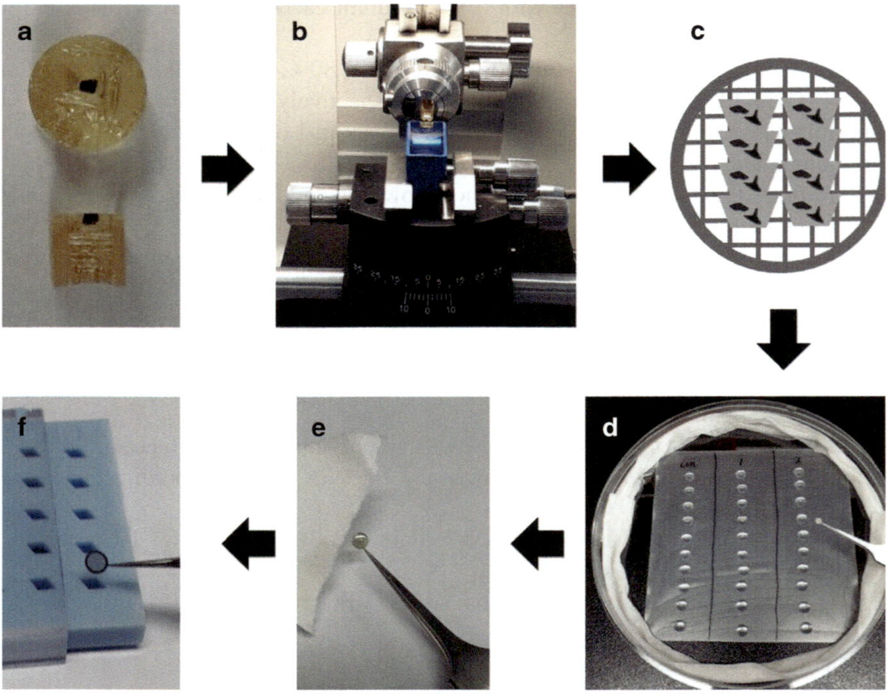

Fig. 5 Process of sample preparation and immuno-staining. (**a**) Samples are embedded in Epon resin. (**b**) Ultrathin sections are cut using an ultra-microtome. (**c**) Sections are mounted on a nickel grid. (**d**) Setup for Immuno-gold labeling. (**e**) Double staining with uranyl acetate and lead citrate. (**f**) Storage in a grid box for TEM observation

Protocol 1

1. Cut 90 nm sections from blocks using an ultramicrotome, and collect them on nickel grids (Fig. 5b and c).
2. Quench free aldehydes with PBS containing 0.1% sodium borohydride and 50 mM glycine for 10 min, with a solution containing 0.05 M glycine for 10 min, or with 0.5 M NH_4Cl for 30 min.
3. Wash grids three times with TBS containing 10% Tween.
   - Before blocking, antigen retrieval can be done by heating. Incubate 1–2 h at 95 °C (Yamashita et al. 2009).
4. Block nonspecific antibody binding sites with PBS containing 1% BSA for 1 h, with Aurion blocking solution for 15 min, or with TBS containing 5% BSA and 5% goat serum for 30 min.
5. Incubate samples with primary antibodies overnight.
   - Incubate grids in 50–100 μL drops of primary antibody in PBS containing 0.1% BSA for 1 h. Perform at 4 °C overnight if necessary.
   - Use a negative control without primary antibody (i.e., no reaction).
   - Use a positive control with primary Ab (i.e., reaction)

6. Wash grids with PBS containing 0.1% BSA for 10 min or wash three times in TBS containing 10% Tween.
7. Block nonspecific antibody binding sites using TBS containing 5% BSA and 5% goat serum for 20 min.
8. Transfer grids to secondary antibody for 1 h (antibody-conjugated with 10 nm gold particles in PBS containing 0.1% BSA).
   - To determine the locations of two different proteins, use multiple secondary antibodies of different origins against respective primary antibodies of differing origins. Double staining should be performed using differing gold sizes for each antibody–antigen complex (Hagiwara et al. 2010).
9. Wash grids once with PBS containing 0.1% BSA for 10 min or three times in TBS containing 10% Tween.
10. Stabilize reactions by incubating in 1% glutaraldehyde in PBS for 5 min.
11. Rinse grids with filtered distilled water (D.W.)
12. Allow grids to air dry.
13. Stain grids with 3% uranyl acetate for 1 min.
14. Rinse grids with filtered D.W.
15. Stain grids with lead citrate for 3 min.
16. Rinse grids with filtered D.W.
17. Observe grids using TEM.

Protocol 2

1. Cut 65–70 nm sections from Lowicryl HM20 blocks using an ultramicrotome, and collect them on nickel grids.
1. Rinse twice in D.W. for 3 min each.
2. Etch sections with 1% periodic acid for 4 min.
   - Some samples can be etched in saturated sodium metaperiodate for 35 min.
3. Rinse samples twice in D.W. for 3 min each.
4. Incubate samples twice in 20 mM Tris HCl buffer (pH 8.2) containing 0.5% BSA, 0.5% normal goat serum, 0.25% Tween 20, and 0.9% NaCl (Tris washing buffer) for 5 min each.
5. React overnight with primary antibody at 4 °C.
   - Dilute primary antibodies to appropriate concentrations in Tris buffer.
6. Incubate samples six times in Tris buffer for 5 min each.
7. React samples with secondary antibody for 2 h at room temperature.
   - Dilute secondary antibody-conjugated colloidal gold particles to appropriate concentrations in Tris buffer.
8. Incubate samples six times in Tris buffer for 5 min each.
9. Rinse samples three times in D.W. for 2 min each.
10. Exclude dust and air dry.
11. Stain samples with 2.5% uranyl acetate in 50% acetone for 20 min.
    - For sections that were etched with saturated sodium metaperiodate, stain with 2.5% uranyl acetate in 50% acetone for 1 h.
12. Rinse sections with D.W. and air dry.
13. Stain sections with lead citrate for 10 min.

- For sections that were etched with saturated sodium metaperiodate, stain with lead citrate for 40 min.

14. Rinse sections with D.W. and air dry.
15. Cover sections with 0.25% Formvar membranes on grids.
16. Observe grids using TEM (example: Fig. 6).

Immuno-negative Staining for Virus or Vesicles (Jung and Mun 2018)

Without sectioning, small size such as vesicle or virus can be applied to immuno-staining on grid.

1. Prepare grids using formavar/carbon film and glow discharge the grids.
2. Fix the particles, such as extracellular vesicles, viruses, and bacteria, with 1 mL of 2% PFA for 5 min.
3. Load 5 μL aliquots of fixed particles onto grids and incubate for 5 min.
4. Rinse grids three times with 100 μL aliquots of PBS for 10 min each.
5. To quench free aldehyde groups, treat grids with 50 μL aliquots of 0.05 M glycine for 10 min.
6. Treat grids with single drops of blocking buffer (PBS containing 1% BSA) for 30 min.
7. Incubate grids with primary antibody in PBS containing 0.1% BSA for 1 h.
8. Wash grids with five separate drops of PBS containing 0.1% BSA for 10 min each.
9. Treat grids for 1 h with single drops of 9–11-nm gold particle-conjugated anti-mouse IgG diluted at 1:100 in PBS containing 0.1% BSA.
10. Wash grids with five separate drops of PBS containing 0.1% BSA for 10 min each.
11. Wash grids with two separate drops of D.W.
12. Stain samples with 2% uranyl acetate.
13. Wash samples with D.W.
14. Observe samples using TEM (example: Fig. 7).

Post-embedding with Cryo-fixation and Low-Temperature Embedding Using Freeze Substitution

In this technique, cryo-fixed samples are processed immediately for low-temperature embedding using a freeze substitution protocol. This low-temperature embedding technique has not only provided better preservation of sensitive antigens (McDonald 1999) but also resulted in superior ultrastructure (Studer et al. 2008) and reduced background staining.

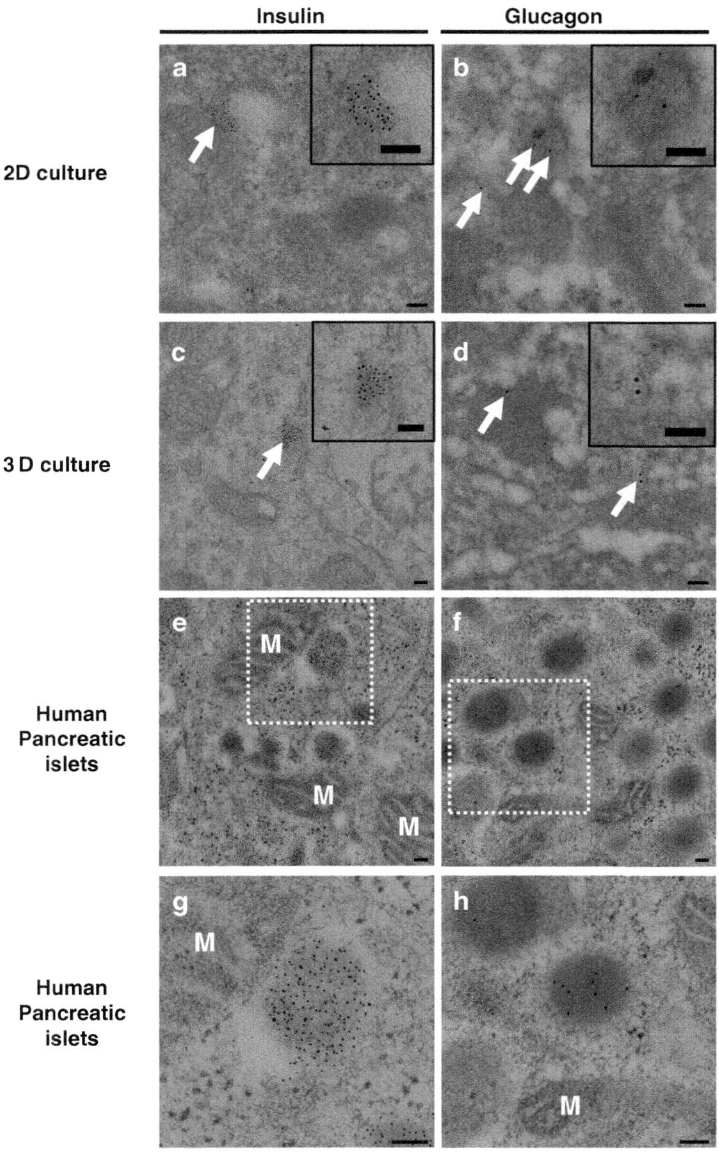

Fig. 6 Transmission electron micrograph of cultured cell and human pancreatic islet tissue. Post-embedding immune-gold labeling was performed with lowicryl HM20 embedding and insulin and glucagon were labeled using a rabbit monoclonal anti-insulin antibody, mouse monoclonal anti-glucagon antibody, 5 nm colloidal gold conjugated goat anti-rabbit IgG secondary antibodies, and 9–11 nm colloidal gold conjugated goat anti-mouse IgG secondary antibodies. Gold particles were labeled with insulin (**a**, **c**, **e**, **g**) and glucagon (**b**, **d**, **e**, **h**), and were localized in the cytoplasm of 2D cultured cells (**a**, **b**) and 3D cultured cells (**c**, **d**). Gold particles are present only on the granules of human pancreatic islets (**e**, **f**, **g**, **h**). *M* mitochondria. Scale bars: 100 nm

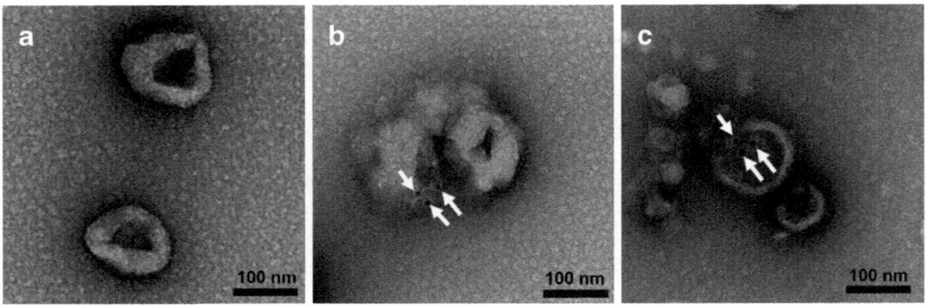

Fig. 7 Immuno-negative staining. (**a**) Representative electron micrograph of exosomes. Negative control of immuno-staining result. (**b**, **c**) Whole-mount immuno-negative staining shows the location of a specific protein. The white arrows show PD-L1 (CD274) expression

Low-Temperature Embedding After Chemical Fixation and Plunge Freezing

1. Fix sample in 0.1 M PB containing 4% PFA and 0.5% glutaraldehyde.
2. Wash samples with 0.1 M PB.
3. Cryo-protect samples with 10%, 20%, and 30% sucrose in 0.1 M PB for 30 min.
4. Replace solutions with fresh 30% sucrose in PB and incubate overnight at 4 °C.
5. Freeze samples at −190 °C.
6. Transfer frozen samples to an automatic freeze substitution (AFS) apparatus.
7. Infiltrate samples with 1.5% uranyl acetate in 100% methanol in the AFS at −90 °C for 30 h.
8. Increase the temperature of the AFS from −90 to −45 for 9 h.
9. Maintain samples in 100% methanol at −45 °C for 2 h.
10. Replace methanol with fresh 100% methanol twice and incubate at −45 °C for 2 h each time.
11. Place gelatin capsules into separate AFS holders and cool in the AFS.
12. Incubate samples with HM20 media in gelatin capsule containing 1:3, HM20:MeOH; 1:1, HM20:MeOH; 3:1, HM20:MeOH; and 100% HM20 for 2 h each.
13. Incubate samples in 100% HM20 overnight and replace the media with fresh media three times.
14. Polymerize with UV light at −45 °C for 24 h.
15. Polymerize by increasing the temperature from −45 °C to 0 °C for 24 h without changing the solution.
16. Polymerize with UV at 0 °C for 36 h without changing the solution.
17. Polymerize with increasing temperature from 0 °C to 20 °C for 5 h without changing the solution.
18. Cut sections of 70–90 nm and load onto nickel grids.
19. Follow the procedures detailed in the immuno-staining of sections and imaging (See section 2-2).

Cryo-fixation and Cryo-ultramicrotomy Using Tokuyasu Techniques

The following procedures were largely developed by Tokuyasu (Tokuyasu and Singer 1976). This technique is considered as a method for minimizing artifacts from ice damage and preserving antigenicity.

1. Immerse fixed tissue fragments for a few min in 10% gelatin in PB at 37 °C.
2. Allow the gelatin to solidify as a thin (0.5 mm) slab at 4 °C and repeat the fixation step for 30 min.
   - Gelatin embedding can be omitted in many cases, but occasionally helps freshly cut sections to withstand surface tension from sucrose drops. Samples embedded in gelatin become harder and darken when fully infiltrated. The optimal gelatin block size is <1 mm^3.
3. Cut tissue or cell pellet in 10% gelatin after primary fixation
4. Immerse the blocks in 2.3 M sucrose for at least 30 min at room temperature or overnight at 4 °C in Eppendorf tubes.
5. Mount blocks on the specimen holder and remove excess cryo-protectant solution on the block using filter paper.
6. Plunge freeze blocks into liquid nitrogen.
   - Samples can be stored in cryo-vials in liquid nitrogen until sectioning. An alternative cryo-protectant, polyvinylpyrrolidone (PVP) can be mixed with sucrose. PVP increases the plasticity of the blocks, allowing cutting of very thin and well-stretched sections.
7. Mount frozen samples in cryo-chamber of the cryo-ultramicrotome and maintain at −90 °C while trimming semi-thin sections, and at −120 °C when performing ultrathin cryo-sectioning.
8. Trim samples for squared block of about 200 × 200 μm. Trim at 50–100 mm/s and take 200–250 nm semi-thin sections.
9. Produce 70–90 nm thin sections using a diamond knife.
10. Retrieve sections using single droplets of 1.15 M sucrose containing 1% methylcellulose.
    - 2.3-M sucrose can be used for section retrieval.
11. Place sections on formvar/carbon-coated grids after thawing.
    - Section quality is considerably dependent on specimen preparation conditions. But the time interval between thawing and attachment of sections can induce ultrastructural damage. The adhesion properties of the support film should also be optimized to eliminate wrinkles. To reduce damage, pick up gelatin embedded samples or tissues with high collagen contents without using methylcellulose (Bos et al. 2011).
12. Follow the procedure for immuno-staining (See section 2-2).

Trouble Shooting

The most important parameter for successful labeling is the specificity of interactions between antibodies and antigens. The specificity of binding can be affected by denaturation and masking of antigens, and by background noise from general physical and chemical properties. High- or low-affinity complexes of antibody and embedding medium are typical sources of background reactions. Hence, to reduce these, low molecular weight background should be eliminated before blocking proteins. Because residual fixatives are aldehydes, amino acids (glycine, lysine), and sodium borohydride (Willingham 1983), hydrosylamine can be used to inactivate free aldehydes. Appropriate blocking steps can be used to eliminate nonspecific reactions occurring on high molecular weight compounds with multiple hydrophobic and negative charge sites. To saturate multipoint hydrophobic areas and highly positive charges with high-affinity protein binding capacities, albumin or normal serum can be used at high concentrations. Appropriate washing and incubation steps can be used to limit background noise from reactions with low molecular compounds that have a single hydrophobic and negative charge. But these procedures must be controlled carefully. To eliminate nonspecific binding of immunoconjugates due to hydrophilic interactions (positive charges), acetylated albumin can be added to blocking solution.

Reactions of specific antigens and antibodies can be affected by antibody concentrations. Thus, antibody concentrations must be optimized in tests of antigen–antibody reactions. Antibodies are best diluted by factors of 10 (1:10, 1:100, 1:1000). Numbers of bound antibodies decrease with higher dilutions and the specificity increases with higher dilutions.

To eliminate the clustering of antibody, Fig. 8 shows trouble shooting. If colloidal gold shows aggregates, larger aggregates can be removed by centrifugation. Because colloidal gold coagulates in NaCl, quantities of protein–gold complexes have to be optimized to prevent NaCl-induced coagulation. Adjustments of pH can

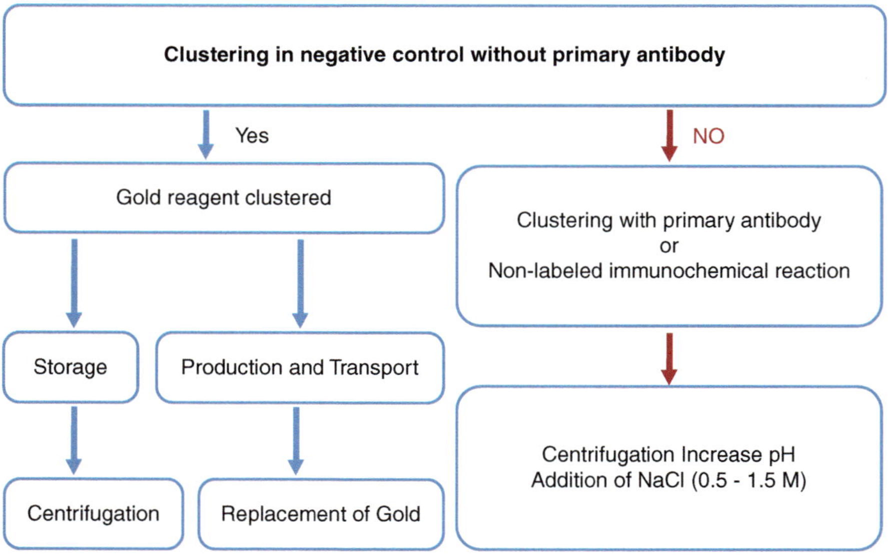

Fig. 8 Trouble shooting for clustering of antibody

prevent protein-induced flocculation and the isoelectric point (PI) of the protein is also important. The absence of aggregates and reactive sites is most important, and it depends on the electrophoretic homogeneity of the protein.

Acknowledgments We thank the Electron Microscopy core facility at the Brain Research Core Facilities in Korea Brain Research Institute and Convergence medicine research center (CREDIT), Asan Medical Center. This manuscript was supported by KBRI basic research program through the Korea Brain Research Institute funded by the Ministry of Science and ICT (21-BR-01-11) and by the National Research Foundation of Korea (NRF) grant funded by the Korea government (MSIT) (No. 2019R1A2C1010634).

References

Acetarin JD, Carlemalm E, Villiger W (1986) Developments of new Lowicryl resins for embedding biological specimens at even lower temperatures. J Microsc 143(Pt 1):81–88

Bendayan M, Zollinger M (1983) Ultrastructural localization of antigenic sites on osmium-fixed tissues applying the protein A-gold technique. J Histochem Cytochem 31(1):101–109

Bendayan M, Nanci A, Kan FW (1987) Effect of tissue processing on colloidal gold cytochemistry. J Histochem Cytochem 35(9):983–996

Bos E, Santanna C, Gnaegi H, Pinto RF, Ravelli RB, Koster AJ, de Souza W, Peters PJ (2011) A new approach to improve the quality of ultrathin cryo-sections; its use for immunogold EM and correlative electron cryo-tomography. J Struct Biol 175(1):62–72

Ferguson DJ, Hughes DA, Beesley JE (1998) Immunogold probes in electron microscopy. Methods Mol Biol 80:297–311

Hagiwara H, Aoki T, Suzuki T, Takata K (2010) Double-label immunoelectron microscopy for studying the colocalization of proteins in cultured cells. Methods Mol Biol 657:249–257

Hainfeld JF, Furuya FR (1992) A 1.4-nm gold cluster covalently attached to antibodies improves immunolabeling. J Histochem Cytochem 40(2):177–184

Hsu SM, Raine L, Fanger H (1981) Use of avidin-biotin-peroxidase complex (ABC) in immunoperoxidase techniques: a comparison between ABC and unlabeled antibody (PAP) procedures. J Histochem Cytochem 29(4):577–580

Jung MK, Mun JY (2018) Sample preparation and imaging of exosomes by transmission electron microscopy. J Visual Exp JoVE (131)

Mandry P, Murray BA, Rieke L, Becke H, Hofler H (1993) Postembedding ultrastructural in situ hybridization on ultrathin cryosections and LR white resin sections. Ultrastruct Pathol 17(2):185–194

McDonald K (1999) High-pressure freezing for preservation of high resolution fine structure and antigenicity for immunolabeling. Methods Mol Biol 117:77–97

Nisman R, Dellaire G, Ren Y, Li R, Bazett-Jones DP (2004) Application of quantum dots as probes for correlative fluorescence, conventional, and energy-filtered transmission electron microscopy. J Histochem Cytochem 52(1):13–18

Piekut DT, Casey SM (1983) Penetration of immunoreagents in Vibratome-sectioned brain: a light and electron microscopic study. J Histochem Cytochem 31(5):669–674

Polishchuk EV, Polishchuk RS (2019) Pre-embedding labeling for subcellular detection of molecules with electron microscopy. Tissue Cell 57:103–110

Roth, J. (1986). Post-embedding cytochemistry with gold-labelled reagents: a review. J Microsc 143(Pt 2), 125–137

Roth J, Bendayan M, Carlemalm E, Villiger W, Garavito M (1981) Enhancement of structural preservation and immunocytochemical staining in low temperature embedded pancreatic tissue. J Histochem Cytochem 29(5):663–671

Singer SJ (1959) Preparation of an electron-dense antibody conjugate. Nature 183(4674):1523–1524

Studer D, Humbel BM, Chiquet M (2008) Electron microscopy of high pressure frozen samples: bridging the gap between cellular ultrastructure and atomic resolution. Histochem Cell Biol 130(5):877–889

Tokuyasu KT, Singer SJ (1976) Improved procedures for immunoferritin labeling of ultrathin frozen sections. J Cell Biol 71(3):894–906

Valentino KL, Crumrine DA, Reichardt LF (1985) Lowicryl K4M embedding of brain tissue for immunogold electron microscopy. J Histochem Cytochem 33(9):969–973

Willingham MC (1983) An alternative fixation-processing method for preembedding ultrastructural immunocytochemistry of cytoplasmic antigens: the GBS (glutaraldehyde-borohydride-saponin) procedure. J Histochem Cytochem 31(6):791–798

Yamashita S, Katsumata O, Okada Y (2009) Establishment of a standardized post-embedding method for immunoelectron microscopy by applying heat-induced antigen retrieval. J Electron Microsc 58(4):267–279

Micro-endoscopy for Live Small Animal Fluorescent Imaging

Bjorn Paulson and Jun Ki Kim

Introduction

Small animal disease models are widely used for the targeted observation of interactive cellular processes in the study of inflammatory disorders, cancers, neural processes, and infectious diseases. Mouse models, in particular, are highly valuable in vivo models of human disease, with similar cardiovascular and immune systems to those of humans, and recently a large variety of genetically engineered mouse models have become available at low cost (Jung et al. 2013). Of particular interest is the development of the many transgenic mouse strains expressing cells tagged with fluorescent proteins such as eGFP and RFP, which express fluorescence upon illumination of specific cell types. These mice allow for the cellular-level analysis of tissues and tumors on the basis of optical fluorescence techniques (Kim et al. 2012a).

In tandem with the development and analysis of these transgenic mouse models, many sophisticated diagnostic systems have been developed for the analysis of these transgenic mouse models. However, trade-offs exist between granularity of the available data in the spatial domain, granularity in the time domain, and spatial extent of the data. At the level of whole-body analysis, common radiological imaging modalities such as computed tomography and magnetic resonance imaging

B. Paulson
Biomedical Engineering Research Center, Asan Institute for Life Sciences, Asan Medical Center, Seoul, Republic of Korea

J. K. Kim (✉)
Department of Convergence Medicine University of Ulsan College of Medicine, Asan Institute for Life Sciences, Asan Medical Center, Seoul, Korea (Republic of)
e-mail: kim@amc.seoul.kr

J. K. Kim et al. (eds.), *Advanced Imaging and Bio Techniques for Convergence Science*, Advances in Experimental Medicine and Biology,
https://doi.org/10.1007/978-981-33-6064-8_8

are available, as well as whole-body fluorescence imaging for mice, but due to their spatial extent, these do not provide data at cellular resolution. At the other extreme, in vitro culture and imaging systems (Rock et al. 2009; You et al. 2002) provide cellular resolution, but many elements that influence the development of pathologies at the cellular level are difficult or impossible to mimic in vitro, such as intercellular signaling cascades, innervation, cellular trafficking, and immune response. Ex vivo histology also provides information about cellular processes; however, the static and sacrificial nature of the imaging increases the cost to attain information about dynamic processes, including cellular trafficking and differentiation (Jung et al. 2013; Kim et al. 2012a, b; Miller 2002; Fan et al. 2010; Fujisaki et al. 2011).

The fluorescence micro-endoscopy technique has been introduced to overcome the limitations of these imaging modalities, combining the high cellular resolution (Ryu et al. 2018) and sensitivity of fluorescence microscopy to monitor epithelial pathologies in vivo, longitudinally, and with minimal invasiveness (Kim et al. 2012a, b; Ryu et al. 2018; Jun et al. 2018; Flusberg et al. 2005; Choi et al. 2014). To date, intravital fluorescence microscopy has been applied for the imaging of diverse cellular targets, from transplanted stem cells (Ryu et al. 2018) and tumor xenografts to immune response and individual neurons in vivo (Jung et al. 2013; Miller 2002; Fan et al. 2010; Fujisaki et al. 2011; Pittet and Weissleder 2011), and also to image cellular processes, including gene expression, cell activation, intercellular interactions, vascular changes, and cell trafficking (Kim et al. 2012a, b), to provide data and discoveries which cannot be easily acquired from other diagnostic modalities, and allows precise inference of the cellular mechanisms behind the pathologies under study (Kim et al. 2012a).

Recent advances in micro-optic and fiber-optic technologies have led to the commercialization of several micro-endoscopic probes from suppliers including Karl Storz (Becker et al. 2007), Mauna Kea Technologies (Hsiung et al. 2008), and Olympus (Dela Cruz et al. 2010), each with their own strengths and limitations. MicroProbe Objectives from Olympus are standard microscope objectives which offer an extension of the imaging plane into tissue via a long (20–25 mm), narrow diameter (1.3–3.5 mm) "stick," and offer high-quality correction for chromatic aberrations, at the expense of a fixed focal plane of the objective and a large overall footprint which restricts tissue access in larger animals (Dela Cruz et al. 2010). The Coloview® endo-microscope from Karl Storz is optimized for small animal colonoscopy, with probes of 10 cm length and narrow diameter (1.9 mm) suitable for tracking tumor progression in the colon on a macroscopic scale (Kim et al. 2012a). Finally, the Cellvizio® probe family by Mauna Kea offers exceptionally miniaturized front-facing probes based on a variety of technologies: microlenses, optical fiber bundles, and gradient index lenses, with diameters ranging from 0.3 to 1.8 mm, field depth from 10 to 60 microns, and resolutions down to 1.4 microns, for a variety of small animal experiments. Herein, we present an alternative, custom-built solution, rigid probes based on gradient index lenses which enable cellular resolution imaging of tissues, optical sectioning, and high stability (Köhler et al. 2018). A side-view extension prism allows a 90° rotation of the imaging path for side-view imaging of luminal organs at cellular resolution. Probes are available with diameters from 0.35 to 1 mm, and maybe fabricated for a range of lengths (20–100 mm) and numerical apertures (NA; 0.4–0.6) (Kim et al. 2012a).

Although several designs have been introduced for the development of flexible micro-endoscopes for human patients, including micro-endoscopes based on coherent fiber bundles (Göbel et al. 2004; Yang et al. 2005), scanning optical fibers (Wu et al. 2009; Lee et al. 2010; Kim et al. 2019), and micro-endoscopes based on chip-on-tip technology (Matz et al. 2017), the need for a flexible probe is less pronounced in small animal micro-endoscopy, since in the small animal disease model it is often more convenient to operate around curved organs than to navigate through their curves, especially in consideration of the fact that those curves are much sharper than in human anatomy. The remainder of this review will concern itself with rigid micro-endoscopic probes built on the recently developed gradient-index (GRIN, also referred to as "graded index") lenses (Kim et al. 2012a).

Geometry and Key Design Parameters for GRIN Probes

Before beginning the manufacture or acquisition of an optical probe, it is recommended to ascertain the optimal design based on the precise organs and tissues for which imaging is desired: the design will vary with practical considerations, such as the target tissue orientation and location relative to the point of entry, the size of the entry point, the desired field of view (FOV), and the desired resolution (Kim et al. 2012a).

Once requirements are specified, the geometry of a GRIN endoscope probe is defined primarily by the GRIN lenses which comprise the probe. These (typically three) lenses are commonly referred to as the coupling lens (CL), the imaging lens (IL), and the relay lens (RL). The CL and IL are high-numerical-aperture (NA) lenses that respectively collimate the input beam with respect to a microscope (or similar assembly) and convey the distal image into the probe. The relay lens (RL) is a longer lens of lower numerical aperture, which connects between the two to extend the beam path and probe length (Kim et al. 2012a).

The probe geometry is also determined by the viewing direction: whether the probe is to observe the target tissue in front or side view. The light propagation and schematic geometries of front-view and side-view GRIN probes are shown in Fig. 1 (Kim et al. 2012a; Jung and Schnitzer 2003). In addition to these three lenses, side-view GRIN micro-endoscopes have a prism mirror affixed to the distal end of the imaging lens. Normally, the front-view probe finds application in the imaging of convex organs or organs in large cavities that can touch the probe tip, such as the pancreas, lymph nodes, kidneys, bladder wall, liver, and heart. Tubular structures, as are found in the epithelium of gastrointestinal and respiratory tracts, are more suitable for imaging by the side-view probe (Kim et al. 2012a).

For gradient index lenses, the length of the lenses is measured in pitch: In forward-view probes, the appropriate lengths for the coupling lens and the imaging lenses are measured in units of pitch: one pitch is the length at which a ray of light passing through the lens will trace out a full sinusoidal path due to the refractive

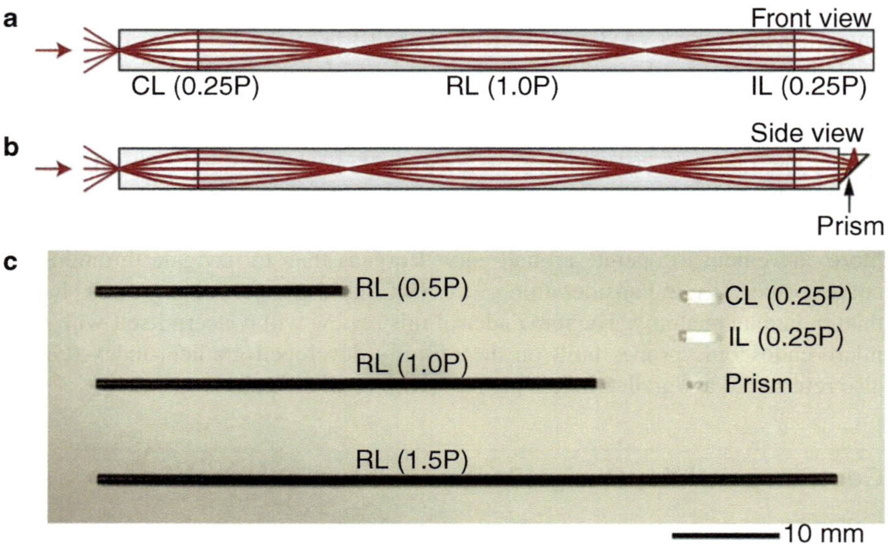

Fig. 1 Schematic and manifestation of gradient index (GRIN) optical probes. (**a**) front- and (**b**) side-view schematics and (**c**) photograph of micro-endoscope probes formed with GRIN collimating lenses (CL), relay lenses (RL), and imaging lenses (IL). The side-view micro-endoscope probe has a prism affixed to the distal end of the imaging lens. Red lines denote a traced optical ray. Reproduced with permission from Kim et al. (2012a), © 2012 Springer Nature

index gradient. The appropriate lengths for the coupling lens and the imaging lenses are 0.25 pitch and 0.23–0.25 pitch, respectively. While the coupling lens serves to collimate the input beam, the imaging lens should focus on the object plane just past the distal end of the probe (Kim et al. 2012a).

For side-view probes, shortening the IL to a pitch of 0.16–0.17 will extend the working distance such that the object plane lies just past the distal end of the prism. This may be achieved either by ordering a custom IL or by polishing down an existing IL. The ideal prism is a glass or quartz cube with the distal end polished at 45° or to reflect light at the desired angle. A metal coating on the polished face results in a second surface mirror. Prisms with edge lengths less than the lens diameter will fit well in stainless steel sheaths or needles used for protecting the optical assembly (Kim et al. 2012a).

For GRIN probes of this design, the field of view (FOV) is proportional to the probe diameter multiplied by the ratio of the relay lens NA to the image lens NA, and thus the diameter should balance the desired FOV with experimental invasiveness. Commercially available GRIN lenses range in diameter from 0.35 to 2 mm but benefit from protective metal sheaths of an additional 0.20–0.30 mm in outer diameter. Although in practice, FOV is limited by GRIN NA to about one-fourth the diameter of the GRIN lens, compositing sequential images from a wider area may overcome this restriction (Kim et al. 2012a).

The length of the probe may be tailored to the application by extending the relay lens in multiples of half the relay lens pitch (Fig. 1c). (Odd multiples of half-pitch result in image inversion.) With increased pitch, optical aberrations (primarily chromatic and spherical) accumulate, resulting in degraded spatial resolution, and the limited stiffness of the probe results in a risk of probe damage due to bending. A length of 20–30 mm is suitable for imaging of the abdominal cavity in mice, as is a length of 60–100 mm for imaging of the gastrointestinal tract (Kim et al. 2012a).

The theoretical aberrations of the GRIN lenses have been calculated (Krishna and Sharma 1996; del Río et al. 2015; Bociort 1994); however, the spherical wavefront error depends on the ion-exchange chemistry of the GRIN fabrication process (Barretto et al. 2009). For this GRIN probe design and ignoring aberrations, the limit of spatial resolution for wavelength λ and imaging lens numerical aperture NA_I is given by 0.6 λ/NA_I in the transverse direction and 1.1 $\lambda/NA_I{}^2$ in the axial direction (Kim et al. 2012a).

The imaging penetration depth by GRIN probes is limited to that of the optical modality, which is in turn bounded by scattering from tissue at the wavelengths used. For confocal fluorescence microscopy in the visible region, the optical penetration depth is restricted to about 100 μm in the majority of soft tissues. GRIN probes are also commonly used in a two-photon fluorescence modality for optogenetics, where they demonstrate penetration depths 2–3 times higher (Kobat et al. 2011), and with new light sources in the 1300 and 1800 nm infrared bands and three-photon processes, penetration depths be improved to the order of millimeters (Smith et al. 2009).

Thus, naive imaging with confocal fluorescent GRIN probes is limited to the dermis, epithelium, or mucosa. By physically injecting the probe into the target tissues, deeper features may be observed in the visible region, while minimizing trauma to surrounding regions (Barretto et al. 2011). Careful planning of surgeries and the incorporation of supplementary technologies such as needles and cannula (to be introduced later in this review) may also mechanically strengthen the probe while reducing surgical trauma (Kim et al. 2010a, 2012a; Barretto et al. 2011).

Fabrication of GRIN Probes

The fabrication of GRIN micro-endoscope probes from GRIN lenses requires some skill in optical polishing and assembly or maybe contracted from a commercial vendor such as GRINTECH GmbH. A suitably equipped laboratory should have two stereoscopes, tweezers, a UV light source and a hotplate for curing thermal epoxies, and a lapping surface and jig for the polishing of lenses. The fabrication process begins with the polishing of the GRIN lens. A lapping tool with an incorporated jig may be used to gently apply steady pressure to the lens while ensuring that the polishing is perpendicular to the axis of the lens.

Completed IL and CL lenses are gently adhered to relay lenses using transparent and nonfluorescent UV-curing optical polymer of matched refractive index, which minimizes back-reflections and aberrations at the surface between GRIN lenses. For

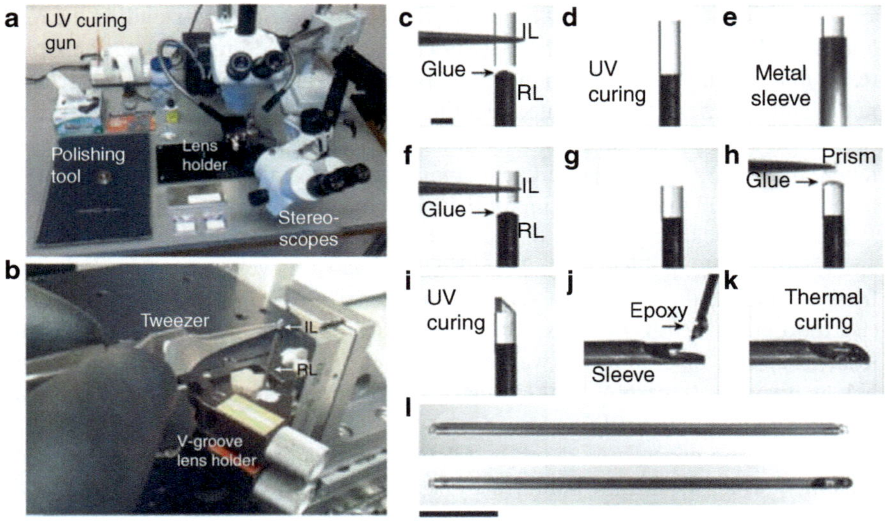

Fig. 2 Assembly of GRIN micro-endoscope probes. Reproduced with permission from Kim et al. (2012a), © 2012 Springer Nature

mechanical stability as well as protection from stray light and moisture, the finalized assembles should be immobilized within a protective stainless steel sheath by heat-sensitive epoxies of high mechanical strength. In Fig. 2a, b, we show typical setups for the lapping and adherence of these processes, and in Fig. 2c–e, we demonstrate the assembly procedure for a front-view probe, with illustrations made at representative steps. Similar steps, with the addition of prism placement onto the IL and heat treatment of the sheathed probe, are depicted for the side-view probe in Fig. 2f–k (Kim et al. 2012a). The protective stainless steel sheaths are available in bulk as tubes of varying quality from a variety of high-gauge medical needle suppliers. A completed probe for side-view micro-endoscopy is shown in Fig. 2i.

Imaging from the GRIN Endo-microscope Probes

For high-quality image capture, the manufactured GRIN optical probes should be integrated into a digital imaging system. High magnifications are not provided by the probe itself, and so it is easier to integrate the endo-microscopic probes with a preexisting or commodity microscope system than to develop a fully customized imaging solution. As shown in Fig. 3 for front-view probes, a mechanical probe mount and lens relay may be assembled to extend the image plane of a confocal microscope to the proximal end of the micro-endoscope probe.

The mechanical mount should guarantee probe stability relative to an objective, as well as fine control of the probe position. Control of the probe is required with single-micron precision perpendicular to the beam path, while in the axial direction

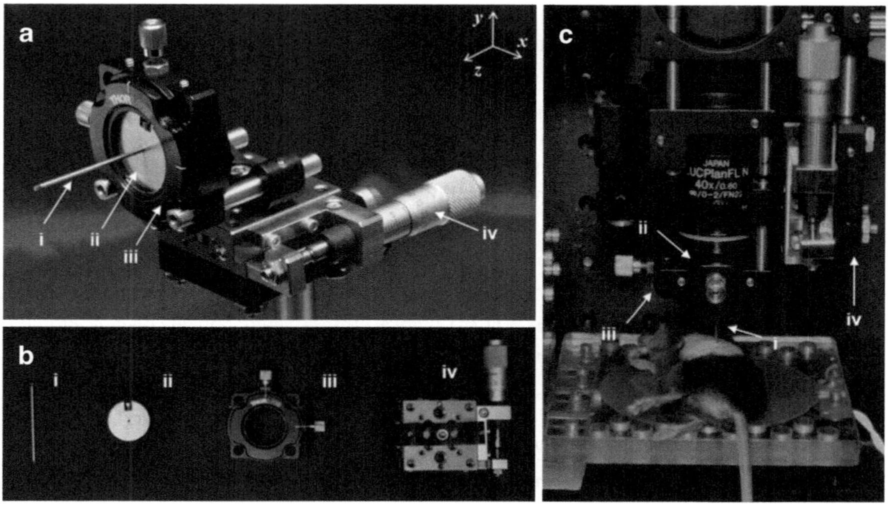

Fig. 3 Assembly for mounting a front-view probe to a confocal microscope. (**a**) Probe mount. (**b**) Individual components. (**c**) The complete assembly, applied for front-view imaging of a mouse model. Adapted with permission from Kim et al. (2012a), © 2012 Springer Nature

equivalent precision should be offered in movement of the microscope objective. For the side-view probe, the ability to rotate the probe 360° with 5° precision and to center the probe with respect to the axis of rotation are required (Kim et al. 2012a). These abilities may be combined into a unified probe mount and relay mount (Kohler et al. 2008).

Applications of GRIN Micro-endoscopes

A. Front-View GRIN Micro-endoscope Applications

Although miniature optical probes offer the benefit of being easily manufactured by the small laboratory, using tools and supplies currently on the market, they are more impressive for their many applications to minimally invasive, cellular resolution imaging in small animal disease models (Fan et al. 2010; Kim et al. 2010a). For the majority of in vivo organs, such as the kidneys (Fan et al. 2010), bladder, and heart, fluorescence cellular imaging is enabled by endo-microscopy systems which incorporate front-view endoscope probes. Side-view probes are well-suited for observing into the folds of the epithelium of tubular tissues as are commonly found in the gastrointestinal and respiratory tracts (Kim et al. 2010a), and will be discussed in the next section.

Longitudinal Renal Imaging with Forward-View GRIN Endoscopes

Through minimally invasive endoscopic microscopy methods, repeated imaging and longitudinal observation become feasible at the cellular scale. As an example of these possibilities, the authors report the observation of cellular motility and longitudinal cellular-resolution tracking of immune response in mouse models of renal transplantation (Kim et al. 2012a; Miyajima et al. 2011). As shown in Fig. 4, over a time-scale of 6 minutes, FoxP3-GFP$^+$ T cells (green) were observed to migrate from the recipient mouse into kidney tissues allotransplanted from a wild-type donor mouse (red autofluorescence). Quantitative analyses of dynamic processes, such as the counting and tracking of individual cells ("Track") are demonstrated in these in vivo measurements (Kim et al. 2012a).

In Fig. 5, a BALB/C wild-type mouse was allografted with a donor kidney from a C57BL/6 mouse with a knock-in MHC class II gene which codes for an eGFP-tagged class II molecule. A front-view GRIN endo-microscope was used repeatedly to observe the fluorescent GFP antigen-presenting cells (APC) from the donor mouse. The renal tissues expressed relatively intense autofluorescence, which reveals kidney nephrons in the cortex. One day after transplantation, a large number of fluorescent dendritic donor APCs are seen (Fig. 5a). After 3 days, most dendritic GFP$^+$ cells had migrated to renal capsules, demonstrating the process of clearance (Fig. 5b). Such repeated imaging and longitudinal imaging would not be directly feasible in more highly invasive observational methods, such as fixed tissue histology.

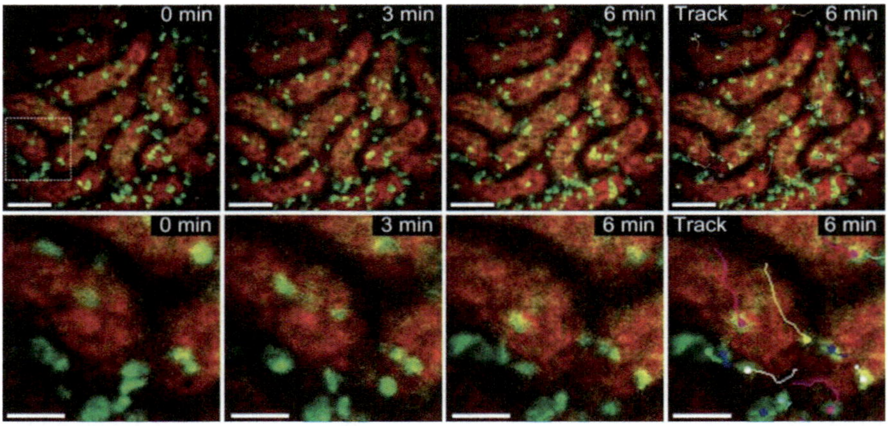

Fig. 4 Micro-endoscopic observations of immune motility in kidney graft allotransplanted from a wild-type donor into a transgenic mouse with FoxP3-GFP$^+$ fluorescent T cells. Green depicts FoxP3-GFP fluorescence while red is autofluorescent kidney capsules. (Bottom) Magnified images of the region shown in the box in the first pane. (Track) Overlaid tracks of individual T cells from time-series images taken every 20 s. Scale bars are 50 μm in the top panels and 20 μm in the lower panels. Reproduced with permission from Kim et al. (2012a), © 2012 Springer Nature

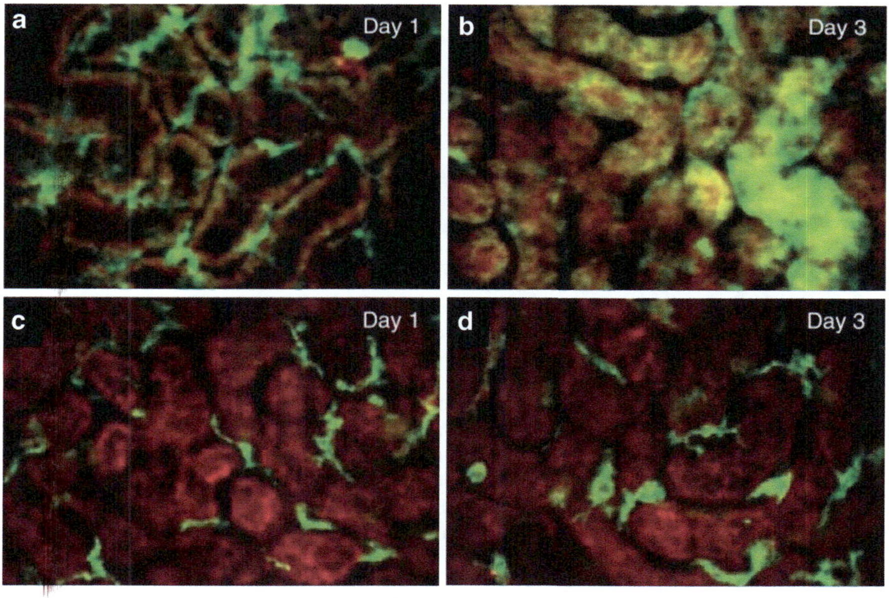

Fig. 5 Longitudinal observations of kidney allografts by front-view GRIN endo-microprobes. Fluorescent MHCII$^+$ cells are in green, while red expresses autofluorescence due to renal tubules. (**a**, **b**) Donor fluorescent cells in kidney graft (**a**) 1 and (**b**) 3 days following transplantation. (**c**, **d**) Controls following sham surgery. Scale bar, 100 μm. Reproduced with permission from Kim et al. (2012a), © 2012 Springer Nature

Synergistic GRIN Micro-endoscopy in Urologic Applications

Small-diameter endoscopic imaging has also been combined with intravital confocal microscopy to enable longitudinal observation of cellular migration in the bladder wall in mouse models of interstitial cystitis/bladder pain syndrome (IC/BPS). This section is adapted from a recently published study (Ryu et al. 2018) in which microscopic imaging was combined with intravital GRIN micro-cystoscopy to track the development of a preclinical stem-cell treatment of IC/BPS (Fig. 6).

Interstitial cystitis/bladder pain syndrome (IC/BPS) is a bladder disorder characterized by chronic inflammation of the bladder epithelium (Propert et al. 2000), resulting in suprapubic pain that is intensified by bladder filling, resulting in urinary frequency, urinary urgency, and reduced quality of life. With unknown causes and no effective treatment (Oravisto 1975), stem cell therapy is one of the few potential treatments (Kim et al. 2016a, b; Song et al. 2015). However, cells did not engraft well in vivo, and undetermined issues remained in assessing their functional integration and the mechanism of treatment. A combination of intravital confocal microscopy and in vivo endo-microscopy was used to image the bladder wall following stem cell treatment, synergizing data from multiple imaging modalities to assess the longitudinal outcome of these treatments in a preclinical mouse model of IC/BPS (Ryu et al. 2018). Previous reports on MSCs have demonstrated beneficial effects on IC/BPS and ketamine-induced cystitis; however, the MSC cells from

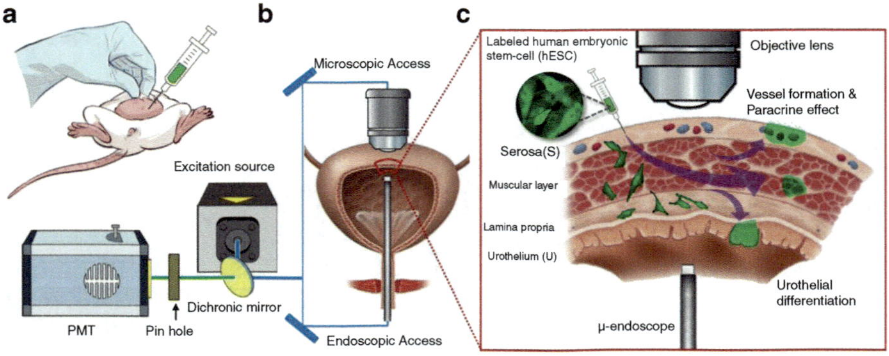

Fig. 6 Longitudinal synergy of microscope and micro-endoscope imaging of transplanted stem cells in a rat model of IC/BPS. (**a**) Injection of hESC-derived M-MSCs into the bladder wall. (**b**, **c**) Synergistic confocal monitoring of transplanted cell migration and observed repair of the denuded urothelial wall. Stem cells are observed to travel from the serosa to the lamina propria and urothelium, where they repair the urothelial wall. Reproduced under CC BY-NC 4.0 license from Ryu et al. (2018), © 2018 Ivyspring International Publishers

previous studies (Song et al. 2015; Kim et al. 2016b) were short-lived and were not able to directly repair damaged tissues. As they did not repair (Liang et al. 2014; Cosenza et al. 2018; Vonk et al. 2018; Wang et al. 2014), MSC-based therapies are still controversial issues (Ryu et al. 2018).

A mouse model of IC/BPS was established by instilling endotoxins into the bladders of female rats weekly for 5 weeks to induce realistic tissue inflammation and chronic urothelial damage (Stein et al. 1996; Birder and Andersson 2018), while a sham group was treated identically with phosphate-buffered saline (PBS). M-mesenchymal stem cells (M-MSCs) were derived from human embryonic stem cells (hESCs) and established with green fluorescence by transfection of a GFP expressing lentivirus (Kim et al. 2018). Compared to other stem cells, hESCs expanding many times and differentiating into therapeutic cells in vitro. They also exhibit high survival, engraftment, and functionality when transplanted in vivo (Kim et al. 2016c). Previous grafts have remained until 6 months post transplantation in an alternative animal model of IC/BPS (Kim et al. 2017).

Fluorescence was confirmed to be stable through long-term (5 week) cultivation and multi-lineage differentiation. M-MSCs were expanded for fewer than ten generations to make sure they maintained multipotency (Ryu et al. 2018; Kim et al. 2016c; Hong et al. 2015), and administered to the bladder urothelium 1 week after the final endotoxin instillation as a single injection from a variety of doses (0.1, 0.25, 0.5, and 1×10^6 cells) via a lower abdominal incision (Ryu et al. 2018; Song et al. 2015, 2014; Kim et al. 2016b, 2017), following confirmation of negligible autofluorescence (Ryu et al. 2018).

Living rat models of IC/BPS were surveyed before transplantation and at regular intervals for up to 42 days post transplantation by in situ confocal microscopy and observation with a micro-endoscopic optical probe. A continuous 488 nm laser was

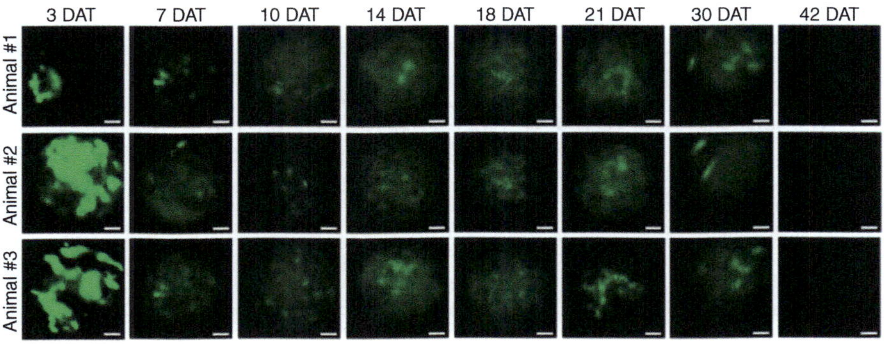

Fig. 7 Longitudinal GRIN probe micro-cystoscopy of engrafted fluorescent stem cells (M-MSCs) in living rats from 3 to 42 days after M-MSC injection (DAT). Microendographs were captured using a 40× objective. Scale bar is 50 μm. Reproduced under CC BY-NC 4.0 license from Ryu et al. (2018), © 2018 Ivyspring International Publishers

used with a scanning galvanometric system to excite GFP for both microscopy and endoscopy, and the signal was detected by filtered photo-multiplicative detectors to quantify emitted light while avoiding most sources of noise. The video capture rate was 30 fps. In situ microscopy was performed via a short incision in the abdomen which exposed the bladder, enabling observation 40× objective lenses mounted on a standard confocal microscope. For micro-endoscopic observation, a probe with 1.2 mm diameter and 5.5 cm length was prepared by affixing GRIN lenses together inside a protective tube, using the procedure above (Kim et al. 2012a). The probe was inserted through the urethra in a consistent manner and focused to a consistent depth within the bladder wall (Fig. 7).

Fluorescence intensity and the focus of the observed fluorescence were recorded in the urothelium by micro-cystoscopy (Fig. 7) and by confocal intravital microscopy (Fig. 8) beginning 3 days after M-MSC transplantation (DAT) and over a 42-day period. Both fluorescence intensity and focus varied over time, and the combination of images from cystoscopy and intravital microscopy allowed survival, migration, and differentiation of the transplanted stem cells to be inferred, with worsening focus indicating migration of cells away from the focal plane and into the lamina propria, while decreased intensity indicated a combination of apoptosis and cellular diffusion across the entire bladder wall. Detailed examination of intravital micro-endoscopic images showed engraftment of M-MSCs into blood vessels of the bladder wall, and subsequent differentiation of MSCs into epithelial, stromal, and perivascular cells appropriate for rebuilding the bladder wall was verified by immunostaining.

Further synthesis of these intravital imaging results with cystometric evaluation demonstrated that although the observed intensity decrease likely included some stem cell apoptosis, therapeutic efficacy as measured by improved bladder voiding parameters indicated that healing of the bladder wall was not limited to the injection site. Careful immunostaining demonstrated that the success of a small, single-point

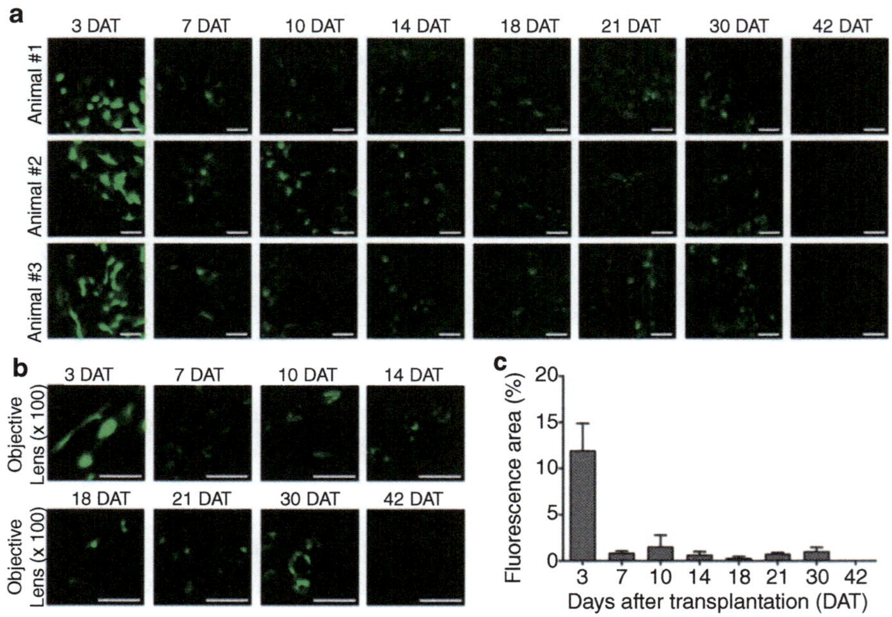

Fig. 8 Intravital microscopy of transplanted stem cells in living rats reveals engraftment from 3 to 42 days after M-MSC injection (DAT). (**a**, **b**) Fluorescent micrographs capture changes in intensity, focus, and morphology over time (**a**, ×40; **b**, ×100; scale bar: 50 μm). (**c**) Fluorescence as a percentage of confocal micrograph area. Reproduced under CC BY-NC 4.0 license from Ryu et al. (2018), © 2018 Ivyspring International Publishers

inoculation of stem cells in decreasing the symptoms of IC/BPS in the rat model and the result was also due to severance of the WNT and IGF signaling pathways (Ryu et al. 2018). In summary, the synthesis of two different intravital imaging techniques allowed accelerated longitudinal tracking of cellular migration, differentiation, and pathogenesis from a preclinical model of advanced stem cell therapy for pathologies of the bladder wall.

Stabilization of Forward GRIN Micro-endoscopes for Cardiac Applications

As seen in the renal imaging with front-view probes, the resolution of endomicroscope probes for cellular time-lapse visualization is often affected by autonomous tissue motion due to the breathing and heartbeats of experimental mice. In cardiac applications, in particular, despite high interest in cellular processes for healing cardiac tissue in mouse models of the heart, the study of cardiac tissue in vivo by microscopy and endo-microscopy has been limited by autonomous tissue movement. Herein is described the development of a minimally invasive tool for immobilizing endoscopes relative to moving tissue, which offers the potential to observe and probe the cellular processes connecting cardiovascular and immune functions in vivo, and thereby markedly accelerate the development and testing of new therapeutic and preventive strategies for cardiac pathologies (Jung et al. 2013).

While the autonomous healing of heart tissue is limited in humans, several cellular processes are known to affect heart health. In humans, the formation of arterial plaques is driven by immune cell infiltration and inflammation (Woollard and Geissmann 2010), and immune cell recruitment is crucial in many processes related to myocardial infarction, including phagocytosis, myofibroblast accumulation, angiogenesis, and collagen deposit (Nahrendorf et al. 2010). Immune cell behavior is also a significant consideration in potential regenerative therapies for atherosclerosis, including stem cell treatments and cellular reprogramming (Jung et al. 2013; Anversa et al. 2006; Chien et al. 2008; Ieda et al. 2010). The present paradigm for cellular observation of cardiac tissues is primarily fixed-tissue histology, which provides only static imagery of completed tissue processes, rather than the cellular movement and interactions which characterize living heart tissue.

Seeing this problem, many research groups have developed means of extracting cellular information from the beating mouse hearts and proxy tissues. Relatively low-movement mouse embryos have been examined, and histology and transplantation procedures have been optimized for minimal tissue trauma (Li et al. 2012; Eriksson 2011; Lucitti et al. 2007). Highly disruptive open heart vasculature imaging (Chilian et al. 1986) and microscopy have been made possible to some extent by image registration (Greenberg and Kerr 2009; Schroeder et al. 2010), temporal gating (Lee et al. 2012; Wiesmann et al. 2003), and motion-dampening windows, glue, and even centimeter-wide suction rings (Looney et al. 2011). Despite these advances, the high invasiveness of open surgery and the resultant visual debris have complicated experimental imaging (Jung et al. 2013).

The requirements for stabilizing cardiac tissue for cellular-level imaging are at first daunting: tissue moving on a range of 1–2 mm must be restricted to sub-micron deviations relative to the imaging system, while also avoiding tissue damage (Looney et al. 2011). In Fig. 9, we present a suction system that achieves minimally

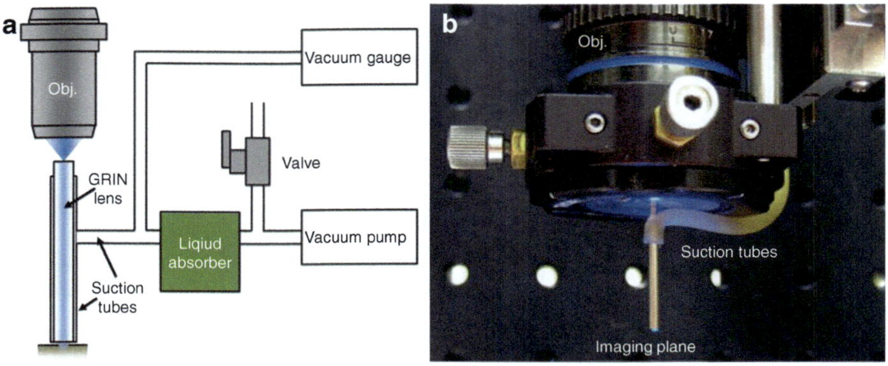

Fig. 9 Schematic and implementation of suction stabilization system for GRIN micro-endoscopes. (**a**) Schematic of the vacuum system, showing liquid absorption system. (**b**) The suction system can be designed to fit flexibly around existing GRIN micro-endoscope setups. Drawn with reference to Jung et al. (2013)

invasive imaging, holding tissues with suction pressure from a narrow tube at non-destructive pressures of 50 to 100 mm Hg. The suction tube was manufactured from two concentric stainless steel tubes of 1.8 mm and 2.9 mm diameter, separated by a 0.45-mm gap. A 20-mm-long gradient index probe was designed as described at the beginning of this chapter, with a protective sheath diameter of 1.25 mm and 1× magnification. As depicted in Figs. 9 and 10, a small, self-sealing incision in the rubber tubing facilitated insertion of the probe into the suction tube, and the tube was joined to a micro-diaphragm pump by way of a vacuum gauge, a liquid trap, and a flow valve. Suction tubes of 1.8 mm and 2.7 mm diameters fit around the endo-microscope probe and were sufficiently narrow to be put through a minimally invasive incision made in the center of the intercostal space (Jung et al. 2013). A longer, flexible suction tube allows resilience to torque and a small tilt angle under translation. Fluorescent dyes and GFP-expressing transgenic mouse models were

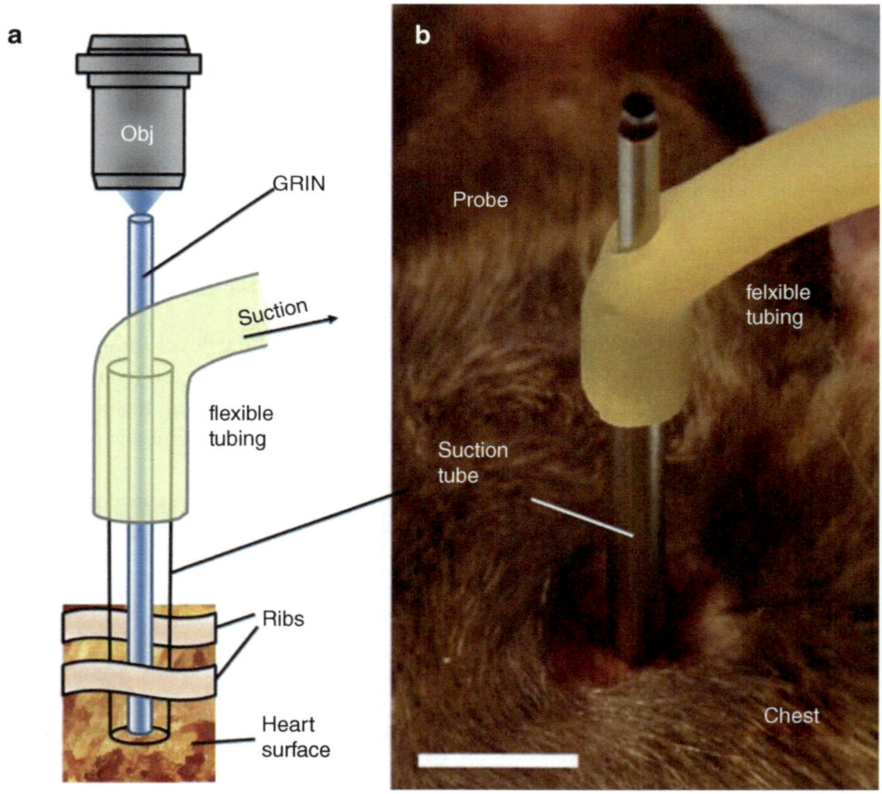

Fig. 10 Motion-stabilization effect of the suction system for cellular endo-microscopy. (**a**) Schematic of the system in use. Lateral motion of cardiac tissue is reduced relative to the probe with suction, and blood flow is maintained in the renal capsule despite suction. (**b**) Image of the system in practice. Scale bars, 50 μm. Image drawn with reference to Jung et al. (2013)

used to image the blood vessels and monocytes, respectively (Jung et al. 2013). To improve the uniformity and noninvasiveness of the suction tube, an O-ring was formed out of UV epoxy at the tube's distal end. The assembled probe was connected to a video-rate laser-scanning fluorescence confocal microscopy system (Kim et al. 2010a; Veilleux et al. 2008) using optical relays as described above. The GRIN probe had a resolution of 1 μm in the focal plane and 10 μm perpendicular to it, an optical penetration depth near 100 μm, and a 250 μm FOV (Kim et al. 2008). As with other micro-endoscopes of this design, the focal plane of the probe can be shifted noninvasively by translating the objective lens (Barretto et al. 2009; Kim et al. 2008).

Anesthetized mice 8- to 10-weeks old were translated on a three-axis stage to align the point of tissue incision with the endo-microscopy platform. The mouse was raised onto the endoscope to achieve a slight tissue pressure without suction before suction was gradually applied, and suction of the tissue was confirmed by slightly lowering the animal stage. Horizontal motion of the stage suction was applied translation of the micro-endoscope probe relative to the stabilized tissue for mosaic imaging.

The relative motion of the probe to the cardiac tissue was quantified by a cross-correlation algorithm (Guizar-Sicairos et al. 2008) and can be seen in Fig. 10b. It was found that leaving a loose connection between the suction and the heart at 50 mm Hg did not cease blood flow while reducing tissue movement to less than 10 μm peak-to-peak. This physical correction for tissue motion enabled further noise correction by image registration, which increased the signal to noise ratio and made possible automated mosaicking (Kim et al. 2010a). As shown in Fig. 10c, application of suction up to 150 mm Hg resulted in no disparity in blood flow, as measured by motion tracking of fluorescently tagged red blood cells and polystyrene beads, and under suction, there was negligible reduction in blood pressure or heart rates, as tracked by a CODA monitor and mouse tail cuff. Furthermore, despite repeated observation over several weeks, there was no sign of tissue damage to the heart: no immune reaction was induced, demonstrating that the suction endo-microscopic imaging procedure is less invasive than open in situ microscopy, utilizing a standard objective or a small objective glued to the pericardium. The minimally invasive nature of suction endo-microscopy enables repeated nonfatal and time-lapse imaging, in contrast to prior techniques, and may be easily combined with commercially obtained microscope platforms (Jung et al. 2013; Kim et al. 2012a).

The suction-assisted micro-endoscope enabled a number of novel observations, including the microvasculature and active immune surveillance of the heart. Of particular interest, rolling monocytes were observed patrolling coronary venules in addition to flowing monocytes. Following an induced infarction, an increase in the frequency of flowing monocytes was seen within 30 min and corresponded with a transient decrease in the observation of circulating monocytes in peripheral vessels. The magnitude of this change suggests monocytes are recruited from the vascular pool and then released from the spleen in the first 60 min after infarction, faster than previously believed (Jung et al. 2013; Swirski et al. 2009).

Side-View GRIN Micro-endoscope Applications

Unlike front-view endoscopes, which are useful for the imaging of organs in cavities and on the rounded surfaces of organs, the side-view micro-endoscope probes based on GRIN lenses are principally applicable to the noninvasive observation of the epithelium of tubular organs in the gastrointestinal tract and airways through natural orifices (Kim et al. 2010a). We survey applications to the colon, esophagus, and trachea which are enabled by the narrow probe diameter, but also look at the application of these probes to minimally invasive injection in the murine brain.

Small Animal Colonoscopy Application

The colon is a particularly relevant target for imaging based on the side-view optical probe, as the probe allows noninvasive imaging of early colorectal tumor formation deep in the crypt bottom. Figure 11 shows endo-micrographs of fluorescently tagged cells and vasculature in the colon of a genetically modified mouse model exhibiting spontaneous carcinogenesis. By knocking out the adenomatous polyposis coli (APC) gene with an adenoviral factor and activating a linked reporter gene for GFP, the formation of fluorescent intestinal tumors can be tracked over several weeks (Kim et al. 2010a). Blood vessels may be observed simultaneously by injection of indocyanine green or tetramethylrhodamine dextran into the bloodstream. In Fig. 11a, early-stage neoplasia is observed with diameters of 50–150 μm after a period of 4 weeks. Observed 2 weeks later, the same neoplasia showed high blood vessel perfusion, with polyp diameters of 200–400 μm (Fig. 11b) (Kim et al. 2012a).

The procedure for side-view imaging of the colon makes use of a horizontal mount and relay mechanism which connects the 1× magnification GRIN endomicroscope probe to a 40× confocal microscope, while allowing probe rotation. Mice are prepared for observation by fasting and a gentle stool removal procedure following anesthetization. Mice are placed on their backs on a translation stage, so that the point of entry may be viewed when the probe is oriented for ventral observation. By gradual translation, the anus is moved around the distal end of the probe.

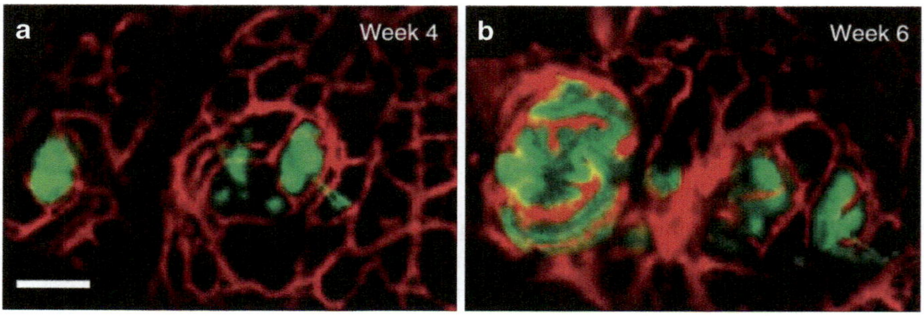

Fig. 11 Time-lapse cellular-level tracking of early colorectal tumor formation in side view. (**a**) Neoplasia expressing GFP in mice 4 weeks after inactivation of *APC* (**b**) The same region shows tumor development and blood vessel perfusion 2 weeks later. Scale bar, 100 μm. Reproduced with permission from Kim et al. (2012a), © 2012 Springer Nature

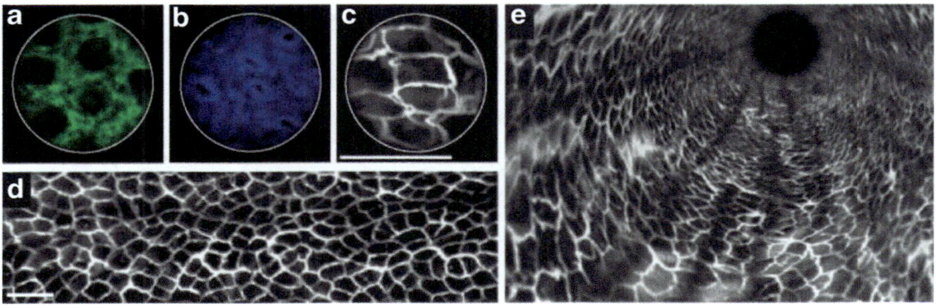

Fig. 12 Potential presentations and imaging targets for side-view microprobes in the descending colon of mice. (**a**–**c**) Images from a single probe field of view (**a**) MHCII$^+$, GFP$^+$ cells (**b**) stained epithelial cells (**c**) vasculature (**d**, **e**) Composite synthesized from many single-FOV images. (**d**) Unwrapped colon wall (**e**) fly-through presentation. Scale bars, 200 μm. Reproduced with permission from Kim et al. (2012a), © 2012 Springer Nature

The probe may be scored for return to consistent locations within the colon, and scanning may be performed to produce image mosaics at 100–200 μm/s. Figure 12a–c depict fluorescently tagged cells in the mucosa of the colon, as well as vasculature visualized by intravenous injection of FITC-dextran. Digital stitching allows a large set of images to be presented as a composite image (Fig. 12d) or as a fly-through rendering (Fig. 12e) (Kim et al. 2010a). The procedure may be adapted for the upper gastrointestinal tract (Kim et al. 2012a; Jun et al. 2018).

Applications to the Upper Digestive Tract for Small Animal Diseased Model

While frequent use is made of mouse models for the study of colorectal cancer, mouse models may also be applied to the simulation of a variety of disorders in the upper digestive tract. Although cancers and related inflammation disorders are common in the esophagus, esophageal passages are also the frequent target of gastrointestinal stenting for the correction of strictures. The pressure induced by these stents on the luminal wall, in turn, may induce a proliferative cellular response (Jun et al. 2017; Werner et al. 2007) which results in complications of restenosis. Here we survey an application of the side-view micro-endoscope to evaluate the safety of a novel self-expanding metal stent (SEMS), observing the proliferation of fibroblast cells and epithelial cell migration over time in a minimally invasive manner. The application of side-view GRIN micro-endoscope probes to the upper digestive tract is interesting for the novel considerations they bring to endo-microscopy in the mouse model. In particular, care must be taken to prevent asphyxiation when accessing target tissue by way of the mouth. Restenosis is not just a concern for the esophagus, but also for the urethra and trachea (Kim et al. 2009, 2007; Park et al. 2012; Song et al. 2003). The results are significant for understanding the cellular processes underlying restenosis in human patients and for motivating new clinical strategies for the mitigation of stent complications (Jun et al. 2018).

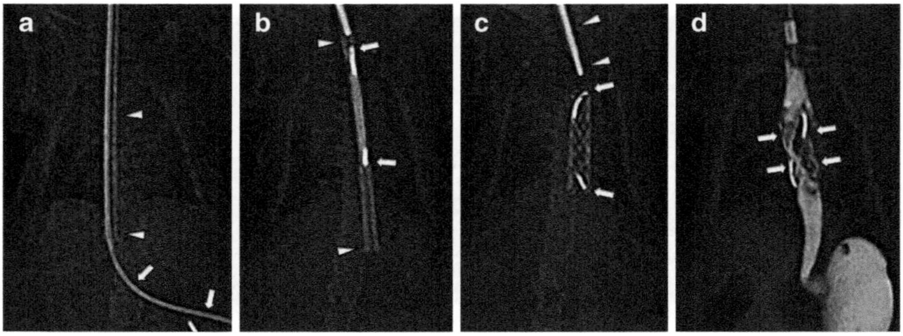

Fig. 13 (**a**, **b**, **c**) Radiographs and esophagography of stent insertion in a mouse model of the proliferative cellular response. (**a**) Guidewire (arrows) and sheath (arrowheads) placement. (**b**) Stent delivery. (**c**) Stent placement. (**d**) Esophagograph captured 8 weeks after stent placement depicting a filling defect (arrows) caused by restenosis. Reprinted from J. Vasc. Interv. Radiol. 29, Jun et al. "In Vivo Fluorescence Microendoscopic Monitoring of Stent-Induced Fibroblast Cell Proliferation in an Esophageal Mouse Model," pages 1756–1763, Copyright 2018, with permission from the Society of Interventional Radiology (SIR)

Fibroblast cells are the primary imaging target for restenosis, as stent-induced proliferative cellular response is characterized by reepithelialization and fibroplasia (Jun et al. 2017), which are modulated by fibroblasts' creation of extracellular matrix and production of collagen (Taylor et al. 2010; Okada et al. 1997; Polyak and Weinberg 2009; Park et al. 2014; Leask and Abraham 2004; Frangogiannis 2006; Lindley and Briegel 2010). Although rat, rabbit, and pig models of human restenosis are available (Jun et al. 2017; Zhu et al. 2017; Feng et al. 2016; Kim et al. 2010b, 2013a; Wang et al. 2015), deeper studies are possible in mice due to their low cost and additional transgenic varieties. Mice were bred expressing fibroblast-specific protein-1 labeled with GFP (Jun et al. 2018; Iwano et al. 2002).

Self-expanding stents for mice were manufactured from threaded nitinol filament with radiopaque markers to assist in stent placement at each end. SEMS were placed under fluoroscopic guidance in ten mice with a pusher catheter (Fig. 13a–c), while ten mice underwent a sham procedure as controls. Fluorescence microendoscopic observations were performed concurrently with fluoroscopy prior to and 2, 4, 6, and 8 weeks after stent placement. Two mice died during the study, one due to breathing difficulties during endoscopy and one due to stent migration (Jun et al. 2018).

As shown in Fig. 14, for fluorescence micro-endoscopic imaging, anesthetized mice were immobilized on a homeostasis-maintaining three-axis translation stage. To ensure a clear airway while facilitating endoscope insertion, the tongue of the mouse was gently pulled out. After insertion of the endoscope into the mid-thoracic esophagus, fluorescence images were captured at 30 Hz using a laser scanning confocal system, including a 488-nm excitation laser and photo-multiplicative detector tubes. Frames were subsequently averaged and filtered to reduce shot noise, and a background normalization procedure was applied to reduce noise from background

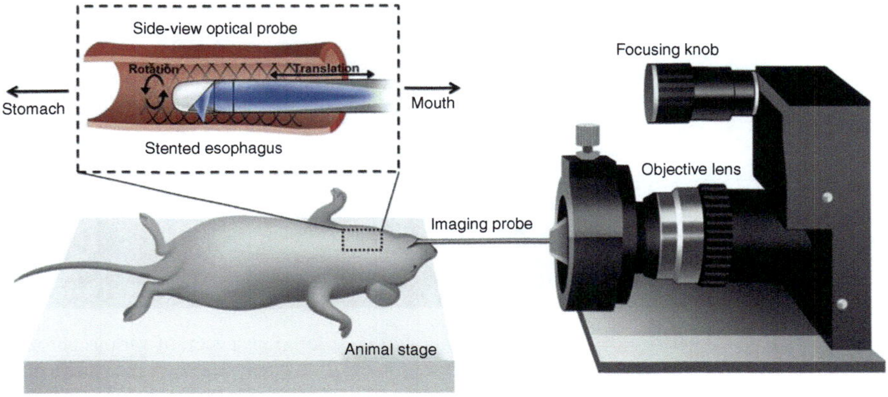

Fig. 14 Application of esophageal fluorescence endo-microscopy to track the cellular response to stenting in vivo. Reprinted from J. Vasc. Interv. Radiol. 29, Jun et al. "In Vivo Fluorescence Microendoscopic Monitoring of Stent-Induced Fibroblast Cell Proliferation in an Esophageal Mouse Model," pages 1756–1763, Copyright 2018, with permission from the Society of Interventional Radiology (SIR)

tissue (Tyack et al. 2014), enabling automated cell counting (Al-Khazraji et al. 2011) and comparison of fluorescence intensities, as a marker of fibroblast health (Jun et al. 2018).

As a result of this imaging procedure, it was possible to capture time-lapse fluorescence micrographs of the GFP fibroblasts, as shown in Fig. 15. In the stenting group, the fibroblast cell density increased immediately, peaking in the first measurement after stent placement before declining gradually (Fig. 15a). The average fibroblast GFP count (Fig. 15b) and average fluorescence intensity (Fig. 15c) were both significantly higher after stenting than in the control group, indicating higher fibroblast health (Jun et al. 2018).

These results are consistent with the standard explanation of a wound-healing process occurring over the first 4 weeks. Within a short period, fibroblasts and epithelial cells migrate to the inflamed tissue, where they undergo highly proliferative cellular responses. Fibroblasts in this situation transform into myofibroblasts, resulting in decreased fluorescence while the lesion becomes a collagen-dense scar (Jun et al. 2018).

Concurrent esophagography and subsequent histology supported the conclusion of the fluorescence monitoring, with a correlated decrease in the luminal diameter after stenting, and increases in tissue hyperplasia, granulation tissue, epithelial tissue, and submucosal fibrosis as measured ex vivo after 8 weeks. In histological samples, markers for fibroblasts, connective tissues, and mesenchymal cells were also increased in the stenting group. These results demonstrate that the inferences made by fluorescence micro-endoscopy are useful for the time-domain monitoring of stent-induced cellular processes in vivo (Jun et al. 2018).

In conclusion, by using minimally invasive micro-endoscopy, insight was gained into the number and behavior of fibroblast GFP cells in a mouse model of

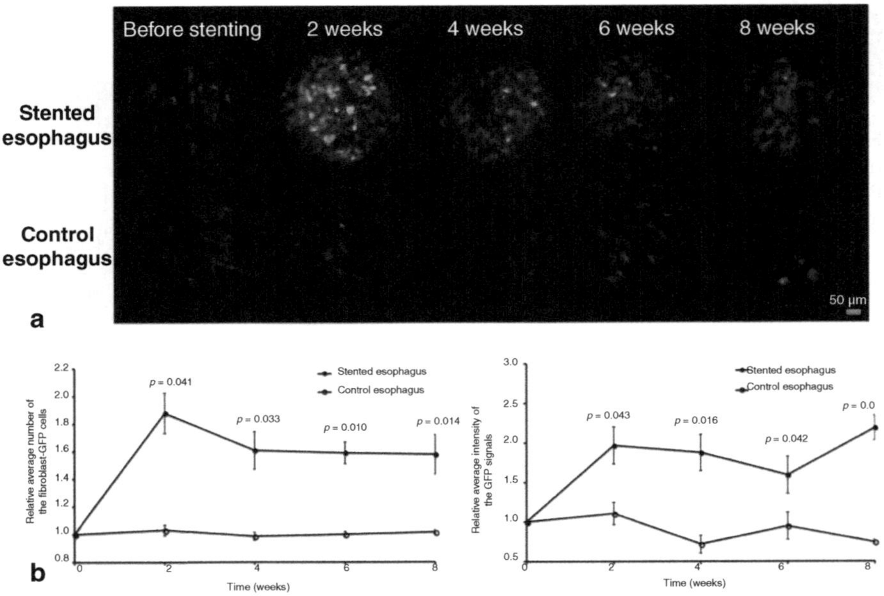

Fig. 15 Fluorescent fibroblasts proliferate in stented mouse esophagi as compared to a control, as shown by longitudinal side-view micro-endoscopy. (**a**) Representative endo-micrographs in vivo before and after stenting. (**b**) Averaged fibroblast counts (**c**) Averaged fibroblast fluorescence intensity. Reprinted from J. Vasc. Interv. Radiol. 29, Jun et al. "In Vivo Fluorescence Microendoscopic Monitoring of Stent-Induced Fibroblast Cell Proliferation in an Esophageal Mouse Model," pages 1756–1763, Copyright 2018, with permission from the Society of Interventional Radiology (SIR)

restenosis. In vivo imaging allowed increased statistical power from a smaller number of animals, by controlling for statistical errors from individual variability through repeated measurement (Kim et al. 2012b; Choi et al. 2014) and supporting radiographic observations and final histological measurements. Gradient-index-based fluorescence micro-endoscopy permits the investigation of cellular and molecular progression of gastrointestinal tract diseases in living small animal models (Jun et al. 2018).

Small-Diameter Endoscopes Enable Respiratory Imaging

Although intravital fluorescence microscopy is a powerful tool for the study of cellular changes during the progression of diseases, mechanistic studies of lung diseases in the airway epithelium of small animal models are limited by probe size and difficulty in accessing the respiratory epithelium without either obstructing breathing or collapsing lungs. In the following section, we summarize the application of a miniature side-view confocal probe to obtain fluorescence cellular-resolution images of the trachea from live transgenic fluorescent reporter mice in vivo. Fluorescent reporters specific to cell types were used to give insight into the cell biology of airway diseases, and a custom small-diameter confocal micro-endoscope was designed to allow simultaneous intubation and observation during repeated in vivo endotracheal observations over 3 weeks following a sulfur dioxide-induced injury. In vivo side-view endo-microscopy offers a minimally invasive method for the investigation of regenerating airways in small animal models of respiratory diseases.

With the recent growth in the accessibility of tools for genetic manipulation and a similarity in cellular organization to the human airway, mice present an ideal model to observe the process of airway regeneration after epithelial damage. In particular, we were interested in observing the processes by which basal stem cells self-renew and differentiate into Clara cells and ciliated cells (Rock et al. 2009), resulting in rapid healing of the airway epithelium after trauma (Rock et al. 2010). A strain of transgenic reporter mice was bred for fluorescence in each of these three major cell types. The CK5-nGFP, B1-EGFP, and FoxJ1-EGFP reporter strains allow visualization of GFP basal, ciliated, and Clara cells, respectively (Fig. 16) (Kim et al. 2012b).

As a model of airway injury and regeneration, we applied SO_2 inhalation, which is widely used due to its induction of rapid sloughing of cells from the suprabasal epithelium in the murine trachea (Kavet and Brain 1974). The practicality of longitudinal endoscopy was verified by the application of doxycycline to induce the generation of GFP^+ Clara cells in the CK5-nGFP mouse model over a 10-day period (Kim et al. 2012b).

Direct imaging and time-lapse visualization are much better suited than ex vivo histology to understanding the regeneration of the epithelium. Imaging of the respiratory system in vivo has been reported by a variety of methods, including open surgery (Kimura et al. 2009), surgical window implantation (Looney et al. 2011), and fluorescence fiber-optic bronchoscopy (Cortez-Retamozo et al. 2008). Our group had previously observed dendrites in the upper airway through a wider, 1.25-mm-diameter side-view GRIN micro-endoscope (Kim et al. 2010a). However, the large probe diameter mandated tracheostomy and ruled out longitudinal imaging. Thus, size remained a barrier to the minimal invasiveness of tracheal imaging (Kim et al. 2012b).

In response to this limitation, GRIN lenses of only 0.35 mm diameter were used to fabricate a new probe with a total outer diameter of 0.61 mm. This thin and semi-rigid probe of length 62 mm was protected in a metal sheath, integrated on a rotating mount, and pushed into the murine airway via an intubation cannula (OD, 1.0 mm; ID, 0.7 mm) so as to not block the flow of air (Fig. 17a) (Kim et al. 2012b).

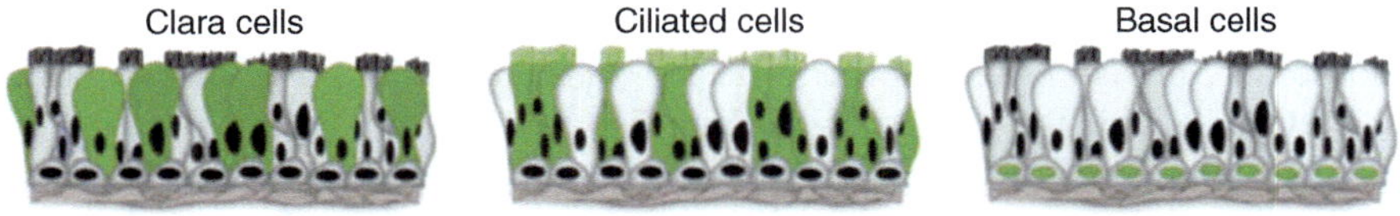

Fig. 16 Sketch of the ciliated, basal, and Clara cells of the tracheal epithelium. Reprinted from Kim et al. (2012b) with permission of the American Thoracic Society. Copyright © 2019 American Thoracic Society

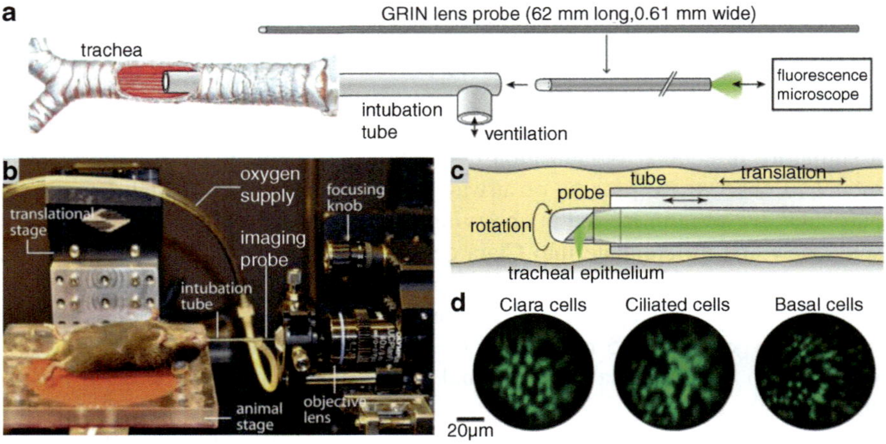

Fig. 17 In vivo GRIN fluorescence endo-microscopy of the trachea. (**a**) Preparation of the GRIN lens imaging through an intubation tube. (**b**) Actual imaging setup. (**c**) Side-view imaging of the tracheal epithelium. (**d**) Resultant in vivo endo-micrographs, showing Clara, ciliated, and basal cells in their respective GFP^+ mouse models. Scale bar, 20 μm. Reprinted from Kim et al. (2012b) with permission of the American Thoracic Society. Copyright © 2019 American Thoracic Society

By adjusting the position of the confocal microscope's objective, the endoscope's focal plane was easily controlled (Kim et al. 2008) (Fig. 17b). Translation of the mouse parallel to the tube and rotation of the probe allowed visualization of an area larger than that of the optical system's native field of view, which was limited to about 6400 μm^2 (Fig. 17c).

Fluorescence images were acquired with an identical image processing workflow to the esophageal imaging described above, with the exception of an image registration algorithm (Zitová and Flusser 2003), which corrected for artifacts of the motion due to respiration (Kim et al. 2012b).

Anesthetized mice were intubated with an endotracheal cannula and connected to an oxygen supply prior to imaging, as shown in Fig. 20. A paraffin film sealed the tube while the narrow-diameter optical probe was pushed in. Imaging sessions lasted approximately 40 minutes, concluding with the slow removal of the micro-endoscope and cannula, and the resumption of autonomous breathing. During in vivo application, the captured images discerned individual cell nuclei, thanks to local confinement of GFP.

Following creation and verification of a new dual-reporter fluorescent Clara cell mouse line, in vivo observation prior to SO_2 inhalation injury revealed uniformly dense Clara cells in the tracheal epithelium (Fig. 18a). One day after exposure, fluorescent Clara cells were completely lost (Fig. 18b). Immunofluorescence of concurrently harvested tissue sections revealed an absence of Clara and ciliated cells while some basal stem cells appeared healthy (Fig. 18c) (Rawlins et al. 2009). One week later and 2 weeks later, fluorescent Clara cells had regenerated, and their population was increasing (Fig. 18d), although the speed of recovery varied between animals (Fig. 18e and f) (Kim et al. 2012b).

Provided cell-type-specific fluorescent reporters from the tracheal epithelium, gradient index micro-endoscopy probes of sufficient miniaturization provide visualization in vivo, over time, and with minimal invasiveness, at the resolution of individual cells. Side-view design enabled the positioning of the target tissue at short focal distances compared to the typical bronchoscopes (Cortez-Retamozo et al. 2008; Thiberville et al. 2007), but an autonomous movement of the respiratory system challenges the imaging system. Through the micro-endoscope imaging system,

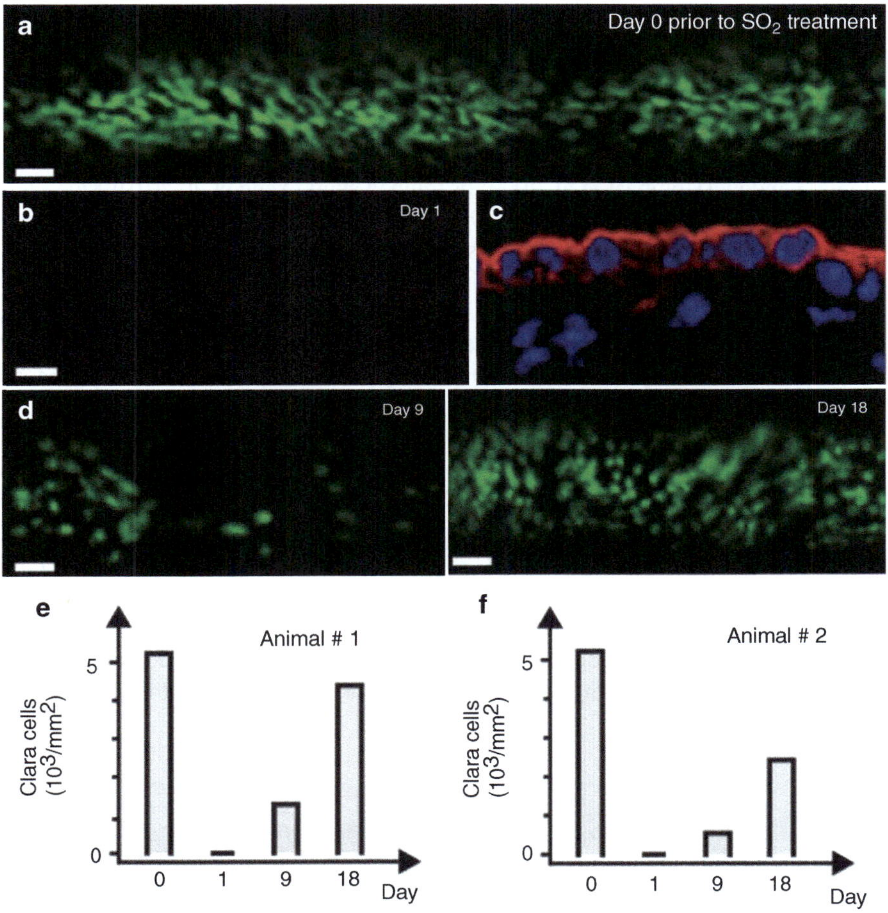

Fig. 18 Micro-endoscopic imaging of Clara cells in vivo following SO_2 injury to the mouse trachea. (**a**) A composite image of the tracheal epithelium pre-injury. (**b**) Complete depletion of Clara cells one day after injury. (**c**) Immunofluorescence of a section obtained one day after exposure reveals basal stem cells. (**d**) Later in vivo images show recovery of Clara cell population. Scale bars, 20 μm. (**e**–**f**) Clara cell densities from two different mice. Reprinted from Kim et al. (2012b) with permission of the American Thoracic Society. Copyright © 2019 American Thoracic Society

Clara cell depletion and recovery were observable in live mice, saving much time and resources over orthodox histological measurement, while reducing errors due to animal variation (Kavet and Brain 1974; Rawlins et al. 2009). The minimally invasive, in vivo observation of cellular processes using fluorescent epithelial reporters in the airway will likely find applications in murine models of asthma and cancer (Kim et al. 2012b).

Small Animal Brain Applications for Minimum Traumatic Injury

A number of significant pathologies involve the brain, including neurodegeneration, stroke, cerebral infection, cerebral inflammation, and metastasis of brain tumors. While clinical treatment of these pathologies is necessarily noninvasive, using radiological brain imaging modalities, the small animal researcher is frequently interested in mechanisms that require cellular-level resolution and prefers techniques less invasive than histology or brain needle biopsy (Feiden et al. 1991; Tilgner et al. 2005). While in situ fluorescence microscopy (Helmchen and Denk 2005) may permit cellular-resolution longitudinal observation of the mouse cerebral cortex via a transparent cranial window (Holtmaat et al. 2009) or thinned skull (Yang et al. 2010), side-view micro-endoscope probes (Kim et al. 2010a) are ideal for observing columns of fluorescent cerebral tissue at imaging depths below the 500–750 μm limit of surface imaging (Kim et al. 2013b). Here is reviewed a narrow 350-μm-diameter side-view micro-endoscope probe and needle-based package which achieves high-resolution, large-volume imaging deep into the small animal brain while minimizing trauma and therefore behavioral side effects. The probe may be inserted into the murine brain via a hole drilled in the cranium (Kim et al. 2013b), or when strengthened with a 22-gauge needle may be directly injected through the murine skull.

The optical probe and needle package were based on standard GRIN endomicroscopic probes, designed with GRIN rod lenses of diameter 350 μm, fashioned into side view probes by attachment of a 0.16-pitch imaging lens and a right-angle prism with 250 μm edges (Kim et al. 2013b). Adding a 22-gauge needle (610 μm diameter) as a protective sheath, as shown in Fig. 19, made it possible for the probe to penetrate the mouse skull without craniotomy.

The optical probe was coupled into a confocal microscope as described above and mounted as shown in Fig. 20, including a translating objective (40×, NA = 0.6)

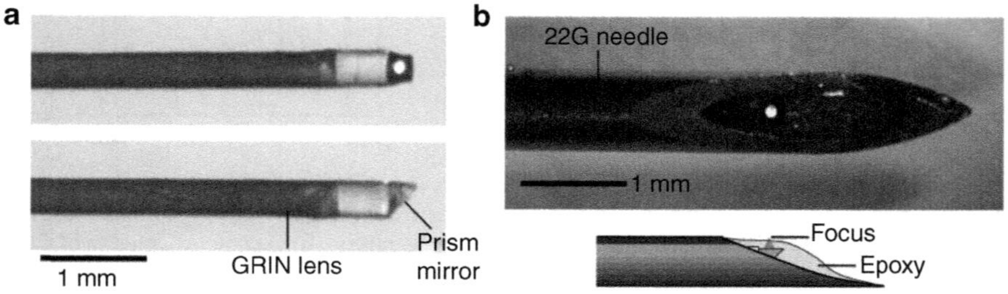

Fig. 19 Photographs of the (**a**) miniaturized and (**b**) stiffened needle side-view probes. Reprinted with permission from Kim et al. (2013c). © 2013 The Optical Society

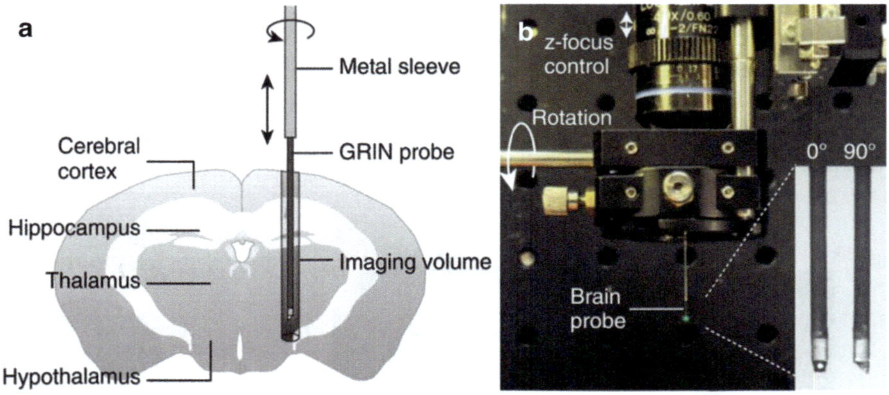

Fig. 20 Small-diameter side-view GRIN endo-microscope probes. (**a**) Application of the probe for deep brain observation. (**b**) The intravital microscopy setup. Reprinted under CC BY 4.0 license from Kim et al. (2013b). Copyright © 2013 SPIE

for adjusting the probe working distance from 0 to 100 μm and a rotational stage for panoramic imaging (Kim et al. 2013b).

As in the preparation of stainless steel prisms, the probe was immobilized relative to the needle by a heat-sensitive epoxy; uniquely, in this application, it is ideal for the epoxy to cover the distal probe tip so as to prevent collection of debris during insertion. The FOV of the 350 μm diameter probe was approximately 90 μm; however, inserting and rotating the needle enabled observation of a much larger tissue area (Kim et al. 2013b, c).

Mouse models were prepared to replicate neuronal degeneration due to intoxication, ischemic stroke, and metastasized melanoma in the brain. Two neural degeneration models (n = 5 each) were created by chronic treatment of methyl-mercury and trimethyltin hydroxide, which are known to cause apoptosis in the cerebellum granule cells (Nagashima et al. 1995; Hara et al. 1997), into Thy1-YFP$^+$ mice, which express yellow fluorescent protein in neuronal cells (Kleinfeld et al. 1998). Injections were performed intraperitoneally every 2 days for 30 days. Cerebral ischemia was modeled by blocking the middle cerebral artery for 1 hour in mice expressing green fluorescent protein in MHC-class-II (Arumugam et al. 2006). Metastatic cancer models were established by craniotomy followed by in situ injection of 2×10^6 B16 melanoma cells modified to express RFP (Kim et al. 2013c). Groups were prepared with and without overexpressed p53 genes, which limits metastasis in melanoma (Kim et al. 2013c; Chudnovsky et al. 2005).

Imaging was performed on anesthetized mice after initial preparation, which consisted of hair removal followed optionally by drill craniotomy (Kim et al. 2013b), after which mice were placed in a homeostasis-maintaining stereotaxic mount on a translational stage. Tissue was moistened with saline, and the stage was raised such that the probe was gradually inserted into the brain. Smooth, low-trauma penetration was facilitated by the sharp distal edge of the prism (or needle), the narrow probe diameter, and the application of force along the probe axis (Kandel 2000). Axial and rotational scanning of the imager were achieved by moving the staged mouse vertically and rotating the probe as described earlier in the chapter (Kim et al. 2013b, c). The setup for this procedure is shown in Fig. 20 (Kim et al. 2013b).

Through axial and rotational composites of a series of images, it becomes possible to used gradient-index-lens-based confocal micro-endoscopes to capture in vivo panoramic and cylindrical sections of fluorescent tissue, including all the layers in the cerebral cortex at cellular resolution, as demonstrated in Fig. 21.

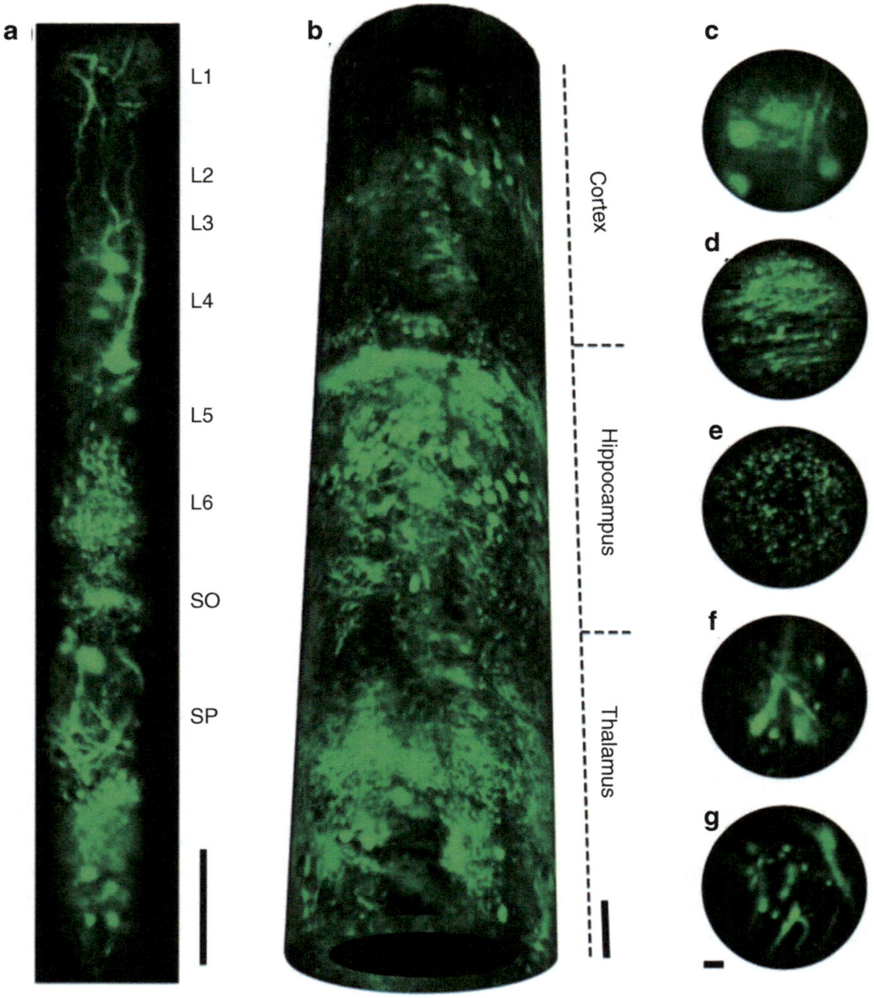

Fig. 21 In vivo composite images of the murine brain captured with a 350-μm side-view probe from Thy1-GFP mouse. (**a**) A composite of neurons and neural cells along a cross-section of the cerebral cortex and hippocampus. (**b**) Cylindrical rendering of captured wide-area deep-brain neuronal images. Scale bars: 100 μm (**c**–**g**) Varieties of cells observed from across the optical section. Reprinted under CC BY 4.0 license from Kim et al. (2013b). Copyright © 2013 SPIE

In Fig. 21a, vertically (axially) scanning the probe during insertion revealed a deep view of the axon structure in the brain tissue encircling the inserted probe, showing the layers of the cortex, and striatum. Adding rotation to the axial translation allowed reconstruction of a cylinder of the brain from the cortex to the thalamus. Individual neurons were observed clearly in the olfactory bulb, hippocampus, thalamus, hypothalamus, midbrain, and medulla, indicating that clear visualization is possible across the entire brain (Kim et al. 2013b).

The resulting side-view GRIN micro-endoscope probes are well suited for optical biopsy. In the murine models of neuronal degeneration, observation by intracerebral endoscopy was performed prior to, after 15 days, and after 30 days of chemical treatment in each model (Fig. 22). The time-lapse endo-micrographs exposed a significant decline in fluorescent granule cell density over time in both treatment groups relative to the control animals (Fig. 22b–e). Behavioral impairment, as measured on mice exempted from endoscopy by the NSS (Beni-Adani et al. 2001), showed onset of symptoms at between 23 and 25 days of treatment

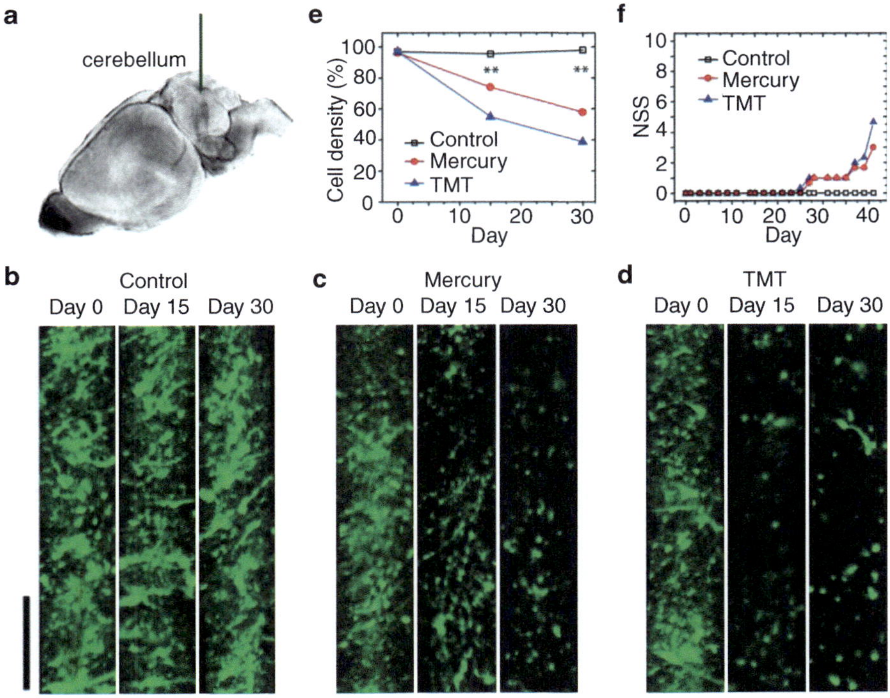

Fig. 22 Longitudinal observation of neuron apoptosis in mouse models of neurodegeneration due to intoxication. (**a**) Schematic of the 22-gauge probe's route through the cerebellum. (**b**, **c**, **d**) Wide-area composite images of fluorescent neurons at days 0, 15, and 30 in mice (**b**) unexposed (**c**) exposed to methyl mercury. (**d**) exposed to trimethyltin hydroxide. (**e**) Cell densities measured in the three cohorts. (**f**) Average neurological severity scores assessed for each cohort ($n = 5$ each). Reprinted with permission from Kim et al. (2013c). © 2013 The Optical Society

(Fig. 22f), demonstrating that in animals, significant cell loss must occur before behavioral deficits are detectable (Kim et al. 2013c).

Similar optical biopsy was applied to the direct imaging of inflammatory cellular response in deep ischemic sites in vivo. One, two, and three days after a surgically induced stroke, the density of inflammation-linked MHC-Class-II-GFP$^+$ fluorescent T cells was observed to increase linearly with time up to 70,000 cells per mm^3 at depths of 1–1.5 mm into the ischemic hemisphere, but not at the symmetric point in the unaffected hemisphere, in agreement with prior findings (Gelderblom et al. 2009). Thus, minimally invasive endoscopic optical biopsy shows promise as a method for understanding the dynamics of inflammation in stroke through optical biopsy, as well as the preclinical assessment of strategies and drugs for the treatment of stroke (Kim et al. 2013c; Storkebaum et al. 2011).

Finally, optical biopsy with GRIN probe was applied to a metastatic melanoma model in the mouse brain, demonstrating a change in cell roundness in the metastasizing, low-p53 variant cells compared to the high-p53 cell variant. While the low-p53 variant proliferated in deep brain tissues, the high-p53 variant disappeared over a period of 3 days (Kim et al. 2013b).

Although tissue damage is unavoidably caused by inserting and removing GRIN microprobes from the brain, the 350-μm diameter optical probe is much less invasive than a similar probe of 1250 μm diameter, as shown in Fig. 23. On the left, brain tissue is shown after one pass of a 350-μm diameter probe. Histology reveals that the wound closed itself due to tissue pressure, in comparison to the clearly hemorrhaging, open wound visible after a single insertion of a 1250-μm probe, as seen on the right (Kim et al. 2013b).

Further histological and behavioral observations with needle-sheathed probes indicated apoptotic tissue damage due to the needle-sheathed probe heals over the

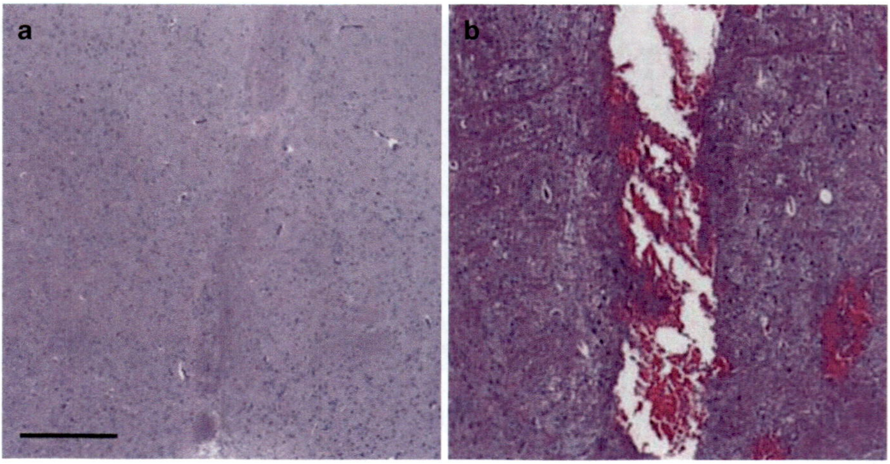

Fig. 23 Trauma received during probe insertion, via stained histology of cerebral tissue (**a**) 350-μm-diameter probe and (**b**) 1250-μm probe. Scale bar: 500 μm. Reprinted under CC BY 4.0 license from Kim et al. (2013b). Copyright © 2013 SPIE

course of a 2-week period, and that the impact of needle insertion into the thalamus or cerebrum on mouse behavior is negligible. Insertion into the cerebellum caused impairment in all mice, but behavior was restored after a 1-week period (Kim et al. 2013c).

In conclusion, the introduction of ultrathin side-view micro-endoscopy probes based on GRIN lenses has resulted in cellular resolution imaging of the brains of living mouse models of neurological disease at depths of more than 10 mm. The probes are minimally invasive and can be inserted with minimal trauma and, for some applications, without causing changes to behavioral patterns. These optical probes allow the researcher to quantify the density of active neurons and immune infiltration longitudinally and in vivo. Further observations of the cellular-level dynamics of the deep brain are promising for the unraveling of disease mechanisms, for developing diagnostic technologies, and for the treatment of neurodegeneration (Kim et al. 2013b, c; Fuhrmann et al. 2010). Further extensions to the imaging method include switching to near-infrared multiphoton excitation for increased imaging depth and developing protocols for the observation of other models based on other animals frequently used in neuroscience and pharmaceutical research.

Conclusion and Future Perspectives

The present review has demonstrated the design, fabrication, and myriad possibilities for the extension and application of endo-microscopic probes based on GRIN lenses. The probes are adaptable, and during the fabrication process may be optimized for front or side views, narrowed for minimal invasiveness, or reinforced for mechanical strength. Suction and gating techniques may be used to correct for large-amplitude autonomous tissue motion, while the combination with other diagnostic techniques and imaging systems may be used to reveal more detail about the behavior of tissues both at the cellular level and at the level of whole organs.

While we have tried to give a representative sampling of the applications of the GRIN micro-endoscope, our breadth has necessarily been limited. Nonfluorescent imaging modalities applying GRIN lens micro-endoscopes have not been discussed, nor have two-photon, three-photon, lifetime, or near-infrared microscopy been discussed. We hope that, when combined with the present techniques, these avenues may serve up a multitude of discoveries for the enterprising researcher, enabling them to approach important the deep unsolved questions concerning cellular-resolution phenomena deep in model animals that concern biology, medicine, and biomedical engineering.

Funding

This work was supported by the National Research Foundation of Korea (NRF) (2019R1A2C2084122, 2020R1F1A1074033) and MRC grant (2018R1A5A2020732) by the Ministry of Science & ICT (MSIT), and by a grant of the Korea

Health Technology R&D Project through the Korea Health Industry Development Institute (KHIDI), funded by the Ministry of Health and Welfare, Republic of Korea (HI18C2391). This work was also supported by grant number 2021IP0017 from the Asan Institute for Life Sciences, Asan Medical Center, Seoul, Korea.

Acknowledgments We thank Mark Myunghwan Choi and Woei-Ming Lee for technical assistance in microscopy; R. Colvin, C. Chase, H. Ploegh, R. Kucherlapati, K. Hung, Thomas Graf, Klaus Ley, and Adam Glick for generous gifts of transgenic mice and mouse models; Dr. Barry Stipp for providing antibodies; and C. Lin, D. Cote, L. Kaplan, M. Ferrari, and G.Y. Koh for discussions which contributed to previous versions of the component manuscripts.

References

Al-Khazraji BK, Medeiros PJ, Novielli NM, Jackson DN (2011) An automated cell-counting algorithm for fluorescently-stained cells in migration assays. Biol Proced Online 13:9

Anversa P, Kajstura J, Leri A, Bolli R (2006) Life and death of cardiac stem cells: a paradigm shift in cardiac biology. Circulation 113:1451–1463

Arumugam TV et al (2006) Gamma secretase–mediated Notch signaling worsens brain damage and functional outcome in ischemic stroke. Nat Med 12:621–623

Barretto RPJ, Messerschmidt B, Schnitzer MJ (2009) In vivo fluorescence imaging with high-resolution microlenses. Nat Methods 6:511–512

Barretto RPJ et al (2011) Time-lapse imaging of disease progression in deep brain areas using fluorescence microendoscopy. Nat Med 17:223–228

Becker C, Fantini MC, Neurath MF (2007) High resolution colonoscopy in live mice. Nat Protoc 1:2900–2904

Beni-Adani L et al (2001) A peptide derived from activity-dependent neuroprotective protein (ADNP) ameliorates injury response in closed head injury in mice. J Pharmacol Exp Ther 296:57–63

Birder L, Andersson K-E (2018) Animal modelling of interstitial cystitis/bladder pain syndrome. Int Neurourol J 22:S3–S9

Bociort F (1994) Imaging properties of gradient-index lenses. Berichte der Technischen Optik, (Koester, 1994)

Chien KR, Domian IJ, Parker KK (2008) Cardiogenesis and the complex biology of regenerative cardiovascular medicine. Science 322:1494–1497

Chilian WM, Eastham CL, Marcus ML (1986) Microvascular distribution of coronary vascular resistance in beating left ventricle. Am J Physiol-Heart Circ Physiol 251:H779–H788

Choi JW, Kim JK, Choi M, Kim YR, Yun SH (2014) In vivo imaging of Lgr5-positive cell populations using confocal laser endomicroscopy during early colon tumorigenesis. Endoscopy 46:1110–1116

Chudnovsky Y, Khavari PA, Adams AE (2005) Melanoma genetics and the development of rational therapeutics. J Clin Invest 115:813–824

Cortez-Retamozo V et al (2008) Real-time assessment of inflammation and treatment response in a mouse model of allergic airway inflammation. J Clin Invest 118:4058–4066

Cosenza S et al (2018) Mesenchymal stem cells-derived exosomes are more immunosuppressive than microparticles in inflammatory arthritis. Theranostics 8:1399–1410

del Río AD, Gómez-Reino C, Flores-Arias MT (2015) Third-order aberrations in GRIN crystalline lens: a new method based on axial and field rays. J Optom 8:77–85

Dela Cruz JM, McMullen JD, Williams RM, Zipfel WR (2010) Feasibility of using multiphoton excited tissue autofluorescence for in vivo human histopathology. Biomed Opt Express 1:1320

Eriksson EE (2011) Intravital microscopy on atherosclerosis in apolipoprotein E-deficient mice establishes microvessels as major entry pathways for leukocytes to advanced lesions. Circulation 124, CIRCULATIONAHA-111

Fan Z et al (2010) In vivo tracking of 'color-coded' effector, natural and induced regulatory T cells in the allograft response. Nat Med 16:718–722

Feiden W, Steude U, Bise K, Gündisch O (1991) Accuracy of stereotactic brain tumor biopsy: comparison of the histologic findings in biopsy cylinders and resected tumor tissue. Neurosurg Rev 14:51–56

Feng Y et al (2016) A comparison of a fully covered and an uncovered segmented biodegradable esophageal stent in a porcine model: preclinical evaluation of degradation, complications, and tissue reactions. Gastroenterol Res Pract 2016:1–7

Flusberg BA et al (2005) Fiber-optic fluorescence imaging. Nat Methods 2:941–950

Frangogiannis N (2006) Targeting the inflammatory response in healing myocardial infarcts. Curr Med Chem 13:1877–1893

Fuhrmann M et al (2010) Microglial Cx3cr1 knockout prevents neuron loss in a mouse model of Alzheimer's disease. Nat Neurosci 13:411–413

Fujisaki J et al (2011) In vivo imaging of Treg cells providing immune privilege to the haematopoietic stem-cell niche. Nature 474:216–219

Gelderblom M et al (2009) Temporal and spatial dynamics of cerebral immune cell accumulation in stroke. Stroke 40:1849–1857

Göbel W, Kerr JND, Nimmerjahn A, Helmchen F (2004) Miniaturized two-photon microscope based on a flexible coherent fiber bundle and a gradient-index lens objective. Opt Lett 29:2521

Greenberg DS, Kerr JND (2009) Automated correction of fast motion artifacts for two-photon imaging of awake animals. J Neurosci Methods 176:1–15

Guizar-Sicairos M, Thurman ST, Fienup JR (2008) Efficient subpixel image registration algorithms. Opt Lett 33:156

Hara A, Niwa M, Yoshimi N, Mori H (1997) Apoptotic cell death in vulnerable subpopulation of cerebellar granule cells. Acta Neuropathol. (Berl.) 94:517–518

Helmchen F, Denk W (2005) Deep tissue two-photon microscopy. Nat Methods 2:932–940

Holtmaat A et al (2009) Long-term, high-resolution imaging in the mouse neocortex through a chronic cranial window. Nat Protoc 4:1128–1144

Hong K-S et al (2015) A porous membrane-mediated isolation of mesenchymal stem cells from human embryonic stem cells. Tissue Eng Part C Methods 21:322–329

Hsiung P-L et al (2008) Detection of colonic dysplasia in vivo using a targeted heptapeptide and confocal microendoscopy. Nat Med 14:454–458

Ieda M et al (2010) Direct reprogramming of fibroblasts into functional cardiomyocytes by defined factors. Cell 149:375–387

Iwano M et al (2002) Evidence that fibroblasts derive from epithelium during tissue fibrosis. J Clin Invest 110:341–350

Jun EJ et al (2017) EW-7197, an activin-like kinase 5 inhibitor, suppresses granulation tissue after stent placement in rat esophagus. Gastrointest Endosc 86:219–228

Jun EJ et al (2018) In vivo fluorescence microendoscopic monitoring of stent-induced fibroblast cell proliferation in an esophageal mouse model. J Vasc Interv Radiol 29:1756–1763

Jung JC, Schnitzer MJ (2003) Multiphoton endoscopy. Opt Lett 28:902–904

Jung K et al (2013) Endoscopic time-lapse imaging of immune cells in infarcted mouse hearts. Circ Res 112:891–899

Eric R Kandel (2000) Principles of neural science. McGraw-Hill, Health Professions Division

Kavet RI, Brain JD (1974) Minireview reaction of the lung to air pollutant exposure. Life Sci 15:849–861

Kim A, Shin D-M Choo M-S (2016a) Stem cell therapy for interstitial cystitis/bladder pain syndrome. Curr Urol Rep 17

Kim A et al (2016b) Mesenchymal stem cells protect against the tissue fibrosis of ketamine-induced cystitis in rat bladder. Sci Rep 6:30881

Kim A et al (2017) Improved efficacy and in vivo cellular properties of human embryonic stem cell derivative in a preclinical model of bladder pain syndrome. Sci Rep 7:8872
Kim DY et al (2019) Lissajous scanning two-photon endomicroscope for in vivo tissue imaging. Sci Rep 9:3560
Kim E-Y, Shin JH, Jung YY, Shin D-H, Song H-Y (2010b) A rat esophageal model to investigate stent-induced tissue hyperplasia. J Vasc Interv Radiol 21:1287–1291
Kim E-Y et al (2013a) IN-1233-eluting covered metallic stent to prevent hyperplasia: experimental study in a rabbit esophageal model. Radiology 267:396–404
Kim JH et al (2007) Benign tracheobronchial strictures: long-term results and factors affecting airway patency after temporary stent placement. Am J Roentgenol 188:1033–1038
Kim JH et al (2009) Temporary metallic stent placement in the treatment of refractory benign esophageal strictures: results and factors associated with outcome in 55 patients. Eur Radiol 19:384–390
Kim JK et al (2012a) Fabrication and operation of GRIN probes for in vivo fluorescence cellular imaging of internal organs in small animals. Nat Protoc 7:1456–1469
Kim JK et al (2012b) In vivo imaging of tracheal epithelial cells in mice during airway regeneration. Am J Respir Cell Mol Biol 47:864–868
Kim JK, Choi JW, Yun SH (2013b) 350-μ m side-view optical probe for imaging the murine brain *in vivo* from the cortex to the hypothalamus. J Biomed Opt 18:050502
Kim JK, Choi JW, Yun SH (2013c) Optical fine-needle imaging biopsy of the brain. Biomed Opt Express 4:2846
Kim JM et al (2016c) Perivascular progenitor cells derived from human embryonic stem cells exhibit functional characteristics of pericytes and improve the retinal vasculature in a rodent model of diabetic retinopathy: hESC-PVPCs improve the retinal vasculature of DR. Stem Cells Transl Med 5:1268–1276
Kim P, Puoris'haag M, Côté D, Yun SH (2008) In vivo confocal and multiphoton microendoscopy. J Biomed Opt 13:010501
Kim P et al (2010a) In vivo wide-area cellular imaging by side-view endomicroscopy. Nat Methods 7:303–305
Kim Y et al (2018) Small hypoxia-primed mesenchymal stem cells attenuate graft-versus-host disease. Leukemia 32:2672–2684
Kimura H et al (2009) Real-time imaging of single cancer-cell dynamics of lung metastasis. J Cell Biochem. n/a–n/a. https://doi.org/10.1002/jcb.22379
Kleinfeld D, Mitra PP, Helmchen F, Denk W (1998) Fluctuations and stimulus-induced changes in blood flow observed in individual capillaries in layers 2 through 4 of rat neocortex. Proc Natl Acad Sci 95:15741–15746
Kobat D, Horton NG, Xu C (2011) In vivo two-photon microscopy to 1.6-mm depth in mouse cortex. J Biomed Opt 16:106014
Kohler A et al (2008) Estimating and correcting Mie scattering in synchrotron-based microscopic Fourier transform infrared spectra by extended multiplicative signal correction. Appl Spectrosc 62:259–266
Köhler M et al (2018) Integrative micro-endoscopic system combined with conventional microscope for live animal tissue imaging. J Biophotonics 11:e201800206
Krishna KSR, Sharma A (1996) Chromatic aberrations of radial gradient-index lenses I Theory. Appl Opt 35:1032
Leask A, Abraham DJ (2004) TGF-β signaling and the fibrotic response. FASEB J 18:816–827
Lee CM, Engelbrecht CJ, Soper TD, Helmchen F, Seibel EJ (2010) Scanning fiber endoscopy with highly flexible, 1 mm catheterscopes for wide-field, full-color imaging. J Biophotonics 3:385–407
Lee S et al (2012) Real-time in vivo imaging of the beating mouse heart at microscopic resolution. Nat Commun 3:1054
Li W et al (2012) Intravital 2-photon imaging of leukocyte trafficking in beating heart. J Clin Invest 122:2499–2508

Liang X, Ding Y, Zhang Y, Tse H-F, Lian Q (2014) Paracrine mechanisms of mesenchymal stem cell-based therapy: current status and perspectives. Cell Transplant 23:1045–1059

Lindley LE, Briegel KJ (2010) Molecular characterization of TGFβ-induced epithelial-mesenchymal transition in normal finite lifespan human mammary epithelial cells. Biochem Biophys Res Commun 399:659–664

Looney MR et al (2011) Stabilized imaging of immune surveillance in the mouse lung. Nat Methods 8:91–96

Lucitti JL et al (2007) Vascular remodeling of the mouse yolk sac requires hemodynamic force. Development 134:3317–3326

Matz G et al (2017) Chip-on-the-tip compact flexible endoscopic epifluorescence video-microscope for in-vivo imaging in medicine and biomedical research. Biomed Opt Express 8:3329

Miller MJ (2002) Two-photon imaging of lymphocyte motility and antigen response in intact lymph node. Science 296:1869–1873

Miyajima M et al (2011) Early acceptance of renal allografts in mice is dependent on Foxp3+ cells. Am J Pathol 178:1635–1645

Nagashima K et al (1995) Apoptotic process of cerebellar degeneration in experimental methylmercury intoxication of rats. Acta Neuropathol (Berl) 91:72–77

Nahrendorf M, Pittet MJ, Swirski FK (2010) Monocytes: protagonists of infarct inflammation and repair after myocardial infarction. Circulation 121:2437–2445

Okada H, Danoff TM, Kalluri R, Neilson EG (1997) Early role of Fsp1 in epithelial-mesenchymal transformation. Am J Physiol-Ren Physiol 273:F563–F574

Oravisto KJ (1975) Epidemiology of interstitial cystitis. Ann Chir Gynaecol Fenn 64:75–77

Park J-H et al (2012) Polytetrafluoroethylene-covered retrievable expandable nitinol stents for malignant esophageal obstructions: factors influencing the outcome of 270 patients. Am J Roentgenol 199:1380–1386

Park J-H et al (2014) Bioreducible polymer–delivered siRNA targeting MMP-9: suppression of granulation tissue formation after bare metallic stent placement in a rat urethral model. Radiology 271:87–95

Pittet MJ, Weissleder R (2011) Intravital imaging. Cell 147:983–991

Polyak K, Weinberg RA (2009) Transitions between epithelial and mesenchymal states: acquisition of malignant and stem cell traits. Nat Rev Cancer 9:265–273

Propert KJ et al (2000) A prospective study of interstitial cystitis: results of longitudinal followup of the interstitial cystitis data base cohort. J Urol 163:1434–1439

Rawlins EL et al (2009) The role of Scgb1a1+ Clara cells in the long-term maintenance and repair of lung airway, but not alveolar, epithelium. Cell Stem Cell 4:525–534

Rock JR et al (2009) Basal cells as stem cells of the mouse trachea and human airway epithelium. Proc Natl Acad Sci 106:12771–12775

Rock JR, Randell SH, Hogan BLM (2010) Airway basal stem cells: a perspective on their roles in epithelial homeostasis and remodeling. Dis Model Mech 3:545–556

Ryu C-M et al (2018) Longitudinal intravital imaging of transplanted mesenchymal stem cells elucidates their functional integration and therapeutic potency in an animal model of interstitial cystitis/bladder pain syndrome. Theranostics 8:5610–5624

Schroeder JL et al (2010) Short communication: subcellular motion compensation for minimally invasive microscopy, in vivo: evidence for oxygen gradients in resting muscle. Circ Res 106:1129–1133

Smith AM, Mancini MC, Nie S (2009) Bioimaging: second window for in vivo imaging. Nat Nanotechnol 4:710–711

Song H-Y et al (2003) Recurrent traumatic urethral strictures near the external sphincter: treatment with a covered, retrievable, expandable nitinol stent—initial results. Radiology 226:433–440

Song M et al (2014) The paracrine effects of mesenchymal stem cells stimulate the regeneration capacity of endogenous stem cells in the repair of a bladder-outlet-obstruction-induced overactive bladder. Stem Cells Dev 23:654–663

Song M et al (2015) Mesenchymal stem cell therapy alleviates interstitial cystitis by activating Wnt signaling pathway. Stem Cells Dev 24:1648–1657

Stein PC, Pham H, Ito T, Parsons CL (1996) Bladder injury model induced in rats by exposure to protamine sulfate followed by bacterial endotoxin. J Urol 155:1133–1138
Storkebaum E, Quaegebeur A, Vikkula M, Carmeliet P (2011) Cerebrovascular disorders: molecular insights and therapeutic opportunities. Nat Neurosci 14:1390–1397
Swirski FK et al (2009) Identification of splenic reservoir monocytes and their deployment to inflammatory sites. Science 325:612–616
Taylor MA, Parvani JG, Schiemann WP (2010) The pathophysiology of epithelial-mesenchymal transition induced by transforming growth factor-β in normal and malignant mammary epithelial cells. J Mammary Gland Biol Neoplasia 15:169–190
Thiberville L et al (2007) In vivo imaging of the bronchial wall microstructure using fibered confocal fluorescence microscopy. Am J Respir Crit Care Med 175:22–31
Tilgner J, Herr M, Ostertag C, Volk B (2005) Validation of intraoperative diagnoses using smear preparations from stereotactic brain biopsies: intraoperative versus final diagnosis—influence of clinical factors. Neurosurgery 56:257–265
Tyack PL, Calambokidis J, Friedlaender A, Goldbogen J, Southall B, Formal Comment on Schorr GS, Falcone EA, Moretti DJ, Andrews RD (2014) First long-term behavioral records from Cuvier's beaked whales (*Ziphius cavirostris*) reveal record-breaking dives. PLoS ONE 9(3):e92633. https://doi.org/10.1371/journal.pone.0092633. PLoS ONE 10, e0142287 (2015)
Veilleux I, Spencer JA, Biss DP, Cote D, Lin CP (2008) *In vivo* cell tracking with video rate multimodality laser scanning microscopy. IEEE J Sel Top Quantum Electron 14:10–18
Vonk LA et al (2018) Mesenchymal stromal/stem cell-derived extracellular vesicles promote human cartilage regeneration *in vitro*. Theranostics 8:906–920
Wang Y, Chen X, Cao W, Shi Y (2014) Plasticity of mesenchymal stem cells in immunomodulation: pathological and therapeutic implications. Nat Immunol 15:1009–1016
Wang Z et al (2015) Nitinol stents loaded with a high dose of antitumor 5-fluorouracil or paclitaxel: esophageal tissue responses in a porcine model. Gastrointest Endosc 82:153.e1–160.e1
Werner S, Krieg T, Smola H (2007) Keratinocyte–fibroblast interactions in wound healing. J Invest Dermatol 127:998–1008
Wiesmann F et al (2003) High-resolution MRI with cardiac and respiratory gating allows for accurate in vivo atherosclerotic plaque visualization in the murine aortic arch. Magn Reson Med 50:69–74
Woollard KJ, Geissmann F (2010) Monocytes in atherosclerosis: subsets and functions. Nat Rev Cardiol 7:77–86
Wu Y, Leng Y, Xi J, Li X (2009) Scanning all-fiber-optic endomicroscopy system for 3D nonlinear optical imaging of biological tissues. Opt Express 17:7907–7915
Yang G, Pan F, Parkhurst CN, Grutzendler J, Gan W-B (2010) Thinned-skull cranial window technique for long-term imaging of the cortex in live mice. Nat Protoc 5:201–208
Yang L, Mac Raighne A, McCabe EM, Dunbar LA, Scharf T (2005) Confocal microscopy using variable-focal-length microlenses and an optical fiber bundle. Appl Opt 44:5928–5936
You Y, Richer EJ, Huang T, Brody SL (2002) Growth and differentiation of mouse tracheal epithelial cells: selection of a proliferative population. Am J Physiol-Lung Cell Mol Physiol 283:L1315–L1321
Zhu Y-Q et al (2017) Technical feasibility and tissue reaction after silicone-covered biodegradable magnesium stent insertion in the oesophagus: a primary study in vitro and in vivo. Eur Radiol 27:2546–2553
Zitová B, Flusser J (2003) Image registration methods: a survey. Image Vis Comput 21:977–1000

Magnetic Resonance Imaging for Drug Development

Jeong Kon Kim

Introduction

Magnetic resonance imaging (MRI) in addition to other imaging modalities is a widely used imaging modality in patient management as well as in animal experiments, and can therefore be effectively applied as a translational tool to bridge the gap between preclinical and clinical research for drug development (Table 1). Anatomical and functional information obtained on high-resolution MRI helps the profiling of drug candidates by demonstrating pathological processes in disease models and the effect of therapeutic agents. Its noninvasiveness allows repeated studies and therefore demonstrates the time-course changes during disease progression and therapeutic intervention. Furthermore, effectively translational MRI can accurately transfer the preclinical results to later clinical experiments, thereby leading to successful clinical trials.

Investigators have introduced a number of MRI biomarkers that can be adequately utilized as disease-and drug-specific parameters. As well as high-resolution T2-weighted images, a variety of MRI techniques, including diffusion-weighted image (DWI), dynamic contrast-enhanced (DCE) MRI and MR spectroscopy (MRS), can provide quantitative information for monitoring the disease severity and therapeutic effects. These MRI biomarkers are theoretically linked to the changes in the molecules, cells and tissues, and have been experimentally validated to adequately indicate the biological processes in both animals and humans. Furthermore, acquisition of multimodal MRI biomarkers in a single session of experiment can enable all-around approaches for testing the effect of treatment.

In this review, commonly used MRI biomarkers will be described in terms of their theoretical and technical backgrounds and potentials to accelerate new drug

J. K. Kim (✉)
Department of Radiology, Asan Medical Center University of Ulsan, Seoul, Republic of Korea

J. K. Kim et al. (eds.), *Advanced Imaging and Bio Techniques for Convergence Science*, Advances in Experimental Medicine and Biology,
https://doi.org/10.1007/978-981-33-6064-8_9

Table 1 Current imaging modalities used for drug development. It is obvious that no single imaging modality is adequate to satisfy all ranges of requirement in this complex field

Modality	Spatial resolution (mm)	Temporal resolution	Clinical imaging	application	advantages	disadvantages
CT	0.05 ~ 0.1	Min	Yes	Anatomical, Functional	High spatial resolution Easy access Useful for bone architecture and cancer diagnosis	Ionizing radiation Poor soft tissue contrast
MRI	0.08 ~ 0.1	s to h	Yes	Anatomical, Functional, molecular	High spatial resolution and soft tissue contrast	High cost Low molecular sensitivity Long scanning time
Ultrasound	0.05	Min	Yes	Anatomical, Functional,	Real-time imaging Low cost	Difficult to image through bone or lungs; microbubbles used for contrast enhancement
SPECT (low energy γ-rays)	1 ~ 2	Min	Yes	Functional	Low spatial and temporal resolution than PET	Ionizing radiation Longer acquisition times and radiation exposure (longer half-lives radioisotopes needed)
PET (low energy γ-rays)	1 ~ 2	Min	Yes	Anatomical, Functional, molecular	High sensitivity (10 ~ 100 times higher than SPECT, nanomolar concentration) Intermediate acquisition times and radiation exposure 3D acquisition	Ionizing radiation High cost Cyclon needed
Bioluminescence	1 ~ 10	s to min	No	Molecular	High sensitivity (picomolar concentration) Transgene-based approach	Light emission prone to attenuation with increased tissue depth
NIRF optical imaging	1 ~ 3	s to min	No	Molecular	High sensitivity (nanomolar concentration)	Excitation and emission light prone to attenuation with increased tissue depth

development. For achieving an understanding about its practical values, actual examples of MRI application for monitoring the treatment effect will be illustrated. Finally, the hurdles to overcome for a wider utilization of MRI biomarker for drug development will be discussed.

Anatomical MRI

Mechanism of Image Contrast

The primary sources of MRI contrast include proton density (PD), longitudinal relaxation time (T1), and transverse relaxation time (T2). PD contrasted is mainly induced by the number of spins (i.e., the number of protons or water in most MRI techniques).

T1 contrast is based on the relaxation time to realign the magnetization from transverse to parallel plane to the main magnetic field, of which the efficiency is influenced by the energy transfer rate between the nuclear spin system and its environment. The T1 relaxation is derived by a dipole–dipole interaction of fluctuating field close to the Larmor frequency in electrons or adjacent protons. As protons in the water molecule tumbles very rapidly, the T1 signal of water is dark on T1-weighted images. Instead, paramagnetic materials, representatively Gd, increase the T1 relaxation rate of adjacent protons, thereby producing high signal intensity on T1-weighted images. As another method to generate T1 contrast, macromolecules are synthesized to have tumbling rate similar to the Lamour frequency may have a rapid T1 relaxation and show high signal intensity on T1-weighted images.

T2-weighted image is the most commonly used MRI for lesion detection and anatomical characterization. T2 contrast is generated by the rate of decays of transverse magnetization, which depends on the difference in the resonance frequency within and between molecules. Inhomogeneity of the intrinsic magnetic field induced by adjacent spins determines the spin–spin relaxation time or transverse relaxation time T2. In addition to the T2 relaxation, field heterogeneity across a voxel causes a difference in resonance frequencies, thereby resulting in a dephasing of the transverse magnetization, i.e., T2* relaxation. This phenomenon is driven by main field inhomogeneities and tissue-related susceptibility-induced field distortions. In spin-echo MRI, this dephasing is cancelled out by the 180-radiofrequency (RF) refocusing the direction of spin.

Application for Drug Development

As diseases generally increase the amount of water in tissue, T2-weighted imaging is the most commonly used MR imaging techniques for lesion detection and volume quantification. Increased amount of water in edematous tissue is recognized by high signal intensity on T2-weighted images. In oncological studies, the change of tumor volume during disease progression or anticancer treatment can be accurately

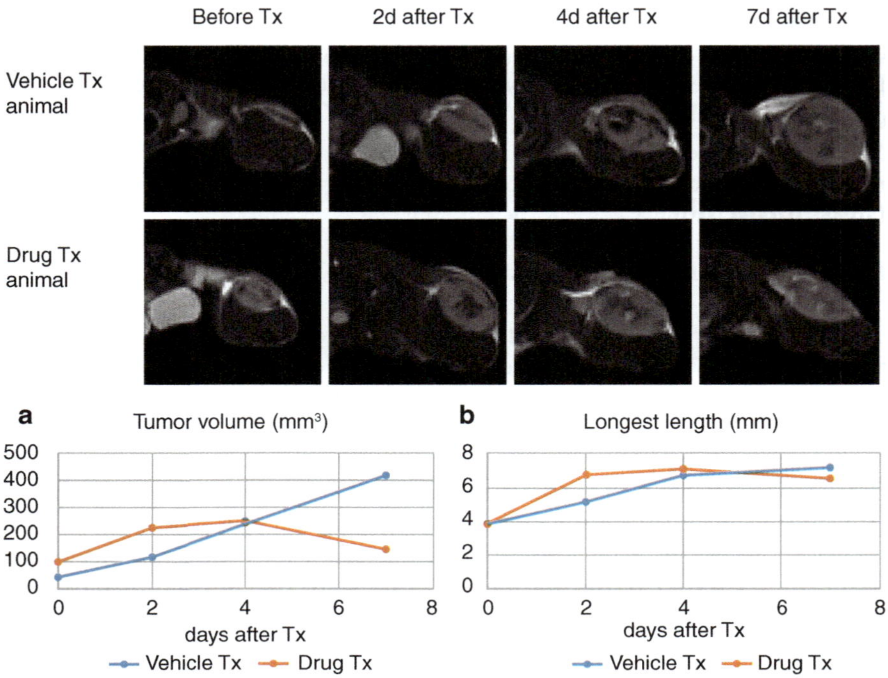

Fig. 1 Tumor volume (**a**) and longest axis length (**b**) measurements before and after vehicle (saline) or drug treatment. Ordinary tumor measurement via longest tumor axis length using caliper rule does not represent true tumor size

measured on T2-weighted images. For volume measurement of tumor, cross-sectional images are more beneficial than caliper-using manual technique because some tumors are not ellipsoid and therefore the visualization of entire tumor boundary on image can provide more accurate volume estimation (Ayers et al. 2010) (Fig. 1). The accuracy of MRI for volume measurement of small tumors on mice was validated with a strong correlation of tumor volume between MRI- and necropsy determined masses. In some tumors, such as brain tumor, which are not apparently identified on CT, MRI is preferred to CT for monitoring anticancer treatment (Shi et al. 1998).

T2-weighted images are also widely used for neurological disorders. In stroke models, high signal intensity on T2-weighted imaging 24 hours after vascular occlusion indicates irreversible ischemic damage. In contrast, the apparent diffusion coefficient (ADC) measured from DWI is an indicator specific for acute stroke but not for irreversible ischemic damage. This structural T2-weighted image can also provide functional information with regard to the ischemic damage in subacute and chronic stages. From multiple acquisitions of T2-weighted images at different echo time, parametric maps of T2 relaxation time can be calculated (Wegener et al. 2006). T2 relaxation time is a direct indicator of water content in each voxel, and therefore can estimate the degree of extracellular fluid accumulation in ischemic tissue (Redwine et al. 2003) (Fig. 2). Increased water accumulation in ischemic brain is

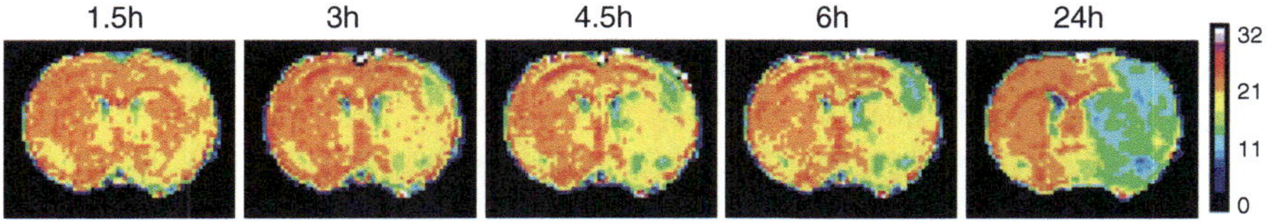

Fig. 2 An example of R2 (=1/T1) changes following middle cerebral artery occlusion, which indicates increasing water accumulation in the ischemic tissue

mainly led by cell death, necrosis, and vasogenic edema. Based on this theoretical background, parametric map of T2 relaxation time can distinguish areas of necrosis from areas suffering selective neuronal death during the subacute phase of stroke, by following the different T2 relaxation times evolution patterns [19]. Therefore, T2-weighted images can be used as a reliable biomarker for testing drugs for reducing ischemic damage or neuroprotection (Redwine et al. 2003).

T2-weighted images are also valuable for assessing cortical atrophy in neurodegenerative disorder. MRI-guided structural analysis of brain can differentiate Alzheimer's from non-Alzheimer's dementia and predict preclinical or prognostic tissue damage. Specifically, structural MRI can accurately visualize the cortical atrophy in the hippocampus and associated entorhinal cortex prior to pervasive progression (Price et al. 1991). Furthermore, the degree of hippocampal atrophy on antemortem MRI has been demonstrated to be correlated with the postmortem severity of neuropathological changes in AD (Jack Jr. et al. 2002). The sensitivity of visual rating of cortical atrophy is greater than 90% for diffuse cortical atrophy and 81%–84% for medial temporal lobe atrophy in clinical studies (Park and Moon 2016). These important anatomical observations can be also achieved in preclinical MRI. In the PDAPP transgenic mouse models of AD, cerebral atrophy can be accurately estimated on MRI (Redwine et al. 2003; Gonzalez-Lima et al. 2001; Weiss et al. 2002). Specifically, these mice showed a significant reduction in the hippocampal volume and a high-degree atrophy/agenesis in the white matter fiber tracts (fornix and corpus callosum). These findings were already noted in 3-month-old animals (Redwine et al. 2003; Gonzalez-Lima et al. 2001; Weiss et al. 2002). Due to early occurrence, these lesions are considered as a neurodevelopmental disorder rather than age-related degenerative changes.

MRI for Detecting Fat

Mechanism of Image Contrast

Chemical shift MRI and fat-suppressed MRI are widely used techniques to identify lipid content. The contrast of chemical shift is based on the theory that the resonance frequency of spin is dependent on the locally experienced magnetic field. Local magnetic field, which consists of externally applied field and magnetism of circulating electrons, is different for protons bound to water molecules and those to fatty acid. Therefore, the difference in resonance precession frequencies between water and fatty acid may generate the contrast between these molecules when specific echo times are applied. Chemical shift-based water–fat separation techniques have been applied to quantify the body fat component. With help of in-house or commercial calculation software, it is feasible to automatically quantify the amount of visceral, intramuscular, and subcutaneous fat (Kob et al. 2015).

Fat-saturated MRI provides signal from tissue component except fat, in which application of radiofrequency focusing to lipid precession frequency can cancel out the lipid signal from the image. In the combination of non-fat-saturated image, this method can quantify the amount of lipid in tissue.

Application for Drug Development

Quantification of fat fraction is an important tool to assess the severity and treatment effect of cachexia, i.e., involuntary loss of muscle and fat mass in patient with chronic disease (Boutin et al. 2015). Cancer cachexia is particularly considered as a crucial prognostic factor in cancer patients. With the combined application of T2-weighted image and chemical shift image, the volume of muscle and fat can be separated measured. This is an important advantage as the loss of skeletal muscle mass is considered as the main component of cancer cachexia (Hudelmaier et al. 2010).

Human brown adipose tissue can also be characterized by chemical shift MRI (Hu et al. 2013). Quantification of brown adipose tissue has been spotlighted for metabolism and obesity. In contrast to white adipose tissue, which stores energy as triglycerides, brown adipose tissue metabolizes fat to generate heat when stimulated. Several works demonstrated that chemical shift MRI is feasible to show distinct signal contrast between brown and white adipose tissues in mice (Branca and Warren 2011; Hu et al. 2010). In a clinical study, Hu et al. demonstrated that fat fractions and T2* relaxation time values jointly derived from chemical shift water-fat MRI are lower in brown adipose tissue than in white adipose tissue due to differences in cellular structures, triglyceride content, and vascularization (Hu et al. 2010).

Fatty liver is a common manifestation of chronic liver disease. Noninvasive detection and quantification of fat is an important requirement, due in large part to the increasing prevalence of nonalcoholic fatty liver disease. Along with frequency-selective MRI and MR spectroscopy, chemical shift MRI is regarded as a useful tool for quantifying the severity of hepatic steatosis. At an echo time when the fat and water signals are in phase, the signals are added constructively; in contrast, at an echo time of out of phase for fat, the signals of fat are cancelled out (Hussain et al. 2005). Fat detection is possible by comparing the signal intensity on the in-phase and out-of-phase images (Fig. 3) (Cassidy et al. 2009). Fat fraction (F) on chemical shift MRI can be calculated as follows:

$$\mathrm{F} = \frac{IP - OP}{2IP}$$

where IP = signal intensity on in-phase image and OP = signal intensity on out-of-phase image.

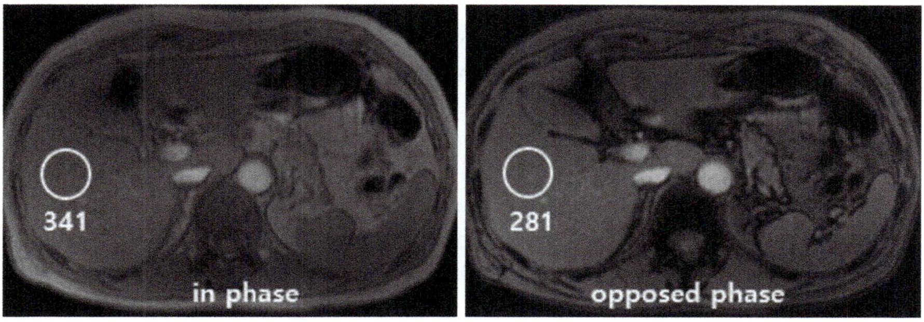

Fig. 3 The difference in signal intensity between in- and opposed-phased MRI indicates the presence of fat in the liver. The fat fraction can be calculated as (341–281)/2*341 = 8.7%

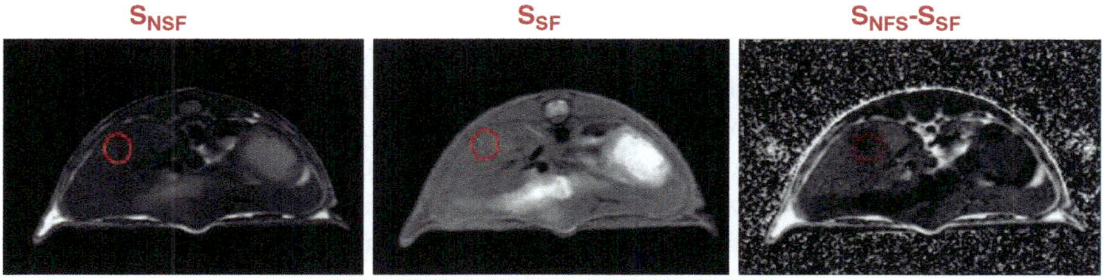

$$\text{fat-fraction signal } (\eta) = \frac{S_{NFS}\text{-}S_{SF}}{S_{NFS}}$$

Fig. 4 The signal intensity from non-suppressed fat images is assumed to represent the sum of water and fat signals (S_{NSF}), the signal intensity from suppressed fat images is assumed to represent water signal (S_{SF}), and the difference in signal intensity between non-suppressed and suppressed fat images is assumed to represent the fat signal (S_{NSF}-S_{SF})

Fat-suppressed pulse sequences separate the liver MR signal into its fat and water components by comparing magnitude images. Two sets of images, one with and the other without fat suppression, are acquired with identical imaging parameters, and comparisons can be made to each other for fat assessment. In the presence of fat, the signal is higher on the non-suppressed fat images than on the suppressed fat images, and the difference in signal intensity is entirely attributed to the presence of fat (Fig. 4).

Diffusion-Weighted Imaging

Theoretical Background

Approximately 70% of human body is composed of water. Brownian motion driven by thermal energy is the main force of diffusion. Pure water at body temperature shows a self-diffusion coefficient of $D \sim 3 \times 10^{-3}$ mm²/s, which leads to ~50 ms root-mean-square displacement at approximately 30 μm. With the dimension of cells (~1–15 μm), water molecules within cells encounter many cellular and subcellular impediments over this 50 ms period. The location of body fluid can be divided into intracellular and extracellular compartments. Extracellular water experiences relatively free diffusion whereas intracellular water has relatively "restricted diffusion." The intracellular diffusivity is influenced by cell organelles, cell size, and cell density influence, and altered by pathologic processes. For instance, highly malignant tumor and acute ischemic damage increase the level of intracellular proportion of water molecules, thereby elevating the intracellular pressure and restricting water diffusivity. Cell death results in the loss of cell membrane integrity, reduction in cell density, and increase in the interstitial space, thereby increasing the diffusivity. Therefore, the diffusivity must respond to cell death that is induced through apoptosis, autophagy, ischemia, cytotoxic agents, radiotherapy, or any other therapeutic approach (Sinkus et al. 2012). These changes are expected to be pharmacodynamic and detectible on DWI before objective volume change.

DWI allows both qualitative and quantitative assessment for the diffusion properties in tissue. This imaging technique exploits the property to diffusion degree in a certain direction and demonstrated micro-architectural detail in tissue. For obtaining DWI, diffusion-sensitization gradients are applied on either side of the 180° refocusing pulse. The parameter of b value, expressed in s/mm², indicates the weight of diffusion, which is proportional to the square of the amplitude and duration of the applied gradient. When diffusion gradient is applied in several directions, the diffusivity in each direction can be estimated. From DWI, apparent diffusion coefficient (ADC), which indicates the degree of diffusion in a direction. ADC is calculated as follows:

$$S_{DWI} = S_{b=0} \times e^{(-b \times ADC)}, \mathrm{ADC} = \ln\left(\frac{S_{b=0}}{S_{DWI}}\right),$$

$$\text{Isotropic ADC}\left(\mathrm{ADC}_i\right) = \frac{\mathrm{ADC}_x + \mathrm{ADC}_y + \mathrm{ADC}_z}{3}$$

where S_{DWI} = signal intensity of diffusion-weighted image with b value and S_0 = signal intensity with b value of 0.

Generally, isotropic ADC is measured. However, in brain white matter that normally has anisotropic water diffusion due to limited molecular movement by axon membranes, diffusion tensor imaging is used for calculating the degree of anisotropic diffusivity.

Application for Drug Development

Oncology

DWI has attracted interest in oncologists and drug developers on the basis of the correlation between ADC and cellular density (Le Bihan et al. 1988). DWI does not use ionizing radiation or require the use of any contrast material injection. Therefore, DWI allows unlimited follow-up, and avoids the costs and risks associated with contrast materials.

Generally, cancer tissue has lower ADC values as high cellularity/cell density in the tumor restricts the diffusivity (Guo et al. 2002; Squillaci et al. 2004). Decreased ADC in neoplastic tissue also depends on the net effect of other alterations such as the composition ratio of free and bound water and the tortuosity in the interstitial space and matrix. Representatively, ADC can differentiate benign from malignant focal hepatic nodules/masses (Taouli et al. 2003) because metastasis and hepatocellular carcinomas have low ADC values whereas benign lesions such as hemangiomas have high ADC values. ADC values have also shown the capability to distinguish the histological subtypes within some tumors. For example, subtypes of breast cancer can be predicted by ADC on the basis of structural differences according to the different histological types such as presence of glandular elements or focal necrosis (Yoshikawa et al. 2008). However, ADC is not specific for cancer cells but just depending on the diffusivity, and therefore similarity of tissue architecture and tissue components between benign and malignant lesions may show overlapped ADC values (Bruegel et al. 2008). In hemorrhagic masses ADC is strongly reduced due to the influence of deoxyhemoglobin or hemosiderin, therefore disturbing accurate differentiation between benign and malignant lesions or between viable and killed tumor tissue (Atlas et al. 2000). Intratumoral necrosis may increase ADC, and thereby treatment-induced necrosis may not be distinguished from the necrosis induced by tumor growth (Schraml et al. 2009). Certain MRI techniques may also influence the ADC value. For example, fat suppression for enhancing the lesion contrast in the fat-containing organs such as bone marrow can also influence the ADC value, particularly at high *b* values (Baron et al. 2010).

Although the interpretation of ADC values for the diagnosis of malignant tissue is complex as described above, the longitudinal change of ADC in tumor growth or chemotherapy can be adequately used as a responsible biomarker for drug development. In conventional treatment study, a decrease in tumor diameter generally serves as a biomarker of therapeutic response. However, this simple measurement does not necessarily indicate the effect of targeted anticancer agents. For example, metastasis from gastrointestinal stromal tumors may become larger even after targeted treatment when cystic degeneration or hemorrhage occur within tumors (Shankar et al. 2005). Similarly, after antiangiogenic treatment, tumor tissue presents altered vasculature and increased necrotic fraction. In these situations, conventional measurement of tumor diameter may be inadequate in assessing the treatment response (Fig. 5).

Following the chemotherapy using cytotoxic as well as targeted agents, there is an early and dynamic change in the micro-architecture of tumor tissue: cell swelling and cell death disintegrate the cell membrane and change the cell density. With widening interstitial space, the freedom of water molecules increases and thereby elevates the ADC in the tumor. These changes are accurately identified with DWI (or ADC) prior to the onset of macroscopic effects such as tumor volume change (Figs. 6 and 7). In rat brain tumor models induced by orthotopically implanted with

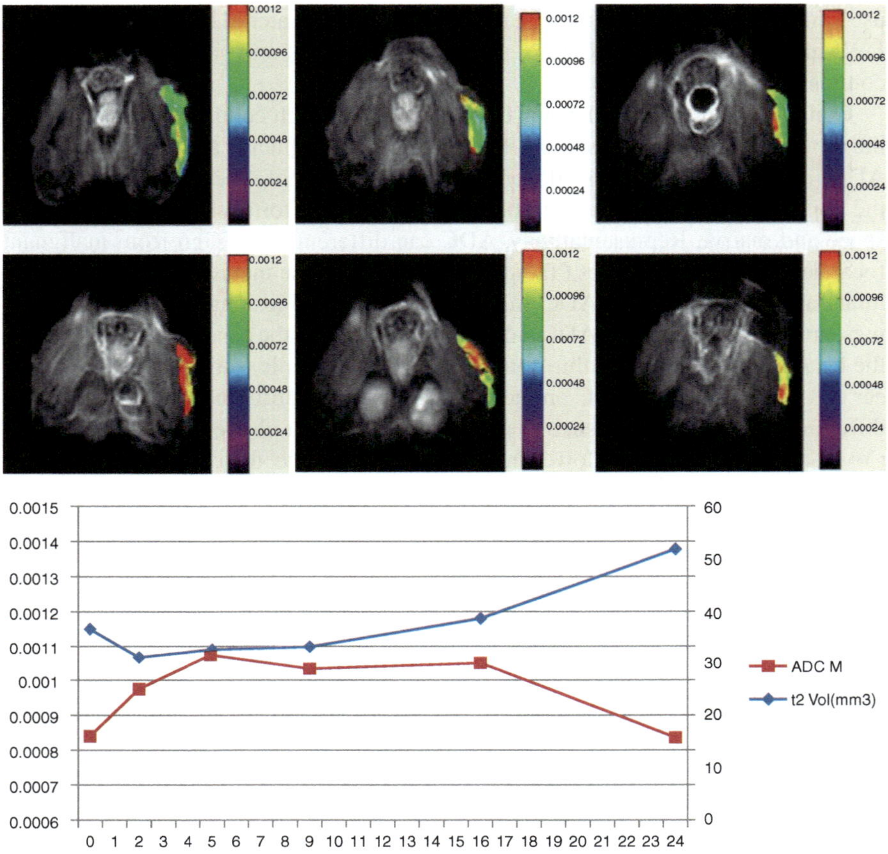

Fig. 5 DWIs are serially obtained in a tumor-bearing mouse. As tumor grows, the ADC decreases due to increasing cell density

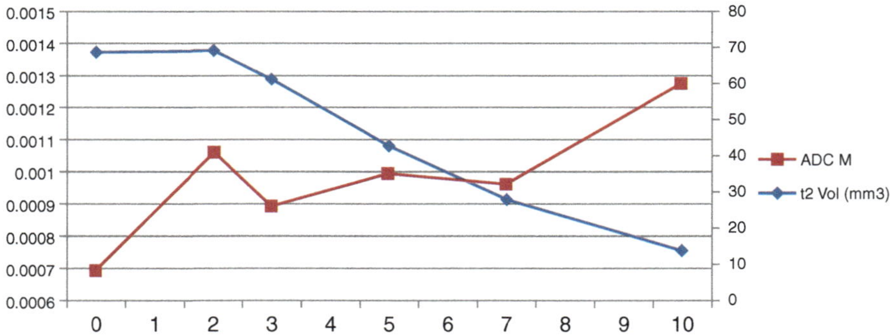

Fig. 6 The changes of ADC and tumor volume after the initiation of. Prior apparent tumor volume reduction, ADC value increases between day 1 and 2

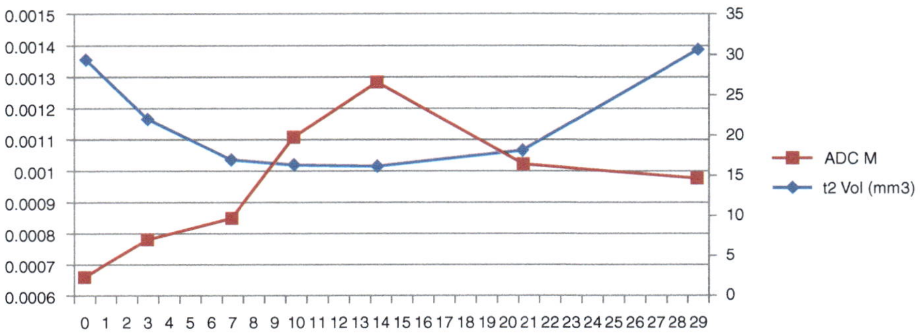

Fig. 7 The changes of ADC and tumor volume in tumor regrowth model. Whereas apparent tumor regrowth is noted between day 21 and 29, significant ADC decrease is noted between day 14 and 21, thereby indicating earlier response of ADC than tumor volume in tumor recurrence

9 L glioma, the influence of yCD/5FC gene therapy was presented by an increase in ADC values by 31% after 8 days (Stegman et al. 2000), which preceded the notification of tumor growth arrest and regression. In animal models of breast cancer, treatment with cyclophosphamide (150 or 300 mg/kg) induced a significant 30% to 40% increase in ADC 2 days after treatment prior to the volumetric response (Zhao et al. 1996). In the study by Theony et al. (Thoeny et al. 2005b), DWI was able to distinguish between non-perfused but viable and necrotic tissues in subcutaneous rhabdomyosarcomas models (Thoeny et al. 2005b). Hence, ADC has capability as a pharmacodynamic biomarker for early detection of treatment response in oncologic drug development (Thoeny et al. 2005a; Hamstra et al. 2007).

In terms of heterogenous treatment response within tumor, several studies demonstrated that the changes in ADC is not uniform across tumors (Hamstra et al. 2004; Moffat et al. 2006); in regions with minimal ADC changes, the tumor was more likely to regrow as compared with those with significant ADC increase. For interpreting this heterogeneity, postprocessing of image data can be used for analyzing the regional ADC changes instead of analyzing the mean ADC in the entire tumor area. In a rodent glioma model, the fraction of therapy-responding tumor was directly correlated with both the therapeutic dosage and the biologic end points (the extent of tumor cell kill, regional necrosis, and animal survival (Moffat et al. 2006). Other studies have extended these observations to diverse histologic tumor types and therapeutic agents (Lee et al. 2006; Lyng et al. 2000; Chinnaiyan et al. 2000), which guided the direction of utilization of DWI in clinical trials (Hamstra et al. 2007).

In order to accept ADC as a biomarker to accurately and reliably determine the therapeutic effect in drug development, it is mandatory to biologically validate the ADC changes. Particularly, histological correlation of ADC change can establish a profound understanding of the impact of the various pathophysiological and therapeutic processes on ADC. Therefore, preclinical animal studies with adequately established models are required to assess the temporal ADC response to the treatment. For this, optimal experimental schedule is highly important and can help to set strategies to translate preclinical results into the clinical trial. In practical aspect, correlation of ADC with cell density, size, and extracellular space may not be easily translated to clinical studies due to tumor heterogeneity. Instead, correlation between ADC and PET is a highly feasible trial to correlate the ADC change with functional biological signals. This early reaction of ADC change in comparison with tumor volume change can be also identified in tumor regrowth.

Neurological Disorder

DWI is a useful MRI biomarker to measure the extent and severity of acute stroke. In acute stroke, cellular edema induced by ischemic damage limits the diffusivity of water molecules and thereby demonstrates decreased ADC. Rapidly after cessation of cerebral blood flow, ADC changes reflect the potentially reversible damages, i.e., energy-dependent processes like cytoplasmic streaming (Harkins et al. 2011) or cell edema due to acidosis (Harris et al. 2000). When the amount of ATP decreases below a critical level of approximately 30% (Obrenovitch et al. 1990), membrane sodium and potassium pumps are damaged, and anoxic depolarization leads to cellular swelling. DWI can show the shift of water molecules between extracellular and intracellular spaces as a change in water diffusivity. Decreased volume of extracellular space further diminishes water diffusion, and a large amount of water flux into the intracellular compartment, where large proteins encumber diffusion, also reduces the water diffusivity. Most lesions with decreased ADC values identified at early time points following ischemic attack become terminal anoxic depolarization and irreversible brain damage in the ischemic core. As such, DWI or ADC map demonstrates the degree of tissue injury in acute stroke with high sensitivity and specificity (Warach 2001).

Based on the abovementioned theoretical background, for evaluating the stroke severity and monitoring the effect of neuroprotective agents, DWI has been applied in many preclinical and clinical trials to test the therapeutic effect of thrombolytic and neuroprotective agents. Particularly, in the combination of perfusion-weighted imaging, PWI–DWI mismatch was been intensively used for identifying the ischemic penumbra (Fig. 8). The area with decreased perfusion and normal ADC is recognized as ischemic penumbra, of which the conservation has been a main

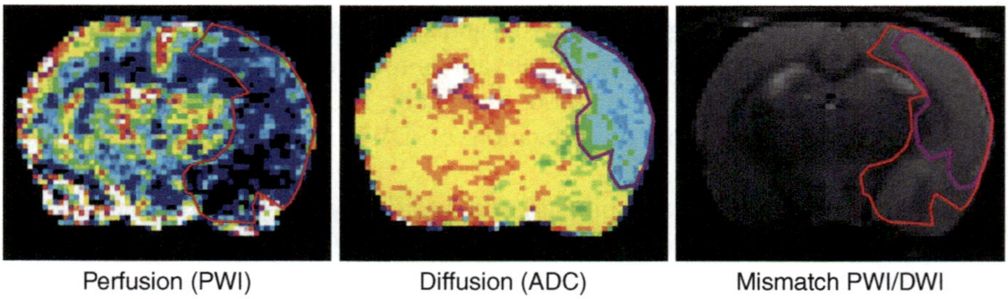

Fig. 8 An example of perfusion–diffusion mismatch. The smaller lesion on diffusion-weighted imaging (DWI) represents the infarct core, meanwhile the much broader area on perfusion-weighted imaging indicates the hypoperfused tissue. The mismatch zone, which contains severely decreased blood flow but favorable ADC value, represents the tissue at risk of infarction (penumbra)

purpose of neuroprotective agents in acute stroke (Mandeville et al. 2017). Although there has been no FDA approved drug as a neuroprotective therapeutic in acute stroke, a number of new candidates are tested in preclinical studies.

Dynamic Contrast-Enhanced MRI

Theoretical Background

With highly specific potential to analyze the vascular characteristics, dynamic contrast-enhanced magnetic resonance imaging (DCE-MRI) has been used as a promising biomarker to assess the hemodynamic changes, particularly in malignant neoplasms. Tumor growth beyond 2 mm must request new vessel formation, i.e., angiogenesis, as passive diffusion alone cannot effectively support the tumor growth (Folkman 1996). Microscopic experiments have demonstrated that tumor angiogenesis increases the blood volume (Weidner 1995). Moreover, new tumor vessels have defective endothelial wall function due to large endothelial gap and loose wall-supporting architectures, and thereby increases the transvascular permeability (McDonald and Baluk 2002). These alterations of tumor vessels eventually accelerates the accumulation of intravenous contrast materials in the voxels and consequently presents rapid and strong contrast enhancement on DCE-MRI (Barrett et al. 2007; Knopp et al. 1999). Based on this theoretical background, DCE-MRI has been accepted as an adequate parameter to display the activity of tumor angiogenesis by profiling the blood volume and transvascular permeability.

Small molecular gadolinium (Gd)-containing contrast agents (such as Gd-DTPA, Gd-BT-DO3A, Gd-BOPTA, and Gd-DOTA) with the molecular weight ranging from 400 to 600 g/mol, are currently used for DCE-MRI. Owing to the small particle size, these agents rapidly move from the intravascular to the extravascular–extracellular space (EES). In order to trace the dynamic change of the tissue or plasma Gd concentration on DCE-MRI, fast and repeated image acquisition is necessary, generally in a temporal resolution of less than 5 secs for each dynamic set. This requirement can be satisfied by utilizing the gradient echo sequence. From the time-intensity curve from DCE-MRI, mathematical modeling can draw very specific parameters describing the nature of tumor vessels.

Quantitative analysis of the DCE-MRI can be classified as non-model-based and pharmacokinetic (PK) model-based approaches. In the non-model-based assessment, the shape of the time–Gd concentration curve ischaracterized by calculating a number of shape-determining factors (Sung et al. 2016) (Table 2). Of them, the blood-normalized initial-area-under-Gd concentration curve is generally recommended as a reliable parameter, while time window to calculate the area is still under controversy (Walker-Samuel et al. 2006).

In the pharmacokinetic model-based approach, several analytical models have been introduced to quantify the blood volume, transvascular permeability, and EES volume (Table 3). Currently, the extended Tofts and Kermode model is most widely used for monitoring the antiangiogenic drugs (Walker-Samuel et al. 2006; Leach et al. 2003).

Table 2 Non-model-based parameters measured from DCE-MRI

Parameter	Definition
Time of arrival (T_0)	Onset time of enhancement curve
Maximum signal intensity (SI_m)	Maximum signal intensity of a given time-varying signal S(t)
Peak enhancement (**ΔS** I)	Difference between SI_m and baseline (SI_0)
Maximum enhancement ratio	SI_m/SI_0
Time-to-peak (T_p)	Time for contrast agent to reach the highest value
Maximum intensity time ratio	Ratio between the peak enhancement(ΔSI) and the time-to-peak (T_P) ($\Delta SI/T_P$)
Wash-in rate	$\Delta SI/(T_p-T_0)$
Wash-out rate	$(SI_{max} - SI_{final})/(T_{max} - T_p)$
Area under curve	Area under the time-concentration curve at a certain time window

Table 3 Pharmacokinetic model-based parameters obtained from DCE-MRI (Nagel et al. 2003; Khalifa et al. 2014; Parker et al. 1998)

Model	Parameter	Definition
Larsson et al. (1990, 1996)	K_{ep}	Transfer constant
	K_i	Bidirectional transfer constant
	F_p	Extraction fraction
	E	Fractional EES volume
Brix et al. (1991)	$K_{ep}(k_{21})$	Contrast agent exchange rate
	K_{el}	Elimination rate
	A^H	Amplitude parameter, corresponds to the EES size
Extended Tofts and Kermode (1999)	K^{trans}	Forward transfer constant
	V_e	Fractional EES volume
	V_p	Fractional plasma volume
Patlak et al. (1983)	K^{trans}	Transfer constant
	V_p	Vascular fraction
Two-compartment exchange (Hwang and Choi 2015)	PS	Permeability
	F_P	Plasma blood flow
	V_p	Vascular fraction
	V_e	Fractional EES volume
Distributed parameter (Sangren and Sheppard 1953)	PS	Permeability
	F_P	Plasma blood flow
	V_p	Vascular fraction
	V_e	Fractional EES volume
	T_c	Mean transit time of the capillary, V_p/F_p

Application for Drug Development

Many antiangiogenic agents have been developed in the last decade. Because their biochemical property, mode of action, and therapeutic response obviously differ from those of traditional cytotoxic drugs, the use of the more therapeutics- and target-oriented imaging biomarker is required (Zweifel and Padhani 2010). In this regard, as angiogenesis-specific indicators, DCE-MRI parameters have attracted strong attentions for monitoring the antiangiogenic effect of new drugs. Furthermore, translation of DCE-MRI results from preclinical to clinical trials enhances its advantages in the drug development.

Monitoring of Antiangiogenic Effect

Approximately 200 or more studies have used the DCE-MRI parameters to monitor the effect of antiangiogenic therapeutics. Early literature demonstrated treatment-driven changes of DCE-MRI parameters after vessel-targeting chemotherapy (Sung et al. 2016), in which a significant decrease of K^{trans} or $IAUGC_{BN}$ was observed much earlier (1–2 days following treatment initiation) than tumor shrinkage. Following phase I trial results have demonstrated that administration of multiple tyrosine kinase receptor inhibitors can reduce the K^{trans} as much as 40–58% (Jiang et al. 2011; Li et al. 2005; Moasser et al. 2007; Rudin et al. 2005; Thomas et al. 2003; Wong et al. 2009; Morgan et al. 2003; Mross et al. 2005). Also, re-increase of DCE-MRI parameter is observed prior to tumor volume re-increase in cases of tumor regrowth (Fig. 9). Other studies presented the therapeutics-induced tumor vessel normalization as the tumor blood volume and permeability became similar to those in the normal organs (Batchelor et al. 2007; Kamoun et al. 2009). Despite such promising results, there still remains controversy and confusion regarding the value of DCE-MRI parameters as the surrogate endpoints because they cannot replace the role of traditional parameter such as tumor volume and patient survival rate. As a potential limitation of DCE-MRI, the baseline values of DCE-MRI parameters vary among studies. This instability limits the application of a common value of DCE-MRI parameter for predicting the treatment outcome. Moreover, an increasing number of studies raise the issue of unsuccessful qualification, which demonstrates only a weak correlation between the DCE-MRI parameters and the clinical outcome (Boxerman and Ellingson 2015; Do et al. 2015; Etxano et al. 2014; Braunagel et al. 2015; Newbold et al. 2009; Ciunci et al. 2014; Dahut et al. 2013;

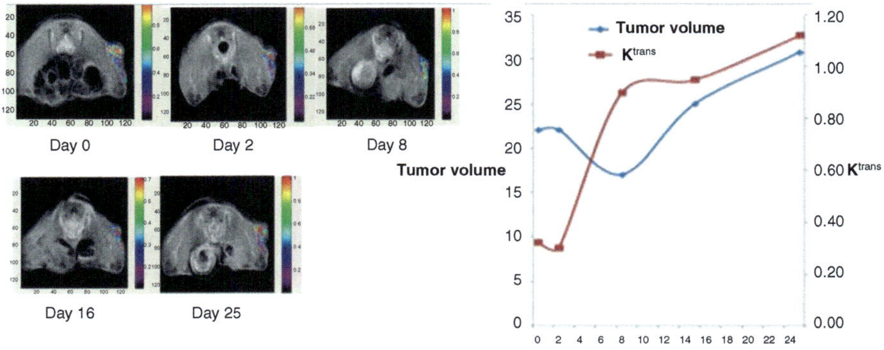

Fig. 9 DCE-MRI serially obtained in a mouse model with tumor regrowth after antiangiogenic treatment. The K^{trans} increases between day 4 and 8, which is prior to tumor regrowth (day 8 and 16)

Flaherty et al. 2015; Guarneri et al. 2015; Horsley et al. 2013). For example, the changes of K^{trans} or $IAUGC_{BN}$ were not significantly correlated with the clinical outcome or tumor volume change (Medved et al. 2004; Dowlati et al. 2005; Rosen et al. 2007; Hahn et al. 2008). Even a negative correlation was exhibited in a study using a tyrosine kinase inhibitor (Xiong et al. 2004).

Predicting the Treatment Outcome

The pre-treatment values of DCE-MRI parameters have been introduced as a baseline parameter to predict the therapeutic response and patient prognosis. A low level of contrast enhancement magnitude or K^{trans} before treatment may indicate a low probability of successful treatment outcome. This suggestion is based on the idea that insufficient blood supply to tumor produces a biologic environment with pool oxygen and nutrient supply and consequently makes tumor cells aggressive and resistant to antiangiogenic treatment (Bernstein et al. 2014). In contrast, a high level of pre-treatment K^{trans} may indicate successful antiangiogenic treatment in many cancers (Zweifel and Padhani 2010; Li and Padhani 2012; Liu et al. 2009; Flaherty et al. 2008). However, such a suggestion to predict treatment outcome has been validated only in a few studies (Flaherty et al. 2008; Ah-See et al. 2008; Johansen et al. 2009). Moreover, it is often contradictory to suggest that effective vascular suppression reduces $IAUGC_{BN}$ and K^{trans} while low pre-treatment $IAUGC_{BN}$ and K^{trans} predict unsuccessful treatment results.

Similarly to the pre-treatment level, the degree of treatment-induced decrease of DCE-MRI parameters early after treatment initiation has been introduced as an indicator to predict the successful treatment outcome. Many studies have shown that the greater reduction of K^{trans} immediately after administration of antiangiogenic drugs may indicate better therapeutic results (Ah-See et al. 2008; Johansen et al. 2009; Padhani et al. 2006). However, the time for determining the early response has been variable between studies.

As described, the pharmacokinetics model-based DCE-MRI parameters have been tested in terms of their adequacy as surrogate endpoints to determine the antiangiogenic effect. Despite the advantages to quantify the blood volume and permeability separately, this approach is often difficult to be standardized in multicenter trials as it requires complex PK modeling and analysis software. By contrast, the non-model-based parameters can be more generally measured by using noncommercial software as it does not require the complex mathematical modeling and curve fitting. $IAUGC_{BN}$ is most commonly applied in addition to or as an alternative to pharmacokinetic parameters (Leach et al. 2003; Engelbrecht et al. 2003; Galbraith et al. 2002). In this context, the reason to use the $IAUGC_{BN}$ has been validated by a simulation study, which demonstrated a strong relationship between the $IAUGC_{BN}$ and the PK parameters, e.g., K^{trans}, V_p, and V_e (Walker-Samuel et al. 2006). To date, no consensus has not been reached regarding which mothed is better for evaluating tumor vasculature. Currently, both K^{trans} and $IAUGC_{BN}$ were chosen as recommended DCE-MRI biomarkers in the Quantitative Imaging Biomarkers Alliance guideline for applying DCE-MRI in clinical trials.

MR Spectroscopy

Theoretical Background

Even though the fundamental mechanism of MR spectroscopy (MRS) is similar to that of MRI, the major difference of MRS from MRI is that MRS is focused for obtaining the chemical shift information in molecules other than water. For example, protons bound to the C2 vs the C1 carbon in ethanol (C(Shi et al. 1998) H3–C(Ayers et al. 2010)H2–OH) show different chemical shifts (Fig. 10). Consequently, these protons in ethanol can be separately identified by MRS. In contrast, MRI is designed to image the water proton at one chemical shift (4.7 parts per million), and therefore cannot differentiate these protons (Hwang and Choi 2015). Given this merit, MRS allows a quantitative assessment of various metabolites and lipids in living tissue, which cannot be separately identified by MRI, by using the different chemical shift between metabolites. In vivo MRS has can measure the relative concentration of molecules typically at mM level either by comparing the intensity ratio of various metabolites or by referring the intensity to an initial value. Absolute quantitation of molecular concentration is limited by many factors such as signal intensity, relaxation properties, and coil loading (Rudin et al. 1995).

Application for Drug Development

^{1}H MRS for Hepatic Lipid Quantification

^{1}H MRS is accepted as a gold standard measurement for quantifying the hepatic lipid content and is often utilized in clinical studies to test hepatic lipid-reducing drugs (Fig. 11). Moreover, as lipid in the hepatocytes and muscle cells liver and muscle cell is reportedly correlated with insulin resistant, MRS has received

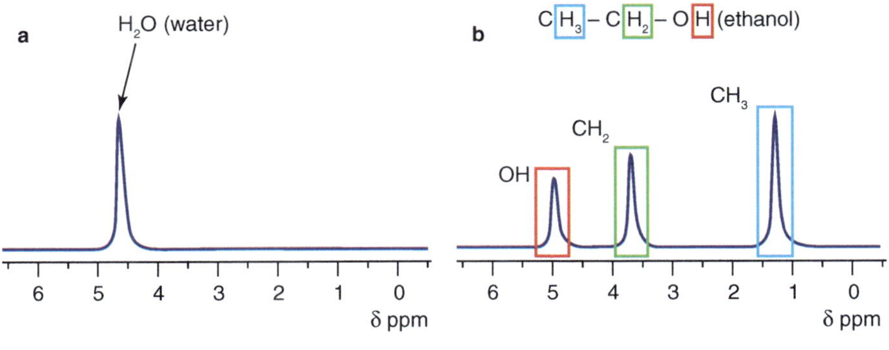

Fig. 10 1H NMR spectrum of the distilled water and ethanol solution. (**a**) Water protons have a chemical shift of 4.7 ppm in living tissues. (**b**) A classic chemical structure of ethanol molecule and its mimetic NMR spectrum. Because different types of protons attached to two carbons and oxygen are resonated at two different chemical shifts, each triplet proton can be discriminated in the spectrum. NMR, nuclear magnetic resonance

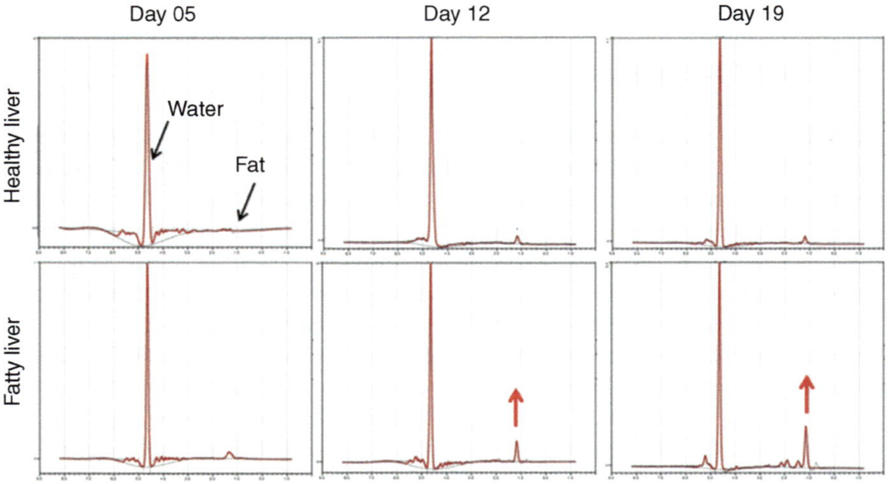

Fig. 11 Comparison of MRS between healthy liver and fatty liver during cholesterol-rich intake. Intensity of lipid peak (arrows) increases in the fatty liver

attention as a predictor for the clinical onset of insulin-resistant DM (Hwang and Choi 2015). Another important benefit of ^{1}H MRS is that it can separately measure the intramyocellular and extramyocellular lipid contents. Many studies have shown a significant correlation between hepatic lipid content and intramyocellular lipid with insulin resistance (Hwang et al. 2007).

^{31}P MRS for Non-insulin Resistance Diabetes Mellitus (NIDD)

^{31}P MRS can detect many phosphorus metabolites, including ATP, PCr, phosphocholine, glycerophosphocholine, glucose-6-phosphate, and others. Glucose-6-phosphate is an important intermediate for glucose oxidation and its lower concentrations in muscles may indicate the severity of NIDDM. These results imply defective glucose transport and/or phosphorylation in insulin-resistant diabetes mellitus patients (Rothman et al. 1992).

^{13}C MRS for Glycogen Detection

^{13}C MRS can detect the C-1 in the liver and muscle glycogen, thereby being utilized to measure the muscular and hepatic glycogen contents during nutrient supply or exercise in metabolic experiments. Natural abundance of hepatic glycogen concentration in human (300–400 mmoles of glucosyl unit per liter of tissue), in humans), ^{13}C MRS can assess the breakdown and synthesis rate of glycogen, which is critical important information for insulin-dependent DM (Hwang et al. 1995). The status of glycogen storage disease can also be quantitatively identified using this method (Roser et al. 1996).

References

Ah-See ML, Makris A, Taylor NJ et al (2008) Early changes in functional dynamic magnetic resonance imaging predict for pathologic response to neoadjuvant chemotherapy in primary breast cancer. Clin Cancer Res 14(20):6580–6589

Atlas SW, DuBois P, Singer MB, Lu D (2000) Diffusion measurements in intracranial hematomas: implications for MR imaging of acute stroke. AJNR Am J Neuroradiol 21(7):1190–1194

Ayers GD, McKinley ET, Zhao P et al (2010) Volume of preclinical xenograft tumors is more accurately assessed by ultrasound imaging than manual caliper measurements. J Ultrasound Med 29(6):891–901

Baron P, Dorrius MD, Kappert P, Oudkerk M, Sijens PE (2010) Diffusion-weighted imaging of normal fibroglandular breast tissue: influence of microperfusion and fat suppression technique on the apparent diffusion coefficient. NMR Biomed 23(4):399–405

Barrett T, Brechbiel M, Bernardo M, Choyke PL (2007) MRI of tumor angiogenesis. J Magn Reson Imaging: JMRI. 26(2):235–249

Batchelor TT, Sorensen AG, di Tomaso E et al (2007) AZD2171, a pan-VEGF receptor tyrosine kinase inhibitor, normalizes tumor vasculature and alleviates edema in glioblastoma patients. Cancer Cell 11(1):83–95

Bernstein JM, Homer JJ, West CM (2014) Dynamic contrast-enhanced magnetic resonance imaging biomarkers in head and neck cancer: potential to guide treatment? A systematic review. Oral Oncol 50(10):963–970

Boutin RD, Yao L, Canter RJ, Lenchik L (2015) Sarcopenia: current concepts and imaging implications. AJR Am J Roentgenol 205(3):W255–W266

Boxerman JL, Ellingson BM (2015) Response assessment and magnetic resonance imaging issues for clinical trials involving high-grade Gliomas. Topics in Magn Reson Imaging: TMRI 24(3):127–136

Branca RT, Warren WS (2011) In vivo brown adipose tissue detection and characterization using water-lipid intermolecular zero-quantum coherences. Magn Reson Med 65(2):313–319

Braunagel M, Radler E, Ingrisch M et al (2015) Dynamic contrast-enhanced magnetic resonance imaging measurements in renal cell carcinoma: effect of region of interest size and positioning on interobserver and intraobserver variability. Investig Radiol 50(1):57–66

Brix G, Semmler W, Port R, Schad LR, Layer G, Lorenz WJ (1991) Pharmacokinetic parameters in CNS Gd-DTPA enhanced MR imaging. J Comput Assist Tomogr 15(4):621–628

Bruegel M, Holzapfel K, Gaa J et al (2008) Characterization of focal liver lesions by ADC measurements using a respiratory triggered diffusion-weighted single-shot echo-planar MR imaging technique. Eur Radiol 18(3):477–485

Cassidy FH, Yokoo T, Aganovic L et al (2009) Fatty liver disease: MR imaging techniques for the detection and quantification of liver Steatosis. Radiographics 29(1):231–260

Chinnaiyan AM, Prasad U, Shankar S et al (2000) Combined effect of tumor necrosis factor-related apoptosis-inducing ligand and ionizing radiation in breast cancer therapy. Proc Natl Acad Sci U S A 97(4):1754–1759

Ciunci CA, Perini RF, Avadhani AN et al (2014) Phase 1 and pharmacodynamic trial of everolimus in combination with cetuximab in patients with advanced cancer. Cancer 120(1):77–85

Dahut WL, Madan RA, Karakunnel JJ et al (2013) Phase II clinical trial of cediranib in patients with metastatic castration-resistant prostate cancer. BJU Int 111(8):1269–1280

Do K, Cao L, Kang Z et al (2015) A phase II study of Sorafenib combined with Cetuximab in EGFR-expressing, KRAS-mutated metastatic colorectal cancer. Clin Colorectal Cancer

Dowlati A, Robertson K, Radivoyevitch T et al (2005) Novel phase I dose de-escalation design trial to determine the biological modulatory dose of the antiangiogenic agent SU5416. Clin Cancer Res 11(21):7938–7944

Engelbrecht MR, Huisman HJ, Laheij RJ et al (2003) Discrimination of prostate cancer from normal peripheral zone and central gland tissue by using dynamic contrast-enhanced MR imaging. Radiology 229(1):248–254

Etxano J, Insausti LP, Elizalde A, Lopez Vega JM, Plazaola A, Martinez P (2014) Analysis of the changes induced by bevacizumab using a high temporal resolution DCE-MRI as prognostic factors for response to further neoadjuvant chemotherapy. Acta Radiol

Flaherty KT, Hamilton BK, Rosen MA et al (2015) Phase I/II trial of Imatinib and Bevacizumab in patients with advanced melanoma and other advanced cancers. Oncologist

Flaherty KT, Rosen MA, Heitjan DF et al (2008) Pilot study of DCE-MRI to predict progression-free survival with sorafenib therapy in renal cell carcinoma. Cancer Biol Ther 7(4):496–501

Folkman J (1996) New perspectives in clinical oncology from angiogenesis research. Eur J Cancer 32A(14):2534–2539

Galbraith SM, Lodge MA, Taylor NJ et al (2002) Reproducibility of dynamic contrast-enhanced MRI in human muscle and tumours: comparison of quantitative and semi-quantitative analysis. NMR Biomed 15(2):132–142

Gonzalez-Lima F, Berndt JD, Valla JE, Games D, Reiman EM (2001) Reduced corpus callosum, fornix and hippocampus in PDAPP transgenic mouse model of Alzheimer's disease. Neuroreport 12(11):2375–2379

Guarneri V, Dieci MV, Bisagni G et al (2015) Preoperative carboplatin-paclitaxel-Bevacizumab in triple-negative breast cancer: final results of the phase II Ca.Pa.Be study. Ann Surg Oncol

Guo Y, Cai YQ, Cai ZL et al (2002) Differentiation of clinically benign and malignant breast lesions using diffusion-weighted imaging. J Magn Reson Imaging: JMRI. 16(2):172–178

Hahn OM, Yang C, Medved M et al (2008) Dynamic contrast-enhanced magnetic resonance imaging pharmacodynamic biomarker study of sorafenib in metastatic renal carcinoma. J Clin Oncol Off J Am Soc Clin Oncol 26(28):4572–4578

Hamstra DA, Lee KC, Tychewicz JM et al (2004) The use of 19F spectroscopy and diffusion-weighted MRI to evaluate differences in gene-dependent enzyme prodrug therapies. Mol Ther 10(5):916–928

Hamstra DA, Rehemtulla A, Ross BD (2007) Diffusion magnetic resonance imaging: a biomarker for treatment response in oncology. J Clin Oncol Off J Am Soc Clin Oncol 25(26):4104–4109

Harkins KD, Galons JP, Divijak JL, Trouard TP (2011) Changes in intracellular water diffusion and energetic metabolism in response to ischemia in perfused C6 rat glioma cells. Magn Reson Med 66(3):859–867

Harris NG, Zilkha E, Houseman J, Symms MR, Obrenovitch TP, Williams SR (2000) The relationship between the apparent diffusion coefficient measured by magnetic resonance imaging, anoxic depolarization, and glutamate efflux during experimental cerebral ischemia. J Cereb Blood Flow Metab 20(1):28–36

Horsley L, Cummings J, Middleton M et al (2013) A phase 1 trial of intravenous 4-(N-(S-glutathionylacetyl)amino) phenylarsenoxide (GSAO) in patients with advanced solid tumours. Cancer Chemother Pharmacol 72(6):1343–1352

Hu HH, Perkins TG, Chia JM, Gilsanz V (2013) Characterization of human brown adipose tissue by chemical-shift water-fat MRI. AJR Am J Roentgenol 200(1):177–183

Hu HH, Smith DL Jr, Nayak KS, Goran MI, Nagy TR (2010) Identification of brown adipose tissue in mice with fat-water IDEAL-MRI. J Magn Reson Imaging: JMRI 31(5):1195–1202

Hudelmaier M, Wirth W, Himmer M, Ring-Dimitriou S, Sanger A, Eckstein F (2010) Effect of exercise intervention on thigh muscle volume and anatomical cross-sectional areas–quantitative assessment using MRI. Magn Reson Med 64(6):1713–1720

Hussain HK, Chenevert TL, Londy FJ et al (2005) Hepatic fat fraction: MR imaging for quantitative measurement and display–early experience. Radiology 237(3):1048–1055

Hwang JH, Choi CS (2015) Use of in vivo magnetic resonance spectroscopy for studying metabolic diseases. Exp Mol Med 47:e139

Hwang JH, Perseghin G, Rothman DL et al (1995) Impaired net hepatic glycogen synthesis in insulin-dependent diabetic subjects during mixed meal ingestion. A 13C nuclear magnetic resonance spectroscopy study. J Clin Invest 95(2):783–787

Hwang JH, Stein DT, Barzilai N et al (2007) Increased intrahepatic triglyceride is associated with peripheral insulin resistance: in vivo MR imaging and spectroscopy studies. Am J Physiol Endocrinol Metab 293(6):E1663–E1669

Jack CR Jr, Dickson DW, Parisi JE et al (2002) Antemortem MRI findings correlate with hippocampal neuropathology in typical aging and dementia. Neurology 58(5):750–757

Jiang F, Albert DH, Luo Y et al (2011) ABT-869, a multitargeted receptor tyrosine kinase inhibitor, reduces tumor microvascularity and improves vascular wall integrity in preclinical tumor models. J Pharmacol Exp Ther 338(1):134–142

Johansen R, Jensen LR, Rydland J et al (2009) Predicting survival and early clinical response to primary chemotherapy for patients with locally advanced breast cancer using DCE-MRI. J Magn Reson Imaging: JMRI. 29(6):1300–1307

Kamoun WS, Ley CD, Farrar CT et al (2009) Edema control by cediranib, a vascular endothelial growth factor receptor-targeted kinase inhibitor, prolongs survival despite persistent brain tumor growth in mice. J Clin Oncol Off J Am Soc Clin Oncol 27(15):2542–2552

Khalifa F, Soliman A, El-Baz A et al (2014) Models and methods for analyzing DCE-MRI: a review. Med Phys 41(12):124301

Knopp MV, Weiss E, Sinn HP et al (1999) Pathophysiologic basis of contrast enhancement in breast tumors. J Magn Reson Imaging: JMRI 10(3):260–266

Kob R, Bollheimer LC, Bertsch T et al (2015) Sarcopenic obesity: molecular clues to a better understanding of its pathogenesis? Biogerontology 16(1):15–29

Larsson HB, Fritz-Hansen T, Rostrup E, Sondergaard L, Ring P, Henriksen O (1996) Myocardial perfusion modeling using MRI. Magn Reson Med 35(5):716–726

Larsson HB, Stubgaard M, Frederiksen JL, Jensen M, Henriksen O, Paulson OB (1990) Quantitation of blood-brain barrier defect by magnetic resonance imaging and gadolinium-DTPA in patients with multiple sclerosis and brain tumors. Magn Reson Med 16(1):117–131

Le Bihan D, Breton E, Lallemand D, Aubin ML, Vignaud J, Laval-Jeantet M (1988) Separation of diffusion and perfusion in intravoxel incoherent motion MR imaging. Radiology 168(2):497–505

Leach MO, Brindle KM, Evelhoch JL et al (2003) Assessment of antiangiogenic and antivascular therapeutics using MRI: recommendations for appropriate methodology for clinical trials. Br J Radiol 76(1):S87–S91

Lee KC, Hall DE, Hoff BA et al (2006) Dynamic imaging of emerging resistance during cancer therapy. Cancer Res 66(9):4687–4692

Li SP, Padhani AR (2012) Tumor response assessments with diffusion and perfusion MRI. J Magn Reson Imaging: JMRI. 35(4):745–763

Li KL, Wilmes LJ, Henry RG et al (2005) Heterogeneity in the angiogenic response of a BT474 human breast cancer to a novel vascular endothelial growth factor-receptor tyrosine kinase inhibitor: assessment by voxel analysis of dynamic contrast-enhanced MRI. J Magn Reson Imaging: JMRI. 22(4):511–519

Liu Y, Bai R, Sun H, Liu H, Zhao X, Li Y (2009) Diffusion-weighted imaging in predicting and monitoring the response of uterine cervical cancer to combined chemoradiation. Clin Radiol 64(11):1067–1074

Lyng H, Haraldseth O, Rofstad EK (2000) Measurement of cell density and necrotic fraction in human melanoma xenografts by diffusion weighted magnetic resonance imaging. Magn Reson Med 43(6):828–836

Mandeville ET, Ayata C, Zheng Y, Mandeville JB (2017) Translational MR neuroimaging of stroke and recovery. Transl Stroke Res 8(1):22–32

McDonald DM, Baluk P (2002) Significance of blood vessel leakiness in cancer. Cancer Res 62(18):5381–5385

Medved M, Karczmar G, Yang C et al (2004) Semiquantitative analysis of dynamic contrast enhanced MRI in cancer patients: variability and changes in tumor tissue over time. J Magn Reson Imaging: JMRI. 20(1):122–128

Moasser MM, Wilmes LJ, Wong CH et al (2007) Improved tumor vascular function following high-dose epidermal growth factor receptor tyrosine kinase inhibitor therapy. J Magn Reson Imaging: JMRI. 26(6):1618–1625

Moffat BA, Chenevert TL, Meyer CR et al (2006) The functional diffusion map: an imaging biomarker for the early prediction of cancer treatment outcome. Neoplasia (New York, NY) 8(4):259–267

Morgan B, Thomas AL, Drevs J et al (2003) Dynamic contrast-enhanced magnetic resonance imaging as a biomarker for the pharmacological response of PTK787/ZK 222584, an inhibitor of the vascular endothelial growth factor receptor tyrosine kinases, in patients with advanced colorectal cancer and liver metastases: results from two phase I studies. J Clin Oncol Off J Am Soc Clin Oncol 21(21):3955–3964

Mross K, Drevs J, Muller M et al (2005) Phase I clinical and pharmacokinetic study of PTK/ZK, a multiple VEGF receptor inhibitor, in patients with liver metastases from solid tumours. Eur J Cancer 41(9):1291–1299

Nagel E, Klein C, Paetsch I et al (2003) Magnetic resonance perfusion measurements for the noninvasive detection of coronary artery disease. Circulation 108(4):432–437

Newbold K, Castellano I, Charles-Edwards E et al (2009) An exploratory study into the role of dynamic contrast-enhanced magnetic resonance imaging or perfusion computed tomography for detection of intratumoral hypoxia in head-and-neck cancer. Int J Radiat Oncol Biol Phys 74(1):29–37

Obrenovitch TP, Scheller D, Matsumoto T, Tegtmeier F, Holler M, Symon L (1990) A rapid redistribution of hydrogen ions is associated with depolarization and repolarization subsequent to cerebral ischemia reperfusion. J Neurophysiol 64(4):1125–1133

Padhani AR, Hayes C, Assersohn L et al (2006) Prediction of clinicopathologic response of breast cancer to primary chemotherapy at contrast-enhanced MR imaging: initial clinical results. Radiology 239(2):361–374

Park M, Moon W-J (2016) Structural MR imaging in the diagnosis of Alzheimer's disease and other neurodegenerative dementia: current imaging approach and future perspectives. Korean J Radiol 17(6):827–845

Parker GJ, Suckling J, Tanner SF, Padhani AR, Husband JE, Leach MO (1998) MRIW: parametric analysis software for contrast-enhanced dynamic MR imaging in cancer. Radiographics 18(2):497–506

Patlak CS, Blasberg RG, Fenstermacher JD (1983) Graphical evaluation of blood-to-brain transfer constants from multiple-time uptake data. J Cereb Blood Flow Metab 3(1):1–7

Price JL, Davis PB, Morris JC, White DL (1991) The distribution of tangles, plaques and related immunohistochemical markers in healthy aging and Alzheimer's disease. Neurobiol Aging 12(4):295–312

Redwine JM, Kosofsky B, Jacobs RE et al (2003) Dentate gyrus volume is reduced before onset of plaque formation in PDAPP mice: a magnetic resonance microscopy and stereologic analysis. Proc Natl Acad Sci U S A 100(3):1381–1386

Rosen LS, Kurzrock R, Mulay M et al (2007) Safety, pharmacokinetics, and efficacy of AMG 706, an oral multikinase inhibitor, in patients with advanced solid tumors. J Clin Oncol Off J Am Soc Clin Oncol 25(17):2369–2376

Roser W, Beckmann N, Wiesmann U, Seelig J (1996) Absolute quantification of the hepatic glycogen content in a patient with glycogen storage disease by 13C magnetic resonance spectroscopy. Magn Reson Imaging 14(10):1217–1220

Rothman DL, Shulman RG, Shulman GI (1992) 31P nuclear magnetic resonance measurements of muscle glucose-6-phosphate. Evidence for reduced insulin-dependent muscle glucose transport or phosphorylation activity in non-insulin-dependent diabetes mellitus. J Clin Invest 89(4):1069–1075

Rudin M, Beckmann N, Mir A, Sauter A (1995) In vivo magnetic resonance imaging and spectroscopy in pharmacological research: assessment of morphological, physiological and metabolic effects of drugs. Eur J Pharm Sci 3(5):255–264

Rudin M, McSheehy PM, Allegrini PR et al (2005) PTK787/ZK222584, a tyrosine kinase inhibitor of vascular endothelial growth factor receptor, reduces uptake of the contrast agent GdDOTA by murine orthotopic B16/BL6 melanoma tumours and inhibits their growth in vivo. NMR Biomed 18(5):308–321

Sangren WC, Sheppard CW (1953) A mathematical derivation of the exchange of a labeled substance between a liquid flowing in a vessel and an external compartment. Bull Math Biophys 15(4):387–394

Schraml C, Schwenzer NF, Martirosian P et al (2009) Diffusion-weighted MRI of advanced hepatocellular carcinoma during sorafenib treatment: initial results. AJR Am J Roentgenol 193(4):W301–W307

Shankar S, van Sonnenberg E, Desai J, Dipiro PJ, Van Den Abbeele A, Demetri GD (2005) Gastrointestinal stromal tumor: new nodule-within-a-mass pattern of recurrence after partial response to imatinib mesylate. Radiology 235(3):892–898

Shi WM, Wildrick DM, Sawaya R (1998) Volumetric measurement of brain tumors from MR imaging. J Neuro-Oncol 37(1):87–93

Sinkus R, Van Beers BE, Vilgrain V, DeSouza N, Waterton JC (2012) Apparent diffusion coefficient from magnetic resonance imaging as a biomarker in oncology drug development. Eur J Cancer 48(4):425–431

Squillaci E, Manenti G, Cova M et al (2004) Correlation of diffusion-weighted MR imaging with cellularity of renal tumours. Anticancer Res 24(6):4175–4179

Stegman LD, Rehemtulla A, Hamstra DA et al (2000) Diffusion MRI detects early events in the response of a glioma model to the yeast cytosine deaminase gene therapy strategy. Gene Ther 7(12):1005–1010

Sung YS, Park B, Choi Y et al (2016) Dynamic contrast-enhanced MRI for oncology drug development. J Magn Reson Imaging: JMRI. 44(2):251–264

Taouli B, Vilgrain V, Dumont E, Daire JL, Fan B, Menu Y (2003) Evaluation of liver diffusion isotropy and characterization of focal hepatic lesions with two single-shot echo-planar MR imaging sequences: prospective study in 66 patients. Radiology 226(1):71–78

Thoeny HC, De Keyzer F, Chen F et al (2005b) Diffusion-weighted MR imaging in monitoring the effect of a vascular targeting agent on rhabdomyosarcoma in rats. Radiology 234(3):756–764

Thoeny HC, De Keyzer F, Vandecaveye V et al (2005a) Effect of vascular targeting agent in rat tumor model: dynamic contrast-enhanced versus diffusion-weighted MR imaging. Radiology 237(2):492–499

Thomas AL, Morgan B, Drevs J et al (2003) Vascular endothelial growth factor receptor tyrosine kinase inhibitors: PTK787/ZK 222584. Semin Oncol 30(3 Suppl 6):32–38

Tofts PS, Brix G, Buckley DL et al (1999) Estimating kinetic parameters from dynamic contrast-enhanced T(1)-weighted MRI of a diffusable tracer: standardized quantities and symbols. J Magn Reson Imaging: JMRI 10(3):223–232

Walker-Samuel S, Leach MO, Collins DJ (2006) Evaluation of response to treatment using DCE-MRI: the relationship between initial area under the gadolinium curve (IAUGC) and quantitative pharmacokinetic analysis. Phys Med Biol 51(14):3593–3602

Warach S (2001) Use of diffusion and perfusion magnetic resonance imaging as a tool in acute stroke clinical trials. Curr Control Trials Cardiovasc Med 2(1):38–44

Wegener S, Weber R, Ramos-Cabrer P et al (2006) Temporal profile of T2-weighted MRI distinguishes between pannecrosis and selective neuronal death after transient focal cerebral ischemia in the rat. J Cereb Blood Flow Metab 26(1):38–47

Weidner N (1995) Intratumor microvessel density as a prognostic factor in cancer. Am J Pathol 147(1):9–19

Weiss C, Venkatasubramanian PN, Aguado AS et al (2002) Impaired eyeblink conditioning and decreased hippocampal volume in PDAPP V717F mice. Neurobiol Dis 11(3):425–433

Wong CI, Koh TS, Soo R et al (2009) Phase I and biomarker study of ABT-869, a multiple receptor tyrosine kinase inhibitor, in patients with refractory solid malignancies. J Clin Oncol Off J Am Soc Clin Oncol 27(28):4718–4726

Xiong HQ, Herbst R, Faria SC et al (2004) A phase I surrogate endpoint study of SU6668 in patients with solid tumors. Investig New Drugs 22(4):459–466

Yoshikawa MI, Ohsumi S, Sugata S et al (2008) Relation between cancer cellularity and apparent diffusion coefficient values using diffusion-weighted magnetic resonance imaging in breast cancer. Radiat Med 26(4):222–226

Zhao M, Pipe JG, Bonnett J, Evelhoch JL (1996) Early detection of treatment response by diffusion-weighted 1H-NMR spectroscopy in a murine tumour in vivo. Br J Cancer 73(1):61–64

Zweifel M, Padhani AR (2010) Perfusion MRI in the early clinical development of antivascular drugs: decorations or decision making tools? Eur J Nucl Med Mol Imaging 37(Suppl 1):S164–S182

Holotomography: Refractive Index as an Intrinsic Imaging Contrast for 3-D Label-Free Live Cell Imaging

Doyeon Kim, Sangyun Lee, Moosung Lee, Juntaek Oh, Su-A Yang, and YongKeun Park

Introduction

Optical images of biological cells and tissues provide invaluable information on the pathophysiology of diseases. Visual diagnosis of blood cells is essential for the diagnosis of various infectious diseases associated with red blood cells (Kaushansky et al. 2016). In pathology and cytology, tissue biopsy and morphological examination of cells, such as checking for abnormally shaped nuclei in the Papanicolaou test, is an essential step for cancer diagnosis (Koss 1989).

The medical diagnostic capabilities of various diseases have evolved with advances in optical imaging technology. In the seventeenth century, Robert Hooke first observed cork cells using his microscope. Since then, various types of microscopes have been developed and likewise the ability to investigate disease-related cellular, and cellular structures have dramatically improved in recent decades. For

D. Kim
Department of Chemistry, Korea Advanced Institute of Science and Technology (KAIST), Daejeon, South Korea
e-mail: faith286@kaist.ac.kr

S. Lee · M. Lee · J. Oh
Department of Physics, KAIST, Daejeon, South Korea
e-mail: prism@kaist.ac.kr; lkaamo@kaist.ac.kr; juntaek.oh@kaist.ac.kr

S.-A. Yang
Department of Biological Sciences, KAIST, Daejeon, South Korea
e-mail: yangsua@kaist.ac.kr

Y. Park (✉)
Department of Physics, KAIST, Daejeon, South Korea

KAIST Institute Health Science and Technology, Daejeon, South Korea

Tomocube Inc., Daejeon, South Korea
e-mail: yk.park@kaist.ac.kr

J. K. Kim et al. (eds.), *Advanced Imaging and Bio Techniques for Convergence Science*, Advances in Experimental Medicine and Biology,
https://doi.org/10.1007/978-981-33-6064-8_10

example, the invention of phase contrast and differential interference contrast microscopes in the mid-twentieth century have accelerated the studies in the field of microbiology and cell biology, because these interferometric microscopy techniques allowed effective visualization of transparent biological cells (Allen et al. 1969; Zernike 1942). The development of fluorescent proteins and fluorescence microscopy have also enabled specific labeling of target molecules or proteins. This breakthrough in technology thus has opened a new era for molecular biology.

Moreover, various super-resolution microscopic techniques have broken the barrier of diffraction-limited optical resolution. The resolution limit of the optical microscope has been extended to the nanometer scale, enabling investigation of biological phenomena at the single-molecule scale (Hell 2007; Huang et al. 2009). More recently, researchers have used adaptive optical approaches for in vivo imaging of biological cells or tissues (Ji et al. 2010; Park et al. 2017a; Yu et al. 2014; 2015).

Over the past several decades, fluorescent protein technology has been widely used to locate specific target molecules and proteins in cells using the molecular specificity of the probe (Specht et al. 2017). This allows effective visualization of specific targets in cells and tissues with very high imaging contrast (Lichtman and Conchello 2005). When combined with the fluorescence correlation spectroscopy technique, the fluorescent probe can also provide information about the physical and chemical information of the surrounding medium (Thompson 2002). In addition, when combined with the Foster resonance energy transfer technique, the intermolecular distance can be accurately measured in the nanometer scale (Roy et al. 2008).

However, the use of fluorescent probes in biological imaging inevitably causes several limitations. It is important to note that fluorescence technology uses exogenous fluorescent molecules as a secondary imaging contrast. Expression or binding of a target molecule of a fluorescent probe in a cell generates various problems (Fei et al. 2011; Hoebe et al. 2007). First, the introduction of an exogenous marker into a cell can affect the intrinsic physiology of the cell, because of possible photodamage and phototoxicity caused by the fluorescent molecule. This issue becomes even more severe when experimenting with neurons or stem cells because these cells are more sensitive to changes in the environment (Braydich-Stolle et al. 2005; Millet et al. 2007). Second, long-term cell imaging may be limited when using fluorescent technologies due to photobleaching of probes (Hoebe et al. 2007). Common fluorescent probes cannot produce strong, continuous fluorescence signals. Most probes photobleach after some time, which means the probes irreversibly lose its fluorescence. Thus maximum period for long-term imaging of live cells is limited to the photobleaching period of fluorescent probes. Third, most fluorescence technologies do not provide quantitative information. The use of fluorescent probes only provides information about the location of the target molecule, but it does not provide quantified information of the mass or concentration of the target molecule.

To complement the limitation caused by the use of exogenous imaging contrast, the use of refractive index (RI), as an intrinsic optical parameter has been exploited recently. All materials have unique RI value, which is correlated with the electrical permittivity of the material. RI is the ratio of the speed of light passing through the specific material to that passing in the vacuum. Conventional phase contrast or differential interference microscopy uses RI values as optical imaging contrast. However, their imaging systems do not provide a one-to-one quantitative mapping of the information about RI distributions in a sample, but only generate high contrast intensity information via interference (Popescu et al. 2008a). Recently, there have been escalating interests in measuring 3-D RI distributions for various applications in biological imaging. Mainly because RI, as the primarily intrinsic optical parameters, provides the possibility for label-free live cell imaging with the capability of providing quantitative information about the sample. Although 3-D RI tomography does not provide molecular specificity in general, some specimen having distinct RI values such as lipid droplets (Kim et al. 2016c) or gold nanoparticles (Kim et al. 2018b; Sung et al. 2018) in the cytoplasm, can be specified and quantified.

Furthermore, 3-D RI tomography provides quantitative imaging capability; cellular dry mass or cytoplasmic concentration can be precisely quantified from the measured RI values, which are inaccessible with fluorescence imaging techniques. Most importantly, 3-D RI tomography does not require the use extracellular agent, and thus, it simplifies sample preparations and is also suitable for long-term live cell imaging (Barty et al. 2000; Lee et al. 2013; Majeed et al. 2016; Popescu 2011). In addition, due to the quantitative and thus reproducible imaging capability, the use of 3-D RI tomography for the disease diagnosis is being actively investigated in combination with other techniques such as microfluidics (Merola et al. 2012; Sung et al. 2014), machine and deep learning algorithms (Jo et al. 2014, 2015, 2017, 2019; Rivenson et al. 2017), and fast imaging processing algorithms (Kim et al. 2013). In this mini review, we introduce the principle of optical techniques that measure 3-D RI tomograms and summarize recent applications for the study of various biological and medical applications.

The Principles of Biological Imaging Using Refractive Index as Imaging Contrast

The RI of material is obtained by measuring the interactions between light and matter. One of the well-known RI measuring techniques includes a refractometer that obtains the average RI value of a solution or a surface plasmonic sensor used to measure the surface RI of metal (Willets and Van Duyne 2007). These refractometer techniques are suitable for measuring a sample with the homogeneous distribution of RI values, such as a transparent solution. However, it is technically challenging to measure a sample with an inhomogeneous distribution of RI values, such as biological cells or tissues (Liu et al. 2016). This is because light refracts and reflects at the interface between the two media with different refractive indexes. When light refraction and reflection occur many times, the coherence summation of these events can be expressed as multiple light scattering (Cheong et al. 1990). This explains why biological tissues appear opaque white, while individual cells seem transparent.

Conventionally, phase contrast or differential inference microscopy have been utilized to exploit RI distributions in samples. When a laser beam passes through a transparent specimen, such as individual biological cells, the laser beam acquires a distorted wavefront or phase information. This occurs because the speed of light passing through a specific part of the sample differs from another part due to inhomogeneous distribution of RI of the sample. Unfortunately, conventional image sensors do not directly measure this wavefront information, because the speed of light is much faster than the capturing ability of an image sensor. Thus, phase contrast or differential inference microscopy exploits the principle of light interference. Significant light interference can be created in an optical imaging system, such as bright-field microscopy, by inserting additional optical components (Lee and Park 2014). This allows the conversion of wavefront information into intensity information that can be measured by an image sensor. This is how one can achieve high imaging contrast when imaging transparent biological cells using phase contrast or differential inference microscopy. Hence, phase contrast or differential inference microscopy enables clear visualization of the boundaries of the cell membrane as well as subcellular organelles (Mann et al. 2005; Smith 1955; Zernike 1942, 1955).

However, these conventional interference microscopes, such as phase contrast or differential interference microscopy, can only provide qualitative information. This is because the relationship between wavefront information and intensity images for these interference microscopes is not straightforward, thus making it difficult to extract quantitative information (Smith 1955; Zernike 1942, 1955). Quantitative phase information can provide valuable information about the sample without using exogenous labeling agents. For example, the measurements of quantitative phase maps of red blood cells can be directly converted into a cell height information (Ikeda et al. 2005; Park et al. 2006; Popescu et al. 2006).

Various quantitative phase imaging (QPI) techniques have been developed and utilized for various research fields (Park et al. 2018a). In particular, Mach–Zehnder or Michelson types of interference microscopic techniques have been extensively utilized. In-line holography techniques simplify the optical setup by removing a reference arm. Quantitative phase microscopy techniques based on the transport of intensity or ptychography have provided enhanced imaging quality with relatively simple instrumentations. Recently, the QPI unit was developed as a filter-type add-on unit, which can be attached to convert a conventional bright-field microscope into a quantitative phase microscope (Lee and Park 2014). The detailed information on QPI and its application to biological studies can be found in elsewhere (Lee et al. 2013; Majeed et al. 2016; Popescu 2011).

The Principle of Measuring 3-D RI Tomography of Cells and Tissues

Even though 2-D QPI techniques provide quantitative and label-free imaging of live cells, they only provide topographic information; i.e., the measured optical phase delay is a coupled parameter of cell height and its RI distribution (Rappaz et al.

2005). Previously, several methods have been suggested in order to decouple the height and RI information in 2-D QPI techniques. Measuring two holographic images obtained with two different extracellular media with different RI values (Rappaz et al. 2005) or illumination with two wavelengths (Rappaz et al. 2008) provides the separation of height and RI from the measured holograms. Alternatively, the mean RI of suspended cells is calculated from measured 2-D optical phase delay images, assuming spherical shapes of cells (Kemper et al. 2007).

In order to measure 3-D RI tomograms of cells, various approaches have been demonstrated (Kim et al. 2016b). Among them, the angle scanning approach has been widely utilized (Fig. 1). First, multiple 2-D holograms of a sample are measured at various angles of illuminations (Figs. 1a-b), from which a 3-D RI tomogram of the sample can be reconstructed via inverse scattering theory. According to the scattering theory, the difference in wavevectors of the incident and scattered lights determine the spatial frequency information of the optical scattering potential of the sample in 3-D Fourier space (Fig. 1c). Finally, a 3-D RI tomogram of the sample (Fig. 1d) is reconstructed by applying 3-D inverse Fourier transform. This technique has been widely known as optical diffraction tomography (ODT) and holotomography (HT). The principle of 3-D RI tomography is very similar to X-ray computed tomography (CT) where multiple 2-D X-ray images of the human body are measured at various illumination angles, and a 3-D X-ray absorptivity tomogram is then retrieved via the inverse scattering theory. Both X-ray CT and laser HT share the same governing equation—Helmholtz equation, the wave equation for a monochromatic wavelength (Kim et al. 2016b).

The history of HT goes back to the late 1960s. After the rise of X-ray CT technology and the invention of lasers, some pioneering physicists realized that the X-ray CT principle could also be applied to a laser. The principles of X-ray CT are based on wave propagation and can be described by the wave equation, except they use different wavelengths of waves. The first theoretical work was presented in 1969

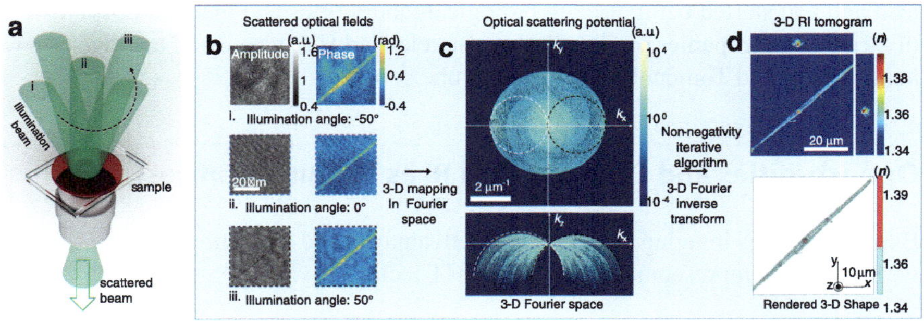

Fig. 1 The schematic of 3-D refractive index tomography. (**a**) The illuminated plane waves and the scattered waves by the sample with distorted wavefronts. Three representative incident and scattered waves are depicted. (**b**) The retrieved 2-D complex optical fields of the light scattered by the sample. (**c**) The synthesized 3-D optical scattering potential in 3-D Fourier space. The regions corresponding to the three illumination angles are specified with distinct colors. (**d**) Three cross-sectional slices and the iso-surface of the reconstructed 3-D refractive index tomogram

(Wolf 1969), and the first experimental demonstration was shown in 1979 (Fercher et al. 1979). However, it seems that at the time many researchers did not realize this HT technique can be applied to biological imaging with significant benefits. The early applications of HT technique had been limited to measuring 3-D shapes of transparent microscopic plastics. In the 2000s, several research groups have revised and employed the HT technique for biological applications (Choi et al. 2007; Kim et al. 2014b; Lauer 2002).

HT technology directly provides the measurements of the 3-D RI distribution of a cell. The reconstruction of 3-D RI tomogram of the specimen is achieved by inversely solving the Helmholtz equation from a set of multiple 2-D optical field images of the sample. This set of 2-D optical field images of the sample containing the 3-D information is generally obtained by varying the illumination angle of a laser impinging onto the specimen (Choi et al. 2007; Kim et al. 2014d; Shin et al. 2015), or rotating the specimen while keeping the light source fixed (Barty et al. 2000; Charrière et al. 2006a; Kuś et al. 2014). For further information, the principle of HT, the detailed procedure with a MatLab code, the phase retrieval algorithms, and the regularization algorithms can be found elsewhere (Debnath and Park 2011; Kim et al. 2014b; Lim et al. 2015) summarize various regularization algorithms used in HT.

From a technical point of view, significant technical advancements have been made recently. For example, the sub-100-nm spatial resolution was achieved using the deconvolution of complex optical field (Cotte et al. 2013). Tomographic RI reconstruction with white-light illumination was presented, which demonstrates a signification reduction of speckle noise and dramatically improved image quality (Kim et al. 2014c). Hyperspectral HT was also demonstrated; it measures 3-D RI tomograms of a sample at various wavelengths using a wavelength-scanning illumination (Jung et al. 2016). The real-time reconstruction and visualization were also demonstrated, which was powered by a graphics processor unit (GPU) (Kim et al. 2013). It is worthy to note that in the mid-2010s, HT technology was commercialized and had started being used in biological laboratories and medical hospitals. As of 2018, two companies provide the commercialized HT systems—Nanolive (www.nanolive.ch) and Tomocube (www.tomocube.com).

Opportunities and Challenges of RI as Imaging Contrast

Exploiting RI as imaging contrasts have advantages and limitations. In this section, we summarize representative features of HT technology.

1. Label free: Because RI is an intrinsic optical parameter of material. No labeling agents or dyes are required for imaging biological cells and tissues. It means 3-D images of live cells can be obtained for a long time as long as physiological conditions are met. In addition, it can save time and cost for sample preparation. This label-free feature might become powerful for some applications where cells

are to be reinjected to human bodies, for example, as in immune therapy (Yoon et al. 2015; 2017) or stem cell therapy (Braydich-Stolle et al. 2005).
2. Quantitative biological imaging: Using HT technology, RI value can be precisely measured. Unlike fluorescence techniques, where the intensity of fluorescence emission highly depends on protocol, and the results are only qualitative, HT technology provides highly reproducible RI value in a quantitative manner. Importantly, the RI value can be directly translated into the protein concentration information (Barer 1953). Furthermore, the dry mass of subcellular structures or cells can also be calculated from RI distributions (Barer 1953; Popescu et al. 2008b; Zangle and Teitell 2014).
3. Potentials in biological and medical applications: From spatial RI distributions, HT technology can retrieve various quantitative parameters. Certain molecules such as lipid (Jung et al. 2018; Kim et al. 2016c) or metal nanoparticles (Kim et al. 2018b; Turko et al. 2013) have distinctly high RI values, which can be addressed by measuring 3-D RI maps. RI values of a solution are linearly proportional to its concentration (Yoon et al. 2017). Local RI values in cells can be converted into cytoplasmic protein concentration (Barer 1952). The integration of RI values over a cell volume can be translated into dry cell mass. Furthermore, these RI values are intrinsic quantitative parameters of live cells, and thus can be utilized for a biophysical marker. For example, 2-D and 3-D RI distributional maps can be utilized in label-free cellular identification (Chalut et al. 2012; Jo et al. 2014, 2015; 2017; Yoon et al. 2017) and long-term growth monitoring (Bettenworth et al. 2014; Chalut et al. 2012). In addition, RI itself can be used as a new marker that represents cellular states or abnormalities (Lenz et al. 2013; Schürmann et al. 2016), and the factors which change the intracellular RI are being actively studied (Ekpenyong et al. 2013; Wang et al. 2011a).

HT technology also has several limitations and challenges:

1. Limited molecular specificity: Although RI values can be precisely measured using HT technology, it is difficult to relate these measured RI to molecular information. This is mainly because proteins have similar RI values regardless of their types. Nonetheless, the spatial distribution RI values can provide limited morphological information about subcellular organelles. For example, nucleus membrane, nucleoli, lipid droplets, and vacuoles can be easily identified from RI distributions because they have distinct RI values different from neighboring environments (Barer 1953; Jung et al. 2018; Kim et al. 2016c; Schürmann et al. 2016). Furthermore, the gradient of RI can also be utilized for further discriminating subcellular structures (Kim and Park 2018).

With these potentials and challenges, measuring RI tomography of various types of cells have become widely adapted for the study of various research topics. Below we highlight recent applications of measuring RI tomograms for various research disciplines.

Microbiology

Observing individual bacteria with conventional optical microscopes is challenging. This is due to a number of reasons. One is the size of most bacteria, which are in the order of micrometers or smaller. In addition, bacterial cytoplasm has refractive indexes similar to that of a medium, and the bacteria are transparent. Therefore, it is difficult to visualize bacteria under a conventional bright-field microscope and a high-resolution microscope is required.

When imaging a bacterial cell using a phase contrast microscope, it is possible to obtain good imaging contrast. In this case, however, only limited morphological information is provided, such as the length, width, and shape of bacterial cells. Although phase contrast microscopy can measure live bacteria without using labels, it has limited capability for the study of microbiology because it provides only 2-D qualitative images. Immunofluorescent labeling technology can provide molecular-specific information inside bacteria, but it can be time consuming and costly to stain the cells. In addition, research is limited due to secondary problems such as photochromism and phototoxicity that occur during the bleaching process. This makes it difficult to observe the bacteria for long periods of time. Traditionally, confocal microscopes and transmission electron microscopes have been used to obtain the internal structure of individual bacteria at high resolution. However, it is difficult to observe live cells for a long time because these techniques require cell staining or fixation.

RI tomographic imaging techniques can solve the problems of these existing imaging techniques. Because it is a noninvasive, label-free method, living bacteria can be observed for a long time without using additional exogenous labeling agents. In particular, by measuring the RI, protein concentration and mass information within the bacteria can be extracted, which has recently led to several studies related to the division of bacterial cells. However, because RI itself does not provide molecular-specific information, an in-depth investigation in the context of molecular biology is significantly limited. In the future, there will be potentials where both 3-D RI tomographic imaging and fluorescence microscope technique are used simultaneously.

Several previous papers had been reported where 3-D RI tomograms of individual bacteria are measured using HT technology. Using Mach–Zehnder interferometer and illumination scanning, 3-D RI tomograms of bacteria extracted from a sample of stool (Lauer 2002) and *E. coli* (Cotte et al. 2013) have been reported. Recently, white-light diffraction tomography was used to image 3-D RI tomogram of *E. coli* (Kim et al. 2014c). More recently, 3-D RI distribution of *Magnetospirillum gryphiswaldense,* a magnetotactic bacterium that produces magnetic particles (magnetosome), were measured (Bennet et al. 2016).

RI information can be exploited to retrieve cellular dry mass and concentration information about individual bacteria. Dry mass refers to the nonaqueous contents inside cells and can be used as an indicator of cellular growth and division. Because RI of cytoplasm is linearly proportional to its concentration, RI tomography represents protein distribution of a cell. Furthermore, the integration of RI over cell volume can also provide information about the dry mass of the cell. Dry mass of a cell can be simply retrieved by measuring 2-D optical phase delay maps and averaging it over cell area because the optical phase delay map of a cell corresponds to the integration of RI differences between nonaqueous contents inside cells (Lee et al. 2013; Popescu et al. 2008b). The cellular dry mass of fission yeast was monitored during the cell cycle with digital holographic microscopy, and the difference in mass production rate between wild-type and mutant fission yeast cells was observed (Rappaz et al. 2009). Using spatial light interference microscopy (SLIM), the dry mass of *Escherichia coli* cells was measured (Mir et al. 2011). In this work, the roles of cell density and morphology in mass regulation had been investigated.

The application of RI-based imaging in microbiology, especially has a strong advantage for the long-term growth monitoring of microbes because RI provides both the morphological information and the quantitative information about cell mass without using exogenous labels. Recent reports show the bacterial species identification using the 2-D optical phase delay maps and artificial intelligence algorithms (Jo et al. 2014, 2015, 2017). More recently, 3D RI tomograms were utilized for evaluating antibacterial activities of the graphene-based film (Kim et al. 2017b).

Hematology

The physical parameters of red blood cells (RBCs) are strongly related to the pathophysiology of various diseases (Suresh 2006). Conventionally, optical microscopy with labeling methods has been used to examine RBC morphology in blood smears; poikilocytosis (e.g., spherocytes, target cells) and blood-borne infectious diseases including malaria are routinely examined. Information about hemoglobin (Hb) in RBCs are of particular importance in laboratory medicine; mean corpuscular Hb concentration (MCHC) and mean corpuscular Hb content (MCH) are extensively examined for medical diagnosis. It is also well known that the deformability of RBCs can be altered by several infectious diseases and genetic disorders (e.g., malaria and sickle cell disease) (Byun et al. 2012; Diez-Silva et al. 2012; Kim et al. 2014a; Mills et al. 2007; Park et al. 2008), with the implications of malfunctions in microcirculation. In clinical hematology, automated blood cell counters based on the complete blood count (CBC) have been utilized to measure the properties of RBCs. Current automated blood cell counter techniques measure the parameters of RBCs, including mean corpuscular volume (MCV), MCHC, MCH, and RBC distribution width (RDW), which serve as the principal and crucial information from which clinicians diagnose abnormalities in RBCs.

The use of 3-D RI tomography in the field of hematology could lead to the simultaneous measurements of various optical parameters of individual RBCs. Figure 2 summarizes the analysis procedure for retrieving the parameters of individual RBCs using 3-D RI tomography, including the volume, surface area, sphericity, Hb content, Hb concentration, and membrane fluctuation, which can be obtained at the single-cell level. Figure 2 summarizes the analytical procedure for retrieving the

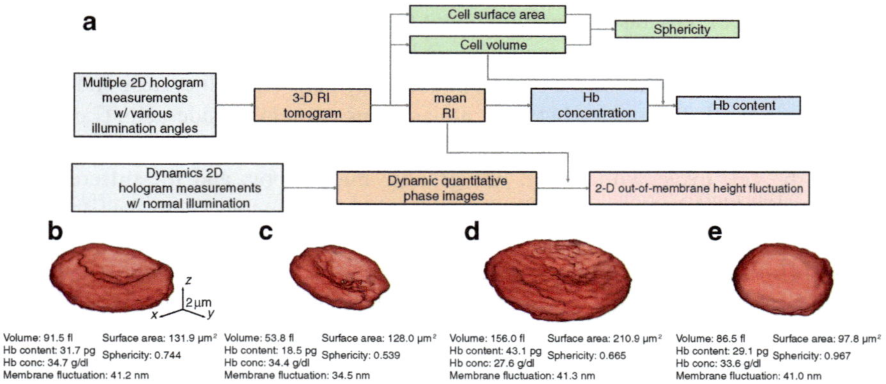

Fig. 2 (**a**) Schematic diagram of the analysis procedure for retrieving the parameters of individual RBCs using 3-D RI tomography. (**b-d**) 3-D rendered isosurfaces of RI tomograms of individual RBCs and retrieved parameters from (**b**) healthy, (**c**) Iron-deficiency anemia, (**d**) reticulocyte, and (**e**) HS red blood cells. Reproduced from Ref. (Kim et al. 2014a) with permission

parameters of individual RBCs using 3-D RI tomography, including volume, surface area, sphericity, Hb concentration, Hb content, and membrane fluctuations.

The measurements of various parameters of RBCs enable single-cell profiling, of which the importance has escalated in recent years (Higgins 2015; Weatherall 2011). Regulation of dynamic cellular systems and related pathophysiology can be better understood when the various parameters of individual cells are simultaneously analyzed in detail. This is because of recent improvements in measurement techniques, which have made single-cell profiling more efficient than before. In addition, data-based research approaches in biology and medicine have also grown rapidly, allowing researchers to explore new perspectives, in addition to existing hypothesis-based research methods. The 3-D RI measurements enable the retrieval of morphological (cell volume, cell surface area, sphericity), biochemical (Hb content, Hb concentration), and biomechanical (dynamic membrane fluctuation) information measured at the individual cell level. This allows correlative analysis, which is not possible with the conventional blood cell counters.

Previously, 2-D QPI techniques have been employed for various applications in hematology. In particular, optical measurements of the parameters of individual RBCs have been widely studied, including malaria-infected human red blood cells (Chandramohanadas et al. 2011; Park et al. 2008), sickle cell diseases (Byun et al. 2012; Shaked 2012), and ATP-dependent fluctuations (Park et al. 2010). RBCs do not have subcellular organelles and exhibit homogeneous RI distributions, which therefore allows measurements of Hb concentration. Thus, 2-D QPI measurements can also provide both the morphological information (cell height map) as well as the biochemical information (Hb contents or dry mass), with prior information about Hb concentration or the RI value of RBC cytoplasm. However, RBCs from an individual with diseases such as malaria infection or sickle cell anemia, the values of Hb concentrations, and thus RI values, vary significantly. Therefore, it is required to directly measure the 3-D RI tomograms of individual RBCs for the systematic study of disease states.

Recently, 3-D RI tomograms of RBCs have been utilized for measuring the parameters of individual RBCs. Lee et al. used common-path diffraction optical tomography (cDOT) (Kim et al. 2014d), and measured 3-D RI tomograms and dynamic membrane fluctuations of RBCs exposed to ethanol (Lee et al. 2015). It was observed that RBCs exposed to an ethanol concentration of 0.1–0.3% v/v becomes more spherical shapes than those of normal cells (Lee et al. 2015). Using the cDOT, the properties of individual RBCs stored with and without a preservation solution, citrate phosphate dextrose adenine-1 (CPDA-1) were reported. In this work, various red blood parameters were analyzed and the results showed that, in the absence of CPDA-1, RBCs undergo a dramatic morphological transformation from discocytes to spherocytes within two weeks (Park et al. 2016). The RBCs also experienced a reduction in cell surface areas and became less flexible. However, RBCs stored with CPDA-1 retained their biconcave shapes and deformability for up to six weeks.

More recently, the RBCs from patients with diabetes mellitus have been systematically measured using 3-D RI tomography and dynamic membrane fluctuations (Lee et al. 2017). Morphologies of the RBCs from diabetic patients were not significantly different from the healthy ones. The deformability of the RBCs from diabetic patients was significantly lower than those of healthy RBCs, which is consistent with the previous literature using ektacytometry or filtration assay. Interestingly, this work reported the negative correlation between dynamic membrane fluctuation and glycated Hb concertation or HbA1c level, for the healthy RBCs; the higher the HbA1c level, the less deformable the cells are. In addition, the alterations in RBCs resulted from the binding of melittin, the active molecule of apitoxin or bee venom, have been studied by measuring 3-D RI maps and dynamic membrane fluctuations (Hur et al. 2017). RBCs from the cord blood of newborn infants and adult mothers or nonpregnant women were also systematically studied using 3-D RI tomography (Shin et al. 2015).

The study of white blood cells using 3-D RI tomography has not been fully exploited, but it will open new applications. More recently, Yoon et al. measured 3-D RI tomograms of mouse lymphocytes and macrophages (Yoon et al. 2015). In this work, the morphological alternations in lymphocytes, caused by lipopolysaccharide which is known to immunologically stimulate lymphocytes, were analyzed.

The applications of 2-D and 3-D RI imaging techniques in hematology has increased. Among them, measurements of membrane fluctuation dynamics and Hb concentration of individual RBCs have been actively investigated due to their correlations with diseases and environmental abnormalities. In addition, because various cellular parameters are simultaneously extracted from individual blood cells, and their correlative analysis can be followed, more previous investigations of cellular alterations associated with various diseases are now accessible with direct experimental approaches.

Infectious Diseases

The visualizations of the structures and dynamism of parasites and host cells are important for the study of parasitic infections. Electron microscopy provides spatial resolutions much higher than optical microscopy, and has been used to visualize the internal structures of parasites. However, the high spatial resolution of electron microscopy comes at the cost of static imaging; it does not provide the time-lapse information of parasitic infections. Optical microscopy techniques have been extensively used for imaging parasites and host cells (Cho et al. 2011). The uses of labeling agents with high molecular specificity have elucidated the molecular biology of various infectious diseases. However, conventional fluorescent labeling techniques only provide qualitative imaging capability, and some parasite is difficult to be labeled.

Recently, 3-D RI tomography techniques have been utilized in the field of infectious diseases. Park et al. have used tomographic phase microscopy and measured 3-D RI maps and dynamic membrane fluctuations of malaria-infected RBCs as a function of various infection stages (Park et al. 2008). *Plasmodium falciparum* parasites invaded into host human red blood cells were visualized from the measurements of 3-D RI tomography. Also, the significantly decreased dynamic membrane fluctuations in the infected RBCs were also reported, indicating the decreased cell deformability (Diez-Silva et al. 2010, 2012). The label-free capability of 3-D RI tomography has been utilized for the study of egress of malaria parasites (Chandramohanadas et al. 2011), which provided a comprehensive body of information on the relationships between biomechanical and biochemical parameters and parasite egress from RBCs infected by malaria-inducing parasites. Kim et al. employed the ODT algorithm to reconstruct 3-D RI tomography of malaria-infected RBCs and it shows better image quality compared to the ones obtained with optical projection algorithm because ODT considers light diffraction inside samples (Kim et al. 2014b). In addition, various morphological information about invading parasites and produced hemozoin structures are obtained and analyzed quantitatively. The evasion mechanisms of *P. falciparum* from host immunity were also studied using holotomography (Tougan et al. 2018).

More recently, *Babesia microti* invaded RBCs were investigated by measuring 3-D RI tomograms at the individual cell level (Park et al. 2015). *B. microti* causes emergency human babesiosis, which shares similar pathophysiology and pathologic symptoms with malaria. In this work, RI information was effectively used for the study of babesiosis, because the RI of *B. microti* parasites is distinct from RBC cytoplasm can thus be clearly visualized, otherwise very difficult to be identified with conventional optical imaging techniques because of the lack of effective labeling agents for *B. microti* parasites. Ekpenyong et al. reported the RI maps of primary murine bone marrow-derived macrophages that were infected by *Salmonella enterica* serovar Typhimurium (Ekpenyong et al. 2013).

3-D RI tomography was also employed for the study of viral infection. Simon et al. used a setup, in which fluorescence confocal microscopy and optical diffraction tomography were combined, and studied human respiratory epithelial carcinoma A549 cells infected with human influenza H3N2 virus (Simon et al. 2010). Interestingly, in the infected cells, the spherical structures with the size of 150–200 nm and with the distinctly high RI values were observed, which were expected to correspond to the buddings of viral particles.

In the field of microbiology, various cases have been studied where parasites can be detected using their distinctive RI values. Also, alterations in infected host cells are quantitative studies by measuring 3-D RI distributions. RI-based imaging techniques can potentially provide advantages over conventional techniques such as chemical assay and fluorescence microscopy, because of its label-free imaging capability and simple sample preparation procedures.

Hepatology

Optical microscopic imaging of hepatocytes has played an important role in hepatology. The structures of cells and subcellular organelles and their dynamics are strongly correlated to the physiology of hepatocytes, and also significantly altered associated with liver-related diseases. Recently, Kim et al. have measured 3-D RI tomograms of human hepatocytes (human hepatocellular carcinoma cell line, Huh-7) were measured at the individual cell level (Fig. 3). In this work, various subcellular structures of hepatocytes are clearly identified using RI values, including cell membrane, nucleus membrane, nucleoli, and lipid droplets (LDs) (Kim et al. 2016c). Also, time-lapse 3-D RI tomograms of hepatocytes were also measured, from which dynamics of individual LDs were quantified.

Among subcellular structures of hepatocytes, lipid droplets (LDs) are of particular interest because they are directly related to the lipid metabolism. LDs consist of

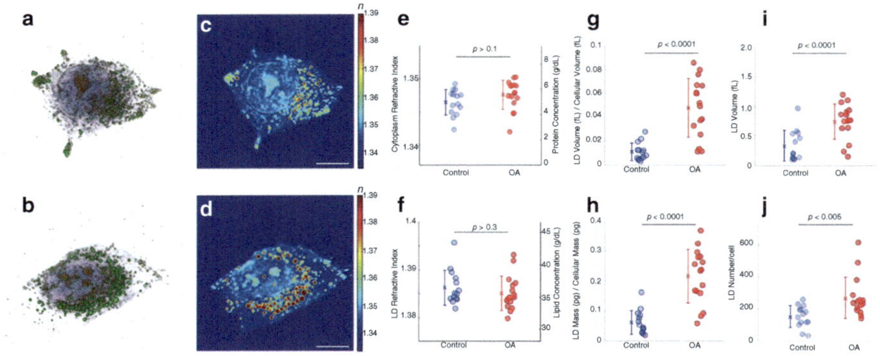

Fig. 3 (**a, b**) 3-D rendered isosurface image of 3-D RI distribution of (**a**) an untreated and (**b**) oleic acid (OA)-treated hepatocyte. (**c, d**) Cross-sectional slice images of 3-D RI distribution of (**c**) the untreated and (**d**) OA-treated hepatocyte. The scale bars indicate 10 μm. **e-j**, Quantitative analysis of (**e**) RI of cytoplasm, (**f**) RI of LDs, (**g**) the ratio of LD volume to cell volume, (**h**) the ratio of LD mass to cellular dry mass, (**i**) volume of individual LDs, and (**j**) number of LDs in untreated and OA-treated hepatocytes. Reproduced from Ref. (Kim et al. 2016c) with permission

a monolayer of phospholipids and associated proteins surrounding a core of neutral lipid and are ubiquitous intracellular organelles storing and supplying lipids in most cell types as well as hepatocytes (Martin and Parton 2006). Recent studies suggest that LDs participate in various pathological roles, such as cancer and diabetes mellitus, and exhibit 3-D motions to regulate lipid storage and metabolism (Welte 2009). However, the detailed process of LDs dynamics including biogenesis, growth, and 3-D subcellular motions are incomplete (Wilfling et al. 2014).

LDs can be effectively visualized exploiting its RI value; the RIs of lipid are significantly higher than protein (Beuthan et al. 1996) and thus LDs can be identified by measuring 3-D RI tomograms. Measuring LDs using the high RI values have several advantages over conventional approaches. Fluorescence techniques for labeling LDs have been widely used (Martin et al. 2005), but the use of fluorescent probes requires for ethanol treatment, which can influence the physiological conditions of LDs such as induced fusions of LDs in live cells (Fukumoto and Fujimoto 2002). Coherent anti-Stokes Raman scattering techniques have been used to visualize LDs in live cells without the use of exogenous labels (Evans et al. 2005; Nan et al. 2003), but it requires highly expensive laser and detection instruments and have the issue of low signal-to-noise ratio for molecules with low concentrations. The recent study demonstrated that the time-lapse 3-D RI distribution of LDs in live hepatocytes could be quantitatively analyzed with ODT (Kim et al. 2016c). The shapes, sizes, and the masses of individual LDs in live hepatocytes were retrieved from the measured 3-D RI tomograms.

One of the direct and important applications of HT technology would be imaging and quantifying individual lipid droplets in hepatocytes. 3-D RI distributions of hepatocytes can provide quantitative imaging and analysis about metabolisms of lipid droplets in live cells. In addition, 3-D RI maps of live cells also provide morphological information, and thus the generations, dynamics, and degradation of lipid droplets can also be studied with the combination with the information about subcellular structures.

Histopathology

RI can potentially serve as an important contrast in histopathology because (i) the use of RI for imaging tissue slides provide imaging contrast for the visualization of anatomical features in tissue slides otherwise invisible under conventional bright-field microscopy and (ii) it does not require for the labeling process, which can save time and cost. In addition, unlike conventional histopathology that relies on staining agents, the use of RI can provide quantitative criteria for pathologies in unlabeled biological tissues.

The RI information, for example, was used to quantify the weight loss in the inflammation-induced colitis (Lenz et al. 2013). Besides the dry mass information, the measurements of optical phase delay maps of a tissue slide could precisely be utilized to extract the scattering parameters such as scattering coefficients (μ_s) and anisotropies (g) (Ding et al. 2011; Wang et al. 2011a, b). This is because light

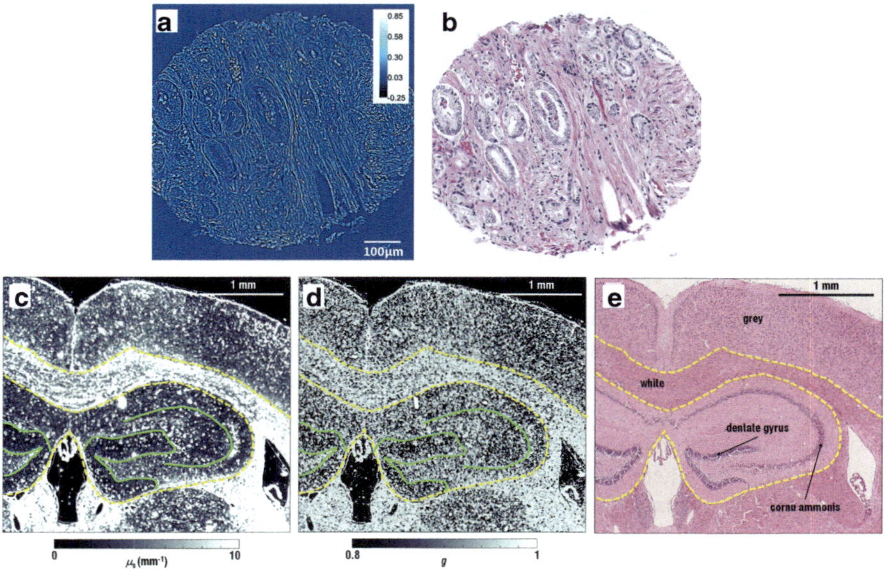

Fig. 4 Applications of RI-based imaging techniques for histopathology and cell dynamics (**a**) 2-D optical phase map of a slide from a biopsy sample obtained from a patient who had a biochemical recurrence of prostate cancer after undergoing radical prostatectomy. (**b**) The adjacent tissue slide of (**a**) labeled with conventional H&E staining. (**c-d**) Scattering parameter (μ_s) map (**c**) and anisotropy (g) map (**d**) of a whole mouse brain tissue slide, which was retrieved from the measured 2-D optical phase map of the slide. (**e**) The adjacent tissue slide is labeled with conventional H&E staining. Reproduced from Refs. (Lee et al. 2016) and (Sridharan et al. 2014) with permission

scattering in tissue is caused by inhomogeneous distributions of RI, and thus as light passes through tissue, it undergoes significantly large events of light refraction and reflection, resulting in complex patterns of multiple light scattering. These scattering parameters have been employed to investigate the morphological alterations in prostate cancers (Figs. 4a-b) (Sridharan et al. 2014), breast cancers (Majeed et al. 2015), and epithelial precancers (Su et al. 2015), which can be potentially utilized for diagnostic purposes.

Recently, the label-free tissue imaging capability by utilizing RI information was adapted to neuroscience. Optical phase delay maps of brain tissues were obtained for whole brain mouse tissue slides (Lee et al. 2016). The maps of scattering parameters (μ_s and g), extracted from the measured optical phase delay images of brain tissue, showed anatomical structures, comparable to the ones obtained with the conventional hematoxylin and eosin (H&E) staining method (Figs. 4c-e). Furthermore, these scattering parameter maps showed a statistical difference between the tissue slides from mice with Alzheimer's disease and the ones from healthy mice, indicating the structural alternations in the tissue.

For the application in histopathology, RI-based imaging of tissues can provide various parameters such dry mass, scattering coefficients, and anisotropies, and these values can be quantitatively analyzed in order to investigate morphological or structural alterations associated with diseases. In the near future, the use of QPI techniques will be further broadened for the diagnosis of various diseases (Kim et al. 2016a). Also, the RI-based imaging techniques do not require complex procedures for sample preparation, which can dramatically reduce the time and cost for histopathological diagnosis (Greenbaum et al. 2014).

Outlook

Here, we presented the principles of quantitative phase imaging techniques that exploit RI as an intrinsic optical imaging contrast for cells and tissues. Although this is a relatively new research field, and only a few topics of the studies have been investigated so far, the research work reviewed here suggests that the use of RI for biological imaging may play an important role in various fields of studies (summarized in Table 1), where label-free and quantitative live cell imaging capability provides benefits. This would significantly enhance our understanding of the pathophysiology of various diseases, which may also lead to the development of novel diagnostic strategies in the future.

The uses of RI as intrinsic imaging contrast for cells and tissues for biological and medical applications have not yet been fully explored; there are still various important issues in the physiology and pathology, which can be addressed by the utilization of RI-based imaging techniques and corresponding analysis methods. Some of the representative cell images in emerging fields are presented in Fig. 5. Many interesting studies which had been performed with 2-D QPI techniques would be readily investigated with 3-D RI tomography techniques because 3-D RI measurements directly provide both RI values and shape information, whereas usual 2-D QPI techniques only provide optical phase maps—a coupled parameter between RI values and the height of a sample. For example, water flux in individual neuron cells (Jourdain et al. 2011), morphologies of tumor cells (Kemper et al. 2006), cell growth and division (Cooper et al. 2013; Kemper et al. 2010; Mir et al. 2011), and cell death (Pavillon et al. 2012) would be studied in more details with 3-D RI tomography.

Label-free live cell imaging capability of 3-D RI tomography will be a strong advantage for its future applications in medicine and biology. For example, it has a potential to be used in the studies of stem and neuron cells that are vulnerable to environmental changes such as the use of fluorescent labeling agents (Braydich-Stolle et al. 2005; Millet et al. 2007). Although 3-D RI tomography does not generally have molecular specificity, it can provide information about the cellular morphology, internal structural changes, and various biophysical parameters. Thus, 3-D RI tomography will be a strong complementary technique to fluorescence techniques. For example, a multimodal approach of using both the 3-D RI tomography and 3-D fluorescence microscopy opens new avenues for the study of cell biology (Kim et al. 2017a, 2018c). Recently, the combination of 3-D RI tomography and structured illumination microscopy for 3-D fluorescence imaging was demonstrated

Table 1 Major research areas in which 2-D and 3-D RI measurement techniques have been employed

Research area	Application	Method	References
Microbiology	Intracellular structures (*Magnetospirillum gryphiswaldense*, *Nannochloropsis oculata*)	3-D RI imaging	Bennet et al. (2016) Jung et al. (2018)
	Cellular growth (fission yeast, *Escherichia coli*)	Dry mass measurement from 2-D optical phase delay map	Rappaz et al. (2009) Mir et al. (2011)
	Bacterial identification (*Escherichia coli*, *Listeria monocytogenes*, *Lactobacillus casei*, *Bacillus subtilis*, *Synechococcus elongatus*)	Machine learning or deep learning with 2-D optical phase delay map	Jo et al. (2014, 2015, 2017)
	Anti-bacterial activity	3-D RI imaging	Kim et al. (2017b)
	Protistology (amoeba, phytoplankton)	3-D RI imaging	Charrière et al. (2006b) Lee et al. (2014)
Hematology	Effect of external stimulus (ethanol, preservation solution, melittin)	2-D optical phase delay map and 3-D RI imaging	Lee et al. (2015) Park et al. (2016) Hur et al. (2017)
	Infectious disease (malaria, babesiosis, typhoid fever, influenza)	2-D optical phase delay map and 3-D RI imaging	Park et al. (2008) Simon et al. (2010) Chandramohanadas et al. (2011) Diez-Silva et al. (2012) Ekpenyong et al. (2013) Kim et al. (2014d) Park et al. (2015) Tougan et al. (2018)
	Chronic human disease (sickle cell anemia, diabetes mellitus)	2-D optical phase delay map and 3-D RI imaging	Byun et al. (2012) Shaked (2012) Lee et al. (2016)
	White blood cell	3-D RI imaging	Yoon et al. (2015) Yoon et al. (2017)
	Learning-based disease diagnosis (iron-deficiency anemia, reticulocytosis, hereditary spherocytosis, diabetes)	2-D optical phase delay map and 3-D RI imaging	Kim et al. (2019)
Hepatology	Lipid droplets in Huh-7 (human hepatocellular carcinoma cell line)	3-D RI imaging	Kim et al. (2016b)

(continued)

Table 1 (continued)

Research area	Application	Method	References
Histopathology	Colitis	2-D optical phase delay map	Lenz et al. (2013)
	Cancer (prostate, breast, and epithelial)	Scattering parameter measurement from 2-D optical phase delay map	Sridharan et al. (2014) Majeed et al. (2015) Su et al. (2015)
	Neurological disorder (Alzheimer's disease and Parkinson's disease)	Scattering parameter measurement from 2-D optical phase delay map and 3-D RI imaging	Lee et al. (2016) Yang et al. (2016)
Cell biology	Membrane fluctuation (red blood cell)	2-D optical phase delay map	Park et al. (2010)
	Cell growth, division, and death	2-D optical phase delay map	Kemper et al. (2010) Mir et al. (2011) Pavillon et al. (2012) Cooper et al. (2013)
	Plant cell study (pollen grain)	3-D RI imaging	Kim et al. (2018b) Park et al. (2018a)

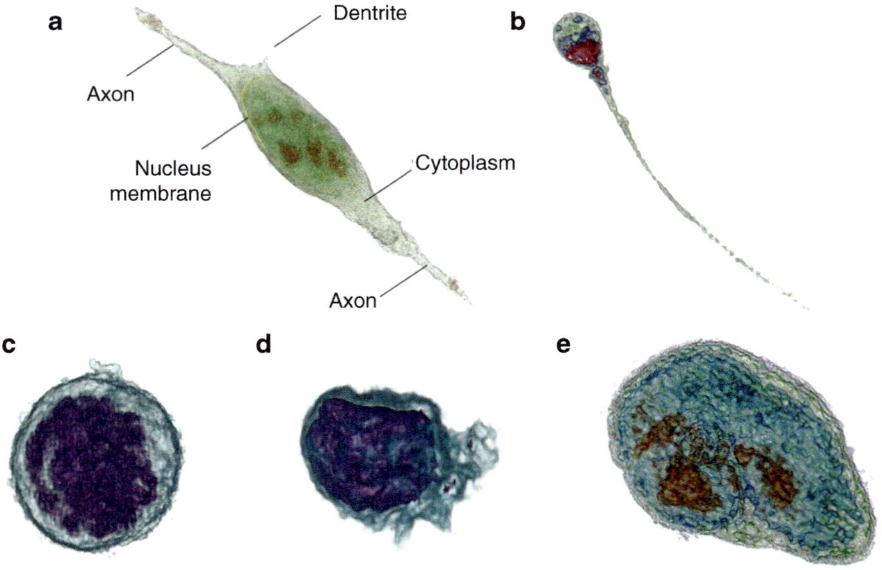

Fig. 5 Emerging fields of research using 3-D RI tomography and representative cell images: (**a**) neuron cell, (**b**) human sperm, (**c**) mouse macrophage, (**d**) LPS-treated mouse macrophage, and (**e**) embryonic stem cells. Image in (**a**) is reproduced from Ref. (Yang et al. 2016) with permission. Images in (**b-e**) are provided by Tomocube Inc.

(Chowdhury et al. 2017; Shin et al. 2018). The specific molecules or proteins can be localized in live cells using 3-D fluorescence microscopy, and then long-term imaging of cell dynamics can be followed. Also, by utilizing 3-D fluorescence images, analyzing quantitative information about subcellular organelles such as concentration or dry mass can be performed. In addition, RI-based imaging has shown promises toward label-free cell identification. Although conventional fluorescence-based cytometry or imaging provides extremely high molecular specificity and signal-to-background ratio, they only provide qualitative information that is highly dependent on protocols, sample conditions, etc. In contrast, RI is an intrinsic optical parameter of material, and thus can serve as a highly reproducible and yet quantitative imaging contrast. When cleverly combined with appropriate algorithms, the 3-D or time-lapsed 3-D RI data obtained from individual cells can be effectively analyzed.

Recently, RI-based imaging data has begun to be exploited in combination with machine learning algorithms (Jo et al. 2019). For example, from the 2-D quantitative phase images of individual bacteria, the genus of various bacteria was distinguished using a machine learning algorithm (Jo et al. 2014, 2015). In addition, non-activated lymphocytes were recently identified and classified from their 3-D RI tomograms (Yoon et al. 2017). Recently, the weapons-grade anthrax spores are optically detected using 2-D QPI techniques and deep learning (Jo et al. 2017). Label-free classification of kinetic cell states was also demonstrated using 2-D QPI and machine learning (Hejna et al. 2017). More recently, parameters of 3-D QPI were exploited for the diagnosis of hematologic diseases with high accuracy (Kim et al. 2019). The combination of RI-based imaging techniques and machine learning will potentially generate strongly synergetic effects for the study of stem cells and neuroscience where the use of exogenous labeling agents is generally avoided (Maxmen 2017).

Finally, yet importantly potential topics in the application would include the 3-D label-free imaging for the study of Protists where 3-D morphometry and spatial movement are important issues. Previously, 3-D RI tomography has shown potentials for the in-depth investigation of amoeba (Charrière et al. 2006b) and phytoplankton (Lee et al. 2014). Also, the imaging of plant cells could be an important topic of study where label-free characterization and quantification of subcellular structures provide useful information (Kim et al. 2018a; Park et al. 2018d).

To facilitate the application of RI-based imaging methods for toxicology and pharmaceutical industry, it is required to achieve a platform to obtain high-throughput data. Toward this direction, several research activities are currently ongoing. Kim et al. presented a method to significantly enhance the speed of 3-D RI tomography employing sparse illumination patterns and a graphics processor unit (GPU) (Kim et al. 2013). Sung et al. demonstrated a method to obtain 3-D RI tomography of cells flowing in a microfluidic channel; without rotating a sample nor illuminating different patterns, optical information to construct a 3-D RI tomogram is acquired from the translations of a sample in the channel (Sung et al. 2014). Also, optical manipulation and 3-D tracking of biological cells in microfluidic channels may also open new applications combined with 3-D RI tomography (Merola et al. 2012).

Developments in new analysis algorithms are also required. Previously, the quantitative imaging capability of RI-based imaging had been limited to static analysis, including cellular dry mass, morphological, and biochemical information. The RI information can also be exploited in analyzing the dynamics of cells and subcellular compartments. Measurements of time-lapse 3-D RI tomograms of cells would accompany with the developments in analysis algorithm (Ma et al. 2016).

New techniques to effectively handle the large size of data are particularly needed to make RI-based imaging approaches even more useful. In order to perform analysis at the individual cellular level and reveal the details of the underlying mechanisms of diseases, at least dozens of cells should be measured in each experimental group. Then a technical issue arises—the sizes of tomogram files are enormously significant. For example, HT-1 series from Tomocube Inc., a commercialized 3-D holographic microscopy system, generate a data file with a 67-Mega voxels for one 3-D RI tomogram; the imaging volume is 84.3 × 84.3 × 41.4 μm with the pixel resolution of 112 × 112 × 356 nm for *x*-, *y*-, and *z*-directions, respectively. With the single-precision (or single) data type, a single tomogram of 67 Mega voxels will have the size of 268 Mbyte. When a 4-D or time-lapse 3-D tomograms the frame number of 512 is recorded, the data size will be approximately 134 Gbyte. More advanced methods to transfer, handle, and store such large-sized data will also be required in the future.

To provide better molecular sensitivity to RI-based imaging, multispectral approaches have been introduced recently (Jung et al. 2014). The use of multiple coherent lasers (Jang et al. 2012; Park et al. 2009) or wavelength-scanning illumination (Jung et al. 2013; Rinehart et al. 2012) enabled measuring the optical phase maps as a function of wavelength, and this information was utilized for molecular imaging because certain molecules have distinct optical dispersion properties, i.e., different RI values as a function of wavelengths. Recently, 3-D RI tomography has been achieved for a large number of wavelengths using a super continuum source and a wavelength-scanning unit (Jung et al. 2016). Besides the technical difficulty, the major challenge in exploiting optical dispersion for molecular specificity is that most molecules present in cells do not exhibit strong optical dispersion except specific molecules such as Hb (Jung et al. 2014). Alternatively, the use of an ultraviolet light source was presented to better visualize chromosomes inside a cell (Sung et al. 2012). More recently, the use of gold nanoparticles (GNPs) has been utilized in RI-based imaging, because the strong light absorption and scattering of GNPs at a resonant wavelength enable high imaging contrast (Kim et al. 2018b; Turko et al. 2013).

Besides the developments in optical instrumentations and correlative imaging strategy, the advancements in reconstruction algorithm are expected to further enhance the quality of 3-D RI tomography. The current reconstruction algorithms are mostly based on the first order approximations—the algorithm only considers single scattered light to inversely solve the wave equations. This assumption is valid for imaging single or up to a few layers of biological cells. However, the reconstruction may fail for highly scattering samples, such as thick biological tissues or embryos. Recently, several methods have been introduced to consider high order scattering in the reconstruction algorithm, including a nonlinear propagation model based on the beam propagation method (Kamilov et al. 2016) and a treatment of an unknown surrounding medium as a self-resonator (Lim et al. 2017). Furthermore, the resolution of the optical imaging system measuring 3-D RI tomography is governed by the numerical aperture of the objective and condenser lenses (Park et al. 2018b). Due to the limited numerical apertures of an imaging system, a fraction of side scattering signals is not collected, resulting into a reduced spatial resolution and inaccurate values of RIs in a reconstructed tomogram—known as a *missing cone* problem. To remedy this issue, several regularization algorithms have been used based on prior information about a sample (Lim et al. 2015). However, the careful use of appropriate prior condition is required and the computing power for the existing regularization algorithms are heavy. The further developments are also expected to more effectively address this missing cone problem.

In addition, optical trapping techniques can be combined with RI-based imaging techniques to revolutionize the way in which biologists approach questions in the field of cell-to-cell interaction and mechanobiology. Toward to this direction, the combination with 3-D RI tomography and optical manipulation techniques are beneficial. The optical tweezer technique has been shown to optically trap spherical particles, which aided to manipulate individual cells. Recently, 3-D RI tomography technique was combined with holographic optical tweezers, which demonstrates the 3-D dynamic imaging of the interaction of an optically trapped particle to a macrophage (Kim et al. 2015). More recently, the 3-D RI maps of biological cells were measured in order to actively control optical wavefront for trapping beams, and the demonstration of stable control of complex shapes objects such as RBCs and dimers was demonstrated (Kim and Park 2017).

From an application point of view, the utilization of microfluidic devices for 3-D RI tomography may open various opportunities. The use of a designed chip can be used to simplify an optical setup for the 3-D RI measurements (Bianco et al. 2017), and to allow the quantification of chemical concentrations of fluids in a microfluidic channel (Park et al. 2017b). The label-free imaging using RI will also have potentials for the cytotoxicity assay (Kwon et al. 2018).

One of the future directions of RI-based imaging would be toward in vivo application. If the use of RI information can be collected in vivo, it would bring significant impacts to the early diagnosis of various diseases. However, because of multiple light scattering in biological tissues deteriorate the delivery of optical information (Yu et al. 2015), the direct application of a simple optical imaging system would not be able to acquire clear images. Recently, 3-D RI tomograms of individual RBCs flowing through microcapillaries (Kim et al. 2016a). However, in this work, the part of a thin mesentery tissue should be placed in 3-D holographic microscopy. Ford et al. demonstrated in vivo phase imaging by collecting *en face* phase gradient images of thick scattering samples (Ford et al. 2012). More recently, Laforest et al. demonstrated in vivo phase imaging of retinal cells using transcleral illumination, which gives a dark field configuration with a high numerical aperture (Laforest et al. 2017). Yet, deep tissue imaging toward applications in dermatology or gastroenterology is technically challenging due to multiple light scattering (Yu et al. 2015).

There are several attempts to overcome multiple light scattering including optical coherence tomography using wavefront shaping (Jang et al. 2013; Park et al. 2018c; Yu et al. 2014; Yu et al. 2016) or accumulating single scattering light in deep tissue imaging (Kang et al. 2015).

Various research results, in which RI was utilized as imaging contrast, are highlighted in this review chapter. Nonetheless, we believe that there are still uncountable potential applications, which are not yet discovered. In order to fully explore the potential and capability of RI-based approaches, interdisciplinary collaborations between biologists, medical specialists, and physicists are crucial. Considering the recent rapid growth of the field and the potentials of the approach, we are optimistic that optical imaging techniques based on RI will play important roles in various topics of studies where label-free and quantitative live cell imaging is important.

Author Contributions

YKP contributed to the overall layout, figures, writing, and revision. All other authors participated in parts of the writing of this chapter and the survey of related literature.

Funding

This work was supported by Tomocube, KAIST Up program, BK21+ program, and the National Research Foundation of Korea (2017M3C1A3013923, 2015R1A3A2066550, 2018 K000396).

Competing Financial Interests

Mr. Doyeon Kim, Mrs. Su-A Yang, and Prof. Park have financial interests in Tomocube Inc., a company that commercializes optical diffraction tomography and quantitative phase imaging instruments.

References

Allen R, David G, Nomarski G (1969) The zeiss-Nomarski differential interference equipment for transmitted-light microscopy. Z Wiss Mikrosk 69:193–221

Barer R (1952) Interference microscopy and mass determination. Nature 169:366–367

Barer R (1953) Determination of dry mass, thickness, solid and water concentration in living cells. Nature 172:1097–1098

Barty A, Nugent K, Roberts A, Paganin D (2000) Quantitative phase tomography. Opt Commun 175:329–336

Bennet M, Gur D, Yoon J, Park Y, Faivre D (2016) A bacteria-based remotely tunable photonic device. Advanced Optical Materials

Bettenworth D, Lenz P, Krausewitz P, Brückner M, Ketelhut S, Domagk D, Kemper B (2014) Quantitative stain-free and continuous multimodal monitoring of wound healing in vitro with digital holographic microscopy. PLoS One 9:e107317

Beuthan J, Minet O, Helfmann J, Herrig M, Müller G (1996) The spatial variation of the refractive index in biological cells. Phys Med Biol 41:369

Bianco V et al (2017) Endowing a plain fluidic chip with micro-optics: a holographic microscope slide. Light: Science Applications 6:e17055

Braydich-Stolle L, Hussain S, Schlager JJ, Hofmann M-C (2005) In vitro cytotoxicity of nanoparticles in mammalian germline stem cells. Toxicol Sci 88:412–419

Byun H et al. (2012) Optical measurement of biomechanical properties of individual erythrocytes from a sickle cell patient Acta Biomaterialia
Chalut KJ, Ekpenyong AE, Clegg WL, Melhuish IC, Guck J (2012) Quantifying cellular differentiation by physical phenotype using digital holographic microscopy. Integr Biol 4:280–284
Chandramohanadas R et al (2011) Biophysics of malarial parasite exit from infected erythrocytes. PLoS One 6:e20869. https://doi.org/10.1371/journal.pone.0020869
Charrière F et al (2006a) Cell refractive index tomography by digital holographic microscopy. Opt Lett 31:178–180
Charrière F et al (2006b) Living specimen tomography by digital holographic microscopy: morphometry of testate amoeba. Opt Express 14:7005–7013
Cheong W-F, Prahl SA, Welch AJ (1990) A review of the optical properties of biological tissues. IEEE J Quantum Electron 26:2166–2185
Cho S, Kim S, Kim Y, Park YK (2011) Optical imaging techniques for the study of malaria. Trends Biotech 30:71–79
Choi W, Fang-Yen C, Badizadegan K, Oh S, Lue N, Dasari RR, Feld MS (2007) Tomographic phase microscopy. Nat Methods 4:717–719
Chowdhury S, Eldridge WJ, Wax A, Izatt JA (2017) Structured illumination multimodal 3D-resolved quantitative phase and fluorescence sub-diffraction microscopy. Biomed Opt Express 8:2496–2518
Cooper KL, Oh S, Sung Y, Dasari RR, Kirschner MW, Tabin CJ (2013) Multiple phases of chondrocyte enlargement underlie differences in skeletal proportions. Nature 495:375–378
Cotte Y et al (2013) Marker-free phase nanoscopy. Nat Photonics 7:113–117
Debnath SK, Park Y (2011) Real-time quantitative phase imaging with a spatial phase-shifting algorithm. Opt Lett 36:4677–4679
Diez-Silva M, Dao M, Han J, Lim C-T, Suresh S (2010) Shape and biomechanical characteristics of human red blood cells in health and disease. MRS Bull 35:382–388
Diez-Silva M et al (2012) Pf155/RESA protein influences the dynamic microcirculatory behavior of ring-stage Plasmodium falciparum infected red blood cells. Sci Rep 2:614
Ding H, Wang Z, Liang X, Boppart SA, Tangella K, Popescu G (2011) Measuring the scattering parameters of tissues from quantitative phase imaging of thin slices. Opt Lett 36:2281–2283
Ekpenyong AE, Man SM, Achouri S, Bryant CE, Guck J, Chalut KJ (2013) Bacterial infection of macrophages induces decrease in refractive index. J Biophotonics 6:393–397
Evans CL, Potma EO, Puoris'haag M, Côté D, Lin CP, Xie XS (2005) Chemical imaging of tissue in vivo with video-rate coherent anti-stokes Raman scattering microscopy. Proc Natl Acad Sci U S A 102:16807–16812
Fei Y et al (2011) Fluorescent labeling agents change binding profiles of glycan-binding proteins. Mol Bio Syst 7:3343–3352
Fercher AF, Bartelt H, Becker H, Wiltschko E (1979) Image-formation by inversion of scattered field data—experiments and computational simulation. Appl Opt 18:2427–2439. https://doi.org/10.1364/Ao.18.002427
Ford TN, Chu KK, Mertz J (2012) Phase-gradient microscopy in thick tissue with oblique back-illumination. Nat Methods 9:1195–1197
Fukumoto S, Fujimoto T (2002) Deformation of lipid droplets in fixed samples. Histochem Cell Biol 118:423–428. https://doi.org/10.1007/s00418-002-0462-7
Greenbaum A, Zhang Y, Feizi A, Chung P-L, Luo W, Kandukuri SR, Ozcan A (2014) Wide-field computational imaging of pathology slides using lens-free on-chip microscopy. Sci Transl Med 6:267ra175–267ra175
Hejna M, Jorapur A, Song JS, Judson RL (2017) High accuracy label-free classification of single-cell kinetic states from holographic cytometry of human melanoma cells. Sci Rep 7:11943
Hell SW (2007) Far-field optical nanoscopy. Science 316:1153–1158
Higgins JM (2015) Red blood cell population dynamics. Clin Lab Med 35:43–57
Hoebe R, Van Oven C, Gadella TJ, Dhonukshe P, Van Noorden C, Manders E (2007) Controlled light-exposure microscopy reduces photobleaching and phototoxicity in fluorescence live-cell imaging. Nat Biotechnol 25:249–253

Huang B, Bates M, Zhuang X (2009) Super resolution fluorescence microscopy. Annu Rev Biochem 78:993

Hur J, Kim K, Lee S, Park H, Park Y (2017) Melittin-induced alterations in morphology and deformability of human red blood cells using quantitative phase imaging techniques. Sci Rep 7:9306

Ikeda T, Popescu G, Dasari RR, Feld MS (2005) Hilbert phase microscopy for investigating fast dynamics in transparent systems. Opt Lett 30:1165–1167

Jang Y, Jang J, Park Y (2012) Dynamic spectroscopic phase microscopy for quantifying hemoglobin concentration and dynamic membrane fluctuation in red blood cells. Opt Express 20:9673–9681

Jang J et al (2013) Complex wavefront shaping for optimal depth-selective focusing in optical coherence tomography. Opt Express 21:2890–2902

Ji N, Milkie DE, Betzig E (2010) Adaptive optics via pupil segmentation for high-resolution imaging in biological tissues. Nat Methods 7:141–147

Jo Y, Cho H, Lee SY, Choi G, Kim G, Min H-S, Park Y (2019) Quantitative phase imaging and artificial intelligence. A Review IEEE Journal of Selected Topics in Quantum Electronics 25:1–14

Jo Y, Jung J, Kim MH, Park H, Kang SJ, Park Y (2015) Label-free identification of individual bacteria using Fourier transform light scattering. Opt Express 23:15792–15805. https://doi.org/10.1364/Oe.23.015792

Jo Y et al (2014) Angle-resolved light scattering of individual rod-shaped bacteria based on Fourier transform light scattering. Sci Rep 4:5090. https://doi.org/10.1038/srep05090

Jo Y et al (2017) Holographic deep learning for rapid optical screening of anthrax spores. Sci Adv 3:e1700606

Jourdain P et al (2011) Determination of transmembrane water fluxes in neurons elicited by glutamate ionotropic receptors and by the cotransporters KCC2 and NKCC1: a digital holographic microscopy study. J Neurosci 31:11846–11854

Jung J, Jang J, Park Y (2013) Spectro-refractometry of individual microscopic objects using swept-source quantitative phase imaging. Anal Chem

Jung J, Kim K, Yoon J, Park Y (2016) Hyperspectral optical diffraction tomography. Opt Express 24:2006–2012. https://doi.org/10.1364/OE.24.002006

Jung J et al (2014) Biomedical applications of holographic microspectroscopy [invited]. Appl Opt 53:G111–G122

Jung J et al (2018) Label-free non-invasive quantitative measurement of lipid contents in individual microalgal cells using refractive index tomography. Sci Rep:8

Kamilov US, Papadopoulos IN, Shoreh MH, Goy A, Vonesch C, Unser M, Psaltis D (2016) Optical tomographic image reconstruction based on beam propagation and sparse regularization. IEEE Transactions on Computational Imaging 2:59–70

Kang S et al (2015) Imaging deep within a scattering medium using collective accumulation of single-scattered waves. Nat Photonics 9:253–258

Kaushansky K, Lichtman MA, Prchal JT, Levi M, Press OW, Burns LJ, Caligiuri MA (2016) Williams hematology. McGraw-Hill Education

Kemper B, Carl D, Schnekenburger J, Bredebusch I, Schäfer M, Domschke W, von Bally G (2006) Investigation of living pancreas tumor cells by digital holographic microscopy. J Biomed Opt 11:034005–034008

Kemper B, Kosmeier S, Langehanenberg P, von Bally G, Bredebusch I, Domschke W, Schnekenburger J (2007) Integral refractive index determination of living suspension cells by multifocus digital holographic phase contrast microscopy. J Biomed Opt 12:054009–054005

Kemper B et al (2010) Label-free quantitative cell division monitoring of endothelial cells by digital holographic microscopy. J Biomed Opt 15:036009–036006

Kim K, Choe K, Park I, Kim P, Park Y (2016a) Holographic intravital microscopy for 2-D and 3-D imaging intact circulating blood cells in microcapillaries of live mice Scientific reports 6

Kim G, Jo Y, Cho H, Min H-S, Park Y (2019) Learning-based screening of hematologic disorders using quantitative phase imaging of individual red blood cells. Biosens Bioelectron 123:69–76

Kim K, Kim KS, Park H, Ye JC, Park Y (2013) Real-time visualization of 3-D dynamic microscopic objects using optical diffraction tomography. Opt Express 21:32269–32278

Kim G, Lee S, Shin S, Park Y (2018a) Three-dimensional label-free imaging and analysis of Pinus pollen grains using optical diffraction tomography. Sci Rep 8:1782

Kim K, Lee S, Yoon J, Heo J, Choi C, Park Y (2016c) Three-dimensional label-free imaging and quantification of lipid droplets in live hepatocytes. Sci Rep 6:36815

Kim D, Oh N, Kim K, Lee S, Pack C-G, Park J-H, Park Y (2018b) Label-free high-resolution 3-D imaging of gold nanoparticles inside live cells using optical diffraction tomography. Methods 136:160–167

Kim K, Park Y (2017) Tomographic active optical trapping of arbitrarily shaped objects by exploiting 3D refractive index maps. Nat Commun 8:15340

Kim T, Park J (2018) Analyzing 3D cell data of optical diffraction tomography through volume rendering. In: Advanced Image Technology (IWAIT), International Workshop on, 2018. IEEE, pp 1–4

Kim K, Park WS, Na S, Kim S, Kim T, Do Heo W, Park Y (2017a) Correlative three-dimensional fluorescence and refractive index tomography: bridging the gap between molecular specificity and quantitative bioimaging. Biomed Opt Express 8:5688–5697

Kim Y, Shim H, Kim K, Park H, Jang S, Park Y (2014a) Profiling individual human red blood cells using common-path diffraction optical tomography. Sci Rep:4

Kim K, Yoon H-O, Diez-Silva M, Dao M, Dasari R, Park Y-K (2014b) High-resolution three-dimensional imaging of red blood cells parasitized by *Plasmodium falciparum* and *in situ* hemozoin crystals using optical diffraction tomography. J Biomed Opt 19:011005–011012

Kim K, Yoon J, Park Y (2015) Simultaneous 3D visualization and position tracking of optically trapped particles using optical diffraction tomography. Optica 2:343–346

Kim K, Yoon J, Shin S, Lee S, Yang S-A, Park Y (2016b) Optical diffraction tomography techniques for the study of cell pathophysiology. Journal of Biomedical Photonics & Engineering 2:020201

Kim T, Zhou R, Mir M, Babacan SD, Carney PS, Goddard LL, Popescu G (2014c) White-light diffraction tomography of unlabelled live cells. Nat Photon 8:256–263. https://doi.org/10.1038/nphoton.2013.350

Kim Y et al (2014d) Common-path diffraction optical tomography for investigation of three-dimensional structures and dynamics of biological cells. Opt Express 22:10398–10407

Kim TI et al (2017b) Antibacterial activities of Graphene oxide–molybdenum disulfide Nanocomposite films. ACS Appl Mater Interfaces 9:7908–7917

Kim YS et al (2018c) Focus: medical technology: combining three-dimensional quantitative phase imaging and fluorescence microscopy for the study of cell pathophysiology. Yale J Biol Med 91:267

Koss LG (1989) The Papanicolaou test for cervical cancer detection: a triumph and a tragedy. JAMA 261:737–743

Kuś A, Dudek M, Kemper B, Kujawińska M, Vollmer A (2014) Tomographic phase microscopy of living three-dimensional cell cultures. J Biomed Opt 19:046009

Kwon S et al (2018) Mitochondria-targeting indolizino [3, 2-c] quinolines as novel class of photosensitizers for photodynamic anticancer activity. Eur J Med Chem 148:116–127

Laforest T, Carpentras D, Kowalczuk L, Behar-Cohen F, Moser C (2017) Quantitative phase imaging of retinal cells arXiv preprint arXiv:170108854

Lauer V (2002) New approach to optical diffraction tomography yielding a vector equation of diffraction tomography and a novel tomographic microscope. J Microsc 205:165–176

Lee K, Park Y (2014) Quantitative phase imaging unit. Opt Lett 39:3630–3633

Lee SY, Park HJ, Best-Popescu C, Jang S, Park YK (2015) The effects of ethanol on the morphological and biochemical properties of individual human. Red Blood Cells PloS one 10:e0145327

Lee S, Park H, Kim K, Sohn Y, Jang S, Park Y (2017) Refractive index tomograms and dynamic membrane fluctuations of red blood cells from patients with diabetes mellitus. Sci Rep 7:1039

Lee K et al (2013) Quantitative phase imaging techniques for the study of cell pathophysiology: from principles to applications. Sensors 13:4170–4191

Lee S et al (2014) High-resolution 3-D refractive index tomography and 2-D synthetic aperture imaging of live phytoplankton. Journal of the Optical Society of Korea 18:691–697
Lee M et al (2016) Label-free optical quantification of structural alterations in Alzheimer's disease. Sci Rep:6
Lenz P et al (2013) Digital holographic microscopy quantifies the degree of inflammation in experimental colitis. Integr Biol 5:624–630
Lichtman JW, Conchello J-A (2005) Fluorescence microscopy. Nat Methods 2:910–919
Lim J, Lee K, Jin KH, Shin S, Lee S, Park Y, Ye JC (2015) Comparative study of iterative reconstruction algorithms for missing cone problems in optical diffraction tomography. Opt Express 23:16933–16948. https://doi.org/10.1364/Oe.23.016933
Lim J, Wahab A, Park G, Lee K, Park Y, Ye JC (2017) Beyond born-Rytov limit for super-resolution optical diffraction tomography. Opt Express 25:30445–30458
Liu P et al (2016) Cell refractive index for cell biology and disease diagnosis: past, present and future. Lab Chip 16:634–644
Ma L et al (2016) Phase correlation imaging of unlabeled cell dynamics. Sci Rep:6
Majeed H, Sridharan S, Mir M, Ma L, Min E, Jung W, Popescu G (2016) Quantitative phase imaging for medical diagnosis. J Biophotonics
Majeed H et al (2015) Breast cancer diagnosis using spatial light interference microscopy. J Biomed Opt 20:111210
Mann CJ, Yu L, Lo C-M, Kim MK (2005) High-resolution quantitative phase-contrast microscopy by digital holography. Opt Express 13:8693–8698
Martin S, Driessen K, Nixon SJ, Zerial M, Parton RG (2005) Regulated localization of Rab18 to lipid droplets EFFECTS OF LIPOLYTIC STIMULATION AND INHIBITION OF LIPID DROPLET CATABOLISM. J Biol Chem 280:42325–42335
Martin S, Parton RG (2006) Lipid droplets: a unified view of a dynamic organelle. Nat Rev Mol Cell Biol 7:373–378
Maxmen A (2017) Machine learning predicts the look of stem cells. https://doi.org/10.1038/nature.2017.21769
Merola F, Miccio L, Memmolo P, Paturzo M, Grilli S, Ferraro P (2012) Simultaneous optical manipulation, 3-D tracking, and imaging of micro-objects by digital holography in microfluidics. IEEE Photonics Journal 4:451–454
Millet LJ, Stewart ME, Sweedler JV, Nuzzo RG, Gillette MU (2007) Microfluidic devices for culturing primary mammalian neurons at low densities. Lab Chip 7:987–994
Mills JP et al (2007) Effect of plasmodial RESA protein on deformability of human red blood cells harboring Plasmodium falciparum. P Natl Acad Sci USA 104:9213–9217. https://doi.org/10.1073/pnas.0703433104
Mir M et al (2011) Optical measurement of cycle-dependent cell growth. Proc Natl Acad Sci 108:13124–13129
Nan X, Cheng J-X, Xie XS (2003) Vibrational imaging of lipid droplets in live fibroblast cells with coherent anti-stokes Raman scattering microscopy. J Lipid Res 44:2202–2208
Park Y, Depeursinge C, Popescu G (2018a) Quantitative phase imaging in biomedicine. Nat Photonics 12:578
Park Y, Diez-Silva M, Popescu G, Lykotrafitis G, Choi W, Feld MS, Suresh S (2008) Refractive index maps and membrane dynamics of human red blood cells parasitized by Plasmodium falciparum. Proc Natl Acad Sci U S A 105:13730–13735. https://doi.org/10.1073/pnas.0806100105
Park G, Han D, Kim G, Shin S, Kim K, Park J-K, Park Y (2017b) Visualization and label-free quantification of microfluidic mixing using quantitative phase imaging. Appl Opt 56:6341–6347
Park J-H, Kong L, Zhou Y, Cui M (2017a) Large-field-of-view imaging by multi-pupil adaptive optics. Nat Methods 14:581
Park H, Lee S, Ji M, Kim K, Son Y, Jang S, Park Y (2016) Measuring cell surface area and deformability of individual human red blood cells over blood storage using quantitative phase imaging. Sci Rep 6:34257
Park Y, Popescu G, Badizadegan K, Dasari RR, Feld MS (2006) Diffraction phase and fluorescence microscopy. Opt Express 14:8263–8268

Park C, Shin S, Park Y (2018b) Generalized quantification of three-dimensional resolution in optical diffraction tomography using the projection of maximal spatial bandwidths. J Opt Soc Am A 35:1891–1898. https://doi.org/10.1364/JOSAA.35.001891
Park Y, Yamauchi T, Choi W, Dasari R, Feld MS (2009) Spectroscopic phase microscopy for quantifying hemoglobin concentrations in intact red blood cells. Opt Lett 34:3668–3670
Park J-H, Yu Z, Lee K, Lai P, Park Y (2018c) Perspective: Wavefront shaping techniques for controlling multiple light scattering in biological tissues: toward in vivo applications. APL Photonics 3:100901
Park Y et al (2010) Metabolic remodeling of the human red blood cell membrane. Proc Natl Acad Sci 107:1289
Park H et al (2015) Characterizations of individual mouse red blood cells parasitized by Babesia microti using 3-D holographic microscopy. Sci Rep 5:10827
Park C et al (2018d) Three-dimensional refractive-index distributions of individual angiosperm pollen grains. Curr Opt Photon 2:460–467
Pavillon N, Kühn J, Moratal C, Jourdain P, Depeursinge C, Magistretti PJ, Marquet P (2012) Early cell death detection with digital holographic microscopy. PLoS One 7:e30912
Popescu G (2011) Quantitative phase imaging of cells and tissues. McGraw-Hill Professional,
Popescu G, Ikeda T, Dasari RR, Feld MS (2006) Diffraction phase microscopy for quantifying cell structure and dynamics. Opt Lett 31:775–777
Popescu G, Park Y, Choi W, Dasari RR, Feld MS, Badizadegan K (2008a) Imaging red blood cell dynamics by quantitative phase microscopy blood cells. Molecules, and Diseases 41:10–16
Popescu G et al (2008b) Optical imaging of cell mass and growth dynamics. Am J Phys Cell Phys 295:C538–C544. https://doi.org/10.1152/ajpcell.00121.2008
Rappaz B, Charrière F, Depeursinge C, Magistretti PJ, Marquet P (2008) Simultaneous cell morphometry and refractive index measurement with dual-wavelength digital holographic microscopy and dye-enhanced dispersion of perfusion medium. Opt Lett 33:744–746
Rappaz B, Marquet P, Cuche E, Emery Y, Depeursinge C, Magistretti PJ (2005) Measurement of the integral refractive index and dynamic cell morphometry of living cells with digital holographic microscopy. Opt Express 13:9361–9373
Rappaz B et al (2009) Noninvasive characterization of the fission yeast cell cycle by monitoring dry mass with digital holographic microscopy. J Biomed Opt 14:034049. https://doi.org/10.1117/1.3147385
Rinehart M, Zhu Y, Wax A (2012) Quantitative phase spectroscopy. Biomed Opt Express 3:958–965
Rivenson Y, Göröcs Z, Günaydin H, Zhang Y, Wang H, Ozcan A (2017) Deep learning microscopy. Optica 4:1437–1443
Roy R, Hohng S, Ha T (2008) A practical guide to single-molecule FRET. Nat Methods 5:507–516
Schürmann M, Scholze J, Müller P, Guck J, Chan CJ (2016) Cell nuclei have lower refractive index and mass density than cytoplasm. J Biophotonics
Shaked NT (2012) Quantitative phase microscopy of biological samples using a portable interferometer. Opt Lett 37:2016–2018. https://doi.org/10.1364/OL.37.002016
Shin S, Kim D, Kim K, Park Y (2018) Super-resolution three-dimensional fluorescence and optical diffraction tomography of live cells using structured illumination generated by a digital micromirror device. Sci Rep 8:9183
Shin S, Kim K, Yoon J, Park Y (2015) Active illumination using a digital micromirror device for quantitative phase imaging. Opt Lett 40:5407–5410
Simon B, Debailleul M, Beghin A, Tourneur Y, Haeberlé O (2010) High-resolution tomographic diffractive microscopy of biological samples. J Biophotonics 3:462–467
Smith F (1955) Microscopic interferometry modern methods of. Microscopy 8:385–395
Specht EA, Braselmann E, Palmer AE (2017) A critical and comparative review of fluorescent tools for live-cell imaging. Annu Rev Physiol 79:93–117
Sridharan S, Macias V, Tangella K, Kajdacsy-Balla A, Popescu G (2014) Prediction of prostate cancer recurrence using quantitative phase imaging. Sci Rep 5:9976–9976
Su J-W et al (2015) Precancerous esophageal epithelia are associated with significantly increased scattering coefficients. Biomed Opt Express 6:3795–3805

Sung Y, Choi W, Lue N, Dasari RR, Yaqoob Z (2012) Stain-free quantification of chromosomes in live cells using regularized tomographic phase microscopy. PLoS One 7:e49502
Sung Y et al (2014) Three-dimensional holographic refractive-index measurement of continuously flowing cells in a microfluidic channel. Physical review applied 1:014002
Sung W et al (2018) Computational modeling and Clonogenic assay for Radioenhancement of gold nanoparticles using 3D live cell images. Radiat Res
Suresh S (2006) Mechanical response of human red blood cells in health and disease: some structure-property-function relationships. J Mater Res 21:1872
Thompson NL (2002) Fluorescence correlation spectroscopy. In: Topics in fluorescence spectroscopy. Springer, pp 337–378
Tougan T et al (2018) Molecular camouflage of Plasmodium falciparum Merozoites by binding of host Vitronectin to P47 fragment of SERA5. Sci Rep 8:5052
Turko NA, Peled A, Shaked NT (2013) Wide-field interferometric phase microscopy with molecular specificity using plasmonic nanoparticles. J Biomed Opt 18:111414–111414
Wang Z, Ding H, Popescu G (2011b) Scattering-phase theorem. Opt Lett 36:1215–1217
Wang Z, Tangella K, Balla A, Popescu G (2011a) Tissue refractive index as marker of disease. J Biomed Opt 16:116017–1160177
Weatherall DJ (2011) Systems biology and red cells. N Engl J Med 364:376–377
Welte MA (2009) Fat on the move: intracellular motion of lipid droplets. Biochem Soc T 37:991–996. https://doi.org/10.1042/Bst0370991
Wilfling F, Haas JT, Walther TC, Farese RV Jr (2014) Lipid droplet biogenesis. Curr Opin Cell Biol 29:39–45
Willets KA, Van Duyne RP (2007) Localized surface plasmon resonance spectroscopy and sensing. Annu Rev Phys Chem 58:267–297
Wolf E (1969) Three-dimensional structure determination of semi-transparent objects from holographic data. Opt Commun 1:153–156
Yang S-A, Yoon J, Kim K, Park Y (2016) Measurements of morphological and biochemical alterations in individual neuron cells associated with early neurotoxic effects in Parkinson's disease using optical diffraction tomography bioRxiv. https://doi.org/10.1101/080937
Yoon J, Jo Y, Kim M-H, Kim K, Lee S, Kang S-J, Park Y (2017) Label-free identification of non-activated lymphocytes using three-dimensional refractive index tomography and machine learning. Sci Rep 7:6654
Yoon J, Kim K, Park H, Choi C, Jang S, Park Y (2015) Label-free characterization of white blood cells by measuring 3D refractive index maps biomedical. Opt Express 6:3865–3875
Yu H, Jang J, Lim J, Park J-H, Jang W, Kim J-Y, Park Y (2014) Depth-enhanced 2-D optical coherence tomography using complex wavefront shaping. Opt Express 22:7514–7523
Yu H, Park J, Lee K, Yoon J, Kim K, Lee S, Park Y (2015) Recent advances in wavefront shaping techniques for biomedical applications. Curr Appl Phys 15:632–641
Yu H et al (2016) In vivo deep tissue imaging using wavefront shaping optical coherence tomography. J Biomed Opt 21:101406–101406
Zangle TA, Teitell MA (2014) Live-cell mass profiling: an emerging approach in quantitative biophysics. Nat Methods 11:1221–1228
Zernike F (1942) Phase contrast, a new method for the microscopic observation of transparent objects. Physica 9:686–698
Zernike F (1955) How I discovered phase contrast. Science 121:345–349

Photoacoustic Tomography Opening New Paradigms in Biomedical Imaging

Joon-Mo Yang and Cheol-Min Ghim

Introduction

Photoacoustic (PA) tomography (PAT), also known as optoacoustic tomography (OAT) or thermoacoustic tomography (TAT), is indeed an invaluable imaging technique that provides new opportunities in biomedical imaging (Xu and Wang 2006a; Wang 2008, 2009; Ntziachristos 2010; Beard 2011; Cox et al. 2012; Wang and Hu 2012; Mehrmohammadi et al. 2013; Yao and Wang 2014a; Nie and Chen 2014; Manohar and Razansky 2016; Weber et al. 2016; Gujrati et al. 2017; Deán-Ben et al. 2017b; Omar et al. 2019). Since its first image demonstration and seminal reports presented by Kruger et al. (Kruger 1994; Kruger and Liu 1994; Kruger et al. 1995) and Oraevsky et al. (Oraevsky et al. 1993; 1994a; Oraevsky et al. 1994b) in the mid-1990s, the technique has greatly advanced, and currently, it is awaiting full maturation as another major clinical imaging technique following conventional mainstream imaging modalities, such as ultrasound (US) imaging, X-ray computed tomography (CT), positron emission tomography (PET), and magnetic resonance imaging (MRI). This evaluation may not be an exaggeration because, among newly emerged imaging techniques after the advent of the aforementioned mainstream imaging techniques, PAT is currently the only imaging modality that permits high-resolution tomographic imaging even in depths of several centimeters. Although other types of imaging techniques (optical or not) (Denk et al. 1990; Helmchen and

J.-M. Yang (✉)
Center for Photoacoustic Medical Instruments, Department of Biomedical Engineering, Ulsan National Institute of Science and Technology, Ulsan, Republic of Korea
e-mail: jmyang@unist.ac.kr

C.-M. Ghim
Department of Physics, School of Natural Science, Ulsan National Institute of Science and Technology, Ulsan, Republic of Korea
e-mail: cmghim@unist.ac.kr

J. K. Kim et al. (eds.), *Advanced Imaging and Bio Techniques for Convergence Science*, Advances in Experimental Medicine and Biology,
https://doi.org/10.1007/978-981-33-6064-8_11

Denk 2005; Drexler and Fujimoto 2015; Boas et al. 2001; Cheney et al. 1999) have also been proposed and even their biomedical utilities have been successfully demonstrated in comparison to the aforementioned mainstream imaging techniques, their effective imaging depths—while conserving a high spatial resolution—do not match that of PAT.

In addition to the high-resolution deep imaging capability, another key feature of PAT is that it can provide a variety of different image contrasts based on the wide selectivity of the wavelength of an excitation laser beam, depending on the endogenous or exogenous contrast agents of interest. Based on this principle, PAT is capable of visualizing various anatomical structures or biological constituents, such as blood vessels (Wang 2009; Beard 2011; Wang and Hu 2012; Deán-Ben et al. 2017b; Omar et al. 2019) and cell nuclei (Yao et al. 2010a), without the aid of any contrast agent (i.e., only using a laser pulse). Moreover, a variety of physiological and functional information, such as the total hemoglobin concentration (Zhang et al. 2006a; Hu et al. 2011), oxygen saturation of hemoglobin (sO_2) (Zhang et al. 2006a; Hu et al. 2011; Yang et al. 2012a), and blood flow (Fang et al. 2007; Yao et al. 2010b), can also be imaged or measured by the endogenous contrast mechanism. In addition, by injecting an appropriate contrast agent, it is possible to produce additional exogenous contrasts, depending on the targeted biomolecule or biomarker of interest (Mehrmohammadi et al. 2013; Nie and Chen 2014; Weber et al. 2016; Gujrati et al. 2017). This wide range of degrees of freedom in contrast creation simply originates from the unique contrast mechanism that utilizes "optical absorption," which forms the core of PAT and makes it versatile. In recent years, the listed imaging capabilities of PAT have been more significantly recognized, as related systems can also be embodied in an integrated form with conventional US imaging, thereby complementing the limitations of US imaging.

After the completion of laying the cornerstones for PAT, such as in terms of the basic theory and instrumentation, major directions for the biological applications of PAT have also been roughly outlined over the previous two decades. Moreover, as has been made evident in recent multiple reports of clinical trials, several areas of clinical applications appear to be promising for their successful translations into actual practice if more specific roles or distinctive contributions are defined. For example, when it comes to biological applications, subjects related to brain imaging and cancer research are currently most popular, and when it comes to clinical applications, breast cancer diagnosis, and endoscopy or minimally invasive imaging applications are dominant in terms of the number of documented papers. Of course, on the basis of the aforementioned versatility of PAT, numerous other notable applications have also been proposed, and completely new concepts or applications that may form a new important area in the future continue to be presented.

In this chapter, we review the technology developments related to PAT in a chronological order for the following seven areas of biological and clinical applications: (1) small animal brain imaging, (2) small animal whole-body imaging, (3) cancer research, (4) breast cancer diagnosis, (5) dermatological applications, (6) endoscopy or minimally invasive imaging applications, and (7) angiography and lymphangiography, which were selected based on their dominancy in terms of the number of publications or their distinctive potentials evaluated subjectively by the authors*. Thus far, a large number of review articles on PAT have been reported (Xu and Wang 2006a; Wang 2008, 2009; Ntziachristos 2010; Beard 2011; Cox et al. 2012; Wang and Hu 2012; Mehrmohammadi et al. 2013; Yao and Wang 2014a; Nie and Chen 2014; Manohar and Razansky 2016; Weber et al. 2016; Gujrati et al. 2017; Deán-Ben et al. 2017b; Omar et al. 2019). However, by reviewing the development histories of the selected areas in chronological order, it would be possible to obtain a certain sense or knowledge with regard to how such application areas have been developed thus far, how fast they have been evolving in recent years, and how they will evolve in the future. Thus, after explaining the basic principles and key imaging features of PAT, we retrace major events achieved over the past two decades and provide rough projections on the future appearances of PAT for the selected areas.

Principles

PAT is often compared to or regarded as a relative of conventional US imaging because it also produces an image by using acoustic waves as a signal mediating medium. However, the main feature of PAT lies in the use of a pulsed laser beam or electromagnetic wave with a duration of ~a few nanoseconds, which induces PA waves within the tissue to be imaged in accordance with the PA effect (Wang 2009; Manohar and Razansky 2016). Thus, unlike conventional US imaging, in which a tomographic image is produced by sending a short pulse of acoustic waves to a target tissue and then detecting some of the acoustic waves reflected from the tissue, PAT produces an image by sending a laser pulse to a target tissue first in order to induce PA waves and then capturing some of the induced PA waves approaching the US sensor that is pre-positioned around the tissue. Due to the distinctive imaging principle, PAT creates an image with a completely different type of image contrast as compared to conventional US imaging, as illustrated in Fig. 1a.

In terms of image contrast that is yielded, US imaging is a technique that maps the magnitude or amplitude of the echo waves that bounce back from each spatial point (but, mostly at a boundary), where the reflection intensity (R_I) is approximately proportional to the square of the spatial variation of the acoustic impedance Z [kg m^{-2} s^{-1} or rayl]—which is defined as the product of tissue density ρ [kg m^{-3}] and acoustic speed c [m s^{-1}] at a particular point (Fig. 1a). Thus, any boundary where acoustic impedance shows a sharp change appears as a hyperechoic region in the acquired image. In contrast, PAT maps the initial pressure distribution p_0 [N m^{-2} or Pa] induced by a deposited laser pulse, which eventually renders information on the spatial distribution of the optical absorption; this is typically represented by the absorption coefficient μ_a [cm^{-1}] and created by the presence of light-absorbing

*This chapter is an opinionated review on the development and prospects of photoacoustic tomography with a view to pedagogical practicality. It has never meant to be comprehensive in scope or authoritative in any way. Rather, we start with fundamental concepts and principles underlying the state-of-the-art technologies, from which selected achievements are reviewed along the chronological progression of the field with an emphasis on the potential breakthrough innovations.

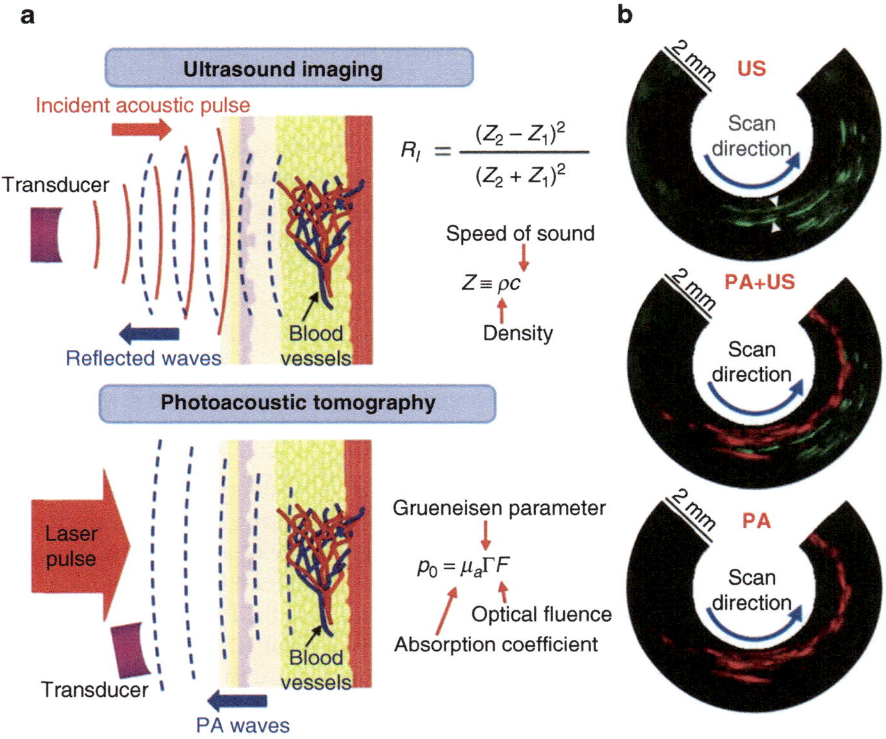

Fig. 1 Comparison of PAT and US imaging. (**a**) Principles of US imaging (upper) and PAT (lower). (**b**) A set of co-registered PA (top), US (bottom), and their combined (middle) images acquired from a rat intestine in vivo using a single-element focused US transducer-based mechanical scanning PA and US dual-modality endoscopic probe (Yang et al. 2012a). These images showcase the contrast difference between the two imaging modalities. During mechanical scanning (see the denoted scanning direction), laser pulses and acoustic pulses were fired radially in turn; subsequent response signals (i.e., PA waves and US echo waves) were captured by a single-element focused US transducer. Note that the US image (top) more clearly visualized the boundary of the rat intestine (see the white arrow heads), whose thickness was less than 500 μm, whereas the PA image visualized the blood vessels distributed inside the intestine wall

biomolecules—that is, chromophores. According to the PAT theory, the initial PA pressure distribution induced at a specific point right after a laser pulse deposition is proportional to the product of the optical fluence and absorption coefficient of the point (Wang 2009; Beard 2011; Cox et al. 2012; Manohar and Razansky 2016). Thus, if the optical fluence is known, which is occasionally assumed to be uniform over an imaged region, the spatial distribution of not only the absorption coefficient but also the gross concentration of chromophores can be recovered in image form.

Due to the explained contrast difference, if, for example, an intestine is imaged using the two techniques, US imaging would more clearly visualize the boundary profile of the intestine wall, whereas PAT would visualize the blood vessels within, as blood is a dominant light absorber (Fig. 1b).

Sending Laser Pulses to Target Tissue in Order to Generate PA Waves

When a target tissue is irradiated by a pulsed laser beam, whether it is focused (Fig. 2a) or broadly (Fig. 2b) illuminated at the surface, the photons carried in the beam are quickly diffused and de-correlated after traveling ~one transport mean free path (l_t') (typically ~1 mm) in the tissue due to the strong scattering nature of biological tissues (Welch and van Gemert 2011). Thus, if we consider the optical fluence $\Phi(r)$ [J/cm^2]—which is defined as the total number of photons (i.e., energy) passing a unit area regardless of the flow directions and over a period of time—at position (r), which is much deeper than the transport mean free path (l_t') from the tissue surface, it shows a broad distribution rather than a localized one (Figs. 2a and b) (Welch and van Gemert 2011). Such a deep region is referred to as a diffusion regime, while the shallow one less than

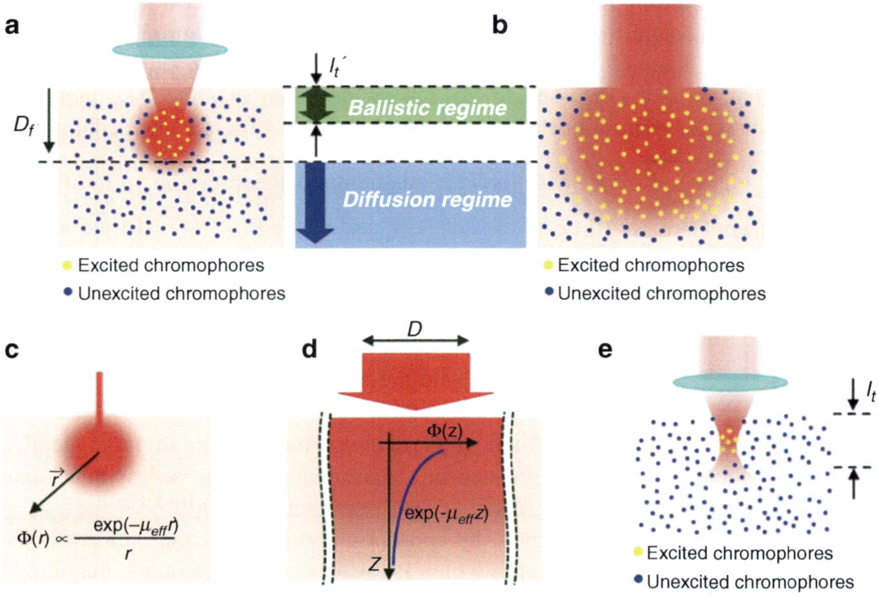

Fig. 2 Depiction of light delivery into a biological tissue and typical optical excitation mechanisms of PAT. Illustrations present the excited chromophores by a focused optical illumination (**a**) and broad optical illumination (**b**). As depicted in (**a**), if the location (D_f) of an aimed optical focus is deeper than one transport mean free path (l_t'), it is impossible to achieve a tight focus by only employing a traditional objective; this is due to the strong light scattering of biological tissues. Note that in (**a**) and (**b**), all the chromophores covered by the optical illumination (i.e., depicted in yellow) generate PA waves. In numerous AR-PAM and PACT systems, a broad optical illumination scheme as that depicted in (**b**) is typically employed. (**c**) Approximate fluence distribution ($\Phi(r)$) for an infinitely narrow beam (i.e., pencil beam), which is incident onto a semi-infinite homogenous medium. (**d**) Approximate fluence distribution ($\Phi(z)$) for a laser beam with an infinitely large beam diameter (i.e., $D \rightarrow \infty$) impinging onto a semi-infinite homogenous medium. (**e**) An illustration depicting excited chromophores (i.e., depicted in yellow) within the focal spot of a focused laser beam. This illumination scheme is typically utilized for OR-PAM

$l_t^{'}$ is referred to as a ballistic regime. In PAT, this implies that all and only the light absorbers included in the illumination zone can generate PA signals in proportion to the optical fluence ($\Phi(r)$) (Fig. 2). Of course, although the PA signal generation in a diffusion regime cannot be localized, PAT can still achieve adequate spatial resolution (the method for achieving this resolution is explained later).

Here, the prediction of exact optical fluence distribution at large depths is not simple if the tissue is complex (i.e., inhomogeneous), and it becomes tougher as the depth of the point under consideration increases because the total number of possible paths of photons that reach the point also increases proportionally. Applying a Monte Carlo simulation (Welch and van Gemert 2011) would be the best option for the prediction, although it is necessary to have a set of data on the optical properties of the tissue. Otherwise, one can use a diffusion approximation (Welch and van Gemert 2011) if the area of interest is located much deeper than the ballistic regime; moreover, the optical properties of the tissue can be assumed to be macroscopically homogenous. According to the diffusion theory, the optical fluence ($\Phi(r)$) at large depths ($>> l_t^{'}$) in a semi-infinite homogenous medium for the illumination with an infinitely narrow beam (Fig. 2c) roughly follows the distribution expressed in Eq. (1); it approaches that expressed in Eq. (2) as the beam diameter (D) increases (Fig. 2d), where μ_{eff} represents the effective attenuation coefficient and is defined as $\mu_{eff} \equiv \sqrt{3\mu_a\left(\mu_a + \mu_s^{'}\right)}$ (for more precise mathematical expression, refer to ref. (Welch and van Gemert 2011).

$$\Phi\left(r\right) \propto \frac{\exp\left(-\mu_{eff} r\right)}{r} \tag{1}$$

$$\Phi\left(z\right) \propto \exp\left(-\mu_{eff} z\right) \tag{2}$$

Thus far, we have discussed the light delivery mechanism in large depths. However, if the targeted area is located in a shallow area—that is, less than ~one transport mean free path—tight focusing of a laser beam can still be achieved, as illustrated in Figure 2e. In this case, by appropriately adjusting the incident pulse energy, one can create a condition in which PA waves with a detectable magnitude are only induced within the focal spot where the optical fluence is the highest. In fact, a related concept is applied to optical resolution PA microscopy (OR-PAM) (Maslov et al. 2008), which is explained later, to achieve a typical optical microscopy-level lateral resolution.

Detecting PA Waves to Produce an Image

Once a short laser pulse is absorbed by chromophores, it causes a temperature rise on account of instant heating, and the thermal energy is converted into PA waves that propagate omni-directionally rather than travel toward a specific direction (e.g., the original laser beam propagation direction); this is because monopole radiation is dominant during the generation of PA waves (Wang 2009). According to the PAT theory, the center frequency of the generated PA waves is inversely proportional to the time duration of the deposited laser pulse. Thus, the shorter the duration, the wider and higher the bandwidth and frequency (refers to ref. (Wang 2009) for the mathematical formalism). Moreover, the propagation of the PA waves can be mathematical modeled by the same wave equation that is typically utilized for US imaging, in which the speed of sound for soft tissues is usually assumed to be constant (~1.55 mm/μs at room temperature). Then, some of the PA waves that approach the US transducer placed near the target object are captured by the transducer and converted into an image.

Due to the involvement of the conventional US technique, although the irradiated laser beam does not focus on a deep tissue region, PAT can still achieve an adequate spatial resolution by ultrasonically resolving the locations of PA signal sources. Of course, considering the time duration of the deposited laser pulse, which is typically in a nanosecond range, the initial PA waves generated immediately after the optical excitation include broad frequency components (sub to hundreds MHz). However, in reality, only a partial frequency range is detected by the employed transducer due to its limited bandwidth characteristics as well as the frequency-dependent acoustic attenuation during wave propagation. Hereafter, several methods for localizing the origins of PA signals are explained, with illustrations in Fig. 3.

The first method is to use a single-element US transducer, in which any type of transducer—that is, either focused or unfocused—could be employed. However, in order to avoid a complex image reconstruction process, like the computed tomographic reconstruction, a focused US transducer is more frequently adopted in numerous PAT systems. For example, although the imaged area is located much deeper than the ballistic regime, such a focused US transducer can easily achieve an adequate lateral resolution by excluding all the PA signals approaching from its off-axial points, which can be attributed to the benefit of the geometrically curved surface (Fig. 3a). Consequently, in this case, the lateral resolution is entirely determined by the acoustic parameters of the employed US transducer and it is known to be inversely proportional to the product of the center frequency and acoustic numerical aperture (NA).

On the other hand, if the imaged area is limited to the ballistic regime (i.e., less than ~1 mm), it is possible to induce PA waves only within the focal spot of a laser beam by appropriately adjusting its pulse energy, as previously mentioned (Fig. 2e). In this case, irrespective of the employed acoustic beam diameter of a US transducer, which is usually wider than that of the focused laser beam, the lateral resolution of the related PAT system is determined by the beam diameter of the optical focus rather than the acoustic beam diameter of the employed transducer (Fig. 3b). Consequently, the optical parameters of the illumination optics, such as the wavelength and optical NA, determine the lateral resolution of the system and, thus, a typical optical microscopy-level lateral resolution (i.e., ~a few micrometers) is achievable by applying this principle.

In fact, the two imaging concepts were applied for the first time in the acoustic-resolution PA microscopy (AR-PAM) and OR-PAM systems reported by Maslov

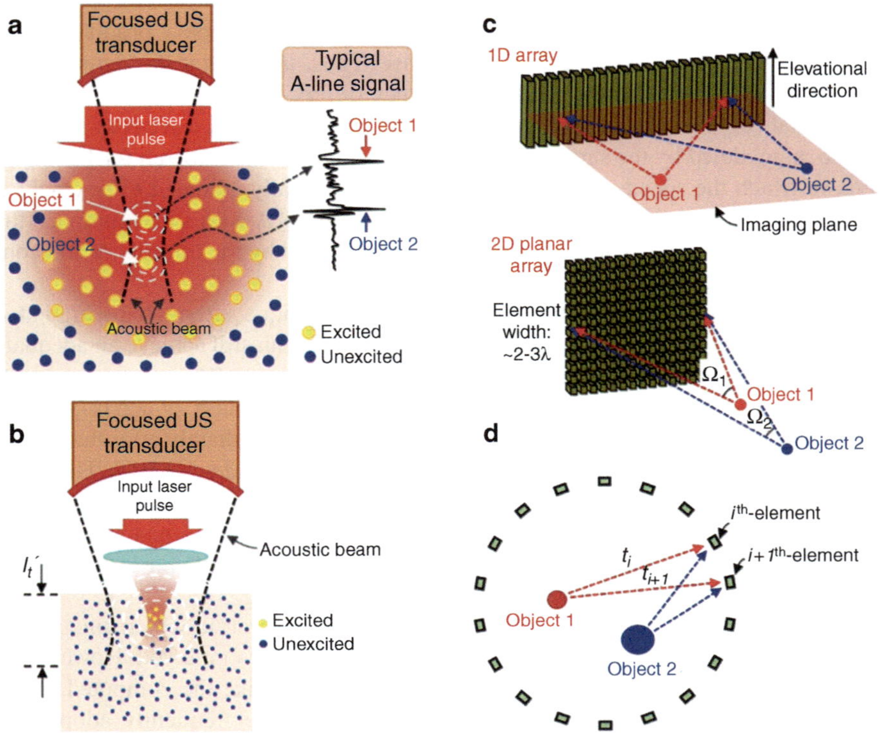

Fig. 3 Typical US signal detection mechanisms in PAT. (**a**) Single-element focused US transducer-based signal detection mechanism typically utilized in AR-PAM. Note that although all the chromophores within the optical illumination zone generate PA signals, only those chromophores located within the acoustic beam axis and the acoustic depth of focus exert a detectable pressure to the US transducer. (**b**) A single-element focused US transducer-based signal detection mechanism typically utilized in OR-PAM. Note that only the chromophores within the optical focus, which is much narrower than the waist of the acoustic beam, generate PA signals, and the optical and acoustic foci must overlap with each other. (**c**) 1D (upper) and 2D (lower) planar array transducer-based signal detection mechanisms, in which optical excitation laser beams are not depicted. Note that Ω_1 is larger than Ω_2. (**d**) Ring array transducer-based signal detection mechanism. t_i denotes the arrival time of the pressure wave originating from Object 1 to i^{th}-element. Note that t_{i+1} is greater than t_i. The depicted signal detection configuration can also be achieved by performing a circular mechanical scanning using a single-element US transducer

et al. of the Wang group in 2005 (Maslov et al. 2005) and 2008 (Maslov et al. 2008), respectively. However, unfortunately, irrespective of whether the optical focusing is employed, the axial resolutions of the AR- and OR-PAM are both determined by the acoustic parameters of an employed US transducer, as the performances are achieved based on the acquired time-resolved signals (see the typical A-line signal presented in Figure 3a). Thus, the axial resolution has an inverse relationship with the product of the center frequency and bandwidth of an employed US transducer—that is, the greater the product, the better the axial resolution.

Although the use of such a focused US transducer can benefit the lateral resolving power, in order to produce an image, both the OR- and the AR-PAM require a mechanical scanning mechanism—for example, raster scanning, because the US transducer can detect only the one-dimensional (i.e., A-line) signal that is localized along the acoustic axis. Consequently, the data acquisition (DAQ) process must be repeated by changing the location of signal detection after each laser pulse firing, which is a time-consuming process. Nevertheless, numerous OR- and AR-PAM systems still adopt the single-element focused US transducer-based mechanical scanning mechanism because they can be implemented more economically compared to an array transducer-based detection mechanism, which is explained next.

The second method for spatially resolving the origins of detected PA signals is to use an array transducer in which a number of piezoelectric elements or its equivalent are arranged in a 1D linear or 2D planar fashion (Fig. 3c). The key advantage of this mechanism is that an array transducer can capture multiple A-line signals simultaneously after firing a single laser pulse only. This implies that it is possible to acquire a two- or three-dimensional image—which is only determined by the spatial configuration of an employed array transducer—at every laser pulse firing process, thereby requiring no additional mechanical scanning process. Consequently, the frame rate for a 2D cross-sectional or 3D volumetric imaging is usually limited by the pulse repetition rate (PRR) of an employed laser system. Nevertheless, it can virtually reach or even exceed the real-time imaging speed of 30 Hz, although the related system cost increases proportionally to the number of channels due to the high cost of the array transducer and parallel DAQ system.

Here, it must be noted that, although a single-element US transducer is mostly employed in a mechanical scanning PAT system, if the employed transducer is unfocused and if its aperture has a size that is slightly larger than the center wavelength of the transducer, the recorded PA data set at multiple different locations renders an almost equivalent effect as that in the case of employing an array transducer with the same center wavelength. Of course, in this discussion, we ignore the fact that the mechanical scanning requires a much larger number of laser pulses as well as the corresponding amount of scanning time. Nevertheless, there is an equivalency because an array transducer could be understood as the spatial iteration of the same type of single-element US transducer. In fact, Wang et al. demonstrated the first in vivo PA imaging for a rat brain by applying a similar concept in 2003 (Wang et al. 2003), even though an array transducer-based PAT system was not available at that time. Figure 3d illustrates the geometric configuration of a ring-type array transducer, although an equivalent data set also could be acquired using a single-element transducer in accordance with the method explained above.

However, irrespective of the method utilized, a sophisticated image reconstruction algorithm must be applied to resolve objects in the imaged area, because each A-line signal is captured at a position by an element in an array transducer or by an unfocused US transducer does not include any information on its angular origin. Only the arrival time of each spike is resolvable in a recorded 1D data set (i.e., no resolving power along a transverse direction because all signals are mixed into a 1D signal). Nevertheless, by employing such an image reconstruction algorithm, an adequate lateral resolution can be achieved, because the presence of an object in a reconstructed image is eventually unveiled by the cumulative projection effect of the PA signals included in a corresponding time point in the A-line signal acquired at each detection position. Note that, as depicted in Figure 3c and d, the arrival time of PA waves originating from a point object (e.g., Object 1) to each transducer element shows a gradual variance due to the corresponding variation of the distance between them (in relation to this principle, it is helpful to refer to ref. (Beard 2011).

In PAT, such an image reconstruction method is usually referred to as a back projection algorithm, which forms the theoretical basis of PA-computed tomography (PACT). Note that this principle is similar to the well-known beam forming process of a phased array transducer in conventional US imaging. In addition to the back projection, other image reconstruction algorithms or concepts—such as time-reversal reconstruction and synthetic focusing (Wang 2009)—are also presented, although the underlying principles are the same.

In the case of a PACT system that employs 1D linear or 2D planar array transducers, as depicted in Figure 3c, it must also be noted that the lateral resolution is usually degraded as the target distance increases because the solid angle subtended by a point object to the entire array elements decreases: in other words, the effective acoustic NA created by the synthetic aperture decreases. The related issue is known as the "limited-view problem," as with the case of conventional US imaging; thus, most of the array transducers with such a limited span area suffer from a similar issue. Thus, in order to achieve a better lateral resolution, the effective acoustic NA must be maximized by increasing the span dimension and the number of array elements, although it is not always possible due to the allowed probe dimension and the correspondingly increasing device cost.

On the other hand, if PA signals are captured by such a ring array transducer or its equivalent—such as the circular-scan 2D PACT imaging, as illustrated in Figure 3d—there is no limited-view problem at least within the detection plane; thus, the in-plane resolution can be almost constant over the imaged area (Li et al. 2017). In this case, like the axial resolution of a single-element US transducer, it approximately has an inverse relationship to the product of the center frequency and bandwidth of the array transducer.

Thus far, we have explained the general principles of PAT. However, before moving to the main part, we briefly review the key imaging features and attainable performances of PAT in terms of the five technical elements—that is, the contrast, imaging depth, resolution, imaging speed, and safety—all of which are importantly evaluated as core factors for an imaging technique to be accepted as a viable biomedical imaging technique.

Contrast

The contrast of an imaging technique is the aspect that represents what kind of image information can be provided by the technique and also accounts for the reason why the technique is necessary as another imaging modality in biomedical imaging. In such an aspect, PAT has a great versatility based on the unique contrast mechanism of the optical absorption and due to which huge possibilities in creating a variety of different image contrasts are provided. In conventional US imaging, for example, available contrast mechanisms are almost fixed and they all basically rely on the reflectivity of acoustic waves. Thus, although the frequency (or wavelength) of an acoustic pulse changes, it does not affect the produced image contrast but only the spatial resolution and penetration depth. However, in PAT, the change in frequency (i.e., wavelength) of a laser beam implies the creation of a different image contrast.

In biological tissues, there are a large number of biomolecules (i.e., chromophores) that absorb light and have their own absorption spectra with different spectral behavior. Figure 4a represents absorption spectra of related examples, such as oxy- or deoxy-hemoglobin (i.e., HbO_2, HbR), DNA and RNA, melanin, lipid, bilirubin, myoglobin (Mb), and water. Thus, if one selects an appropriate laser wavelength considering the absorption peaks of a specific chromophore of interest, it is possible to selectively visualize the spatial distribution of the chromophores by using only the laser pulse. Indeed, this is the key principle spanning the endogenous contrast mechanism of PAT; thus far, actual imaging demonstrations for the mentioned chromophores and other biological constituents, such as methemoglobin, cytochrome *c*, and glucose, have been presented in multiple papers (Yao et al. 2010a; Zhang et al. 2006a, 2011, 2013; Hu et al. 2011; Yang et al. 2012a; Viator et al. 2004; Allen et al. 2012; Wilson et al. 2014; Zhou et al. 2012; Lin et al. 2016; Xu et al. 2010; Tang et al. 2015a; Kottmann et al. 2012; Pleitez et al. 2013).

Notably, since hemoglobin is the core composition of a blood vessel, its presence is usually revealed as a form of vasculature image when a PA imaging is performed for in vivo biological tissue. Moreover, if one harnesses the different absorption properties of oxy- and deoxy-hemoglobin (Fig. 4a), it is also possible to acquire a map of oxygen saturation of hemoglobin (sO_2) based on multiwavelength imaging (Fig. 4b), which constitutes the core of PAT functional imaging. In a similar manner, there have also been reports that demonstrated the visualization of other anatomic structures or biological constituents, such as cell nuclei (Yao et al. 2010a), tissue hydration (Xu et al. 2010), lipid (Allen et al. 2012; Wilson et al. 2014), and nerve (Matthews et al. 2014; Li et al. 2016). In addition, it is also possible to acquire additional physiological and functional information such as blood flow (Fang et al. 2007; Yao et al. 2010b), temperature (Larina et al. 2005; Shah et al. 2008; Pramanik and Wang 2009; Landa et al. 2019), and metabolism (Wang 2008; Hu et al. 2011; Pleitez et al. 2020; Ntziachristos et al. 2019).

The above examples are cases that are all achievable by the endogenous contrast mechanism. However, by using (i.e., administrating) an appropriate contrast agent to target or label the specific molecule or biomarker of interest (Mehrmohammadi et al. 2013; Nie and Chen 2014; Weber et al. 2016; Gujrati et al. 2017), it is also possible to produce additional exogenous image contrasts for numerous important anatomical structures and markers, such as the lymphatic system (Yang et al. 2012a; Song et al. 2008), amyloid-β (Hu et al. 2009b), integrin $\alpha_v\beta_3$ (De la Zerda et al. 2008), epidermal growth factor receptor (EGFR) (Hudson et al. 2014), and reporter

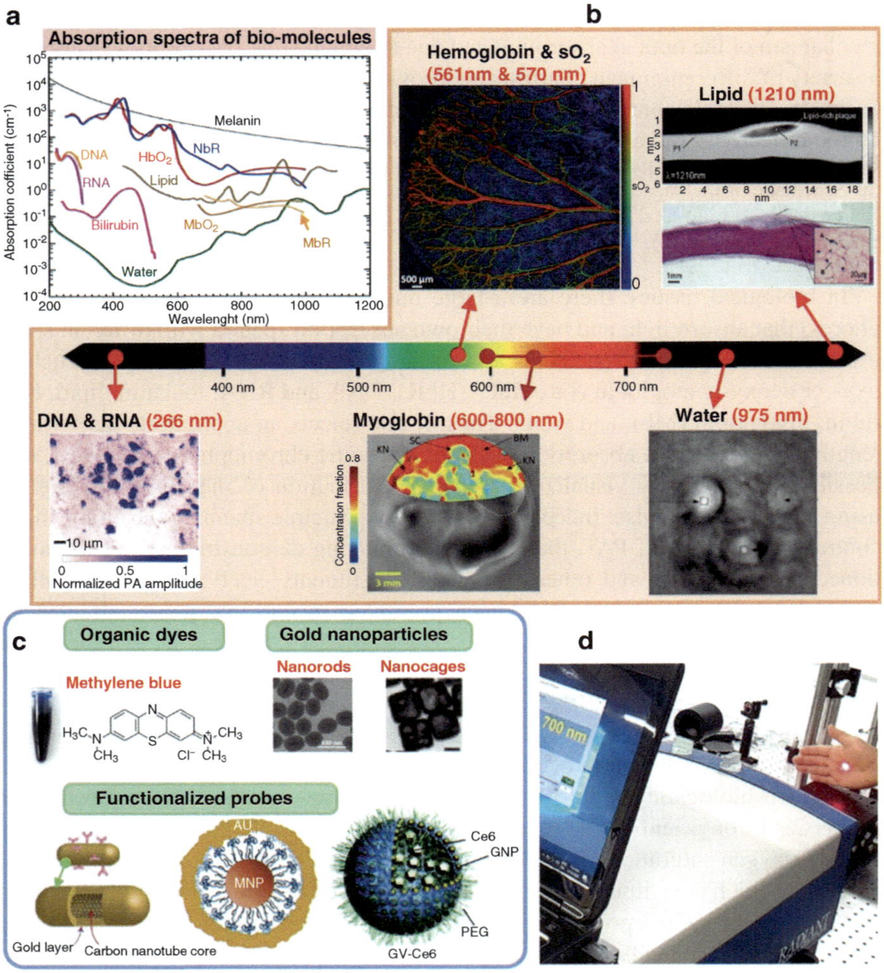

Fig. 4 Examples of available image contrasts in PAT. (**a**) Absorption spectra of various chromophores (Yao and Wang 2014a). (**b**) Examples of PAT images acquired based on the endogenous contrast mechanism (Yao et al. 2010a; Hu et al. 2011; Allen et al. 2012; Lin et al. 2016; Xu et al. 2010). The hemoglobin and sO_2 maximum amplitude projection (MAP) image were acquired from a mouse ear using dual-wavelength laser beams. (**c**) Examples of contrast agents applicable for PAT to produce exogenous image contrasts (Chen et al. 2011; Yang et al. 2007; Kim et al. 2009; Jin et al. 2010; Lin et al. 2013). Note that the presented images are very few examples of PA contrast agents developed thus far (for more detailed information, refer to refs. (Mehrmohammadi et al. 2013; Nie and Chen 2014; Weber et al. 2016; Gujrati et al. 2017). (**d**) A photo showing a human hand directly exposed to a 700-nm laser beam with an energy of 20 mJ and a beam diameter of ~6.5 mm (operating at the PRR of 20 Hz), which yields an approximate optical fluence of 60 mJ/cm^2 for a single laser pulse on the surface of the skin (note that although this value is three times greater than that of the ANSI safety limit, there was no damage to the skin)

genes (Brunker et al. 2017). Thus far, a variety of contrast agents or molecular probes have been developed for the purpose (Mehrmohammadi et al. 2013; Nie and Chen 2014; Weber et al. 2016; Gujrati et al. 2017), and Figure 4c presents a few examples of related contrast agents, which can be roughly classified as organic dyes, nanoparticles (Chen et al. 2011; Yang et al. 2007), and functionalized probes (Kim et al. 2009; Jin et al. 2010; Lin et al. 2013; Huynh et al. 2015). Related to the physiological or functional imaging of PAT, there were also reports that demonstrated a pH mapping in phantoms (Chatni et al. 2011) and tumors (Jo et al. 2017) by using a pH-sensitive fluorescent dye (SNARF-5F carboxylic acid). Moreover, in 2017, Knox et al. of the Chan group presented a novel N-oxide-based probe (named HyP-1) developed for PAT imaging of hypoxia in ischemia and cancer (Knox et al. 2017).

Of course, it might be a general situation that there are several different types of chromophores that are distributed over a region of interest rather than only a single type that exists independently. Although such a situation occurs, PAT can differentiate the spatial distribution of each chromophore by applying the spectral unmixing technique in which multiple wavelength imaging is performed to separate out the necessary information only. As presented, PAT has a wide range of freedom in creating contrasts, which must be importantly evaluated as the core aspect of how PAT can complement conventional US imaging if the two techniques are combined in a single imaging system.

Imaging Depth

Adding to the unique contrast mechanism, PAT's deep imaging capability is another important aspect that highlights its technical value. Although PAT is openly classified as an optical imaging modality owing to the use of a pulsed laser illumination, its imaging depth can reach up to several centimeters, thereby overtaking the well-known, ballistic optical imaging modalities—such as confocal microscopy (CFM) or two-photon microscopy (TPM) (Denk et al. 1990; Helmchen and Denk 2005) and optical coherence tomography (OCT) (Drexler and Fujimoto 2015)—by also being able to employ diffused light as an energy deliverer to deep tissues.

Thus, if deep imaging is the major goal of a PAT system, like the case of a PACT system, more care must be given to achieve an effective light energy delivery to deep tissues because only those chromophores covered by an optical illumination can generate PA waves, as explained earlier (Figs. 2a and b). Of course, employing a US transducer with a better sensitivity is also important. However, the maximum imaging depth of a PAT system is more affected by the light illumination parameters because, due to the highly scattering and light-absorbing nature of biological tissue, the optical fluence distribution ($\Phi(r)$) of photons within a biological tissue shows a sharper decay than the amplitude of the acoustic waves propagating in biological tissue. For example, acoustic waves with a center frequency of 3 MHz have an acoustic penetration depth of approximately 2.9 cm for soft tissue, like muscle, whereas photons with a wavelength of 720 nm, at which the penetration is highest, have an optical penetration depth of only 0.57 cm (here, the penetration depth was defined as $1/\mu_{eff}$).

Of course, a deeper imaging could be achieved by exciting a target surface with a higher laser pulse energy. However, the achievable maximum imaging depth does not have a simple linear relationship to the input energy; moreover, the associated safety issue, which is presented later, must be considered. Thus far, a penetration depth of ~4 cm has been demonstrated in actual female breasts and a male prostate in vivo by broadly illuminating a target region while regulating the maximum permissible exposure (MPE)—that is, 20 mJ/cm^2 (Kruger et al. 2010, 2013; Lin et al. 2018; Kothapalli et al. 2019).

Resolution

As explained earlier, spatial resolutions of typical AR-PAM and PACT are mostly determined by the three acoustic parameters—that is, center frequency, bandwidth, and acoustic NA (or viewing angle)—of an employed US transducer because the two techniques employ a broad optical illumination to achieve deep imaging. In accordance with the principle, the first AR-PAM system achieved an axial resolution of ~15 μm and a lateral resolution of ~45 μm by employing a 50-MHz transducer with an acoustic NA of 0.44 and a fractional bandwidth of 70% (Maslov et al. 2005), which provides a rough estimation for the resolution performance of other AR-PAM systems. In the case of a circular-scan PACT system, implemented based on either a single-element transducer or ring array transducer, its in-plane resolution exhibits an isotropic and quasi-uniform performance (Li et al. 2017; Razansky et al. 2011), provided that the total viewing angle created by an entire scan is greater than ~180° (Razansky et al. 2011) and is typically at a hundreds-of-micrometers level.

The lateral resolution of a typical OR-PAM system is in the range of sub-microns to a few microns, whereas the axial resolution has a larger value, like a few ten microns (in general, >15 μm) (Hu et al. 2011; Maslov et al. 2008; Zhang et al. 2010). Thus, in order to overcome this limitation, several new concepts—such as the use of an even higher frequency US transducer (e.g., 125 MHz) (Zhang et al. 2012) and the capitalization of the Grueneisen effect (Wang et al. 2014a; Ma et al. 2016)—have been proposed since 2012, which eventually further dragged down the axial resolution to 2.3–12.5 μm. Similarly, in order to push the lateral resolution limit of OR-PAM, a few studies considered the nonlinear effects in the PA wave generation (Danielli et al. 2014; Yao et al. 2014) or harnessed the Grueneisen effect along with the use of dual-laser pulses with different wavelengths, etc. (Shi et al. 2019).

Moreover, since 2015, multiple reports have been presented to achieve a PA super-resolution imaging even in an optical diffusion regime, in which the spatial resolution is known to be limited by the acoustic diffraction of an employed US transducer (Conkey et al. 2015; Chaigne et al. 2016; Chaigne et al. 2017; Murray et al. 2017; Vilov et al. 2017; Deán-Ben and Razansky 2018). Eventually, sub-acoustic spatial resolutions were achieved by applying the concepts of exploiting a PA feedback for wave front optimization (Conkey et al. 2015), multiple speckle illumination (Chaigne et al. 2016), temporal PA signal fluctuations caused by the dynamics of flowing absorbers (Chaigne et al. 2017), optical speckle pattern-aided structured illumination (Murray et al. 2017), PA signal localization based on injected microbeads (Vilov et al. 2017; Deán-Ben and Razansky 2018), etc.

Imaging Speed

As with conventional US imaging, PAT is classified as a real-time imaging modality. Although a single-element US transducer-based mechanical scanning PAT system requires a relatively longer scanning time due to the involvement of a point or raster scanning mechanism, an array transducer-based state-of-the-art PAT system—in which a simultaneous signal capturing is possible—can easily achieve a real time (~30 Hz) or even higher imaging speed provided that an employed laser system permits corresponding PRR (Li et al. 2017; Deán-Ben et al. 2017a). For example, in a report presented by Deán-Ben et al. of the Razansky group in 2017 (Deán-Ben et al. 2017a), a 100-Hz frame rate was achieved even for performing volumetric imaging by employing a 510-element hemispherical matrix array transducer.

In the case of OR-PAM, although there were a few exceptions (Song et al. 2011; Li et al. 2013; Xia et al. 2013), most related embodiments have been embodied based on single-element US transducer-based mechanical scanning mechanisms, which inherently require a relatively longer scanning time than that of an AR-PAM system, as it requires much denser scanning steps. In fact, in the first OR-PAM system reported in 2008 (Maslov et al. 2008), in which a stepper motor was employed, its B-scan imaging could be performed only at ~1.3 Hz for a 1-mm scan range with a step size of 1.25 μm. However, in the demonstration presented in ref. (Lan et al. 2018), a B-scan imaging speed as high as 900 Hz was achieved for a much wider scan range (12 mm) by utilizing a fast rotating hexagonal mirror-based mechanical scanning mechanism.

As shown, the imaging speed of a mechanical scanning PAM system is mostly affected by the employed scanning mechanism; thus, the speed could be further increased by employing a more advanced scanning mechanism. However, it must be noted that there is a fundamental limitation on the maximum imaging speed. Unlike pure optical imaging techniques, PAT requires a much longer gate time to capture all the PA signals necessary to cover a desired imaging depth. Note that the speed of sound in biological tissues is merely ~1.55 mm/μm, whereas the tissue penetration time of a fired laser beam is almost negligible. Consequently, if the distance between the detector and a tissue sample increases or the total depth to be imaged increases, a corresponding amount of time is required to wait for all the arrivals of necessary signals. Nevertheless, it is true that the theoretical imaging speed of PAT could be approximately twice as fast than that of US imaging because, unlike US imaging, PAT only involves one-wave propagation of acoustic waves.

Safety

In the inception of PAT, there was a concern regarding the safety issue, as PAT utilizes a nanosecond level pulsed laser radiation in which its instant peak power is extremely high, although its average power is within a reasonable range. In fact, one may also encounter a situation in which a laser beam emitted from a laser head generates a strong sound when it directly hits the surface of a metallic object, such as optical table. Although the sound is rather frightening, like that of linagraph paper, it is not a general case but a rather exceptional one, because metal is a conductive material in which numerous free electrons are included and the incident laser beam is coherent and polarized. Thus, unlike such a conductive material, no damage occurs when it is incidental on a biological tissue surface, as it is closer to a dielectric material than a conductive material; moreover, a laser beam incident on tissue is quickly diffused and de-correlated, losing its original polarization after traveling a few transport mean free paths. Figure 4d presents a human hand directly exposed to an unexpanded 760-nm laser beam with a single pulse energy of 20 mJ and beam diameter of 6 mm (operating at 20 Hz PRR). More importantly, it must also be noted that the energy of a single photon in the visible range, where PAT is typically performed, is in the range 1.6–3.3 eV, which is much less than that of X-ray radiation and, thus, makes PAT fall into a nonionizing imaging modality.

In conclusion, if the intensity of the laser beam is well regulated within a safety limit, it does not cause any tissue damage. Currently, related safety for a nanosecond laser is guided by the ANSI standards in terms of the MPE and the mean irradiance (Laser Institute of America 2007). Thus, in the spectral region of 400–700 nm, the MPE on the tissue surface must not exceed 20 mJ/cm^2, whereas it becomes looser as the wavelength increases and can go up to even 100 mJ/cm^2 at 1050 nm. In addition, if the same area is exposed to a laser beam for over 10 s, the mean irradiance must not exceed 200 mW/cm^2 in the visible and 1 W/cm^2 in the NIR. Nevertheless, complying with the safety guideline, a PA signal detection even from a depth of ~4 cm has been demonstrated in an actual human subject (Kruger et al. 2010; 2013; Lin et al. 2018; Kothapalli et al. 2019), as previously mentioned.

Biological Applications

Reviewing the initial papers of PAT, it is evident that the PAT technology originally emerged to mainly target several clinical applications, such as breast imaging, dermatological applications, and endoscopy (Oraevsky et al. 1994b; Esenaliev et al. 1997; Oraevsky et al. 1997). Although the related research outcomes reported to date have been very successful, most of them are still ongoing, whereas PAT has already successfully demonstrated its utility in various biological applications, as many of the systems developed for these applications were already commercialized. Thus, in this section, we introduce the development histories of PAT biological applications in three main subjects: (1) brain imaging, (2) small animal whole-body imaging, and (3) cancer research.

Brain Imaging

The brain imaging application of PAT was the subject where the in vivo imaging capability of PAT was first demonstrated in 2003 by Wang et al. of the Wang group (Wang et al. 2003). Currently, the related research activities account for one of the most active and rapidly advancing research subjects of PAT. Although the earlier works reported in the mid-1990s by Kruger et al. and Oraevsky et al. demonstrated the feasibility of PAT's in vivo imaging, their studies were limited to phantom or ex vivo animal studies (Kruger 1994; Kruger and Liu 1994; Kruger et al. 1995; Oraevsky et al. 1993; Oraevsky et al. 1994a; Oraevsky et al. 1994b). However, after the in vivo imaging capability of PAT was clearly demonstrated (Wang et al. 2003), the entire area of PAT, as well as the brain imaging application, has begun to grow in earnest. In addition, the report presented several prototypical experimental demonstrations that suggested that PAT could be further developed as a new tool for both functional and structural imaging.

Figure 5a depicts the system that they utilized for the experiment. They guided a 532-nm laser beam from a Q-switched Nd:YAG laser to the central zone of a rat brain, where the scanning axis of a single-element US transducer (3.5 MHz) was also set to perform a rotational scanning around the animal head. Thus, after the firing of each laser pulse, the transducer detected PA waves at a certain distance apart from the brain and achieved a full 360° scanning that included 240 steps of the employed stepper motor (~16 min was required for one full scanning due to a signal averaging at the given 10-Hz pulse PRR of the employed laser system).

By performing a circular mechanical scanning under the orthogonal configuration between the laser illumination and acoustic detection, they could acquire the first in vivo image of a rat brain. In particular, after applying physical stimulations to left and right whisker of a rat, they successfully mapped the corresponding hemodynamic responses of the rat brain via the intact scalp and skull, as depicted in Figure 5b. At that time, the removal of both or at least one of the scalp and skull layers was typical in numerous optical microscopies. Moreover, they were also able to capture a blood volume increase in accordance with the alteration of the oxygen level from hypoxia to hyperoxia in the inhaled gas.

Although the experiments were performed in a water tank to provide acoustic matching between the brain and US transducer, the study was the first case that demonstrated the functional brain imaging capability of PAT by applying the circular-scanning PACT concept. After the in vivo demonstration, the same PACT concept presented in Figure 5a, or its modified versions, were continuously applied to most of the PAT brain imaging studies reported by the Wang group until ~2010, except one case that imaged a Rhesus monkey brain ex vivo by replacing the Q-switched laser source with a 3-GHz microwave source, as reported by Y. Xu et al. in 2006 (Xu and Wang 2006b). The following account describes several notable outcomes reported in the period.

In 2004, the first in vivo imaging demonstrator, Wang et al. presented another report that applied several exogenous contrast agents, such as indocyanine green (ICG) polyethylene glycol (Wang et al. 2004b), and Y. Wang et al. applied

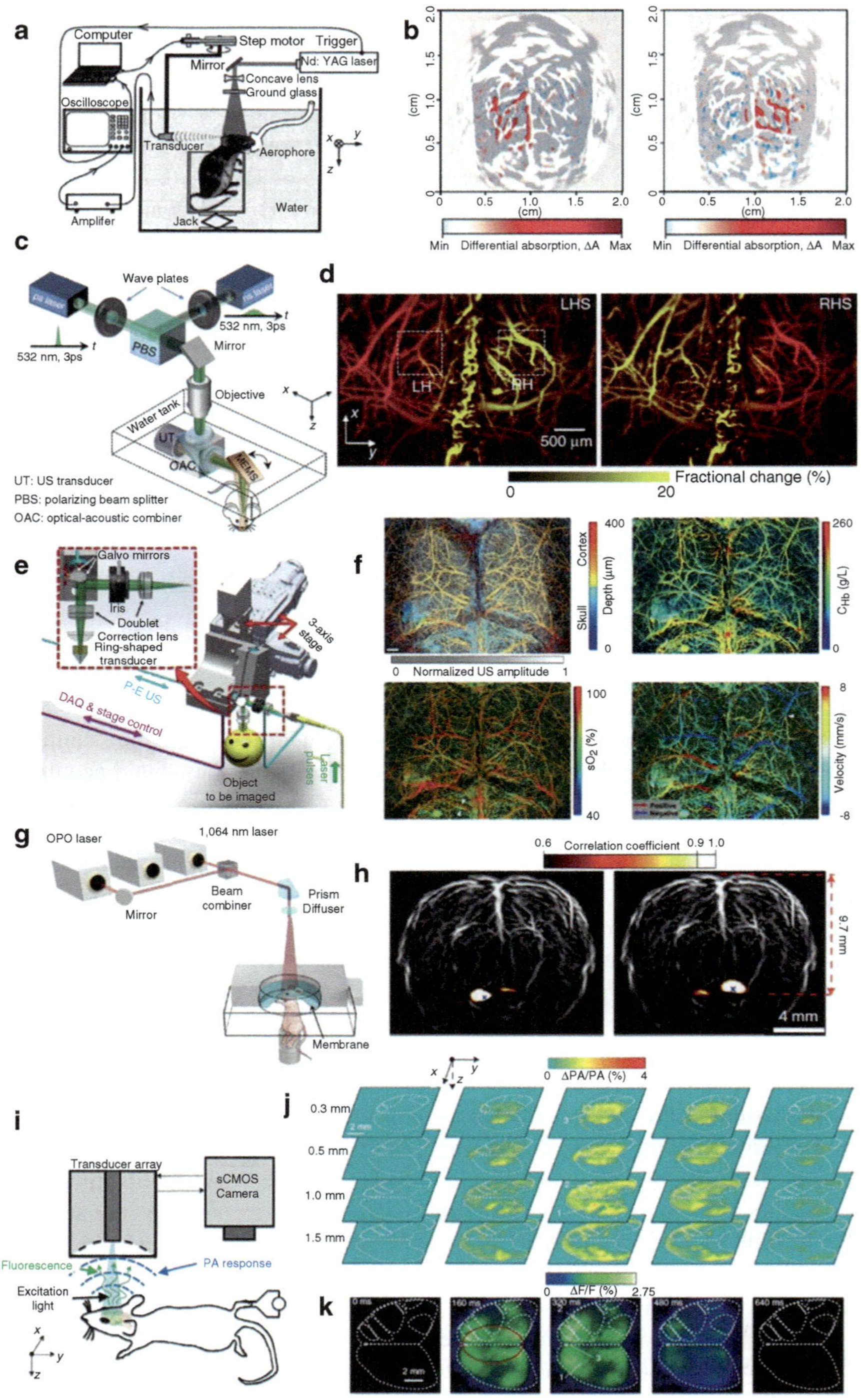
a
Computer
Step motor
Trigger
Nd: YAG laser
Mirror
Concave lens
Ground glass
Oscilloscope
Transducer
Aerophore
Amplifer
Jack
Water
b
(cm)
Min Differential absorption, ΔA Max
c
Wave plates
532 nm, 3ps
PBS
Mirror
Objective
Water tank
UT
OAC
MEMS
UT: US transducer
PBS: polarizing beam splitter
OAC: optical-acoustic combiner
d
LHS
RHS
LH
RH
500 μm
Fractional change (%)
0
20
e
Galvo mirrors
Iris
Doublet
Correction lens
Ring-shaped transducer
3-axis stage
P-E US
DAQ & stage control
Object to be imaged
Laser pulses
f
Cortex
Skull
Depth (μm)
400
0
Normalized US amplitude
C_{Hb} (g/L)
260
sO_2 (%)
100
40
Velocity (mm/s)
8
-8
g
OPO laser
1,064 nm laser
Mirror
Beam combiner
Prism
Diffuser
Membrane
h
0.6 Correlation coefficient 0.9 1.0
9.7 mm
4 mm
i
Transducer array
sCMOS Camera
Fluorescence
PA response
Excitation light
j
0 ΔPA/PA (%) 4
0.3 mm
0.5 mm
1.0 mm
1.5 mm
2 mm
k
0 ΔF/F (%) 2.75
0 ms
160 ms
320 ms
480 ms
640 ms
2 mm

Fig. 5 PAT small animal brain imaging systems and produced images. (**a**, **b**) A circular-scan PAT system (a) for transdermal and transcranial rat brain imaging and corresponding PACT images (**b**) depicting the cerebral hemodynamic changes in response to left- and right-side whisker stimulation (Wang et al. 2003). (**c**, **d**) MEMS scanner-based fast scan OR-PAM brain imaging system (**c**), and corresponding OR-PAM images (**d**) depicting fractional PA amplitude changes (shown in yellow) in response to left hindlimb stimulation (LHS) and right hindlimb stimulation (RHS), superimposed on the vascular image (shown in red) (Yao et al. 2015b). LH, left hemisphere; RH, right hemisphere. (**e**, **f**) Schematic of a US-aided multi-parametric PAM system (**e**) and corresponding PAM images (**f**) of the mouse brain through the intact skull (Ning et al. 2015). Left top image: Depth-encoded skull vasculature (indicated by a cool color) and cortical vasculature (indicated by a warm color) separated by the ultrasonically determined skull (in gray). Other images (clockwise): Simultaneously acquired high-resolution maps of C_{Hb}, sO_2, and CBF (both speed and direction), respectively. Scale bar: 0.5 mm. (**g**, **h**) Schematics of a SIP-PACT system (**g**), and corresponding PACT image (**h**) depicting a rat whole-brain vasculature in the coronal plane and seed-based functional connectivity analyses of the thalamus region on both sides of the brain (Li et al. 2017). (**i**) Bimodal PA and fluorescence imaging setup for noninvasive imaging of the GcaMP6f brain in vivo (Gottschalk et al. 2019). (**j**) Noninvasive imaging of somatosensory-evoked rapid calcium transients in the GcaMP6f brain in vivo. Four-dimensional photoacoustically recorded brain activation maps (ΔPA/PA) in response to the current stimulus at $t = 0$ at the right hind paw. Slices across the entire brain are shown to be located at different depths from the brain surface. Functional regions of the brain are indicated as dashed white lines. (**k**) Temporal sequence of fluorescence recorded brain activation maps ($\Delta F/F$) in response to the stimulus at $t = 0$

poly(ethylene glycol)-coated nanoshells (Wang et al. 2004a), in order to increase the signals from the cortical vasculature. In the next year (2005), Ku et al. reported the first PAT-based tumor imaging result using the rat mammary adenocarcinoma cell line (Br7-C5) and successfully visualized angiogenesis around the tumor tissues developed in a cerebral hemisphere after the inoculation (Ku et al. 2005). Thereafter, in 2006, Wang et al. demonstrated the feasibility of PAT-based sO_2 mapping, which subsequently played a central role in the functional imaging capability of PAT throughout the entire PAT area (Wang et al. 2006). By using 584- and 600-nm wavelengths, they successfully visualized the sO_2 change in the brain cortex according to the artificial modulation of the oxygen level in the inhaled gas (i.e., between hyperoxia and normoxia or hypoxia and normoxia). Further, in 2008, Li et al. demonstrated the first tumor molecular imaging by using an ICG derivative contrast agent, IRDye800-c(KRGDf), which was developed to target integrin $\alpha_v\beta_3$ overexpressed in human U87 glioblastomas in nude mouse brains (Li et al. 2008c).

Meanwhile, in the late 2000s, two new PAT technical platforms—called the AR- and OR-PAM—were presented in the field (Maslov et al. 2005, 2008) and began to be applied to the brain imaging application as well. Since the techniques could provide much finer vasculature images at also an increased imaging speed compared to the conventional PACT, research progress of the brain application could be boosted and also widened in terms of the spectrum. The three reports presented by Stein et al. (Stein et al. 2009a, b) and Li et al. (Li et al. 2009) in 2009 were the first cases that applied the AR-PAM technique to the brain imaging study, in which more detailed views of the vasculature in a brain cortex (Stein et al. 2009a, b) and solid brain tumor (Li et al. 2009) were presented, respectively. In addition, in the same year, Hu et al. (Hu et al. 2009a) presented the first report that applied the OR-PAM technique for brain imaging. In the paper, they presented a transcranial mouse brain imaging result that achieved a much finer visualization of the vasculature with a spatial resolution of ~5–10 μm, which enabled a capillary-level visualization, and also demonstrated an sO_2 map for the single capillary-level blood vessels distributed in a local brain cortex.

In addition to the two technical platforms, there was another important technical platform presented in 2009, which thereafter has continuously had a great impact on the brain imaging study thus far. It was the 512-element-based full-ring array transducer-based PACT system (Gamelin et al. 2009), constructed by Gamelin et al. of the Zhu group, and was an upgraded version of the previous partial ring system reported in 2008 (Gamelin et al. 2008). Since the array system could capture all the PA waves diverging from a source along the detection plane simultaneously, unlike the previous mechanical scanning PACT system, it enabled cross-sectional images with a frame rate of up to 8 Hz, which was almost 200 times faster than that of the mechanical scanning PACT system (Wang et al. 2003). Due to the emergence of the array technique, most of the later PACT-based brain imaging studies were replaced with it; finally, the array transducer-based PACT, as well as the AR- and OR-PAM, became the three major technical platforms for brain imaging as well as other applications.

Based on the techniques, PAT brain imaging studies began to show a more active progress in the beginning of the 2010s, as the related concepts were popularized among other groups. Although, consequently, related research outcomes were also presented from other groups, most key achievements reported until 2015 appeared to be only from the Wang group. In particular, in the period between 2010 and 2015, the Wang group also presented several notable techniques and methods, such as the flow speed measurement based on the Doppler principle (Yao et al. 2010b), visualization of a single-RBC blood flow based on high-speed OR-PAM (Wang et al. 2013), glucose metabolic imaging of a mouse brain based on 2-NBDG (Yao et al. 2013), and investigation of the functional connectivity of a rat brain based on PACT (Nasiriavanaki et al. 2014). Although some of these techniques were not directly related to the brain imaging application, they made the related research spectrum expand further.

Overall, entering the 2010s, PAT techniques have been steadily improved in terms of the imaging speed, imaging depth, sensitivity, and functional or molecular imaging capability, thereby resulting in various research outcomes. However, among numerous reports, here, we introduce several recent key papers that well represented the imaging capabilities of the state-of-the-art PAT brain imaging systems.

First, in the report presented by Yao et al. of the Wang group in 2015, the authors showcased how fast a brain imaging could be performed using a state-of-the-art OR-PAM system (Yao et al. 2015b). Although the imaging speed of a mechanical scanning OR-PAM is generally lower than that of an array transducer-based PAT system, when they are compared in terms of a volumetric imaging, it can do much better in terms of a B-scan imaging if an adequate scanning mechanism is employed.

As demonstrated in the report, they achieved a B-scan imaging as fast as 400-Hz over a ~ 3 mm scanning range, which was ~200-times faster than that of the second-generation OR-PAM system, by employing a fast-oscillating MEMS mirror scanner that reflected and steered both the excitation laser beam and the detection acoustic axis (Fig. 5c).

Most of all, due to the greatly improved imaging speed, they were able to reproduce the brain hemodynamic response imaging experiment presented in ref. (Wang et al. 2003) at a more detailed (~3 μm spatial resolution) and sensitive level. As depicted in Figure 5d, which are transcranial PA maximum amplitude projection (MAP) images, they successfully recorded the contralateral hemodynamic responses in the somatosensory regions of a mouse brain according to the electrical stimulations alternatingly applied to the left and right hindlimbs in a much faster imaging speed. Moreover, it was also notable that the system's sensitivity was sufficient to capture the increase in blood volume in the ipsilateral somatosensory region, which might be a phenomenon that hemodynamically represented the functional interconnection of blood vessels distributed in the two hemispheres.

In addition to this, for the first time in the paper, they presented a new method that mapped the sO_2 of a mouse brain in action in vivo based on the single-wavelength, pulse-width-based sO_2 measurement. Unlike conventional dual-wavelength-based sO_2 mapping (Zhang et al. 2006a; Hu et al. 2011; Yang et al. 2012a), this new method featured the use of two laser pulses with a different pulse length (nano- and pico-second) but with the same wavelength of 532 nm, thereby capitalizing the different absorption saturation effect of both oxy- and deoxy-hemoglobin at the two pulse lengths. Since the method is not affected by the unknown fluence information typically occurring in the conventional sO_2 measurement method, it also could have an impact on numerous other applications. Overall, the demonstrated high-speed, high-resolution, label-free functional hemodynamic imaging capability suggests that PAT could be a new powerful tool for understanding the neurovascular coupling phenomenon (i.e., hemodynamic response associated with neuronal activity).

The B-scan imaging speed demonstrated in the study could be understood as a kind of speed limit achievable by the MEMS scanner-based scanning mechanism. However, in a report presented by Lan et al. of the Yao group in 2018 (Lan et al. 2018), an even faster brain imaging was demonstrated by adopting a different scanning mechanism that employed a fast rotating hexagonal mirror. In the schematic presented in Figure 5c, the MEMS mirror performed a mechanical scanning based on its mechanical oscillation, in which acceleration and deceleration were involved. However, in the new system (Lan et al. 2018), they steered a coaxially aligned laser beam and acoustic waves by employing a hexagonal mirror that constantly rotated only in one direction with a rather fast rotational speed of 150 Hz. Since the hexagonal mirror was actuated by a DC motor with no directional change, it was possible to achieve a B-scan imaging speed that was ~two times faster than the MEMS scanner-based method (Yao et al. 2015b)—actually, it was an area scanning speed that was ~10 times faster when considering the scanning range augmentation widened up to ~12 mm.

Thus far, we have introduced the two papers that showcased a technical extreme in terms of B-scan imaging speed. As presented, OR-PAM can visualize cortical vasculature based on the endogenous contrast mechanism, because PAT is very sensitive to hemoglobin absorption. Although the presented examples were mostly focused on related image productions, it is also well known in the PAT field that from the acquired vasculature image, one can extract quantitative information on the following five anatomical and physiological parameters—tissue volume, vessel diameter, total hemoglobin concentration (C_{Hb}), sO_2, and cerebral blood flow (CBF) of a blood vessel distributed in a mouse brain cortex (Wang 2008; Hu and Wang 2010). Since ~2008, the Wang group has been solidifying the related technique formalisms to establish a method to quantitatively measure how much oxygen is consumed in a specific or entire area of a brain cortex—that is, the cerebral metabolic rate of oxygen ($CMRO_2$) (Wang 2008; Hu and Wang 2010). Although the group presented multiple reports related to the issue, here, we introduce one of the recent reports presented by Ning et al. of the Hu group in 2015, as it discussed a more accurate measurement method for the mentioned parameters (Ning et al. 2015).

Unlike previous reports (Hu et al. 2011; Yao et al. 2015b), in the study, they paid attention to the optical and acoustic wave distortion effect caused by the skull, which eventually hindered the accurate mapping of the aforementioned physiological data set. Thus, in order to address the issue, they constructed a new version of the OR-PAM system that was combined with the conventional US imaging function (Fig. 5e) and performed an OR-PAM imaging by applying a dynamic optical focusing to a target tissue during its scanning based on the skull contour information provided by the US imaging. Consequently, based on the method, they could acquire much clearer and more precise data on C_{Hb}, sO_2, and CBF simultaneously and also at the same spatial scale of 10 μm, as illustrated in Figure 5f. Although the imaging speed had to be lowered due to the intervening of the dynamic focusing process and actual estimation of $CMRO_2$ in a brain cortex was not exemplified, the achievement was considered important progress in the practical implementation of the related goal.

Apart from the presented cortical imaging, PAT also can be utilized for small animal whole-brain imaging. In the paper by Li et al. of the Wang group in 2017, they demonstrated an ~11 mm image penetration, which can potentially cover the entire volume of an adult rat brain (Li et al. 2017). Based on their advanced PACT theory and technology termed "single-impulse panoramic PACT" (SIP-PACT), they demonstrated the small animal whole-brain imaging capability of PAT by using the 512-element full-ring array system (5 MHz) (Fig. 5g), which was upgraded from its earlier version (Gamelin et al. 2009). As previously mentioned, the prototype of the 512-element full-ring array transducer-based PACT technology was originally presented by the Zhu group (Gamelin et al. 2009). However, the Wang group continuously improved the technology in terms of the illumination optics, DAQ hardware, and image reconstruction algorithm. In particular, by developing a simultaneous one-to-one mapped DAQ system engaged with the array transducer and a half-time dual-speed-of-sound universal back-projection algorithm, they achieved a 2D panoramic acoustic detection completable within 50 μs and 125 μm isotropic in-plane resolution within a field of view (FOV) of ~16 mm in diameter for a single laser pulse.

The performance of the system was surprising. As illustrated in Figure 5h, they could acquire detailed vasculature images of a whole rat brain with a penetration extended up to 11 mm. Moreover, by analyzing the spontaneous hemodynamic responses between 16 contralateral regions distributed in the coronal plane (~bregma −2.16 mm) of the rat brain, they found a clear correlation (i.e., functional connectivity) between the corresponding regions across the left and right hemispheres, which was consistent with previous research on both fMRI and power Doppler US. Notably, they were able to observe a left–right correlation even at the thalamus regions located at 9.7 mm depth from the surface. Previously, a report presented by Nasiriavanaki et al. of the same group provided a similar PACT-based functional connectivity analysis study for the first time (Nasiriavanaki et al. 2014). However, this new study performed related analysis over a more global brain area and also with a much improved spatial resolution.

Indeed, the demonstrated in vivo high-resolution deep brain imaging was a great achievement that far exceeded the capabilities of previous PAT studies, which were mostly limited to visualizing only cortical vasculatures (Hu et al. 2009a, 2011; Wang et al. 2003, 2013; Gamelin et al. 2009; Gamelin et al. 2008; Yao et al. 2013; Nasiriavanaki et al. 2014; Yao et al. 2015b; Hu and Wang 2010; Ning et al. 2015). Since the deep brain area is recently recognized as being important in relation to understanding the origin of memory and numerous neurological diseases, such as Alzheimer's and Parkinson's, while functional MRI (fMRI) and power Doppler US are only currently available tools for such deep brain functional studies, the presented demonstration is a good example showcasing the importance of PAT for such an application.

Although the presented real-time and label-free brain vasculature and hemodynamics imaging capability of PAT is the core technical contribution to neuroscience, it has also been a long-standing question whether PAT has the ability to directly map a neuronal activity because it is undeniably the major function of the brain. In the paper by Gottschalk et al. of the Razansky group in 2019, a result that answered this question was presented (Gottschalk et al. 2019).

By using a mouse model that expressed a genetically encoded calcium indicator GCaMP6f, the group demonstrated the real-time and brain-wide mapping capability of PAT for the calcium dynamics in vivo for the first time. Figure 5i illustrates the imaging system that they developed and customized to perform the study. First, the system featured the dual-mode imaging capability, presenting conventional planar fluorescence and three-dimensional (3D) PA images simultaneously or sequentially. In addition, it performed a rather fast volumetric PA imaging, as high as 100 Hz, over the entire imaging volume (~2 cm^3) based on the 512-element spherical matrix transducer array with a 5-MHz central frequency and produced a PA image with an almost isotropic 3D spatial resolution of ~150 μm.

Surprisingly, with the system, they were able to successfully acquire the first snapshots of the spatiotemporal propagation profiles of the calcium transient waves based on the PA imaging function, as illustrated in Figure 5j. Notably, the calcium signals could be photoacoustically differentiated even in the presence of strong hemoglobin background absorption, and the GCaMP6f's responses were evoked by the electric stimulations applied to the right hind paw (one pulse with a 50-ms duration at every 20 s). A set of sequentially acquired planar fluorescence images also visualized the wave propagation (Fig. 5k). However, it was less clear compared to the PA imaging in terms of the initial cortical localization and depth-wise entanglements of the waves around the corresponding somatosensory region and thereafter propagation. Moreover, in the PA imaging mode, it was observed that the peak response appeared after a delay of ~360 ms from the stimulation, which is in accordance with known knowledge; clear bilateral activations were detectable at a depth of approximately 1 mm.

In the presented PA images (Fig. 5j), in order to effectively differentiate the calcium waves, they presented the results in a differential mode normalized by a reference image (i.e., ΔPA/PA); this is because when the peak excitation wavelength of GCaMP6f (488 nm) was applied to the GCaMP6f-expressing mouse for a PA excitation, the acquired PA images mostly exhibited vascular contrasts rather than GCaMP6f's contrasts (note that hemoglobin was still the dominant absorber at the wavelength). In addition, the PA imaging experiment was performed at a relatively lowered frame rate of 25 Hz, which, however, was still sufficient for capturing the calcium spikes that typically disappeared after ~0.8 s; moreover, the pulse duration of the electric stimulation was set at as short as 50 ms to minimize background hemodynamic responses.

Indeed, the reported results are great achievements that evidence that PAT is capable of providing 3D neuronal activity with unparalleled spatiotemporal resolution and sensitivity over conventional planar fluorescence imaging and also that no other currently available imaging modality can replace the ability of PAT in terms of the versatile contrast, speed, resolution, and FOV, all of which are critical for studying neurodynamics.

Thus far, we have introduced several representative examples showcasing the current forefronts of the PAT brain imaging study. Although it is true that these are currently the dominant research directions highlighting the core advantages of PAT in neuroscience, a couple of notable papers presenting a new approach have been reported in recent years (Tang et al. 2015b, 2016; Cao et al. 2017). Unlike the presented conventional brain imaging studies (Xu and Wang 2006b; Wang et al. 2004a, b; Ku et al. 2005; Wang et al. 2006, 2013; Li et al. 2008c, 2009; Stein et al. 2009a, b; Hu et al. 2009a; Gamelin et al. 2009; Gamelin et al. 2008; Yao et al. 2013, 2015b; Nasiriavanaki et al. 2014; Hu and Wang 2010; Ning et al. 2015; Gottschalk et al. 2019), in which an animal was imaged in a resting and restrained state, the new approach is related to imaging an awake (Cao et al. 2017) or moving animal (Tang et al. 2015b, 2016). Figure 6 depicts the related examples that imaged a freely behaving mouse brain based on a wearable device (Fig. 6a) and a head-restrained OR-PAM system (Fig. 6c), which provided a lower degree of freedom compared to the first.

Actually, such an imaging approach was first proposed by researchers in the CFM and TPM field to understand how neural circuits produce animal behavior by directly monitoring related cellular events that occur in small animal brain in situ and real time (Wilt et al. 2009). Currently, in the reported PA images (Fig. 6b and d), meaningful signals directly reflecting the animal's behavior were hardly captured

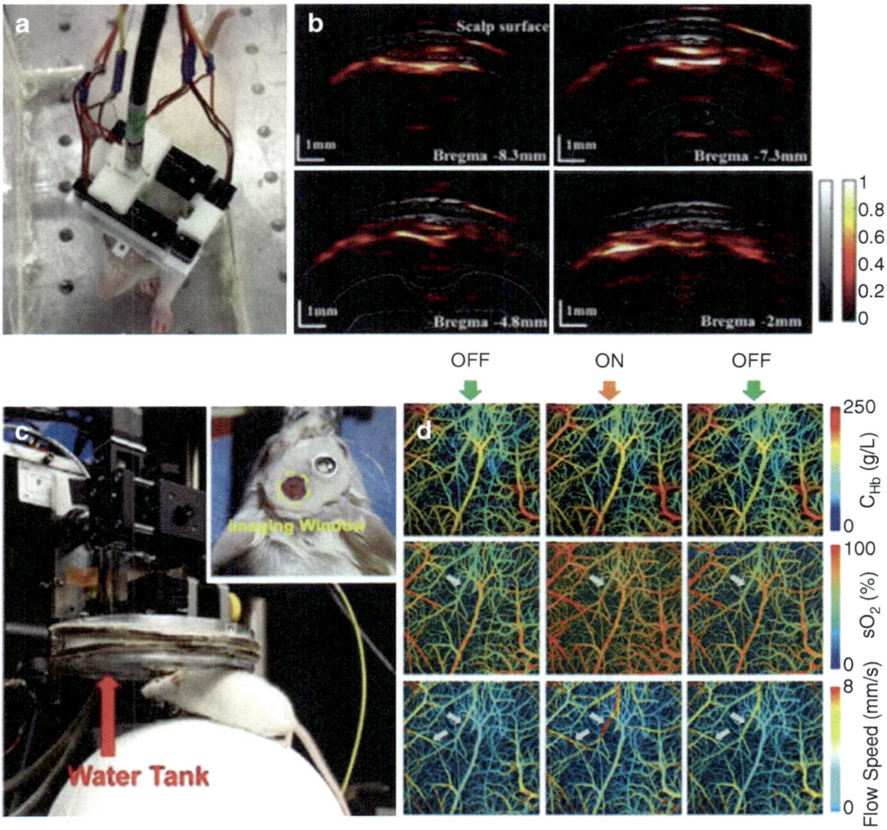

Fig. 6 Wearable and head-mounted PAT systems and produced images. (**a**, **b**) A wearable PA imaging device (**a**) for imaging the brain of a freely behaving mouse and acquired B-scan image (**b**) (Tang et al. 2016). (**c**) Photo of a head-restrained PAM system (Cao et al. 2017). The inset shows the photo of a mouse brain with a thinned skull window and a nut attached using dental cement. (**d**) Head-restrained PAM of cerebral C_{Hb}, sO_2, and blood flow speed in the absence (OFF) and presence (ON) of isoflurane. The white arrows in the second and third rows highlight the isoflurane-induced changes in the average sO_2 values of the draining veins and blood flow speed

because there was a tricky technical issue related to stable light delivery and US detection that must be overcome. The first paper only presented a 2D cross-sectional image of blood vessels, and the second study was also limited to several hemodynamic parameter changes according to the variation of inhaled anesthesia gas (isoflurane). Nevertheless, it is an undeniably important research direction to ultimately answer the mentioned question; in this regard, it is expected that PAT could make a valuable contribution to the study because PAT can provide real-time image information on the hemo- and neurodynamics of a brain in action at unprecedented imaging depth compared to other optical neuroimaging modalities.

In this section, we reviewed several notable research outcomes reported in the PAT brain imaging application. As presented, PAT has a unique advantage for the brain imaging application in terms of the label-free visualization of vasculature and real-time hemodynamic imaging. Again, it is the only method that can provide five anatomic and physiological parameters—that is, the tissue volume, vessel diameter, concentration of hemoglobin, sO_2, and blood flow—quantitatively based on a single imaging system (Wang 2008; Hu and Wang 2010; Yao and Wang 2014b). In addition, by optimizing necessary optical illumination and acoustic detection parameters, it is also possible to image the deep brain area of a small animal (Li et al. 2017). Further, global neuronal dynamics also can be imaged in real time with the aid of existing calcium indicators (Gottschalk et al. 2019). Based on the listed strengths, it is highly expected that PAT could make a crucial contribution to understanding the complex mechanism of a whole brain's in vivo dynamics as well as its neurovascular coupling, thereby complementing conventional neuroimaging modalities.

Small Animal Whole-Body Imaging

The feasibility of PAT-based small animal whole-body imaging and molecular imaging was first demonstrated in the two papers reported by Kruger et al. in 2003, in which several TAT image slices of a mouse's head and body trunk were well presented (Kruger et al. 2003a; b). However, earnest research interest appeared after Brecht et al. of the Oraevsky group reported the first 3D whole-body image of a mouse in vivo in 2009 (Brecht et al. 2009). As the paper presented a detailed 3D image of internal organs visualized noninvasively, the related application became one of the major areas of PAT and also became associated with numerous biological or preclinical studies in conjunction with various disease models like cancer and models like gene expression.

Needless to say, such a small animal-based whole-body imaging approach is a valuable methodology for both preclinical and fundamental biological studies, as it may provide us with a better understanding or new findings on numerous important biological processes and deeper insights on disease development mechanisms on a whole-body scale. In addition, it could enable the testing of a custom-designed disease model or a treatment effect of a new drug or therapeutic method. Although MRI, X-ray, and US imaging are the major imaging tools presented in related studies, there are numerous unique benefits of PAT. Thus, in this section, we introduce several notable achievements that showcase the current capabilities of PAT for small animal whole-body imaging, focusing on the papers reported from 2009 onward.

Figure 7a depicts a photo of the system, with which the Oraevsky group acquired the first small animal whole-body image in vivo (Brecht et al. 2009) and that was embodied in the form of an array transducer-based PACT. In particular, the system features the use of a concave arc-shaped array transducer comprising 64 piezo-elements with a center frequency of 3.1 MHz (−6 dB bandwidth: >80%). However, by adding additional rotational scanning with 150 steps to the animal along its mounting axis, it was possible to create a virtual spherical array with 9600 detection positions, thereby achieving a spatial resolution of ~0.5 mm. In the reported imaging demonstration, although the PRR of their laser system was 10 Hz, a complete scanning time for producing a whole-body 3D imaged required ~8 min due to the signal averaging over 32 laser pulses.

Figure 7b depicts the exquisite performance of the system, revealed through the 3D whole-body imaging of two mice. Both images were acquired using a 755-nm excitation, but each was from a female (left) and male (right) mouse, respectively. In the left image, both kidneys, as well as the spleen and a partial lobe of the liver, were clearly imaged. In addition, the ovarian vessels, as well as the inferior vena cava and its bifurcation into the femoral veins, were visible. Similar structures were also imaged in the right image. However, the kidneys appeared obscured, whereas the spine—in which numerous small vessels are distributed—was visualized more clearly than in the left image.

In 2011, Razansky and Ntziachristos et al. reported another type of whole-body imaging system that employed a similar arc-shaped 1D curved array transducer with 64 elements (5 MHz center frequency) (Fig. 7c) (Razansky et al. 2011). However, their transducer was different from that of the Oraevsky group in that it had an additional focusing capability along the longitudinal direction (i.e., *z*-direction) owing to its spherically concaved surface along the mentioned direction, which resulted in a geometrical focal length of 4 cm. In addition, they utilized a different animal positioning strategy that placed an animal inside a horizontally arranged holder to avoid direct contact of the animal with the surrounding matching medium, and illuminated the animal at 10 different positions to more uniformly excite the trunk during PA imaging, which eventually enabled a simultaneous DAQ. Notably, although the geometrical spanning of the array transducer only covered the animal by 172°, no rotational scanning of the transducer around the animal's trunk nor the rotation of the animal itself was added. Only the animal was translated along the *z*-axis to image different body sites. Nevertheless, an in-plane resolution of 150 μm was guaranteed, while the elevation resolution—whose value was determined by the aforementioned geometric focusing—was ~800 μm.

Such horizontal placement provided for a more comfortable imaging condition for the animal, and the DAQ system that they constructed enabled a cross-section imaging speed as fast as <1 ms/frame, although the actual imaging speed demonstration was imposed by the PRR of the employed laser system to ~10 Hz. Unlike the system of the Oraevsky group, whose imaging was mostly limited to the trunk, this system was capable of imaging even the thoracic and head regions in a 2D fashion (Fig. 7d) by simply positioning the area of interest along the horizontal direction (i.e., *z*-axis); moreover, it also featured a real-time imaging capability by aligning the imaging plane of the arc-shaped transducer across the body trunk (Fig. 7c). Further, in order to demonstrate the system's fast B-scan imaging capability, they imaged the perfusion of ICG in vivo after its injection into a mouse kidney (Fig. 7e) and also indicated the dynamic imaging of the respiratory motion at a frame rate of 10 Hz. In addition to this paper, there also were two notable preceding reports that showcased the system's more detailed imaging capability for renal and cardiovascular systems (Buehler et al. 2010; Taruttis et al. 2010). Moreover, in 2012, Taruttis et al. of the group also demonstrated its more comprehensive imaging capability for the pharmacokinetics of an injected drug (Taruttis et al. 2012).

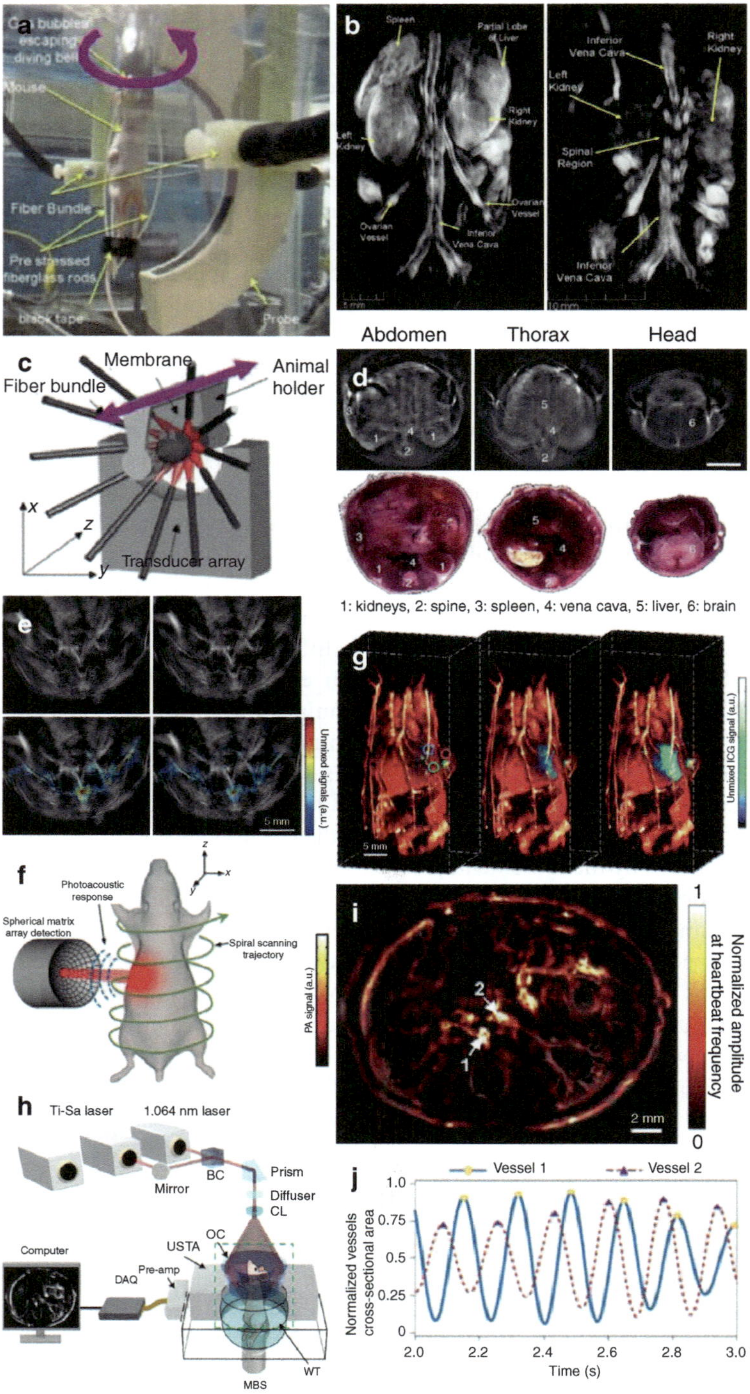
a
Fiber Bundle
Pre stressed fiberglass rods
Probe
b
Spleen
Partial Lobe of Liver
Left Kidney
Right Kidney
Ovarian Vessel
Inferior Vena Cava
Spinal Region
5 mm
10 mm
c
Membrane
Fiber bundle
Animal holder
Transducer array
x
z
y
Abdomen
Thorax
Head
d
1: kidneys, 2: spine, 3: spleen, 4: vena cava, 5: liver, 6: brain
e
Unmixed signals (a.u.)
5 mm
f
Photoacoustic response
Spherical matrix array detection
Spiral scanning trajectory
PA signal (a.u.)
g
Unmixed ICG signal (a.u.)
5 mm
h
Ti-Sa laser
1,064 nm laser
Mirror
BC
Prism
Diffuser
CL
OC
USTA
Computer
DAQ
Pre-amp
WT
MBS
i
Normalized amplitude at heartbeat frequency
2 mm
j
Vessel 1
Vessel 2
Normalized vessels cross-sectional area
Time (s)

Fig. 7 PAT small animal whole-body imaging systems and produced images. (a, b) An arc-array transducer-based PACT system (**a**), and corresponding 3D PA images (**b**) acquired from nude mice using a 755-nm laser excitation (Brecht et al. 2009). (**c**, **d**) A 64-element (5 MHz) curved array transducer-based PACT system (**c**) in which an animal could be translated horizontally over a 15-cm range to acquire a 3D image and corresponding PA images (**d**) showing the cross-sections (upper row) acquired at the abdominal, thoracic, and head levels using a 750-nm laser excitation (Razansky et al. 2011). Below photos are corresponding ex vivo cryoslices images. (**e**) In vivo tracking of systemic ICG injection into a mouse. The upper left image depicts a single-wavelength (810 nm) PA image of the lower abdomen before intravenous injection of 40 nmol of ICG, and the upper right image depicts the same region 2 min after administering the injection. The lower two images depict spectrally unmixed ICG signals (in color), superimposed onto a single-wavelength image, captured at 2 min (left) and 4 min (right) after the injection. (**f**) Schematic of the SVOT system (Deán-Ben et al. 2017c). (**g**) Spectrally unmixed images (green) for three representative instants following tail vein injection of 100 nmol of ICG at 0 s. The kinetic images were superimposed onto the whole-body PA anatomical reference (red) acquired at 800 nm wavelength. (**h**) Schematics of the SIP-PACT system for trunk imaging (Li et al. 2017). BC, beam combiner; CL, conical lens; MBS, magnetic base scanner; OC, optical condenser; USTA, (full-ring) US transducer array; and WT, water tank. (**i**) Heartbeat encoded arterial network mapping overlaid on the anatomical image. (**j**) A graph showing the relative phase delay between the temporal variations of the cross-sectional areas of the two blood vessels 1 and 2 marked in (**i**)

The presented imaging demonstrations (Figs. 7d and e) were all limited to a 2D recording of the dynamics. However, in 2017, Deán-Ben et al. of the Razansky group reported a new type of small animal whole-body imaging system that enabled a much clearer 3D visualization of the organs and vasculature inside a body trunk (Fig. 7f). As included in the technical term "spiral volumetric optoacoustic tomography (SVOT)" used in the paper (Deán-Ben et al. 2017c), the system also added a mechanical scanning motion with a helical trajectory to an array transducer (4 MHz, 256-elements) that itself was capable of yielding an isotropic spatial resolution of ~200 μm over an ~1 cm^3 imaging volume—that is, FOV (note that this spherical array transducer is a similar one to that employed in Figure 5i, although this paper was reported earlier). Since the laser system employed in the study enabled a PRR of 100 Hz, the imaging system was eventually able to cover a ~ 100 cm^3 imaging volume in merely one second by including additional mechanical scanning; moreover, the system could also perform a dynamic imaging of an organ with a temporal resolution of 10 ms (~50 ms for multispectral imaging) if it was set to continuously record image data in a stationary position.

This system led to a very striking result that has not ever been reported thus far. As illustrated in Fig. 7g, a real-time tumor perfusion of the ICG solution (i.e., a contrast agent) injected via the tail vein of a tumor-bearing mouse was clearly visualized with a frame rate of 20 Hz. Furthermore, an interesting phenomenon of a long delay of ~1.26 s between the appearance of the contrast agent in a major lateral vein versus the tumor area was also observed. The researchers ascribed the extended time delay to the obstruction caused by the tumor growth. Then, the perfused drug flowed into the liver of the animal and was finally revealed as a PA signal decay due to its clearance.

In the paper (Deán-Ben et al. 2017c), their demonstrations were not limited to the experiment. By imaging the flow dynamics of an ICG bolus injected in a similar manner, the real-time motions of a cardiac system and the bolus' appearance in each ventricle could also be successfully observed (using an independent animal). Further, the renal artery infiltration of Alexa Fluor 750 (AF750) after five-minute post-injection and its filtration in the kidney cortex and subsequent excretion were successfully visualized. In the demonstrations, the lowered frame rate was because of the multi-wavelength imaging requirement for a spectral unmixing, in which five different laser wavelengths were used. Nevertheless, the imaging speed was sufficient for visualizing the contrast agent's kinetics in a living biological system. Undoubtedly, the showcased model was the best imaging performance ever achieved in biomedical PAT, in terms of the effective combination of the imaging speed, image quality, and spectral unmixing capability, and was an invaluable achievement that indicated a new paradigm change in small animal whole-body imaging.

Although their sizes are much smaller, zebrafish and *Drosophila* are also regarded as important models for such body-wide imaging research, and the Ntziachristos and the Razansky groups also reported numerous interesting results based on the two models (Razansky et al. 2009; Ma et al. 2009, 2012; Omar et al. 2017). However, we have not introduced these applications in this chapter because the body dimensions of zebrafish and *Drosophila* are much smaller than those of rodents and, thus, can be imaged by using other optical imaging techniques as well. In addition, we mostly focused on array transducer-based imaging systems to highlight PAT's real-time imaging capability for whole-body imaging.

Meanwhile, by employing the same US detection system utilized for the SIP-PACT-based brain imaging presented in Figure 5g in the previous section, Li et al. of the Wang group in 2017 also demonstrated its applicability to small animal whole-body imaging in the same paper (Li et al. 2017). However, for the experiment, they modified the related illumination optics, as illustrated in Fig. 7h to more efficiently irradiate the trunk zone of a small animal.

First, as with the case of the SVOT system presented in Fig. 7f, this system also could image the entire cross-section of a small-animal body and, thus, visualize the major organs in the thoracic (heart and lung) and abdominal cavities (liver, spleen, kidney, caecum, and intestine). However, the authors of the SIP-PACT system presented an interesting result that highlighted the high-speed and high-resolution imaging capability of the system, which was sufficiently advanced to image a body trunk with an isotropic in-plane resolution of 125 μm at a 50-Hz frame rate.

As depicted in Fig. 7i, they were able to selectively map the artery network in a whole-body cross-sectional image by analyzing the temporal correlation of the vessel diameter changes for all the imaged blood vessels according to the heartbeat frequency. The change in vessel diameter in the arterial network is a natural phenomenon caused by instant blood ejection from the heart; the pressure waves then propagate along the arterial tree, dilating the aortic wall of the arterial tree, as it is directly connected to the heart. Figure 7j depicts the changes in vessel area according to time for the two blood vessels highlighted by arrows 1 and 2 in Fig. 7i and in which a steady phase delay was clearly revealed. Indeed, this was a great achievement that photoacoustically captured the pulse wave propagating through the arterial network for the first time. Based on the result, they suggested that such a capability could provide a noninvasive and direct diagnostic information for chronic coronary artery disease and chronic renal disease.

Here, it must be noted that the 50-Hz imaging speed of the system, which was also subjected to the PRR of the employed laser, was relatively slower than that of the SVOT system and the speed benefit was only available in the 2D imaging mode. Nevertheless, vessel diameters could be better resolved in a cross-sectional image (Fig. 7i) because the in-plane spatial resolution (125 μm) was better than that of the SVOT system due to the use of 512-transducer elements with the 5-MHz center frequency.

In addition to the imaging demonstration, in the paper, they also presented another interesting result that observed different sO_2 response characteristics of individual organs according to the artificially manipulated oxygen challenge (i.e., hypoxia) via inhaled gas. Although related data are not presented here, they observed a relative sO_2 decrease in most organs—such as the brain, liver, and kidney—and a sO_2 increase in a few organs, like the cecum. Based on the observed results, they presented a plausible interpretation that there may exist an adjustment mechanism for the animal's whole-body metabolic activity to survive during hypoxia.

As is evident, the presented research examples showcase that PAT can offer numerous unique opportunities in small animal-based future biological and clinical studies. Among a large number of outstanding technical aspects, this is attributed to the nonionizing radiation-based, high-resolution, real-time, and safe longitudinal imaging capability of PAT. Indeed, this could provide researchers great benefits in terms of versatility and research freedom when designing a relevant experimental model and conducting studies. Although X-ray CT and MRI also have been utilized for such research, X-ray CT requires the use of contrast agents and is always accompanied by numerous concerns caused by its ionizing radiation. Further, MRI suffers from motion artifacts, as a much longer scanning time is necessary to achieve a similar level of spatial resolution for vasculature as that with PAT.

All the presented systems (Fig. 7) had to be accompanied by an additional mechanical scanning mechanism due to the limited number of detection channels in an employed array transducer. However, in the near future, it will be eventually possible to image faster biological processes over the whole-body region without the effect of motion artifacts. Then, more delicate experiments associated with disease models and pharmacodynamics studies can be performed. In this section, the presented examples were mostly focused on the whole-body and label-free imaging capability of PAT; thus, no disease model was involved. However, examples presented in the next section reveal what additional research scenarios can be provided by PAT.

Cancer Research

When an imaging technique is utilized for cancer imaging, there are two important needs. From the perspective of biology, it can promote understanding of underlying mechanisms related to tumor genesis, development, metastasis, and associated microenvironments (Weissleder and Pittet 2008; Fukumura et al. 2010; Amornphimoltham et al. 2011; Brown et al. 2019). On the other hand, if the technique is also applicable clinically—that is, to humans—it is of particular importance in identifying the tumor location (i.e., visualization) in relation to cancer screening and staging and also to use the technique for guiding other diagnostic or therapeutic procedures (Weissleder et al. 2016; Yun and Kwok 2017). Thus, irrespective of whichever application is eventually considered, it is evident that, among the performance factors listed in "Principles" section, contrast is the most important element to effectively visualize cancerous tissues as well as the spatial distribution of the contrast agent administrated to visualize them.

In terms of this aspect, PAT has immense potential as it can create a variety of different image contrasts based on excellent spectroscopic imaging ability. As already mentioned in "Principles" section, thus far, numerous studies have demonstrated that, compared to conventional US imaging, PAT can provide much clearer image contrasts for cancerous tissues even using the endogenous contrast mechanism. In addition, in order to create additional exogenous image contrasts for tumor, a variety of dedicated contrast agents for PAT have been suggested, along with successful demonstrations of their biocompatibility and functionalities, such as targeting and therapeutic functions (Mehrmohammadi et al. 2013; Nie and Chen 2014; Weber et al. 2016; Gujrati et al. 2017).

Since 2005, in which the first PAT-based in vivo tumor imaging experiment was presented (Ku et al. 2005), the largest number of papers have been reported in relation to cancer application (for example, among all the papers published in 2019, over 50% expressed their relationship to this application). In particular, it is evident that related publications showed a sharper increase around 2014, probably due to the global spread of the PAT technique along with the inflow of more researchers and that they presented various approaches and goals, such as the ones based on various kinds of tumor models (xenograph, orthotopic, etc.), and also the ones that aimed at a drug test or the validation of a developed contrast agent to target specific biomarkers (Mehrmohammadi et al. 2013; Taruttis et al. 2015). Among many, particularly the studies associated with contrast agent development for actual clinical use accounted for a considerable portion of the reports (Mehrmohammadi et al. 2013; Nie and Chen 2014; Weber et al. 2016; Gujrati et al. 2017). Of course, although many of them successfully demonstrated their feasibility for actual clinical use, it is currently true that, as a completely new type of chemical developed for PAT, they require more rigorous and systematic studies because the toxicity and accompanying clearance issues of an injected contrast agent must be completely resolved before the translation. Thus, in this section, we present several possible research scenarios of PAT in relation to biological or preclinical cancer research, mostly focusing on those papers that well represented the distinctive imaging features of PAT over existing imaging modalities for the purpose.

First, PAT can provide quantitative information on numerous functional and physiological parameters associated with tumor development in an animal model based on an employed single PAT imaging system only. As already mentioned in "Brain Imaging" section, PAT is a unique imaging modality that can simultaneously image anatomical and physiological parameters such as the tissue volume, vessel cross-section (diameter), concentration of hemoglobin, sO_2, and blood flow. Thus, the first statement represents that related parameters also can be utilized for understanding the tumor microenvironment. In 2011, Yao et al. of the Wang group presented a report that actually demonstrated such a capability of PAT and also showcased the importance of the other parameter termed the metabolic rate of oxygen (MRO_2) (Yao et al. 2011), which can be calculated based on the measurements of the aforementioned parameters and is also typically regarded as a more useful index compared to other oxygenation indexes of tissue—such as sO_2 and partial oxygen pressure (pO_2)—because it directly reflects the rate of oxygen consumption instead of the static oxygen concentration (Buxton and Frank 1997).

In the study, in order to observe a significant change in the aforementioned parameters, they designed a series of experiments that imaged nude mouse ears representing four different physiological and pathological conditions—that is, hyperthermia, cryotherapy, melanoma, and glioblastoma—by employing the first OR-PAM system reported in 2008 (Fig. 8a)(Maslov et al. 2008). As mentioned earlier, the system featured the greatly improved spatial resolution (5 μm for transverse and 15 μm for axial) compared to any other PAT system that was available at that time. They chose the mouse ear model because it can be treated as a well-defined biological system for the in vivo validation of such a quantitative imaging approach and its thickness can be covered by the penetration depth (~700 μm) of the OR-PAM system.

Among the results acquired from the aforementioned four conditions, Figure 8b–d illustrates the cases of a control (upper row: before inoculating the tumor) and U87 human glioblastoma (GT)-bearing mouse ear (lower row: day 7 after the xenotransplantation of the tumor), respectively, and Figure 8e represents the calculated physiological parameters for the two cases. After 7 days of tumor inoculation, it was evident that numerous new blood vessels appeared in the intratumoral region due to angiogenesis (see the lower image of Fig. 8c). Moreover, there was a significant increase in the volumetric flow rate, while the oxygen extraction fraction (OEF) decreased (see Fig. 8e); here, the OEF is defined as $(sO_{2in} - sO_{2out})/sO_{2in}$ and represents the proportion of O_2 molecules that cross the capillary wall (the subscripts *in* and *out* denote feeding and draining vessels, respectively).

However, there was a more notable finding that—rather counterintuitively—the early-stage glioblastoma expressed a hyperoxic state instead of a hypoxic state, although there was a significant increase in MRO_2 (i.e., hypermetabolism) compared to the control data (see Fig. 8e). Here, the hyperoxic state of the tumor region could be conjectured indirectly by the analyzed sO_2 value of the draining vein, but also could be determined by directly analyzing the sO_2 value of intratumoral vasculature in the lower image of Figure 8c; Figure 8f represents the result acquired from the direct analysis. It must be noted that a similar tendency was also observed in the result of the melanoma tumor imaging experiment performed in the study (but not presented here). Based on this result, the researchers suggested that the MRO_2 parameter could be a more reliable measure for early cancer detection if related imaging techniques are applied to clinics.

Currently, the presented imaging experiments are only limited to tumors that are developed in a shallow tissue—that is, mouse ear—where OR-PAM (that is, the ballistic imaging version of PAT) can be applied and where the region of interest is

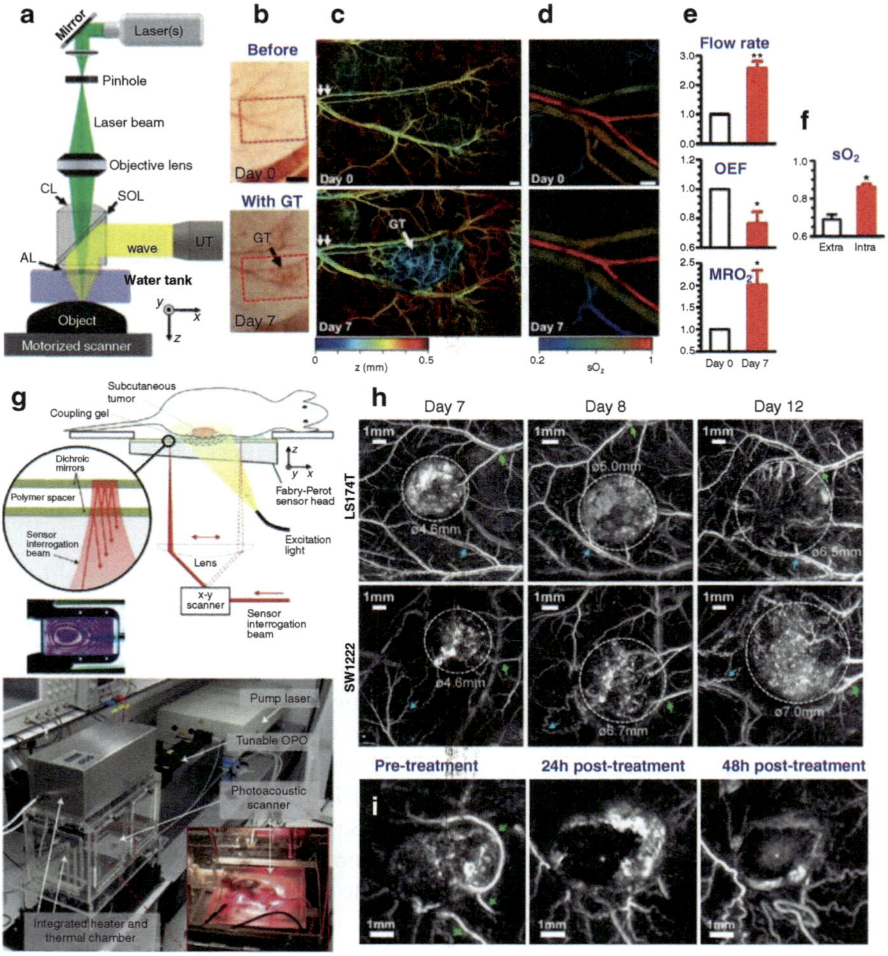

Fig. 8 Examples of PAT cancer imaging applications. (**a**) Schematic of an OR-PAM system (Yao et al. 2011). CL: correction lens; AL: acoustic lens; SQL: silicone oil layer; UT: US transducer. (**b**) Photos of a representative mouse ear before (day 0) and 7 days after the xenotransplantation of U87 glioblastoma tumor cells. Scale bar: 2 mm. (**c**) OR-PAM images (584 nm) of microvasculature in the tumor region, which correspond to the dashed boxes in (**b**) and *z* is coded by colors. Scale bar: 250 μm. GT: glioblastoma tumor. (**d**) Metabolic PAM images of sO_2 in the artery–vein pair (i.e., double arrows in (**c**)) that supports the tumor region. The two images were acquired at days 0 and 7, respectively. Scale bar: 100 μm. (**e**) OR-PAM quantification of volumetric blood flow rate, OEF and MRO_2 7 days after the tumor xenotransplantation, normalized by the values of day 0. (**f**) Comparison of the averaged sO_2 values in the intra- and extra-tumoral vasculatures. (**g**) Schematic and photo of the Fabry-Pérot polymer film US sensor-based PAT system (Laufer et al. 2012a). The thickness of the polymer spacer in a natural state is 40 μm. (**h**) PA *x–y* MAP images showing the development of two types of human colorectal tumor (LS174T, SW1222) and the surrounding vasculature between day 7 and day 12 post-inoculation. The dashed lines indicate the tumor margins. (**i**) PA images (640 nm) showing the effect of a vascular disrupting agent (OXi4503) on the blood vessel network of a tumor (LS174T). These are *x–y* MAP images through the center of the tumor before treatment, 24 h after treatment, and 48 h after treatment

well-defined because the quantitative measurement of sO_2 and flow speed is currently challenging with the optical diffusion regime. Nevertheless, there is sufficient possibility for the related imaging approach to be further extended to deeper tissue regions, as multiple studies have presented several feasible methods for solving such issues (Cox et al. 2012; Deán-Ben et al. 2015; Perekatova et al. 2016; Tzoumas et al. 2016; Hochuli et al. 2019; van den Berg et al. 2015, 2016; Brunker and Beard 2016; Bücking et al. 2018). If successful, PAT could be a useful and powerful tool for the pathophysiological study of cancers.

Second, PAT is an excellent tool for studying the longitudinal observation of an angiogenesis as well as the therapeutic response to anticancer treatment at unprecedented depth scales. Angiogenesis is known as a hallmark of cancers (Hanahan and Weinberg 2011; Pavlova and Thompson 2016; Weis and Cheresh 2011), and PAT has demonstrated its excellence for such a vasculature visualization. Although previous multiple studies demonstrated PAT's angiogenesis imaging capability, all these were limited to shallowly seated tumors, mostly based on OR-PAM or AR-PAM. However, in 2012, Laufer et al. of the Beard group demonstrated a much larger-scale visualization of angiogenesis with an acoustic diffraction-limited spatial resolution over a depth extended to ~10 mm (Laufer et al. 2012a). By using a Fabry-Pérot (FP) polymer film US sensor-based PAT system (Fig. 8g), they imaged two different types of human colorectal adenocarcinoma xenografts, LS174T and SW1222, over 12 days after the inoculation of the tumor cells into the flanks of nude mice. Briefly, the system also employed a typical wavelength-tunable optical parametric oscillator (OPO) system whose output beam with a pulse duration of 7 ns and a PRF of 50 Hz was guided to a target object via an optical fiber. However, unlike other PAT systems that used conventional piezoelectric sensors, it featured the use of the FP polymer sensor that detected generated PA waves based on the interference of an interrogation laser beam (1550 nm) reflected from the two different reflection surfaces formed by dichroic mirror coating, whose distance (or the thickness of the polymer spacer, see black circle in Figure 8g) was changed according to the varied intensity of incoming PA waves. Of course, due to the dichroic nature, the two reflection surfaces were transparent to the PA excitation laser beam over a wavelength range of 590–1200 nm.

Thus, in the sensing mechanism, the beam diameter of the interrogation laser beam determines the aperture size of the detector and also can virtually form a point detector, which can receive PA waves approaching over a much wider angle than those of conventional piezoelectric sensors. In addition, compared to the conventional piezoelectric sensor-based PAT systems, the concept of the presented system also has an advantage that it can exhibit an excellent axial-resolving ability due to its exceptionally broad bandwidth characteristics and it also allows one to achieve a very simple, but more efficient and high energy light delivery mechanism into a target object through the entire area of the transparent sensor (i.e., full-field illumination over the entire imaging zone). Consequently, the measured transverse and axial resolutions of the system were as high as 81 μm and 51 μm, respectively, which could also be conserved up to an imaging depth of 10 mm.

As revealed by the tumor imaging experimental results presented in Figure 8h, a neovascularization was commonly observed in the two tumor models as they grew, and estimated tumor sizes (volumes) also increased by approximately 50% per day; however, they exhibited somewhat different features. Compared to the SW1222 tumor (bottom row), the LS174T tumor (upper row) showed more strongly varying contrast over the entire tumor region—that is, heterogeneous vascular distribution—particularly toward the tumor center (days 7 and 8). Interestingly, from days 8 to 12, a noticeable decrease in the optical absorption was observed in the central zone, which could be attributed to a reduced blood content or possibly tumor necrosis, thereby creating an avascular region. The result was generally consistent with the findings of previous studies that were performed on the basis of vascular corrosion casting methods (Folarin et al. 2010). In contrast, in the case of the SW1222 tumor (bottom row), the development of a more homogeneous vasculature and an increase in the tortuosity of the normal vasculature surrounding the tumor was also observed (see the blue arrows in the bottom row), which was also consistent with previous studies on this type of tumor (Folarin et al. 2010).

The authors also performed another interesting experiment that illustrated the potential of PAT for evaluating the response of tumor vasculature to a therapeutic vascular disrupting agent. By using OXi4503 (combretastatin A1 diphosphate)—which is known to selectively block and destroy tumor vasculature and, thus, result in extensive central necrosis—they visualized the response of LS174T tumor after 24 h and 48 h of the administration of the drug. Note that the imaging time was determined based on the documented time course of the effects of the drug.

Surprisingly, the effect of the drug administration appeared very clearly. As illustrated in Figure 8i, the contrast of the tumor core in the 24 h image was much reduced compared to the pre-treatment image due to the disintegration of the vasculature caused by the drug; on the other hand, the tumor rim exhibited a strong image contrast, which suggested an increase in hemoglobin concentration caused by the reperfusion of blood vessels. However, after 48 h, the tumor core began to show a slight and diffusely distributed increase in the signal intensity. The authors interpreted that this might indicate an accumulation of hemoglobin due to the reperfusion of vessels in the rim that had partially disintegrated during treatment.

Thus far, in PAT researches, there have been numerous other reports of the application of PAT for such a tumor imaging study. However, as clearly evidenced in the two cases, it is undoubted that the safe, high-resolution, deep tissue imaging capabilities of PAT will provide an unprecedented research platform that enables one to longitudinally observe the progress of tumor development.

Third, PAT can also be used with a variety of contrast agents developed for molecular imaging. The two papers presented were both based on endogenous contrast mechanisms to visualize tumor vasculature. However, PAT also can be employed as a new type of molecular imaging modality (Mehrmohammadi et al. 2013; Nie and Chen 2014; Weber et al. 2016; Gujrati et al. 2017). Currently, related research is very actively growing because, compared to existing imaging modalities, PAT can provide wider options in selecting related probes based on its unique optical absorption-based contrast mechanism. Of course, it is more desirable to visualize a tumor based on the endogenous contrast mechanism. However, this is not always possible because a large number of tumors still do not sufficiently express such chromophores that can absorb light. Thus, in order to visualize such tumors more effectively, thus far, a variety of contrast agents, such as small molecule dye- or nanostructure-based targeting probes, have been developed (Mehrmohammadi et al. 2013; Nie and Chen 2014; Weber et al. 2016; Gujrati et al. 2017). Of course, a number of these contrast agents, particularly earlier versions, were employed only to boost PA signals. However, in recent years, more sophisticatedly engineered contrast agents have been developed to target specific biomarkers—such as $\alpha_v\beta_3$ integrin (De la Zerda et al. 2008), epidermal growth factor receptor (EGFR) (Hudson et al. 2014), human epidermal growth factor receptor (HER) 2 (Li et al. 2008b), and CD44 (Swierczewska et al. 2012)—which are known to be overexpressed in numerous types of cancers.

Indeed, a great amount of effort has been made in relation to the development of contrast agents for PAT; this led to a vast number of documented papers that cannot be sufficiently covered here. Thus, in the latter part of this section, we introduce only several core papers related to reporter gene (RG) imaging—which is regarded as one of the forefronts of molecular imaging—to exemplify what kind of RG imaging approaches are possible in PAT (for more general information on PAT contrast agents, refer to refs. (Mehrmohammadi et al. 2013; Nie and Chen 2014; Weber et al. 2016; Gujrati et al. 2017).

The main advantage of in vivo RG imaging is that it enables a direct visualization of the location and level of gene expression, promoter activity, protein–protein interaction as well as more complex biological behaviors, like cell growth dynamics, by selectively encoding the genes of interest (Weber et al. 2016). In PAT, approximately two types of RGs based on enzymes and fluorescent proteins have been utilized thus far (Nie and Chen 2014; Weber et al. 2016; Brunker et al. 2017; Laufer et al. 2013).

In 2007, Li et al. of the Wang group reported the first PAT RG imaging result based on lacZ (Li et al. 2007), which is one of the most widely utilized RGs (Ghim et al. 2010). It is known that the lacZ gene encodes the enzyme β-galactosidase, which is an *Escherichia coli* enzyme responsible for lactose metabolism. In the study, by locally administrating X-gal—which consists of galactose linked to a substituted indole—to a target point of interest (i.e., near the inoculated 9 L gliosarcoma tumor expressing lacZ gene [Fig. 9a]), the authors could photoacoustically visualize the presence of the lacZ expression (Fig. 9b). It was photoacoustically possible because according to the explained role, if lacZ is expressed, the X-gal's glycosidic linkage is cleaved by β-galactosidase, a process that produced a stable dark-blue product—that is, chromophore—with a strong absorption in a red region. Note that although β-galactosidase and X-gal alone are colorless, the final product—which can be produced only when they coexist—is dark blue (Fig. 9a); thus, it can be visualized by PAT. In Fig. 9c, we present one of the core results of another report from the authors (Li et al. 2008a) in which gliosarcoma cells expressing the lacZ gene under a rat scalp and also its neighboring vasculature were clearly visualized by a dual-wavelength AR-PAM.

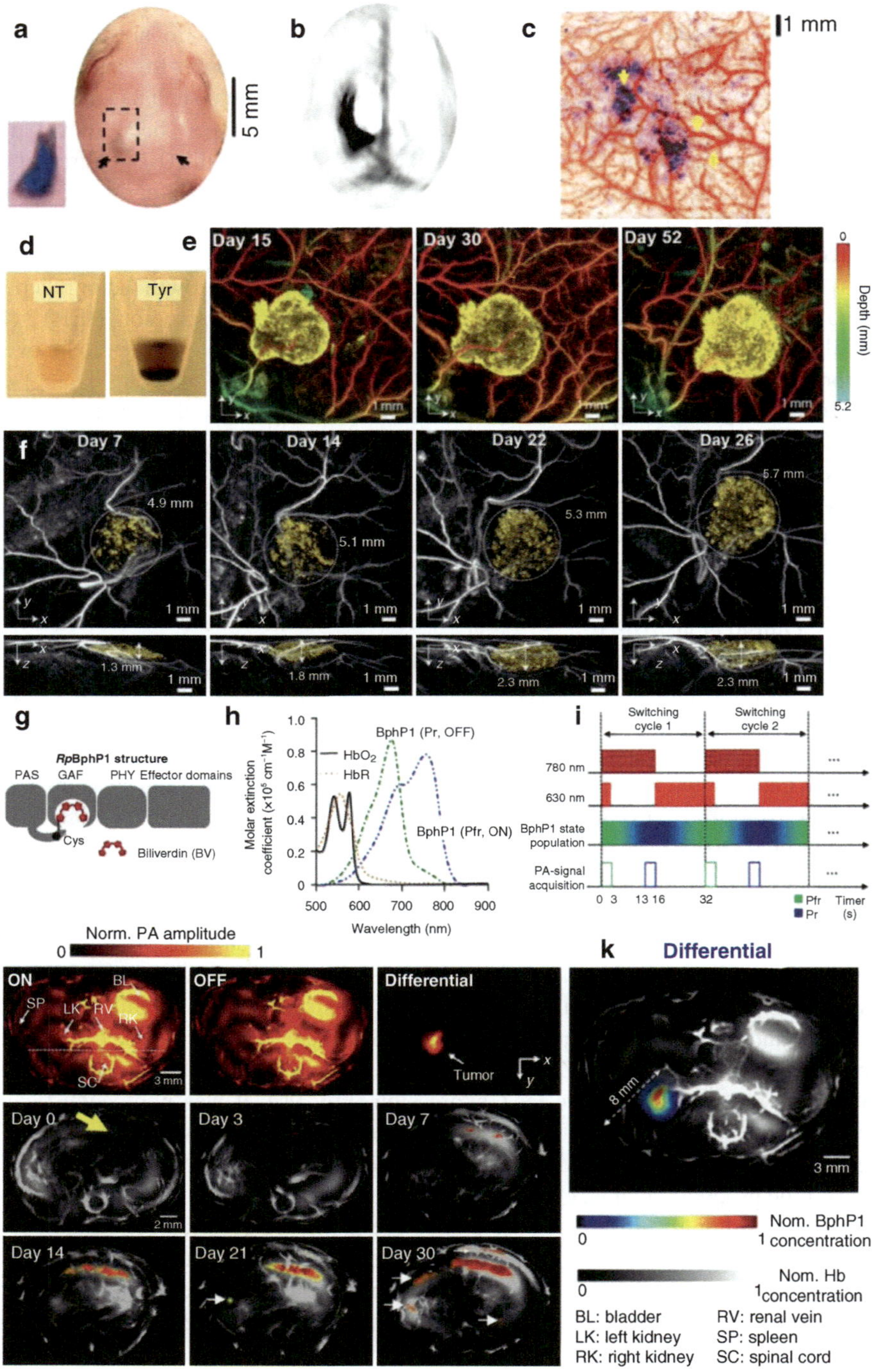

a
5 mm
b
c
1 mm
d
NT
Tyr
e
Day 15
Day 30
Day 52
Depth (mm)
0
5.2
f
Day 7
Day 14
Day 22
Day 26
4.9 mm
5.1 mm
5.3 mm
5.7 mm
1.3 mm
1.8 mm
2.3 mm
2.3 mm
1 mm
g
RpBphP1 structure
PAS GAF PHY Effector domains
Cys
Biliverdin (BV)
h
BphP1 (Pr, OFF)
BphP1 (Pfr, ON)
HbO2
HbR
Molar extinction coefficient (x10^5 cm^-1M^-1)
Wavelength (nm)
i
Switching cycle 1
Switching cycle 2
780 nm
630 nm
BphP1 state population
PA-signal acquisition
Pfr
Pr
Timer (s)
j
Norm. PA amplitude
ON
OFF
Differential
BL
SP
LK
RV
RK
SC
Tumor
3 mm
k
Differential
8 mm
3 mm
l
Day 0
Day 3
Day 7
Day 14
Day 21
Day 30
2 mm
Nom. BphP1 concentration
Nom. Hb concentration
BL: bladder
LK: left kidney
RK: right kidney
RV: renal vein
SP: spleen
SC: spinal cord

Fig. 9 Examples of representative PAT-based reporter gene imaging studies. (**a**, **b**) Photo showing an imaged rat head (**a**) and corresponding PAT image (**b**) acquired at 650 nm after the injection of X-gal (Li et al. 2007). In (**a**), the inset depicts the underside of the rat's scalp excised after sacrificing the animal. (**c**) In vivo images of lacZ-marked tumor by dual-wavelength AR-PAM (Li et al. 2008a). This is a merged pseudo-colored image of two PAM images acquired at 584 nm and 635 nm, which depict blood vessels (red) and tumor (blue), respectively. Arrows indicate the feeding vessels of the tumor. (**d**) Photos showing the pigmentation of 293 T cell pellets expressing Tyr (Jathoul et al. 2015). Left: non-transduced (NT); right: transduced with Tyr. Artificial expression of Tyr caused the pigmentation of the otherwise-transparent tumor cells. (**e**) Serial longitudinal in vivo PA MAP images (600 nm) of Tyr-expressing K562 cells after subcutaneous injection of 5 million K562 cells into the flank of a nude mouse (vasculature is color-coded for depth; K562 cells are false-colored yellow). (**f**) In vivo PA images (640 nm) of Tyr-expressing 293 T cells acquired at different times after the inoculation of 5 million 293 T cells. Dashed circles provide an indication of the increase in diameter. (**g**) Domain structure of BphP1 (Yao et al. 2015a). (h) Molar extinction spectra of HbO_2, HbR, Pfr (ON-state) BphP1, and Pr (OFF-state) BphP1. (**i**) Time sequences of the photo-switching cycle of BphP1. The 780-nm light was used for PA imaging and switching off of the protein. The PA signals were acquired only with 780-nm illumination. (**j**) In vivo whole-body PACT images of the kidney region of a nude mouse, acquired 1 week after injection of ~10^6 BphP1-expressing U87 cells into the left kidney. The differential image clearly shows the tumor in the left kidney. (**k**) An overlay of the U87 tumor (in color) in the left kidney and the blood dominated OFF-state image (in grayscale). Hb, hemoglobin. (**l**) Whole-body PACT images of the liver region of a nude mouse acquired repeatedly over 30 days after the injection of BphP1-expressing U87 cells into the right liver lobe. Differential signals (in color) are overlaid on top of the structural signals from the blood (in grayscale). The white arrows in the images for days 21 and 30 indicate secondary tumors due to metastasis

Here, an important thing that must be noted is that a single β-galactosidase molecule can cleave multiple X-gal molecules and produce a large number of blue products. Although this could have a positive implication that the method could enable the detection of a low-level expression of lacZ because the related process could be understood as a kind of signal amplification, the downside is that a 1:1 stoichiometric mapping between the amount of detected blue products and the level of RG expression is fundamentally impossible. Moreover, another big disadvantage of this method is that it requires the local injection of an exogenous chromogenic substance (X-gal) into the region of interest and this process may not lead to a systemic delivery.

Another well-known enzymatic approach in PAT is the tyrosinase (Tyr)-based reporter system (Jathoul et al. 2015). Since Tyr plays the role of catalyzing the formation of the dark brown pigment eumelanin from cellular tyrosine, the mechanism can be utilized for an artificial PAT contrast creation by genetically expressing it. By creating a novel retroviral vector that co-expressed Tyr and a surface marker gene (dCD34) in the two types of human cancer cell models (K562 and 293T), Jathoul et al. of the Beard group successfully demonstrated in vivo longitudinal imaging of their growths in the flanks of nude mice over the longest period (52 days) ever achieved. They co-expressed the dCD34 gene to facilitate the easy selection of the tumor cells during the long-term cultures of the cell lines and the Tyr gene to create a PAT image contrast for the inoculated tumors, as mentioned earlier. Figure 9d illustrates the photos of 293 T cell pellets without (left) and with (right) expressing Tyr. For the experiment, they utilized the FP US sensor-based PAT system, as depicted in Figure 8g. Figure 9e and f depicts the longitudinal imaging results of the growth of K562 and 293 T tumor cells.

Previously, several studies also have employed Tyr for PAT (Paproski et al. 2011; Krumholz et al. 2011; Stritzker et al. 2013; Qin et al. 2013; Paproski et al. 2014). However, the greatest advancement of the presented report was that they achieved a greatly extended longevity of the Tyr-expressing tumor in living mice, which extended up to >7 weeks (Fig. 9e). Moreover, although the inoculated tumor xenograft was genetically engineered, it was evident that the entire volume of the 293 T tumor cells increased by 38% over a period of 26 days (Fig. 9f). Basically, such achievements were possible because these studies developed a novel retroviral vector that enabled a stable transduction in which the cell genome was permanently altered so that the changes were passed to progeny cells, thereby resulting in the continuous production of PA image contrast caused by the steady increase in tumor cell populations. As long-lived high contrast is essential for numerous practical biological applications, like visualizing cell growth as presented, the study suggests that retroviral-mediated expression of Tyr can provide high expression levels and also be well tolerated with little effect on cell viability.

In addition to the biological implications, the report has another important technical implication—high-resolution three-dimensional imaging of the genetically labeled cells was achieved at depths approaching 1 cm with unprecedented image quality, which is not possible with other optical imaging modalities even now.

As presented, the Try-based RG system exhibits a high detection sensitivity based on the intrinsic signal amplification mechanism. Moreover, unlike the lacZ-based enzymatic approach, the tyrosinase RG system does not require an exogenously administered substrate and, thus, related approaches could address potential toxicity concerns. However, the introduced two enzymatic reporter systems are basically less desirable because they are regarded as indirect methods—which depend on substrate availability—and, thus, limit quantification. In addition, melanin's featureless broad absorption spectrum makes it difficult to be effectively discriminated from intrinsic background signals generated by hemoglobin (Weber et al. 2016; Brunker et al. 2017; Laufer et al. 2013).

Due to the limitations of the enzymatic reporters, a new form of approaches based on nonenzymatic reporters—such as fluorescent proteins, nonfluorescent chromoproteins, and photoswitchable nonfluorescent proteins—has been actively explored in recent years and is attracting increased attention (Weber et al. 2016; Brunker et al. 2017; Laufer et al. 2013). As a direct method expressing RGs, such a protein-based approach enables a 1:1 mapping of the expression level of the protein of interest and also enables one to take advantage of using various proteins already available in the class.

In this context, there was a report that exploited eGFP (enhanced green fluorescent protein) and mCherry—well-known fluorescent proteins—for a PAT-based RG imaging study (Razansky et al. 2009). However, the limited penetration capability of light at the absorption peaks (488, 587 nm) of the two proteins made them applicable only to the two small size organisms, *Drosophila melanogaster* pupae and adult zebrafish. Subsequently, in order to overcome the issue related to the undesirable spectral properties of these fluorescent proteins, more red-shifted fluorescent proteins, like iRFPs, began to be employed for PAT since 2012 (Filonov et al. 2012). Obviously, the iRFPs are better suited for related applications because they are derived from bacterial phytochrome photoreceptors (BphPs), which are generally known to exhibit relatively low fluorescence quantum yield and a high intrinsic extinction coefficient compared to GFP-like proteins as well as to have ideal spectral properties that absorb far-red light due to the incorporation of biliverdin IXα (BV).

In the related class, iRFP 713 (also simply called iRFP) with a peak absorption of 690 nm was the first one to be utilized for the related approach, followed by two variants iRFP 670 and iRFP 720 that have peak absorptions of 645 nm and 703 nm, respectively (Krumholz et al. 2014). However, although the two papers were those that employed BphPs for PAT for the first time, the developed probes did not relate to such an optical switching mechanism of BphP, as demonstrated by the following paper.

In 2015, Yao et al. of the Wang group reported an important achievement that harnessed the reversible optical switching mechanism of BphP1 to enhance the detection sensitivity of PACT at large depths as well as to achieve a super-resolution OR-PAM imaging (Yao et al. 2015a). BphP1 is a nonfluorescent BphP from the bacterium *Rhodopseudomonas palustris* (BphP1 is a short name for *Rp*BphP1). Interestingly, as with all canonical natural BphPs, it also exhibits very interesting photochromic behavior: it undergoes reversible Pfr → Pr photoconversion (i.e., *trans* to *cis* conformation) upon 730–790 nm light irradiation and Pr → Pfr photoconversion (i.e., *cis* to *trans* conformation) upon 630–690 nm light irradiation. Here, the two states Pr and Pfr are named according to the meaning "pigment red absorbing" and "pigment far-red absorbing," respectively; the two conformational states are the result of the photoisomerization of BV. Notably, due to the photoconversion, the two states exhibit different absorption spectra, where the Pr state has a dominant absorption at 670–700 nm (the Pr state) and the Pfr state at 740–780 nm.

It is known that BphPs consist of two key elements: a photosensory core module (PCM) and an output effector domain (Fig. 9g). The PCM is further divided into the three protein domains PAS, GAF, and PHY; these are connected with an α-helix linker, and inside a chromophore-binding pocket of the GAF domain, photoisomerization of the BV—which is covalently attached to the cysteine (Cys) residue of the PAS domain—leads to the two conformational states called Pfr and Pr. In the unbound state, switching does not affect the absorption properties of BV even though photoisomerization occurs; however, when covalently bound to the protein barrel, the two conformational states exhibit the distinct absorption spectra, as illustrated in Fig. 9h, in which the Pfr state is called "the ON state" and the Pr state is referred to as "the OFF state." The Pfr state is also regarded as "the ground state," because although BphP1 is photoswitched to the PrF state, it naturally relaxes back to the Pfr state with a half-life of ~210 s.

In the report, the authors utilized the interesting on/off switching mechanism of BphP1for the in vivo differential imaging of BphP1-expressing U87 glioblastoma cells injected into the left kidney of a nude mouse 1 week ahead. According to the time sequences presented in Fig. 9i, the animal was irradiated with 630 nm and 780 nm and then the PA images were acquired. Here, it is important to note that the PA images were acquired only with 780-nm excitation, although the times sequence shows that during the PA signal acquisition period, the 630-nm light was also fired—that is, in each switching cycle, the 630-nm illumination (16 s) was not for PA signal generation but only for inducing the Pr → Pfr photoconversion, whereas the 780-nm illumination (16 s) was for PA signal generation as well as the Pfr → Pr photoconversion. Consequently, they could acquire the two sets of 780 nm-excited PA images of BphP1-expressing glioblastoma in the two different states (ON and OFF) of BphP1.

Since the two PA image data sets were at the same 780-nm wavelength, it was possible to accurately extract only the tumor signal by simply subtracting the two data sets pixel-wise (Fig. 9j). Figure 9k illustrates the extracted tumor image overlaid on the OFF-state PA image that depicts blood vasculature. Indeed, the showcased differential imaging concept is a very important demonstration that illustrates another technical scope of PAT in relation to contrast creation. Even though a local optical fluence information is completely unknown due to the inhomogeneity of intervening complex biological tissues and also the PA signal change at the two different states is obscured by the overwhelming background vascular signals, it was possible to accurately separate only the reporters' signals that came from large depths approaching ~10 mm. Based on the same method, they were also able to successfully visualize the growth and metastasis of the glioblastoma tumor—which was originally inoculated into the right liver lobe of a mouse—for over 30 days (Fig. 9l), and also the mouse brain with a glioblastoma located at ~3 mm beneath the scalp surface (not presented here). Moreover, by applying a similar principle, they achieved a super-resolution PA imaging with a lateral/axial resolution of ~141 nm/400 nm (not presented here).

After the first demonstration was presented, the same group as well as other groups also presented more advanced photoswitchable probes—such as DrBphP (Li et al. 2018a) and sGPC2 (Chee et al. 2018)—which exhibited more improved performance and/or other optical characteristics. As the concept provides a new research paradigm in PAT-based RG imaging, it goes without saying that this is likely to lead to numerous important new findings and follow-up technical advances.

In this section, several notable examples of PAT related to cancer research have been introduced. As evident, related examples have a close relation to the previous two sections. In short, based on the multi-scale, multi-contrast, real-time RG imaging capability of PAT, various forms of cancer research can be realized not only in the illustrated areas but also over the entire brain and whole-body areas of small animals. Moreover, although not sufficiently covered in this section, the relatively wider options in creating endogenous contrasts as well as the great flexibility in engineering dedicated contrast agents for PAT is also likely to have a profound impact in clinics by presenting viable solutions against issues of toxicity and insufficient targeting ability of conventional contrast agents developed for other imaging modalities.

Other Notable Biological Applications

In relation to the small animal whole-body imaging capability of PAT, there is another important application that needs to be noted. The examples presented in the previous section mostly showcased how PAT could be utilized for disease models-based preclinical studies, especially focusing on cancer application. However, PAT is also expected to open a new research paradigm in developmental biology based on the safe and real-time, deep imaging capabilities, which however cannot be provided simultaneously in conventional imaging modalities (Ripoll et al. 2015; Laufer et al. 2012b).

As depicted in Fig. 10, Laufer et al. of the Beard group demonstrated longitudinal visualization of organ and vascular genesis during embryo development in a pregnant mouse (Laufer et al. 2012b). Very surprisingly, numerous details of several major organs—such as heart, liver, umbilical cord, and many other major vessels—were clearly visualized even in the noninvasive in vivo imaging, which was performed via the abdominal wall of the mother mouse. Further, in another experiment performed ex vivo within one hour after the sacrifice of the mother mouse, since no

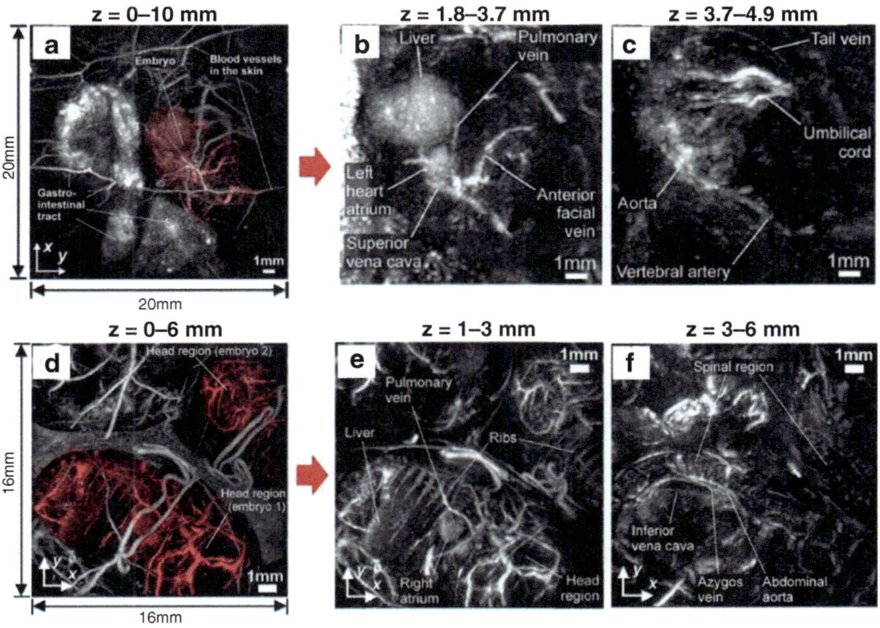

Fig. 10 In situ PAT images (640 nm) of the abdomen of pregnant female mice (Laufer et al. 2012b). (**a**) In vivo MAP of a 3D image data set of the abdomen of a pregnant mouse (E15.5) containing a single embryo. (**b**) MAP for z = 1.8 to 3.7 mm, (**c**) MAP for z = 3.7 to 4.9 mm. (**d**) MAP of the complete 3D image data set (z = 0 to 6 mm) obtained from the abdomen of a pregnant mouse containing two embryos (shaded red). (e) MAP for z = 1 to 3 mm. (f) MAP for z = 3 to 6 mm

motion artifact was involved, the PAT system enabled the visualization of a large number of additional details related to smaller vessels, like Azygos vein, and not only related to major organs, such as the spinal cord, ribs, and right atrium.

Although the presented demonstration was based on the FP US sensor-based PAT system, which has not yet sufficiently advanced to provide real-time imaging, the authors presented a feasible projection that related imaging approach could also be applied to such applications as the longitudinal imaging studies of transgenic mouse embryos, in which specific genetic knockouts induce heart and vascular malformations. Since all such studies are currently performed ex vivo based on MRI or optical projection tomography—which require extensive preparation of the excised embryos and also may involve the draining of blood and the use of contrast agents, such as gadolinium or fluorescent dyes—it is highly expected that the new technical addition provided by PAT could have profound impact on various developmental biological studies (Laufer et al. 2012b).

Clinical Applications

The key benefit of the clinical application of PAT is that its related systems are compatible with conventional US imaging functions. Although a number of previous biological applications also demonstrated the capability of combined PA and US imaging, since the inherent integration benefit—apart from the functional and molecular imaging capability—was more pronounced in clinical applications, it must be importantly evaluated as the core technical aspect that could make PAT more easily acceptable in clinics. Needless to say, US imaging already occupies a solid position in clinics as many of its related procedures are regarded as gold standards. Thus, PAT's compatibility with US complements the weaknesses of US and doubles the value of the technology.

Thus far, the clinical applications of PAT have been pursued over various application areas. However, considering the number of publications and likelihood of success, we selected several major application areas and made the following subjective categorization: 1) breast cancer diagnosis, 2) dermatological application, and 3) endoscopy and minimally invasive imaging application. In addition to those, we allotted a fourth topic called PA angiography and lymphangiography because several studies related to this topic have reported very promising results.

Breast Cancer Diagnosis

The application of PAT for breast cancer imaging has been a major topic of interest since the inception of the technique in the 1990s (Esenaliev et al. 1997; Oraevsky et al. 1999; Kruger et al. 1999a, b); it has played a leading role in PAT development, as numerous important breakthroughs associated with imaging performance, such as imaging depth, were achieved in relation to this application. However, it must be noted that although over 30% of all PAT papers pursuing clinical applications disclosed research objectives addressing this application, thus far, real images of human breasts have been mainly reported by two PAT pioneers and several other research groups that emerged later; In other words, most of the research from other groups were limited to animal-based studies associated with disease models or contrast agent developments. This feature can probably be explained by the fact that related study requires a long technical accumulation and a number of expensive system elements, such as high-power laser, array transducer, and multichannel DAQ. Thus, in this section, we introduce related technological developments, focusing on the results reported by several major research groups.

The first case that applied PAT to breast cancer diagnosis was presented by Oraevsky and Kruger in the 1990s (Esenaliev et al. 1997; Oraevsky et al. 1999; Kruger et al. 1999a, b), and thereafter, the two researchers made many important initial contributions to the development of clinical breast imaging systems up to 2013. Meanwhile, in 2004, Manohar and Steenbergen et al. also appeared as new players and began reporting their work on system development (Manohar et al. 2004; Manohar et al. 2005). In short, until 2013, these were the three main researchers reporting their works on system developments for practical clinical translation.

The first paper on real human breast image production appears to be the one reported by Oraevsky et al. in 1999 (Oraevsky et al. 1999). In this work, they presented the first human breast image ex vivo and also reported the important observation that PA image contrasts between tumors and normal tissues were as great as ~300% and also much greater than those of conventional ultrasonography, MRI, and X-ray mammography images, which were acquired concurrently. Although it was performed ex vivo, they conjectured that the strong contrast was attributed to the dense microvascularity of the tumors because the imaging was performed within the first 10–20 min after harvesting tumor-bearing breast samples from a radical mastectomy. In the report, they named the system the laser optoacoustic imaging system (LOIS), and it was equipped with a 12 element-based linear array transducer (0.05–3 MHz, polyvinylidene difluoride (PVDF)) and 1064-nm laser system.

In the next year, the group presented another report on a second-generation breast imaging system named LOIS-2 (Andreev et al. 2000). The system was also constructed based on a PVDF-based array transducer. However, the number of detection channels in the array transducer was greatly increased to 32, and the array had an arc-shape rather than the previous linear shape. It was an important modification, and its particular geometric features have been adopted in many other PAT systems from other groups.

Almost one decade later in 2009, the group reported the latest version of the arc-array transducer-based LOIS along with the results of a pilot clinical study obtained from 27 patients (Ermilov et al. 2009). It appears to be the first clinical study that involved such a large number of subjects.

Figure 11a shows a schematic of the latest system in which the number of the detection channels of the arc-array transducer was further increased to 64 (Fig. 11b) from the previous 32, and the laser illumination was provided from one side by busing an optical fiber bundle. The updated configuration made it possible to acquire a single slice breast image in vivo in a fashion similar to craniocaudal or mediolateral projection, with a spatial resolution that was better than 0.5 mm. To excite the breast

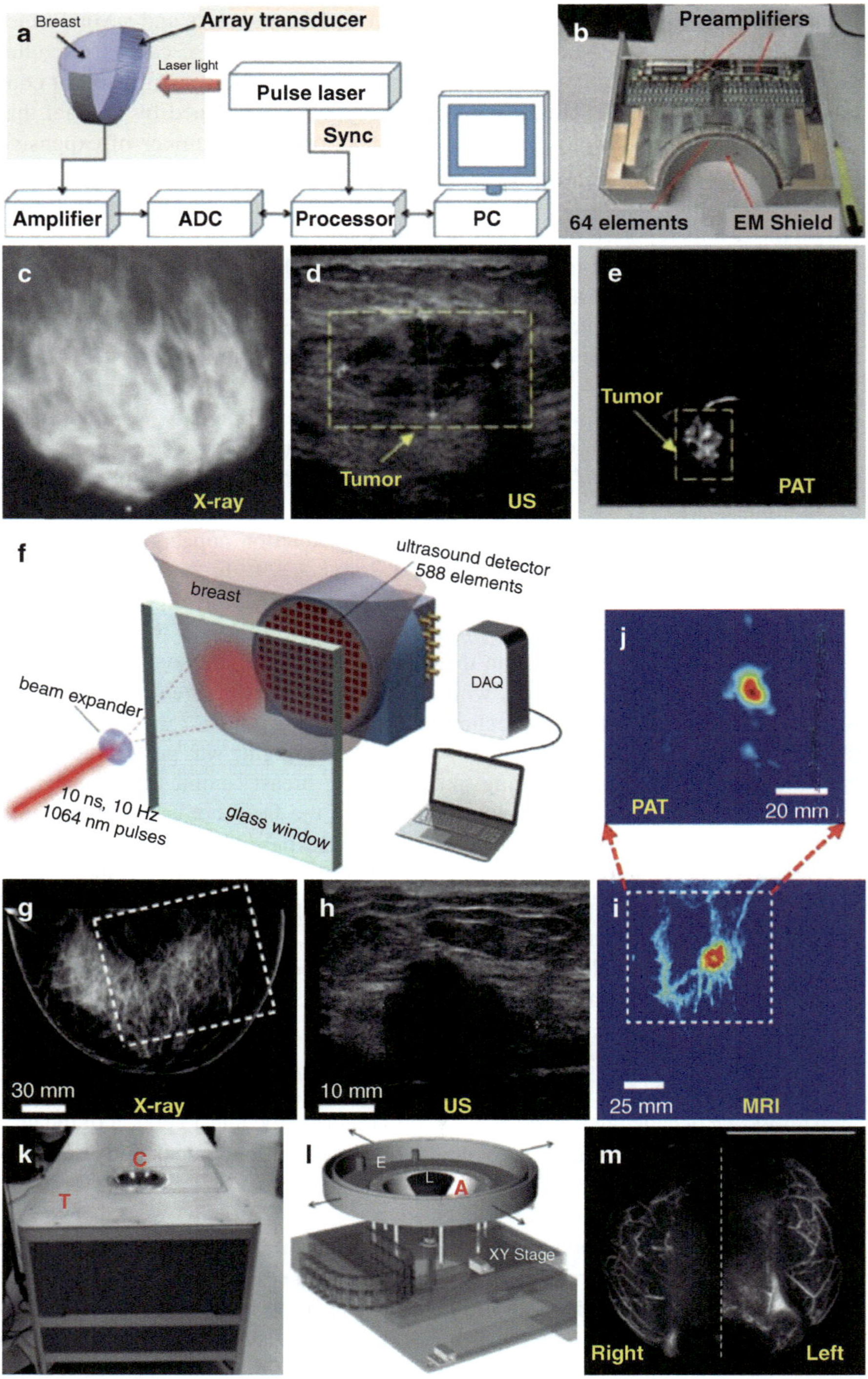
a
Breast
Array transducer
Laser light
Pulse laser
Sync
Amplifier
ADC
Processor
PC
b
Preamplifiers
64 elements
EM Shield
c
X-ray
d
Tumor
US
e
Tumor
PAT
f
ultrasound detector
588 elements
breast
beam expander
DAQ
10 ns, 10 Hz
1064 nm pulses
glass window
j
PAT
20 mm
g
30 mm
X-ray
h
10 mm
US
i
25 mm
MRI
k
C
T
l
E
L
A
XY Stage
m
Right
Left

Fig. 11 PAT breast imaging systems and imaging results from related pilot clinical studies performed based on the systems. (**a**, **b**) Schematic (**a**) of a 64-element arc-shaped array transducer-based breast imaging system (LOIS-64) and a photo (**b**) of the employed array transducer (Ermilov et al. 2009). (**c**–**e**) A set of imaging results acquired from a malignant tumor based on conventional X-ray mammography (**c**), US (**d**), and LOIS-64 (**e**). High contrast of the object in the PA image (**e**) implies the advanced angiogenesis indicative of a malignant tumor. (**f**) Schematic of the 2D 588-elements US array transducer-based PA breast imaging system (Heijblom et al. 2015). (g–j) A set of imaging results acquired from a patient with IDC (grade 3 of 19 mm) based on conventional X-ray mammography (**g**), US (h), MRI (**i**), and PA MAP image (**j**). The X-ray mammogram (**g**) did not show any abnormalities, whereas the US image (h) revealed a highly suspect image contrast for the 22-mm palpable mass. The white boxes in (**g**) and (**i**) indicate the FOV of the PA image presented in (**j**). (k) Photo of the hemispherical detector array (HDA)-based PAT breast imaging system depicting the exam Table (T) and the breast positioning cup (C); the HDA is located below this (Kruger et al. 2013). (l) Schematic depicting the HDA (A) mounted on an XY-translational stage. (m) Medial-lateral MAP image of the right- and left-breasts of a healthy volunteer with known mammographic breast density (heterogeneously dense) and brassiere cup size (D cup), which was acquired by the system (**k**)

tissue, a 757-nm laser beam with a single pulse energy of approximately 380 mJ was employed, and it was uniformly expanded over a 70-mm diameter area on the breast surface, which yielded an optical fluence of 10 mJ/cm^2.

In addition to the system result, they also accomplished a very promising outcome from the clinical study: They successfully visualized 18 malignant tumors out of 20 (based on the number confirmed by biopsy) using LOIS, excepting those cases in which PA images could not be obtained due to technical issues. Importantly, among the 18 visualized tumors using LOIS, five were not seen on X-ray mammography images, whereas only one tumor was seen on the mammography image. Figure 11c–e shows an example in a breast cancer case that eventually turned out to be a poorly differentiated infiltrating ductal carcinoma grade 3/3 (according to the biopsy). As shown in Figure 11c, the tumor could barely be localized on the mammography image acquired from the radiologically dense breast, whereas it was clearly visualized in the PA image (Fig. 11e). The US image (Fig. 11d) also helped to identify a 23 × 15 mm size tumor located at the depth of ~21 mm; however, the tumor margins were not clear. Furthermore, the tumor contrast in the US image was lower than that in the PA image, a result that was attributed to the strong signal generation by the associated microvasculature.

In 2004, Manohar et al. of the Steenbergen group in the Netherlands reported a different type of breast imaging system that was designed to image an entire breast based on parallel plate geometry (Manohar et al. 2004). Since then, the group has made continuous reports on the progress of related system development until the mid-2010s. In 2007, they also reported their first human breast imaging result obtained by the system from 13 subjects (Manohar et al. 2007); however, here we introduce their later report published in 2015 (Heijblom et al. 2015) since it was more comprehensive.

Figure 11f depicts the latest version of the related system reported in 2015 as described earlier; its basic configuration and key elements have remained almost the same since the first published paper (Manohar et al. 2004), except the increase in the number of detection channels of the engaged DAQ system. The main feature of the system is the use of a 2D US detector array consisting of 588 elements distributed over an 85-mm diameter area (pitch: 3.175 mm). As a result, they could achieve a relatively larger FOV (90 × 85 mm^2) compared to those of other groups. The size of each element was 2 × 2 mm^2, and a vast number of detection channels could be created by forming corresponding electrodes on the surface of a 110-μm thick PVDF film, which yielded a center frequency of ~1 MHz. For breast imaging, the target was gently pressed between the disc-shaped transducer and a glass window and then photoacoustically excited by a 1064-nm laser beam (~350 mJ) over a 35-mm^2 illumination area.

With this system, they performed a pilot clinical test that imaged 29 human subjects with symptomatic breast malignancies. Then, they compared the acquired PA images with conventional X-ray and US images and also with MRI and vascular staining histopathological images for subsets of cases. Surprisingly, they acquired a promising result that successfully identified lesions in suspect breasts at the expected locations in 28 out of 29 cases. Moreover, they also made the crucial observation that the PA contrast of the lesions was independent of the conventional mammographically estimated breast density, whereas there was a significant drop in lesion contrast in X-ray mammography for high-density breasts compared to low-density breasts.

In addition, a large number of other important findings and conclusions were reported in the paper. Among these are a set of results that correspond to the aforementioned observation and are presented in Figure 11g–j. The data were obtained from a subject with a past history of infiltrating ductal carcinoma (IDC), which was also confirmed by histopathological analysis in the study. However, as shown in Figure 11g, no abnormalities were observed in the X-ray image, whereas the US image (Fig. 11h) revealed the presence of an unsharply delineated, irregular, hypoechoic mass that was 22 mm, which was highly suspected of malignancy as per the US evaluation. Thus, they decided to proceed to an MRI and PA examination, from which they were able to identify that the PAT image (Fig. 11j) could accurately delineate the highly vascularized region, as with the MRI image (Fig. 11i). Although the individual blood vessels were not visualized in their PA images due to the relatively poor spatial resolution (~3.5 mm) of the system, they attributed the acquired high contrast production to the increased vascularity inside the tumors as it could be confirmed by the immunohistochemistry analysis based on CD31 staining.

As mentioned earlier, Kruger also pioneered PAT since the beginning by making many initial crucial contributions to instrumentational and theoretical works (Kruger 1994; Kruger and Liu 1994; Kruger et al. 1995). In particular, he reported the first prototype of PAT image reconstruction algorithms (Kruger et al. 1995). Furthermore, for breast imaging applications, he suggested the hemispherical detector array (HDA)-based signal detection concept for the first time (Kruger et al. 1999a, b). Among many of his important papers, we mention only three papers.

First, in his 2000 paper (Kruger et al. 2000), Kruger et al. reported the first in vivo human breast images acquired from five subjects. Notably, they acquired the images by employing a 434-MHz, high power (25 kW peak power) radio wave generator rather than a typical pulsed laser. Thus, the related concept was termed thermoacoustic tomography by the researchers. Although the images were poorer than those in his subsequent papers, this paper was the first to present in vivo breast images over the entire PAT area. Approximately one decade later, Kruger et al. disclosed a new version of a breast imaging system that was constructed based on a pulsed laser, and they published two studies in 2010 (Kruger et al. 2010) and 2013 (Kruger et al. 2013), in which a different scanning mechanism and a different number of transducer elements were applied. Although the former paper was also technically important, we focus here on the second paper (Kruger et al. 2013) because it presented 3D breast vasculature images with superior quality compared to any other system reported so far at that time.

Figure 11k depicts the system reported in the paper. Basically, the overall system configuration and scanning mechanism were inherited from its former version (Kruger et al. 2010). However, in this study (Kruger et al. 2013), they greatly improved the spatial resolution and FOV by employing a 512-elements array transducer (2 MHz) and performing a spiral scan, which differed from the former 128-element (5 MHz)-based rotational scanning mechanism (Kruger et al. 2010). The main advantage of the spiral scanning method lay in its flexibility in setting the lateral FOV, which can be set arbitrarily, exceeding the demonstrated size in the paper (i.e., 24 cm in diameter), while conserving the uniform spatial resolution of ~0.42 mm over the imaged volume. The schematic presented in Figure 11l was the key module that enabled the spiral scanning.

Figure 11m depicts one of the representative images acquired from four subjects. Surprisingly, vasculature details even at a depth of ~4 cm were clearly visualized using a laser pulse energy of ~280 mJ (756 nm) irradiated to the breast over a 60-mm diameter area and then collecting PA data over the described detection points. Previously, the researchers also claimed a similar imaging depth. However, the visualized vasculature was not as detailed and global as the presented image. Indeed, the paper represented a great achievement, demonstrating that the developed system could image breasts up to the DD cup size, which accommodates 90% of all women in the United States.

Meanwhile, in 2014, a Japanese group teamed with Canon Inc. reported the results of a clinical study that was conducted using their first prototype named PAM-01 (Kitai et al. 2014). The overall configuration of their first version was similar to that of the Netherlands group (Heijblom et al. 2015) in terms of the imaging concept, where an imaged breast was placed between two parallel plates. However, they applied a different illumination scheme that adopted dual-side illumination, in which another illumination was provided around the detector made of a 15 × 23 matrix array (1 MHz). Consequently, their imaging resolution was much better (~2 mm) than that of the Netherlands group and could also successfully visualize breast cancer lesions in 20 out of 27 subjects. More comprehensive results were reported in their other paper published in 2015 (Fakhrejahani et al. 2015). After the study, the group also reported another improved version of the system (named PAM-02). It featured an advanced imaging performance capable of providing PA and US images simultaneously with an improved spatial resolution, and it also produced a sO_2 image (Asao et al. 2016).

However, in the most recent paper published in 2017, with Toi as a lead author (Toi et al. 2017), the group reported an achievement that could be considered one of the most significant advancements in the practical clinical translations of the PAT breast imaging applications. In the report, they also released their new imaging setup (named PAI-03) and reported another pilot clinical study result acquired from

22 subjects with malignancies based on the new system. However, in the new study, they could acquire much clearer 3D vasculature images than the previous one (Asao et al. 2016) because they changed their system concept to the HDA-based breast imaging concept that was originally proposed by Kruger et al. (Fig. 11k).

Figure 12a and b–c respectively show the new PAT system developed for the study and a representative imaging result (PA and MRI images) acquired from a subject. As shown in the PA (Fig. 12b) and MRI (Fig. 12c) images acquired from the same area of a normal breast, the PAT system could depict detailed 3D vasculature and blood vessel branching structures along with morphological characteristics much better than conventional MRI based on standard contrast-enhanced imaging. The superior imaging performance was also evident in tumor regions, as shown in a

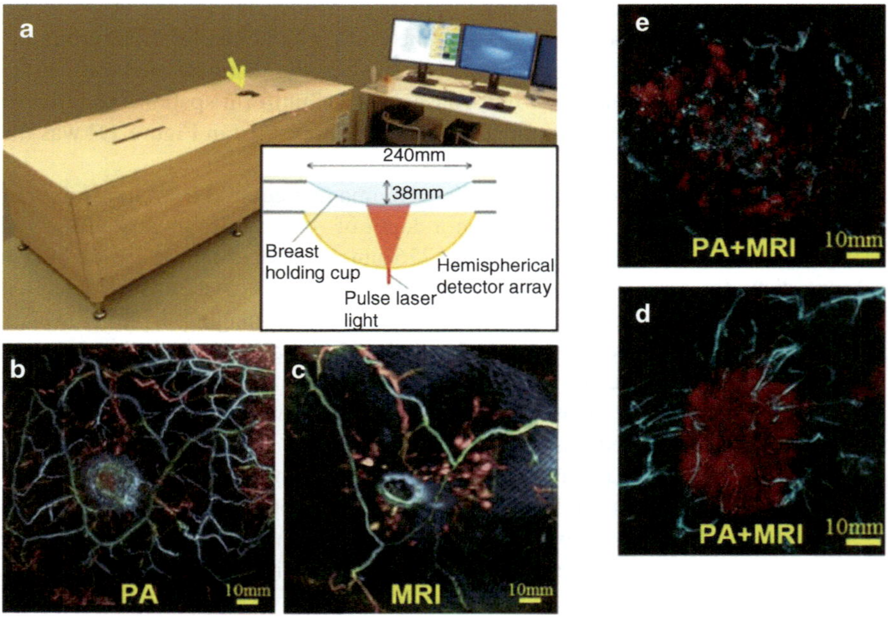

Fig. 12 PAT breast imaging systems and imaging results from related pilot clinical studies performed based on the systems. (**a**) Photo of the HDA-based PAT breast scanner (PAI-03) developed by Toi et al. (Toi et al. 2017). (**b**) PA image acquired from a normal breast. (**c**) Standard contrast-enhanced MRI image acquired from the same area of (**b**). (d) PA and MRI fusion image acquired from a 40-year-old woman with IBC (tumor is 47 mm in diameter). (**e**) Another PA and MRI fusion image acquired from a 46-year-old woman with DCIS; this reveals a trend toward increased identification of spotty intra-tumoral signals. (**f**) Photo of the dual-modality PA and US imaging system embodied by modifying a clinical US system (Garcia-Uribe et al. 2015). (**g**) Photo of the imaging probe, in which two fiber bundles were added on the two sides of a clinical linear array US probe. (**h**) An illustration showing the PA and US image-guided fine needle aspiration biopsy of an SLN. (**i**) Co-registered PA and US image of an SLN. (**j**) Schematic of the 512-element full-ring array transducer-based PAT breast imaging system (Lin et al. 2018). (**k**) Representative results were acquired from two breast cancer patients with radiographically dense breasts. Images from the left to right columns: X-ray mammography images; depth-encoded PA images; PA MAP image presented in sagittal planes marked by white dashed lines in the depth-encoded PA images; PA images showing the contrast-boosted tumors after applying the vessel density calculation algorithm (tumors are identified by green circles, and background images in grayscale are the MAP of vessels that lie deeper than the nipple)

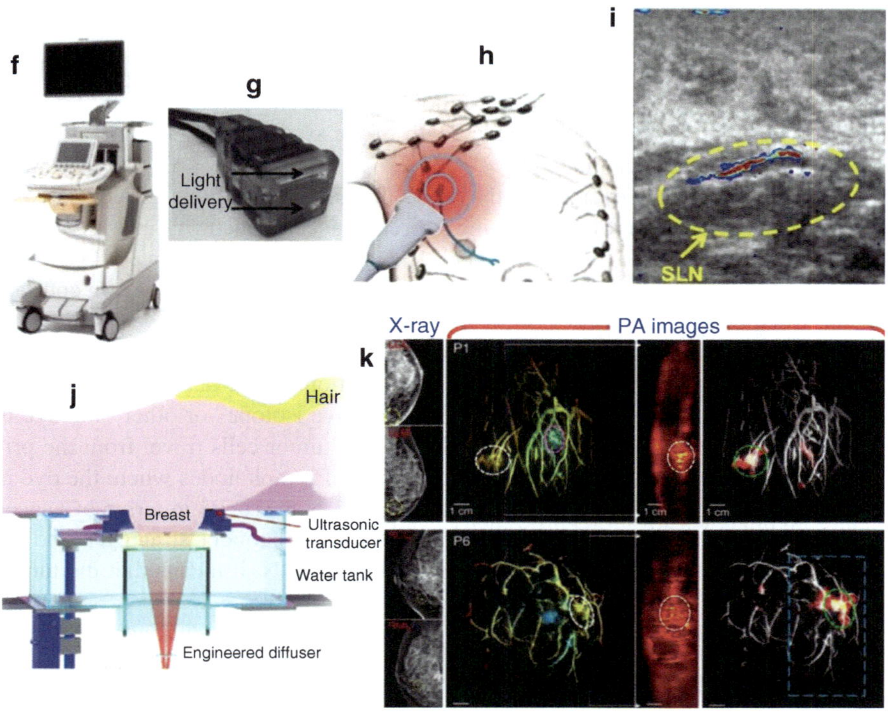

Fig. 12 (continued)

representative image (Fig. 12d) acquired from another subject with invasive breast carcinoma (IBC). Such high-quality vasculature images could be acquired because they adopted the HDA-based signal detection concept in which 512 US transducers (2 MHz) were engaged. Based on the system presented in Figure 12a, PA spatial resolutions were 0.57 mm along the horizontal plane and 0.37 mm along the gravitational direction.

In addition to substantial improvement in the imaging performance, they also reported many interesting findings obtained from the 22 subjects, including five cases of ductal carcinoma in situ (DCIS) and 17 cases of IBC and general conclusions derived from them. Among them, the notable ones were as follows: (1) More blood vessels were observed in the tumor-bearing breast than the contralateral side in most cases; (2) peritumoral vasculature was detected in 86% of all cases; (3) in IBC cases, most tumor-related blood vessels were centripetally directed toward the tumor (Figure 12d corresponds to a related example), and 93% of centripetal blood vessels appeared to be disrupted or rapidly narrowed at the tumor boundary; (4) a significant difference was observed between DCIS and IBC in terms of the vessels' centripetal vascularity; IBC cases tended to have a centripetal blood vessel structure (Fig. 12d), whereas there was a trend toward increased identification of intratumoral spotty signals among the five DCIS cases (Fig. 12e). In addition, although there was only one case, they observed one interesting phenomenon: Fine intratumoral blood vessels were visualized after chemotherapy, whereas no significant change in tumor size was detected in US imaging.

The reported results represent an important achievement highlighting the diagnostic value of PAT. According to the authors, with the employed laser pulse energy of ~200 mJ, the maximally imageable depth of the system was ~30 mm, although this is somewhat underestimated.

Another notable achievement in the breast imaging application of PAT related to the sentinel lymph node (SLN) mapping in breast cancer, which was investigated by the Wang group. Prior to the study in 2008, the group also demonstrated PAT-based SLN mapping in a rat model for the first time by using methylene blue as a contrast agent (Song et al. 2008).

In the lymphatic system, SLN is defined as the first node that drains from a primary tumor. From a practical perspective, its location can be determined based on the finding that an injected agent such as radioactive isotopes or other blue dye is taken up into the node in a manner similar to how tumor cells travel from the primary tumor. Thus, the first lymph node or cluster of lymph nodes where the dye is taken up represent the SLN. In clinical settings, SLN is crucial in terms of cancer staging because it guides the patient prognosis and treatment strategy based on biopsy results: that is, if tumor cells are found in an SLN, it means that the tumor has already metastasized from the primary tumor.

To demonstrate PAT's advantage in finding an SLN and guiding the related biopsy, Garcia-Uribe et al. of the Wang group developed a dual-modality PA and US imaging system by modifying a clinical US imaging system (Fig. 12f) and implementing an imaging probe (Fig. 12g), and they reported their first pilot clinical study result in 2015 (Garcia-Uribe et al. 2015). In the study, methylene blue dye with a peak absorption wavelength of 665 nm was employed as the contrast agent to visualize the SLNs stained by it (Fig. 12h), and a laser wavelength (667 nm) near the peak wavelength was chosen.

Excepting two cases in which the system malfunctioned, 16 women who had pathologically proven breast cancer and thus were scheduled to have an axillary lymph node dissection participated in the study and underwent SLN biopsy under the guidance of the developed system. Figure 12i shows a typical co-registered PA and US image in which the colored area corresponds to the SLN visualized by PAT imaging. Under the guidance of the system, they performed percutaneous needle biopsies after deploying a single titanium marker clip via a needle for 13 cases. Among them, the presence of the marker clip in the surgically removed SLN specimen could be confirmed in six cases.

Beginning with the SLN imaging project, the Wang group more aggressively pursued breast imaging application, and in 2018 the group reported a new type of imaging system that significantly improved spatiotemporal resolution by employing a full-ring 512-element ultrasonic transducer array (2.25 MHz) (Fig. 12j) (Lin et al. 2018). Although many of the aforementioned studies also demonstrated the imaging of a whole breast (Kruger et al. 2013; Toi et al. 2017), the imaging speeds of their systems were relatively slow due to the limited capacity of the employed DAQ or scanning mechanism. However, this new system enabled full 3D imaging of a breast within the approximate scanning time of a single breath hold (~15 s) by employing vast numbers of DAQ channels that were mapped to all the elements one by one. In the reported version, the scanning time was a kind of minimum time required to acquire a complete volumetric data set by adding additional mechanical translational motion to the array transducer along the elevational direction under the limited PRR (10 Hz) of the employed laser system. However, 2D cross-sectional imaging speed itself was as fast as 10 Hz.

As with the Kruger (Kruger et al. 2013) and Asao (Toi et al. 2017) groups' systems, their new imaging system was also able to provide a vasculature image of a full breast in 3D, and it was also able to depict vasculature details in a tumor region as fine as 255-μm along the transducer array plane; however, the image resolution along the elevational direction was relatively worse (~5.6 mm) due to the large extent (5 mm) of the transducer elements. Based on the performance, they were able to acquire high-quality 3D vasculature images from eight subjects (one healthy, seven cancerous). Among these, representative results acquired from two breast cancer patients with radiographically dense breasts are presented in Figure 12k.

Although the X-ray images (left-most ones) did not show any distinguishable image contrast, the PA images clearly visualized tumor regions. In addition, by applying the vessel density (local vessel number/local area) calculation algorithm they developed, they were able to further boost the image contrasts of the tumors (right-most ones). Moreover, they disclosed a new concept that differentiated tumor regions based on the compliance difference of tumor and normal tissues with respect to a mechanical deformation (such as a breathing motion as mentioned in the paper) by using the acquired PA images. In terms of the imaging depth, their system's performance was comparable with those of previous studies (Kruger et al. 2010, 2013) and reached ~4 cm.

In this section, several representative system configurations developed for breast imaging have been introduced. As shown, some of the earlier reported imaging systems suffered from either limited FOV or scanning speed, limited spatial resolution, and insufficient imaging depth. However, the two systems that were more recently reported (Lin et al. 2018; Toi et al. 2017) have sufficiently advanced to delineate numerous vascular details at much improved imaging speeds and depths. Consequently, they could successfully illustrate the core benefit of PAT for dense breast tissues for which conventional X-ray mammography does not work effectively.

Considering the speed with which related technology has progressed, we expect that PAT could attain its meaningful position in the near future since it can provide detailed vasculature associated with tumor development without the involvement of contrast agents and ionizing radiation. As angiogenesis is a well-known hallmark of cancer, PAT could be more useful than MRI in future breast cancer diagnosis and management, when considering such technical aspects, such as excellency in vasculature visualization, real-time imaging capability, better accessibility to the system with virtually no limitation to the frequency of use, and versatile compatibility with various contrast agents. Although clinically approved dedicated functional contrast agent is not available yet, we anticipate that it will be realized in the near future and thus greatly increase the utility of PAT in clinics.

To more successfully realize this goal, however, further improvement in imaging depth remains necessary. Moreover, in the case of the presented systems, equipping them with conventional US imaging function, which is technically possible, would be a desirable development. We imagine a future PAT system capable of whole breast imaging based on a single shot of laser pulse and capable of providing co-registered US images in real time.

Dermatological Applications

Currently, the application of PAT for dermatological uses is not large as compared to the breast imaging application, as the former only emerged in earnest in the late 2000s. However, several papers reported in recent years have demonstrated its promise in terms of clinical success. Although Oraevsky et al. also mentioned the dermatological application of PAT in an earlier 1990s study (Oraevsky et al. 1994b), full-fledged research outcomes could only be reported after the AR-PAM concept was first introduced by Maslov et al. of the Wang group in 2005 (Fig. 13a) (Maslov et al. 2005). Before 2005, image quality of the PACT systems that were typically available at that time were not good enough to visualize detailed morphologies in the skin. However, due to the substantial resolution improvement provided by the AR-PAM technique, vascular details distributed in a cutaneous tissue could be very clearly visualized at an axial resolution of ~15 μm and a lateral resolution of ~45 μm (Maslov et al. 2005), which represents an approximately 10-fold improvement compared to previous PACT images.

After reporting on the AR-PAM concept, researchers in the Wang group first attempted to showcase its applicability to dermatology by imaging various skin diseases or conditions. For example, at approximately the same time in 2006, Zhang et al. (Zhang et al. 2006a) and Oh et al. (Oh et al. 2006) in the Wang group reported the first PA image feature of an inoculated B16 melanoma tumor, which is a representative murine cancer cell line frequently utilized for human skin cancer studies.

Fig. 13 Representative systems developed for the dermatological applications of PAT and produced images. (**a**) Photo of an AR-PAM system (Maslov et al. 2005). (**b**) The first AR-PAM-based sO_2 image was acquired from a subcutaneous section of rat skin in vivo (Zhang et al. 2006a). (**c**) In vivo AR-PAM image of a subcutaneously inoculated B16-melanoma in an immunocompromised nude mouse, acquired using 584 nm and 764 nm (Zhang et al. 2006a). (**d**) Photo of a nevus located on the forearm of a subject (Favazza et al. 2011). (**e**) Corresponding AR-PAM image (570 nm) delineating the nevus and surrounding blood vessels. (**f**) Schematic of the OR-PAM system developed for cuticle imaging (Hsu et al. 2016). (**g**) Photo showing the imaged area. (**h**) Wide-field PA image of cuticle capillaries depicted with normalized PA amplitude. (**i**) Selected time-lapse images of single-RBC sO_2. (**j**) Time-averaged images (~10 s) of all time-lapse frames of sO_2 imaging. (**k**) Photo of the portable UB-RSOM system (Aguirre et al. 2017). (**l**) Schematic depicting the operation of UB-RSOM. The transducer was raster-scanned parallel to the skin surface, acquiring the PA A-line signals generated by laser illumination (532 nm) from two fixed fiber bundles. Note that the focal point of the transducer was kept above the skin surface. (**m, n**) UB-RSOM images of psoriatic (**m**) and healthy (**n**) skin. The histology images were acquired based on hematoxylin and eosin staining to compare with the corresponding UB-RSOM cross-sections

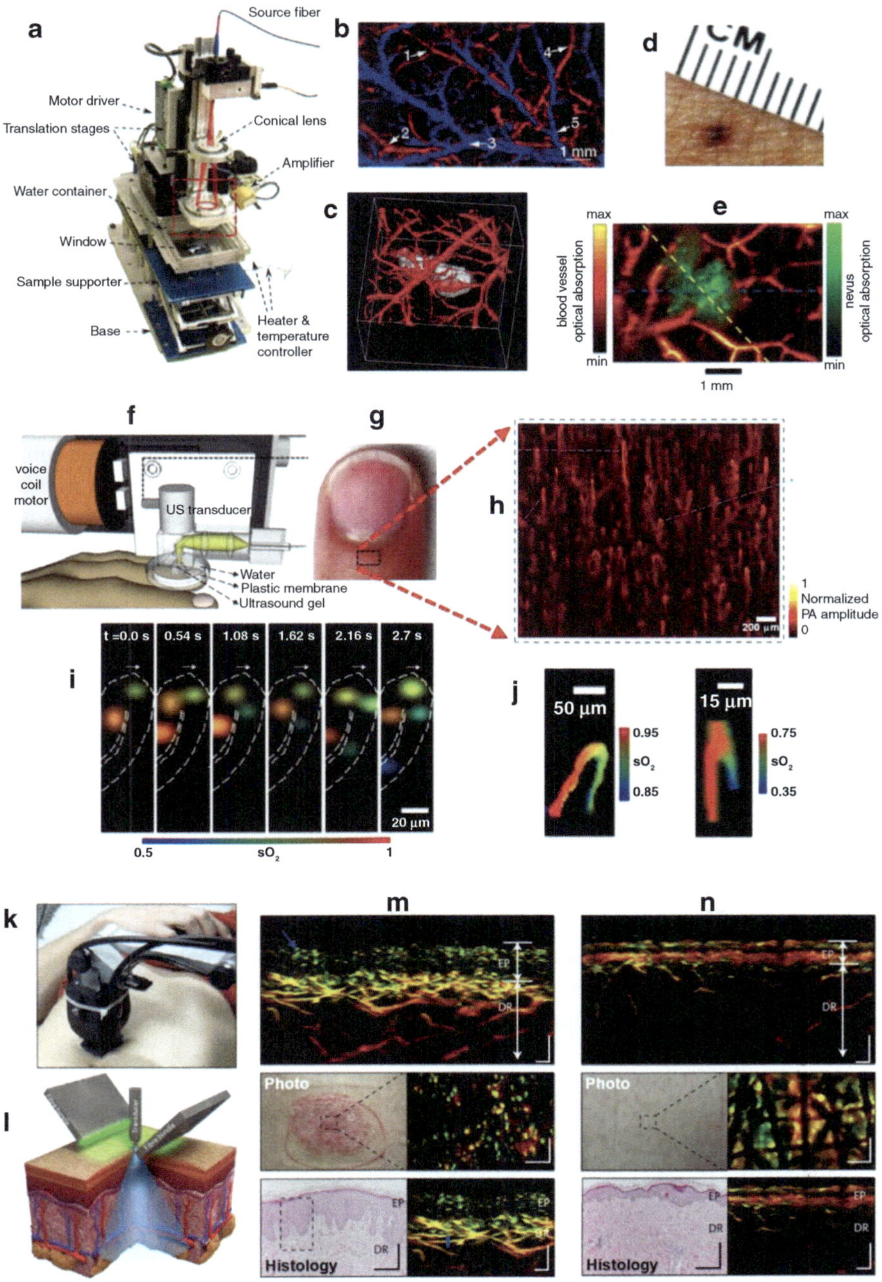
a
Source fiber
Motor driver
Translation stages
Conical lens
Amplifier
Water container
Window
Sample supporter
Base
Heater &
temperature
controller
b
1 mm
c
d
CM
e
max
blood vessel
optical absorption
min
max
nevus
optical absorption
min
1 mm
f
voice
coil
motor
US transducer
Water
Plastic membrane
Ultrasound gel
g
h
200 μm
1
Normalized
PA amplitude
0
i
t =0.0 s
0.54 s
1.08 s
1.62 s
2.16 s
2.7 s
20 μm
0.5
sO2
1
j
50 μm
0.95
sO2
0.85
15 μm
0.75
sO2
0.35
k
l
m
n
EP
DR
Photo
Histology

In particular, the former paper (Zhang et al. 2006a) presented a sO_2 map of blood vessels distributed in rat skin (Fig. 13b), which is referred to as the first experimental demonstration of the functional (i.e., sO_2) imaging capability of PAM. Moreover, they also presented the first PAM image of human palmar vessels distributed at a ~ 3 mm depth. Although the melanoma tumor image presented in the paper (Fig. 13c) was not from an actual human subject, but from an animal (nude mouse), the range of demonstrations presented in the paper presented an important event announcing the beginning of the dermatological application of PAM. In the same year, Zhang et al. presented another report that demonstrated the technique's diagnostic potential in evaluating the depth of a thermal burn. For this, they artificially induced a thermal burn on porcine skin and imaged it ex vivo (Zhang et al. 2006b).

Approximately 5 years later in 2011, Favazza et al. of the Wang group presented a report that demonstrated more direct potential for the AR-PAM technique for in vivo dermatological applications by imaging a nevus developed in real human skin (Fig. 13d) (Favazza et al. 2011). Clinically, nevus is known as a benign tumor of melanocytes, which should be accurately differentiated from melanoma in order to avoid unnecessary biopsies. However, many pigmented lesions still display some of the clinical characteristics of melanoma, and their discrimination still relies heavily on clinical suspicion based on evaluation parameters such as asymmetry, border irregularity, color, size (lateral dimensions), and change (evolution, lesion growth). Thus, if a lesion raises sufficient concern, it is biopsied and histologically analyzed, and a definitive diagnosis is rendered.

To demonstrate related applicability, they imaged a volunteer with nevus, and effectively visualized the affected areas and surrounding microvasculature using a 570-nm laser excitation wavelength as shown in Figure 13e. In addition, by analyzing the depth-resolved PA images, they also demonstrated that the PAM system could accurately identify the thickness, depth, and volume of the lesion, all of which could be confirmed using a histological analysis performed on the excised tissue.

In addition to the nevus imaging experiment, they also performed a normal cutaneous imaging experiment, and they found an interesting result; namely, a large difference in the vascular patterns between the forearm and palm. Based on the demonstrated high-resolution vasculature imaging capability of PAM, they suggested that the PAM technique could serve as a new tool for characterizing and assessing cutaneous malignancies or other skin disorders, such as melanoma and hemangioma in situ, and also for diagnosing and assessing systemic diseases that affect the microvasculature, such as diabetes, renal disease, hypertension, and many other cardiovascular diseases.

Meanwhile, in 2008, the OR-PAM technique, which is another version of PAM that was invented to provide an optical-resolution PA image, was first introduced to the PAT field by Maslov et al. of the Wang group (Maslov et al. 2008). Although only relatively few researchers have investigated the dermatological application of the OR-PAM technique, it has immense potential as it possesses a unique functional imaging capability in terms of microcirculation. According to the literature, most OR-PAM-based research to date have focused on other research subjects, such as brain or cancer imaging studies that are based on animal models rather than human subjects.

Of the several papers that demonstrated the dermatological application of OR-PAM, Hsu et al. of the Wang group in 2016 showcased more direct potential for dermatological application (Hsu et al. 2016). Using a fast voice-coil scanning-based OR-PAM system (Fig. 13f), which was an improved version from the original system presented in ref. (Wang et al. 2013) in terms of the dual-wavelength imaging capability, they investigated the vascular structures and other functional parameters (sO_2) of human cuticles for the first time in vivo (Fig. 13g). They chose the cuticle region because of its ability to present microcirculation well, and they imaged nine healthy subjects.

The cellular-level and high-resolution imaging capability of the system (3 μm lateral and 15 μm axial) made it possible for the detailed structures of the imaged capillary loops to be clearly visualized in the MAP image (Fig. 13h). Moreover, the high-speed imaging capability providing 100 B-scan frames enabled the tracing of a single RBC flow in a capillary loop along with the record of its sO_2 change in real time (Fig. 13i). Surprisingly, by averaging multiple sO_2 images acquired in such a manner for ~10 s (Fig. 13j), they discovered that more oxygen was released in the curved cuticle tip than in other regions of the capillary loop. Furthermore, they observed that the RBC flow speed decreased in the tip region. Based on further analysis, they concluded that in addition to the decreased RBC flow speed, other factors, such as the drop of the partial oxygen pressure in the tip region, drove RBCs to release more oxygen in the tip region.

As the cuticle region serves as a window representing the status of microcirculation, monitoring such functional parameters at the fundamental level of the physiology of oxygen transport can potentially help physicians define clinical standards for early-stage diagnosis and evaluation of perivascular diseases, such as Raynaud's phenomenon and systemic scleroderma, before capillaries undergo observable changes in morphology.

Meanwhile, in 2017, Aguirre et al. of the Ntziachristos group also reported a research outcome that was important enough to rekindle interest in this application area (Aguirre et al. 2017). By using a raster scan optoacoustic mesoscopy implemented in ultra-broadband (10–180 MHz) detection mode (UB-RSOM) (Fig. 13k and l), they visualized detailed vasculature of normal and psoriatic skins with a spatial resolution as high as 15 μm. However, it should be noted that such a high spatial resolution could be sustained up to a 2-mm depth, and it was not achieved by applying the OR-PAM concept, but by employing a custom-made focused ultrasonic transducer (f = 3 mm, NA: 0.5, $LiNbO_3$) with the ultra-broad detection bandwidth (170 MHz) and a laser system with an ultra-short pulse duration of 0.9 ns. Furthermore, to effectively utilize the detected broad frequency components, they applied a frequency band equalization technique that they developed, and they finally produced images by applying a dedicated image reconstruction algorithm that was based on a beam forming concept.

Figure 13m and n depicts the representative imaging results of psoriatic and healthy tissues, respectively. As shown in the three UB-RSOM images presented in the left column, elongation and dilation of capillary loops (visualized in green), thickening of the epidermal layer due to prominent skin acanthosis, and increased vessel diameter and vascularization in the dermal layer were clearly observed compared to healthy skin (right column). Furthermore, the layered structure observed in the healthy skin disappears in psoriasis, and coronal UB-RSOM images close to the surface of the psoriatic skin were markedly different from the corresponding images of healthy skin. Observation of the histology images revealed that the overall features provided by the label-free UB-RSOM demonstrated good agreement with them.

Indeed, the presented results are remarkable achievements that demonstrate the potential of PAT for the crucial issue of dermatological application. However, apart from its dermatological implications, it must be noted here that although they produced high-quality images over a large volume of tissue (~4 × 2 × 1 mm^3) with a high spatial resolution (lateral/axial: 18.4/4.5 μm) that remained mostly consistent over the entire volume, the scanning time was merely ~70 s, which was a rather short scanning time compared to that required by large scanning points like 266 × 135 scanning points. This result could be explained by the synergetic effect of the employed ultra-broad band acoustic detection, frequency band equalization, and dedicated image reconstruction techniques, and it would have probably been impossible if the conventional OR-PAM scanning concept was applied.

Considering this aspect, we expect that the presented imaging concept could provide a more effective solution to making a seamless connection between OR- and AR-PAM techniques; that is, visualizing those structures lying around the boundary between ballistic and diffusion regimes without drastic change of the spatial resolution. Although several researchers have attempted to address the issue, unsatisfactory resolution zones remain because they constructed related systems by simply merging OR- and AR-PAM concepts.

Currently, only a relatively small number of studies have explored this domain of application. However, continuous attention to this area may be needed because interest in beauty and healthcare has steadily increased recently. In addition, it should be noted that after the paper, the Ntziachristos group also recently presented a report that applied artificial intelligence (AI) to skin vasculature analysis (Moustakidis et al. 2019).

Endoscopy or Minimally Invasive Imaging

Another important clinical application of PAT is minimally invasive imaging, and continuous research has been made in this area in recent years. As already demonstrated by many studies based on precedented endoscopic methods, such as endoscopic OCT (Tearney et al. 1997; Yun et al. 2006; Gora et al. 2013) and confocal endoscopy (Kiesslich et al. 2004, 2006, 2007), the optical approach is known to be superior to conventional endoscopic ultrasound (EUS) (Shami and Kahaleh 2010; Akahoshi and Bapaye 2012) in terms of disease specificity, and as an optical imaging modality, PAT-based minimally invasive imaging applications have also gained validity based on their unique imaging capability that provides rich functional and molecular information at clinically relevant depths. The importance of related applications is evident from the recent increase in the number of related publications: Currently, it forms the second largest clinical application area after breast cancer application.

Reviewing the literature pertaining to this application area revealed several initial papers that were published in the 1990s and that addressed gastrointestinal (GI) tract (Oraevsky et al. 1997) or intravascular imaging applications (Chen et al. 1993; Beard et al. 1998). Thereafter, two more studies reported in 2001 and 2005 pursued the same applications (Viator et al. 2001; Henrichs et al. 2005). However, although these reports also recognized the importance of related research, earnest research activities arose in the latter part of the 2000s, as Sethuraman et al. of the Emelianov group reported the first PA image acquired from a rabbit aorta ex vivo in 2007 (Sethuraman et al. 2007). In the paper, not only did they present the aorta image, they also proposed a conceptual idea on the related imaging device to be embodied in an integrated PA and US imaging design. Indeed, it was an important event that reignited attention because it illustrated how the PA and US images differed and what synergetic effect could be derived through integrated dual-modality imaging, more details of which are presented in the next section titled "Intravascular photoacoustic imaging."

Unlike conventional EUS and endoscopic OCT probes, a miniaturized PA imaging probe requires the integration of optical and acoustical elements in a single imaging probe, which creates a challenge and thus slows down the technology's progress. Nevertheless, the minimally invasive imaging approach would ultimately represent one of the most important clinical application directions of PAT because the typical imaging depth of PAT is less than ~5 cm; however, there are many important clinical issues that can be sufficiently covered by this depth. Moreover, in clinics, there are already well-established US-based endoscopic tools, such as EUS, transesophageal echocardiography (TEE), intravascular ultrasound (IVUS), transrectal ultrasound (TRUS), and transvaginal ultrasonography, which can be potentially integrated with their PA counterparts.

Related to this fact, minimally invasive imaging application of PAT has been pursued for the three main endoscopic divisions: GI, intravascular, and urogenital imaging applications. Thus, in this section, we introduce major achievements related to these areas.

Gastrointestinal Tract Imaging

After categorizing documented reports according to their application areas, it was obvious that compared to the intravascular imaging application to be presented next, relatively fewer studies related to Gastrointestinal (GI) application have been reported. Nevertheless, GI application is a crucial endoscopic area because it forms the largest endoscopic clinical application, and there are also many important diseases, such as various GI cancers, where PAT can be effectively used. Although PAT has already demonstrated its potential for critical clinical issues, such as breast cancer detection and dermatological issues that develop near the body surface, most other major diseases arise in internal organs, where PA imaging from outside the body may not be applicable.

Recognizing the importance of GI application, Yang et al. from the Wang group undertook related technology developed and reported their first achievement of a miniaturized imaging probe in 2009 (Yang et al. 2009). In the paper, they demonstrated a physical embodiment with a 4.2-mm diameter and ~ 4-cm long imaging probe by applying a micromotor-based built-in scanning mechanism, and they also showcased its endoscopic imaging capability by imaging excised tissues ex vivo. The key feature of the imaging probe was the use of a scanning mirror that was customized to reflect both the laser beam and acoustic waves during constant rotation (~4 Hz). Notably, by applying the scanning mirror concept, it was possible to transmit laser pulses and electric signals at the proximal unit via static cables. Compared to previous endoscopes (Oraevsky et al. 1997; Chen et al. 1993; Beard et al. 1998; Viator et al. 2001; Henrichs et al. 2005; Sethuraman et al. 2007), the main advancement of the probe lay in its embodiment as a fully encapsulated form, in which all necessary optical and acoustic detection mechanisms and scanning mechanisms were integrated, thus enabling the first endoscopic imaging.

After the 2009 report, the authors also successfully implemented the advanced version with a 3.8-mm diameter and demonstrated the first case of in vivo endoscopic imaging applicability, using rabbits and rats (Yang et al. 2012a). Figure 14a depicts the imaging probe and imaging concept presented in the paper. Compared to the previous study (Yang et al. 2009), they demonstrated a co-registered, simultaneous, dual-wavelength PA and US endoscopic imaging concept for the first time, and they also greatly improved the lateral resolution by employing a focused US transducer.

In addition to the technical advancement, the paper presented several notable endoscopic demonstrations that, for the first time among optical endoscopic techniques, detected PA signals even from major organs, such as the lung and trachea located around the esophagus, and also successfully mapped sO_2 for the two major blood vessels; that is, the aorta and caudal vena cava, which directly branched out from the heart of a rabbit (Fig. 14b). Moreover, by using Evans blue as a contrast

Fig. 14 PA endoscopes developed for GI tract imaging applications and produced images. (**a**) An illustration depicting the principle of simultaneous dual-wavelength PA and US endoscopic imaging (Yang et al. 2012a). (**b, c**) Representative images were acquired from a rabbit esophagus (b) and rat colorectum (c) **in vivo**. (**d**) Schematic and photo of the dual-modality PA and LUS endoscopic probe (Tsyboulski et al. 2014). (**e**) Conventional US (middle), PA, and LUS image of a tissue-mimicking phantom with embedded glass spheres. The PA and LUS images were acquired according to the presented concept. (**f**) Photos of the optical-resolution PA endomicroscopic probe and its optical illumination and acoustic detection unit (Yang et al. 2015). (**g**) Schematic of the optical illumination and acoustic detection unit, which corresponds to the lower photo in (**f**). The inserted photo represents the optical illumination unit comprising an optical fiber and GRIN lens. (**h**) Label-free **in vivo** optical-resolution PA endomicroscopy image acquired from a rat colorectum. (**i**) Schematic and photo showing the distal end of the 2.5-mm outer diameter catheter (torque coil)-based PA and US dual-mode endoscopic probe capable of full angular FOV imaging (Li et al. 2018b). (**j**) Corresponding 3D PA and US images acquired from a rat colorectum in vivo

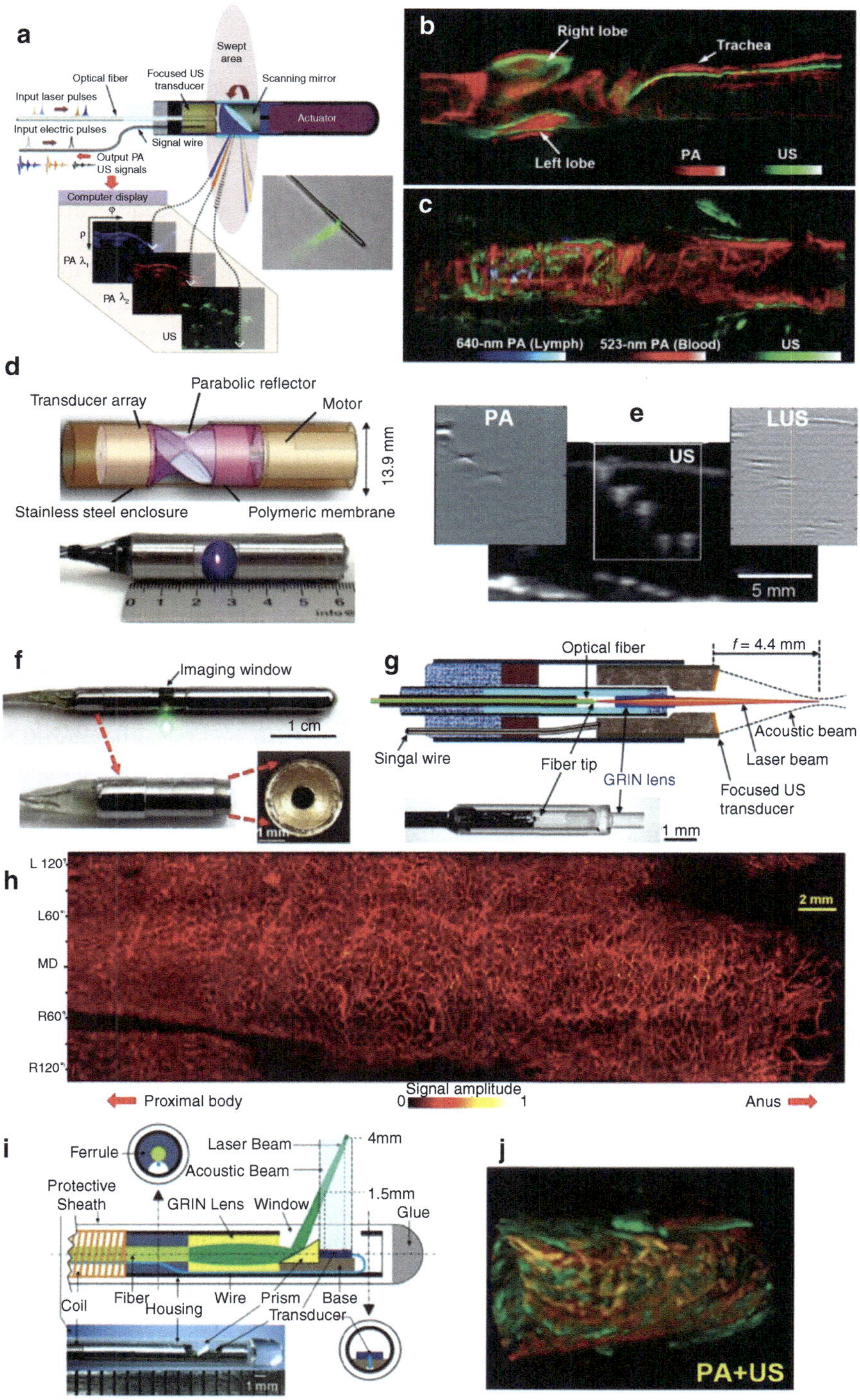
a
Swept area
Optical fiber
Focused US transducer
Scanning mirror
Input laser pulses
Actuator
Input electric pulses
Signal wire
Output PA
US signals
Computer display
PA λ1
PA λ2
US
b
Right lobe
Trachea
Left lobe
PA
US
c
640-nm PA (Lymph)
523-nm PA (Blood)
US
d
Parabolic reflector
Transducer array
Motor
13.9 mm
Stainless steel enclosure
Polymeric membrane
e
PA
US
LUS
5 mm
f
Imaging window
1 cm
1 mm
g
Optical fiber
f = 4.4 mm
Acoustic beam
Laser beam
Singal wire
Fiber tip
GRIN lens
Focused US transducer
1 mm
h
L 120°
L60°
MD
R60°
R120°
2 mm
Proximal body
Signal amplitude
0
1
Anus
i
Ferrule
Laser Beam
4mm
Acoustic Beam
Protective Sheath
GRIN Lens
Window
1.5mm
Glue
Coil
Fiber
Housing
Wire
Prism
Base
Transducer
1 mm
j
PA+US

agent, they successfully visualized the lymph nodes and vessels distributed around the descending colon of a rat (Fig. 14c). Although the image demonstrations were performed in animals (i.e., rabbits and rats), it must be noted that they were all achieved using transesophageal or trans-enteric imaging, which suggests that PAE technology could be further developed to be utilized in a manner similar to current clinical TEE or EUS probes.

After the in vivo demonstration, the group reported two advanced versions of a PAE probe, a 2.5-mm diameter probe (Yang et al. 2012b) and a catheter-based PA endoscope (Yang et al. 2014), to be used for a clinical study that evaluated its utility for Barrett's esophagus diagnosis. Compared to the previous version (Yang et al. 2012a), the two probes featured a unique technical aspect: They were embodied in a narrower diameter or a fully flexible form to be used for real human GI tract imaging via the instrument channel of a clinical video endoscope. However, they turned out to be insufficient for use in the clinical study due to multiple technical issues, such as the long rigid distal section (Yang et al. 2012b) or the fragileness of the signal wire near the distal connection point (Yang et al. 2014).

While the Wang group developed the endoscopic probes, two notable endoscopes were also reported by other groups in 2010 (Yuan et al. 2010) and 2014 (Tsyboulski et al. 2014). Unlike those reported by the Wang group, in which a single-element US transducer was employed to develop the probes and which mostly aimed for a mini-probe, the two probes featured the use of an array transducer that served to develop them into a large probe.

The first was the 64-element array transducer (5 MHz)-based PAE probe reported by Yuan et al. in the Xing group in 2010 (Yuan et al. 2010), and it was developed for human colorectal cancer diagnosis. With this probe, they imaged an excised human colorectal tissue, including a tumor mass and claimed that their probe was capable of discriminating the tumor region with adequate contrast. However, although the overall device concept may have been feasible for practical clinical use, the large probe size (~3 cm in diameter and ~ 9.5 cm in length) needed to be further reduced. Furthermore, the fragile glass-based probe casing needed to be replaced with other materials.

The second probe was the 8-element array transducer-based PA and laser US endoscope developed by Tsyboulski et al. of the Oraevsky group in 2014 (Fig. 14d) (Tsyboulski et al. 2014). The probe was developed for esophageal and colon cancer diagnosis and harnessed the dual contrast benefit provided by PA and US imaging. However, interestingly, they applied a laser pulse-induced acoustic pulsing principle, namely "laser ultrasound (LUS)," instead of applying the conventional piezoelectric transducer-based acoustic pulsing method to perform US imaging. As per the concept, an LUS pulse is generated by a tubular polymeric membrane affixed around a rotating parabolic reflector in accordance with the same PA principle caused by some of the absorbed laser energy when a laser beam passes through the membrane. Of course, the membrane was acoustically transparent. Then, additional mechanical rotation was also provided to the parabolic mirror to achieve radial scan imaging. Compared to the previous probe (Yuan et al. 2010), there was a technical advancement in terms of the size (1.4 cm in diameter and 6 cm in length) (Tsyboulski et al. 2014). However, their image demonstration was limited to a proof-of-concept study based on nonbiological samples (Fig. 14e).

As shown, the two endoscopic probes were representative examples that were embodied as GI endoscopes employing an array transducer. However, their image demonstrations were all limited to ex vivo imaging, and no follow-up studies in relation to the probes were reported until now.

Meanwhile, in 2015, Yang et al. of the Wang group presented another report that demonstrated optical-resolution PA endomicroscopy in vivo for the first time (Yang et al. 2015). Figure 14f depicts the implemented imaging probe encapsulated in a stainless steel housing with a 3.8-mm outer diameter. As shown, its overall appearance and mechanical structure inherited those of the previous AR-PAE probe (Yang et al. 2012a). However, they achieved an OR-PAM level lateral resolution (~10 μm) by installing a 0.5-mm outer diameter gradient-index (GRIN) lens (0.20 pitch) inside the central hole of a ring-shape single-element focused US transducer (f = 4.4 mm, 42 MHz, $LiNbO_3$) and by guiding the necessary laser beam via a single-mode optical fiber (SMF) (Figs. 14f and g). The embodiment was key in attaining the best overlap between the optical focus and the acoustic focus to achieve maximum sensitivity, and the in vivo image presented in Figure 14h that was acquired from a rat colorectum shows that the alignment was successful.

The in vivo demonstration demonstrated that even a PAE probe with such a small diameter could provide a detailed vasculature map of a GI tract with a spatial resolution comparable to that of OR-PAM. However, for such a mini-PAE probe to be successfully translated into clinical usage, embodying related imaging probes in a fully flexible form over the entire probe section remains a key issue because it is typical that related imaging procedures are performed via the instrument channel of a standard clinical video endoscope. The paper reported by Li et al. of the Song group in 2018 eventually addressed the issue (Li et al. 2018b).

Unlike the previous built-in micromotor-based rotational scanning mechanism (Yang et al. 2009, 2012a, b, 2015), they applied the torque coil-based proximal actuation mechanism that is frequently utilized in conventional EUS mini-probes (Shami and Kahaleh 2010; Akahoshi and Bapaye 2012) or endoscopic OCT probes (Tearney et al. 1997; Yun et al. 2006; Gora et al. 2013). As already mentioned, the mechanism features include the use of a torque coil, which is commercially available and manufactured for the purpose of transmitting mechanical torque very smoothly from one end to the other, even in a bent condition. Furthermore, in such a torque coil, a hollow tubular hole is present, which enables a necessary signal wire and optical fiber for PA imaging to be placed inside. Consequently, leaving aside its great flexibility, applying this mechanism enabled related imaging probes to be embodied without limited angular FOV issues, which usually occur in the micromotor-based scanning mechanism (Yang et al. 2009, 2012a, b, 2015).

Based on these principles, the researchers implemented a 2.5-mm outer diameter catheter probe (Fig. 14i) that could potentially fit the 2.8-mm instrumental channel of a clinical video endoscope, and they demonstrated its in vivo applicability by imaging a rat colorectum (Fig. 14j). In the reported system, a slip ring-based optical–electrical rotary joint was also embodied at the connection point between the flexible catheter and the proximal driving unit. Furthermore, it can be seen that a GRIN lens unit was installed inside the scanning tip as in the case of ref. (Yang et al. 2015). However, it should be noted that the achieved lateral resolution was not as great as that of ref. (Yang et al. 2015) because the employed optical fiber was not an SMF, but a multimode optical fiber (core dia.: 105 μm). Consequently, the attained PA transverse resolution was ~250 μm at the maximum SNR distance where the optical beam and the acoustic beam overlapped each other.

In short, the presented mechanism can be understood as an integration of an EUS and an OCT mini-probe, and it should also be noted that it has been the most frequently applied general mechanism in many intravascular PA imaging devices developed for cardiovascular applications, not only the GI applications. Related examples are introduced in the next intravascular imaging application section.

Previously, Yang et al. also adopted such a torque coil-based proximal actuation mechanism once in their catheter-based PAE probe (Yang et al. 2014). However, as the signal wire of an employed US transducer was not placed inside the torque coil but on a side of the plastic sheath, which was meant only to provide a static connection between the catheter and the proximal driving unit, the signal wire was easily damaged during its introduction into the instrument channel of a video endoscope. Eventually, such a design turned out to be undesirable. In this aspect, the configuration demonstrated in ref. (Li et al. 2018b) appeared to be more suitable for translation into practical clinical use. However, the unaddressed technical issue of related optical–electrical rotary joints being embodied based on a single-mode optical fiber remained. Although the related embodiment is not a simple issue due to the key requirement of delivering uniform optical energy during its rotation, it must be superior to the presented example in terms of the achievable spatial resolution.

All the PAE probes presented until now employed the conventional piezoelectric material-based US detection mechanism. Although this mechanism is more dominant, PA signals can also be detected by applying pure optics-based US detection mechanisms (Wissmeyer et al. 2018). Several mechanisms, such as FP polymer film (Beard et al. 1998; Zhang and Beard 2011; Ansari et al. 2018), micro-ring resonator (Hsieh et al. 2012), and a CCD camera-based US detection mechanism (Nuster et al. 2014), have been proposed so far. Among them, however, the first mechanism has been the most successful in terms of endoscopic applicability. Thus, in this section, we introduce a recent 2018 report by Ansari et al. (Ansari et al. 2018) from the Beard group as an example.

Figure 15a depicts their imaging probe that applied the FP polymer film-based US sensing mechanism and was developed for various laparoscopic approaches or interventional procedures, such as surgery, laser therapy, and needle biopsy. Since 1998, the group has pursued related technology development (Beard et al. 1998). However, in the presented paper (Ansari et al. 2018), they achieved high-resolution 3D PA imaging for the first time by implementing the optical sensor in a 2D array form. As shown in the schematic, the imaging probe was embodied as a forward viewing endoscope in which 50,000 optical fibers that acted as array elements were included and encapsulated in a 76-mm long rigid housing. Although the array consisted of such a vast number of detection elements, surprisingly, the achieved probe diameter was as small as 3.2 mm, a size that can numerically fit the standard instrument channel with a 3.7-mm diameter. Indeed, realizing such a high-density US array within such a small diameter was possible because they utilized the optical detection mechanism. This would have never been possible if the conventional

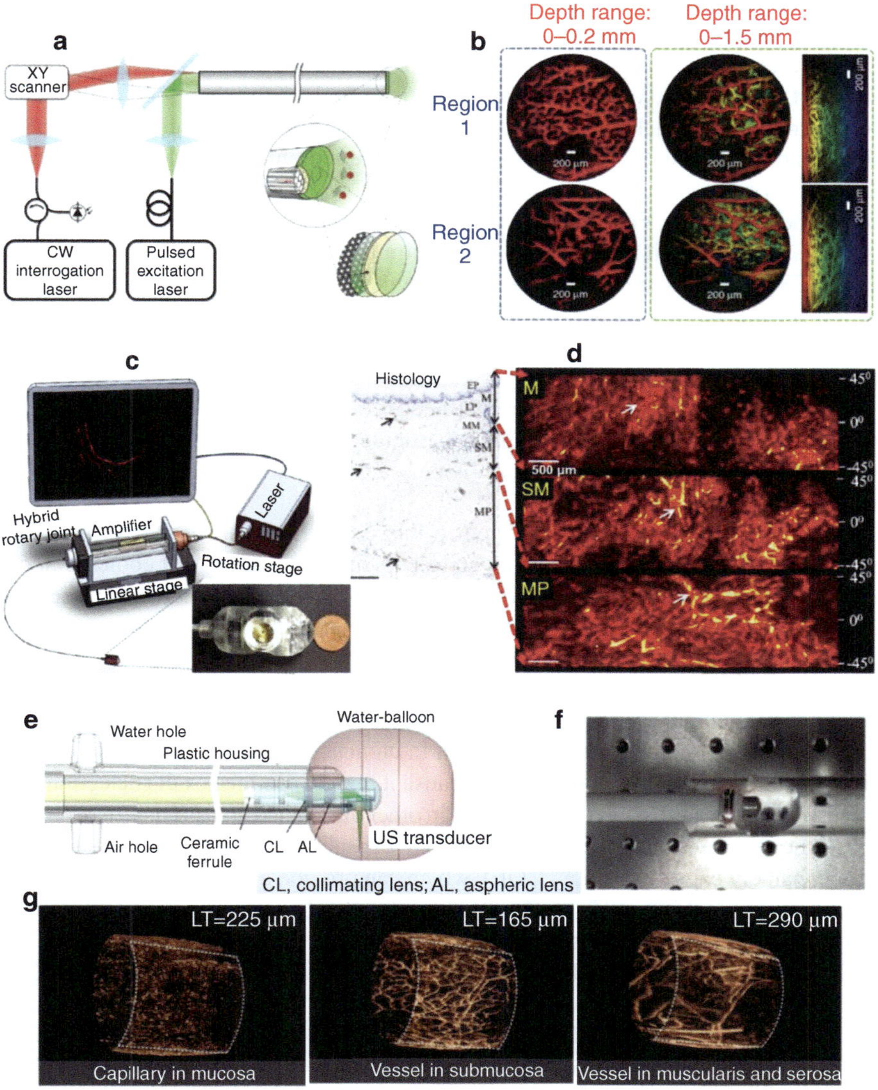

Fig. 15 PA endoscopes developed for GI tract imaging applications and produced images. (**a**, **b**) Schematic of an FP polymer US sensor-based forward viewing PA endoscopic probe (**a**) and corresponding 590-nm PA MAP images (**b**) acquired from an ex vivo duck embryo at two different regions (upper/lower rows) and depth ranges (left/right columns) (Ansari et al. 2018). The lateral and axial resolutions were ~ 45–70 µm and ~ 31 µm, respectively, for depths between 1 and 7 mm. (**c**) A schematic depicting the concept of the capsule-based PA endoscopic probe. The inset presents a picture of the implemented capsule (He et al. 2019). (**d**) A histology image (left) showing blood vessels at different esophageal layers revealed by anti-CD31 immunostaining, and radial-MAP images of the mucosa (M), submucosa (SM), and muscularis propria (MP) layers of the esophageal wall, respectively. (**e**) Schematic of the shape-adapting panoramic PA endomicroscopic probe (Xiong et al. 2019). (**f**) Photograph of the probe. (**g**) 3D PA images acquired from a rabbit colorectum in vivo. Each image depicts a vascular network in mucosa (left), submucosa (middle), and muscularis and serosa (right), respectively

piezoelectric method was applied. They validated its performance by imaging a duck embryo ex vivo (Fig. 15b).

The main shortcoming of this mechanism, however, is that it requires the installation of an additional complex optical system around the proximal part of the probe to interrogate the thickness changes of the FP polymer film attached at the other distal end, according to the pressure variation exerted onto the surface. In short, an interrogation beam should scan individual fibers one by one in order to address different points on the FP sensor film, a process that currently cannot be achieved quickly. In addition, it has the fundamental limitation of being incapable of conventional US pulse-echo imaging because it does not provide a reverse energy transduction mechanism. Nevertheless, attention should be paid to this technology because such an optical sensor exhibits the highest detection sensitivity and wide bandwidth characteristics (f_{-3dB} = 34 MHz).

Another notable technical issue that needs to be discussed relates to the target distance adjustment mechanism. This is particularly important when one designs an OR-PAE probe whose best resolution is only achieved at a specific distance called working distance.

As a representative example related to this issue, He et al. of the Ntziachristos group in 2019 reported an interesting endoscopic probe concept that could be applicable to a single-element transducer-based mechanical scanning mechanism (He et al. 2019). Although the aforementioned narrow diameter catheter has the advantage that it can be used via the standard instrument channel of a video endoscope, it is still true that if it is applied to imaging a human esophagus with a relatively larger diameter by directly contacting the esophageal wall, it can provide PA signals only from a limited angular portion where it contacts the wall. In clinical contexts, such a reduced effective angular FOV issue is undesirable because it greatly diminishes the detection possibility of cancerous tissues that may exist in other unscanned areas. Thus, to overcome the issue, the authors developed a tethered capsule-like optoacoustic endoscopic probe, namely capsule OE (i.e., COE), which can provide a full 360° angular FOV.

As shown in Figure 15c, it featured a capsule-shaped structure (18.6 mm in diameter and 20 mm in length) attached at the tip of a catheter. Its internal structure and components that were placed inside the catheter section were almost similar to those of the catheter-based probe (Fig. 14i), in which a 400-μm core diameter optical fiber was installed inside the conduit of the 3.2-mm diameter torque coil to deliver laser pulses. However, adding the capsule with a large diameter enabled them to employ a large diameter US transducer (30 MHz, PVDF), which performed a rotational scanning inside the capsule while receiving a mechanical torque from the torque coil. Consequently, the transducer, torque coil, optical fiber, and illumination optics placed around the transducer rotated together.

By attaching such a large diameter capsule, multiple benefits can be derived. These include the following: It gently expands an esophageal wall so that the wall surface is naturally positioned at an optimal working distance, where the lateral resolution is highest; no additional acoustic matching medium, excepting the body fluid, is necessary; and a room for installing a preamplifier, which is crucial for achieving a better SNR, can be secured very close to the transducer. Simply speaking, it provides an equivalent effect to using a medical balloon. Although the presented image demonstration was limited to an ex vivo swine esophagus imaging (Fig. 15d), the presented concept could become practically useful due to the explained advantages.

In addition to this paper, another report addressed a technical issue related to acoustic coupling between a PAE probe and an imaged tissue. Although the presented capsule idea could also resolve this issue, it may impose a somewhat strong constraint on target tissues. However, in the 2018 report by Xiong et al. of the Yang group (Xiong et al. 2019), the issue was approached differently by integrating a water balloon to their rigid-type PAEM probe developed for colorectal cancer detection (Figs. 15e and f). Since the water balloon was very soft, it provided a gentler physical contact to target tissues by naturally changing its shape than that of the capsule-based method. In addition, it also enabled a boundary recognition to extract the holonomic vascular networks layer by layer, as shown in the rabbit colorectum image (Fig. 15g). As water balloons are frequently utilized in conventional EUS technology, the presented method could also be a key element in future clinical PAE. In relation to this, it could also be useful to refer to the adjustable focusing concept reported by Guo et al. in the Chen group (Guo et al. 2019).

In this section, we summarized PAE technology developments, focusing on several key papers that implemented a PAE probe with similar probe embodiment (i.e., mini probes or large diameter stand-alone probes) or typical technical features (i.e., single-element transducer- or array transducer-based probes) that are already applied in conventional EUS technology (Shami and Kahaleh 2010; Akahoshi and Bapaye 2012). Although the presented papers made their advancements close to the benchmarked EUS technology, only animal-based esophagus imaging or lower GI tract imaging has been reported so far, and no disease model-related or clinical study results have been reported yet (excepting the partially signals captured from a rectal wall as included in the human prostate image to be presented in a next section).

Intravascular Photoacoustic Imaging

The motivation for Intravascular photoacoustic (IVPA) imaging is to apply the excellent spectroscopic imaging capability of PAT for better screening and characterization of atherosclerotic plaques that build up in the cardiovascular system. In particular, vulnerable plaques that develop in the coronary artery are known to cause acute coronary artery syndromes, such as angina pectoris, sudden cardiac death, and myocardial infarction, all of which are major causes of death in many Western countries.

Presently, X-ray-based coronary angiography is regarded as the gold standard for examining stenosis, in which a radiocontrast agent is injected using the catheter introduced via the femoral or radial artery, and then its flow is displayed based on X-ray projection imaging. IVUS is also being increasingly used to delineate wall thickness and to determine calcification levels. In addition, a somewhat newly emerging technique called intravascular OCT has proven its clinical utility by providing unprecedented detailed anatomical image features, such as for vasa-vasorum (Yun et al. 2006), based on high-speed and high-resolution imaging capability. However, none of the mentioned imaging techniques can provide spectroscopic image information as rich as those by PAT. Consequently, more research efforts have been made in this area of application. Therefore, related research outcomes have also demonstrated an increase in recent years.

Although the idea of applying PAT for intravascular imaging was also suggested in earlier reports (Chen et al. 1993; Beard et al. 1998; Henrichs et al. 2005), earnest research activities relating to it emerged in the latter part of the 2000s; that is, after the Emelianov group reported the first PA and US image acquired from a rabbit aorta in 2007 (Sethuraman et al. 2007). The group paid their attention to structure and composition, rather than the degree of stenosis, which is a more important factor associated with acute clinical events, and thus PAT could be a valuable complement for better disease characterization and risk stratification.

Figure 16a and b shows the experimental setup and the first IVPA and IVUS images acquired from a normal rabbit aorta. As illustrated, the system was set up to provide co-registered PA and US images by incrementally rotating an excised aortic tissue sample, while a commercial IVUS catheter remained static inside the sample, and the necessary laser beam was delivered from the outside. Although the study did not address implementing a real intravascular imaging probe, it was the first one presenting how the IVPA image appeared and what kind of complementary image information could be provided over conventional single-mode IVUS images.

In 2008, the same group reported another notable demonstration using the same system (Sethuraman et al. 2008). As mentioned earlier, the key rationale for using PAT for this specific application was the excellent spectroscopic imaging capability that has the potential to discriminate plaque regions. To demonstrate feasibility, they imaged a normal and diseased atherosclerotic aorta, harvested from a rabbit subjected to 0.15% cholesterol chow for 10 months, using multiple laser wavelengths (from 680 to 900 nm), and they then spectroscopically differentiated plaque regions by applying a finite difference analysis method.

As shown in Figure 16c, the atherosclerotic (left) and normal (right) aorta images exhibited quite different slope patterns. Since the two images were processed from two PA images acquired at 680 nm and 900 nm, where lipid and collagen type I are known to manifest a large variation in the absorption, but with a different sign of the

Fig. 16 Systems and probes developed for IVPA imaging applications and produced images. (**a**) A commercial IVUS imaging catheter-based experimental setup for combined IVPA and IVUS imaging (Sethuraman et al. 2007). (**b**) IVUS (left) and IVPA (right) B-scan images acquired from a normal rabbit artery ex vivo. (**c**) The spectroscopic (first derivative) IVPA images of the atherosclerotic (left) and control (right) aorta calculated at 900 nm using a finite difference approach (Sethuraman et al. 2008). The reference image for evaluating the first derivative was obtained at 680 nm. In (**b**) and (**c**), the scale bars represent 1 mm. (**d**) Schematic (upper) and photo (lower) of the integrated IVUS and IVPA imaging catheter utilized for the first in vivo demonstration (Karpiouk et al. 2012). (**e**) First in vivo IVUS and IVPA images acquired from a rabbit's thoracic artery, at two different locations, in which a deployed stent was visualized. (**f**) Schematic showing the distal tip of the 1.25-mm diameter catheter probe and its peripheral systems (Jansen et al. 2011). (**g**) Histology and ex vivo IVPA and IVUS images acquired from a human coronary artery with an advanced atherosclerotic plaque. The histology (Oil Red O stain) reveals the presence of a lipid-rich plaque (*) as well as a calcified area (Ca). Lu, lumen; Pf, peri-adventitial fat. The two IVPA images were acquired using 1210 nm and 1230 nm, in which the lipid absorption was high and low, respectively. (**h**) Photo of the 1.1-mm diameter intravascular optical-resolution PA catheter probe showing its flexibility (Bai et al. 2014). (**i**) Schematic (upper) and photo (lower) of the distal end. (**j**) Schematic presenting the peripheral systems of the catheter probe. The 3D scanner consists of an optical–electric rotary joint (ROJ), a step motor, and a motorized pullback stage. (**k**) Three-dimensional cut-away PA image of a stent deployed in a plastic tube

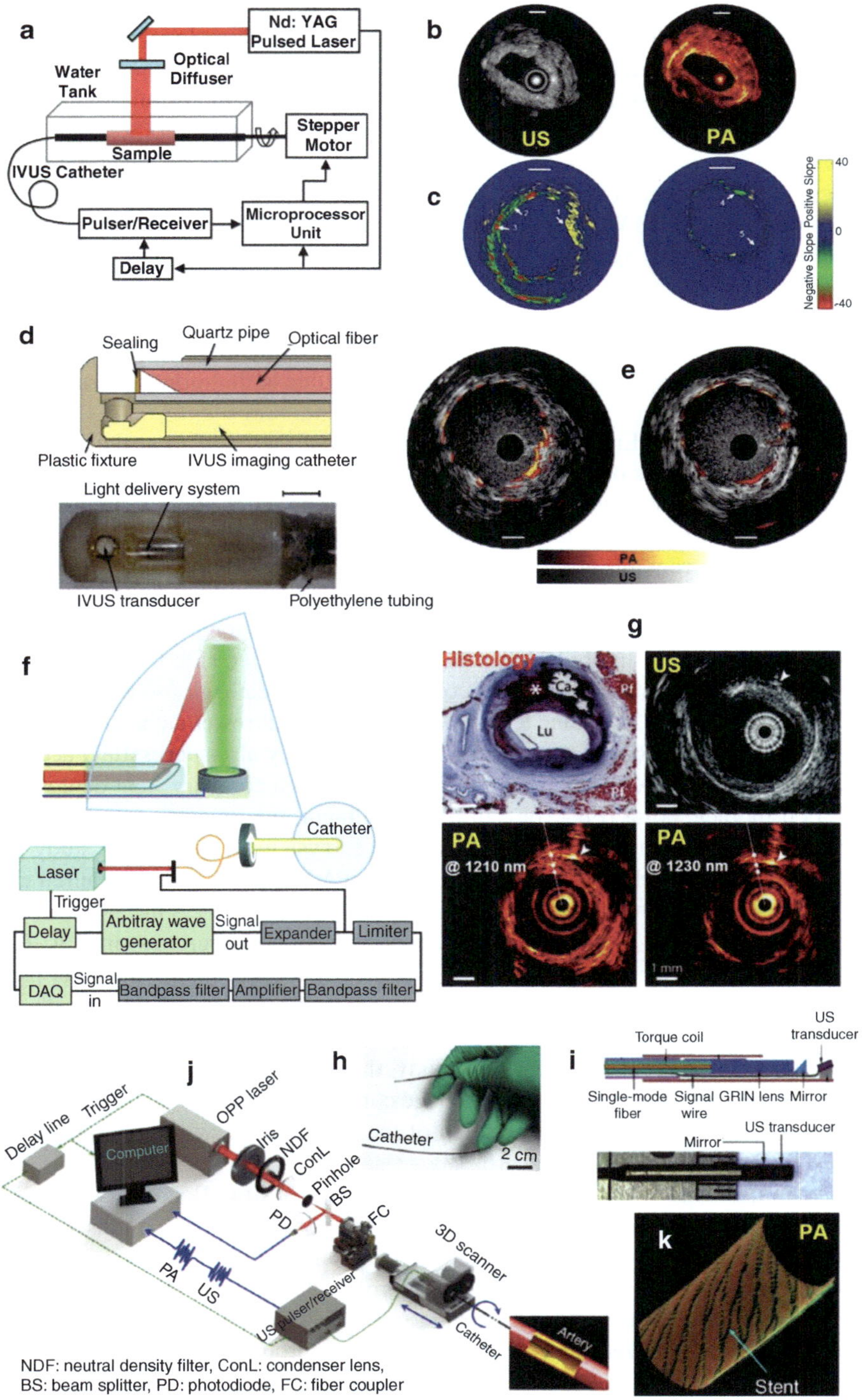
a
Nd: YAG Pulsed Laser
Optical Diffuser
Water Tank
Stepper Motor
Sample
IVUS Catheter
Pulser/Receiver
Microprocessor Unit
Delay
b
US
PA
c
Negative Slope Positive Slope
40
0
-40
d
Sealing
Quartz pipe
Optical fiber
Plastic fixture
IVUS imaging catheter
Light delivery system
IVUS transducer
Polyethylene tubing
e
PA
US
f
Catheter
Laser
Trigger
Delay
Arbitray wave generator
Signal out
Expander
Limiter
DAQ
Signal in
Bandpass filter
Amplifier
Bandpass filter
g
Histology
Ca
Lu
US
PA @ 1210 nm
PA @ 1230 nm
1 mm
j
Delay line
Trigger
Computer
OPP laser
Iris
NDF
ConL
Pinhole
BS
PD
FC
3D scanner
PA
US
US pulser/receiver
Catheter
Artery
NDF: neutral density filter, ConL: condenser lens,
BS: beam splitter, PD: photodiode, FC: fiber coupler
h
Catheter
2 cm
i
Torque coil
US transducer
Single-mode fiber
Signal wire
GRIN lens
Mirror
Mirror
US transducer
k
PA
Stent

slope, it can be seen that Region 1, in which a positive slope appeared, is suspected to be a lipid-rich region, while Region 3, which has a negative slope, is suspected to be a collagen type I-rich region.

In addition to the structural and spectroscopic imaging demonstrations, the group also reported the first IVPA-based molecular imaging result in 2009 (Wang et al. 2009). It is known that various biomarkers are present during the development of atherosclerotic lesions (Wang et al. 2010b). By using gold nanoparticles (Au NPs) as contrast agents, Bo Wang et al. from this group identified macrophages in atherosclerotic plaques (Wang et al. 2009). Macrophages are known to play a crucial role in the development and vulnerability of plaques. As such, Au NP has multiple advantages: small size on a nanometer level, good biocompatibility, and high volumetric absorption coefficient. Notably, when it is attached to other particles, it exhibits different optical absorption characteristics due to the surface plasmon resonance coupling effect.

As presented, the Emelianov group made most of the initial contributions demonstrating what kinds of actual benefits can be derived by the IVPA imaging technique if it were to be translated into clinical use. Although the demonstrations were promising, realizing a real imaging catheter remained a key issue. To address this, Karpiouk et al. from this group undertook related imaging device development, and they reported a prototype with a ~ 5-mm diameter in 2010 (Karpiouk et al. 2010). In their first outcome, the achieved probe diameter was somewhat large because they embodied it by simply adding a 600-μm core diameter, side-firing, and angle-polished optical fiber to the side of a commercial IVUS catheter (40 MHz, 3.6 F).

However, in 2012, the same team achieved an even narrower catheter with a 2.2-mm diameter (Fig. 16d) (Karpiouk et al. 2012), and they also successfully acquired the first in vivo IVPA (1064 nm) and IVUS images of a coronary stent deployed in a rabbit thoracic aorta (Fig. 16e). In the experiment, the imaging speed was very slow (~0.1 Hz) due to the PRR of the employed system. However, the study presented a valuable technical baseline—that such PA images could be acquired even in the presence of luminal blood using a laser pulse energy of 2.5 mJ.

Multiple research outcomes were also reported from other groups since 2010. In 2011, Jansen et al. in the van Soest group reported an even smaller diameter probe of 1.25 mm (Fig. 16f) along with the first ex vivo PA images of the investigated atherosclerotic human coronary artery samples that ranged between 715 nm and 1800 nm (Jansen et al. 2011). In their study, they demonstrated experimentally for the first time that an adequate level of image contrast for lipids could be created near 1210 nm, where a prominent harmonic of the C—H stretch vibrational mode signaled the presence of lipids (Fig. 16g).

A similar result was also reported by Allen et al. of the Beard group in 2012 (Allen et al. 2012). They investigated the spectral behavior of PA contrasts for atherosclerotic human aorta samples that ranged from 740 nm to 1400 nm, thereby reconfirming the contrast prominence of approximately 1210 nm. In addition, they performed a similar under-blood imaging test and discovered that a plaque region could still be visualized and differentiated, even through a 2.8-mm thick blood medium by using tens of mJ of laser pulse energy, which, however, was broadly expanded over the sample (exact fluence was unspecified). The under-blood imaging test result increases the chance of actual clinic translation for the IVPA technique.

After related technical bases were established as described above, multiple groups began focusing on developing an applicable IVPA catheter for actual coronary artery imaging and reported related results. In 2014, Bai et al. in the Song group reported a 1.1-mm diameter probe (Fig. 16h) (Bai et al. 2014), whose size included its tubing thickness, implying that the IVPA technique could now achieve a typical thickness level comparable with those of existing clinical IVUS probes. Furthermore, the report presented two notable achievements: The device was embodied based on an OR-PAM concept, which thereby achieved a lateral resolution of 19.6 μm (Fig. 16i), and a slip-ring-based electrical and optical rotary junction at the proximal unit was demonstrated for the first time (Fig. 16j). It should be noted that the mechanism is commonly applied in most IVPA imaging systems reported thereafter. As the rotary joint was a key element that enabled a continuous rotational scanning unlike previous systems (Karpiouk et al. 2010, 2012; Jansen et al. 2011), they could acquire the first volumetric image from a stent as shown in Figure 16k.

In the next year, the same group also reported an even smaller probe at a diameter of 0.9 mm (Li et al. 2015); however, a quasi-focusing concept rather than an optical focusing concept was applied. Moreover, in the paper, they demonstrated a considerable advancement in the A-line acquisition rate, as high as 1 kHz; this was possible due to the effect of a rotary joint use and not merely the performance of the employed laser system. Although the two size achievements were technically considerable, the image demonstrations of the two papers were all limited to ex vivo experiments, presumably due to the still insufficient imaging speed.

To successfully translate the IVPA technique into clinical use, achieving a real-time imaging speed is also crucial because it is desirable to perform related imaging procedures within a few cardiac cycles. However, the imaging speeds of previous studies were all limited by either the low PRR or the availability of required wavelength from a commercial laser system. In general, the IVPA technique requires a laser wavelength of 1.2 or 1.7 μm, where the optical absorption of lipid generates strong PA signals. Unfortunately, however, existing commercial OPO systems do not provide a sufficient energy and pulse repetition rate at these two wavelengths. Recognizing this issue, the Cheng group has been reporting related achievements since 2014 (Wang et al. 2014b; Hui et al. 2015, 2017; Cao et al. 2016; Cao et al. 2018), presenting new types of catheter designs.

First, in 2014, Wang et al. of the Cheng group reported a custom-built barium nitrite [$Ba(NO_3)_2$] Raman laser system capable of providing a 1197-nm laser beam with a pulse energy as high as 2 mJ at a 2-kHz PRR, along with an integrated IVPA and IVUS catheter with a 2.9-mm outer diameter (Wang et al. 2014b). Although the probe diameter was relatively larger than those of previous reports, mainly due to the use of a large diameter ring transducer, embodiment levels for other parts were similar to them, and most notably, they demonstrated an A-line acquisition rate that was twice as fast as ref. (Li et al. 2015). Afterward in 2015, the group presented another report related to laser source development (Hui et al. 2015) that featured a potassium titanyl phosphate (KTP)-based OPO with an output pulse energy of up to

2 mJ at a PRR of 500 Hz and a wavelength of 1724 nm. The high energy achievement at this wavelength should be particularly noted because it was very close to the 1730-nm wavelength where the first overtone excitation of the methylene group CH_2 could be applied. For this reason, the 1730-nm wavelength has also been utilized in many later IVPA imaging studies.

In addition, in a 2016 paper reported and led by Cao et al., the Cheng group presented a new probe design that applied a collinear alignment concept between optical illumination and acoustic detection (Cao et al. 2016). In all previous studies that adopted a misaligned configuration between the optical illumination and acoustic detection (Karpiouk et al. 2010, 2012; Jansen et al. 2011; Bai et al. 2014; Li et al. 2015; Wang et al. 2014b; Hui et al. 2015), a significant signal drop occurred when an object location was beyond the overlapped position. However, in the new study (Cao et al. 2016), they solved this issue by configuring the illumination optics and acoustic sensor as depicted in Figure 17a. In fact, the probe concept was applied to the experiment reported in 2017 in which a 16-Hz B-scan imaging speed was demonstrated for the first time along with the use of an improved KTP-based OPO laser providing a 2-kHz PRR (Hui et al. 2017).

Would it be eventually possible to image an actual human coronary artery in vivo using such an IVPA technique? This may be the most important question when the related technology is pursued. In 2017, the van Soest group presented results showing the feasibility of this concept (Wu et al. 2017).

In the paper (Wu et al. 2017), Wu et al. acquired the first in vivo coronary artery images from a pig using their 1.3-mm diameter catheter probe composed of a 100-μm core diameter, angle-polished multimode optical fiber installed in a torque coil and a 40-MHz US transducer (Fig. 17b). Notably, the size included a sheath made of polyethylene (PE). As shown in Figure 17c, the imaging probe and all necessary peripheral systems were set up in a portable trolley to enable combined IVPA and IVUS imaging in real time in the experimental room. With the system, they could

→

Fig. 17 Systems and probes developed for IVPA imaging applications and produced images. (**a**) Schematic (upper) and photo (lower) of the 1.6-mm outer diameter IVPA and IVUS catheter with a collinear overlap between optical and acoustic waves (Cao et al. 2016). (**b**) Schematic (left) and photo (right) of the integrated IVPA and IVUS imaging catheter with a 1.3-mm outer diameter, whose measure included a polyethylene (PE) sheath. (**c**) A photo depicting the peripheral systems placed in a portable trolley for easy transportation (Wu et al. 2017). (**d**) An X-ray image showing the catheter and the two deployed stents (marked by two yellow boxes) with lipid target mounted on a balloon catheter. Note that the artificially created lipid target was located between the two stents. (**e**) Representative combined IVPA and IVUS B-scan image. The yellow-dashed area represents the artificially created lipid target. (**f**) Photograph of the 1.6-mm outer diameter IVPA and IVUS catheter with a quasi-collinear overlap between optical and acoustic waves (Cao et al. 2018). Note that the diameter measure included a polymer sheath. (**g**) A schematic illustrating the IVPA and IVUS imaging as well as the pullback C-scan. (**h**) A photograph exemplifying a related proximal actuation unit in which the optical rotary joint, electric slip ring, and mechanical pullback stage were included (this photograph was adopted from a different report (ref. Hui et al. 2017) presented by the same group). (**i**) Reconstructed 3D PA and US images acquired from a rabbit aorta over a pullback length of 20 mm. The image data were acquired at 4 fps and a pullback speed of 0.25 mm/s

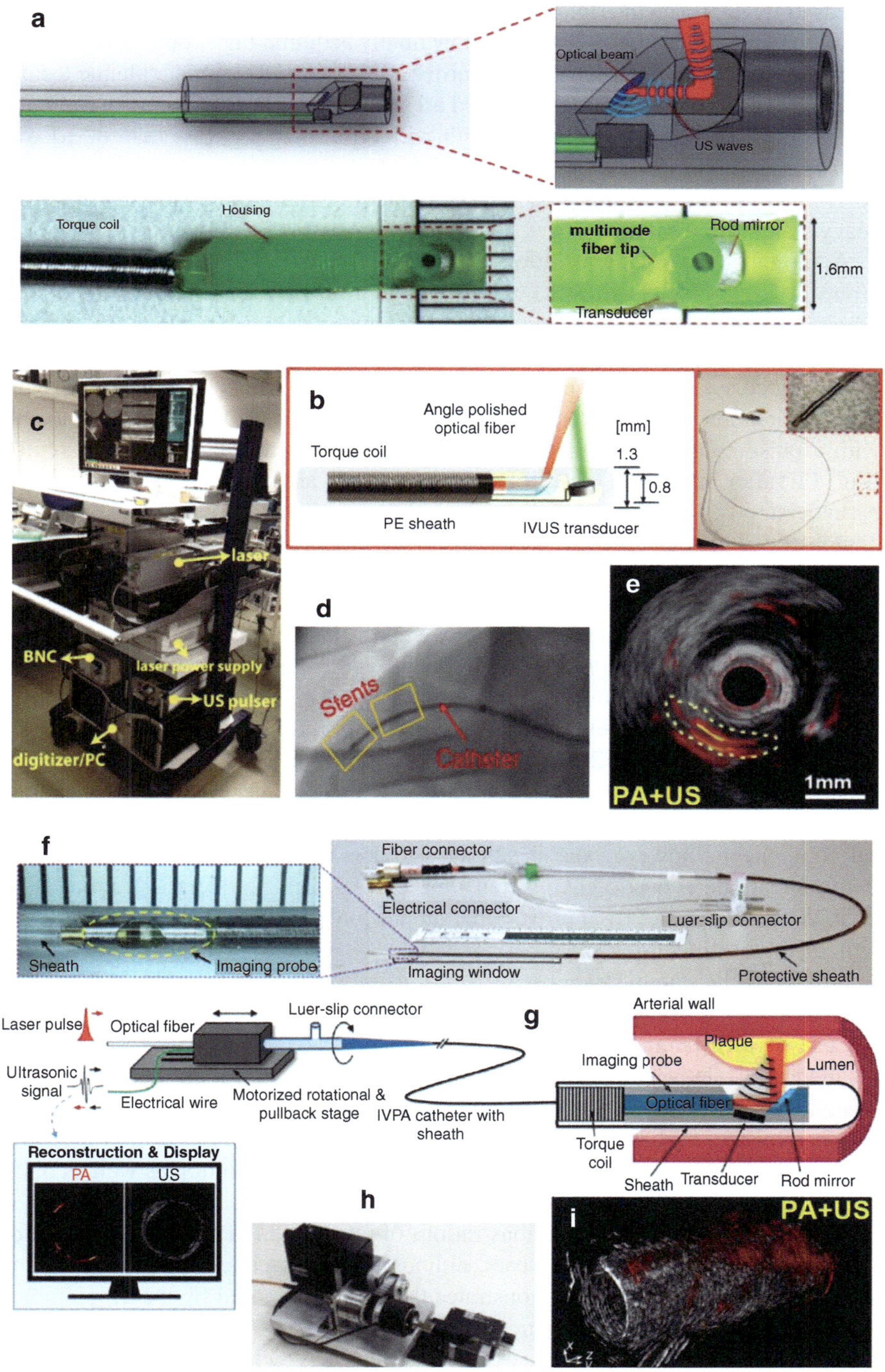
a
Optical beam
US waves
Torque coil
Housing
multimode fiber tip
Rod mirror
1.6mm
Transducer
c
laser
BNC
laser power supply
US pulser
digitizer/PC
b
Angle polished optical fiber
Torque coil
[mm]
1.3
0.8
PE sheath
IVUS transducer
d
Stents
Catheter
e
PA+US
1mm
f
Fiber connector
Electrical connector
Luer-slip connector
Sheath
Imaging probe
Imaging window
Protective sheath
Laser pulse
Optical fiber
Luer-slip connector
g
Arterial wall
Plaque
Lumen
Imaging probe
Ultrasonic signal
Electrical wire
Motorized rotational & pullback stage
IVPA catheter with sheath
Optical fiber
Torque coil
Sheath
Transducer
Rod mirror
Reconstruction & Display
PA
US
h
i
PA+US

successfully approach the right coronary artery (Fig. 17d) via the carotid artery with the aid of a guiding catheter (6 F), and they finally acquired high contrast PA images (Fig. 17e) from a lipid region that was artificially created to mimic a plaque lesion: The lipid was introduced into the imaged area prior to the imaging procedure. In the experiment, a 1720-nm laser wavelength with a pulse energy less than 80 μJ was utilized, and the frame rate of the B-scan, which was composed of 250 A-lines, was 20 Hz. During the imaging procedure, the artery was continuously flushed using heavy water-based saline.

A similar in vivo demonstration was also reported by Cao et al. of the Cheng group in 2018 (Cao et al. 2018), whose imaging result was, however, based on a rabbit model and performed using a 1.6-mm diameter catheter capable of achieving 16 Hz B-scan imaging speed (Fig. 17f). Figure 17g and h is the schematic of the implemented system and a photo of the rotary junction, respectively. In the in vivo imaging demonstration, they utilized a slightly different laser wavelength (1730 nm) with a pulse energy of 100 μJ at the output, and imaged a thoracic aorta region (Fig. 17i) after introducing the probe via the femoral artery. The imaged area might be understood as being less challenging compared to the coronary artery imaging demonstrated by the van Soest group (Wu et al. 2017). However, it should be noted that it was performed without flushing the blood.

Indeed, the presented in vivo IVPA imaging demonstrations along with the 3D pullback imaging were remarkable achievements that photoacoustically mimicked the current IVUS imaging procedure frequently performed clinically. From a technical perspective, it was possible because the torque coil and scanning tip of the two probes were all adequately sheathed in polymer tubing, unlike earlier probes (Karpiouk et al. 2010, 2012; Jansen et al. 2011; Bai et al. 2014; Li et al. 2015; Wang et al. 2014b; Cao et al. 2016; Hui et al. 2015, 2017). Needless to say, the sheathing is essential to protect the fragile coronary artery during the fast rotation of mechanical components; however, the sheathing also acts as an obstacle that deteriorates both optical and acoustic energy transmission. As such, the demonstrations (Cao et al. 2018; Wu et al. 2017) proved that an adequate SNR could be achieved even with the involvement of a sheath.

Although the demonstrations were promising, an unaddressed technical issue remained. Unlike ref. (Bai et al. 2014), the two demonstrations were not based on the use of a SMF and thus did not achieve optical focusing, which eventually resulted in a relatively poorer spatial resolution production than that of ref. (Bai et al. 2014). Although the spatial resolution may not be a critical part of IVPA application, applying such an optical focusing along with the use of a SMF or narrower diameter optical fiber would greatly improve the spatial resolution as well as SNR, even when using lower pulse energy.

As presented, real in vivo demonstrations of integrated IVPA and IVUS imaging were presented only in recent years, although the first concept was proposed in 2007, a decade earlier. This demonstrates the challenge posed by the development of this technology. Although some other technical issues related to the diameter, flexibility, and SNR still exist, we expect that the IVPA technique will be ultimately translated into clinical use because no fundamental limitation of the principle exists (van Soest et al. 2017; Sowers and Emelianov 2018). In relation to the intravascular multi-modality imaging, it should also be considered to realize a tri-modality imaging of PA, US, and OCT. In 2011, related development targeting ovarian cancer diagnosis was addressed for the first time (Yang et al. 2011). However, the demonstration was limited to ex vivo imaging only because the required mechanical scanning mechanism was not effectively embodied. If the concept were to be realized in the form of an intravascular imaging catheter, there could be much great synergetic effect because each imaging modality possesses different strengths. So far, integration of only two imaging modalities (i.e., PA and US or OCT and US) has been demonstrated due to this challenge (Li et al. 2020).

Urogenital Imaging

Another crucial minimally invasive imaging application of PAT is the imaging of urogenital systems; thus far, prostate cancer (Kothapalli et al. 2019; Yaseen et al. 2010; Wang et al. 2010a; Bell et al. 2016; Horiguchi et al. 2016; Horiguchi et al. 2017) and numerous gynecologic cancers, such as endometrial (Miranda et al. 2018; Li et al. 2014), cervical (Qu et al. 2018), and ovarian (Yang et al. 2011; Kumavor et al. 2013) cancers, have been the major targets of application for capitalizing PAT as a new tool for the screening of these diseases as well as for guiding other procedures in relation to disease management. Among them, the prostate imaging application has been the most representative subject with the longest history of investigation because the organ can be relatively easily accessed via anal introduction. Currently, TRUS, which is also performed based on the same procedure, is regarded as one of the most representative tools for the diagnosis and management of prostate cancer. However, its clinical usage is mostly limited to biopsy guidance due to insufficient contrast. Oraevsky et al. recognized the importance of developing TRUS-like PAT probes and have been investigating this since the inception of PAT in addition to the application of PAT for breast imaging.

As a result of their efforts, in 2010, Yaseen et al. of the Oraevsky group presented a laser OA imaging system developed for prostates (LOIS-P), with an in vivo canine prostate imaging experiment that was designed to demonstrate efficacy for image-guided biopsy (Yaseen et al. 2010). In the system, a commercial 128-channel TRUS probe (Fig. 18a) was equipped with their OA imaging platform, including a signal amplification and DAQ function previously developed for the breast application (Ermilov et al. 2009). For the in vivo demonstration, they artificially created a bloody lesion, which simulated a cancerous tumor with high blood content, by surgically exposing the prostate of a dog and inserting an abrasive brush into the prostate via a 17G hypodermic needle (Fig. 18b). The exposed prostate was then placed on the colon and photoacoustically imaged using a 757-nm laser beam that approached the acoustic detection plane from an orthogonal direction. As shown in Figure 18c, in which two OA images acquired before (upper) and after (lower) the lesion induction are presented, only the OA image could successfully visualize the induced lesion, whereas the US image (Fig. 18d) did not show any change. Although the study demonstrated the feasibility of PAT-based prostate imaging, the developed

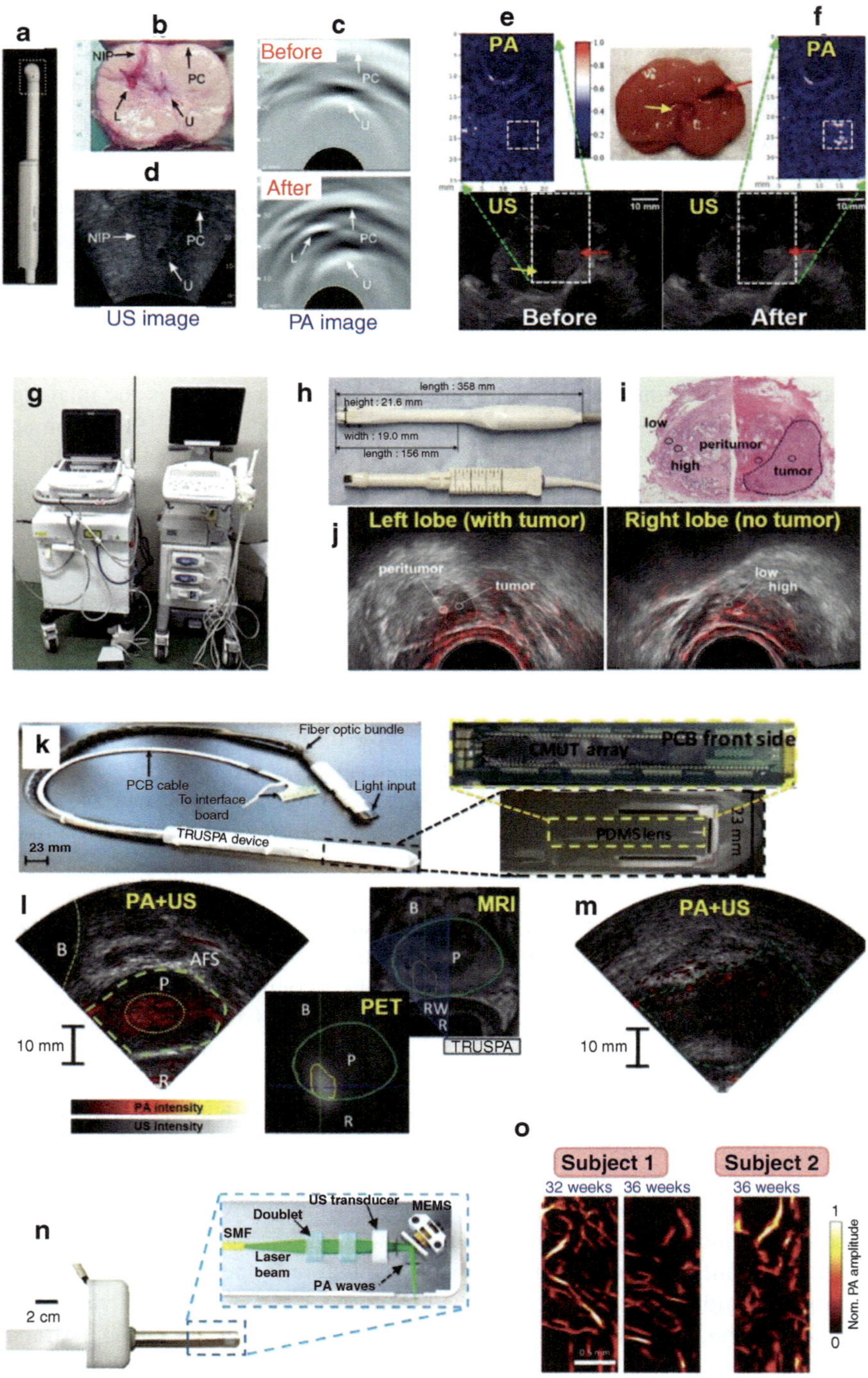
a
b
NIP
PC
L
U
d
US image
c
Before
After
PA image
e
f
PA
US
10 mm
Before
After
g
h
length : 358 mm
height : 21.6 mm
width : 19.0 mm
length : 156 mm
i
low
high
peritumor
tumor
j
Left lobe (with tumor)
Right lobe (no tumor)
k
Fiber optic bundle
PCB cable
To interface board
Light input
TRUSPA device
23 mm
CMUT array
PCB front side
PDMS lens
l
PA+US
B
AFS
P
R
10 mm
PA intensity
US intensity
MRI
PET
RW
TRUSPA
m
n
2 cm
US transducer
Doublet
MEMS
SMF
Laser beam
PA waves
o
Subject 1
Subject 2
32 weeks
36 weeks
Nom. PA amplitude
1
0

Fig. 18 Representative outcomes reported in relation to the PAT-based urogenital tract imaging applications. (**a**) Commercial 128-channel endocavity US probe utilized for the dog prostate imaging experiment (Yaseen et al. 2010). (**b**) Photo of the sliced dog prostate, in which the needle insertion path (NIP), induced lesion (L) with blood, prostate capsule (PC), and urethra (U) are marked by an arrow. (**c**) PA images were acquired in vivo from the same dog prostate before (upper) and after (lower) the lesion was induced. The induced bloody lesion (L) is only visible in the lower image. (**d**) An US in vivo image acquired from the same dog prostate after surgery using a conventional US imaging system. (**e, f**) In vivo PA (upper) and US (lower) dual-modality mages of a canine prostate acquired before (**e**) and after (**f**) the generation of a lesion by injecting 0.1 ml of blood (Wang et al. 2010a). The photograph in the center represents the imaged prostate. In all images, red arrows indicate the plastic cannula inserted into the prostate for the blood injection, and yellow arrows indicate the urethra. Note that, although the PA and US images were co-registered, the presented PA images only correspond to the dashed rectangles in the US images. (**g**) Photo of the PAT system (left) developed for the human prostate imaging study and a conventional US imaging system (right); the two machines were presented together for an appearance comparison (Horiguchi et al. 2017). (**h**) Photograph of the developed TRUS-type PA probe (upper) and conventional TRUS probe (lower). (**i**) Histopathology image of the whole-mount specimen acquired from a subject. (**j**) Combined PA (red color map) and US (gray color map) image of the index tumors (left panel) and normal prostatic tissues (right panel) acquired from the subject. (**k**) Photograph of the 23-mm diameter TRUSPA imaging probe (Kothapalli et al. 2019). (**l**) Combined PA (red color map) and US (gray color map) in vivo image of the prostate in a patient with prostate cancer. In this image, the yellow contour indicates the suspicious region with a distinct PA contrast. The next two rectangular panels depict PET and MRI images acquired ahead of the TRUSPA imaging and utilized as references for the TRUSPA imaging. The blue-shaded triangular region in the MRI image roughly corresponds to the TRUSPA image. R: rectum, RW: rectal wall, B: bladder, AFS: anterior fibromuscular stroma, and P: prostate (green contour) (**m**) 2-min post-ICG TRUSPA image showing an enhanced PA contrast (red color map). (**n**) Photo and schematic of the transvaginal fast-scanning optical-resolution PAE probe (Qu et al. 2018). The probe features a custom-designed MEMS scanning mirror that permits a fast B-scan imaging speed of 250 Hz and has an encasing tube that is 20 cm in length and 20 mm in diameter. (**o**) Corresponding in vivo images (532 nm) acquired from the first pregnant woman at 32 and 36 weeks of gestation and the second pregnant woman at 36 weeks of gestation

system did not possess a combined US imaging function. It should be noted that the presented US image (Fig. 18d) was acquired using another commercial US system.

However, in the same year (2010), Wang et al. reported a more advanced imaging system that addressed this technical issue (Wang et al. 2010a). By modifying a commercial US system combined with a P4–1 phased array probe (128 chs), they successfully implemented a PA imaging function so that co-registered PA and US images could be provided by the single system. However, for this system, the necessary illumination optics remained unintegrated into the imaging probe, as with the previous system (Yaseen et al. 2010), and it also presented a limitation to producing these images by alternatively switching the related units on an off. However, there was a great improvement in the resolution, and also the contrast features in the PA and US images could be better differentiated due to the realized image co-registration capability. As with ref. (Yaseen et al. 2010), they also performed a similar invasive canine prostate imaging experiment (Fig. 18e) and demonstrated that the lesion artificially induced by the injection of small amounts of blood (as low as 0.025 ml) was detected only by PA imaging (Fig. 18f).

Although the latter study achieved dual-mode PA and US imaging successfully, the imaging probe was still unintegrated with a necessary illumination optic for PA imaging. However, in the paper reported by Bell et al. of the Zemp group in 2016, they addressed the issue by presenting a notable probe design (Bell et al. 2016). Unlike the previous two cases (Yaseen et al. 2010; Wang et al. 2010a), they implemented a custom-designed transrectal imaging probe by integrating a conventional 192-element linear array transducer and multiple optical fibers, which were distributed around the array transducer facing the scanning direction of the transducer to perform side-scanning. It should be noted here that the probe embodiment was not based on such a simple addition of illumination optics to an existing clinical US probe as in the next case. Although the image demonstration was limited to in vivo human skin imaging, they achieved a 10-Hz frame rate, even with the vast number of detection channels.

In 2016, as these three PAT groups conducted the delineated developments, Horiguchi et al. suddenly appeared in the field reporting two radical advancements that involved the first pilot clinical study results acquired from real human subjects with prostate cancer (Horiguchi et al. 2016; Horiguchi et al. 2017).

First, in the paper published in 2016 (Horiguchi et al. 2016), they reported the results of investigating whether PAT could help visualize periprostatic neurovascular bundles (NVBs) based on microvessels, which are known to be closely co-localized with NVBs and thus expected to reveal their presence in the form of PA signals. They raised the question because in urology there has been a crucial need to preserve NVBs around the prostate to preserve sexual potency after a radical prostatectomy (RP). First, to perform the study, they developed an integrated PA and US imaging system as well as two types of imaging probes, a conventional linear array-type US probe and a TRUS-type probe, which could be connected to the console. Then, they enrolled seven patients who were undergoing non-nerve-sparing retropubic RP, as they already had clinically proven organ-confined prostate cancer, and performed an in situ prostate imaging intraoperatively (using the TRUS-type imaging probe) as well as an ex vivo imaging with the resected specimens (using the linear array-type US probe).

Through the in vivo imaging experiments performed intraoperatively using 756 nm, they found that PAT was capable of locating NVBs using microvessels as the landmark of the NVBs and that it was also better at identifying NVBs than US alone. Importantly, the derived conclusion could also be identified and validated by experiments performed ex vivo, using the resected non-nerve-sparing RP specimens. As with the intraoperative imaging experiments, strong PAI signals were observed on the posterior and posterolateral surface of the prostate in most patients, and the location of the strong PAI signals was consistent with the location of coexisting microvessels and nerve fibers that could be identified by the immunohistochemistry analysis of the sample treated through double immunostaining with CD31 (for microvessels) and S-100 (for nerves). Furthermore, the location and extent of NVBs were more clearly visible in PAI images than in images by US alone.

The serial experiments were an important demonstration that suggested that PAT could be helpful in nerve-sparing radical prostatectomy. Thus, based on this experience, they proceeded to a subsequent 2017 study (Horiguchi et al. 2017) that investigated whether the developed PAT system could describe the map of microvessels that arose from angiogenesis in prostate cancer and thus whether it could be utilized for detecting prostate cancer based on angiogenesis.

For the study, they recruited three patients who had undergone prostate biopsy and subsequent RP, and they acquired combined PA and US images just before the procedures by using the imaging system and probe (Fig. 18g and h), which was improved from the previous version (Horiguchi et al. 2016). Then, they retrospectively reviewed the acquired PA images and compared the most relevant PA images with the CD34-immunostained histopathological images of the RP specimens over four selected areas; that is, areas with high- and low-intensity PAI signals in normal prostatic tissue and areas peripheral to and inside of an index tumor, based on the correlation of PA signal intensity with three microvascular parameters; namely, microvascular density (MVD), total vascular area (TVA), and total vascular length (TVL), where MVD was defined by the number of vessels per unit area.

First, from the analysis of the normal prostate tissue, they found that in all three patients, the PA signal intensity, TVA, and TVL in areas with high-intensity PA signals were significantly higher than those in areas with low-intensity PA signals. This was indirect evidence suggesting that PA appearance could describe the distribution of microvasculature in prostatic tissue correctly. Next, from the comparative analysis on the two areas (i.e., peripheral to and inside of an index tumor), they found that the PA signal intensity, TVA, and TVL in the periphery of the index tumors were significantly higher than those inside the index tumors for all three subjects. However, MVDs in the index tumor and peripheral tumor were comparable, with the exception of one subject.

In addition to these findings, it was also observed that all index tumors showed a ring-like PAI appearance consisting of a peripheral area of high signal intensity completely or partially surrounding an area with low signal intensity, which was consistent with the results acquired from a previous breast cancer imaging study (Heijblom et al. 2015), and those tumors were recognized as hypoechoic regions on TRUS imaging. Figure 18i and j respectively shows representative histopathological and combined PA-US images acquired from a subject in which the aforementioned general tendency appeared.

Although the report was limited by the fact that the number of recruited subjects was small and the possibility of a spatial mismatch between the chosen histology and PA image, it demonstrated that PAT had potential in visualizing microvasculature in prostate cancer. Moreover, they demonstrated for the first time that PA signals could be detected at depths greater than 20 mm in actual human subjects.

In 2019, Kothapallie et al. reported another notable achievement in PAT prostate imaging application (Kothapalli et al. 2019). Unlike the other aforementioned groups' systems, in which commercial piezoelectric sensor-based commercial US detectors were employed, they developed an integrated TRUS and PA imaging probe (named TRUSPA) by compactly integrating a lab-made capacitive micromachined US transducer (CMUT) array (64 elements, 5 MHz) and illumination optics into a ~ 23-mm diameter imaging probe (Fig. 18k). They used it to perform an animal experiment on mice with subcutaneous prostate cancer, and they also demonstrated its actual clinical applicability by imaging human subjects with prostate cancer (n = 20).

Among the numerous imaging results, Figure 18l and m is notable. First, Figure 18l depicts in vivo TRUSPA and PET-MRI imaging results from a patient with proven advanced prostate cancer as evidenced by preoperative PET-MRI that was followed by PET-MRI contrast-targeted biopsy (with MRI-TRUS fusion) of the prostate using a conventional TRUS device. As shown, the TRUSPA device displayed a distinct PA contrast from the right peripheral base of the prostate, which was not present when other prostatic regions of this patient was scanned, a result which was in agreement with PET-MRI results.

In the paper, they also demonstrated first-in-man contrast-enhanced TRUSPA imaging (n = 10) by intravenously administrating FDA-approved ICG solution (2.5 mg/ml) in the dose range of 5 to 75 mg (n = 10). As shown Figure 18m, an example acquired from a patient with biopsy-proven cancer in the left peripheral base of the prostate and captured at 2 min post-injection using an 800-nm laser beam, a PA contrast enhancement was clearly observed after the injection of ICG (25-mg dose). It should be noted that the dose was quite lower than the average dose (10 mg/kg) of MRI contrast agents in humans.

In addition to the encouraging results of the pilot clinical study, their achievement presented several important technical implications. First, unlike other groups' systems (Bell et al. 2016; Horiguchi et al. 2016; Horiguchi et al. 2017), the compact integration was not based on the simple addition of fiber optic illumination units to existing TRUS probes, but on a customized design and fabrication targeting the specific application. Consequently, they could detect PA signals from the dorsal vascular complex located even at ~40 mm depth from the rectal wall of a subject. Second, they achieved the highest imaging speed—10 frames per second (fps), even for the simultaneous B-mode presentation of PA, US, and combined PA-US images—among all array transducer-based, miniaturized PAT probes manufactured in-house. Lastly, they introduced the novel CMUT technology, which is known to be superior to conventional piezoelectric sensors in terms of bandwidth, SNR, and ease of fabrication in forming a 1D or 2D array, to PAT research.

In addition to the prostate imaging application, the report from Qu et al. of the Wang group in 2018 is also notable for its female genital system imaging (Qu et al. 2018). In 2015, Li et al. of the same group also reported a urogenital PA endoscope working in an AR mode for a similar purpose (Li et al. 2014). However, in the new study, they developed a transvaginal fast-scanning optical-resolution PAE probe to understand vascular relationships during cervical remodeling in pregnancy (Fig. 18n). The development of such an imaging probe would be crucial for understanding abnormal delivery issues, such as premature labor. In this regard, the proposed working hypothesis was also reasonable because there were studies reporting that cervical remodeling during pregnancy was associated with increased vascularity. Considering the application, they developed the probe so that it could exhibit a performance such as 250-Hz for the B-scan imaging speed over a 3-mm scanning range—which was faster than the previous device (Li et al. 2014) by a factor of 10—and 3.1 μm and 46.5 μm for the lateral and axial resolutions, respectively.

With the probe, they imaged two pregnant women and analyzed five images acquired from three different areas (i.e., the sublingual mucosa, the uterine body, and the ectocervix) based on the two vascular parameters: (1) the microvessel density (the number of vessels per unit area) and (2) the total microvascular area (the percentage of area occupied by blood vessels). Notably, although the number of enrolled subjects was too small, they arrived at the tentative conclusion that microvessel density was a more promising parameter for identifying the progress of cervical remodeling compared to total microvascular area, whereas the total microvascular area was a more discriminatory parameter for classifying the type of tissue. Figure 18o represents representative PA vasculature images acquired from the ectocervix of the two subjects. It can be seen that the blood vessels of Subject 2 had a smaller aspect ratio than those of Subject 1 at the same gestational age (36 weeks).

As we already briefly mentioned in the previous IVPA imaging section, there was also another notable achievement demonstrating the tri-modality imaging of PA, US, and OCT reported by Yang et al. of the Zhou group in 2011 (Yang et al. 2011). Although the demonstration was limited to human ovarian tissue imaging ex vivo, further improvement of the technique would have a significant impact on urology in relation to laparoscopic or image-guided surgery applications.

In this section, we reviewed the three active domains of minimally invasive imaging applications for PAT. To summarize, the IVPA application has been the most active area of application because cardiovascular diseases are the leading cause of death in Western countries, and in recent years it is in vivo coronary artery imaging capability has been successfully demonstrated based on a swine model (Cao et al. 2018; Wu et al. 2017). In addition, the prostate imaging of PAT has advanced enough to apply to actual human subjects (Kothapalli et al. 2019; Horiguchi et al. 2016; Horiguchi et al. 2017). Although related imaging probes do not require a higher miniaturization and also have the advantage of relatively easy access to the target area compared to the other two applications—GI endoscopy and IVPA imaging—it was a great achievement that demonstrated image penetration of up to 4 cm in an actual human subject (Kothapalli et al. 2019), which suggests that a similar penetration could also be achieved in GI endoscopy. In short, if a PAE probe were to be realized in such a large form as with a conventional EUS probe (Shami and Kahaleh 2010; Akahoshi and Bapaye 2012), it may be able to provide images of peri-luminal organs.

Overall, the minimally invasive application of PAT continues to face many challenges related to the required imaging probe fabrication because unlike the existing EUS or optical endoscopic probes, a PAE probe requires the integration of both optical and acoustic elements in a small space. Nevertheless, considering the aforementioned fundamental penetration limit of PAT, we expect that this application could ultimately be one of the most important clinical application areas of PAT.

PA Angiography and Lymphangiography

Angiography refers to a technique utilized to visualize the vascular network (i.e., vasculature), a specific blood vessel of interest, or its internal structure in relation to a diagnostic or interventional purpose. Further, lymphangiography refers to a similar procedure but one that is performed to visualize lymph vessels or lymph nodes. For this, US imaging, X-ray imaging (either planar or CT), and MRI have been utilized thus far. Although MRI-based angiography can provide high-quality vasculature images, its clear visualization still requires the injection of a gadolinium-based contrast agent. Moreover, it requires a rather long scanning time, which makes the technique inapplicable for a real-time guidance of other medical procedures; thus, it is mostly utilized for pre- or after-treatment scanning purposes.

On the other hand, US imaging has the advantage of being capable of visualizing large blood vessels without the aid of a contrast agent and also does not have any safety concerns. However, in order to visualize a blood vessel clearly, there must be a sufficient level of blood flow because it relies on the Doppler principle and also requires the use of a contrast agent when it visualizes microvasculature. Due to the limitations of the two techniques, X-ray-based angiography has been a dominant tool and is regarded as a gold standard for numerous important clinical procedures; however, it also requires the injection of an iodine-based contrast agent and involves concern related to radiation hazards caused by X-ray bombardment when it is utilized for real-time imaging. Representative applications include the visualization of cerebral circulation, vascular network in a cancerous organ, and guiding a catheter for interventional cardiology.

As presented in the previous sections, PAT has already demonstrated its superiority of visualizing blood vessels (even microvasculature) without using any contrast agent and also the lymphatic system if it could be properly stained by a contrast agent injected into or around the target region of interest. Since blood vessels and lymphatic systems are the core parts of the circulatory system and also the major pathways of cancer metastasis (Steeg 2006; Alitalo 2011), addition of PAT to clinical angiography would have a significant impact on medicine.

Since the inception of PAT, various results that visualized a superficial vasculature have been steadily reported. Actually, such images were already good examples that indicated the angiographic imaging capability of PAT. However, in order to make PAT applicable to real clinical angiography, as with X-ray, its penetration depth must be extended to larger depths. In such aspects, research outcomes showing a more clinically relevant level (i.e., organ size) of visualization of vasculature began to be reported since the mid-2000s, as the energy level of available laser sources had greatly increased at the time. The breast vasculature images presented in "Breast Cancer Diagnosis" section correspond to related examples. However, in recent years, more encouraging results that visualized the detailed vasculature of various human body sites have been reported. Thus, in this section, although the related reports are relatively fewer than those related to other applications, we introduce several notable research outcomes that reveal the current status and future potential of PA angiography.

First, a report showcased the feasibility of PAT-based in vivo human carotid artery (CA) and jugular vein imaging for the first time (Fig. 19a), which was presented by Dima et al. of the Ntziachristos group in 2012 (Dima and Ntziachristos 2012). In clinic, as with coronary artery imaging, the CA imaging has also been an important subject in relation to atherosclerosis diagnosis, as it is a representative aspect in assessing the possibility of cerebral hemorrhage and ischemic stroke. In the report, they successfully visualized these vessels located 12–18 mm beneath the skin by using a curved array US transducer (5 MHz, 64 elements) with a radius of a curvature of 40 mm and applying a side optical illumination irradiated over a 4 cm × 0.5 cm area with ~26 mJ at 800 nm (Fig. 19b). The imaged features were CA and jugular vein because they could be confirmed by a conventional duplex US that provided combined anatomical and color Doppler US images (Fig. 19c).

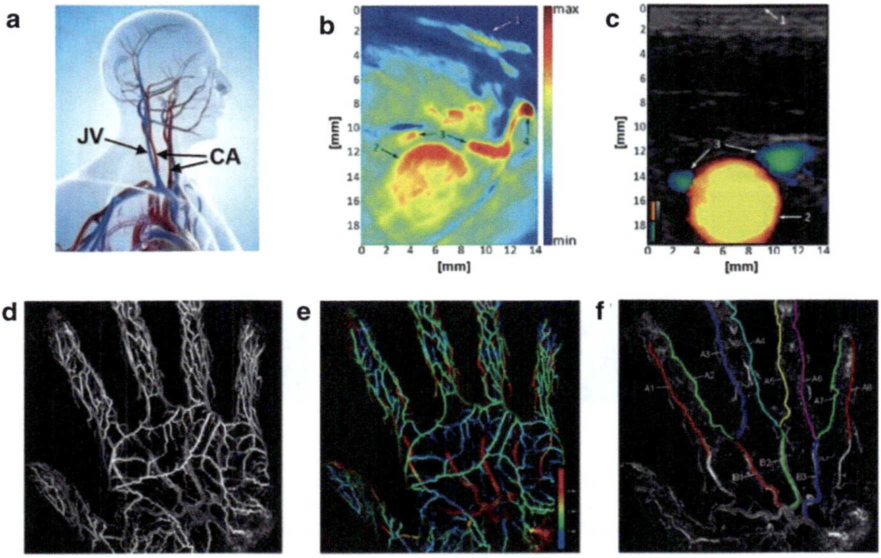

Fig. 19 Examples of reported outcomes showing the feasibility of clinical PA angiography. (**a**) Anatomic diagram presenting the location of the carotid artery (CA) and jugular vein (JV). (**b**) PA cross-sectional image presenting the cardio-vasculature of a female volunteer at the lower neck level; this image was acquired using the developed system (Dima and Ntziachristos 2012). (**c**) Corresponding directional duplex US image (i.e., an image acquired using a different machine) indicating positive flow in yellow/red and negative flow in green/blue. Annotations: 1. Skin surface; 2. Common carotid; 3. Internal jugular vein; 4. External jugular vein. (**d**) PA MAP image (795 nm) showing a human palmar vasculature (Matsumoto et al. 2018). The image was produced using a complete 3D data set acquired by PAI-03. (e) Depth-encoded PA image. (f) Superimposed image of the binarized colored lines extracted from the common and proper palmar digital arteries over an PA background image (gray) after removing the vein signals near the skin's surface. (**g**) Schematics illustrating the postures of a subject during PA imaging (Nagae et al. 2019). (**h**) PA images obtained from various body sites. ⑦ is an image of whole anterolateral thigh (ALT) image. The pink square in ⑦ indicates the corresponding area of the S-factor image presented alongside. In the S-factor image, the two large blood vessels appear to be an artery and vein in the ALT, and the yellow arrow indicates blood vessels that are considered perforators

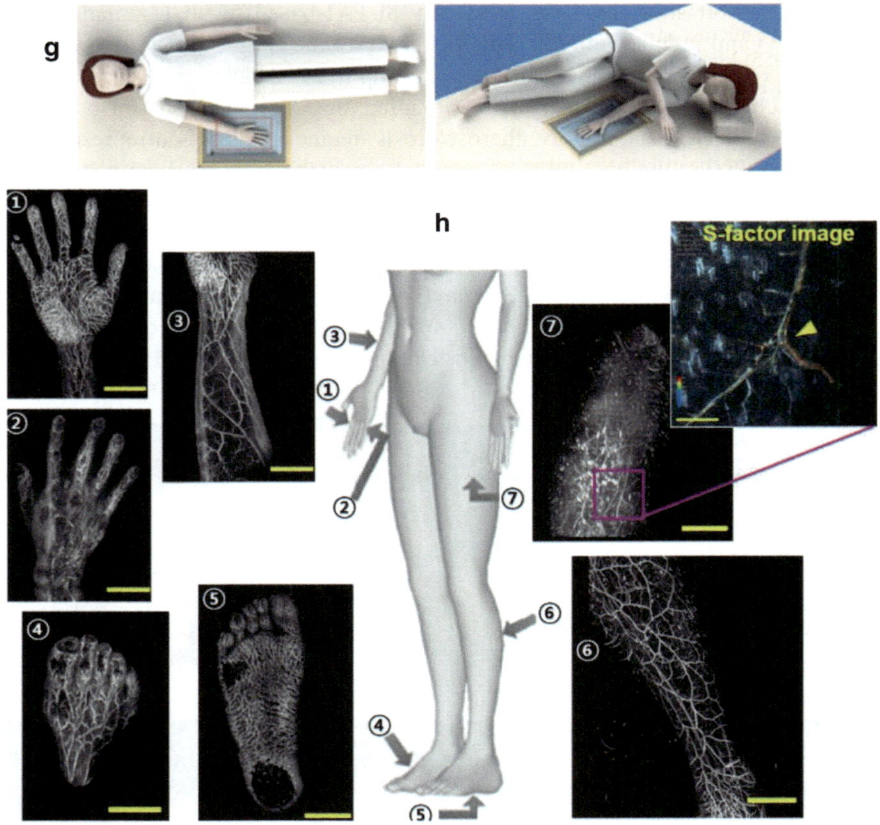

Fig. 19 (continued)

Since the report, the CA imaging has continued to generate interest. For example, in 2014, Kruizinga et al. of the van Soest group presented a report that investigated CA samples ex vivo by using laser wavelengths that ranged from 1130 nm to 1250 nm (Kruizinga et al. 2014); they found that the atherosclerotic plaques accumulated in a CA wall also could be photoacoustically visualized by the same 1210-nm laser wavelength that was frequently applied in numerous IVPA imaging studies (Jansen et al. 2011; Li et al. 2015), as presented earlier. Moreover, in the report, they also presented a different strategy that photoacoustically imaged CA by applying internal illumination from the pharynx and detecting the generated PA signals externally from a neck surface using an US detector. In addition to the report, there were several follow-up studies that demonstrated (1) a more accurate visualization of anatomical features distributed around the thyroid as well as the CA using a PAT probe custom-designed to better fit the human neck (Dima and Ntziachristos 2016), (2) the first in vivo 2D real-time mapping of the blood flow (US) as well as the oxygen state (PA) of the CA (Mercep et al. 2018), and 3) the first 3D real-time visualization of the structures around the CA (Ivankovic et al. 2019).

In addition to the CA imaging, several research outcomes that demonstrated the potential of PAT for clinical angiography began to be reported in very recent years. Among these, the two papers reported by a Japanese team provided more strong evidence in this regard.

First, in 2018, Matsumoto and Asao et al. (Matsumoto et al. 2018) presented a rather impressive achievement that visualized the detailed vasculature distributed in human hands for the first time by using their third prototype of the HDA-based PAT system (named as "PAI-03"). As mentioned in "Breast Cancer Diagnosis" section, the employed system was originally developed for the breast imaging application (Fig. 12a); however, they improved the previous version for this new study. By using a 795-nm laser wavelength, they imaged palmar blood vessels (Fig. 19d and e) and analyzed their morphology, particularly focusing on the curvature parameters of the common and proper palmar digital arteries. However, for the analysis, it was necessary to remove the signals from superficial veins first to better distinguish the targeted arteries (Fig. 19f). In this manner, they analyzed the 3D palmar vasculature PA data acquired from 23 healthy subjects (12 females and 11 males) in an age range of between 22 and 59 years; interestingly, they found that the level of the tortuousness increased with age regardless of the sex.

After this report, in 2019, the research team also presented another notable research result that visualized vasculatures distributed in several body sites of a human subject (Nagae et al. 2019). The recruited subject was a male in his 50s, and the imaging experiment was performed using their fifth prototype (named "PAI-05"), which was further improved from the previous version in terms of the number of detection channels (now 1024-chs), scanning pattern (rectangular spiral), and the ease in positioning a target area. Thus, it was possible to position various body sites—such as hand, foot, arm, and thigh—on the scanner (as depicted in the illustrations presented in Figure 19g) and acquire nice vasculature images from the mentioned areas (as shown in Figure 19h), which were the first-of-its-kind in PAT. Moreover, by performing a dual-wavelength PA imaging (756 nm and 797 nm), they also demonstrated the sO_2 mapping capability of the developed system (see the magnified color image extended from the anterolateral thigh image).

Thus far, such a detailed vasculature analysis of human extremities has not been facilely attempted in clinic due to the limited availability of conventional imaging modalities and the cost issue. However, this has now become possible, as PAT can provide vasculature images with unprecedented image quality. Based on the presented demonstrations, the authors commented that the information provided on vascular morphology and related parameters, like arterial tortuousness, would be important information representing individuals' lifestyles, medical history, and disease symptoms. Moreover, apart from the evident medical implications, the new angiographic imaging technique could also extend into the fields of healthcare and beauty, for example, to treat issues such as varicose veins.

Then, could PAT eventually replace the conventional X-ray-based clinical angiography? Of course, it would be difficult to replace this due to its relatively lower penetration depth. However, it may be still possible for some areas or at least it may be realized in such a form as to complement or aid conventional procedures. As evident from the examples discussed in other sections, there was a case in which PAT detected signals from a maximum depth of ~5-cm, even in actual human tissue imaging in vivo. In particular, the real-time and label-free safe nonionizing imaging capability of PAT would make the expectation rather feasible. Moreover, PAT has the unique imaging ability of distinguishing between a vein and an artery, which could make a critical contribution to interventional cardiology by aiding related procedures, like the arterial introduction of a catheter.

Thus far, we have discussed the potential of PAT for clinical angiography. However, in comparison, there are fewer studies involving PAT-based lymphangiography. This is because in order to visualize the lymphatic system, it is necessary to administrate an appropriate contrast agent, like methylene blue. Nevertheless, here we briefly review related progress because the lymphatic system is another important circulatory system in addition to the vascular system and is also another area in which PAT can function effectively.

As previously mentioned, the first PAT-based lymphatic system imaging was demonstrated by Song et al. of the Wang group in 2008 in relation to the SLN mapping in a rat (Fig. 20a) (Song et al. 2008). Thereafter, a number of papers were further documented in the category, particularly in relation to the development of a contrast agent that enabled more effective visualization by increasing its drainage efficiency into the lymphatic system or by increasing its optical detection sensitivity. For example, in 2010, Pan et al. investigated the size dependence of gold nanobeacons (GNBs) for SLN detection and found that smaller GNBs were noted trafficking through the lymphatic system and accumulating more efficiently in the lymph nodes in comparison to larger GNBs (Pan et al. 2010). In addition, in 2009, Kim and Zharov et al. demonstrated a dual-modality PA and photothermal mapping of lymphatic endothelial cells—which line the internal surface of lymphatic vessels—by using golden carbon nanotubes (GNTs) conjugated with an antibody specific to the lymphatic endothelial hyaluronan receptor-1 (Kim et al. 2009). As the GNTs were engineered to provide an enhanced PA contrast as high as ~10^2-fold, they were able to successfully visualize lymphatic vessels in mice even using an extremely low laser fluence of a few mJ/cm^2 (Fig. 20b and c).

Although the mentioned papers demonstrated the potential of PAT for the imaging of the lymphatic system, they were all limited to proof-of-concept studies based on animal models. However, in 2015, in addition to the aforementioned report by Alejandro et al. of the Wang group (Garcia-Uribe et al. 2015), Stoffels et al. of the Klode group also presented human SLN imaging results acquired from their first pilot clinical study performed in relation to melanoma metastasis (Stoffels et al. 2015). Notably, by using ICG as a contrast agent, they could reliably map SLNs in 20 patients in vivo, with a 100% concordance with conventional ^{99m}Tc-marked SLN lymphoscintigraphy. Moreover, they detected SLNs located even up to a depth of 5 cm—whose measure was based on a natural state and whose location was identified by SPECT/CT—by applying pressure during the PAT image acquisition to reduce the distance between the SLNs and the surface of the skin (Fig. 20d). Figure 20e represents a PA image displaying a spectrally unmixed ICG signal in an SLN overlaid on a structural image acquired at 800 nm. Overall, from the in vivo study, they were able to photoacoustically identify SLNs without any false negative and with 100% sensitivity and 48% specificity.

Although the report was also significant in terms of the first-in-human study related to melanoma, it was mostly focused on mapping a lymph node rather than

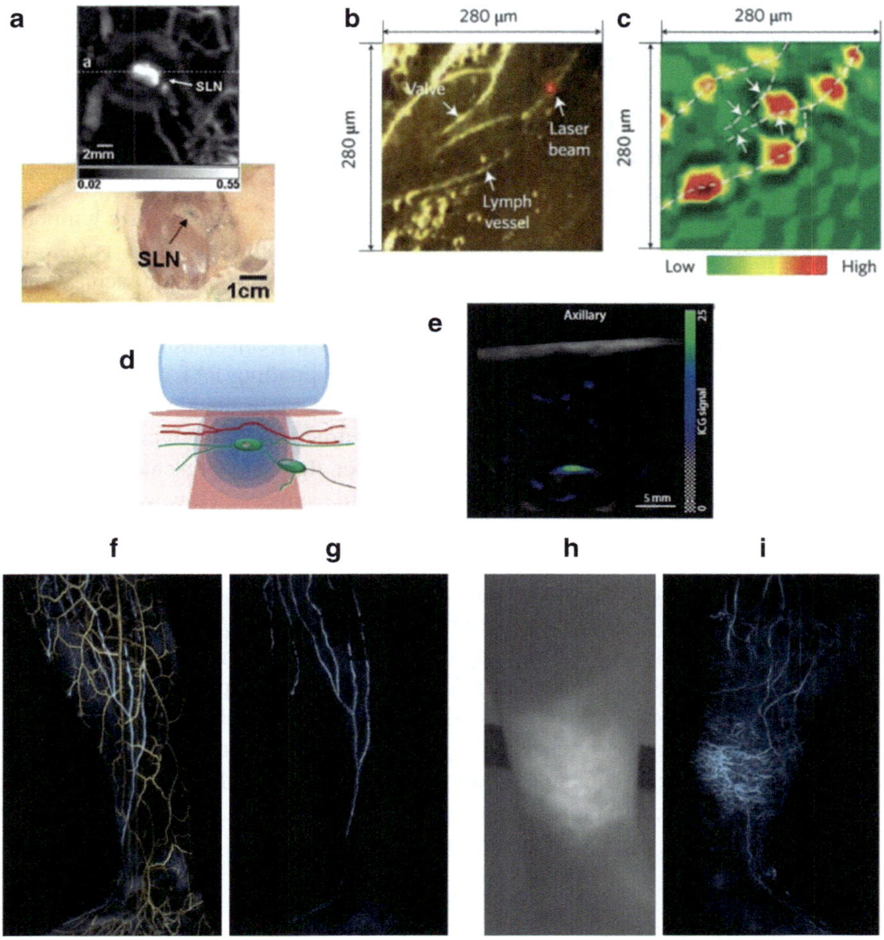

Fig. 20 Representative reports demonstrating preclinical lymphatic system imaging and the feasibility of clinical PA lymphangiography. (**a**) PA MAP image (upper) acquired 52-min post-injection, in which the stained SLN was visualized, and photo (lower) with skin removed after the PA imaging (Song et al. 2008). (**b**) Conventional microscope image showing the imaged area in a fragment of mouse mesentery (Kim et al. 2009). (**c**) Corresponding PA image (850 nm) acquired in vivo at 60 min after the administration of antibody-GNTs, which was engineered to target the lymphatic system. Dashed white lines indicate the lymphatic wall and valve. Arrows indicate valve leaflets. (**d**) A schematic illustrating the PA detection of a SLN indicated by ICG in conjunction with a handheld PA imaging probe (Stoffels et al. 2015). (**e**) Representative preoperative PA image of axillary SLN, with ICG signal (green) overlaid on a single-wavelength (800 nm) background image (gray). (**f**) The medial-side view of the PA angiography and lymphangiography of the right lower leg of a woman in her thirties without any past medical history; this image depicts both lymphatic vessels and venules (Kajita et al. 2019). (**g**) PA image showing lymphatic vessels only. (**h**) The medial-side view of the conventional fluorescence lymphography of the left lower leg of a 47-year-old woman with secondary lymphedema. (**i**) The PA lymphangiography of the same part, depicting only lymphatic vessels

visualizing lymphatic vasculature. However, in 2019, Kajita et al. presented an impressive result that achieved the first human lymph vessel imaging by using ICG as a contrast agent (Kajita et al. 2019). For the study, they employed the same PAT system called "PAI-05" utilized for the human angiography imaging presented in Figure 19g and imaged 20 healthy subjects and 30 lymphedema patients by administering ICG subcutaneously to the dorsal aspect of each foot or hand, as done in fluorescence lymphography.

Surprisingly, as depicted in Figure 20f, the PAT system could visualize lymphatic vessels and their neighboring venules with a spatial resolution as fine as 0.2 mm; Figure 20g depicts only lymphatic vessels. Indeed, this is a significant achievement that acquired a PAT-based human lymphatic vessel image for the first time. In order to showcase the superiority of PAT, they also presented a conventional fluorescence lymphographic image (Fig. 20h) along with a corresponding PAT image (Fig. 20i) obtained from a lymphedema patient. As illustrated, only the PAT system could clearly differentiate the dense interconnecting 3D structures of the lymphatic vessels distributed around the leg of the patient. In the paper, they only presented representative images obtained from healthy and lymphedema subjects. However, they stated that it would possible to analyze morphological patterns with the acquired lymphatic data and running course of each lymphatic vessel, along with their comparisons to accompanying venules, which would be reported elsewhere.

As presented, only the three papers involved human subjects and demonstrated more direct feasibility of PAT-based lymphangiography (Garcia-Uribe et al. 2015; Stoffels et al. 2015; Kajita et al. 2019). Although the related application is relatively more challenging because there is a fundamental issue regarding how to effectively deliver a contrast agent to the lymph system, PAT is expected to make a crucial contribution to clinical lymphangiography because it is an undeniable fact that once a targeted lymphatic structure is successfully stained, PAT can provide a much clearer contrast on it compared to conventional US imaging.

Other Notable Clinical Applications

Although the aforementioned clinical application areas are dominant in terms of the number of publications, there remain numerous other important applications that must be noted. Among these, PAT-based circulating tumor cell (CTC) detection as well as an intraoperative imaging application appear to be promising and of particular importance.

The first subject was mostly pioneered by the Zharov group in relation to the PAT-based flow cytometry technique development (Galanzha and Zharov 2012), and the result presented in Figure 20c could be understood as one of the related outcomes. In particular, in 2009, Galanzha et al. of the Zharov group also reported another notable article that presented a novel method to increase the possibility of CTC detection in the bloodstream (Galanzha et al. 2009). By applying a magnetic enrichment concept along with the use of dual chemical agents—that is, GNTs and magnetic nanoparticles—which were engineered to target human breast cancer cells, they could significantly increase the PA signal intensity generated by the magnetically captured tumor cells (e.g., ~88-fold increase in an in vivo mouse model).

Thereafter, similar approaches also began to be addressed by other groups. For example, in 2010, Jin et al. reported iron oxide and gold-coupled core-shell nanoparticles that could be utilized as a multifunctional contrast agent for multi-modality imaging—for example, MRI as well as the magnetomotive PAT (Jin et al. 2010). In addition, in the paper reported by He et al. of the Wang group in 2016 (He et al. 2016), they demonstrated an imaging-based single melanoma tumor cell detection in a bloodstream by applying the OR-PAM technique (Figs. 21a–c) as well as its on-the-spot laser killing.

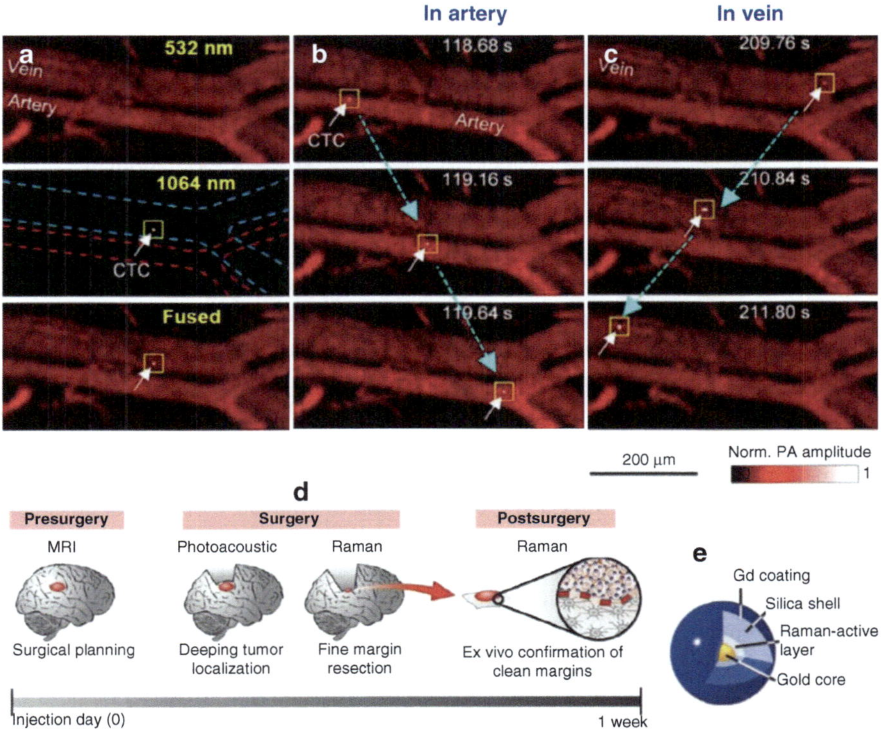

Fig. 21 PAT-based CTC detection and intraoperative imaging applications. (**a**) Snapshots showing single CTCs travelling in vasculature: 532-nm laser-induced (top), 1064-nm laser-induced (middle), and fused (bottom) flow cytography images (He et al. 2016). In the 1064-nm image, the white arrow and yellow square indicate the detected CTC; the red and blue dashed lines delineate the artery and vein boundaries, respectively. (**b**) Three fused snapshots spanning ~1 s revealing a single CTC traveling in the artery. (**c**) Three fused snapshots spanning ~2 s revealing a single CTC traveling in the vein. The times labeled in (**b**) and (**c**) are relative to CTC injection. (d) A diagram illustrating the strategy of the triple-modality MPR imaging-based brain tumor surgery (Kircher et al. 2012). (**e**) Approximate structure of the developed MPR nanoparticle. (**f**) A mouse model-based in vivo demonstration of the proposed strategy. Images from the left to the right columns are 2D axial MRI, PA, and Raman images, respectively. The post-injection images (lower row) of all the three modalities showed clear tumor visualization (dashed boxes outline the imaged area). (**g**) An in vivo surgical demonstration of a sequentially resected a brain tumor. Quarters of the tumor were sequentially removed (upper row) and intraoperative Raman imaging was performed after each resection step (lower row) until the entire tumor had been removed, as assessed by visual inspection. (**h**) The proposed imaging concept for the PAT-guided pituitary tumor surgery (left) and coronal view of the anatomy surrounding the sphenoid bone (right) (Lediju Bell et al. 2015a; Delman 2009)

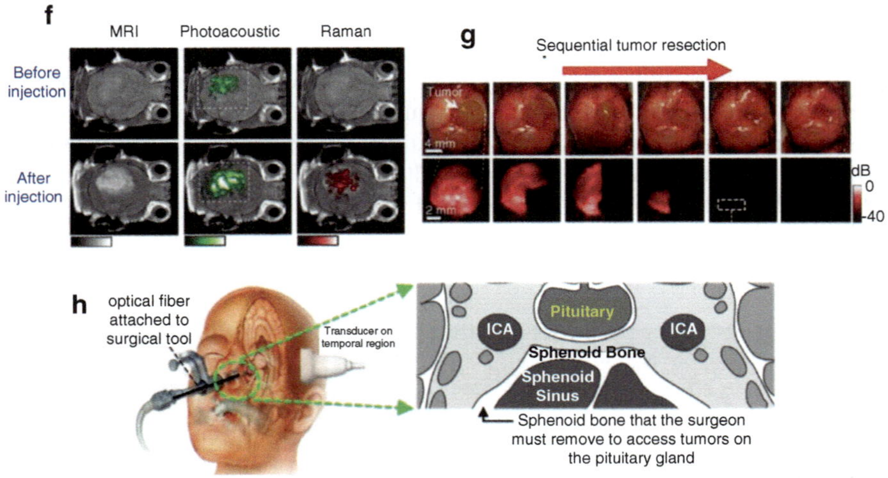

Fig. 21 (continued)

As presented, a key issue in CTC detection is to increase detection sensitivity or possibility. Surprisingly, Galanzha et al. of the Zharov group in 2019 (Galanzha et al. 2019) demonstrated their technique's in vivo clinical applicability by successfully detecting CTCs in the skins of 27 out of 28 patients with melanoma, with a detection sensitivity as low as 1 CTC/liter of blood. Moreover, they also successfully showcased the feasibility of laser illumination-based CTC eradication even in a human subject. As CTCs are regarded as potential indicators for tumor metastasis, in vivo detection of these CTCs is a critical issue in clinics for early diagnosis and treatment of cancer.

In relation to the intraoperative imaging application of PAT, the development of a dedicated miniaturized imaging probe, such as laparoscopic or handheld probes, is rather urgent and promising because there are numerous unmet needs in clinics—for example, although an MRI can provide a clear view of a tumor region, it is not directly applicable for an actual surgery. In this regard, in 2012, Kircher et al. of the Gambhir group proposed a feasible strategy that could more effectively remove a brain tumor based on the synergetic effect derived from the combined use of the three imaging modalities—MRI, PAT, and Raman (Kircher et al. 2012). As per the strategy, 1) MRI is performed for preoperative macroscopic delineation of the brain tumor; 2) PAT for intraoperative high-resolution and deep 3D imaging; and 3) Raman for intraoperative high sensitivity, high specificity, and high-resolution surface imaging of tumor margins (Fig. 21d). In order to make the strategy more effective in actual clinical settings, they also developed a dedicated contrast agent called the MRI-PA-Raman (MPR) nanoparticle (Fig. 21e), which could be utilized as a common contrast agent for the three mentioned imaging techniques (Fig. 21f). Figure 21g depicts an actual photo of the dissection of a brain tumor according to the strategy.

In addition to the brain imaging application, other reports also suggested PAT as a guiding tool for other surgical procedures, such as those related to prostate tumor

(Lediju Bell et al. 2015b) and pituitary tumor (Lediju Bell et al. 2015a). Figure 21h depicts a related example that addressed a PAT-guided pituitary tumor surgery based on an endonasal transsphenoidal approach.

Summary and Outlook

In this chapter, we reviewed the development history of PAT, mostly focusing on several key application areas that have been actively investigated over the previous two decades. However, as new applications and methodologies continue to be presented even currently, like the several examples that we briefly mentioned above, it is also highly expected that a completely new field of application or new embodiments that are beyond our current expectations could continuously emerge. This is probably due to the versatile contrast creation capability of PAT, which is based on the unique optical contrast mechanism termed "light absorption," and the distinct system embodiment feature that uses the conventional US imaging technique.

Of course, although in this review, we mainly highlighted the positive aspects of PAT, it is an undeniable fact that the imaging depth of PAT is relatively lower than that of existing mainstream imaging modalities, such as the MRI, CT, and US imaging techniques. However, it is also true that no other biomedical imaging technique, among all those that have emerged after current clinical mainstream imaging modalities, can provide such high-resolution images as PAT at depths reaching several centimeters in vivo (thus far, signal detection in a depth of ~5 centimeters has been recorded as the maximum in PAT); this is probably the core aspect that accounts for why PAT is currently the most rapidly growing technique in the biomedical imaging field compared to other imaging techniques, thereby indirectly suggesting the importance of PAT technology. Most of all, the unique anatomical, functional, and molecular imaging capability—which is not possible with the conventional mainstream imaging modalities but affordable in real time by PAT—will make a significant contribution to biomedicine in the future by enabling numerous breakthroughs and also paving the way for new medical imaging paradigms.

Further, the technical aspect that PAT can be easily integrated with the conventional US imaging technique—which is currently regarded as one of the three core tomographic imaging modalities in clinics—must also be importantly evaluated because based on the integration, PAT could provide a feasible solution or new innovation for numerous scattered issues in the current clinical application areas of US imaging, for which conventional methods or procedures face limitations. Simply speaking, the emergence of PAT has enabled color visibility in the US imaging technique, thereby enabling better identification and characterization of suspect tissues; this event could probably be likened to a previous historic event like when black-and-white TVs were replaced with color TVs in the previous century.

As discussed above, in biological application, the role and contribution of PAT for small animal whole-brain and whole-body imaging is expected to be particularly unique and, thus, be ultimately successful in the near future. In the clinical application, the breast cancer diagnostic application appears to be in a state that is closest to the final clinical translation compared to other application areas. In addition, new light is currently being shed on the dermatological applications of PAT due to the recent innovations that have been presented. Moreover, the endoscopy and minimally invasive imaging applications of PAT are also expected to be successfully translated into clinics because there are numerous attractive clinical issues and also a corresponding number of efforts being made. Moreover, based on the unprecedented image quality for visualization of the vascular and lymphatic systems, which are the two core circulation systems, there is also a high expectation of the development of new forms of applications and contributions of PAT in clinical angiography and lymphangiography.

Although only several core application areas were mainly introduced in this chapter, there are numerous other notable and interesting application areas that could become more significant in the future. For example, in a case in which AI was applied to skin vascular analysis (Moustakidis et al. 2019), the incorporation of smart AI technology into PAT would make a critical contribution to clinical application by enabling early detection of disease symptoms and more accurate characterization of suspicious tissues in dermatological and numerous other applications areas, based on the rich spectroscopic and functional information on the vasculature and associated biomarkers provided by PAT.

Of course, however, additional efforts must be made for successful clinical translations. First, more precise and clear protocols and criteria on related imaging procedures and quality assurance must be established. Moreover, numerous technical standards necessary for related system embodiments—such as the one related to a light illumination parameter as well as for the quantification of related imaging performances, such as imaging depth and resolution—must be established. In addition, it will be necessary to further lower the cost of implementing related systems and also improve overall imaging performance. In particular, further improvements in imaging depth will be most critical, as current imaging depth is still insufficient to achieve broader clinical applications. Although this may not be essential for certain applications, such as the IVPA imaging and dermatological applications, significant improvements in the image depth will ensure the successful clinical translation of PAT. The results presented in this chapter represent the achievements in the development of PAT over the past ~25 years. We look forward to the full maturation of PAT in the next 25 years, when it becomes a commonly used imaging technique—for example, in hospitals.

Acknowledgments This work was supported by the U-K Brand Research Fund (1.200032.01) of UNIST (Ulsan National Institute of Science and Technology) and Basic Science Research Program through the National Research Foundation (NRF) of Korea funded by the Ministry of Education (2015R1D1A1A01059361).

References

Aguirre J, Schwarz M, Garzorz N et al (2017) Precision assessment of label-free psoriasis biomarkers. Nat Biomed Eng 1:0068

Akahoshi K, Bapaye A (2012) Practical handbook of endoscopic ultrasonography. Publisher. Springer Japan, Tokyo

Alitalo K (2011) The lymphatic vasculature in disease. Nat Med 17:1371–1380

Allen TJ, Hall A, Dhillon AP et al (2012) Spectroscopic photoacoustic imaging of lipid-rich plaques in the human aorta in the 740 to 1400 nm wavelength range. J Biomed Opt 17:061209. https://doi.org/10.1117/1.JBO.17.6.061209

Amornphimoltham P, Masedunskas A, Weigert R (2011) Intravital microscopy as a tool to study drug delivery in preclinical studies. Adv Drug Deliv Rev 63:119–128

Andreev V, Karabutov A, Solomatin S et al (2000) "Optoacoustic tomography of breast cancer with arc-array transducer", Proc SPIE 3916. Biomedical Optoacoustics:36–47

Ansari R, Zhang EZ, Desjardins AE et al (2018) All-optical forward-viewing photoacoustic probe for high-resolution 3D endoscopy. Light Sci Appl 7:75

Asao Y, Hashizume Y, Suita T et al (2016) Photoacoustic mammography capable of simultaneously acquiring photoacoustic and ultrasound images. J Biomed Opt 21:116009

Bai X, Gong X, Hau W et al (2014) Intravascular optical-resolution photoacoustic tomography with a 1.1 mm diameter catheter. PLoS One 9:e92463

Beard P (2011) Biomedical photoacoustic imaging. Interface Focus 1:602–631

Beard PC, Pérennès F, Draguioti E, Mills TN (1998) Optical fiber photoacoustic-photothermal probe. Opt Lett 23(15):1235–1237

Bell KL, Harrison T, Usmani N et al (2016) "Integrated transrectal probe for translational ultrasound-photoacoustic imaging", Proc SPIE 9708, Photons Plus Ultrasound: Imaging and Sensing 2016, 97080A

Boas DA, Brooks DH, Miller EL et al (2001) Imaging the body with diffuse optical tomography. IEEE Signal Process Mag 18(6):57–75

Brecht HP, Su R, Fronheiser M et al (2009) Whole-body three-dimensional optoacoustic tomography system for small animals. J Biomed Opt 14:064007. https://doi.org/10.1117/1.3259361

Brown E, Brunker J, Bohndiek SE (2019) Photoacoustic imaging as a tool to probe the tumour microenvironment. Dis Model Mech 12:dmm039636

Brunker J, Beard P (2016) Velocity measurements in whole blood using acoustic resolution photoacoustic Doppler. Biomed Opt Express 7(7):2789–2806

Brunker J, Yao J, Laufer J et al (2017) Photoacoustic imaging using genetically encoded reporters: a review. J Biomed Opt 22(7):070901

Bücking TM, van den Berg PJ, Balabani S et al (2018) Processing methods for photoacoustic Doppler flowmetry with a clinical ultrasound scanner. J Biomed Opt 23(2):1–8

Buehler A, Herzog E, Razansky D et al (2010) Video rate optoacoustic tomography of mouse kidney perfusion. Opt Lett 35(14):2475–2477

Buxton RB, Frank LR (1997) A model for the coupling between cerebral blood flow and oxygen metabolism during neural stimulation. J Cereb Blood Flow Metab 17(1):64–72

Cao Y, Hui J, Kole A et al (2016) High-sensitivity intravascular photoacoustic imaging of lipid-laden plaque with a collinear catheter design. Sci Rep 6:25236

Cao Y, Kole A, Hui J et al (2018) Fast assessment of lipid content in arteries in vivo by intravascular photoacoustic tomography. Sci Rep 8(1):2400

Cao R, Li J, Ning B et al (2017) Functional and oxygen-metabolic photoacoustic microscopy of the awake mouse brain. NeuroImage 150:77–87

Chaigne T, Arnal B, Vilov S et al (2017) Super-resolution photoacoustic imaging via flow-induced absorption fluctuations. Optica 4(11):1397–1404

Chaigne T, Gateau J, Allain M et al (2016) Super-resolution photoacoustic fluctuation imaging with multiple speckle illumination. Optica 3(1):54–57

Chatni MR, Yao J, Danielli A et al (2011) Functional photoacoustic microscopy of pH. J Biomed Opt 16:100503

Chee RKW, Li Y, Zhang W et al (2018) In vivo photoacoustic difference-spectra imaging of bacteria using photoswitchable chromoproteins. J Biomed Opt 23:1–11

Chen QX, Davies A, Dewhurst RJ et al (1993) Photo-acoustic probe for intra-arterial imaging and therapy. Electron Lett 29(18):1632–1633

Chen YS, Frey W, Kim S et al (2011) Silica-coated gold nanorods as photoacoustic signal nanoamplifiers. Nano Lett 11(2):348–354

Cheney M, Isaacson D, Newell JC (1999) Electrical impedance tomography. SIAM Rev 41(1):85–101

Conkey DB, Caravaca-Aguirre AM, Dove JD et al (2015) Super-resolution photoacoustic imaging through a scattering wall. Nat Commun 6:7902
Cox B, Laufer JG, Arridge SR et al (2012) Quantitative spectroscopic photoacoustic imaging: a review. J Biomed Opt 17:061202
Danielli A, Maslov KI, Garcia-Uribe A et al (2014) Label-free photoacoustic nanoscopy. J Biomed Opt 19(8):086006
De la Zerda A, Zavaleta C, Keren S et al (2008) Carbon nanotubes as photoacoustic molecular imaging agents in living mice. Nat Nanotechnol 3(9):557–562
Deán-Ben XL, Fehm TF, Ford SJ et al (2017c) Spiral volumetric optoacoustic tomography visualizes multi-scale dynamics in mice. Light Sci Appl 6:e16247
Deán-Ben XL, Gottschalk S, Mc Larney B et al (2017b) Advanced optoacoustic methods for multiscale imaging of in vivo dynamics. Chem Soc Rev 46:2158–2198
Deán-Ben XL, López-Schier H, Razansky D (2017a) Optoacoustic micro-tomography at 100 volumes per second. Sci Rep 7:6850
Deán-Ben XL, Razansky D (2018) Localization optoacoustic tomography. Light-Sci Appl 7:18004
Deán-Ben XL, Stiel AC, Jiang Y et al (2015) Light fluence normalization in turbid tissues via temporally unmixed multispectral optoacoustic tomography. Opt Lett 40(20):4691–4694
Delman BN (2009) Imaging of pediatric pituitary abnormalities. Endocrinol Metab Clin N Am 38(4):673–698
Denk W, Strickler J, Webb W (1990) Two-photon laser scanning fluorescence microscopy. Science 248(4951):73–76
Dima A, Ntziachristos V (2012) Non-invasive carotid imaging using optoacoustic tomography. Opt Express 20:25044–25057
Dima A, Ntziachristos V (2016) In-vivo handheld optoacoustic tomography of the human thyroid. Photo-Dermatology 4:65–69
Drexler W, Fujimoto JG (2015) Optical coherence tomography technology and applications, 2nd edn. Springer International Publishing, Cham. Imprint: Springer
Ermilov SA, Khamapirad T, Conjusteau A et al (2009) Laser optoacoustic imaging system for detection of breast cancer. J Biomed Opt 14(2):024007. https://doi.org/10.1117/1.3086616
Esenaliev RO, Karabutov AA, Tittel FK et al (1997) "Laser optoacoustic imaging for breast cancer diagnostics: limit of detection and comparison with x-ray and ultrasound imaging", Proc SPIE 2979, Optical Tomography and Spectroscopy of Tissue: Theory, Instrumentation, Model, and Human Studies II, 71–82
Fakhrejahani E, Torii M, Kitai T et al (2015) Clinical report on the first prototype of a photoacoustic tomography system with dual illumination for breast cancer imaging. PLoS One 10:e0139113
Fang H, Maslov K, Wang LV (2007) Photoacoustic Doppler effect from flowing small light-absorbing particles. Phys Rev Lett 99:184501
Favazza CP, Jassim O, Cornelius LA et al (2011) In vivo photoacoustic microscopy of human cutaneous microvasculature and a nevus. J Biomed Opt 16(1):016015. https://doi.org/10.1117/1.3528661
Filonov GS, Krumholz A, Xia J et al (2012) Deep-tissue photoacoustic tomography of a genetically encoded near-infrared fluorescent probe. Angew Chem Int Ed Engl 51(6):1448–1451
Folarin AA, Konerding MA, Timonen J et al (2010) Three-dimensional analysis of tumour vascular corrosion casts using stereoimaging and micro-computed tomography. Microvasc Res 80(1):89–98
Fukumura D, Duda DG, Munn LL et al (2010) Tumor microvasculature and microenvironment: novel insights through intravital imaging in pre-clinical models. Microcirculation 17:206–225
Galanzha EI, Menyaev YA, Yadem AC et al (2019) In vivo liquid biopsy using cytophone platform for photoacoustic detection of circulating tumor cells in patients with melanoma. Sci Transl Med 11(496):eaat5857
Galanzha EI, Shashkov EV, Kelly T et al (2009) In vivo magnetic enrichment and multiplex photoacoustic detection of circulating tumour cells. Nat Nanotechnol 4:855–860
Galanzha EI, Zharov VP (2012) Photoacoustic flow cytometry. Methods 57(3):280–296

Gamelin J, Aguirre A, Maurudis A et al (2008) Curved array photoacoustic tomographic system for small animal imaging. J Biomed Opt 13(2):024007
Gamelin J, Maurudis A, Aguirre A et al (2009) A real-time photoacoustic tomography system for small animals. Opt Express 17(13):10489–10498
Garcia-Uribe A, Erpelding TN, Krumholz A et al (2015) Dual-modality photoacoustic and ultrasound imaging system for noninvasive sentinel lymph node detection in patients with breast cancer. Sci Rep 5:15748
Ghim CM, Lee SK, Takayama S, Mitchell RJ (2010) The art of reporter proteins in science: past, present and future applications. BMB Rep 43(7):451–460
Gora MJ, Sauk JS, Carruth RW et al (2013) Tethered capsule endomicroscopy enables less invasive imaging of gastrointestinal tract microstructure. Nat Med 19:238–240
Gottschalk S, Degtyaruk O, Mc Larney B et al (2019) Rapid volumetric optoacoustic imaging of neural dynamics across the mouse brain. Nat Biomed Eng 3:392–401
Gujrati V, Mishra A, Ntziachristos V (2017) Molecular imaging probes for multi-spectral optoacoustic tomography. Chem Commun (Camb) 53(34):4653–4672
Guo Z, Ye Z, Shao W et al (2019) Miniature probe for optomechanical focus adjustable optical resolution photoacoustic endoscopy. arXiv e-prints, arXiv:1909522
Hanahan D, Weinberg RA (2011) Hallmarks of cancer: the next generation. Cell 144(5):646–674
He H, Stylogiannis A, Afshari P et al (2019) Capsule optoacoustic endoscopy for esophageal imaging. J Biophotonics 12(10):e201800439
He Y, Wang L, Shi J et al (2016) In vivo label-free photoacoustic flow cytography and on-the-spot laser killing of single circulating melanoma cells. Sci Rep 6:39616
Heijblom M, Piras D, Brinkhuis M et al (2015) Photoacoustic image patterns of breast carcinoma and comparisons with magnetic resonance imaging and vascular stained histopathology. Sci Rep 5:11778
Helmchen F, Denk W (2005) Deep tissue two-photon microscopy. Nat Methods 2(12):932–940
Henrichs PM, Meador JW, Fuqua JM et al (2005) "Atherosclerotic plaque characterization with optoacoustic imaging", Proc SPIE 5697, Photons Plus Ultrasound: Imaging and Sensing 2005: The Sixth Conference on Biomedical Thermoacoustics, Optoacoustics, and Acousto-optics, 217–223
Hochuli R, An L, Beard PC et al (2019) Estimating blood oxygenation from photoacoustic images: can a simple linear spectroscopic inversion ever work? J Biomed Opt 24(12):1–13
Horiguchi A, Shinchi M, Nakamura A et al (2017) Pilot study of prostate cancer angiogenesis imaging using a photoacoustic imaging system. Urology 108:212–219
Horiguchi A, Tsujita K, Irisawa K et al (2016) A pilot study of photoacoustic imaging system for improved real-time visualization of neurovascular bundle during radical prostatectomy. Prostate 76(3):307–315
Hsieh BY, Chen SL, Ling T et al (2012) All-optical scanhead for ultrasound and photoacoustic dual-modality imaging. Opt Express 20:1588–1596
Hsu HC, Wang L, Wang LV (2016) In vivo photoacoustic microscopy of human cuticle microvasculature with single-cell resolution. J Biomed Opt 21(5):56004. https://doi.org/10.1117/1.JBO.21.5.056004
Hu S, Maslov K, Tsytsarev V et al (2009a) Functional transcranial brain imaging by optical-resolution photoacoustic microscopy. J Biomed Opt 14:040503
Hu S, Maslov K, Wang LV (2011) Second-generation optical-resolution photoacoustic microscopy with improved sensitivity and speed. Opt Lett 36:1134–1136
Hu S, Wang LV (2010) Neurovascular photoacoustic tomography. Front Neuroenerg 2:10
Hu S, Yan P, Maslov K et al (2009b) Intravital imaging of amyloid plaques in a transgenic mouse model using optical-resolution photoacoustic microscopy. Opt Lett 34(24):3899–3901
Hudson SV, Huang JS, Yin W et al (2014) Targeted noninvasive imaging of EGFR-expressing orthotopic pancreatic cancer using multispectral optoacoustic tomography. Cancer Res 74(21):6271–6279
Hui J, Cao Y, Zhang Y et al (2017) Real-time intravascular photoacoustic-ultrasound imaging of lipid-laden plaque in human coronary artery at 16 frames per second. Sci Rep 7(1):1417

Hui J, Yu Q, Ma T et al (2015) High-speed intravascular photoacoustic imaging at 1.7 μm with a KTP-based OPO. Biomed Opt Express 6(11):4557–4566

Huynh E, Leung BY, Helfield BL et al (2015) In situ conversion of porphyrin microbubbles to nanoparticles for multimodality imaging. Nat Nanotechnol 10(4):325–332

Ivankovic I, Mercep E, Schmedt CG et al (2019) Real-time volumetric assessment of the human carotid artery: handheld multispectral optoacoustic tomography. Radiology 291:45–50

Jansen K, Van Der Steen AF, Van Beusekom HM et al (2011) Intravascular photoacoustic imaging of human coronary atherosclerosis. Opt Lett 36:597–599

Jathoul AP, Laufer J, Ogunlade O et al (2015) Deep in vivo photoacoustic imaging of mammalian tissues using a tyrosinase-based genetic reporter. Nat Photonics 9(4):239

Jin Y, Jia C, Huang SW et al (2010) Multifunctional nanoparticles as coupled contrast agents. Nat Commun 1:41

Jo J, Lee CH, Kopelman R et al (2017) In vivo quantitative imaging of tumor pH by nanosonophore assisted multispectral photoacoustic imaging. Nat Commun 8:471

Kajita H, Oh A, Urano M et al (2019) Photoacoustic lymphangiography. J Surg Oncol 121:48–50

Karpiouk AB, Wang B, Amirian J et al (2012) Feasibility of in vivo intravascular photoacoustic imaging using integrated ultrasound and photoacoustic imaging catheter. J Biomed Opt 17(9):96008. https://doi.org/10.1117/1.JBO.17.9.096008

Karpiouk AB, Wang B, Emelianov SY (2010) Development of a catheter for combined intravascular ultrasound and photoacoustic imaging. Rev Sci Instrum 81:014901

Kiesslich R, Burg J, Vieth M et al (2004) Confocal laser endoscopy for diagnosing intraepithelial neoplasias and colorectal cancer in vivo. Gastroenterology 127:706–713

Kiesslich R, Goetz M, Vieth M et al (2007) Technology insight: confocal laser endoscopy for in vivo diagnosis of colorectal cancer. Nat Clin Pract Oncol 4:480–490

Kiesslich R, Gossner L, Goetz M et al (2006) In vivo histology of Barrett's esophagus and associated neoplasia by confocal laser endomicroscopy. Clin Gastroenterol Hepatol 4:979–987

Kim JW, Galanzha EI, Shashkov EV et al (2009) Golden carbon nanotubes as multimodal photoacoustic and photothermal high-contrast molecular agents. Nat Nanotechnol 4:688–694

Kircher MF, De La Zerda A, Jokerst JV et al (2012) A brain tumor molecular imaging strategy using a new triple-modality MRI-photoacoustic-Raman nanoparticle. Nat Med 18:829–834

Kitai T, Torii M, Sugie T et al (2014) Photoacoustic mammography: initial clinical results. Breast Cancer 21:146–153

Knox HJ, Hedhli J, Kim TW et al (2017) A bioreducible N-oxide-based probe for photoacoustic imaging of hypoxia. Nat Commun 8:1794

Kothapalli SR, Sonn GA, Choe JW et al (2019) Simultaneous transrectal ultrasound and photoacoustic human prostate imaging. Sci Transl Med 11(507):eaav2169

Kottmann J, Rey JM, Luginbühl J et al (2012) Glucose sensing in human epidermis using mid-infrared photoacoustic detection. Biomed Opt Express 3(4):667–680

Kruger RA (1994) Photoacoustic ultrasound. Med Phys 21:127–131

Kruger RA, Kiser WLJR, Reinecke DR et al (2003a) Thermoacoustic computed tomography using a conventional linear transducer array. Med Phys 30(5):856–860

Kruger RA, Kiser WL, Reinecke DR et al (2003b) Thermoacoustic molecular imaging of small animals. Mol Imaging 2(2):113–123

Kruger RA, Kopecky KK, Aisen AM et al (1999b) Thermoacoustic CT with radio waves: a medical imaging paradigm. Radiology 211(1):275–278

Kruger RA, Kuzmiak CM, Lam RB et al (2013) Dedicated 3D photoacoustic breast imaging. Med Phys 40:113301

Kruger RA, Lam RB, Reinecke DR et al (2010) Photoacoustic angiography of the breast. Med Phys 37:6096–6100

Kruger RA, Liu P (1994) Photoacoustic ultrasound: pulse production and detection of 0.5% Liposyn. Med Phys 21:1179–1184

Kruger RA, Liu P, Fang Y et al (1995) Photoacoustic ultrasound (PAUS)—reconstruction tomography. Med Phys 22(10):1605

Kruger RA, Miller KD, Reynolds HE et al (2000) Breast cancer in vivo: contrast enhancement with thermoacoustic CT at 434 MHz-feasibility study. Radiology 216:279–283
Kruger RA, Reinecke DR, Kruger GA (1999a) Thermoacoustic computed tomography--technical considerations. Med Phys 26:1832–1837
Kruizinga P, Van Der Steen AF, De Jong N et al (2014) Photoacoustic imaging of carotid artery atherosclerosis. J Biomed Opt 19:110504
Krumholz A, Shcherbakova DM, Xia J et al (2014) Multicontrast photoacoustic in vivo imaging using near-infrared fluorescent proteins. Sci Rep 4:3939
Krumholz A, Vanvickle-Chavez SJ, Yao J et al (2011) Photoacoustic microscopy of tyrosinase reporter gene in vivo. J Biomed Opt 16(8):080503
Ku G, Wang X, Xie X et al (2005) Imaging of tumor angiogenesis in rat brains in vivo by photoacoustic tomography. Appl Opt 44(5):770–775
Kumavor PD, Alqasemi U, Tavakoli B et al (2013) Co-registered pulse-echo/photoacoustic transvaginal probe for real time imaging of ovarian tissue. J Biophotonics 6(6–7):475–484
Lan B, Liu W, Wang Y et al (2018) High-speed widefield photoacoustic microscopy of small-animal hemodynamics. Biomed Opt Express 9(10):4689
Landa FJO, Penacoba SR, de Espinosa FM et al (2019) Four-dimensional optoacoustic monitoring of tissue heating with medium intensity focused ultrasound. Ultrasonics 94:117–123
Larina IV, Larin KV, Esenaliev RO (2005) Real-time optoacoustic monitoring of temperature in tissues. J Phys D Appl Phys 38(15):2633–2639
Laser Institute of America (2007) American National Standard for safe use of lasers. In: ANSI Z136, vol 1–2007. American National Standards Institute, Inc., New York, NY
Laufer J, Jathoul A, Pule M et al (2013) In vitro characterization of genetically expressed absorbing proteins using photoacoustic spectroscopy. Biomed Opt Express 4:2477–2490
Laufer J, Johnson P, Zhang E et al (2012a) In vivo preclinical photoacoustic imaging of tumor vasculature development and therapy. J Biomed Opt 17:056016. https://doi.org/10.1117/1.JBO.17.5.056016
Laufer J, Norris F, Cleary J et al (2012b) In vivo photoacoustic imaging of mouse embryos. J Biomed Opt 17:061220. https://doi.org/10.1117/1.JBO.17.6.061220
Lediju Bell MA, Guo X, Song DY et al (2015b) Transurethral light delivery for prostate photoacoustic imaging. J Biomed Opt 20:036002
Lediju Bell MA, Ostrowski AK, Li K et al (2015a) Localization of transcranial targets for photoacoustic-guided endonasal surgeries. Photo-Dermatology 3:78–87
Li Y, Chen J, Chen Z (2020) Multimodal intravascular imaging technology for characterization of atherosclerosis. J Innov Opt Heal Sci 13(1):2030001
Li Y, Gong X, Liu C et al (2015) High-speed intravascular spectroscopic photoacoustic imaging at 1000 A-lines per second with a 0.9-mm diameter catheter. J Biomed Opt 20(6):065006
Li Y, Lin R, Liu C et al (2018b) In vivo photoacoustic/ultrasonic dual-modality endoscopy with a miniaturized full field-of-view catheter. J Biophotonics 11(10):e201800034
Li G, Maslov KI, Wang LV (2013) Reflection-mode multifocal optical-resolution photoacoustic microscopy. J Biomed Opt 18(3):030501
Li ML, Oh JT, Xie X et al (2008c) Simultaneous molecular and hypoxia imaging of brain tumors in vivo using spectroscopic photoacoustic tomography. P IEEE 96(3):481–489
Li R, Phillips E, Wang P et al (2016) Label-free in vivo imaging of peripheral nerve by multispectral photoacoustic tomography. J Biophotonics 9(1–2):124–128
Li L, Shemetov AA, Baloban M et al (2018a) Small near-infrared photochromic protein for photoacoustic multi-contrast imaging and detection of protein interactions in vivo. Nat Commun 9(1):2734
Li ML, Wang JC, Schwartz JA et al (2009) In-vivo photoacoustic microscopy of nanoshell extravasation from solid tumor vasculature. J Biomed Opt 14(1):010507
Li PC, Wang CR, Shieh DB et al (2008b) In vivo photoacoustic molecular imaging with simultaneous multiple selective targeting using antibody-conjugated gold nanorods. Opt Express 16(23):18605–18615
Li C, Yang JM, Chen R et al (2014) Urogenital photoacoustic endoscope. Opt Lett 39:1473–1476

Li L, Zemp RJ, Lungu G et al (2007) Photoacoustic imaging of lacZ gene expression in vivo. J Biomed Opt 12:020504. https://doi.org/10.1117/1.2717531
Li L, Zhang HF, Zemp RJ et al (2008a) Simultaneous imaging of a lacZ-marked tumor and microvasculature morphology in vivo by dual-wavelength photoacoustic microscopy. J Innov Opt Health Sci 1:207–215
Li L, Zhu L, Ma C et al (2017) Single-impulse panoramic photoacoustic computed tomography of small-animal whole-body dynamics at high spatiotemporal resolution. Nat Biomed Eng 1:0071
Lin L, Hu P, Shi J et al (2018) Single-breath-hold photoacoustic computed tomography of the breast. Nat Commun 9:2352
Lin J, Wang S, Huang P et al (2013) Photosensitizer-loaded gold vesicles with strong plasmonic coupling effect for imaging-guided photothermal/photodynamic therapy. ACS Nano 7(6):5320–5329
Lin L, Yao J, Li L et al (2016) In vivo photoacoustic tomography of myoglobin oxygen saturation. J Biomed Opt 21(6):61002. https://doi.org/10.1117/1.JBO.21.6.061002
Ma R, Distel M, Deán-Ben XL et al (2012) Non-invasive whole-body imaging of adult zebrafish with optoacoustic tomography. Phys Med Biol 57(22):7227–7237
Ma J, Shi J, Hai P et al (2016) Grueneisen relaxation photoacoustic microscopy in vivo. J Biomed Opt 21(6):66005
Ma R, Taruttis A, Ntziachristos V et al (2009) Multispectral optoacoustic tomography (MSOT) scanner for whole-body small animal imaging. Opt Express 17:21414–21426
Manohar S, Kharine A, Van Hespen JC et al (2004) Photoacoustic mammography laboratory prototype: imaging of breast tissue phantoms. J Biomed Opt 9:1172–1181
Manohar S, Kharine A, Van Hespen JC et al (2005) The Twente photoacoustic mammoscope: system overview and performance. Phys Med Biol 50:2543–2557
Manohar S, Razansky D (2016) Photoacoustics: a historical review. Adv Opt Photonics 8(4):586–617
Manohar S, Vaartjes SE, Van Hespen JC et al (2007) Initial results of in vivo non-invasive cancer imaging in the human breast using near-infrared photoacoustics. Opt Express 15:12277–12285
Maslov K, Stoica G, Wang LV (2005) In vivo dark-field reflection-mode photoacoustic microscopy. Opt Lett 30(6):625–627
Maslov K, Zhang HF, Hu S et al (2008) Optical-resolution photoacoustic microscopy for in vivo imaging of single capillaries. Opt Lett 33:929–931
Matsumoto Y, Asao Y, Yoshikawa A et al (2018) Label-free photoacoustic imaging of human palmar vessels: a structural morphological analysis. Sci Rep 8:786
Matthews TP, Zhang C, Yao DK et al (2014) Label-free photoacoustic microscopy of peripheral nerves. J Biomed Opt 19:16004
Mehrmohammadi M, Yoon SJ, Yeager D et al (2013) Photoacoustic imaging for cancer detection and staging. Curr Mol Imaging 2:89–105
Mercep E, Deán-Ben XL, Razansky D (2018) Imaging of blood flow and oxygen state with a multi-segment optoacoustic ultrasound array. Photo-Dermatology 10:48–53
Miranda C, Barkley J, Smith B (2018) Intrauterine photoacoustic and ultrasound imaging probe. J Biomed Opt 23(4):1–9
Moustakidis S, Omar M, Aguirre J et al (2019) Fully automated identification of skin morphology in raster-scan optoacoustic mesoscopy using artificial intelligence. Med Phys 46(9):4046–4056
Murray TW, Haltmeier M, Berer T et al (2017) Super-resolution photoacoustic microscopy using blind structured illumination. Optica 4(1):17–22
Nagae K, Asao Y, Sudo Y et al (2019) Real-time 3D photoacoustic visualization system with a wide field of view for imaging human limbs. F1000Research 7:1813
Nasiriavanaki M, Xia J, Wan H et al (2014) High-resolution photoacoustic tomography of resting-state functional connectivity in the mouse brain. Proc Natl Acad Sci U S A 111:21–26
Nie L, Chen X (2014) Structural and functional photoacoustic molecular tomography aided by emerging contrast agents. Chem Soc Rev 43:7132–7170
Ning B, Sun N, Cao R et al (2015) Ultrasound-aided multi-parametric photoacoustic microscopy of the mouse brain. Sci Rep 5:18775

Ntziachristos V (2010) Going deeper than microscopy: the optical imaging frontier in biology. Nat Methods 7(8):603–614

Ntziachristos V, Pleitez MA, Aime S, Brindle KM (2019) Emerging technologies to image tissue metabolism. Cell Metab 29(3):518–538

Nuster R, Slezak P, Paltauf G (2014) High resolution three-dimensional photoacoustic tomography with CCD-camera based ultrasound detection. Biomed Opt Express 5(8):2635

Oh JT, Li ML, Zhang HF et al (2006) Three-dimensional imaging of skin melanoma in vivo by dual-wavelength photoacoustic microscopy. J Biomed Opt 11(3):34032

Omar M, Aguirre J, Ntziachristos V (2019) Optoacoustic mesoscopy for biomedicine. Nat Biomed Eng 3:354–370

Omar M, Rebling J, Wicker K et al (2017) Optical imaging of post-embryonic zebrafish using multi orientation raster scan optoacoustic mesoscopy. Light Sci Appl 6:e16186

Oraevsky A, Andreev V, Karabutov A et al (1999) "Laser optoacoustic imaging of the breast: detection of cancer angiogenesis", Proc SPIE 3597, Optical Tomography and Spectroscopy of Tissue III, 352–363

Oraevsky AA, Esenaliev RO, Karabutov AA (1997) "Laser optoacoustic tomography of layered tissue: signal processing", Proc SPIE 2979, Optical Tomography and Spectroscopy of Tissue: Theory, Instrumentation, Model, and Human Studies II, 59–70

Oraevsky AA, Jacques SL, Esenaliev RO et al (1994a) Time-resolved optoacoustic imaging in layered biological tissues. Advances in optical imaging and photon migration 21:161–165

Oraevsky AA, Jacques SL, Esenaliev RO et al (1994b) "Laser-based optoacoustic imaging in biological tissues", Proc SPIE 2134, Laser-Tissue Interaction V; and Ultraviolet Radiation Hazards, 122–8

Oraevsky AA, Jacques SL, Tittel FK (1993) "Determination of tissue optical properties by piezo-electric detection of laser-induced stress waves", Proc SPIE 1882, Laser-Tissue Interaction IV, 86–101

Pan D, Pramanik M, Senpan A et al (2010) Near infrared photoacoustic detection of sentinel lymph nodes with gold nanobeacons. Biomaterials 31:4088–4093

Paproski RJ, Forbrich AE, Wachowicz K et al (2011) Tyrosinase as a dual reporter gene for both photoacoustic and magnetic resonance imaging. Biomed Opt Express 2(4):771–780

Paproski RJ, Heinmiller A, Wachowicz K et al (2014) Multi-wavelength photoacoustic imaging of inducible tyrosinase reporter gene expression in xenograft tumors. Sci Rep 4:5329

Pavlova NN, Thompson CB (2016) The emerging hallmarks of cancer metabolism. Cell Metab 23(1):27–47

Perekatova V, Subochev P, Kleshnin M et al (2016) Optimal wavelengths for optoacoustic measurements of blood oxygen saturation in biological tissues. Biomed Opt Express 7(10):3979–3995

Pleitez MA, Khan AA, Soldà A et al (2020) Label-free metabolic imaging by mid-infrared optoacoustic microscopy in living cells. Nat Biotechnol 38(3):293–296

Pleitez MA, Lieblein T, Bauer A et al (2013) In vivo noninvasive monitoring of glucose concentration in human epidermis by mid-infrared pulsed photoacoustic spectroscopy. Anal Chem 85(2):1013–1020

Pramanik M, Wang LV (2009) Thermoacoustic and photoacoustic sensing of temperature. J Biomed Opt 14:054024

Qin C, Cheng K, Chen K et al (2013) Tyrosinase as a multifunctional reporter gene for photoacoustic/MRI/PET triple modality molecular imaging. Sci Rep 3:1490

Qu Y, Li C, Shi J et al (2018) Transvaginal fast-scanning optical-resolution photoacoustic endoscopy. J Biomed Opt 23(12):1–4. https://doi.org/10.1117/1.JBO.23.12.121617

Razansky D, Buehler A, Ntziachristos V (2011) Volumetric real-time multispectral optoacoustic tomography of biomarkers. Nat Protoc 6:1121–1129

Razansky D, Distel M, Vinegoni C et al (2009) Multispectral opto-acoustic tomography of deep-seated fluorescent proteins in vivo. Nat Photonics 3(7):412

Ripoll J, Koberstein-Schwarz B, Ntziachristos V (2015) Unleashing optics and optoacoustics for developmental biology. Trends Biotechnol 33:679–691

Sethuraman S, Aglyamov SR, Amirian JH et al (2007) Intravascular photoacoustic imaging using an IVUS imaging catheter. IEEE Trans Ultrason Ferroelectr Freq Control 54(5):978–986
Sethuraman S, Amirian JH, Litovsky SH et al (2008) Spectroscopic intravascular photoacoustic imaging to differentiate atherosclerotic plaques. Opt Express 16(5):3362–3367
Shah J, Park S, Aglyamov S et al (2008) Photoacoustic imaging and temperature measurement for photothermal cancer therapy. J Biomed Opt 13:034024
Shami VM, Kahaleh M (eds) (2010) Endoscopic ultrasound, 1st edn. Humana Press, Totowa, NJ. Imprint: Humana
Shi J, Wong TTW, He Y et al (2019) High-resolution, high-contrast mid-infrared imaging of fresh biological samples with ultraviolet-localized photoacoustic microscopy. Nat Photonics 13:609
Song L, Maslov K, Wang LV (2011) Multifocal optical-resolution photoacoustic microscopy in vivo. Opt Lett 36(7):1236–1238
Song KH, Stein EW, Margenthaler JA et al (2008) Noninvasive photoacoustic identification of sentinel lymph nodes containing methylene blue in vivo in a rat model. J Biomed Opt 13:054033. https://doi.org/10.1117/1.2976427
Sowers T, Emelianov S (2018) Exogenous imaging contrast and therapeutic agents for intravascular photoacoustic imaging and image-guided therapy. Phys Med Biol 63(22):22TR01
Steeg PS (2006) Tumor metastasis: mechanistic insights and clinical challenges. Nat Med 12:895–904
Stein EW, Maslov KI, Wang LV (2009a) Noninvasive, in vivo imaging of blood-oxygenation dynamics within the mouse brain using photoacoustic microscopy. J Biomed Opt 14(2):020502
Stein EW, Maslov KI, Wang LV (2009b) Noninvasive, in vivo imaging of the mouse brain using photoacoustic microscopy. J Appl Phys 105(10):102027
Stoffels I, Morscher S, Helfrich I et al (2015) Metastatic status of sentinel lymph nodes in melanoma determined noninvasively with multispectral optoacoustic imaging. Sci Transl Med 7:317ra199
Stritzker J, Kirscher L, Scadeng M et al (2013) Vaccinia virus-mediated melanin production allows MR and optoacoustic deep tissue imaging and laser-induced thermotherapy of cancer. Proc Natl Acad Sci U S A 110(9):3316–3320
Swierczewska M, Choi KY, Mertz EL et al (2012) A facile, one-step nanocarbon functionalization for biomedical applications. Nano Lett 12(7):3613–3620
Tang J, Dai X, Jiang H (2016) Wearable scanning photoacoustic brain imaging in behaving rats. J Biophotonics 9:570–575
Tang J, Xi L, Zhou J et al (2015b) Noninvasive high-speed photoacoustic tomography of cerebral hemodynamics in awake-moving rats. J Cereb Blood Flow Metab 35:1224–1232
Tang M, Zhou Y, Zhang R et al (2015a) Noninvasive photoacoustic microscopy of methemoglobin in vivo. J Biomed Opt 20(3):036007
Taruttis A, Herzog E, Razansky D et al (2010) Real-time imaging of cardiovascular dynamics and circulating gold nanorods with multispectral optoacoustic tomography. Opt Express 18:19592–19602
Taruttis A, Morscher S, Burton NC et al (2012) Fast multispectral optoacoustic tomography (MSOT) for dynamic imaging of pharmacokinetics and biodistribution in multiple organs. PLoS One 7(1):e30491
Taruttis A, van Dam GM, Ntziachristos V (2015) Mesoscopic and macroscopic optoacoustic imaging of cancer. Cancer Res 75(8):1548–1559
Tearney GJ, Brezinski ME, Bouma BE et al (1997) In vivo endoscopic optical biopsy with optical coherence tomography. Science 276:2037–2039
Toi M, Asao Y, Matsumoto Y et al (2017) Visualization of tumor-related blood vessels in human breast by photoacoustic imaging system with a hemispherical detector array. Sci Rep 7:41970
Tsyboulski D, Conjusteau A, Oraevsky A (2014) "Dual modality optoacoustic and laser ultrasound endoscopy system", Proc SPIE 8943, Photons Plus Ultrasound: Imaging and Sensing 2014, 89432S https://doi.org/10.1117/12.2044424
Tzoumas S, Nunes A, Olefir I et al (2016) Eigenspectra optoacoustic tomography achieves quantitative blood oxygenation imaging deep in tissues. Nat Commun 7:12121

van den Berg PJ, Daoudi K, Steenbergen W (2015) Review of photoacoustic flow imaging: its current state and its promises. Photo-Dermatology 3(3):89–99

van den Berg PJ, Daoudi K, Steenbergen W (2016) Pulsed photoacoustic flow imaging with a handheld system. J Biomed Opt 21(2):26004

van Soest G, Marcu L, Bouma BE et al (2017) Intravascular imaging for characterization of coronary atherosclerosis. Curr Opin Biomed Eng 3:1–12

Viator JA, Komadina J, Svaasand LO et al (2004) A comparative study of photoacoustic and reflectance methods for determination of epidermal melanin content. J Invest Dermatol 122(6):1432–1439

Viator JA, Paltauf G, Jacques SL et al (2001) "Design and testing of an endoscopic photoacoustic probe for determination of treatment depth after photodynamic therapy," Proc SPIE 4256, Biomedical Optoacoustics II, 16–27

Vilov S, Arnal B, Bossy E (2017) Overcoming the acoustic diffraction limit in photoacoustic imaging by the localization of flowing absorbers. Opt Lett 42(21):4379–4382

Wang LV (2008) Prospects of photoacoustic tomography. Med Phys 35(12):5758–5767

Wang LV (2009) Photoacoustic imaging and spectroscopy. CRC, Boca Raton

Wang LV, Hu S (2012) Photoacoustic tomography: in vivo imaging from organelles to organs. Science 335:1458–1462

Wang X, Ku G, Wegiel MA et al (2004b) Noninvasive photoacoustic angiography of animal brains in vivo with near-infrared light and an optical contrast agent. Opt Lett 29(7):730–732

Wang P, Ma T, Slipchenko MN et al (2014b) High-speed intravascular photoacoustic imaging of lipid-laden atherosclerotic plaque enabled by a 2-kHz barium nitrite raman laser. Sci Rep 4:6889

Wang L, Maslov K, Wang LV (2013) Single-cell label-free photoacoustic flowoxigraphy in vivo. Proc Natl Acad Sci U S A 110:5759–5764

Wang X, Pang Y, Ku G et al (2003) Noninvasive laser-induced photoacoustic tomography for structural and functional in vivo imaging of the brain. Nat Biotechnol 21:803–806

Wang X, Roberts WW, Carson PL et al (2010a) Photoacoustic tomography: a potential new tool for prostate cancer. Biomed Opt Express 1:1117–1126

Wang B, Su JL, Karpiouk AB et al (2010b) Intravascular photoacoustic imaging. IEEE J Sel Top Quantum Electron 16:588–599

Wang X, Xie X, Ku G et al (2006) Noninvasive imaging of hemoglobin concentration and oxygenation in the rat brain using high-resolution photoacoustic tomography. J Biomed Opt 11(2):024015

Wang Y, Xie X, Wang X et al (2004a) Photoacoustic tomography of a nanoshell contrast agent in the in vivo rat brain. Nano Lett 4(9):1689–1692

Wang B, Yantsen E, Larson T et al (2009) Plasmonic intravascular photoacoustic imaging for detection of macrophages in atherosclerotic plaques. Nano Lett 9:2212–2217

Wang L, Zhang C, Wang LV (2014a) Grueneisen relaxation photoacoustic microscopy. Phys Rev Lett 113(17):174301

Weber J, Beard PC, Bohndiek SE (2016) Contrast agents for molecular photoacoustic imaging. Nat Methods 13:639–650

Weis SM, Cheresh DA (2011) Tumor angiogenesis: molecular pathways and therapeutic targets. Nat Med 17(11):1359–1370

Weissleder R, Pittet MJ (2008) Imaging in the era of molecular oncology. Nature 452:580–589

Weissleder R, Schwaiger MC, Gambhir SS et al (2016) Imaging approaches to optimize molecular therapies. Sci Transl Med 8(355):355ps16

Welch AJ, van Gemert MJC (2011) Optical-thermal response of laser-irradiated tissue. Springer; 2 edition (January 15, 2011)

Wilson KE, Bachawal SV, Tian L et al (2014) Multiparametric spectroscopic photoacoustic imaging of breast cancer development in a transgenic mouse model. Theranostics 4(11):1062–1071

Wilt BA, Burns LD, Wei Ho ET et al (2009) Advances in light microscopy for neuroscience. Annu Rev Neurosci 32:435–506

Wissmeyer G, Pleitez MA, Rosenthal A et al (2018) Looking at sound: optoacoustics with all-optical ultrasound detection. Light Sci Appl 7:53

Wu M, Springeling G, Lovrak M et al (2017) Real-time volumetric lipid imaging in vivo by intravascular photoacoustics at 20 frames per second. Biomed Opt Express 8(2):943–953

Xia J, Li G, Wang L et al (2013) Wide-field two-dimensional multifocal optical-resolution photoacoustic-computed microscopy. Opt Lett 38(24):5236–5259

Xiong K, Wang W, Guo T et al (2019) Shape-adapting panoramic photoacoustic endomicroscopy. Opt Lett 44(11):2681

Xu Z, Li CH, Wang LV (2010) Photoacoustic tomography of water in phantoms and tissue. J Biomed Opt 15(3):036019. https://doi.org/10.1117/1.3443793

Xu M, Wang LV (2006a) Photoacoustic imaging in biomedicine. Rev Sci Instrum 77:041101

Xu Y, Wang LV (2006b) Rhesus monkey brain imaging through intact skull with thermoacoustic tomography. IEEE Trans Ultrason Ferroelectr Freq Control 53:542–548

Yang JM, Chen R, Favazza C et al (2012b) A 2.5-mm diameter probe for photoacoustic and ultrasonic endoscopy. Opt Express 20:23944–23953

Yang JM, Favazza C, Chen R et al (2012a) Simultaneous functional photoacoustic and ultrasonic endoscopy of internal organs in vivo. Nat Med 18:1297–1302

Yang JM, Li C, Chen R et al (2014) Catheter-based photoacoustic endoscope. J Biomed Opt 19:066001

Yang JM, Li C, Chen R et al (2015) Optical-resolution photoacoustic endomicroscopy in vivo. Biomed Opt Express 6(3):918–932

Yang Y, Li X, Wang T et al (2011) Integrated optical coherence tomography, ultrasound and photoacoustic imaging for ovarian tissue characterization. Biomed Opt Express 2(9):2551–2561

Yang JM, Maslov K, Yang HC et al (2009) Photoacoustic endoscopy. Opt Lett 34:1591–1593

Yang X, Skrabalak SE, Li ZY et al (2007) Photoacoustic tomography of a rat cerebral cortex in vivo with au nanocages as an optical contrast agent. Nano Lett 7(12):3798–3802

Yao J, Kaberniuk AA, Li L et al (2015a) Multiscale photoacoustic tomography using reversibly switchable bacterial phytochrome as a near-infrared photochromic probe. Nat Methods 13:67–73

Yao J, Ki M, Zhang Y et al (2011) Label-free oxygen-metabolic photoacoustic microscopy in vivo. J Biomed Opt 16:076003. https://doi.org/10.1117/1.3594786

Yao J, Maslov KI, Shi Y et al (2010b) In vivo photoacoustic imaging of transverse blood flow by using Doppler broadening of bandwidth. Opt Lett 35:1419–1421

Yao DK, Maslov K, Shung KK et al (2010a) In vivo label-free photoacoustic microscopy of cell nuclei by excitation of DNA and RNA. Opt Lett 35:4139–4141

Yao J, Wang LV (2014a) Sensitivity of photoacoustic microscopy. Photo-Dermatology 2:87–101

Yao J, Wang LV (2014b) Photoacoustic brain imaging: from microscopic to macroscopic scales. Neurophotonics 1(1):1877516

Yao J, Wang L, Li C et al (2014) Photoimprint photoacoustic microscopy for three-dimensional label-free subdiffraction imaging. Phys Rev Lett 112(1):014302

Yao J, Wang L, Yang JM et al (2015b) High-speed label-free functional photoacoustic microscopy of mouse brain in action. Nat Methods 12:407–410

Yao J, Xia J, Maslov KI et al (2013) Noninvasive photoacoustic computed tomography of mouse brain metabolism in vivo. NeuroImage 64:257–266

Yaseen MA, Ermilov SA, Brecht HP et al (2010) Optoacoustic imaging of the prostate: development toward image-guided biopsy. J Biomed Opt 15:021310. https://doi.org/10.1117/1.3333548

Yuan Y, Yang S, Xing D (2010) Preclinical photoacoustic imaging endoscope based on acousto-optic coaxial system using ring transducer array. Opt Lett 35:2266–2268

Yun SH, Kwok SJJ (2017) Light in diagnosis, therapy and surgery. Nat Biomed Eng 1:0008

Yun SH, Tearney GJ, Vakoc BJ et al (2006) Comprehensive volumetric optical microscopy in vivo. Nat Med 12:1429–1433

Zhang EZ, Beard PC (2011) "A miniature all-optical photoacoustic imaging probe", Proc SPIE 7899, Photons Plus Ultrasound: Imaging and Sensing 2011, 78991F

Zhang HF, Maslov K, Stoica G et al (2006a) Functional photoacoustic microscopy for high-resolution and noninvasive in vivo imaging. Nat Biotechnol 24:848–851
Zhang HF, Maslov K, Stoica G et al (2006b) Imaging acute thermal burns by photoacoustic microscopy. J Biomed Opt 11(5):054033
Zhang C, Maslov K, Wang LV (2010) Subwavelength-resolution label-free photoacoustic microscopy of optical absorption in vivo. Opt Lett 35:3195–3197
Zhang C, Maslov K, Yao J et al (2012) In vivo photoacoustic microscopy with 7.6-μm axial resolution using a commercial 125-MHz ultrasonic transducer. J Biomed Opt 17(11):116016
Zhang X, Zhang HF, Puliafito CA et al (2011) Simultaneous in vivo imaging of melanin and lipofuscin in the retina with photoacoustic ophthalmoscopy and autofluorescence imaging. J Biomed Opt 16(8):080504
Zhang C, Zhang YS, Yao DK et al (2013) Label-free photoacoustic microscopy of cytochromes. J Biomed Opt 18(2):20504
Zhou Y, Zhang C, Yao DK et al (2012) Photoacoustic microscopy of bilirubin in tissue phantoms. J Biomed Opt 17(12):126019

Imaging Motion: A Comprehensive Review of Optical Coherence Tomography Angiography

Woo June Choi

Introduction

Living system, from the simplest single-celled organism to multicellular organism, is an astonishing entanglement of multi-scale dynamics. In the single-celled organism, for instance, the interior of the eukaryotic cell is a crowded, heterogeneous, and highly dynamic environment that undergoes continuous agitation caused by the activity of molecules and other nonequilibrium cellular processes (Brangwynne et al. 2008; Bressloff and Newby 2013). The subtle and random choreographies of cytoplasmic molecules and organelles are directly associated with the functional efficiency and the overall metabolic state of the cells, which often influence the physiological conditions of the cells. Malfunction of molecular motors, for instance, is believed to be responsible for incorrect distribution of chromosomes, one of the main signatures of Down syndrome (Nicklas 1997), and aberrant dynamics of the actin cytoskeleton is a distinct characteristic of invasive and metastatic cancer cells (Li et al. 2007). On the larger scale of organism, however, the vascular circulatory system permits blood to continually circulate blood cells (such as red blood cells (RBCs)) through branching arteries, to reach and transverse the microscopic capillary vessels in all parts of the body. This uninterrupted movement of the blood is critical to maintain the supply of oxygen and nutrients to the cells in the brain and other organs, and the delivery of waste for excretion from all body tissues for vital functions in normal living (Segal 2005). Disorders in the circulation are intimately involved with various vascular diseases (such as peripheral arterial diseases (PAD)) or other serious health conditions (such as ischemic stroke).

W. J. Choi (✉)
School of Electrical and Electronics Engineering, College of ICT Engineering, Chung-Ang University, Seoul, Republic of Korea
e-mail: cecc78@cau.ac.kr

J. K. Kim et al. (eds.), *Advanced Imaging and Bio Techniques for Convergence Science*, Advances in Experimental Medicine and Biology,
https://doi.org/10.1007/978-981-33-6064-8_12

This wide range of dynamics of the intracellular organelles or cellular particles are a primary basis of phenomenon of life, which could thus be a critical readout of indicating the healthy and diseased states of living single cells and even human organ tissues in vivo. Advances in optical imaging for the past few decades have revolutionized the field of cell biology, biomedicine, and clinics, enabling to devise methodologies for visualizing and quantifying the dynamics at high temporal and spatial resolution. The quantitative imaging of dynamics would thus aid in building a better understanding of the underlying mechanisms of cellular states and functions, and opening avenues for practical biomedical applications.

In this chapter, hence, we introduce optical imaging technologies that have been recently developed for visualizing the biological dynamics. Among the dynamics imaging modalities, in particular, we review optical coherence tomography angiography (OCTA) as an emerging technique to delineate the microcirculation within vascularized tissue beds in vivo or intracellular dynamics using the dedicated endogenous motion-contrast manners, instead of using artificial exogenous dyes or probes as contrast sources. The chapter starts with basic concept and principle of OCTA and followed by various algorithms of OCTA to achieve the motion-contrast imaging, highlighting the advantages and disadvantages of each of the algorithms. Next, it covers examples of clinical researches using OCTA and finally followed by discussion and outlook of OCTA in the translational research.

Optical Coherence Tomography Angiography (OCTA)

Optical coherence tomography (OCT) is a noninvasive interferometric imaging technology, capable of providing high-resolution, high-speed, depth-resolved cross-sectional images of highly scattering samples, such as biological tissue, and is being gained much attention for both medical and nonmedical imaging applications (Huang et al. 1991). OCT measures echo time-delayed backscattered light signals (i.e., reflectance) at the boundaries between multiple structural layers within the sample through the principle of low coherence interferometry (Drexler and Fujimoto 2007). A collection of the reflectances along optical axis exhibits internal structures of the sample in depth at a micrometer-scale resolution (1–10 μm). By step-wise dislocating either the beam position or the sample itself along lateral (*x* or *y* or both) direction, a cross-sectional image (tomography) or volumetric tomographic image of the sample is displayed (Drexler and Fujimoto 2007). Since its introduction in 1991 (Huang et al. 1991), there have been continuous advances in OCT technologies, in particular, advent of Fourier domain OCT (FDOCT) have led to remarkable technical advances in OCT imaging performance in terms of imaging speed and imaging sensitivity along with developments in lasers, optical devices, and components (Klein et al. 2013; Drexler 2004; Choi and Wang 2015; Klein and Huber 2017). Today, several implementations of OCT extension have been proposed to observe the structure as well as other characteristic features caused by the interaction of light with the sample such as polarization (de Boer et al. 2017) and attenuation (Wang et al. 2017). These additional improvements in OCT methodologies have allowed for combination of tissue anatomy with functional information of the living tissues, prompting wide-spread clinical acceptance of OCT to date.

Furthermore, enhanced processing of an OCT signal has offered information on blood flow in the tissue in vivo. Early work has harnessed the Doppler phase shift caused by backscattering from erythrocyte (red blood cells, RBCs) motion within blood vessel lumen to measure blood flow velocity and delineate vessels, called Doppler OCT. Though Doppler OCT has been widely used, however, its flow sensitivity is too low to detect slow blood flow (0.1–0.9 mm/s) within capillary vessels in human skin (Stucker et al. 1996) and blood flow in the vessels perpendicular to the orientation of OCT probe beam is hardly detectable, making it difficult to visualize whole microcirculation, particularly within the human skin. In order to achieve the flow sensitivity enhancement, algorithms utilizing phase variance, sequential scan subtraction, and speckle variance have been developed and successfully implemented to enhance microvasculature in OCT images. These functional OCT strategies are commonly named as OCT angiography (OCTA) (Wang 2010). OCTA is a type of motion-contrast technique for imaging blood perfusion in the vascularized tissue beds. OCTA exploits scattering dynamics of the time-course OCT signals, which is mainly contributed by the random motion of RBCs through the vessel lumen. By assessing the dynamics, OCTA is able to isolate the dynamic scattering component (due to blood perfusion) from the static scattering component (due to surrounding tissues), enabling precise location of the patent blood vessels (Wang 2010). This capability to image the functional blood vessels of OCTA has pushed the recent OCT development into a new height of being able to noninvasively assess the microcirculation in healthy and diseased states, permitting preclinical and clinical applications of OCTA to biomedicine. In a few years since invention, particularly, its clinical translation to ophthalmology has been first realized to examine various retinal vascular abnormalities and diseases such as retinal arterial venous occlusion, retinitis pigmentosa, choroidal neovascularization, glaucoma, uveitis, age-related macular degeneration (AMD), and diabetic retinopathy (DR) (Tan et al. 2018), representing a major contribution toward advancing the conventional ophthalmic diagnosis using current standard-of-care methods, including fluorescein angiography (FA) and indocyanine green angiography (ICG) (Spaide and Cooney 2015a).

In this section, we review concept and principle of OCTA and several OCTA algorithms for microcirculation imaging, and followed by its biomedical applications in the next section.

System Setup

Basically, OCTA is achieved on a standard FD-OCT platform with unique scan protocols and data processing algorithms dedicated to blood flow mapping. Fig. 1 shows a typical layout of 1.3 μm spectral domain OCT setup, one of FD-OCT systems (another one is termed of swept-source OCT (SS-OCT) (Gora et al. 2009)), which is similar to that commonly used in previous OCTA works (Li et al. 2018).

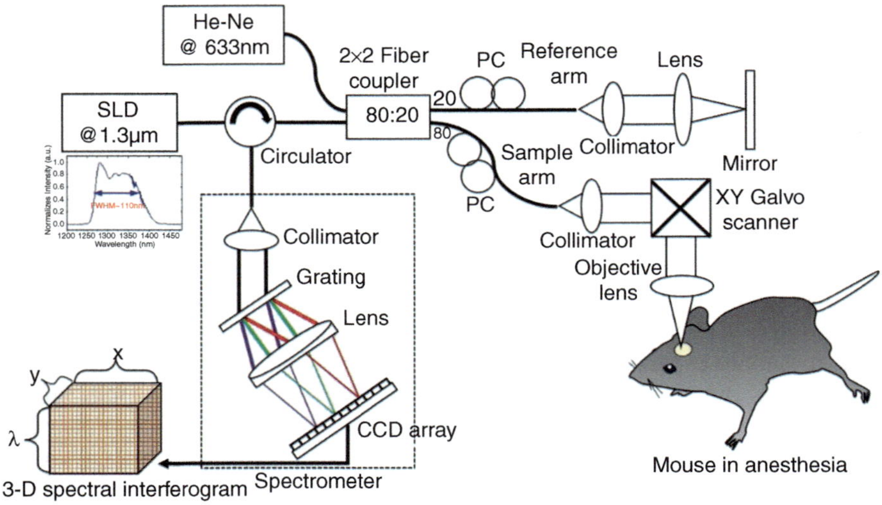

Fig. 1 Schematic of the standard spectral domain OCT system used in OCTA study to image the blood perfusion, where PC represents the polarization controller and CCD is the charged coupled device. The He-Ne laser diode emitting light at 633 nm was used as an aiming beam during imaging

Here we briefly describe its main parameters. The system used a superluminecent diode (SLD) with a central wavelength of 1310 nm and a spectral bandwidth (>100 nm) as a broad bland light source, which provided an axial imaging resolution of ~7 μm in air (~5 μm in tissue). The light from the SLD was coupled into a fiber-based Michelson interferometer, via an optical circulator. In the reference arm, the light was delivered onto a stationary mirror; and in the sample arm, the light was focused into a sample via an objective lens. The zero delay line of the system was set at ~0.5 mm above the focus spot of sample beam. With an 18-mm focal length of the objective lens in the sample arm, the size of the focus spot on the sample was ~7 μm, i.e., the lateral resolution of the system was ~7 μm. The light backscattered from the sample and reflected from the reference mirror were recombined by a 2 × 2 optical fiber coupler, creating interference signals, and then the interference signals were routed to a home-built high-speed spectrometer via the optical circulator. The spectrometer consisted of a collimator of 30 mm focal length, a 1200-lines/mm transmission grating, an achromatic lens with a 100-mm focal length and a 14-bit, a 1024 pixels InGaAs line scan camera. The maximum line scan rate of the camera was ~92 KHz. In the spectrometer, therefore, the interference signals through the grating are spatially dispersed in wavelength and thus the spectral interferogram is finally projected onto the array of charge-coupled device (CCD) for storage in computer. This spectrometer setup had a designed spectral resolution of 0.141 nm, which gave a measured imaging depth of ~3.0 mm on each side of the zero-delay line. The signal-to-noise ratio was measured to be ~100 dB with an incident light power onto the sample at ~3 mW.

Concept and Principle

The basic concept of OCTA is to use the variation in the OCT signal caused by moving scatters in tissue such as red blood cells (RBCs), as a source of contrast to image blood flow (Zhang et al. 2015). Figure 2 conceptually illustrates how OCTA works. The cartoon (left) in Fig. 2 describes one A-scan (blue arrow) through the skin tissue involving one blood vessel in function where RBCs are flowing in random through the vessel lumen. If you would repeat the A-scan at the same location and record OCT signals at two positions of the depth profile—one is backscattered from inside the vessel lumen and another one is backscattered from surrounding tissue outside the vessel—over time, you would be able to observe that the OCT signal backscattered from the structural tissue remains steady, while the OCT signal backscattered from the flowing blood changes over time due to constant motion of RBCs which are tumbling and tank-treading in shear flow (Dupire et al. 2012) through the vessel lumen as a graph in Fig. 2. The arbitrary oscillation in OCT signal over time is the result of random dynamics in backscattering from the biological particles that should be observed in the vessel lumen. The temporal changes of OCT signal in subsequent scan caused by the moving particles can differentiate the moving particles from the static tissue by analyzing the extent of the signal fluctuation, providing the opportunity to visualize the microvasculature without the need of exogenous dyes. The motion-contrast may be generated by any moving particles (e.g., particles in tissue fluid or lymphatic vessel), however, the predominant movement in typical tissue beds is from the RBCs in blood.

Repeated B-scans at the same position along the *x*-direction produce a series of OCT signals in cross section at different time points. For example, Fig. 3(a) shows

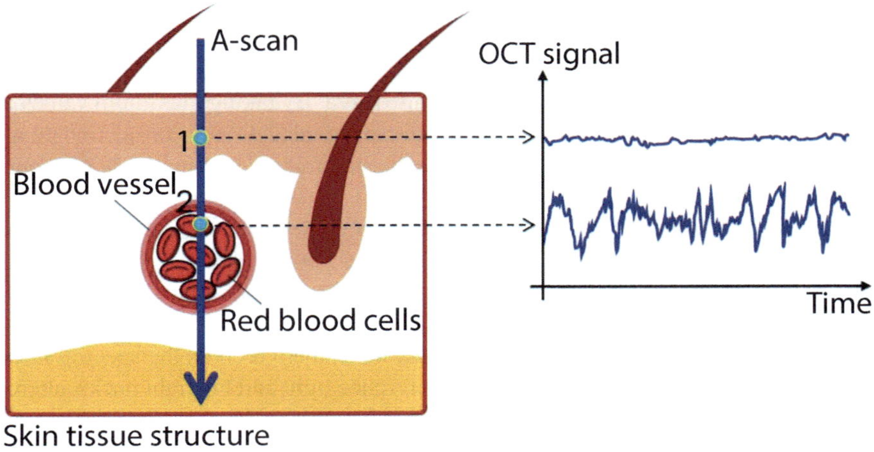

Fig. 2 Illustration of the concept of OCTA. OCT signals are sampled from two points in the A-scan through skin tissue, where a pixel 1 is located at the static tissue, and an another pixel 2 is located within a functional blood vessel. Dynamic changes in the OCT signals for pixel 2 can be observed over time while the OCT signals from pixel 1 remain steady over time

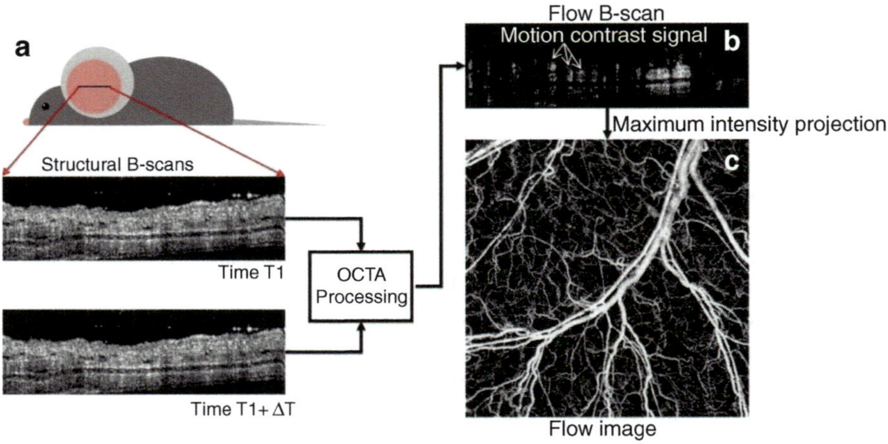

Fig. 3 The OCTA signal processing scheme. (**a**) Two adjacent OCT images of mouse pinna obtained at the same location but with a time gap ΔT. (**b**) An OCT angiography image obtained by using OCTA processing with the OCT images in (**a**). The intensities (bright signals) in (**b**) indicate motion-contrasts corresponding to red blood cell flow in the blood vessel lumens. (**c**) A *en-face* projection view of 3-D OCTA images obtained at the scanned area representing a vessel network of the mouse pinna

two OCT images of a live mouse pinna obtained at time T1 and T1 + ΔT, respectively, where ΔT is an elapsed time to complete one B-scan. Now, we know that OCT signals from the blood vessels in the mouse pinna should be different at T1 and T1 + ΔT because of motion of RBCs flowing through the vessel lumen. By performing the OCTA processing with the ensemble of OCT images, one flow B-scan is produced with all of the OCT signals from the stationary tissue excluded. The motion-contrast signals in the flow B-scan localize the blood vessels (Fig. 3b). As repeating the OCTA protocol along the *y*-direction, the flow B-scans at different locations would be obtained within the scanned area. By taking maximum values of depth profiles in the flow B-scans, one two-dimensional (2-D) flow projection image in *x*-*y* en face orientation is produced, showing the microvessel network of mouse pinna (Fig. 3c).

In specific, the theoretical analysis of OCTA begins with a spectral interferogram captured by each pixel of the CCD camera of the SD-OCT system (Wang and An 2009). It is assumed that the wavenumbers of the broadband light source are from k_0 to $k_0 + \Delta k$, where $k_0 = 2\pi/\lambda_0$, and these wavenumbers cover 1024 pixels of the line scan camera, respectively. As a consequence, the camera records the spectral interference fringe signal formed between the reference light and the light backscattered from within sample, which can be written as a function of k_j:

$$I\left(k_j\right) = S\left(k_j\right)\left\langle \left| E_{\mathrm{R}} \exp\left(i2k_j r\right) + \int_{-\infty}^{\infty} a(z) \exp\left\{i2k_j\left[r + n(z)z\right]\right\} dz \right|^2 \right\rangle \quad j = 1,2,3 \ldots 1024 \qquad (1)$$

where $i = \sqrt{-1}$, $\langle\rangle$ is the time average, k_j is the wavenumber of the light captured by the jth pixel of the CCD camera, $I(k_j)$ is the light intensity captured by the jth detector, and $S(k_j)$ is the spectral density of the light source at k_j, E_R is the light reflected from the reference mirror. $a(z)$ is the amplitude (reflectivity) of the light backscattered at depth z. r and $n(z)$ are the optical path length for the light traveled in the reference arm and the refractive index of tissue at depth z, respectively. However, the real living tissue is often optically heterogeneous, which means that $a(z)$ and $n(z)$ can be time-variable (e.g., due to extracellular fluid flowing between tissue cells or intracellular motion). Thus, Eq. (1) can be rewritten as:

$$I\left(k_j,t\right) = S\left(k_j\right)\left\langle\left|E_R \exp\left(i2k_j r\right) + \int_{-\infty}^{\infty} a(z,t)\exp\left\{i2k_j\left[r + n\left(z,t\right)z\right]\right\}dz\right|^2\right\rangle \quad j = 1,2,3...1024 \tag{2}$$

where t is the timing when one of A-lines in B-scan was captured. Because the light backscattered from the sample is quite weak compared to the light reflected from the reference mirror, the self cross-correlation between the light backscattered from different positions within the sample itself is not considered. Also, the DC background signals are excluded in Eq. (2) because of no contribution to useful OCTA signals. In these cases, Eq. (2) can be written as:

$$I\left(k_j,t\right) = 2S\left(k_j\right)E_R \int_{-\infty}^{\infty} a\left(z,t\right)\cos\left(2k_j n\left(z,t\right)z\right)dz \quad j = 1,2,3...1024 \tag{3}$$

Let us assume that there is one patient blood vessel within a tissue and its central position is at a depth z_1, in which the blood cells as scattering particles move toward the incident beam at a velocity v. Then, the amplitude of the light backscattered from these blood cells passing through the vessel position z_1 will be fluctuated due to the random arrangement of the blood cells and the frequency of the backscattered light will be also modulated because of their velocity. Thus, Eq. (3) must be modified to include the effect of the random motion of blood cells on the interference signal as:

$$I\left(k_j,t\right) = 2S\left(k_j\right)E_R\left[\int_{-\infty}^{\infty} a\left(z,t\right)\cos\left[2k_j n\left(z,t\right)z\right]dz + a(z_1,t)\cos\left[2k_j n\left(z_1,t\right)\left(z_1 - vt\right)\right]\right] \tag{4}$$

Eq. (4) represents two backscattering signal components involved in the spectral interferogram; backscattering signals from the static tissue with the amplitude of $a(z, t)$ (1st term) and the backscattering from the moving particles (blood cells) with the amplitude of $a(z_1, t)$ and a velocity of v at the position z_1 (2nd term). Since the changes in the amplitude and refractive index in the heterogeneous tissue over time are subtle, therefore, we can assume $a(z, t)$ and $n(z, t)$ to be time-constant variables as a function of depth z. In doing so, the final spectral interference signal now becomes:

$$I\left(k_j,t\right)=2S\left(k_j\right)E_{\mathrm{R}}\left[\int_{-\infty}^{\infty}a\left(z\right)\cos\left[2k_jn\left(z\right)z\right]dz+a(z_1,t)\cos\left[2k_jn\left(z_1,t\right)\left(z_1-vt\right)\right]\right] \tag{5}$$

Eq. (5) indicates that the construction of the backscattering and refractive index in the tissue (1st term) will be constant over time and does not affect the 2nd term resulted from the blood flow. Fourier transform upon wavelength k to Eq. (5) produces a complex based OCT signal in depth (A-line):

$$\tilde{I}\left(z,t\right)=FT^{-1}\left\{I\left(k_j,t\right)\right\}\Big|_k=A\left(z,t\right)\exp\left[i\phi\left(z,t\right)\right] \tag{6}$$

where $\tilde{I}\left(z,t\right)$ is the complex form of OCT signal: $\tilde{I}\left(z,t\right)=a+ib$ (a, b are real numbers) and $\phi(z,t)$ is the phase of the analytic signal involving the velocity of particles in motion. Amplitude $A(z,t)$ of the $\tilde{I}\left(z,t\right)$ is obtained by taking an absolute of Eq. (6). In order to obtain the optical scattering signals by the moving scatters, at the very least, an ensemble of two more time-varying B-scans consisting of OCT A-lines (Eq. (6)) is necessary, from which the static signal component can be filtered out while preserving the dynamic scattering components (blood flow). A variety of OCTA methods based on this high-pass filtering technique have been reported to image the blood flow within the perfused vessels in tissue beds which will be described in Section 2.3 in detail.

However, the phase difference (phase shift) between the inter-A lines with the time interval ΔT is then evaluated with Kasai's autocorrelation method (Kasai et al. 1985):

$$\Delta\phi\left(z,t\right)=\tan^{-1}\left[\frac{Im\left[\tilde{I}\left(z,t\right)\cdot\tilde{I}^{*}\left(z,t+\Delta T\right)\right]}{\mathrm{Re}\left[\tilde{I}\left(z,t\right)\cdot\tilde{I}^{*}\left(z,t+\Delta T\right)\right]}\right]=k_0 2n\left(z_1,t\right)v\Delta T \tag{7}$$

Therefore, in Eq. (7), the velocity of the flow signal v can be directly written as:

$$v=\frac{\Delta\phi\left(z,t\right)\lambda_0}{4\pi n\left(z_1,t\right)\Delta T} \tag{8}$$

It is noted again that the velocity in Eq. (8) is parallel to OCT beam (i.e., it becomes an axial (z-direction) component of flow velocity of RBC). For typical laminar blood flow in the large vessels or a single file flow of RBCs in capillary networks perpendicular to the OCT beam, therefore, it is quite difficult to measure the flow velocities using the phase shift in Eq. (7). The maximum detectable flow velocity in Eq. (8) is determined by the time interval ΔT for the phase shift of 2π. If flow velocity in a capillary is $v \leq 100$ μm/s, then it would require $\Delta T \geq \sim 4.7$ ms, which is the time spacing between adjacent B-scans.

Algorithms

In order to image the blood flow, over the last decade, a variety of high-pass filtering techniques have been introduced and applied to OCTA measurements. The different OCTA methods harness the different components of OCT signal such as amplitude, intensity, and phase to create the dynamic-contrasts representing the extent of blood perfusion. In usual, the OCTA methods reported can be categorized into three types: phase-signal-based OCTA algorithms, intensity-signal-based OCTA algorithms, and complex-signal-based OCTA algorithms.

Phase-Signal-Based OCTA Algorithms

OCTA algorithms utilizing the phase variation as an angiographic contrast have been reported. The phase of OCT signal is altered by the moving RBCs because of changes in OPD, which is mainly induced by axial (z-axis) displacement of RBCs in motion. Since the even small displacement reflects the detectable change in phase, the phase-signal-based approach takes advantage to detect the larger dynamic range of flow velocities. A few methods have been proposed to obtain the difference in phases between the sequential pixels, adjacent A-scans (Makita et al. 2006; Zhang and Chen 2015), or adjacent B-scans (Fingler et al. 2018), achieved by different scan protocols. Especially, the latter called phase variance OCT (pvOCT) (Schwartz et al. 2014) measures the phase variance between the adjacent B-scans, where a time interval between the inter B-scans is significantly increased. The increased time interval induces the larger displacement of moving particles during the time interval, and as a result, the detection sensitivity to the slower flow (e.g., capillary flow) can be dramatically increased. The flow signal in pvOCT is expressed as the following equation (Schwartz et al. 2014):

$$pvOCT(x,z) = \frac{1}{N-1}\sum_{i=1}^{N-1}\left[\Delta\Phi_i(x,z) - \frac{1}{N-1}\sum_{i=1}^{N-1}\Delta\Phi_i(x,z)\right]^2$$
$$\Delta\Phi_i(x,z) = \Phi_{i+1}(x,z,t+\Delta T) - \Phi_i(x,z,t) \quad (9)$$

where N represents the repetition number of B-scans acquired at the same location, $\Phi_i(x,z,t)$ and $\Delta\Phi_i(x,z,t)$ indicate the phase value and phase difference in the *i-th* B-scans at lateral location x, depth position z, and time t, ΔT is the time interval between two consecutive B-scans, and i is the index of the *i-th* B-scan. A main drawback of the phase-signal-based approach is that it is prone to perturbation, contaminating the real flow signals with the phase noise caused by the OCT laser source (Yun et al. 2004) or the bulk motion of the tissue. The phase stabilization has been attempted to remove the phase noise using computational methods on post-processing (Song et al. 2017) or developing phase-stable light sources (Chen et al. 2016).

Intensity-Signal-Based OCTA Algorithms

In order to overcome the drawback (vulnerable to phase noise) of phase-signal-based OCTA algorithms, OCT researchers have proposed intensity-signal-based OCTA algorithms using the change in amplitude or intensity of OCT signal to contrast the particle motion. Currently, these OCTA approaches are the most preferably being used for OCTA studies because of being insensitive to the phase noise, saving the trouble for the phase compensation. The intensity-signal-based OCTA algorithms include three representative methods: speckle variance OCT (svOCT) (Mariampillai et al. 2008), correlation mapping OCT (cmOCT) (Enfield et al. 2011), and split-spectrum amplitude decorrelation (SSADA) (Jia et al. 2012).

Speckle Variance OCT (svOCT)

OCT signals involve speckles, a form of constructive or destructive interference generated by superposition of multiply scattered lights through the heterogeneous tissue. It has been shown that the speckles in OCT signals play a dual role; both as a source of noise and as a carrier of information about tissue microstructure and flow. However, a basic concept of classical laser speckle imaging was to use the time dependency of laser speckles to quantify the particle flow (Boas 2010). In OCT regime, Barton et al. (Barton and Stromski 2005) in 2005 have first coined this concept to TD-OCT with a hypothesis that the OCT signal can be treated as a speckle and the strength of speckle coming from flow would become different from that of the static tissue. By measuring the variance in OCT signal amplitudes taken at the adjacent four-pixels in one OCT cross-section, the authors have been able to successfully visualize the flow image of an in vitro tube phantom and in vivo rodent skin (Barton and Stromski 2005). This was the first application of using speckle analysis in OCT images to obtain flow signals. In 2008, Mariampillai et al. have introduced the concept of speckle variance on SD-OCT (Mariampillai et al. 2008). The B-scans are repeated with a constant time interval and then the inter B-frame (inter B-scans) speckle variance signal can be computed using the following equation (Mariampillai et al. 2008):

$$svOCT(x,z) = \frac{1}{N}\sum_{i=1}^{N}\left(I_i(\mathrm{x,z}) - I_{mean}\right)^2 \qquad (10)$$

where N represents the repetition number of B-scans acquired at the same location, $I_i(x,z)$ denotes the intensity value in the *i-th* B-scans at lateral location x, depth position z, and I_{mean} is the average of the intensity value over the same set of pixels. Since the speckle variance approach is readily suited to both SD- and SS-OCT implementation because of its immunity to phase noise, allowing for a wide range of application of biomedical imaging of retina (Xu et al. 2015), skin (Markowitz et al. 2016), spinal cord (Cadotte et al. 2012), and embryo (Grishina et al. 2017).

Correlation Mapping OCT (cmOCT)

Another intensity-signal-based OCTA algorithm has been introduced by Enfield et al. in 2011 (Enfield et al. 2011). In the proposed method, the authors have evaluated the degree of similarity (resemblance) of the OCT speckle intensities between adjacent B-scans to obtain flow signal, that is, by calculating the correlation of inter-frames, the flow signal can be extracted from the correlogram. As particle flow regions characterize relatively lower correlation than static regions, it is possible to isolate the microvasculatures from the surrounding static tissues by estimating the calculated correlation values with a given threshold (Enfield et al. 2011). In specific, the flow signals can be acquired by calculating cross-correlation between two of $M \times N$ grid consisting of neighboring pixels taken at the same transverse location from the inter B-frames which is expressed as (Enfield et al. 2011):

$$cmOCT\left(x,z\right) = \sum_{p=0}^{M}\sum_{q=0}^{N} \frac{\left[I_A\left(x+p,z+q\right)-\overline{I_A\left(x,z\right)}\right]\left[I_B\left(x+p,z+q\right)-\overline{I_B\left(x,z\right)}\right]}{\sqrt{\left[I_A\left(x+p,z+q\right)-\overline{I_A\left(x,z\right)}\right]^2+\left[I_B\left(x+p,z+q\right)-\overline{I_B\left(x,z\right)}\right]^2}} \tag{11}$$

where $I_{\#}(x, z)$ (# is A or B) indicates the intensity value at an arbitrary pixel in the grids from frame A and frame B, and M and N is the grid size. $\overline{I_{\#}\left(x,z\right)}$ (# is A or B) is the grid mean value. This grid is then shifted across the entire XZ image and a 2-D correlation map (a correlogram) is generated. The resulting correlation map contains values on the range from −1 to +1, where the value toward +1 indicates a positive correlation and vice versa. 0 means no correlation at all. The grid size used is arbitrarily chosen for optimal image quality with the trade-off of processing time; larger grids require greater processing times and pixel bining, resulting in the loss of smaller vessels in the flow image.

The background noise gives weak correlation values, comparable to the correlation values in the flow region. This flow signal ambiguity has been effectively resolved by masking the correlation image with a structural mask, a region that contains structural information in the OCT image (Enfield et al. 2011). Further masking with a flow image obtained using other intensity-signal-based OCTA algorithm has markedly enhanced the flow image quality (Choi et al. 2014).

Split-Spectrum Amplitude-Decorrelation Angiography (SSADA)

Although the speckle variance or correlation mapping method mentioned above does not suffer from phase noise artifacts, the methods are still sensitive to the global movement of biological samples in axial direction (e.g., breathing, heartbeat, involuntary tissue bulk motion). To overcome this limitation, a new correlation technique termed "split-spectrum amplitude-decorrelation angiography" (SSADA) has been proposed by Jia et al. in 2012 (Jia et al. 2012).

A main key of the method is to lower axial resolution in OCT imaging; by extending the height of resolution cell, the depth-resolving power is degraded, which can dampen the axial decorrelation sensitivity caused by the axial motion noise (Jia et al. 2012). The authors have achieved the lowered axial resolution on purpose by splitting full OCT spectrum into several narrower bands. With the split four sub-bands, the original 2-D spectral interferogram is subdivided into four new interferograms with smaller bandwidth. This strategy of splitting the spectrum can reduce the predominant bulk motion noise in the axial dimension without sacrificing the flow signal. Inter B-scan decorrelation was computed using the spectral interferograms separately and then averaged using the following equation (Jia et al. 2012):

$$SSADA(x,z) = 1 - \frac{1}{N-1}\frac{1}{M}\sum_{n=1}^{N-1}\sum_{m=1}^{M}\frac{A_{nm}(x,z)A_{(n+1)m}(x,z)}{\left[\frac{1}{2}A_{nm}(x,z)^2 + \frac{1}{2}A_{(n+1)m}(x,z)^2\right]} \tag{12}$$

where x and z are lateral and depth indices of the B-scan images, and n and m denotes the B-scan slice index and the index of the split spectrum, respectively. N indicates the number of repeated B-scans taken at the same position and M is the number of split-spectrums. $A_{nm}(x,z)$ and $A_{(n+1)m}(x,z)$ are the adjacent amplitude frames obtained from the *m-th* split spectral interferogram. By increasing the number M, the decorrelation signal-to-noise ratio (SNR) can be improved without increasing the scan acquisition time.

SSADA has been shown to improve the SNR of flow detection and connectivity of microvascular networks when compared to other amplitude decorrelation algorithms (Gao et al. 2015). This split-spectrum method is particularly optimal for imaging of retinal flow prone to the pulsatile bulk motion in the axial direction, but it is not preferred for imaging the cerebral flow of head-fixed rodent brain which is immune to the bulk motion.

Complex-Signal-Based OCTA Algorithms

Optical Microangiography

The third category utilizes both the phase and the intensity (amplitude) information of the OCT signal for flow imaging. The representative of the complex-signal-based OCTA algorithms is optical microangiography (OMAG), which has been developed and named by Wang et al. in 2007 (Wang et al. 2007) and later renovated to the current implementation by Wang in the same group in 2010 (An et al. 2010). In this hybrid OCTA method, the authors have included both the phase and the amplitude components of the OCT signal in the flow signal calculation in order to pursue the sensitivity enough to detect the slow flow that is only detectable by using the phase component. The flow signal is calculated by absolute of differential operation on the complex OCT signals in the adjacent B-scans which is formulated as (An et al. 2010):

$$OMAG(x,z) = \frac{1}{N-1}\sum_{i=0}^{N-1}\left|C_{i+1}(x,z) - C_i(x,z)\right| \tag{13}$$

where N indicates the repetition number of B-scans at the same transverse location, and $C_i(x,z)$ denotes the complex signal (having both amplitude and phase values) in *i-th* B-scans at lateral location x and depth position z. The possible phase noise induced by the axial bulk motion of the sample can be diminished by using a phase compensation (An et al. 2010). The technique is theoretically sensitive enough to image the very slow blood flows at ~4 μm/sec (An et al. 2010), which is sufficient to image particle Brownian motion (Li and Raizen 2013). The flow image quality can be improved by increasing the number of repetitions of B-scan, compromising the scanning time and the improvement in the SNR is saturated over 12-times repetitions (Wang et al. 2016).

Translational Applications of OCTA

Since most OCT studies have been intensively conducted in ophthalmology and many ophthalmic OCT machines are currently commercially available, OCTA has been rapidly translated to clinical ophthalmology within a short period of time since its first demonstration. Dye-based angiography such as fluorescein angiography (FA) and indocyanine green angiography (ICGA) is a gold standard to evaluate blood flow in retinal and choroidal vessels in the ocular tissue. Despite high sensitivity rates in detecting vascular pathology, however, FA and ICGA still face several challenges of intravenous dye injection, long acquisition and waiting times, and the potential side-effects of the dyes. The invasive, time-consuming nature of FA and ICGA impedes their practical use for the routine following of treatment responses. The dye-based technology also does not reveal the retinal tissue structure. OCTA, on the other hand, is quick, safe, and noninvasive tool that combines angiographic visualization of the exact site, size, and extent of vascular lesions, even those that may be missed by FA, together with cross-sectional representation of the anatomical structure of retina. Moreover, OCTA angiograms can be segmented into en face slabs to evaluate the different retinal vascular layers (e.g., superficial inner retina, deep inner retina, outer retina, and choriocapillaris) in isolation (Fig. 4). Therefore, OCTA as a complimentary tool to FA and ICGA can offer the convenience and practicality required for the frequent follow-up of the diseased eyes in clinics.

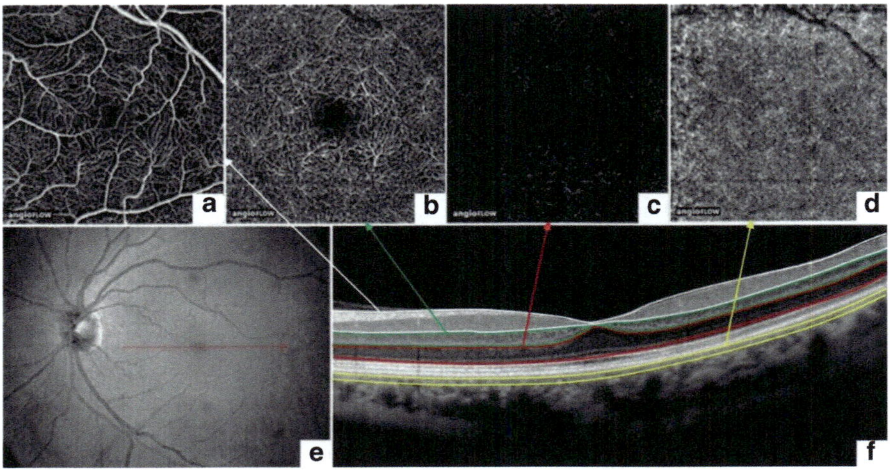

Fig. 4 OCT angiograms (3 mm × 3 mm) of a normal eye of a 56-year-old Caucasian man using the Angiovue which are made by a mean projection of OCTA dataset segmented along specific depth ranges of retina displayed in its OCT crossection (**f**): (**a**) Superficial inner retina, (**b**) Deep inner retina, (**c**) Outer retina (avascular region), and (**d**) Choriocapillaris. (**e**) En face view of retinal OCT image. Red solid line indicates the location of (**f**) (This figure was reprinted from ref. (de Carlo et al. 2015) with permission

With clinical translation of OCTA, many pilot and case studies have been performed using home-built OCTA systems or commercialized ones available throughout the world (Table 1) and have demonstrated its ability to visualize characteristic alterations in retinal and choroidal vascular morphology occurred in various ocular diseases such as proliferative diabetic retinopathy (PDR) (Khadamy et al. 2018; Thompson et al. 2019; Hwang et al. 2015; Sandhu et al. 2018) (Fig. 5(top), retinal vein occlusion (Chen and Wang 2017; Lee et al. 2016; Zhang et al. 2018), retinitis pigmentosa (Spaide et al. 2015b; Chen et al. 2015), polypoidal choroidal vasculopathy (PCV) (Fig. 5(Bottom) (Srinivasan and Radhakrishnan 2014; Zhang et al. 2016, 2017), choroidal neovascularization caused by age-related macular degeneration (AMD) (Qin et al. 2011; Blatter et al. 2012), which have been comparable and correlated with the results by FA and ICGA.

Furthermore, recent papers have reported wide-field OCTA imaging of retina, which has been achieved either with motion tracking through a real-time auxiliary line scan ophthalmoscope (LSO) (Lee et al. 2016) or using montage scanning protocols (Zhang et al. 2018) that is clinically feasible to image functional retinal vasculature in patients, with a coverage of more than 60 degrees of retina while still preserving high definition and resolution. This wide-field imaging would provide more detailed microvasculature in the peripheral regions, where it is believed to have some early defects taking place before hitting the central vision (Zhang et al. 2018). Even though OCTA is not able to detect leakage in the retina, OCTA may

Table 1 Details of main commercially available ophthalmic OCTA systems

Commercial product name					
	AngioVue*	AngioPlex*	SSOCT Angio*	Spectralis OCTA	AngioScan
Company	OptoVue	Zeiss	Topcon	Heidelberg	Nidek
Algorithm	SSADA	OMAG	OCTARA	Amplitude decorrelation	Modified OMAG
A-scans/sec	70,000	68,000	100,000	85,000	53,000
Scan time	3 s	3.8 s			>40 s
Image dimensions	3 × 3 mm, 4.5 × 4.5 mm, 6 × 6 mm, 8 × 8 mm, 6 × 10 mm (montage)	3 × 3 mm, 6 × 6 mm	3 × 3 mm, 4.5 × 4.5 mm, 6 × 6 mm	10° × 10°, 15° × 15°, 20° × 20°, 15° × 30°	3 × 3 mm, 6 × 6 mm, 9 × 9 mm, 12 × 9 mm (montage)
Eye tracking	IR video tracking	FastTrack	SMARTTrack	Active eye tracking	SLO fundus tracking
Motion correction	Motion correction technology	Motion correction software to remove artifacts	No active motion correction software		

Data sourced from company web pages and product brochures
*These systems were FDA-cleared as of 2019

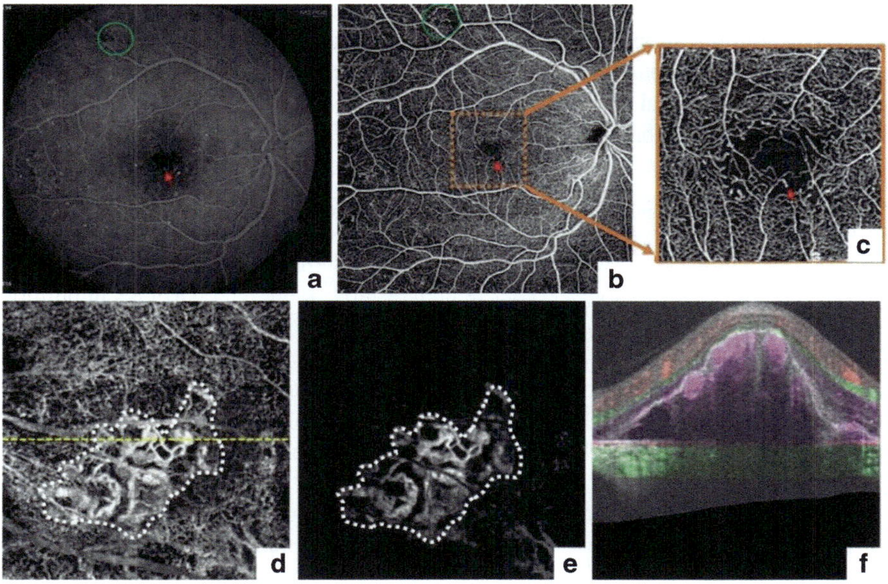

Fig. 5 Top row: OCT angiograms using OMAG method from a 33-year-old patient (male) diagnosed with proliferative diabetic retinopathy (PDR). (**a**) Early phase fluorescein angiography (FA) image of the patient and a corresponding OCT angiography image (12 mm × 12 mm) (**b**). (**c**) Zoom into a 3 mm × 3 mm area centered at the foveolar from (**b**). Microaneurysms, the earliest visible changes of DR, are observed as indicated by the red arrows and green ovals in both FA and OCTA images. Bottom row: OCT angiogram (3 mm × 3 mm) from a polypoidal choroidal vasculopathy (PCV) eye. (**d**) En face OCTA image within outer retinal layer where projection shadow artifacts are significant across the image that would impede the image interpretation. (**e**) After projection shadow artifact removal, a real vasculature (white dotted line enclosure) is much clear to see, exhibiting a polyps and a branching vascular network (BVN). (**f**) Color composite cross-sectional OCTA taken at a yellow dotted line in (**d**) shows that the abnormal vasculature was in the space between the RPE and the Bruch's membrane. This figure was reprinted from ref. (Chen and Wang 2017) with permission

still provide further insightful information to help understand the pathophysiology of retinal diseases. Having OCTA as one of the clinical standard routines may help clinicians and physicians monitor the progression of diseases and guide therapeutic treatment at its earliest possible time point.

However, multiple OCTA image artifacts may affect image quality. Motion artifacts due to blinking or gross eye movement appears as horizontal or vertical black and white lines, respectively. Motion correction techniques published could compensate for these motion artifacts but can create vessel doubling, stretching of the image, quilting, or loss of detail (Spaide et al. 2015b; Chen et al. 2015). Inherently, furthermore, shadow-graphic projection artifacts (Srinivasan and Radhakrishnan 2014), false flow effect due to tail-like artifacts created by dynamic scattering of moving blood cells in the superficial vasculature interfere with the interpretation of deeper vascular network in OCTA image. A series of recent attempts for removing the shadow artifacts (Zhang et al. 2016, 2017) have been reported and have demonstrated the minimization of the artifacts on OCTA images of the deeper retinal layers.

Since OCTA's debut in ophthalmology, researchers have also been trying to expand its translational applications in dermatology because skin is an easy site to hand and access OCT beam. Table 2 gives a brief overview of dermatological OCTA

Table 2 Chronological review of dermatological OCTA studies

Year	Last author	Location	Feature	References
2010	R. K. Wang	Palm	First demonstration	Opt. Express (An et al. 2010)
2011	M. Leahy	Forearm	–	Biomed. Opt. Express (Enfield et al. 2011)
2011	R. K. Wang	Antecubital area	Psoriasis	Laser Surgery Med. (Qin et al. 2011)
2012	R. Leitgeb	Forehead, forearm, cheek	Eczema, BCC	Biomed. Opt. Express (Blatter et al. 2012)
2013	D. Sampson	Forearm, thigh	Scars	J. Biomed. Opt. (Liew et al. 2013)(Gong et al. 2014)
2014	R. K. Wang	Hand	Laser burn	J. Biomed. Opt. (Choi et al. 2014)
2014	R. K. Wang	Oral cavity, nasal cavity, lip	–	Biomed. Opt. Express (Choi and Wang 2014)
2015	R. K. Wang	Face	Acne	Laser Surgery Med. (Baran et al. 2015)
2015	R. K. Wang	Finger	Inflammation	J. Biophotonics (Wang et al. 2015)
2015	A. Tosti	Nail	Psoriasis, 16 patients	Skin Appendage Disord. (Aldahan et al. 2017)
2016	R. K. Wang	Forehead, arm, chest, areola	Nevi, acne, 7 day	Skin Res. Technol. (Baran et al. 2016a), Sei. Rep. (Baran et al. 2016b)
2016	R. McLaughlin	Forearm	Heat stress, 8 subjects	J. Appl. Physiol. (Carter et al. 2016)
2016	R. Leitgeb	Palm, finger	–	Biomed. Opt. Express (Chen et al. 2016)
2016	R. McLaughlin	Abdomen, arm, thigh	Burn scar, 7 patients for 7 days	Biomed. Opt. Express (Gong et al. 2016b), J. Biophotonics (Gong et al. 2016a)
2016	G. Jemec	Various	35 subjects	Microvasc. Res. (Themstrup et al. 2016)
2016	J. Welzel	Various	Atlas of diseases	Dermatology (Ulrich et al. 2016)
2017	R. McLaughlin	Various	Scar under treatment, 7 patients 20 weeks	J. Biophotonics (Es'haghian et al. 2017)
2017	G. Pellacani	Face	Acne, 31 patients, 15 for 60 days	J. Eur. Acad. Dermatol. Venereol. (Manfredini et al. 2017)
2017	R. K. Wang	Various	15 subjects	Skin Res. Technol. (Men et al. 2017)

Table 2 (continued)

Year	Last author	Location	Feature	References
2017	J. Welzel	Various	Atlas of diseases	Dermatol. Ther. (Schuh et al. 2017)
2017	M. Ulrich	Various	18 AK, 12 BD, 24 SCC patients	J. Eur. Acad. Dermatol. Venereol. (Themstrup et al. 2017)
2018	R. K. Wang	Residual limb	Reactive hyperaemia	Skin Res. Technol. (Baran et al. 2018)
2018	R. K. Wang	Oral cavity	Monitoring of wound healing	Lasers Med. Sei. (Wei et al. 2018)
2018	S. Matcher	Various	Eczema for 32 patients	Biomed. Opt. Express (Byers et al. 2018)
2018	R. K. Wang	Various	Atlas of diseases	Laser Surgery Med. (Deegan et al. 2018a)
2018	R. K. Wang	Hand	Wound healing for 44 days	Quant. Imaging Med. Surg. (Deegan et al. 2018b)
2018	G. Jemec	Various	280 subjects	Exp. Dermatol. (Lindsø Andersen et al. 2018)
2018	R. K. Wang	Upper arm	Blood collection, 19/21 subjects	Laser Surgery Med. (Men et al. 2018)
2018	G. Jemec	Forearm	Histamine induced wheals, 10 subjects	Skin Res. Technol. (Olsen et al. 2018)
2018	G. Jemec	Various	13AK, 22 BCC, 17 SCC and 81 patients for subtypes of BCC	Skin Res. Technol. (Sigsgaard et al. 2018), Exp. Dermatol. (Themstrup et al. 2018)

papers up to now. As we can see from the table, in the first 5 years of OCTA's application in dermatology, most of the initial publications have been limited to some preliminary or pilot studies, which have applied OCTA to skin lesions for testing its angiographic performance. Since 2016, however, development of a clinical prototype OCTA system designed for skin imaging and release of commercial dermatological OCTA systems (VivoSight OCT, Michelson Diagnostics Ltd.) have boosted functional and cohort studies of OCTA in clinical settings, providing a new insight into investigating diseased skin. Hence, clinical OCTA researches have been intensively reported and demonstrated its potential and promise as a diagnostic tool in dermatology for assessing microcirculatory and structural feature in human skin diseases such as acne (Manfredini et al. 2017), eczema (Byers et al. 2018), psoriasis (Deegan et al. 2018a), chronic graft-versus-host-disease (cGvHD) (Deegan et al. 2018a), skin cancers (Sigsgaard et al. 2018; Themstrup et al. 2018), burn scars (Gong et al. 2016a, b; Es'haghian et al. 2017), and monitoring the wound healing of oral cavity (Wei et al. 2018) and cutaneous skin tissues (Deegan et al. 2018b).

Although OCTA provides a fast, reliable, and noninvasive way of skin vasculature imaging, the imaging depth of OCTA in skin is limited to <1 mm, preventing its application in skin diseases that affect deeper cutaneous vessels. Moreover, OCTA alone is incapable of extracting functional parameters such as oxygen saturation (sO_2) level and the metabolic rate, valuable information in assessing cutaneous vascular diseases. Multimodal approaches for combining OCTA with photoacoustic tomography (PAT) may fill in these shortcomings in a complementary way (Liu et al. 2016; Liu and Drexler 2019).

Discussion and Outlook

OCTA has yet to achieve broadly accepted (e.g., it has not been adopted in clinical neurology because of small field of view (>3 mm^2) and imaging depth (~1 mm) insufficient to penetrate through a human skull) and definitive clinical indications, but it is surely on the path to becoming a revolutionary tool in the diagnosis and the management of retinal vascular diseases in ophthalmology. While commercial ophthalmic OCTA devices have only been released and available for a just few years, clinical adoption of the commercial systems has been rapid, despite several controversial and logistical hurdles, including the lack of reimbursement and questions regarding the role of OCTA, FA, and ICGA.

The current main limitation of OCTA in clinical applications for retinal diseases is the field of view. In diabetic retinopathy (DR), for instance, more evidence has shown that the early changes in blood flow or vessel morphology happen in the peripheral region, and may not be noticed if the scanning area only covers 6 × 6 mm^2 (approximately 20° × 20°) area of central fovea. A wide field of view imaging system or a robust vascular image montaging algorithm is critical to enable the visualization of functional vascular networks in the peripheral region (with a viewing angle of at least ~100°). Another limitation is wavelength of light source. Using a longer wavelength of 1050 nm, as opposed to 800 nm range wavelength for ophthalmology applications, is also required. Incident light beams with longer wavelength can penetrate deeper, reach deeper tissue, and therefore may be able to provide flow information in the deeper tissue, such as the choriocapillaris and choroid. SS-OCTA systems (e.g., PLEX® Elite 9000) with longer wavelengths of coherent light will likely overcome these limitations although these systems are still very early in their clinical evolution. The same is true for dermatology applications where a central wavelength range of 1680 nm as opposed to the current use of 1300 nm wavelength should provide more opportunity for the light to penetrate deeper into the dermis.

Given the speed with which clinical interest is evolving around OCTA, it will not be long before all of these limitations will be overcome. It is expected that OCTA will stand to change the practice of ophthalmology and dermatology in the next 10 years as profoundly as OCT has changed it in the past 10 years.

References

Aldahan AS, Chen LL, Fertig RM, Holmes J, Shah VV, Mlacker S, Hsu VM, Nouri K, Tosti A (2017) Vascular features of nail psoriasis using dynamic optical coherence tomography. Skin Appendage Disord 2(3–4):102–108

An L, Qin J, Wang RK (2010) Ultrahigh sensitive optical microangiography for *in vivo* imaging of microcirculations within human skin tissue beds. Opt Express 18(8):8220–8228

Baran U, Choi WJ, Wang RK (2016a) Potential use of OCT-based microangiography in clinical dermatology. Skin Res Technol 22(2):238–246

Baran U, Li Y, Choi WJ, Kalkan G, Wang RK (2015) High resolution imaging of acne lesion development and scarring in human facial skin using OCT-based microangiography. Lasers Surg Med 47(3):231–238

Baran U, Qin W, Qi X, Kalkan G, Wang RK (2016b) OCT-based label-free in vivo lymphangiography within human skin and areola. Sci Rep 6:21122

Baran U, Swanson E, Sanders JE, Wang RK (2018) OCT-based microangiography for reactive hyperaemia assessment within residual limb skin of people with lower limb loss. Skin Res Technol 24(1):152–155

Barton JK, Stromski S (2005) Flow measurement without phase information in optical coherence tomography images. Opt Express 13(14):5234–5239

Blatter C, Weingast J, Alex A, Grajciar B, Wieser W, Drexler W, Huber R, Leitgeb RA (2012) In situ structural and microangiography assessment of human skin lesions with high-speed OCT. Biomed Opt Express 3(10):2636–2646

Boas DA (2010) Laser speckle contrast imaging in biomedical optics. J Biomed Opt 15(1):011109

Brangwynne CP, Koenderink GH, MacKintosh FC, Weltz DA (2008) Cytoplasmic diffusion: molecular motors mix it up. J Cell Biol 183(4):583–587

Bressloff PC, Newby JM (2013) Stochastic models of intracellular transport. Rev Mod Phys 85(1):135–196

Byers RA, Maiti R, Danby SG, Pang EJ, Mitchell B, Carré MJ, Lewis R, Cork MJ, Matcher SJ (2018) Sub-clinical assessment of atopic dermatitis severity using angiographic optical coherence tomography. Biomed Opt Express 9(4):2001–2017

Cadotte DW, Mariampillai A, Cadotte A, Lee KKC, Kiehl TR, Wilson BC, Fehlings MG, Yang VXD (2012) Speckle variance optical coherence tomography of the rodent spinal cord: *in vivo* feasibility. Biomed Opt Express 3(5):911–919

Carter HH, Gong P, Kirk RW, Es'haghian S, Atkinson CL, Sampson DD, Green DJ, McLaughlin RA (2016) Optical coherence tomography in the assessment of acute changes in cutaneous vascular diameter induced by heat stress. J Appl Physiol 121(4):965–972

Chen Z, Liu M, Minneman M, Ginner L, Hoover E, Sattmann H, Bonesi M, Drexler W, Leitgeb RA (2016) Phase-stable swept source OCT angiography in human skin using an akinetic source. Biomed Opt Express 7(8):3032–3048

Chen FK, Viljoen RD, Bukowska DM (2015) Classification of image artifacts in optical coherence tomography angiography of the choroid in macular diseases. Clin Exp Ophthalmol 44(5):388–399. https://doi.org/10.1111/ceo.12683

Chen CL, Wang RK (2017) Optical coherence tomography based angiography [invited]. Biomed Opt Express 8(2):1056–1082

Choi WJ, Reif R, Yousefi S, Wang RK (2014) Improved microcirculation imaging of human skin *in vivo* using optical microangiography with a correlation mapping mask. J Biomed Opt 19(3):036010

Choi WJ, Wang RK (2014) In vivo imaging of functional microvasculature within tissue beds of oral and nasal cavities by swept-source optical coherence tomography with a forward/side-viewing probe. Biomed Opt Express 5(8):2620–2634

Choi WJ, Wang RK (2015) Swept-source optical coherence tomography powered by a 1.3-μm vertical cavity surface emitting laser enables 2.3-mm-deep brain imaging in mice *in vivo*. J Biomed Opt 20(10):106004

de Boer JF, Hitzenberger CK, Yasuno Y (2017) Polarization sensitive optical coherence tomography – a review [invited]. Biomed Opt Express 8(3):1838–1873

de Carlo TE, Romano A, Waheed NK, Duker JS (2015) A review of optical coherence tomography angiography (OCTA). Int J Retina Vitreous 1:5

Deegan AJ, Talebi-Liasi F, Song S, Li Y, Xu J, Men S, Shinohara MM, Flowers ME, Lee SJ, Wang RK (2018a) Optical coherence tomography angiography of normal skin and inflammatory dermatologic conditions. Lasers Surg Med 50(3):183–193

Deegan AJ, Wang W, Men S, Li Y, Song S, Xu J, Wang RK (2018b) Optical coherence tomography angiography monitors human cutaneous wound healing over time. Quant Imaging Med Surg 8(2):135–150

Drexler W (2004) Ultrahigh-resolution optical coherence tomography. J Biomed Opt 9(1):47–74
Drexler W, Fujimoto JG (2007) State-of-the-art retinal optical coherence tomography. Prog Retin Eye Res 27(1):45–88
Dupire J, Socol M, Viallat A (2012) Fully dynamics of a red blood cell in shear flow. Proc Natl Acad Sci U S A 109(51):20808–20813
Enfield J, Jonathan E, Leahy M (2011) In vivo imaging of the microcirculation of the volar forearm using correlation mapping optical coherence tomography (cmOCT). Biomed Opt Express 2(5):1184–1193
Es'haghian S, Gong P, Chin L, Harms K-A, Murray A, Rea S, Kennedy BF, Wood FM, Sampson DD, McLaughlin RA (2017) Investigation of optical attenuation imaging using optical coherence tomography for monitoring of scars undergoing fractional laser treatment. J Biophotonics 10(4):511–522
Fingler J, Readhead C, Schwartz DM, Fraser SE (2018) Phase-contrast OCT imaging of transverse flows in the mouse retina and choroid. Invest Ophthalmol Vis Sci 49(11):5055–5059
Gao SS, Liu G, Huang D, Jia Y (2015) Optimization of the split-spectrum amplitude-decorrelation angiography algorithm on a spectral optical coherence tomography system. Opt Lett 40(10):2305–2308
Gong P, Es'haghian S, Harms KA, Murray A, Rea S, Kennedy BF, Wood FM, Sampson DD, McLaughlin RA (2016a) Optical coherence tomography for longitudinal monitoring of vasculature in scars treated with laser fractionation. J Biophotonics 9(6):626–636
Gong P, Es'haghian S, Harms KA, Murray A, Rea S, Wood FM, Sampson DD, McLaughlin RA (2016b) In vivo label-free lymphangiography of cutaneous lymphatic vessels in human burn scars using optical coherence tomography. Biomed Opt Express 7(12):4886–4898
Gong P, McLaughlin RA, Liew YM, Munro PR, Wood FM, Sampson DD (2014) Assessment of human burn scars with optical coherence tomography by imaging the attenuation coefficient of tissue after vascular masking. J Biomed Opt 19(2):21111
Gora M, Karnowski K, Szkulmowski M, Kaluzny BJ, Huber R, Kowalczyk A, Wojtkowski M (2009) Ultra high-speed swept source OCT imaging of the anterior segment of human eye at 200 kHz with adjustable imaging range. Opt Express 17(17):14880–14894
Grishina O, Wang S, Larina IV (2017) Speckle variance optical coherence tomography of blood flow in the beating mouse embryonic heart. J Biophotonics 10(5):735–743
Huang D, Swanson EA, Lin CP, Schuman JS, Stinson WG, Chang W, Hee MR, Flotte T, Gregory K, Puliafito CA, Fujimoto JG (1991) Optical coherence tomography. Science 254(5035):1178–1181
Hwang TS, Jia Y, Gao SS, Bailey ST, Lauer AK, Flaxel CJ, Wilson DJ, Huang D (2015) Optical coherence tomography angiography features of diabetic retinopathy. Retina 35(11):2371–2376
Jia Y, Tan O, Tokayer J, Potsaid B, Wang Y, Liu JJ, Kraus MF, Subhash H, Fujimoto JG, Hornegger J, Huang D (2012) Split-spectrum amplitude decorrelation angiography with optical coherence tomography. Opt Express 20(4):4710–4725
Kasai C, Namekawa K, Koyano A, Omoto R (1985) Real-time two-dimensional blood flow imaging using an autocorrelation technique. IEEE Trans Sonics Ultrason 32(3):458–464
Khadamy J, Aghdam KA, Falavarjani KG (2018) An update on optical coherence tomography angiography in diabetic retinopathy. J Ophthalmic Vis Res 13(4):487–497
Klein T, Huber R (2017) High-speed OCT light sources and systems. Biomed Opt Express 8(2):828–859
Klein T, Wieser W, Reznicek L, Neubauer A, Kampik A, Huber R (2013) Multi-MHz retinal OCT. Biomed Opt Express 4(10):1890–1908
Lee QZCS, Chao J, Chen C-L, Zhang T, Sharma U, Zhang A, Liu J, Rezaei K, Pepple KL, Munsen R, Kinyoun J, Johnstone M, van Gelder RN, Wang RK (2016) Wide-field optical coherence tomography based microangiography for retinal imaging. Sci Rep 6:22017
Li Y, Choi WJ, Wei W, Song S, Zhang Q, Liu J, Wang RK (2018) Aging-associated changes in cerebral vasculature and blood flow as determined by quantitative optical coherence tomography angiography. Neurobiol Aging 70:148–159

Li J, Lykotrafitis G, Dao M, Suresh S (2007) Cytoskeletal dynamics of human erythrocyte. Proc Natl Acad Sci U S A 104(12):4937–4942

Li T, Raizen MG (2013) Brownian motion at short time scales. Ann Phys 525(4):281–295

Liew YM, McLaughlin RA, Gong P, Wood FM, Sampson DD (2013) *In vivo* assessment of human burn scars through automated quantification of vascularity using optical coherence tomography. J Biomed Opt 18(6):61213

Lindsø Andersen P, Olsen J, Friis KBE, Themstrup L, Grandahl K, Mortensen OS, Jemec GBE (2018) Vascular morphology in normal skin studied with dynamic optical coherence tomography. Exp Dermatol 27(9):966–972

Liu M, Chen Z, Zabihian B, Sinz C, Zhang E, Beard PC, Ginner L, Hoover E, Minneman MP, Leitgeb RA, Kittler H, Drexler W (2016) Combined multi-modal photoacoustic tomography, optical coherence tomography (OCT) and OCT angiography system with an articulated probe for *in vivo* human skin structure and vasculature imaging. Biomed Opt Express 7(9):3390–3402

Liu M, Drexler W (2019) Optical coherence tomography angiography and photoacoustic imaging in dermatology. Photochem Photobiol Sci 18(5):945–962

Makita S, Hong Y, Yamanari M, Yatagi T, Yasuno Y (2006) Optical coherence angiography. Opt Express 14(17):7821–7840

Manfredini M, Greco M, Farnetani F, Ciardo S, de Carvalho N, Mandel VD, Starace M, Pellacani G (2017) Acne: morphologic and vascular study of lesions and surrounding skin by means of optical coherence tomography. J Eur Acad Dermatol Venereol 31(9):1541–1546

Mariampillai A, Standish BA, Moriyama EH, Khurana M, Munce NR, Leung MKK, Jiang J, Cable A, Wilson BC, Vitkin IA, Yang VXD (2008) Speckle variance detection of microvasculature using swept-source optical coherence tomography. Opt Lett 33(13):1530–1532

Markowitz O, Schwartz M, Minhas S, Siegel DM (2016) Speckle-variance optical coherence tomography: a novel approach to skin cancer characterization using vascular patterns. Dermatol Online J 22(4):13030/qt7w10290r

Men SJ, Chen CL, Wei W, Lai TY, Song SZ, Wang RK (2017) Repeatability of vessel density measurement in human skin by OCT-based microangiography. Skin Res Technol 23(4):607–612

Men S, Wong JM, Welch EJ, Xu J, Song S, Deegan AJ, Ravichander A, Casavant B, Berthier E, Wang RK (2018) OCT-based angiography of human dermal microvascular reactions to local stimuli: implications for increasing capillary blood collection volumes. Lasers Surg Med 50(9):908–916

Nicklas RB (1997) How cells get the right chromosomes. Science 275(5300):632–637

Olsen J, Birch-Johansen FH, Themstrup L, Holmes J, Jemec GBE (2018) Dynamic optical coherence tomography of histamine induced wheals. Skin Res Technol 24(4):592–598

Qin J, Jiang J, An L, Gareau D, Wang RK (2011) *In vivo* volumetric imaging of microcirculation within human skin under psoriatic conditions using optical microangiography. Laser Surg Med 43(2):122–129

Sandhu HS, Eladawi N, Elmogy M, Keynton R, Helmy O, Schaal S, El-Baz A (2018) Automated diabetic retinopathy detection using optical coherence tomography angiography: a pilot study. Br J Ophthalmol 102(11):1564–1569

Schuh S, Holmes J, Ulrich M, Themstrup L, Jemec GBE, de Carvalho N, Pellacani G, Welzel J (2017) Imaging blood vessel morphology in skin: dynamic optical coherence tomography as a novel potential diagnostic tool in dermatology. Dermatol Ther(Heidelb) 7(2):187–202

Schwartz DM, Fingler J, Kim DY, Zawadzki RJ, Morse LS, Park SS, Fraser SE, Werner JS (2014) Phase-variance optical coherence tomography: a technique for noninvasive angiography. Ophthalmology 121(1):180–187

Segal SS (2005) Regulation of blood flow in the microcirculation. Microcirculation 12(1):33–45

Sigsgaard V, Themstrup L, Theut Riis P, Olsen J, Jemec GB (2018) In vivo measurements of blood vessels' distribution in non-melanoma skin cancer by dynamic optical coherence tomography – a new quantitative measure? Skin Res Technol 24(1):123–128

Song S, Xu J, Men S, Shen TT, Wang RK (2017) Robust numerical phase stabilization for long-range swept-source optical coherence tomography. J Biophotonics 10(11):1398–1410

Spaide RF, Klancnik JM, Cooney MJ (2015a) Retinal vascular layers imaged by fluorescein angiography and optical coherence tomography angiography. JAMA Ophthalmol 133(1):45–50
Spaide RF, Fujumoto JG, Waheed NK (2015b) Image artifacts in optical coherence tomography angiography. Retina 35(11):2163–2180
Srinivasan VJ, Radhakrishnan H (2014) Optical coherence tomography angiography reveals laminar microvascular hemodynamics in the rat somatosensory cortex during activation. NeuroImage 102(Pt2):393–406
Stucker M, Baier V, Reuther T, Hoffmann K, Kellam K, Altmeyer P (1996) Capillary blood cell velocity in human skin capillaries located perpendicularly to the skin surface: measured by a new laser Doppler anemometer. Microvasc Res 52(2):188–192
Tan ACS, Tan GS, Denniston AK, Keane PA, Ang M, Milea D, Chakravarthy U, Cheung CMG (2018) An overview of the clinical applications of optical coherence tomography angiography. Eye 32(2):262–286
Themstrup L, de Carvalho N, Nielsen SM, Olsen J, Ciardo S, Schuh S, Nørnberg BM, Welzel J, Ulrich M, Pellacani G, Jemec GBE (2018) In vivo differentiation of common basal cell carcinoma subtypes by microvascular and structural imaging using dynamic optical coherence tomography. Exp Dermatol 27(2):156–165
Themstrup L, Pellacani G, Welzel J, Holmes J, Jemec GBE, Ulrich M (2017) In vivo microvascular imaging of cutaneous actinic keratosis, Bowen's disease and squamous cell carcinoma using dynamic optical coherence tomography. J Eur Acad Dermatol Venereol 31(10):1655–1662
Themstrup L, Welzel J, Ciardo S, Kaestle R, Ulrich M, Holmes J, Whitehead R, Sattler EC, Kindermann N, Pellacani G, Jemec GBE (2016) Validation of dynamic optical coherence tomography for non-invasive, in vivo microcirculation imaging of the skin. Microvasc Res 107:97–105
Thompson IA, Durrani AK, Patel S (2019) Optical coherence tomography angiography characteristics in diabetic patients without clinical diabetic retinopathy. Eye(Lond) 33(4):648–652
Ulrich M, Themstrup L, de Carvalho N, Manfredini M, Grana C, Ciardo S, Kästle R, Holmes J, Whitehead R, Jemec GB, Pellacani G, Welzel J (2016) Dynamic optical coherence tomography in dermatology. Dermatology 232(3):298–311
Wang RK (2010) Optical microangiography: a label free 3D imaging technology to visualize and quantify blood circulations within tissue beds *in vivo*. IEEE J Sel Top Quantum Electron 16(3):545–554
Wang RK, An L (2009) Doppler optical micro-angiography for volumetric imaging of vascular perfusion *in vivo*. Opt Express 17(11):8926–8940
Wang H, Baran U, Wang RK (2015) In vivo blood flow imaging of inflammatory human skin induced by tape stripping using optical microangiography. J Biophotonics 8(3):265–272
Wang RK, Jacques SL, Ma Z, Hurst S, Hanson SR, Gruber A (2007) Three dimensional optical angiography. Opt Express 15(7):4083–4097
Wang H, Magnain C, Sakadzic S, Fischl B, Boas DA (2017) Characterizing the optical properties of human brain tissue with high numerical aperture optical coherence tomography. Biomed Opt Express 8(12):5617–5636
Wang RK, Zhang A, Choi WJ, Zhang Q, Chen CL, Miller A, Gregori G, Rosenfeld PJ (2016) Wide-field optical coherence tomography angiography enabled by two repeated measurements of B-scans. Opt Lett 41(10):2330–2333
Wei W, Choi WJ, Wang RK (2018) Microvascular imaging and monitoring of human oral cavity lesions in vivo by swept-source OCT based angiography. Lasers Med Sci 33(1):123–134
Xu J, Han S, Balaratnasingam C, Mammo Z, Wong KS, Lee S, Cua M, Young M, Kirker A, Albiani D et al (2015) Retinal angiography with real-time speckle variance optical coherence tomography. Br J Ophthalmol 99(10):1315–1319
Yun SH, Tearney GJ, de Boer JF, Bouma BE (2004) Motion artifacts in optical coherence tomography with frequency-domain ranging. Opt Express 12(13):2977–2998
Zhang J, Chen Z (2015) *In vivo* blood flow imaging by a swept laser source based Fourier domain optical Doppler tomography. Opt Express 13(19):7449–7457

Zhang M, Hwang TS, Campbell JP, Bailey ST, Wilson DJ, Huang D, Jia Y (2016) Projection-resolved optical coherence tomographic angiography. Biomed Opt Express 7(3):816–828
Zhang Q, Rezaei KA, Saraf SS, Chu Z, Wang F, Wang RK (2018) Ultra-wide optical coherence tomography angiography in diabetic retinopathy. Quant Imaging Med Surg 8(8):743–753
Zhang A, Zhang Q, Chen CL, Wang RK (2015) Methods and algorithms for optical coherence tomography-based angiography: a review and comparison. J Biomed Opt 20(10):100901
Zhang Q, Zhang A, Lee CS, Lee AY, Rezaei KA, Roisman L, Miller A, Zheng F, Gregori G, Durbin MK, An L, Stetson PF, Rosenfeld PJ, Wang RK (2017) Projection artifact removal improves visualization and quantitation of macular neovascularization imaged by optical coherence tomography angiography. Ophthalmol Retina 1(2):124–136

Spectral Reflectometry in Biomedical Imaging and Sensing

Junhwan Kwon, Yongjae Jo, and Myunghwan Choi

Introduction

A soap bubble blower has long been a beloved toy for kids because of the iridescent colors of the bubbles. Their luminous colors dynamically change when seen from different angles. These visually fascinating colors are found in multiple areas such as in oil slicks in gasoline stations and the cockpit window in airplanes (Fig. 1a-c). These examples are also commonly observed in nature including the patches of the peacock butterfly, the gloss of the buttercup flower, and most famously, the feathers of peacocks (Fig. 1d-f). These iridescent colors in natural plants and animals often provide evolutionary benefits by attracting mates or startling predators.

Since it is prevalent in nature, physicists have continually attempted to explain the underlying physical mechanism of iridescence. Although, the earliest study reported by Robert Hooke in 1665 postulated that the iridescence of peacock feathers might be from thin alternating layers. Most of these early studies tried to explain the iridescence by surface colors produced by special dyes or pigments. In 1919, Lord Rayleigh proposed the concept of structural color which states that iridescence can be formed by microscopic structures without dyes or pigments (Kinoshita et al. 2008). In 1925, Ernest Merritt first described the theory of thin-film interference as a physical mechanism of iridescence (refer to the next section, "the principle of spectral reflectometry" for details). Subsequently, experimental evidences were provided by electron micrographs on the peacock feathers and morpho butterfly wings, revealing multilayered nanoscale structures (Kinoshita et al. 2008).

[a]These authors equally contributed to this work.

J. Kwon · Y. Jo
Department of Biomedical Engineering, Sungkyunkwan University, Suwon, South Korea

M. Choi (✉)
School of Biological Sciences, Seoul National University, Seoul, South Korea

J. K. Kim et al. (eds.), *Advanced Imaging and Bio Techniques for Convergence Science*, Advances in Experimental Medicine and Biology,
https://doi.org/10.1007/978-981-33-6064-8_13

Fig. 1 Examples of thin-film interference. (**a**) A thin membrane of a soap bubble. (**b**) An oil slick floating on water. (**c**) Airbus windshield coated with an indium tin oxide film. (**d**) Blue wing patches of the peacock butterfly. (**e**) Gloss of buttercup flower. (**f**) Breast feathers of the bird of paradise. The photographs are permitted for reuse with a license under the CC 3.0

Theoretical understanding of thin-film interference allowed the designing of artificial thin films to have iridescent colors in a controlled manner. Recent advances in nanofabrication techniques, such as chemical vapor deposition and physical vapor deposition, have led to the generation of artificial multilayered thin films. Currently, a variety of optical filters and mirrors designed to transmit or reflect various wavelengths are commercially available and are serving as an essential building block for modern microscopy. Antireflective coating on lenses or glasses is also widely utilized to reduce reflection and improve transmission.

In reverse, spectroscopic analysis of iridescent reflected color can provide nanoscale information on multilayered transparent thin films. This technique, named spectral reflectometry, acquires the spectrum of reflected light (intensity versus wavelength), and a theoretical model is fit to the spectral data to extract the thicknesses of the thin films (Hlubina et al. 2008). For fitting, optical constants (refractive index and absorbance) for each layer should be provided a priori. However, if the optical constants are unknown, spectral ellipsometry can be used instead (Fujiwara et al. 2007). Spectral reflectometry is widely used in silicon industries to inspect the quality of thin-film coatings on a silicon wafer (Hlubina et al. 2006).

Recently, spectral reflectometry has gained considerable interest in biology and medicine because it provides nanoscale information without exogenous labeling and does not require a phototoxic light dose. Its applications can be broad, ranging from molecular sensing to in vivo imaging. In this chapter, we comprehensively discuss the relevant topics on spectral reflectometry in biomedicine from the principle to applications.

Physical Principle

Thin-film interference is a natural optical phenomenon in which light waves reflected at the interfaces of multilayered thin films interfere with one another, either constructively or destructively, to redistribute the reflected and transmitted waves. Depending on the thicknesses and refractive indices of the thin films, interference patterns in the spectral domain (i.e., iridescence) can be observed with a broadband input source. Inversely, structural information, such as thicknesses of thin films, can be obtained at nanoscale precision by analyzing the reflectance spectrum. This optical measurement technique is called "spectral reflectometry" or "spectral interferometry."

Thin-film interference can be explained by reflection and interference of light waves. Reflection at the interfaces of distinct dielectric thin films can be described using the Fresnel equation, according to which reflectance intensity (R) depends on the polarization (s, p), incident angle (θ_i), and refractive indices (n_i). The Fresnel equation is described as

$$R_s = \left| \frac{n_1 \cos\theta_i - n_2 \cos\theta_t}{n_1 \cos\theta_i + n_2 \cos\theta_t} \right|^2$$

$$R_p = \left| \frac{n_1 \cos\theta_t - n_2 \cos\theta_i}{n_1 \cos\theta_t + n_2 \cos\theta_i} \right|^2$$

where n_1 and n_2 are the refractive indices of the two media and θ_t is the refracted angle.

Next, interferences among the reflected and transmitted light waves should be considered, which can be formulated by the transfer matrix method (Lu et al. 2014; Jo et al. 2018). This method is based on the continuity conditions for the electromagnetic fields in the Maxwell equation. Thus, if the input electromagnetic field is known at the beginning of a layer, the field at the end of the layer can be derived from a simple matrix operation. If the input electric and magnetic fields at position z are $E(z)$ and $H(z)$, the resultant fields after propagating a stack of thin films with distance L can be calculated by matrix operation with a system matrix M as follows:

$$\begin{pmatrix} E(z+L) \\ H(z+L) \end{pmatrix} = \begin{pmatrix} M_{11} & M_{12} \\ M_{21} & M_{22} \end{pmatrix} \cdot \begin{pmatrix} E(z) \\ H(z) \end{pmatrix} = M \cdot \begin{pmatrix} E(z) \\ H(z) \end{pmatrix}$$

The system matrix of the thin films can be represented as the product of the matrices representing each individual layer (M_i):

$$M = M_N \cdot M_{N-1} \cdot \ldots \cdot M_2 \cdot M_1$$

where N is the number of layers. The reflectance spectrum (R) can then be obtained from the two entities of the system matrix:

$$R = \left| \frac{M_{21}}{M_{11}} \right|^2$$

A more detailed derivation and formulas can be found in previous literatures on thin-film interference (e.g., Chap. 9 of "Optics" by Eugene Hecht). More generalized wave simulations can also be performed by scalar or vector diffraction theories (Jo et al. 2018).

Optical Labeling

As biomolecules exist in a complex mixture in a living system, technologies to label target molecules of interest have played a critical role in biomedical research. Among the various optical labeling techniques proposed, the current gold standard is fluorescence labeling because it offers high specificity and sensitivity with a variety of readily available fluorophores. However, the number of simultaneously observable colors (i.e., multiplexing capacity) is typically limited to several entities due to the broad spectral emission in fluorescence (Resch-Genger et al. 2008).

Various approaches have been proposed to improve the multiplexing capacity, among which spectral reflectance has recently gained interest as an optical label (Fu et al. 2016). Since iridescent colors from structured dielectric materials are highly sensitive to nanoscale geometry and material compositions, the reflectance spectrum can serve as an optical barcode for highly multiplexed labeling (Jo et al. 2018). An additional compelling benefit of spectral reflectance compared with fluorescence is that the structural color is from reflection, so it is nearly invulnerable to photobleaching, which provides practically everlasting labeling or infinite readouts.

One famous example is the use of photonic crystals, which refer to periodic nanostructures made of dielectric materials. Due to the periodicity in dielectric properties, they exhibit photonic bandgaps in which light wave propagation is forbidden (Wu et al. 2019). Thus, when white light impinges on a photonic crystal, the light waves corresponding to the photonic bandgap are reflected, resulting in an iridescent sheen. By varying the periodicity in the photonic crystal, the bandgap can be tuned to generate various colors. Previously, photonic crystals were fabricated in rectangular shapes by low-throughput photolithography (Cunin et al. 2002), but recently, they are being produced as colloidal particles by self-assembly for high-throughput production (Shang et al. 2015). These colloidal self-assembly methods provide facile tuning of the structural periodicity for multiplexed labeling of more than 10 entities. These colloidal particles have been applied in various biological fields, for example for observing circulating tumor cells (Zheng et al. 2014), biodistribution of pathogenic bacteria, cell–material interactions, and enzyme cascade reactions (Xu et al. 2018; Liu et al. 2013; Wang et al. 2018).

Higher degree multiplexing could be attained by spatial barcoding, such as photo-patterned fluorescent microspheres (Braeckmans et al. 2003), reflective metallic barcodes, and embossed 3D barcodes (Nicewarner-pen et al. 2001; Gómez-Martínez et al. 2013). However, these approaches require a relatively large volume for the labels, thus restricting their utility in biological systems. Alternatively, color-coding has been adapted to obtain high-degree multiplexing with smaller volumes. For example, Kim et al. demonstrated a photonic ball encapsulating several kinds of photonic crystals, and Zhao et al. developed multicolored Janus particles with spatially varying wettability (Zhao et al. 2012). Lee et al. integrated barcoding and color coding to dramatically enhance the multiplexing capacity (Lee et al. 2010) by printing photonic crystals on binary codes (Fig. 2a, b). In this work, the photonic bandgap was tunable using an external magnetic field so that various colors could be easily generated with a single material (Fig. 2c). The multiplexing capacity in this approach can be logarithmically increased (multiplexing capacity = $2^{\text{color} \times \text{bits}}$).

Recently Jo et al. reported an interferometric optical probe, named reflectophore (Jo et al. 2018). Contrary to most previous approaches, off-the-shelf dielectric microspheres, such as polystyrene or silicon beads, can function as reflectophores without further modification. The minute differences in the sizes of reflectophores form the individual labels so that readout precision for the size determines multiplexing capacity, which is mediated by self-interference. Focused light incident onto the center of a microsphere causes thin-film interference, which is geometrically analogous to interference on a Fabry–Pérot cavity (Fig. 3a). By analyzing the phase and frequency of the measured spectrum, the diameter can be determined with a readout precision of ~1 nm (Fig. 3b, c). This precision provides a multiplexing capacity of ~1000 in a size span of 1 μm, and this multiplexing capacity was used to demonstrate the longitudinal tracing of multiple migrating cancer cells (Fig. 3d).

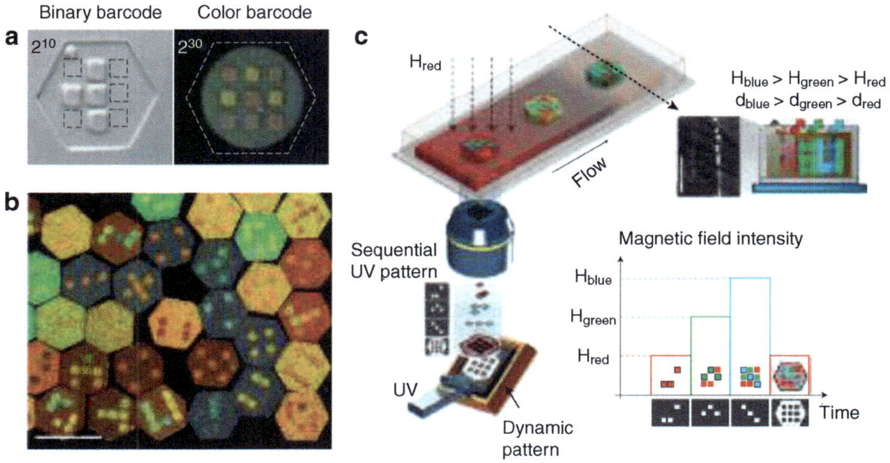

Fig. 2 Color-barcoded magnetic particles based on photonic crystals. (**a**) 10-bit binary markers (left) and 30-bit color barcodes (right). (**b**) Various magnetic color barcodes. Scale bar, 500 μm. (**c**) Schematics for microfluidic particle production. Colors on the particles were tuned by modulating the magnetic field. Reproduced from Ref. Lee et al. (2010) with permission

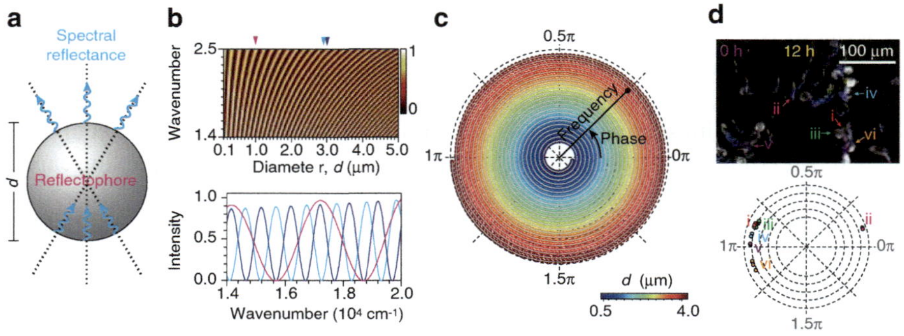

Fig. 3 Reflectophore. (**a**) Self-interference in reflectophore. (**b**) Simulated reflectance spectra with different diameters. (**c**) Phasor representation of the simulated spectrums. (**d**) Long-term tracking of migrating cancer cells labeled with reflectophores. Gray color indicates the Mitotracker fluorescence, and pseudo-colored traces are time series of cellular movements. Reproduced from Ref. Jo et al. (2018) with permission

Biosensors

Because spectral reflectance is highly sensitive to structural change, it has been adapted to sense specific molecules or environmental changes, representative examples of which include multilayered stimuli-responsive hydrogels and colloidal photonic crystals. Various types of stimuli-responsive hydrogels are available that can swell or shrink in response to specific external physicochemical stimuli. The physical changes modify the photonic bandgap, which can be readily obtained by spectral detection (Burgess 2013). Various sensors based on spectral reflectance have been fabricated, such as thermometers (Liu et al. 2015), pH meters (Hu et al. 2012), photodetectors (Zhang et al. 2014), glucose meters (Nakayama et al. 2003), and DNA sensors (Zhao et al. 2010a, b) (Fig. 4a, b).

Among the various sensors, mechanical sensors have gained interest because of their exquisite sensitivity and utility in studying cellular physiology (Zhang and Wang 2018; Park et al. 2018). For measuring intracellular forces, Gómez-Martínez et al. fabricated a multilayered Fabry–Pérot sensor, which is tiny enough to be introduced into a single cell. The sensor has an internal vacuum cavity, which is compressed with respect to the intracellular pressure, resulting in a spectral shift (Fig. 4c, d). Using this sensor, they found that intracellular pressure is maintained even under a strong osmotic shock.

Sensors based on thin-film interference inevitably suffer from orientation sensitivity as the Fresnel reflection is dependent on the incident angle (Gao et al. 2015; Park et al. 2018; Zhang and Wang 2018). On the contrary, reflectophores based on dielectric microspheres are free from the orientation issue due to their spherical geometry (Jo et al. 2018). Using biotin-coated reflectophores, adhesion of streptavidin proteins onto the reflectophore surface could be readily detected (Fig. 5a, b). In addition, using a reflectophore made of a nematic liquid crystal, the local electric field could be sensed (Fig. 5c).

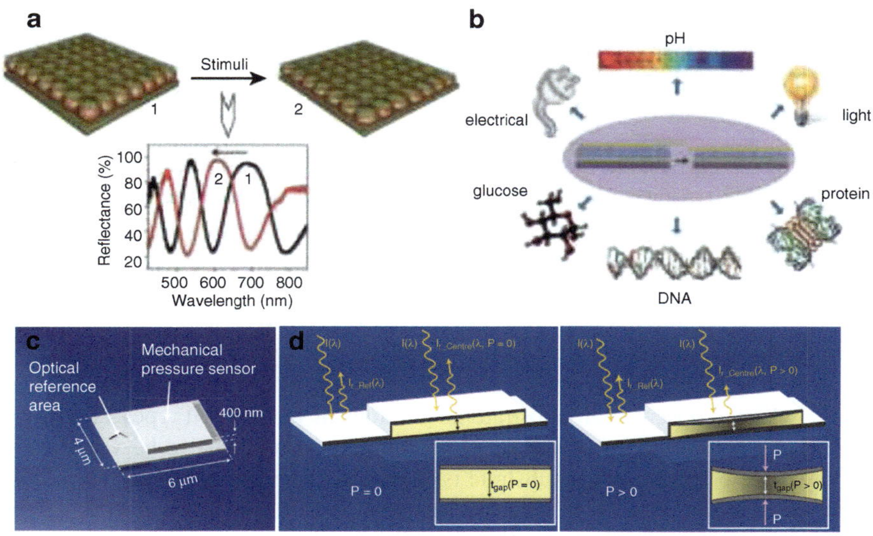

Fig. 4 Sensors based on spectral reflectometry (**a**) Stimuli-responsive etalon colorimetric sensors. Stimuli-responsive hydrogel microspheres are aligned between the gold plates. (**b**) Various kinds of external stimuli for sensing. (**c**) Intracellular pressure sensor chip manufactured by a patterning process. (**d**) Schematics of the working principle of the intracellular pressure sensor. Images are reproduced from Refs. Gao et al. (2015) and Gómez-Martínez et al. (2013) with permission

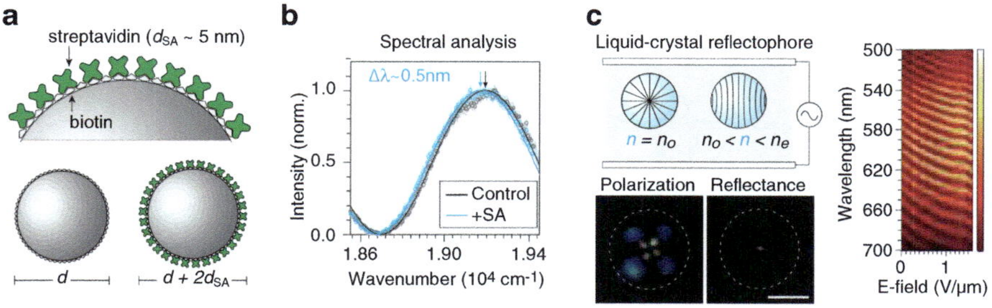

Fig. 5 Microenvironmental sensing. (**a**) Schematic illustration for molecular sensing. (**b**) Spectral reflectance measurements on control and streptavidin bound reflectophores. The arrows correspond to spectral peaks. (**c**) Measurement of the electric field with a self-assembled liquid crystal reflectophore. Reproduced from Ref. Jo et al. (2018) with permission

Imaging Cell–Substrate Interactions

Living cells dynamically interact with the external physicochemical environment. Various types of optical microscopies, such as phase-contrast and fluorescence microscopy, have been developed to study cellular dynamics; however, certain interactions with spatial scales smaller than the optical diffraction limit (~200 nm) cannot be studied via conventional microscopy. For example, cells cultured in a dish make focal contacts on the surface and also exert forces on the dish. These dynamic cell–substrate interactions typically occur at the nanoscales and thus cannot be studied using conventional microscopes. Interference microscopy based on thin-film interference provided a solution to this problem with its diffraction-unlimited precision.

One of the earliest applications of interference microscopy (Verschueren 1985; Hwang et al. 2015) was for the measurement of the distance between a cultured cell and a glass substrate (Fig. 6a). The cell–substrate distance, which is the gap between the cell membrane and the glass substrate, is typically in the order of 100 nm. Although this thickness could not be resolved by conventional diffraction-limited microscopes, an interference reflection microscope (IRM) could generate a sensitive contrast in reflectivity by constructive or destructive interference (Fig. 6b, c). Because of its simple optics, nanoscale sensitivity, and noninvasiveness, IRM has been applied to observe cell-substrate interactions in various cell types.

Cellular forces involved in cell–substrate interactions can also be visualized if structural deformation in the substrate is detectable. The cellular force exerted on a substrate was first visualized by culturing cells on a thin silicon substrate (Harris and Patricia Wild 1980) (Fig. 7). Because of the high elasticity of thin-layered silicon, even minute cellular forces generated wrinkles on the substrate, which can provide a qualitative cellular force map in a conventional microscope. Quantitative approaches based on a micropillar array or a substrate labeled with fluorescent beads (i.e., traction-force microscopy) were subsequently introduced and widely utilized in the mechanobiology field (Munevar et al. 2001).

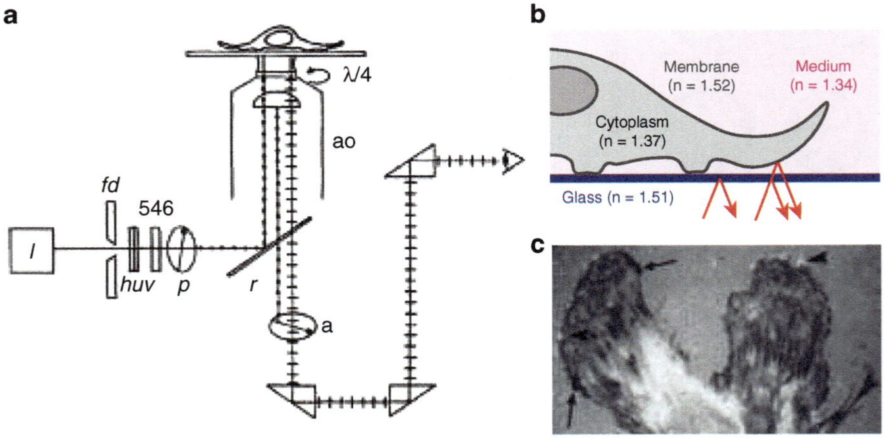

Fig. 6 One of the earliest designs of interferometric microscopy for observing cell–substrate interaction. (**a**) An optical layout to measure interference patterns on cultured cells. (**b**) Illustration for cultured cell morphology and optical indices. (**c**) Cultured mouse embryo cells imaged by interferometric microscopy. The white fringe at the edge of the cell is a result of the close contact area (arrowhead). The black fringe is a result of the destructive interference in the case of the thin lamellar cytoplasm (arrows). Images reproduced from Ref. Verschueren (1985) with permission

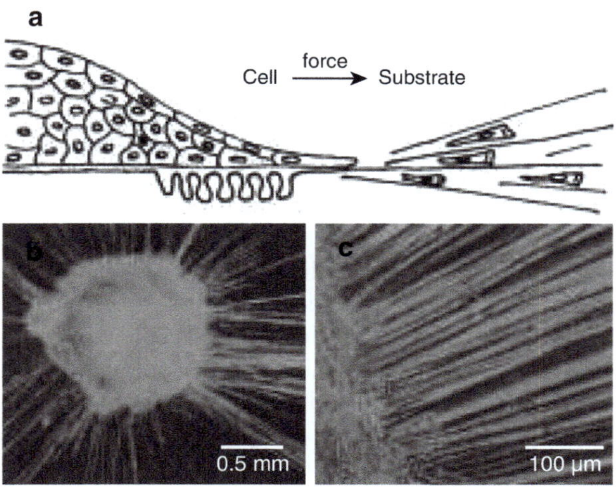

Fig. 7 Wrinkles resulting from cell–substrate interaction. (**a**) A schematic side-view of distorting and wrinkling of the elastic silicone substrate during outgrowing of cells. (**b**) Chick heart fibroblast imaged by dark-field microscopy, (**c**) Higher magnification on the edge of the cell in (**b**). Images reproduced from Ref. Harris and Patricia Wild (1980) with permission

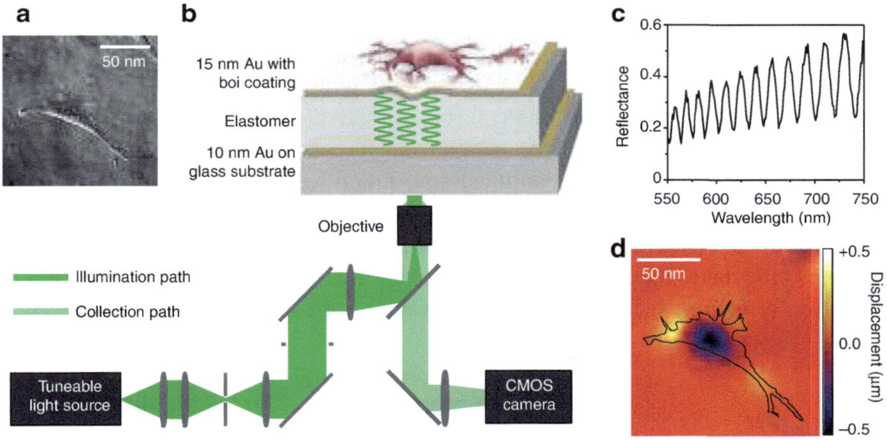

Fig. 8 ERISM configuration and demonstration (**a**) Cultured cell imaged by phase-contrast microscopy. (**b**) Illustration of ERISM using a biological sample. (**c**) Representative acquired reflectance spectrum at one point. (**d**) Deformation map on elastic substrate acquired by spectrum. Images reproduced from Ref. Kronenberg et al. (2017) with permission

The previous quantitative approaches for measuring cell–substrate interaction had a poor resolution along the axial direction, restricting their utility only to shear forces. To overcome this limitation, thin-film interference has recently been adapted by culturing cells on an elastic micro-cavity and measuring its deformation by interference microscopy (Kronenberg et al. 2017). This technique, named elastic resonator interference stress microscopy (ERISM), quantitatively measures the nanoscale deformation in a micro-cavity by using a broadband light source and via spectroscopic analysis. By measuring the axial displacement at the nanoscale, ERISM can provide an exquisite force sensitivity of ~1 pN along a vertical axis (Fig. 8).

Imaging Neuronal Structures

Imaging has played a pivotal role for novel discoveries in neuroscience by providing structural, functional, and molecular information (Y Cajal and Swanson 1995) (Tian et al. 2009) (Shi et al. 1991). So far, the major player has been fluorescence microscopy with a variety of available fluorophores, but label-free microscopes, such as nonlinear harmonic microscopies, are slowly finding specific roles in neuroscience studies (Lim et al. 2014a, b; Mascaro et al. 2015). Recently, interference microscopy based on spectral reflectance has been suggested to be useful for studies on myelinated axons (Schain et al. 2014; Kwon et al. 2017).

Several years ago, Schain et al. first reported that myelinated axons in the mammalian nervous system exhibit efficient spectral reflectance because of their multilayered thin-film cytoarchitectures (Schain et al. 2014) (Fig. 9). Using spectral confocal reflectance microscopy (SCoRe), they could image living myelinated axons using labels. SCoRe uses several discrete laser lines to obtain colorful iridescence from the myelinated axons and also worked on a human brain explant (Fig. 9b, c). Independently, Mascaro et al. showed that a near-infrared femtosecond laser typically used in two-photon microscopy provides a similar reflectance contrast in the myelinated axons (Mascaro et al. 2015). Later, Fei et al. showed that SCoRe imaging of subcortical layers (~1.3 mm depth) is feasible with an infrared light source of ~1700 nm (Xia et al. 2018).

Kwon et al. further revealed that spectral analysis of the iridescence from myelinated axons enables nanoscale quantifications of their subcellular structures (Fig. 10) (Kwon et al. 2017). This technique, termed spectral reflectometry (SpeRe), is based on thin-film interference occurring at multilayered subcellular structures. SpeRe acquires reflectance spectra from the geometric centers of an axon fiber and extracts spatial information, such as axon diameter and g-ratio (the ratio of the inner axon

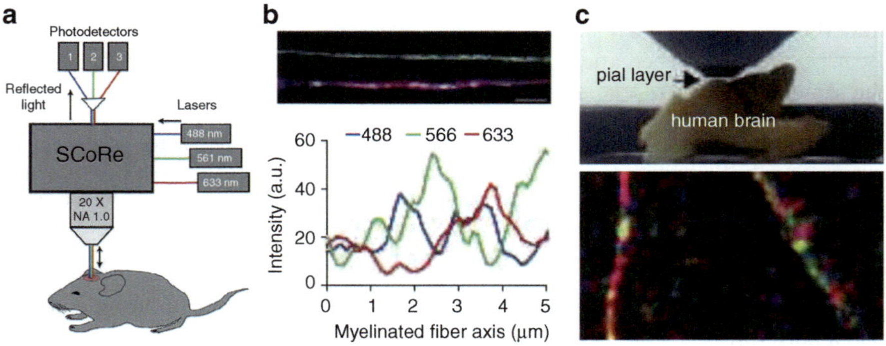

Fig. 9 Vertebrate nervous imaging using SCoRe. (**a**) Illustration of in vivo SCoRe imaging of mouse cortex. (**b**) Representative SCoRe images with colorful patterns of the sciatic nerve (above) and spectrum along the myelinated axon (below). (**c**) Demonstration of SCoRe imaging on paraformaldehyde-fixed human postmortem human brain explant (above) and representative SCoRe image (below). Images reproduced from Ref. Schain et al. (2014) with permission

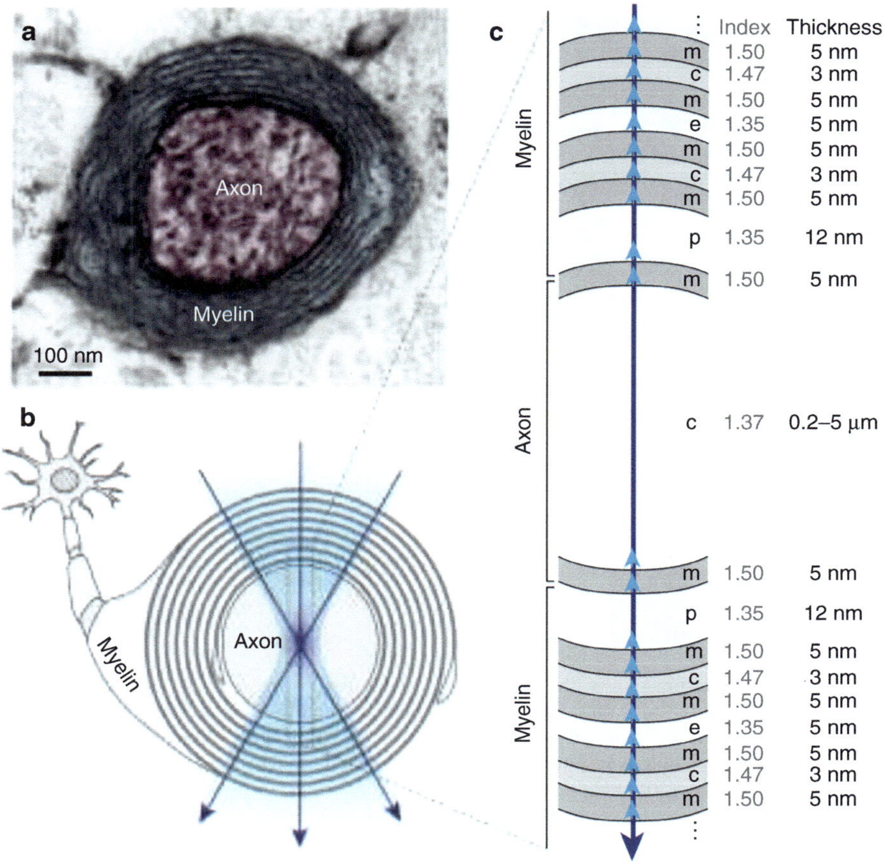

Fig. 10 Myelin morphology and optical structure. (**a**) Myelinated axon in mouse brain imaged by electron microscopy. (**b**) Cross-sectional view of a myelinated axon, with an interrogating focused beam. (**c**) Physiological thin-film model and structure scheme with corresponding refractive indices and thicknesses. Images reproduced from Ref. Kwon et al. (2017) with permission

diameter to the total outer diameter), based on a simulated database. This approach allows nanoscale reconstruction of a 3-dimensional map of myelinated axons and tracking of physiological osmotic swelling of the myelin (Fig. 11).

Because SpeRe provides label-free nanoscale information, it has great potential for clinical applications, such as for diagnosing demyelinating diseases or traumatic axonal injury. To demonstrate its applicability in diagnosing axonal injury, Kwon et al. performed longitudinal SpeRe imaging on the brain cortex of a live mouse before and after introducing a mild compressive injury. They found that the injured axons often showed enlarged axonal diameter, and the degree of swelling could be quantified based on spectral periodicity (Fig. 11).

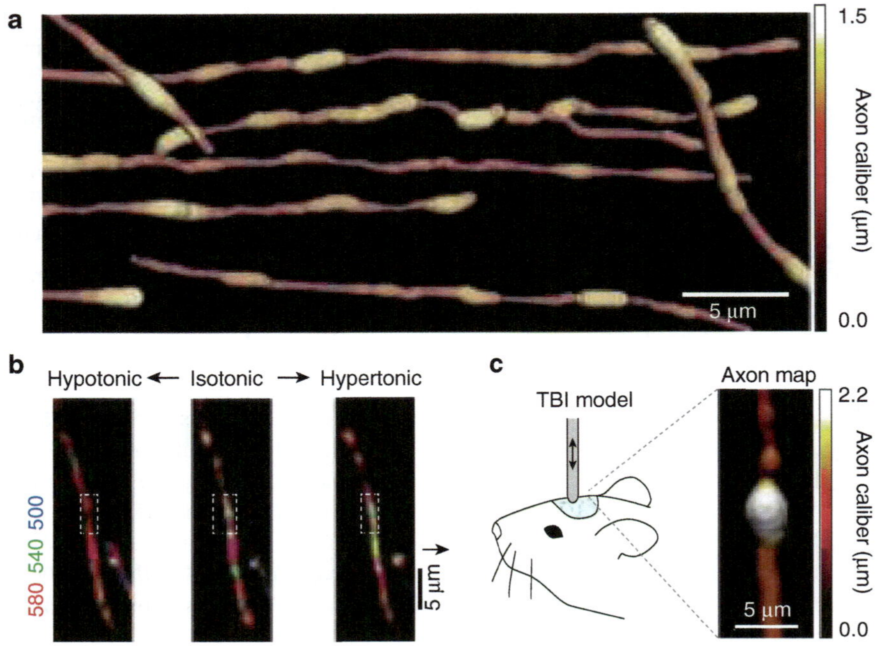

Fig. 11 Nanoscale imaging with SpeRe in a living tissue and animal. (**a**) A volumetric 3-dimensional view of the axon rendered using axons extracted from the reflectance spectrum of the myelinated axon. (**b**) Nanoscale sensing during osmotic challenges with representative SpeRe images. (**c**) Axonal pathology in a mechanical injury model detected by SpeRe. Images reproduced from Ref. Kwon et al. (2017) with permission

Clinical Imaging

A tear film on the cornea is composed of three layers, lipid, aqueous, and mucin. The lipid layer is the outermost layer produced by the meibomian gland (Mishima and Maurice 1961) and plays important physiological roles including lubrication of the ocular surface and prevention of aqueous layer evaporation (Yokoi et al. 2008). Thus, the thickness of the lipid layer is a valuable diagnostic indicator for dry eye syndrome or other dysfunctions of the meibomian glands (Bai et al. 2017). Its physiological thickness is typically in the order of a hundred nanometers, and it forms a thin film floating on the aqueous layer of ~50 μm (Khamene et al. 2000). Additionally, because of the fragility of the tear film, its thickness should be measured in situ without physical perturbation.

The tear film was first measured by Mishima et al., who instilled saline fluorescein on the cornea. However, this technique failed to be widely adapted due to limited measurement precision and invasiveness. The first noninvasive measurement was demonstrated in 1989 by Doane, using optical interferometry. Although, Doane successfully measured physiological dynamics of the tear film, the technique required a dry reference area or color clues that were cumbersome for wide adaptation. In 1998, Fogt and King-Smith first introduced spectral reflectometry for tear film measurement, which successfully solved issues associated with invasiveness, precision, and simplicity. By offering noncontact, nanoscale quantification in situ, spectral reflectometry attracted great interest in the field of ophthalmology and was rapidly adapted into the clinic (Fig. 12).

Conventional approaches have estimated the thickness of the lipid layer from the ratio of intensities reflected at different wavelengths, based on the fact that the reflectance spectrum from thin films produces sinusoidal functions of the wavenumber (Doane 1989). Typically, a color camera with red, green, and blue channels is

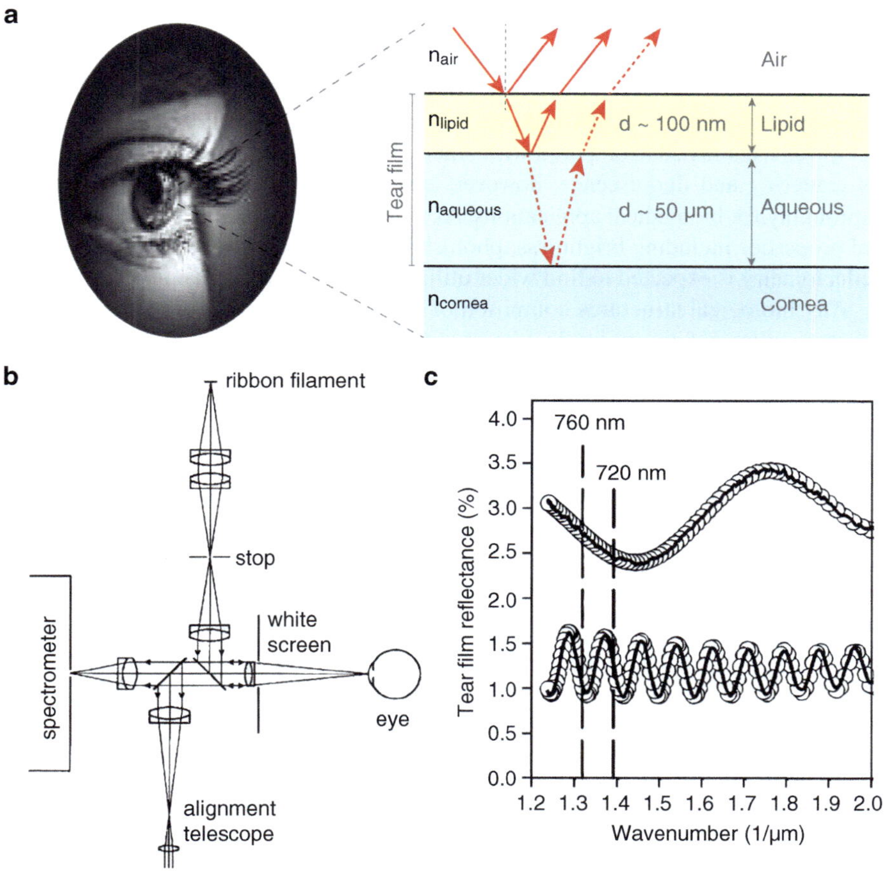

Fig. 12 Spectral reflectometry on a human tear film. (**a**) A schematic representation of the two-layer model of a tear film on a human cornea. The photograph on the left is permitted for reuse with a license under CC 3.0. (**b**) The first spectral reflectometric setup was designed by Fogt and King-Smith in 1998. (**c**) A representative spectrograph of human tear films measured by the optic setup is shown in (**b**). The solid lines represent the sinusoidal fit to the measured data points. Images reproduced from Ref. Fogt and King-smith (1998) with permission

used to image the colorful interference fringe, and the ratio between channels is fitted to a sinusoidal function to estimate the thickness of the lipid layer. Recent approaches utilize a spectrometer to minimize the measurement errors produced by the fitting with three colors. A number of optical designs have also been proposed to image the interference fringes by minimizing the background from the cornea, such as introducing a polarization-dependent optical isolator to remove back-reflected signals from the cornea (Josephson 1983; Hamano 1979; Khamene et al. 2000). For versatile uses in the clinic, compact and flexible spectral reflectometers based on fiber optics are also available.

Outlook

Beginning with the basic physical understanding of beautiful iridescence observed in peacock feathers, spectral reflectometry has impacted multiple fields from the silicon industry to biomedicine. Theoretical understanding of thin-film interference has been well established. Compared with other popular optical contrast techniques such as scattering and fluorescence, however, spectral reflectance is yet to be explored thoroughly for biomedical applications. As it has unique advantages in terms of optical properties including brightness, photostability, and nanoscale precision, spectral reflectometry is expected to find wider utilization for biomedicine in the near future.

Any biological structures composed of thin dielectric layers are potential areas for applications of spectral reflectometry for biomedical imaging. So far, it has been successfully applied to the imaging of myelinated axons for neuroscience, thus providing their nanoscale dynamics in the living state, and tear films for ophthalmology, serving as a clinical diagnostic tool for dry eyes and other dysfunctions of the meibomian glands. Compared with other imaging modalities such as fluorescence, spectral reflectometry has unique features. Most notably, detection is sensitive to the angle of incidence, and a geometric structural model including refractive indices should be estimated for precise quantification. Thus, correlative imaging with other structural imaging modalities can be complementary. Especially for dynamic imaging of myelinated axons, faster acquisition may be achieved by utilizing spatial sparsity of the sample. Data analysis based on geometric models may also be improved by employing recent deep learning algorithms.

For biomedical sensing, any dielectric material can be adopted as the optical probe for spectral reflectometry, called a reflectophore (Jo et al. 2018). With advances in biomaterials, various types of biocompatible hydrogels have become readily available for biomedical uses (Zhang et al. 2017). Hydrogels can be further engineered to change their physical size or optical properties in response to chemical or physical milieu (e.g., pH, temperature, and electric field). Since the physical changes in size or refractive indices can be robustly measured by spectral reflectometry, reflectophores made with these stimulus-responsive materials can be utilized as biomedical sensors. Incorporation of multilayered structures with distinct sensors will allow multiplexed sensing. Moreover, introducing surface chemistry will enable targeted loading on specific organs, cell subtypes, or subcellular organelles.

Adoption of rapidly expanding knowledge in material science and chemistry will pave the way toward the wide utilization of reflectophores as biomedical sensors.

Author Contributions All the authors designed the layout and prepared the manuscript. JK mainly wrote the sections on imaging; YJ wrote the sections on labeling and sensing; and MC wrote the remaining sections.

Funding This work was supported by Basic Science Research Program through the National Research Foundation of Korea (NRF) funded by the Ministry of Education (2018R1D1A1B07042834); and by the National Research Foundation of Korea(NRF) funded by the Ministry of Science, ICT & Future Planning (2019M3A9E2061789, 2020M3C1B8016137).

Competing Financial Interests The authors declare competing financial interests: all the authors are the co-inventors of the patents for SpeRe (Kwon et al. 2017) and reflectophore (Jo et al. 2018).

References

Harris AK, Patricia Wild DS (1980) Silicone rubber substrata : a new wrinkle in the study of cell locomotion. Science 208(4440):177–180

Mascaro A et al (2015) Label-free near-infrared reflectance microscopy as a complimentary tool for two-photon fluorescence brain imaging. Biomed Opt Express 6(11):4410–4483

Bai Y, Nichols JJ, MPH OD (2017) Progress in retinal and eye research advances in thickness measurements and dynamic visualization of the tear fi lm using non-invasive optical approaches. Prog Retin Eye Res 58:28–44

Braeckmans K et al (2003) Encoding microcarriers by spatial selective photobleaching. Nat Mater 2(3):169–173

Burgess IB (2013) J Mater Chem C:6075–6086. https://doi.org/10.1039/c3tc30919c

Cunin F et al. (2002). Biomolecular screening with encoded porous-silicon photonic crystals, 39–41

Nakayama D, Takeoka Y, Watanabe M, Kataoka K (2003) Simple and precise preparation of a porous gel for a colorimetric glucose sensor by a templating technique. Angew Chem Int Ed Engl 42(35):4197–4200

Doane MG (1989) An instrument for in vivo tear film interferometry. Optometry Vision Sci 66(6):383–388

Fogt N, King-smith PE (1998) Interferometric measurement of tear film thickness by use of spectral oscillations. J Opt Soc Am A Opt Image Sci Vis 15(1):268–275

Fu F et al. (2016) Cells cultured on Core-Shell photonic crystal barcodes for drug screening

Fujiwara H (2007) Spectroscopic ellipsometry: principles and applications. John Wiley & Sons

Gao Y, Li X, Serpe MJ (2015) Stimuli-responsive microgel-based etalons for optical sensing, 44074–44087

Gómez-Martínez R et al (2013) Silicon chips detect intracellular pressure changes in living cells. Nat Nanotechnol 8(7):517–521

Hamano H (1979) Bio-differential interference microscope observation on anterior segment of the eye. First report: observations of precorneal tear film J Jpn CL Soc 21:229–238

Hlubina P, Ciprian D, Luňáňek J, Lesňák M (2006) Dispersive white-light spectral interferometry with absolute phase retrieval to measure thin film. Opt Express 14(17):7678–7685

Hlubina P, Luňáček J, Ciprian D, Chlebus R (2008) Spectral interferometry and reflectometry used to measure thin films. Applied Physics B 92(2):203–207

Hu L, Serpe MJ (2012) Color modulation of spatially isolated regions on a single poly(N- isopropylacrylamide) microgel based etalon. J Mater Chem 22(17):8199-8202

Hwang J, Kim S, Heo J, Lee D, Ryu S (2015) microscopy, 23(5), 770–772

Jo Y et al. (2018) Microsphere-based interferometric optical probe. Nature Communications, 9, 4577(2018), 1–10

Josephson JE (1983) Appearance of the preocular tear film lipid layer. Optom Vis Sci 60(11):883–887

Khamene A, Member S, Negahdaripour S, Member S, Tseng SCG. (2000) A spectral-discrimination method for tear-film lipid-layer thickness estimation from fringe pattern images, 47(2), 249–258

Kinoshita S (2008) Structural colors in the realm of nature. World Scientific p 3–6

Kronenberg NM et al. (2017) Long-term imaging of cellular forces with high precision by elastic resonator interference stress microscopy, 19(7)

Kwon J et al (2017) Label-free nanoscale optical metrology on myelinated axons in vivo. Nat Commun 8(1):1832

Lee H, Kim J, Kim H, Kim J, Kwon S (2010) Colour-barcoded magnetic microparticles for multiplexed bioassays, 9(August), 745–749

Lim H, Sharoukhov D, Kassim I, Zhang Y, Salzer JL (2014a) Label-free imaging of Schwann cell myelination by third harmonic generation microscopy

Lim H et al (2014b) Label-free imaging of Schwann cell myelination by third harmonic generation microscopy. Proc Natl Acad Sci 111(50):18025–18030. JOUR

Liu C, Yao C, Zhu Y, Ren J, Ge L (2015) Sensors and actuators B : chemical dually responsive one dimensional photonic crystals with reversible color changes. Sensors & Actuators: B Chemical 220:227–232

Lu L, Joannopoulos JD, Soljačić M (2014) Topological photonics. Nat Photonics 8(11):821–829

Liu W, Shang L, Zheng F, Lu J, Qian J, Zhao Y (2013) Photonic crystal encoded microcarriers for biomaterial evaluation, 88–93

Mishima S, Maurice DM (1961) The oily layer of the tear film and evaporation from the corneal surface. Exp Eye Res 1(1):39–45

Munevar S, Wang YL, Dembo M (2001) Traction force microscopy of migrating normal and H-ras transformed 3T3 fibroblasts. Biophys J 80(4):1744-1757

Nicewarner-pen SR et al. (2001) Submicrometer metallic barcodes, 294 (October), 137–142

Park TH et al (2018) Block copolymer structural color strain sensor. NPG Asia Materials:328–339

Resch-Genger U, Grabolle M, Cavaliere-Jaricot S, Nitschke R, Nann T (2008) Quantum dots versus organic dyes as fluorescent labels. Nat Methods 5(9):763

Schain AJ, Hill RA, Grutzendler J (2014) Label-free in vivo imaging of myelinated axons in health and disease with spectral confocal reflectance microscopy. Nat Med 20(4):443–449

Shan-Rong Shi MEK, KLK (1991) Antigen Retrieval in Formalin-fixed, Tissues: An Enhancement Method for Immunohisto-chemical Staining Based on Microwave Oven Heating of Tissue Sections

Shang L, Fu F, Cheng Y, Wang H, Liu Y, Zhao Y, Gu Z (2015) Photonic crystal microbubbles as suspension barcodes

Tian L, Hires SA, Mao T, Huber D, Chiappe ME, Chalasani SH et al (2009) Imaging neural activity in worms, flies and mice with improved GCaMP calcium indicators. Nat Methods 6(12):875–881

Verschueren H (1985) Interference reflection microscopy in cell biology: Methodology and applications, 301, 279–301

Wang H, Zhao Z, Liu Y, Shao C, Bian F, Zhao Y (2018) Biomimetic enzyme cascade reaction system in microfluidic electrospray microcapsules, 1–8

Wu Y, Zhang K, Yang B (2019) Ordered hybrid micro/nanostructures and their optical applications, 1800980, 1–30

Xia F, Wu C, Sinefeld D, Li B, Qin Y, Xu C (2018) In vivo label-free confocal imaging of the deep mouse brain with long-wavelength illumination. Biomed Opt Express 9(12):6545-6555

Xu Y, Wang H, Luan C, Liu Y, Chen B, Zhao Y (2018) Biosensors and bioelectronics Aptamer-based hydrogel barcodes for the capture and detection of multiple types of pathogenic bacteria. Biosens Bioelectron 100(July 2017):404–410

Y Cajal SR, Swanson N (1995) Histology of the nervous system of man and vertebrates (no. 6). Oxford University Press, New York

Yokoi N et al. (2008) Rheology of tear film lipid layer spread in normal and aqueous tear–deficient dry eyes, 1–6

Zhang QM, Li X, Islam MR, Wei M, Serpe MJ (2014) Light switchable optical materials from azobenzene crosslinked poly(N-isopropylacrylamide)-based microgels, 6961–6965

Zhang YS, Khademhosseini A (2017) Advances in engineering hydrogels. Science, 356(6337):eaaf3627

Zhang R, Wang Q (2018) Flexible mechanochromic photonic crystals : routes to visual sensors and their mechanical properties, (1), 3182–3199

Zhao BY, Zhao X, Gu Z (2010a) Photonic Crystals in Bioassays, 2970–2988

Zhao BY, Zhao X, Tang B, Xu W, Li J (2010b) Quantum-dot-tagged bioresponsive hydrogel suspension array for multiplex label-free DNA detection, 976–982

Zhao Y, Xie Z, Gu H, Jin L, Zhao X, Wang B, Gu Z (2012) Multifunctional photonic crystal barcodes from microfluidics, (July), 1–7

Zheng F, Cheng Y, Wang J, Lu J, Zhang B, Zhao Y, Gu Z (2014). Aptamer-functionalized barcode particles for the capture and detection of multiple types of circulating tumor cells, 7333–7338

Label-Free Raman Spectroscopic Techniques with Morphological and Optical Characterization for Cancer Cell Analysis

Sanghwa Lee and Jun Ki Kim

Introduction

Current cancer detection and diagnosis is accomplished through various processes such as clinical symptoms, medical imaging, biopsy, pathological analysis, and surgery. Histopathologic analysis, especially including immunohistochemical staining, can be relied on at the molecular level, such as structure and concentration, but has disadvantages such as invasiveness, cost, and time-consuming (Rehman et al. 2007). There is a diagnostic method using spectroscopy as an approach to overcome the disadvantages of the pathological analysis method. Such vibration spectroscopy diagnosis is noninvasive and has advantages such as not destroying biological material and requiring no additional labeling (Oshima et al. 2010; Singh et al. 2012; Gautam et al. 2015). Detection of molecular vibrational spectra such as infrared (IR) and Raman spectroscopy has been widely applied in examining chemical and molecular changes in biological samples (Haka et al. 2005, 2009; Saha et al. 2011), and the spectra can fingerprint biomolecule composition without the need for additional labeling or sample separation (Kamemoto et al. 2010; Abramczyk et al. 2012).

Recently, Raman spectroscopy has been reassessed as a measurement method for biological samples of the breast, lungs, stomach, and other organs (Larraona-Puy et al. 2009; Sato et al. 2009; Kirsch et al. 2010; Feng et al. 2011; Kawabata et al.

S. Lee
Biomedical Engineering Research Center, Asan Institute for Life Sciences, Asan Medical Center, Seoul, Republic of Korea

J. K. Kim (✉)
Department of Convergence Medicine University of Ulsan College of Medicine, Asan Institute for Life Sciences, Asan Medical Center, Seoul, Korea (Republic of)
e-mail: kim@amc.seoul.kr

J. K. Kim et al. (eds.), *Advanced Imaging and Bio Techniques for Convergence Science*, Advances in Experimental Medicine and Biology,
https://doi.org/10.1007/978-981-33-6064-8_14

2011; Larraona-Puy et al. 2011). Furthermore, Raman spectroscopy has been studied for the verification of biological samples of various sizes, such as tissue (Kawabata et al. 2011; Ishigaki et al. 2016), cells (Carvalho et al. 2015; Smith et al. 2016; Talari et al. 2015), bacteria (Liu et al. 2011; Zhou et al. 2014), and protein (Lyng et al. 2007; Maiti et al. 2004; Rygula et al. 2013). The Raman method is based on the inelastic scattering of photons incident on the sample to be measured. When a single energy laser beam is injected into the sample, the energy of a part of the incoming photons is changed by the molecular vibration of the sample. This change is called Stokes shift and the spectral representation of Stokes-shifted photon energy distribution is a Raman spectroscopic signal. Since the initial energy is diverted by various molecular vibration modes, individual peaks in the spectrum include structural information of molecules. Thus, the Raman measurement method that incorporates this chemical information is a great tool for identifying biomaterials such as cancer tissues and cells (Li et al. 2010; Neugebauer et al. 2010; Shapiro et al. 2011; Abramczyk et al. 2011; McQueenie et al. 2012; Marro et al. 2014).

In this study, we compared and analyzed the Raman signal according to the cell surface position through the synchronization between the optical image with high spatial resolution and the measurement result of surface roughness at the nanometer level as shown in Fig. 1. The topography of the cells provided quantitative thickness and quantity indicating the relationship between local Raman spectrum and intensity (Lee et al. 2018). By introducing local RGB intensity in optical measurements, we have identified additional factors that normal cells can use to differentiate cancer cells. The optical microscope shows an image of transmitted light, and the intensity of each pixel in RGB represents an internal event such as absorption and scattering. We compared the optical signals according to the differences in the intracellular heights. As a result, the inelastic scattering Raman and the transmission type optical microscope confirmed that there was rare local variation in the cancer cells, unlike the normal cells.

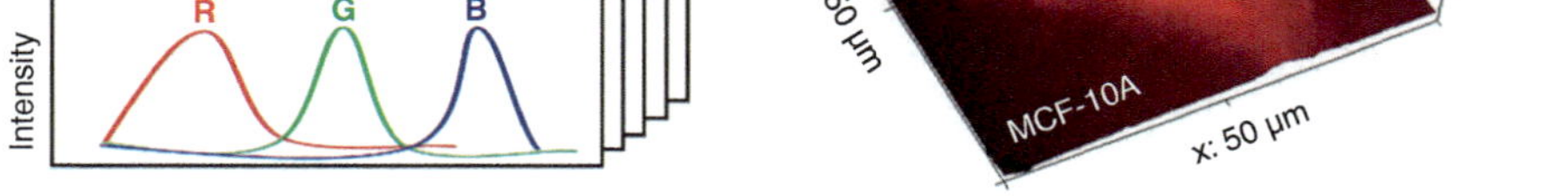

Fig. 1 Schematic diagram showing the examination of cell line using spectroscopy, tomography, and optical signals. Reprinted from Spectrochimica Acta Part A: Molecular and Biomolecular Spectroscopy 205, Lee et al., “Local-dependency of morphological and optical properties between breast cancer cell lines,” p. 132–138, Copyright 2018, with permission from Elsevier

Raman Measurement Mechanism and System Configuration

Raman method is a spectroscopic analysis that measures the inelastic scattering of the photons of an incident single-wavelength laser. In 1928, Raman spectroscopy was carried out to investigate the effect of light, which is different from the energy of light incident upon reflection or scattering. The emission of these other energies is determined by the energy exchange with the molecule vibration that forms between the light and matter, and the difference can be expressed as wavenumber (cm^{-1}) or wavelength (nm) as shown below:

$$\Delta E = hc\kappa = h\frac{c}{\lambda} = h\nu$$

where ΔE is the energy difference, h is Planck's constant, c is the speed of light, κ is the wavenumber, ν is the frequency, and λ is the wavelength. From this representation of quantum mechanical light, it is possible to explain the difference between the Rayleigh scattering, the Stokes (anti-Stokes) in Raman scattering, and the fluorescence as shown in Fig. 2a. These inelastic scattering and fluorescence signals sometimes interfere with each other (Fig. 2b). Most biomaterials appear as spectra with multiple peaks because of inelastic scattering and emission of fluorescence signals in various forms. On the other hand, these peaks have the advantage of being able to specify a sample like a fingerprint because it provides a parameter, such as the position of the peak and the relative intensity between the various peaks, in a narrow form.

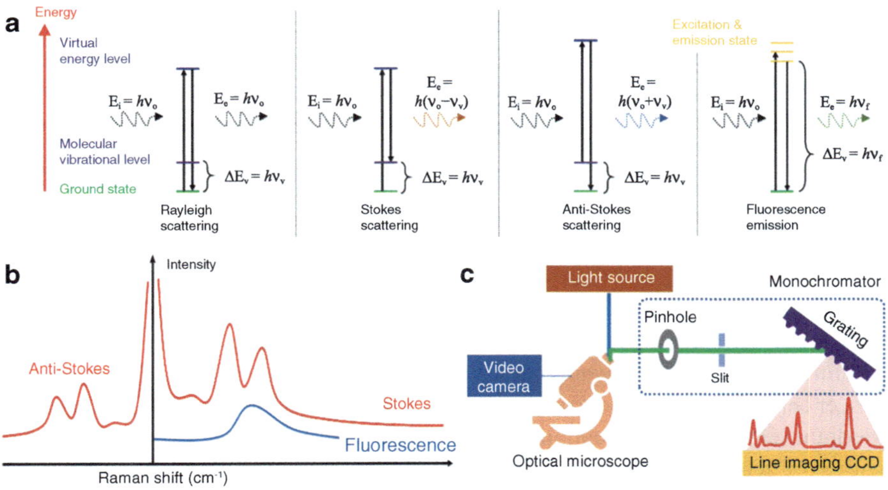

Fig. 2 (**a**) Schematic illustrating the interaction of incident light with materials such as Rayleigh scattering, Raman scattering and fluorescence to show the Raman spectroscopic mechanism. (**b**) Schematic representation of the spectrum formed by Raman inelastic scattering and fluorescence. (**c**) Conventional Raman system configuration. © 2020 the authors

The configuration of a typical Raman system consists of a confocal microscope, a laser source, a monochromator, and a detector as shown in Fig. 2 (c). The Raman signal is obtained by a spectroscope measuring the intensity of an inelastically scattered photon in a sample from a short-wavelength laser (785 nm). 514 nm, 632 nm, and 785 nm, it is possible to improve the specificity of the sample by installing the Raman apparatus of the wavelength variable system because the signal is different depending on the wavelength of the laser used (Synytsya et al. 2014). Since hemoglobin in blood has a large absorption spectrum in the wavelength range of 200–700 nm, it can be seen that the wavelength selection of the incident light is an important issue according to the characteristics of the target sample. By enlarging the scan area through Raman mapping, it is possible to cover a wide range of areas, thereby increasing measurement sensitivity and reliability. In mapping, it is possible to raise the spot size of the beam by reducing the magnification of the lens for a close scan of the area of interest. As a practical experience of our Raman system setting, the spot size is increased from 1.92 um to 3.58 um when switching from ×50 lens (0.5NA) to ×10 lens (0.25NA). Recently, the Raman spectroscopy system has been introduced into the endoscope or made into a portable device to broaden its application range.

Cell Culture and Sample Pretreatment

A variety of cell lines have been used for in vitro models in breast cancer studies at the cellular level. These cell lines have demonstrated advantages such as ease of handling, high homogeneity, and unlimited self-replicating potential (Burdall et al. 2003). Raman spectra have shown not only differences in biological composition between cell lines but also biological processes such as cell differentiation and apoptosis. Therefore, by selecting a few kinds of normal cells and cancer cells related to breast cancer and acquiring the difference between Raman and optical signals, it is possible to secure the appropriateness of Raman and the transmission optical analysis method. Three cells of MCF-10A, MDA-MB-453, and MDA-MB-231 were examined by Raman spectroscopy in order to determine the differences between normal cells and cancer cells, and these cells were selected because they play an important role in cancer metastasis.

For securing these cell lines, one human breast cancer cell line (MDA-MB-453) and two human breast epithelial cell lines (MCF-10 and TNBC) were purchased from ATCC (Manassas, VA), and the same TNBC cell line and human breast cancer cell line (MDA-MB-231) were purchased from Korea Cell Bank. The human breast cancer cell lines were cultured using Dulbecco's modified Eagle's Minimal Essential Medium (DMEM; Life Technologies, Inc., Grand Island, NY, USA) supplemented with 10% fetal bovine serum (FBS; Hyclone Laboratories, Logan, UT, USA) and 1% penicillin–streptomycin solution (Life Technologies, Inc.). The human breast epithelial cell line was grown using mammary epithelium growth medium, MEGM BulletKit (Lonza Group Ltd. Basel, Switzerland), except for GA-1000 (gentamicin–amphotericin B mix), with the addition of 100 ng/ml cholera toxin from

Sigma-Aldrich (St. Louis, MO). All of the cell cultures were maintained on a humidified 5% CO_2 incubator at 37 °C.

All cells were cultured on gold-coated substrates in Petri dishes until they reached 70 ~ 80% confluency to reduce Raman noise from bare glass. In order to synchronize the results of various measurement devices, it is necessary to fix the cells and measure Raman and optical images. Therefore, phosphate-buffered saline (PBS) was removed after 24 h, and cells were fixed using 4% paraformaldehyde for 20 minutes at 37 °C and 5% humidity in an incubator. Fixed cells were washed three times in PBS for 10 minutes each time. Because dried PBS produces Raman noise signals, the cells were rinsed with distilled water approximately 10 times to remove the PBS before the Raman measurements.

Raman Measurement Conditions and Analysis of Spectral Results

Raman line profiles were measured using a SENTERRA confocal Raman system (Bruker Optics, Billerica, MA, USA), and a 785-nm diode laser with 100 mW power was used for excitation. The laser beam was focused on the fixed cells through a 50 × (0.4 NA) objective with an approximate spot size of 2.4 μm. Raman lines were scanned through a raster scan at approximately 2.5 μm of step which refers to the distance between laser beam centers. The spectral range of each point was measured twice in the range from 400 to 2500 cm^{-1}. A spectral resolution was 5 cm^{-1} and an integration time was 30 sec at room temperature. Before the Raman measurements, the Raman spectrum was calibrated by measuring a silicon crystal. Following the Raman measurements, the spectrum was preprocessed by a Savitzky–Golay smoothing filter. The autofluorescence background was removed by a third-order polynomial fitting to make the spectrum better suited for an analysis.

The Raman spectra obtained for the normal cell line and the breast cancer cell lines showed major peaks in phenylalanine, amides I and III, CH_2 deformation, CH_2 wagging/twisting, and S-S bonding vibration regions as shown in Fig. 3. Each spectrum was previously smoothed and baseline corrected, and the spectra obtained from each cell line showed visual differences including standard deviations. The concentrations of amide III, CH_2 deformation vibration, CH_2 wagging/twisting, amide I, and so on were different in breast cell lines from normal cell lines. The difference in Raman spectra can be distinguished by peak position, relative intensity between peaks, and absolute intensity of a particular peak, and these factors relate to the chemical and quantitative values. The biochemical species such as amide III, CH_2 deformation vibration, CH_2 wagging/twisting, amide I were labeled from the peak position. The CH_2 deformation and amide I regions showed similar intensities in two cancer cell lines, but the MDA-MB-231 spectrum had a higher relative amount of CH_2 wagging/twisting. In particular, the Raman spectrum for MDA-MB-231 had higher CH_2 wagging/twisting intensity than another cancer cell. This difference is greater than the error bars in situations where the intensities of other peaks, such as S-S bonding and CH_2 deformation vibration, are similar. These

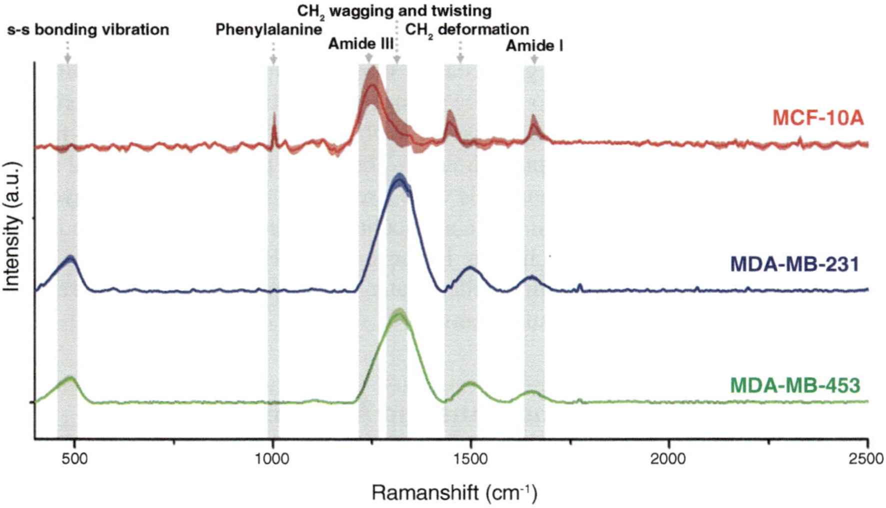

Fig. 3 Averaged Raman spectra of the normal breast cell line (**a**) MCF-10A and the breast cancer cell lines (**b**) MDA-MB-231 and (**c**) MDA-MB-453. The standard deviations for each point are represented by shading around the solid line. Reprinted from Spectrochimica Acta Part A: Molecular and Biomolecular Spectroscopy 205, Lee et al., "Local-dependency of morphological and optical properties between breast cancer cell lines," p. 132–138, Copyright 2018, with permission from Elsevier

features seem to indicate the different chemical characteristics between the cancer cells and normal cells, which is a factor in the statistical analysis.

Principal Component Analysis (PCA) of Raman Spectroscopic Signals

Analysis of multiple samples can be accessed through statistical analysis. Principal component analysis (PCA) is a method used to analyze a large number of data signals such as individual peaks of Raman signal. It is an algorithm that projects complex data with multiple dimensions into a small number of principal components (PC). Since many data have many dimensions, it is difficult to directly classify the data. The PCA transforms these high-dimensional data into a low-dimensional space and transforms the data so that it can be easily classified.

The remaining Raman spectrum reflects the conformational characteristics of the samples in addition to the background noise. In the Raman spectrum, the x-axis corresponds to the wavelength of light, and the wavelength corresponding to each spectrum can be an independent variable. Since the intensity according to these variables is an input value for analysis, signal processing including smoothing and baseline

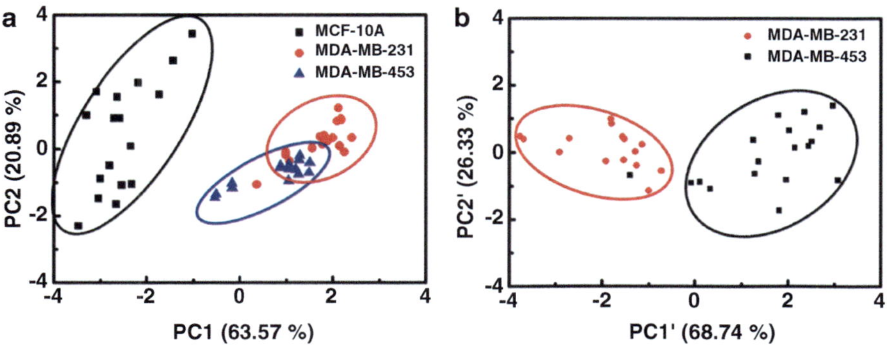

Fig. 4 Principal component analysis results for (**a**) MCF-10A, MDA-MB-231, and MDA-MB-453 and (**b**) MDA-MB-231 and MDA-MB-453. Reprinted from Spectrochimica Acta Part A: Molecular and Biomolecular Spectroscopy 205, Lee et al., "Local-dependency of morphological and optical properties between breast cancer cell lines," p. 132–138, Copyright 2018, with permission from Elsevier

correction was necessary for accurate data analysis of each sample. We evaluated the spectral differences between the normal and cancer cell lines using PCA. In this study, PCA was conducted as an attempt to group useful information from the pre-processed Raman spectra and to highlight the differences between MCF-10A, MDA-MB-231, and MDA-MB-453. All analyses were conducted using the XLSTAT 2016 software (Microsoft, Redmond, WA, USA).

From the PCA of the spectrum obtained from the normal and cancer cell lines, it became possible to further distinguish between three cell lines as shown in Fig. 4a. It also became possible to distinguish between cancer cells which were difficult to distinguish on the spectrum of Fig. 3. Figure 4a shows a clear separation between the normal breast cells and breast cancer cells in the PCA plot, where the first principle component explained 63.57% of the variance, and the second component explained 20.89%. Additionally, while the breast cancer cells clustered near each other in Fig. 4a, the groups show clear separation from normal cells. Figure 5b shows an additional PC analysis on just the cancer cells, which clearly separates the two cell types, with the first principle component explaining 68.74% of the variance and the second component explaining 23.33%. The Raman spectral variables that were responsible for the discrimination on PC space were the phenylalanine, amide I, amide III, CH_2 deformation, CH_2 wagging/twisting, and S-S Bonding vibration peaks. Thus, components in the highly confined spectrum including the above chemical constitution alone occupy a large portion as a factor that differentiates each cell line. This statistical analysis method provides the criteria for discrimination between normal and cancer cells. The validity of the diagnosis can be confirmed by peak reference for multiple dimensions.

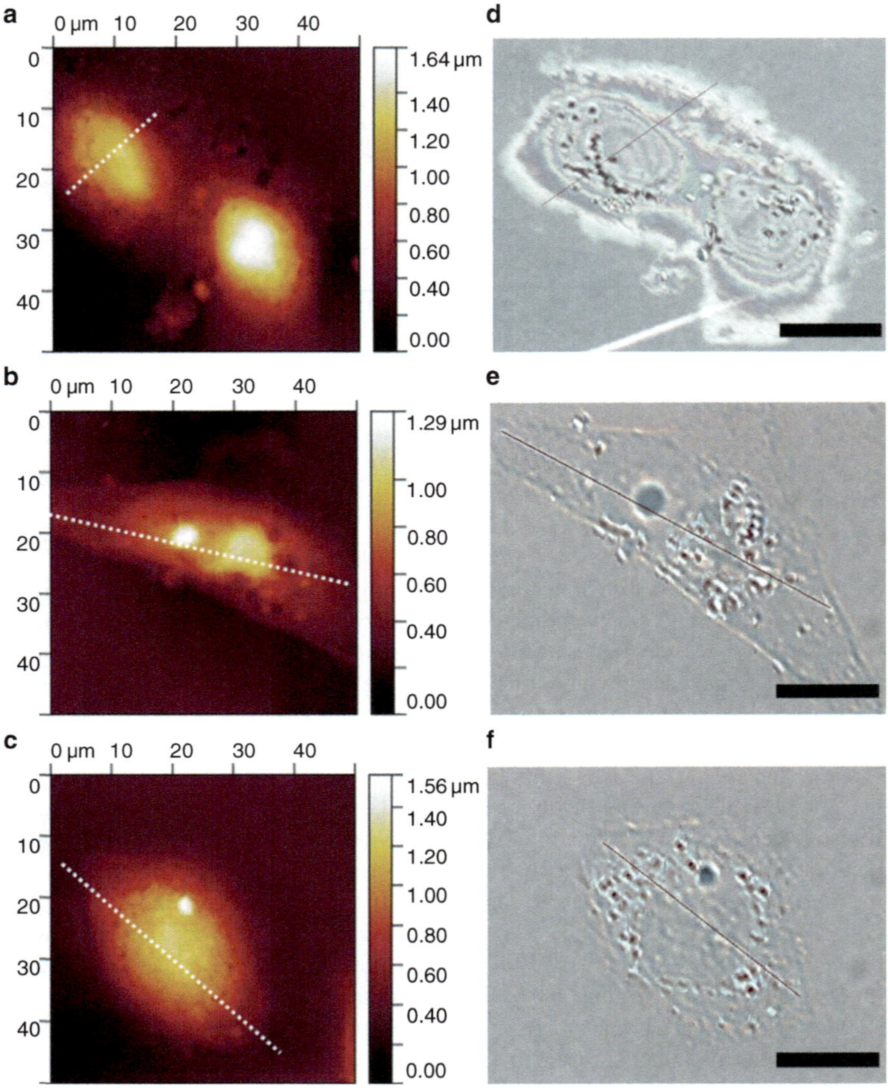

Fig. 5 Synchronizing the topographic and optical images of the three cell lines. AFM images of the normal cell (**a**) MCF-10A and cancer cell lines (**b**) MDA-MB-231 and (**c**) MDA-MB-453 on the left column. Transmission optical images are displayed in the right column to match the measured lines, where each black scale bar is 20 μm. Reprinted from Spectrochimica Acta Part A: Molecular and Biomolecular Spectroscopy 205, Lee et al., "Local-dependency of morphological and optical properties between breast cancer cell lines," p. 132–138, Copyright 2018, with permission from Elsevier

Atomic Force Microscopy for Comparative Analysis of Local Raman Signals and Cellular Morphology

Atomic force microscopy (AFM) has been widely used to study biological phenomena such as the cytotoxicity of cellular systems (Lee et al. 2013) and the mechanical properties of cells and molecules (Haase and Pelling 2015; Bao and Suresh 2003). Furthermore, AFM has been applied to investigate Young's modulus indicating cellular stiffness, which properties can in turn differentiate between nonmalignant and breast cancer cells (Li et al. 2008; Calzado-Martin et al. 2016). From the AFM results, it is now well-known that cancerous cells are remarkably softer than normal cells, and there is a potential as an analytical tool that can give information on morphology and partial classifications together. In this work, Raman spectroscopy was analyzed by morphological localization through synchronization with AFMs having high spatial resolution and optical microscope. The topography of depending on the in-plane direction of the substrate gives local quantitative thicknesses and volumes that reveal the relationships between local Raman spectra and intensities.

Topography and surface roughness data were acquired using an atomic force microscope (TT-AFM, AFM workshop) in a contact mode at room temperature. Normal and cancer cells fixed on a gold-coated substrate of approximately 15 nm roughness (RMS value) were mounted on the AFM x-y motor stage for scanning. A scan area of 50 μm × 50 μm was constructed using 256 scan lines and a scanning rate of 0.2 Hz. After the samples were prepared to synchronize Raman and AFM, the rinsed cells were dried and alignment markers were printed on them before the loading Raman stage. The individual cells measured by Raman spectroscopy were pre-checked with distance from the alignment marker and magnitude of the objective lens. Prior to approaching and positioning the cantilever, alignment markers on the sample and a crossbar on the AFM optical image were matched to position the cells and rotate the substrate. The magnitudes of the AFM topography and the Raman optical image were synchronized by a scale calibration target on AFM and Raman.

AFM topology and optical images are shown in Fig. 5, indicating matched line profiles of height and Raman spectroscopy. The individual AFM topologies, shown in the left column of (a) MCF-10A, (b) MDA-MB-231, and (c) MDA-MB-453 had 50 μm × 50 μm scan areas covering one or two cells. The minimum data value was shifted such that the height of each cell varied from 0 to 1.5 μm. There were no significant differences in height between cell types or species. The optical image is displayed in the right column corresponding to each AFM topogram column. The optical MCF-10A image showed a whitish cell boundary, a Newton ring with a contour like a rainbow, and blackhead-like points. Most blackheads corresponded to hollows on the AFM topography, but this was not always the case. For the invasive cancer cells MDA-MB-231 and MDA-MB-453, images transmitted from 3e and f showed blackhead-like points, black spots larger than blackheads, and cell boundaries less clear than MCF-10A (See Fig. 3d). The blackheads on the optical images were highly distributed around the hollows shown on the morphology images, but the black spots were at the highest peak in each cell. Note that these spots appear in approximately one out of ten invasive cancer cells. Thus, the whitish boundaries can be a significant indicator of whether or not a cell is invasive, but two cancer cell lines have barely visible boundaries as shown in Fig. 5 on the right column.

To investigate the comparative factors among noninvasive and invasive breast cell lines, line measurements of height and Raman spectroscopy were performed along the white dotted lines on the AFM topography and the red line on the optical image, as shown in Fig. 6.

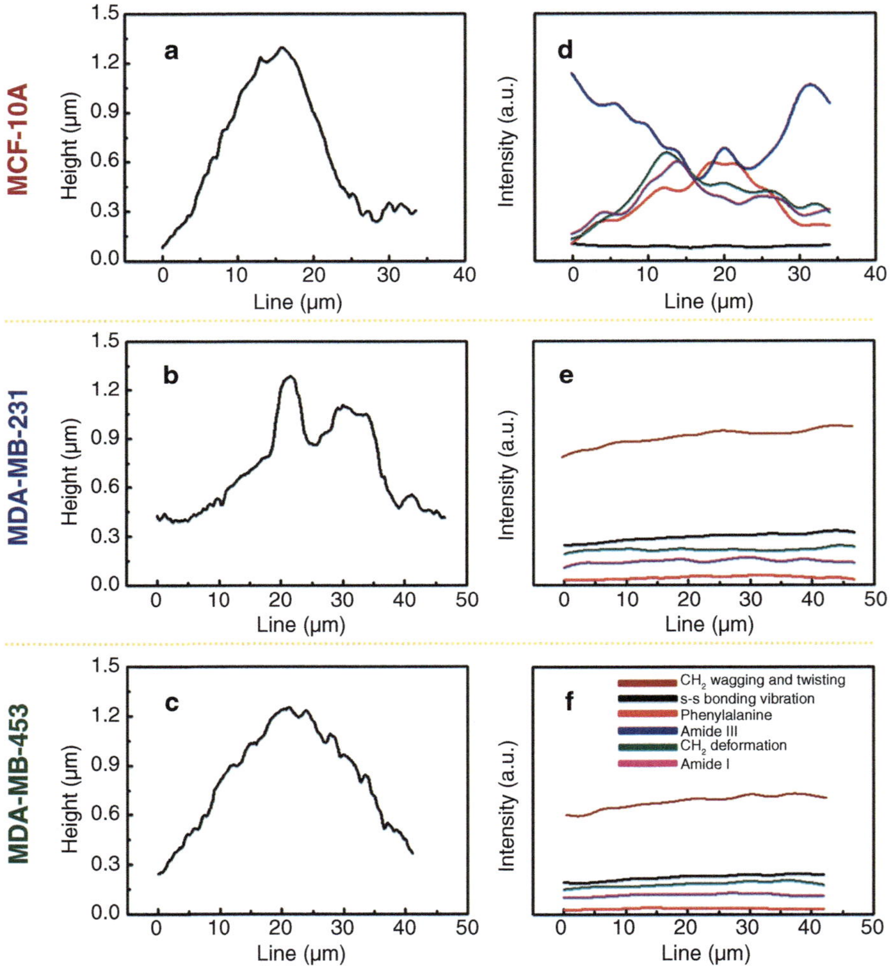

Fig. 6 Comparative results of extracted line morphology (**a-c**) and Raman line scan (**d-f**) for matching local variations. Raman line profiles were plotted for phenylalanine, amide I, amide III, CH_2 deformation, CH_2 wagging/twisting, and S-S bonding vibration. The inset in f shows the corresponding colors. Adapted from Spectrochimica Acta Part A: Molecular and Biomolecular Spectroscopy 205, Lee et al., "Local-dependency of morphological and optical properties between breast cancer cell lines," p. 132–138, Copyright 2018, with permission from Elsevier

In the AFM results, all cross-sectional intensity height profiles were convex, and thus it was expected that Raman would depend on radial differences in cell morphology. Each profile covered one cell and was 34 μm long for MCF-10A (as shown in Fig. 6a and d), 47 μm long for MDA-MB-231 (as shown in Fig. 6b and e), and 42 μm long for MDA-MB-453 (as shown in Fig. 6c and f). To confirm Raman differences in cell morphology, we plotted the local differences in Raman signal, phenylalanine, amide I, amide III, and CH_2 deformation, CH_2 wagging/twisting, and S-S bonding vibration peaks for individual cell lines synchronized with AFM. With the normal MDF-10A cells, signals of phenylalanine, CH_2 deformation, and amide I looked convex as in AFM morphology, but the signal for amide III was concave. These results indicate that the relative proportions of structural components in normal cells vary radially but not in the invasive cell line. The order of downward peak intensity for the invasive cells was maintained with CH_2 wagging/twisting, S-S bonding, CH_2 deformation, amide I, and phenylalanine as shown in Figs. 6 e and f.

Comparative Analysis of Raman Signal and RGB Transmission

By introducing local RGB intensity in optical measurements, we further identified factors that can be used to distinguish normal cells from cancer cell lines. An optical microscope shows the image of the transmitted light, and the intensity of each pixel of RGB indicates internal events such as absorption and scattering. Although we found that the difference in height within each cell was about 1.5 μm in the AFM results, the results of the inelastic scattering Raman and transmitted light optical microscope confirmed that there is no local variation in cancer cells, unlike normal cells.

To obtain RGB profile, optical microimages were observed using an optical microscope (BX51, Olympus) with mercury (reflection) and halogen (transmission) white light sources, respectively. A combined 50 × objective lens and a second lens of 2 × magnification were used. To reduce chromatic aberrations and blurred image edges on the reflected image and the transmitted image, an extended focus imaging calculation was performed. The resolution of all images was 1360 × 1024 pixels, and snap images were acquired within 1 sec. For calibration of a white balance, white paper was initially mounted, and subsequently, reflection optical images and RGB profiles were acquired. For acquiring transmission images, white balance was performed on the bare gold surface.

Notable observation from the transmission images was the difference in absorption behavior between the normal and cancer cells. When the transmitted optical images were acquired after light scattering and absorption through the cells, the individual red, green, and blue (RGB) differences gave information associated with scattering and absorption in the cells. At shorter wavelengths in incident white light, Rayleigh scattering is more common, so there is a difference in transmission between RGB. Thus, the RGB sequence on black spots in MDA-MB-231 was as shown in Fig. 7b, where the range from 18 to 22 μm appeared to indicate scattering. In contrast, the relative ratios of RGB from MCF-10A in Fig. 7a showed a fringe

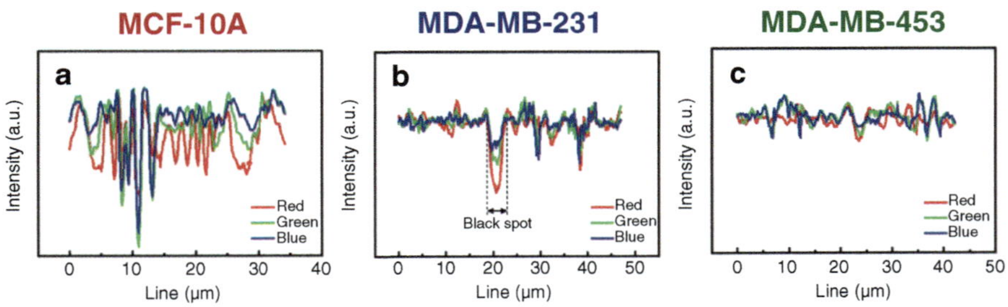

Fig. 7 RGB intensity for (**a**) MCF-10A, (**b**) MDA-MB-231, and (**c**) MDA-MB-453. Adapted from Spectrochimica Acta Part A: Molecular and Biomolecular Spectroscopy 205, Lee et al., "Local-dependency of morphological and optical properties between breast cancer cell lines," p. 132–138, Copyright 2018, with permission from Elsevier

Table 1 The individual deviations and means in RGB intensity between applied cell lines

Standard deviations (Mean values)	MCF-10A	MDA-MB-231	MDA-MB-453
Red	28.1 (194.0)	10.3 (213.2)	8.1 (213.5)
Green	33.0 (209.8)	10.1 (215.2)	11.0 (217.8)
Blue	28.8 (215.9)	9.2 (214.3)	9.5 (217.2)

Reproduced with permission from Ref. (Lee et al. 2018)

with no red-green-blue sequence from 14 to 25 μm; the fringe came from the shape of Newton's ring as shown in Fig. 5d, which indicated different transmission properties between the normal and cancer cells. By introducing standard deviations for RGB intensity, the intensity was compared quantitatively. The individual deviations in RGB intensity for invasive cell lines were much smaller than those for noninvasive cell lines as shown in Figs. 7 and in Table 1. The standard deviations of RGB in MCF-10A were 28.1, 33.0, and 28.8 where the means were 194.0, 209.8, and 215.9, respectively. For the MDA-MB-231 with absorption information at the black spots removed, the standard deviations for RGB intensity were 10.3, 10.1, and 9.2 with RGB means of 213.2, 215.2, and 214.3. In total, the standard deviations of red, green, and blue intensity in the noninvasive cell lines were nearly triple those of the invasive cells as summarized in Table 1.

Additionally, the local relation between morphology, Raman, and RGB deviation was significant in distinguishing between the normal and cancer cell lines. Since Raman spectroscopy is a technique for the detection of conformational structures, and if the optical transmission is measured with RGB profiling in the radial direction of the cell, the characterization factor that distinguishes the normal and the cancer can be found locally. In the case of the MCF-10A in this Raman study, the position-dependent projected signal difference in the radial direction seemed to be due to the cell organelles. However, in malignant cells of MDA-MB-231 and MDA-MB-453, the spectral difference did not appear to be dependent on the intracellular location. Additionally, since there is little difference in the degree of transmission between the red, green, and blue color components across the cells, it is suggested that RGB and Raman characteristics offer promising data for understanding the mechanism of cancer cell development, diagnosis, and treatment.

Conclusion

In summary, Raman spectra were acquired for the normal cell line (MCF-10A) and cancer cell lines (MDA-MB-231 and MDA-MB-453). The major peaks of Raman spectra are in the regions for phenylalanine, amides I and III, CH_2 deformation, CH_2 wagging/twisting, and S-S bonding vibration. These peaks were used as the main spectrum markers of cell classification. When those peaks were used as statistical variables, PCA on spectral differences allowed for further discrimination between cancer and normal cells. Although the breast cancer cells of MDA-MB-231 and MDA-MB-453 clustered close to each other when the PCA results were calculated for all three types of cells containing MCF-10A. Each group was well differentiated when only the breast cancer cells were recalculated. The study of local differences of Raman and RGB transmission with the image synchronization in this cell unit suggests the methodology of label-free and nondestructive field study and has potential for further application to other cell lines. When the results obtained in this study are extended to tissue and tumor size, these results will be the basis for establishing clinical diagnostic criteria and endoscopy applications.

Funding

This work was supported by the National Research Foundation of Korea (NRF) (2018R1D1A1B07048562, 2019R1A2C2084122) and MRC grant (2018R1A5A2020732) by the Ministry of Science & ICT (MSIT), and by a grant of the Korea Health Technology R&D Project through the Korea Health Industry Development Institute (KHIDI), funded by the Ministry of Health & Welfare, Republic of Korea (HI18C2391). This study was also supported by a grant (2019-7212) from the Asan Institute for Life Sciences, Asan Medical Center, Seoul, Korea.

References

Abramczyk H et al (2011) The label-free Raman imaging of human breast cancer. J Mol Liq 164(1–2):123–131

Abramczyk H et al (2012) Raman 'optical biopsy' of human breast cancer. Progress in Biophysics & Molecular Biology 108(1–2):74–81

Bao G, Suresh S (2003) Cell and molecular mechanics of biological materials. Nat Mater 2(11):715–725

Burdall SE et al (2003) Breast cancer cell lines: friend or foe? Breast Cancer Res 5(2):89–95

Calzado-Martin A et al (2016) Effect of actin organization on the stiffness of living breast cancer cells revealed by peak-force modulation atomic force microscopy. ACS Nano 10(3):3365–3374

Carvalho LFCS et al (2015) Raman micro-spectroscopy for rapid screening of oral squamous cell carcinoma. Exp Mol Pathol 98(3):502–509

Feng SY et al (2011) Study on gastric cancer blood plasma based on surface-enhanced Raman spectroscopy combined with multivariate analysis. Science China-Life Sciences 54(9):828–834

Gautam R et al (2015) Raman and mid-infrared spectroscopic imaging: applications and advancements. Curr Sci 108(3):341–356

Haase K, Pelling AE (2015) Investigating cell mechanics with atomic force microscopy. J R Soc Interface 12(104)

Haka AS et al (2005) Diagnosing breast cancer by using Raman spectroscopy. Proc Natl Acad Sci U S A 102(35):12371–12376

Haka AS et al (2009) Diagnosing breast cancer using Raman spectroscopy: prospective analysis. J Biomed Opt:14(5)

Ishigaki M et al (2016) Diagnosis of early-stage esophageal cancer by Raman spectroscopy and chemometric techniques. Analyst 141(3):1027–1033

Kamemoto LE et al (2010) Near-infrared micro-Raman spectroscopy for in vitro detection of cervical cancer. Appl Spectrosc 64(3):255–261

Kawabata T et al (2011) Near-infrared multichannel Raman spectroscopy with a 1064 nm excitation wavelength for ex vivo diagnosis of gastric cancer. J Surg Res 169(2):E137–E143

Kirsch M et al (2010) Raman spectroscopic imaging for in vivo detection of cerebral brain metastases. Anal Bioanal Chem 398(4):1707–1713

Larraona-Puy M et al (2009) Development of Raman microspectroscopy for automated detection and imaging of basal cell carcinoma. J Biomed Opt 14(5)

Larraona-Puy M et al (2011) Discrimination between basal cell carcinoma and hair follicles in skin tissue sections by Raman micro-spectroscopy. J Mol Struct 993(1–3):57–61

Lee YJ et al (2013) Label-free and quantitative evaluation of cytotoxicity based on surface nanostructure and biophysical property of cells utilizing AFM. Micron 49:54–59

Lee S et al (2018) Local-dependency of morphological and optical properties between breast cancer cell lines. Spectrochim Acta A Mol Biomol Spectrosc 205:132–138

Li QS et al (2008) AFM indentation study of breast cancer cells. Biochem Biophys Res Commun 374(4):609–613

Li Y et al (2010) Research on the Raman spectral character and diagnostic value of squamous cell carcinoma of oral mucosa. J Raman Spectrosc 41(2):142–147

Liu TY et al (2011) Functionalized arrays of Raman-enhancing nanoparticles for capture and culture-free analysis of bacteria in human blood. Nat Commun 2

Lyng FM et al (2007) Vibrational spectroscopy for cervical cancer pathology, from biochemical analysis to diagnostic tool. Exp Mol Pathol 82(2):121–129

Maiti NC et al (2004) Raman spectroscopic characterization of secondary structure in natively unfolded proteins: alpha-synuclein. J Am Chem Soc 126(8):2399–2408

Marro M et al (2014) Dynamic molecular monitoring of retina inflammation by in vivo Raman spectroscopy coupled with multivariate analysis. J Biophotonics 7(9):724–734

McQueenie R et al (2012) Detection of inflammation in vivo by surface-enhanced Raman scattering provides higher sensitivity than conventional fluorescence imaging. Anal Chem 84(14):5968–5975

Neugebauer U et al (2010) Towards detection and identification of circulating tumour cells using Raman spectroscopy. Analyst 135(12):3178–3182

Oshima Y et al (2010) Discrimination analysis of human lung cancer cells associated with histological type and malignancy using Raman spectroscopy. J Biomed Opt 15(1)

Rehman S et al (2007) Raman spectroscopic analysis of breast cancer tissues: identifying differences between normal, invasive ductal carcinoma and ductal carcinoma in situ of the breast tissue. J Raman Spectrosc 38(10):1345–1351

Rygula A et al (2013) Raman spectroscopy of proteins: a review. J Raman Spectrosc 44(8):1061–1076

Saha A et al (2011) Raman spectroscopy: a real-time tool for identifying microcalcifications during stereotactic breast core needle biopsies. Biomed Opt Express 2(10):2792–2803

Sato H et al (2009) Raman study of brain functions in live mice and rats: a pilot study. Vib Spectrosc 50(1):125–130
Shapiro A et al (2011) Raman molecular imaging: a novel spectroscopic technique for diagnosis of bladder cancer in urine specimens. Eur Urol 59(1):106–112
Singh B et al (2012) Application of vibrational microspectroscopy to biology and medicine. Curr Sci 102(2):232–244
Smith R, Wright KL, Ashton L (2016) Raman spectroscopy: an evolving technique for live cell studies. Analyst 141(12):3590–3600
Synytsya A et al (2014) Raman spectroscopy at different excitation wavelengths (1064, 785 and 532 nm) as a tool for diagnosis of colon cancer. J Raman Spectrosc 45(10):903–911
Talari ACS et al (2015) Raman spectroscopic analysis differentiates between breast cancer cell lines. J Raman Spectrosc 46(5):421–427
Zhou HB et al (2014) SERS detection of bacteria in water by in situ coating with Ag nanoparticles. Anal Chem 86(3):1525–1533

Diagnostic and Therapeutic Nanomedicine

Jinmyoung Joo

Introduction

Since the discovery of X-ray technology for medical imaging, many noninvasive methodologies have been invented and successfully applied in a wide range of research fields from clinical diagnostics to cell biology and drug discovery. The biomedical imaging is taking advantage of significant advances in electronics, information technology, and more recently, nanotechnology. Nanomedicine is a new area of interdisciplinary study dealing with the application of nanotechnology to healthcare and medicine. Emerging advances in nanometer-scaled materials (nanomaterials) have opened a new era of diagnostics and therapeutics, which is called as a nanomedicine, and made a broad and significant impact on medical science and healthcare applications (Kim et al. 2010; Theis et al. 2006). Research in translational nanomedicine has already led to the development of clinical products in a wide range of diagnostic and/or therapeutic imaging agents, in vitro diagnostics, and medical devices (Min et al. 2015; Etheridge et al. 2013). Because nanomaterials are similar in scale to biological elements and systems while they can be intentionally engineered to have various useful functions, nanotechnology has the significant potential for medical applications. The field of nanomedicine aims to use the unique properties including physical and chemical characteristics of nanomaterials owing to their dimensional scales for the diagnosis and treatment of diseases at the molecular level (Fig. 1).

The properties of nanomaterials are typically transitional between molecular and bulk regimes, and unique features of the nanomaterials extend the area of their uses in medicine. Those unique properties of nanoscale materials exist not only for

J. Joo (✉)
Department of Biomedical Engineering, Ulsan National Institute of Science and Technology, Ulsan, South Korea
e-mail: jjoo@unist.ac.kr

J. K. Kim et al. (eds.), *Advanced Imaging and Bio Techniques for Convergence Science*, Advances in Experimental Medicine and Biology,
https://doi.org/10.1007/978-981-33-6064-8_15

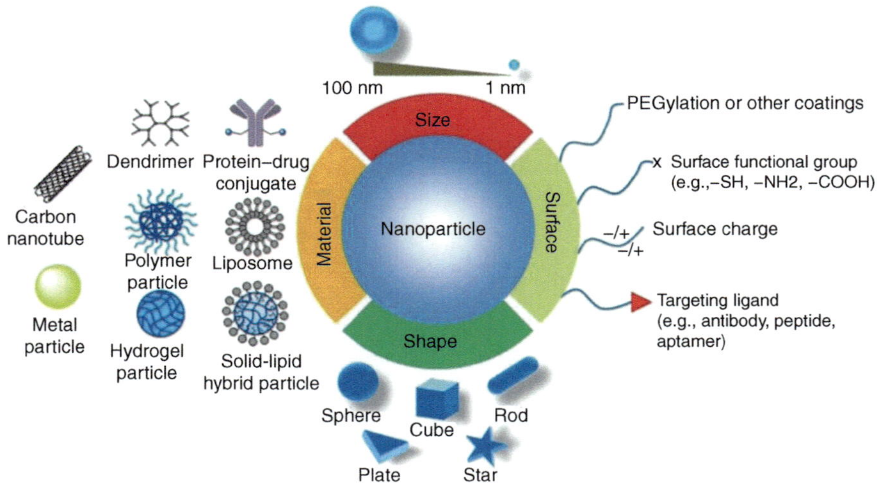

Fig. 1 Schematic illustration of nanomedicine with precisely engineered physicochemical properties for diagnostic and therapeutic applications (Sun et al. 2014). The nanoparticles used for medical applications can be readily synthesized from either soft (organic and polymeric) or hard (inorganic) materials, with controlled size and compositions/structures being engineered to employ unique signatures such as luminescence, magnetic momentum, and drug loading with a variety of configurations for diagnostic and/or therapeutic applications. The physicochemical properties of the nanoparticles can also be finely tuned by tailoring their chemical compositions, sizes, shapes, structures, morphologies, and surface properties

artificially synthetic materials but also for natural products. Nanoparticles (typically, 1–100 nm in size) have widely demonstrated with attractive characteristics exploited for the development of new diagnostics and treatment strategies. Furthermore, given the broad scope of nanomedicine, it is expected to involve all aspects of medicine eventually. Nanomedicine developed in bench side as a proof-of-concept can be a general strategy to be translated to the bedside (clinics) and be a part of conventional clinical practice assuming all aspects of medical requirements and needs are satisfied, including safety, regulatory, and ethical requirements (Pelaz et al. 2017). It is expected that nanomedicine will lead to the development of better medical devices, drugs, and other biomedical applications for early diagnosis and/or treatment in a wide range of diseases with improved efficiency and specificity. In addition, personalized therapeutics upon patients' stage of diseases will enhance the quality of life. In this chapter, overview of materials and technology belonging to nanomedicine is thoroughly presented.

Nanoparticles are a type of nanoscale materials and key components of nanomedicine, thus various nanoparticles have been developed for medical purposes over a decade. Of those, many nanoparticles enable to replicate certain functions of globular biological components, and support to fix malfunctions occurred in biological processes (Kotov 2010). For instance, lipid micelles are one of the early generations of nanoparticles while still widely used for therapeutic carriers in drug delivery systems (Torchilin 2007). Advances in materials science have provided tremendous adoptions of biocompatible nanomaterials derived from polymers, proteins, nucleic acid, and many types of organic composites including carbon nanotubes and graphene (Zhang et al. 2014a; Lee et al. 2007; Guo 2010; Baker and Baker 2010; Yu et al. 2005; Baughman et al. 2002; Nurunnabi et al. 2014). Inorganic materials, in particular, such as mesoporous silica, superparamagnetic iron oxide, quantum dots, plasmonic gold, and upconverting nanoparticles are also widely applied for diagnosis and therapeutics as a translational medical platform (Vallet-Regi et al. 2018; Colombo et al. 2012; Parak et al. 2005; Sperling et al. 2008; Wilhelm et al. 2015). These nanoparticles have unique features of optical, electronic, and magnetic characteristics with size- and shape-dependent manner, in which methods of synthesis, purification, and characterization critically determine the physicochemical properties (Sapsford et al. 2011; Kovalenko et al. 2015; Schaak and Williams 2012). It is very important that small changes in size and shape significantly affect those properties. Therefore, precision fabrication and synthesis are necessary to prepare the nanoparticles with narrow size distributions and shape similarity in order to achieve the specific function of interest and to correlate observed functions with particular characteristics of nanoparticles in the biological environment. It is also crucial for injectable diagnostic or therapeutic nanomedicine to understand their ultimate fates and in vivo characteristics as being administrated into the body.

Extensive efforts have been conducted to synthesize and functionalize nanoparticles with precisely controlled size, shape, and surface chemistry. This is particularly very important for nanomedicine, because the biodistribution and interaction of nanoparticles with proteins and other biological species are strongly size- and surface-dependent. Thus, heterogeneous nanoparticles will distribute in very different ways and may exhibit undesired effects causing toxicity. In this regard, the precisely controlled property of nanoparticles will surely reduce the potential uncertainty of interactions between nanoparticles and biological components in the body. However, one should note that nanoparticles are certainly heterogeneous with distinct subpopulations even after a careful synthetic process. This is the reason why the synthesis of uniform-sized nanoparticles has attracted great attention not only in the field of chemistry and materials science, but also in nanotechnology and medicine. Separation and purification process of those heterogenous nanoparticles are another field of interest, particularly for the medical applications in consideration with large-scale production and reproducible formation. Microscopic imaging provides to obtain structural information such as size and shape, and recent advancement in instrumentation enables to analyze property-determining features such as surface functionalization and elemental composition (Mulvaney et al. 2016). Therefore, the characterization of nanoparticles should be a part of starting point prior to use in human, and must be an important focus for nanomedicine applications. As mentioned, it is also required to develop new and improved methods of nanoparticle separation and purification to produce optimal samples for medical applications and for studying behavior as the nanoparticles are administered and flowed inside the body (which is important to design optimal nanoparticle formulations with reduced cytotoxicity for medical use) (Melamed et al. 2016; Saha et al. 2016).

One of the main advantages of nanoparticles for medical uses is that they can be designed and employed as a contrast agent for bioimaging. Optical or magnetic signals originated from the nanoparticle can provide contrast at the pathological site inner body after administration, and noninvasively report the information about the local tissue microenvironment related to certain diseases. Anatomical fine structures of organs are readily visualized, and tissues are also labeled with certain biomarkers for detection and treatment of disease. Local read-out of concentration of target molecules helps to analyze the early stage of disease effectively inside the body. Functional nanoparticles also enable to analyze both tissue and liquid biopsy extracted from the body for rapidly detecting biomarkers and molecular alterations in vitro. Simultaneous analysis of multiple biomarkers will improve diagnostic precision and allows to make an early decision for post-diagnostic processes such as chemotherapy and surgery. Lastly, multifunctional nanoparticles that can simultaneously diagnose, treat, and even monitor therapeutic efficacy are being engineered as a theranostic nanomedicine (Chow and Ho 2013).

Diagnostic Imaging in Nanomedicine

Early diagnosis is highly desirable to decide effective and decisive treatment in the clinic, where accurate identification of biological species is crucial. Diagnostic imaging techniques are probably the most intuitive and useful methods, which also represent one of the most investigated field of applications for nanoparticles in nanomedicine. There are still large demands for innovative contrast agents in diagnostic imaging, and the increasing requirement leads to the development of new diagnostic modalities and associated imaging agents. Considering the bioimaging modalities, magnetic resonance imaging (MRI) and luminescence imaging are the most common tools, which are widely used in laboratories and clinics. X-ray, computed tomography (CT), and positron emission tomography (PET) are other imaging modalities for whole-body visualization, providing additional information on metabolic alterations and soft tissues, which is also applicable area of diagnostic nanomedicine. Imaging resolution and measuring time are depending on the working principle of each imaging modality, thus appropriate imaging methods should be chosen along with the investigational subject (Wolfbeis 2015; Kobayashi et al. 2010).

First of all, the luminescence imaging has shown a huge step forward with the discovery of semiconducting quantum dots and upconverting nanocrystals. These nanoparticles show unique luminescent emission characteristics: the emission spectra can be readily tuned in the range from visible to infrared region by simply varying the size or elemental compositions, along with intensity and/or lifetime modulation (Martynenko et al. 2017; Kim et al. 2004; Wen et al. 2018; Zhu et al. 2019). The luminescence characteristics are significantly useful and have many advantageous points compared to conventional organic dyes. Although early generation of emissive quantum dots suffers from potential toxicity to be applied to human due to release of heavy metal ions such as cadmium, biocompatible, and environment-friendly emission nanocrystals based on silicon and carbon are newly suggested as an alternative luminescent probe while retaining all the unique optical properties (Park et al. 2019; Li et al. 2020).

On the other hand, magnetic nanoparticles have attracted great attention as a clinically relevant diagnostic imaging agent. Magnetic nanoparticles have a high potential usage for MRI imaging because of their biocompatibility, versatility, and high diagnostic performance upon applying an external magnetic field. One of the first nanomedicine that has been approved by the US Food and Drug Administration (FDA) is magnetic iron oxide nanoparticles for MRI-based diagnosis of hepatic lesion (Ros et al. 1995). There have been huge advancements from the first iron oxide nanoparticle MRI agent, as the field of nanomaterials grows tremendously, and still in one of the most attractive research area of nanomedicine. Nowadays, topics on developing magnetic nanoparticles as MRI contrast agents aim to produce controlled size, shape, composition, and crystallinity for the enhancement of magnetic momentum and corresponding performance. The magnetic nanoparticles act as an actuator upon external magnetic field, and the vibration characteristics are strongly affected by those factors. Recent studies of magnetic nanoparticles for MRI contrast agents include the formation of core@shell structure for the locally confined magnetic field in the core and heterogenous nanostructures for multiplexed imaging. Indeed, co-localization of signals deriving from the multiple imaging modalities (i.e., MRI and fluorescence imaging) enables to provide more accurate diagnostic imaging. Furthermore, spatiotemporal improvement in the visualization can be achieved through high-resolution signals came from the utilized magnetic nanoparticles.

Considering CT imaging, metallic nanoparticles (typically noble metal such as gold, platinum, or silver) play a crucial role, leading to a significant enhancement of the CT signals, along with enhanced tissue accumulation and retention. In addition, certain adverse effects such as the contrast-induced nephropathy can be avoided by using the nanoparticles over conventional means (Pasternak and Williamson 2012). In this regard, the metal nanoparticles, a form of nanomedicine, are also seriously considered as a valid alternative to traditional contrast agents.

In the next sections, a brief introduction of each technique is presented, explaining the physical principles behind it, followed by detailed examples of the diagnostic nanomedicine for the applications in each imaging modality.

Luminescence Imaging

Luminescence is typically defined as the emission of light from certain substances either molecules or materials, when excitation energy fits on or higher than the specific electromagnetic spectrum of a band structure. Luminescence imaging is one of the most versatile and widely used visualization modality in cellular and preclinical studies because the emission of the fluorescent probes such as organic dyes and quantum dots with sufficient energy for excitation are readily detectable at high resolution with simple instrumentation such as an optical microscope or even naked eye. High sensitivity, high spatial resolution, and ease of use are major features of the luminescence imaging. Fluorophores are commonly tailored to specific applications, and a varied range of emissive fluorophores including organic dyes and quantum dots are developed over a few decades. However, many traditional organic fluorophores typically suffer from aggregation-caused quenching or photobleaching due to decomposition upon light irradiation, thereby limiting design schemes where specific interactions promote fluorophore localization and degradation in biological environments (Ma et al. 2016). On the other hand, increased aggregation of organic dyes often yields an enhanced fluorescent signal, called aggregation-induced emission, which makes intensified luminescent signals at the site of accumulation (Mei et al. 2014). The aggregation-induced emission is particularly useful for long-term tracking of target cells or biological components over a wide time window, for instance, metastatic tumor tracking, drug delivery, and stem cell differentiation (Ding et al. 2013a, 2014). Once organic dyes form nanoscale aggregates (or clusters), which are further functionalized with targeting moieties (antibody, peptide, aptamer, etc.) (Field et al. 2015), the binding specificity to particular receptors allows active homing of the diseased tissues and cells (Liang et al. 2015; Kwok et al. 2015). Integration of an RGD-based peptide with those nanoaggregates, for example, has led to integrin-specific fluorescence signals for tumor cell targeting, image-guided drug delivery, live monitoring of therapeutic effect as well (Zhang et al. 2014b; Xue et al. 2014).

Over the past decades, two important classes of luminescent probes—semiconductor quantum dots and upconverting nanocrystals—have been thoroughly studied and considered as an alternative to common organic fluorophores for luminescence imaging. These inorganic nanoparticles have robust characteristics and superior optical properties including large Stokes shift, high quantum yield, and photostability. Moreover, their high surface-to-volume ratio allows facile and stable surface functionalization for targeting moiety conjugation. All these features lead to the development of these inorganic fluorophores as versatile in vitro and in vivo tools for biological labeling and imaging in the field of biomedical science and engineering.

Quantum Dots

Since the late 1990s, quantum dots (QDs) have been applied for luminescence labeling of biological components and cells (Chan and Nie 1998; Bruchez et al. 1998). Colloidal QDs are semiconductor nanocrystals, which have a few nanometer scales in size, and composited with elements from Groups II–VI, III–V, or IV–VI of the periodic table (Medintz et al. 2005). The unique optical properties arise from their peculiar electronic structure and dimensions close to or smaller than the exciton Bohr radius (Aldakov and Reiss 2019). In this size scale, typically referred to as quantum-confined domain, the electrons and holes are physical confined in discrete energy bands, resulting in the creation of an electron–hole pair (exciton). Since the bandgap extended, absorbance of energy in the QDs increases over the conventional bulk level (Fig. 2a). The radiative recombination of the exciton then generates the

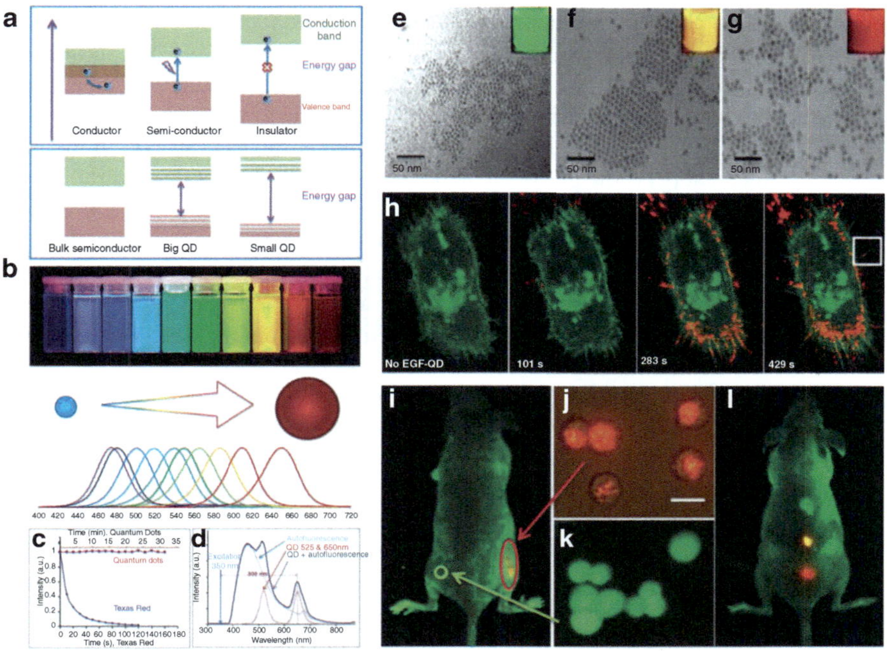

Fig. 2 (**a**) Electronic structure of bulk conductor, semiconductor, and insulator materials (top panel) and semiconductor nanoparticles (bottom panel) (Zrazhevskiy et al. 2010). Bulk semiconductor materials have a fully populated valence band and empty conduction band separated by a relatively small bandgap. When energy exceeding the bandgap is supplied, valence band electrons acquire enough energy to populate conduction band and enable electric current flow. In nanoparticles, valence and conduction bands split into discrete energy levels, with the energy gap between closest possible valence and conduction levels increasing with decreasing particle size, corresponding to increased degree of confinement of charge carriers. (**b**) Size-tunable luminescence emission profile enables precise control over the probe color via varying the nanoparticle size (Zrazhevskiy et al. 2010). (**c**) QDs have superior photostability compared to conventional organic dyes, enabling real-time and long-term monitoring of luminescent probes and accurate quantitative analysis, while quick photobleaching/photodecomposition of organic dyes limits such applications (Gao et al. 2005). (**d**) Capability of absorbing high-energy (UV-blue) light without damaging the probe and emitting fluorescence with a large Stokes shift enables efficient separation of the QD signal over the fluorescent background (Gao et al. 2005). (**e-g**) Transmission electron microscopy (TEM) images of CdSe/ZnS (core/shell) quantum dots with different size and emission color as referred in (**b**): (**e**) small size (left, green), (**f**) middle size (middle, yellow), and (**g**) large size (right, red) (Baig et al. 2008). (**h**) Retrograde transport of EGF-QDs on filopodia: Labeling of erbB/HER transmembrane receptors with QD-EGF probes. Continuous observation of QDs in live cells as a function of time enabled monitoring of receptor heterodimerization, probe endocytosis, and QD-EGF retrograde transport along cell filopodia (Lidke et al. 2004). (**i-k**) In vivo imaging of implanted QD-tagged tumor cells: (**j**) Bright QD tags enable visualization of tumor cells through skin with a noninvasive whole-animal fluorescence imaging, (**k**) whereas organic dye signal is not clearly distinguishable from tissue autofluorescence. (**l**) Luminescence imaging of subcutaneously implanted QD-loaded microbeads shows the potential for multiplexed in vivo cell detection and tracking (Gao et al. 2004)

emission, called photoluminescence, which shows size-dependent characteristics as a consequence of the quantum confinement effect (Yu et al. 2017). The confinement of excitons in such a small-sized domain of QDs corresponds to increased bandgap, indicating higher energy emission for smaller QDs while lower energy emission for larger-sized QDs. For instance, 2 nm-sized CdSe QDs shows emission in the blue region (higher energy), on the other hand, 7 nm-sized CdSe QDs exhibit lower energy emission in the red to near infrared region (closer to the bulk bandgap energy of CdSe) (Fig. 2b, e–g). Moreover, high quantum efficiency driven by superior molar extinction coefficients is another interesting characteristic of QDs. Bright emission and high stability to photobleaching are other advantages of QDs over conventional organic dyes, and those are attractive features for long-term imaging and single-molecule imaging in living cells (Clarke et al. 2010; Pinaud et al. 2010; Chang et al. 2008). Large Stokes shifts and narrower emission bandwidth are also important features of QDs, leading to getting rid of possible misleading caused by unfeasible emission shoulder peaks (Fig. 2c, d). In addition to broad absorption spectra that allow the excitation of multiple QDs at a single wavelength, multiplexed imaging of biological components enables simultaneous detection of biological components in situ (Fig. 2h). Due to these significant performances as a luminescence probe, QDs have been widely employed in biological studies, not only from single-molecule tracking to cell labeling, but also from gene delivery to in vivo imaging (Fig. 2i–l) (Wang et al. 2013a; Ho and Leong 2010). However, early generation of several QDs such as cadmium selenide (CdSe) are cytotoxic toward cells because of heavy metal ions, which is huge barrier for FDA approval and limits clinical uses (Derfus et al. 2004; Silva et al. 2014; Kirchner et al. 2005).

Biocompatible and biodegradable QDs have been reported for safe administration of bright QDs inside the body with reduced risk of harmful toxicity (McHugh et al. 2018; Zhou et al. 2015; Ma et al. 2010). For instance, semiconducting silicon nanocrystals are a promising candidate for bioimaging since nanocrystalline silicon has intrinsic photoluminescence while readily degrades in physiological environment and cleared from the body via urine (Park et al. 2009; 2019). Carbon-based nanoparticles (CNPs) are one particular type of luminescent probes for bioimaging. Similar to silicon nanocrystals, CNPs possess the advantageous characteristics of QDs over organic dyes: high photostability, easily tunable emission, and large two-photon excitation cross sections (Du et al. 2019). Moreover, CNPs are water soluble, and can be functionalized via facile conjugation chemistry. Synthetic steps are also relatively safe, and the CNPs are economically produced on a large scale as well. The CNPs have attracted large attention for intracellular sensing and targeting because of promising cell permeability and biocompatibility (Baker and Baker 2010; Sahu et al. 2012; Li et al. 2010). Subjected to a systematic assessment of potential toxicity including acute-, subacute-, and genotoxicity, the CNPs are much safer for biological applications than traditional heavy metal-containing QDs. Therefore, increasing number of studies have been reported for CNPs as a bioimaging agent for clinical uses including neuronal signaling and cancer targeting in vivo.

Upconverting Nanoparticles

Upconverting nanoparticles (UCNPs) consist of a host inorganic matrix doped with luminescent rare earth elements such as lanthanide, which constitutes another type of bioimaging probes. The UCNPs showed quite similar physical characteristics with the QDs, including high photostability, significant thermal and chemical stability, and sharp emission bandwidth (Wang et al. 2011). However, they typically show low luminescence intensity though luminescence quantum yield is relatively high, because of the low absorption of the parity-forbidden Ln^{3+} 4f–4f transitions (Fig. 3a–h) (Wisser et al. 2015). The unique luminescence property contributed by the upconversion mechanism of the photon is that long-wavelength radiation (lower energy) such as infrared or near-infrared can be converted to short-wavelength emission (higher energy) such as visible to UV region. The upconversion process involves an anti-Stokes mechanism, which is a nonlinear optical phenomenon consisting of the excitation of lower electronic levels and in the following emission at higher electronic levels. In principle, fundamental mechanism of the upconversion process is initiated with two or more photons adsorption, followed by excited-state absorption, energy transfer, and photon avalanche (Park et al. 2015; Eliseeva and Bunzli 2010). In this unique energy transfer process, the host inorganic lattice determines a spatial distribution of the dopants to improve the luminescence efficiency while reducing unwanted energy loss. The sensitizer ions act as energy donators while the activator species are radiation emitters upon appropriate energy excitation. Therefore, a number of inorganic host materials have been developed such as $NaYF_4$, $NaYbF_4$, and LaF_3, with various ionic dopants such as Er^{3+}, Tm^{3+}, or Yb^{3+} (Wu et al. 2015). It should also be noted that there are even more complex mechanisms of working involved beyond this unique nonlinear optical property, derived by the combination of these components in UCNPs. Accordingly, the optical characteristics of UCNPs can be easily modulated by controlling doping contents and concentration. For instance, intense red luminescence can be achieved by using indirect excitation of the Eu^{3+} cations via the vanadate anions. The incorporation of Bi^{3+} into the inorganic host matrix results in a shift of the original absorption band toward longer wavelengths (corresponding to the vanadate), yielding the UCNPs excitable by UV and visible lights (Escudero et al. 2016).

Besides the efficient upconverting emission lanthanide-doped UCNPs also have significant features: resistance to photoblinking and photobleaching. These characteristics are particularly very important for bioimaging since most of the conventional organic dyes suffer from stability issues upon long-term imaging and sensitive detection of low abundant biological targets (Fig. 3i–k). Moreover, the upconversion mechanism is based on the use of NIR excitation, thus it can overcome the concerns related to radiation absorbance and substantial autofluorescence of tissues in vivo. It is worth reminding that in vivo imaging is more challenging if compared to in vitro processes, due to the diffusion and absorption of the light by the tissues and to the autofluorescence emitted from intrinsic molecules. The near-infrared region of the spectrum offers deeper photon penetration and reduced scattering in comparison to the ultraviolet excitation source. In addition, some imaging setups require multiplexing analysis, crucial for the simultaneous recording of multiple

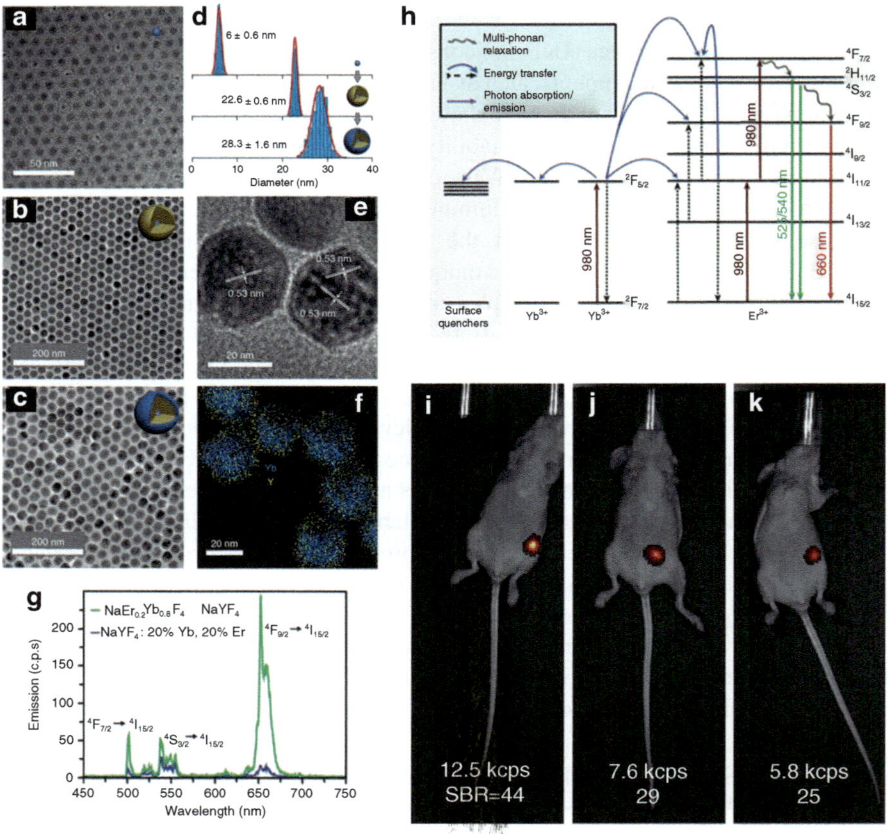

Fig. 3 (**a–f**) Structural characterizations of core/shell/shell $NaYF_4/NaYb_{0.98}F_4:Er_{0.02}/NaYF_4$ UCNPs: (**a–c**) TEM images of core $NaYF_4$ (**a**), $NaYF_4/NaYb_{0.98}F_4:Er_{0.02}$ (**b**) and $NaYF_4/NaYb_{0.98}F_4:Er_{0.02}/NaYF_4$ (**c**). (**d**) Size distributions of the UCNPs at different stages during core/shell/shell structure synthesis. The red lines represent Gaussian fits. (**e**) High-resolution TEM image, and (**f**) corresponding elemental mapping of the UCNPs obtained via energy dispersive X-ray spectroscopy (EDX) (Liu et al. 2018). (**g**) Emission spectra of single core/shell $NaYb_{0.8}Er_{0.2}F_4$ UCNPs and $NaYF_4$: 20% Yb^{3+}, 20% Er^{3+} UCNPs under 2×10^6 W cm^{-2} 980 nm laser excitation (Tian et al. 2018). (**h**) Multiphoton energy absorption, transfer, surface loss, and emission in Yb^{3+}/Er^{3+} UCNPs (Tian et al. 2018). (**i-k**) Deep-tissue imaging of the UCNPs: 12-nm core/shell $NaEr_{0.6}Yb_{0.4}F_4$ (8-nm core with 2-nm shell) UCNPs was injected into mammary fat pads 3–4 mm below the skin, and then imaged at (**i**) 2 hr., (**j**) 4 hr., and (**k**) 6 hr. after injection. Laser intensity is 0.1 W cm^{-2} focused at or near the injection site, and emission is from the Er^{3+} $^4S_{3/2}$ band (530–550 nm). SBR: signal-to-background ratio (Tian et al. 2018)

events or for the real-time labeling of different compartments and biological targets (Frangioni 2003). In this regard, the UCNPs-based visualization systems enable multimodal imaging, due to the possibility of doping with appropriate ions, which can be exploited in MRI, CT, or PET (i.e., Gd^{3+}, Lu^{3+}, and $^{153}Sm^{3+}$, respectively) (Qiu et al. 2013).

Environmental-Responsive Hybrid Nanoparticles

Besides using nanoparticles as contrast agents for bioimaging, luminescent fluorophores that have microenvironment-responsive emission characteristics are suggested for active diagnosis of diseases in the body. For example, pH-sensitive fluorophores attachment onto the nanoparticles can report the information of nanoparticle location whether in extracellular region (neutral) or endosome/lysosome (acidic) upon local microenvironment (Fig. 4a–d) (Semmling et al. 2008; Hartmann et al. 2015). This is particularly very useful to understand the cellular trafficking of nanoparticles, and gives a critical information for uptake mechanism and kinetics of nanoparticles. Regarding the pH-sensitive detection, activatable nanoparticles formed with self-assembled covalent composition of fluorescent organic dyes and crosslinking agents (i.e., tertiary amine) have been reported to undergo a dramatic transition within a narrow range of pH changes (0.3 pH unit or less) (Zhou et al. 2012; Wang et al. 2014). The pH-induced transition can lead to rapid and complete dissociation of the composites, resulting in change of the dye's fluorescent state from a self-quenched "off" to a brightly emissive "on" state. This nonlinear response to local pH changes is very sensitive, thus enables to target the acidic microenvironment in tumors. It is an important feature to address the heterogeneity of solid tumors, indicating a potential breakthrough to solve the challenges for diagnostic and therapeutic strategies based on molecular recognition and receptor targeting (Fig. 4e). By targeting the indication of acidic microenvironment in tumors, tremendous opportunities to diagnose and treat a broad range of solid tumors have been claimed. In addition, the microenvironment-sensitive detection can be improved by using multiple emission probes, and even more open up potential uses of fluorescent nanoparticle couples by leading to bright emission signals amplified from energy transfer between two neighboring fluorophores. To determine the local concentration of particular analytes in the vicinity of the nanoparticles, target-sensitive fluorophores can be also combined to report the extent of each analyte (Riedinger et al. 2010; Kantner et al. 2015).

Magnetic Resonance Imaging

T_1 Contrast Agent

Magnetic resonance imaging (MRI) is a noninvasive technique based on nuclear magnetic resonance to visualize inner parts of the body. In principle, the working mechanism of MRI is based on the interaction of nuclei with surrounding molecules (typically, water for clinical imaging) in the external magnetic field, thus MRI possesses unlimited penetration depth into the body while it has no need for ionizing radiation (James and Gambhir 2012). Magnetic relaxation of the hydrogen nuclear spin of water molecules in the tissue is recorded with high spatial resolution and contrast, thus clinically relevant to monitor anatomical information and functionality of organs. However, MRI suffers from relatively lower sensitivity in comparison to optical modalities (i.e., fluorescence imaging), leads to longer acquisition time and to use higher amounts of contrast agents which might cause harmful side effect.

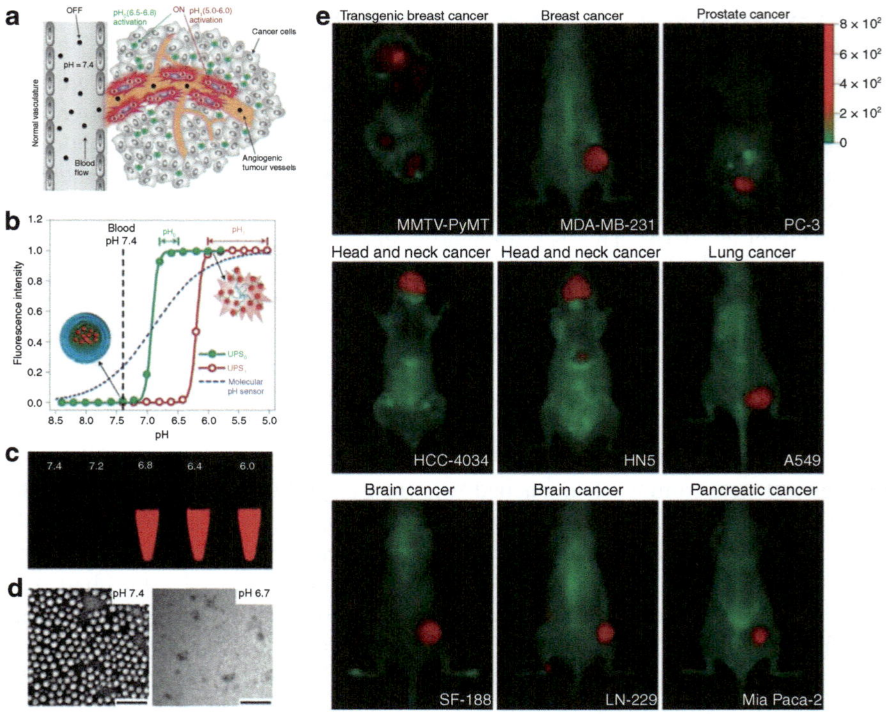

Fig. 4 (**a**) A schematic illustration of the tumor microenvironment-specific imaging using pH-sensitive nanoparticles: The nanoparticles stay "OFF" state at neutral pH (~7.4) during blood circulation. Upon reaching tumors, the nanoparticles are turned "ON" by acidic extracellular pH_e (6.5–6.8) in the tumor milieu, or endocytic organelles (pH_i, 5.0–6.0) in the tumor endothelial cells after receptor-mediated endocytosis (Wang et al. 2014). (**b**) Normalized fluorescence intensity as a function of pH for two types of ultra pH-sensitive nanoparticles (UPS_e and UPS_i). At neutral pH (~7.4), both nanoparticles stay silent (OFF state). At pH below their transitions (that is ~6.9 (UPS_e) and ~ 6.2 (UPS_i), respectively), the nanoparticles can be activated as a result of micelle dissociation (shown as schematic images). The blue dashed line simulates the pH response of a small molecular pH sensor with a pK_a of 6.9 based on the Henderson–Hasselbach equation. For the pH-sensitive nanoparticles, the pH response ($\Delta pH_{10-90\%}$) is extremely sharp (<0.25 pH unit between ON/OFF states) with >100-fold signal amplification. In contrast, small molecular pH sensors require 3 pH units for a comparable signal change (Wang et al. 2014). (**c**) Fluorescence imaging of UPS_e–Cy5.5 nanoparticles solution in different pH buffers ($\lambda_{ex}/\lambda_{em}$ = 675/710 nm) (Wang et al. 2014). (**d**) TEM images of UPS_e nanoparticles at pH 7.4 and 6.7 (scale bars, 100 nm) (Wang et al. 2014). (**e**) The nanoparticles enable to target both acidic pH_e and tumor vasculature with broad tumor specificity: The nanoprobes show broad tumor imaging specificity and efficacy in ten different tumor models of different cancer types (breast, prostate, head and neck, lung, brain, and pancreatic cancers) and organ sites (Wang et al. 2014)

Applying higher magnetic fields (typically, larger than 5 T) could give higher resolution with enhanced signal-to-noise ratio and rapid scanning, but this also raises concerns on safety for human body (Formica and Silvestri 2004; Hartwig et al. 2009). A hyperpolarized MRI, developed by polarizing the nuclear magnetic moments far beyond thermal equilibrium, has been known to improve the sensitivity of particular nuclei magnetization and enables real-time quantitative analysis (Leawoods et al. 2001).

To enhance the sensitivity and image quality, contrast agents have been widely studied and applied over the past decades. Many MRI contrast agents typically shorten the relaxation time of protons, which is located in the surrounding regions, resulting in improved morphological details and specific structures (Na et al. 2009). The MRI contrast agents is generally classified as T_1 and T_2 agents. The T_1 contrast agents, generally based on gadolinium (Gd) chelates, are paramagnetic ionic metals with unpaired electrons in the outer orbital. These ions respond and generate strong magnetic dipoles, inducing rapid T_1 relaxation of neighboring protons followed by positive contrast enhancement. Conventional T_1 contrast agents are based on small molecular chelating complexes with paramagnetic ions (e.g., Gd^{3+}, Mn^{2+}, or Fe^{3+}). However, these chelating complexes can circulate the blood vessel and rapidly clear through the renal pathway, resulting in limited acquisition of high-resolution images of desired sites. Nanoparticle-based MRI agents is suggested as they have advantageous features over those conventional chelating compounds (Johnson et al. 2011). Tailoring the size, shape, and composition, the nanoparticles can be modulated for controlled circulation time, specific targetability, and magnetic resonance characteristics (Fig. 5a–e). In this regard, the nanoparticle-based contrast agents allow to safely obtain physiological information more in detail complementing anatomical images.

On the other hand, superparamagnetic iron oxide nanoparticles can increase the rate of transverse T_2 decay due to inhomogeneity in the magnetic field. As a result, the accelerated transverse relaxation process (with shorter T_2 and higher r_2 values) causes an enhanced negative contrast (Mahmoudi et al. 2011). A number of formulations based on superparamagnetic iron oxide nanoparticles have been developed as contrast agents for MRI and several products, such as Resovist® (also known as Cliavist®), Combidex®, Endorem®, and Sinerem®, have been approved by FDA for clinical uses. Among them, Endorem® is one of the first commercially available compounds, consisting of an aqueous colloidal suspension of superparamagnetic iron oxide nanoparticles (Laurent et al. 2008). The iron oxide nanoparticles were synthesized by a coprecipitation method, and further functionalized with a biopolymer dextran. The superparamagnetic core is 10–20 nm in diameter and the final hydrodynamic size upon dextran coating increases to 120–200 nm. These biocompatible forms of nanoparticles have been tested for clinical uses in the diagnosis of liver lesions. Indeed, the macrophages that are localized in the healthy region of the liver can readily uptake the iron oxide nanoparticle very quickly after intravenous administration (less than 5 min) as they recognize the foreign substances. This leads to a decrement of T_2 value in the healthy region, consequently, the abnormal lesions can be recognized through MRI. In contrast, damaged region is not stained with the nanoparticle owing to the absence of macrophages, thus the MR imaging contrast is highly enhanced (superior signal-to-noise ratio). However, this is limited to only the case of liver diseases diagnostics because the selective sequestration of iron oxide

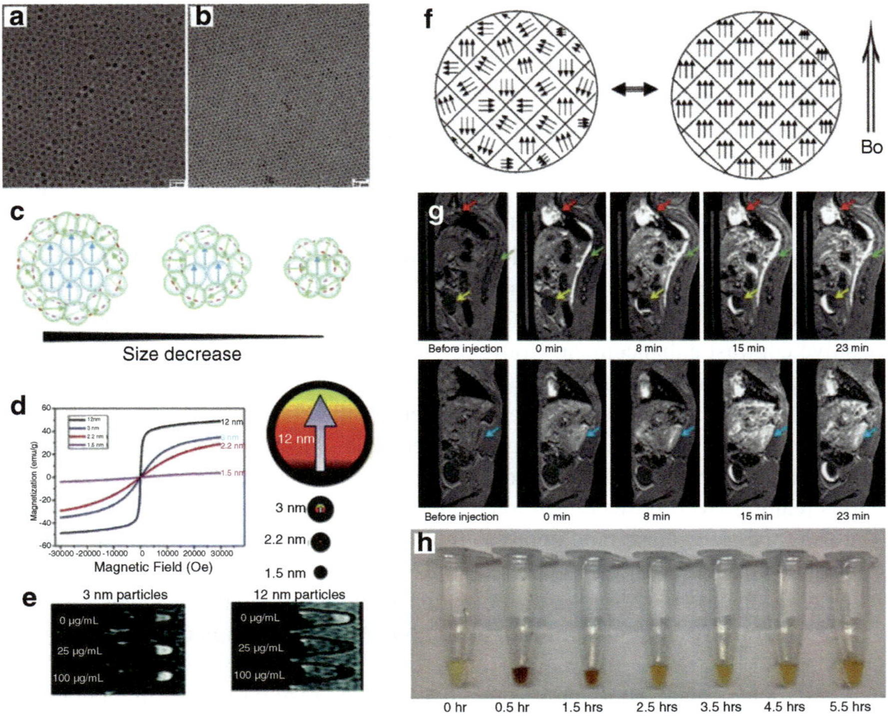

Fig. 5 (**a, b**) TEM images of superparamagnetic iron oxide nanoparticles obtained via thermal decomposition of iron oleate precursor: (**a**) ~7.0 nm and (**b**) ~ 5.5 nm in diameter, respectively (Wei et al. 2017). (**c–e**) Size effects of iron oxide nanoparticles: (**c**) Schematic illustration depicting the increased surface effects on magnetic momentum, (**d**) magnetization of different sized iron oxide nanoparticles, and (**e**) MRI phantom images of 3 nm and 12 nm iron oxide nanoparticle (Bao et al. 2018). (**f**) Crystal lattices structure of superparamagnetic iron oxide nanoparticles in the absence of an external magnetic field: the orientation of the magnetic domains is random. An external magnetic field (B_0) causes the magnetic domains of the nanoparticle crystal to reorient according to B0, which is reversible after the B_o is removed (Wang and Idee 2017). (g) T_1-weighted MR images of a mouse injected with iron oxide nanoparticles at 7 T: (Upper panel) one sagittal slice showing the heart (red arrow), the vena cava (green arrow), and the bladder (yellow arrow), and (Bottom panel) another sagittal slice showing the kidney (blue arrow). It should be noted that the heart (red arrow), the vena cava (green arrow), and the bladder (yellow arrow) do not show appreciable positive contrast before the injection of iron oxide nanoparticles. After injection, the heart and the vena cava display high positive contrast immediately after injection. At 8 min after injection, the bladder displays some positive contrast, indicating excretion of urine containing the nanoparticles (Wei et al. 2017). (h) Urine samples from mice taken at different time points post-injection showing renal clearance of the nanoparticles in vivo. With the increase of time post-injection, the positive contrast region of the bladder increases, corresponding to the accumulation of urine with the iron oxide nanoparticles (Wei et al. 2017)

nanoparticles to the macrophages in the healthy liver is driven by the low colloidal stability in vivo.

Recent advancements in nanoparticle synthesis and surface treatment methods have opened great potential for wide range of biomedical applications. Ultrasmall-sized iron oxide nanoparticles with narrow size distribution have been subjected to clinical trials since they are considered to enhance the physiological stability of nanoparticles, especially in biological fluids (Fig. 5g, h). Increasing the circulation time of the nanoparticles is crucial for specific targeting of the diseases over unavoidable liver uptake. However, iron oxide nanoparticles have been associated with certain toxic effects needed to be taken into account for safe uses in clinic (Khawaja et al. 2015; Singh et al. 2010). Other than iron oxide nanoparticles, researchers have also struggled to develop functional nanoparticles for enhanced MRI measurement and safe administration. Efforts have tailored the size, shape, and compositions of inorganic nanoparticles, thus controlling the circulation time, targeting efficiency, and magnetic properties to meet the biological requirements for optimized imaging. For example, melanin nanoparticles are suggested as a tumor-targeting MRI agent by chelating with Fe^{3+} ions and subsequent conjugation with cyclic arginine–glycine–aspartic acid (RGD) peptides. These nanoparticles feature the unique characteristics (served as T_1 contrast agents) for targeted MRI of U87MG glioblastoma (Fan et al. 2014; Qin et al. 2013).

T_2 Contrast Agent

As T_2 contrast agents, superparamagnetic iron oxide nanoparticles establish a substantial dipolar field perturbed locally to shorten proton relaxation of the surrounding tissues (Fig. 6). Magnetic momentum arisen by external magnetic fields is strongly dependent on the size and crystalline anisotropy of the magnetic domain. In this regard, for example, oxtapod-shaped iron oxide nanoparticles exhibit a significantly enhanced transverse relaxation behavior compared to spherical nanoparticles. Differences in effective radius and inhomogeneity of the local magnetic field contribute to the dramatic increment of MRI signals, thus particularly increases the detection sensitivity to diagnose microstructured tissues such as early state of cancers (Zhao et al. 2013). On the other hand, due to the negative contrast effect for T_2 transverse relaxation, it is still a major challenge to distinguish whether the region of signals is induced by iron oxide nanoparticles from low-level background magnetic resonance signals or originated from adjacent tissues such as bone, vasculature, fat, hemorrhage, and blood clots.

As a combination of paramagnetic and superparamagnetic properties, T_1/T_2 dual-mode contrast agents have been also suggested with formation of core-shell nanoparticles (Ali et al. 2011), because they help validating reconstruction and visualization of the detected MR signals in more accurate and reliable analysis. However, potential interference of MR read-out occurs inevitably in those hybrid structures as integrated in close proximity. For instance, the magnetic field induced by the T_2 contrast material perturbs the relaxation process of paramagnetic T_1 contrast materials, resulting in undesirable quenching of the T_1 signal. Therefore, the hybrid core-shell structured nanoparticles should reduce the magnetic coupling by introducing a

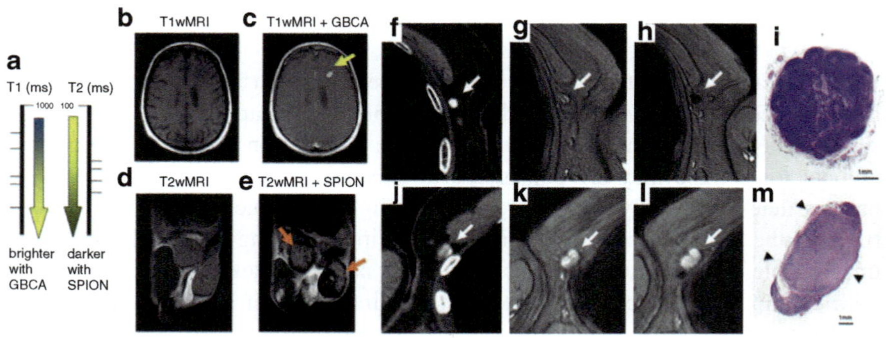

Fig. 6 (**a**) Two major classes of MRI contrast agents: Paramagnetic gadolinium-based contrast agents (GBCA) are considered T_1-positive contrast agents, by decreasing the spin-lattice T_1 relaxation time, they produce bright T_1 images. Superparamagnetic iron oxide (SPION) causes negative T_2 contrast, thus deceased spin–spin T_2-relaxation time produces the darkening of T_2-weighted images. (**b–e**) Comparison of T_1 and T_2 contrast MR imaging. (**b**) pre- and (**c**) post-GBCA T_1-weighted MRI on a brain metastasis in a melanoma patient (15 min post-injection); (**d**) pre- and (**e**) post-SPION T_2-weighted MRI on inflamed mouse mammary gland tumors (24 h post-injection) (Serkova 2017; Zhou and Lu 2013; Anchordoquy et al. 2017). (**f-m**) MR imaging obtained with FDA-approved Resovist® as a contrast agent, demonstrating metastasis negative and positive lymph nodes. (**f**) CT lymphography demonstrates a sentinel node (arrow); (**g**) corresponding node is identified on pre-SPION-contrast T_2-weighted axial MRI (arrow), showing high signal intensity; (**h**) after administration of SPION, the node shows strong SPION enhancement and was diagnosed as benign (arrow); (**i**) histologic findings confirmed it accordingly. (**j**) CT lymphography demonstrates a sentinel node (arrow); (**k**) corresponding node is identified on pre-SPION-contrast T_2-weighted axial MRI (arrow), showing high signal; (**l**) after administration of SPION as a contrast agent, the node shows no SPION enhancement and is diagnosed as malignant (arrow); (**m**) histologic findings confirmed it as malignant. This node is almost entirely replaced by metastatic tissue (arrowheads) (Motomura et al. 2011)

separating layer between the superparamagnetic and paramagnetic domains. Such magnetical decoupling provides superior MRI contrast in both T1 and T2 mode, and further enables to adopt "AND logic gate" to reduce the susceptible artifacts, resulting in self-confirmation of the measured images to enhance the accuracy of the MRI (Shin et al. 2014; Choi et al. 2010). In contrast to core-shell structures, dumbbell-like nanostructures that combine T_1 and T_2 contrast agents together in a single formulation with nanoparticles as a bridge spatially separate two contrast agents at a sufficient distance to reduce their magnetic interference or coupling (Cheng et al. 2014a). Unlike core-shell structures, these dumbbell-like nanostructures allow both T_1 and T_2 contrast agents to be directly exposed to neighboring environment with negligible magnetic properties compromising. Moreover, both surfaces of T_1 and T_2 contrast agents can be functionalized with each chemical or biologics for specific targeting and enhanced stability.

Smart Nanomedicine for MRI

Besides simply being a contrast agent for MRI (Tassa et al. 2011; Hu et al. 2006), advanced formulations of MR-responsive nanoparticles that have additional functions such as tumor treatment have been widely investigated over the past decades.

These are another type of nanomedicine, a similar platform to the hybrid luminescent nanoparticle (discussed in previous Section Environmental-Responsive Hybrid Nanoparticles), actively responsive to tumor microenvironmental changes to instigate the diagnostic and therapeutic mechanisms (Lee et al. 2012; Kleijn et al. 2011). In this case, microenvironmental stimuli such as pH, temperature, metabolites, ions, and proteases trigger the extra functions (Delcea et al. 2011). As the tumor microenvironment regulates tumor progression and metastatic spread of cancer, the responsive smart nanoparticles are capable of reporting physicochemical properties of the tumor microenvironment, thus diagnostically visualize the early stage of cancer through MRI. In other applications, pH-responsive nanomedicine containing MRI contrast agents is of particular interest since the pH is an important physiological parameter as dysregulation can be a cancer marker. For instance, pH-responsive ^{19}F-probe for detecting pH transitions in the range from 5.5 to 7.4 have been demonstrated to actively target and visualize through MR signals (Huang et al. 2013). Fluorinated polymers with tertiary amines form nanoparticles capable to determine the local pH as a consequence of the amine protonation and corresponding disassembly of the micellar polymer that activates ^{19}F-MRI signals to be sensitively detected.

Another class of smart nanomedicine for MRI is thermosensitive probes for detecting temperature changes in different tissues (Zhu et al. 2015), as temperature differences between tissues is also one of the common signatures in tumors. A combined nano-formulation containing temperature-sensitive liposomal ^{1}H chemical exchange saturation transfer (CEST) (Rivlin et al. 2013) and ^{19}F magnetic resonance (MR) contrast agent is a potential carrier for MRI-guided drug delivery system (Langereis et al. 2009). In this platform, the liposomal nanoparticles contain both a chemical shift agent and a fluorinated compound (Fig. 7). Upon reaching the melting temperature of the lipid membrane, the nuclear magnetic contrast (driven

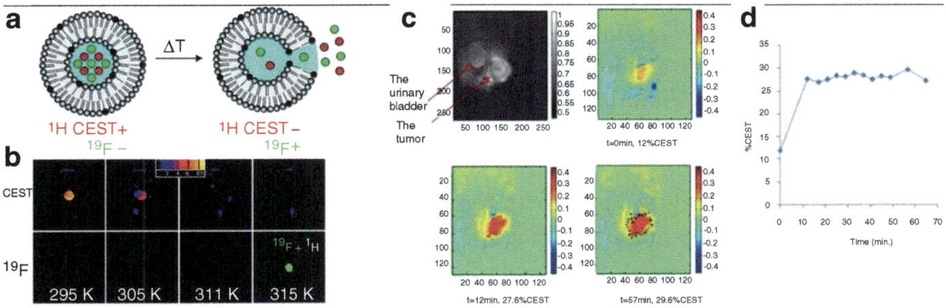

Fig. 7 (**a**) Schematic illustration depicting ^{1}H liposomal CEST and ^{19}F MR contrast agent (Langereis et al. 2009), and (**b**) corresponding MR images of temperature-sensitive liposomes on a clinical 3.0 T MRI scanner. The CEST signal (color scale in percent) vanished at $T \geq 311$ K while the fluorine signal appeared at 315 K (overlay with the ^{1}H image for colocalization and clarity) (Langereis et al. 2009). (**c-d**) CEST MRI kinetic measurements in the tumor at different times following injection of 2-fluoro-2-deoxy-D-glucose (FDG) (Rivlin et al. 2013). (**c**) The T_2-weighted image identifies the location of the tumor prior to administration of FDG. The marked ROI was used for the %CEST calculation (**d**) (Rivlin et al. 2013)

by CEST) vanishes, due to the release of the chemical shift agent while the paramagnetic shift agent induces appearance of the ^{19}F-MRI signal. This hybrid structure allows demonstrating dual functionality including "on" and "off" switch system for simultaneous diagnosis and signaling.

Redox reactions participate in many biological processes, and play crucial roles in major chemical mechanisms in organisms. Therefore, abnormal redox reactions are implicated in various conditions of diseases including liver damage and human immunodeficiency virus (HIV). Nanoparticle-based MRI agents can be also used to diagnose tumor microenvironment through redox reaction tracking. For example, once the Gd^{3+} complexes were bound to the surface of ^{19}F-containing nanoparticles via disulfide bonds, closer distance between Gd^{3+} and ^{19}F reduces the transversal T_2 signal by the paramagnetic relaxation enhancement effect.

Computed Tomography

X-ray radiation has led to a huge breakthrough for noninvasive anatomical imaging of the human body in high resolution. Classical X-ray imaging, in general, utilizes the tissue-specific attenuation of X-ray transmission to generate contrast in the radiographs. Differences in the electron density is a key origin to obtain a distinguished contrast of each tissue. For instance, bones have much higher electron density than soft tissues, thus generate more contrast to clarify the skeletal structures. In order to enhance the low contrast of soft tissue and clearly visualize it, high electron density elements such as iodine or barium are used to develop contrast agent, which is administrated to the blood vessel and allows to image the soft tissues including gastrointestinal (GI) tract and tumors. Computed tomography (CT) is a whole-body imaging technique using X-ray as a radiation source, and widely used in medicine. X-rays emitted from source generator pass through the body, and the penetrated radiation reaches the detector to record the signals. The X-ray source and the detector (scanner) are synchronized in motion and rotate around the body to acquire 3D anatomical images. Each 2D planar images are reconstructed through computer-assisted algorithms to build up inner parts of the body. Since the X-rays are produced outside the body and modulated by the electron density of each tissue, the large photon flux improves superior signal-to-noise ratio, leading CT to outperform other 3D imaging modalities in terms of spatial resolution. Therefore, CT imaging is widely used in clinical diagnostics owing to the simplicity of the instruments and rapid image generation, and relatively lower cost.

To obtain an accurate and rapid reconstruction of 3D anatomical images, much effort have been made to develop innovative contrast agents, particularly based on nanoparticle formulations (Fig. 8) (Cormode et al. 2014). Hypersensitivity of many patients to iodinated compounds is another reason for new CT contrast agents (Shilo et al. 2012). Nanoparticles provide several advantages in terms of sensitivity, stability, and functionality compared to conventional CT contrast agents. Firstly, gold nanoparticles are engaged to develop highly sensitive CT imaging with low abundance of contrast agent administration. Thanks to the high atomic number and

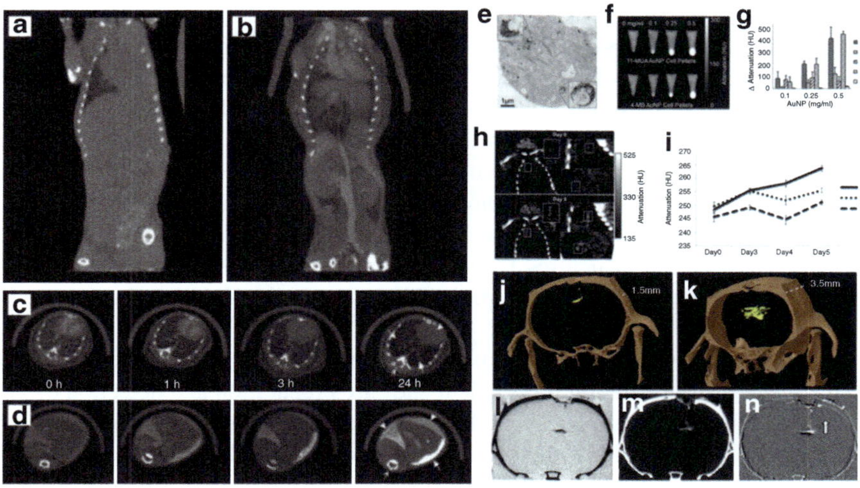

Fig. 8 (**a–d**) Serial in vivo imaging of vasculature: (Rabin et al. 2006) (**a, b**) Coronal curved multiplanar reformatted reconstructions of a live BALB/c mouse before (**a**) and after (**b**) intravenous administration of bismuth sulfide PVP nanoparticles (BPNPs) as a contrast agent, showing the vascular enhancement of large vessels and the heart. (**c, d**) Serial CT scans of a live BALB/c mouse following tail-vein injection of BPNPs suspension (0.228 M bismuth). (**c**) Transverse slices through the heart (arrowhead), the lungs (solid arrow), and a vertebra (open arrow). (**d**) Transverse slices including the liver (arrowhead), the spleen (solid arrow), and a vertebral body (open arrow) (Rabin et al. 2006). (**e**) TEM image of a monocyte cell after 24 h incubation with gold nanoparticles (black dots) (Chhour et al. 2016). (**f**) CT images of pellets of monocytes that had been labeled with 11-MUA (up) or 4-MB (bottom) coated gold nanoparticles (Chhour et al. 2016). (**g**) Quantification of CT attenuation for each gold nanoparticle formulation (Chhour et al. 2016). (**h**) CT scans of an atherosclerotic mouse before (day 0) and after (day 5) injection with gold nanoparticles-labeled monocytes. The boxed area indicates the aortic region of interest (Chhour et al. 2016). (**i**) Average CT attenuation in the aortas of mice (Chhour et al. 2016). (**j-k**) 3D in vivo volume-rendered micro-CT scans of rat brains acquired one month after injection with (**j**) gold nanoparticle-labeled hMSCs and (**k**) free gold nanoparticles (Betzer et al. 2014). Note that the distinctive migration behavior of the gold nanoparticle-labeled hMSC was detected while free gold nanoparticles spread and scattered into many regions of the brain. The result supports the hypothesis that hMSCs navigate and home to regions of the brain involved in depression. (**l-n**) Coronal brain slice (**l**) 1 week post-transplantation, (**m**) 3 week post-transplantation, and (**n**) Overlay of (**l**) and (**m**) showing the migration pattern of the cells (Betzer et al. 2014)

electron density (79 and 19.32 g/cm^3) of gold over iodine, the base element for typical CT contrast agent (53 and 4.9 g/cm^3), respectively), gold nanoparticles show superior attenuation coefficients, indicating potential uses as contrast agents for X-ray imaging, CT, and micro-CT (Bao et al. 2014; Lin et al. 2009). Once these gold nanoparticles are further functionalized with targeting molecules such as folic acid, they can exhibit specific accumulation to the folic acid receptor-positive cells (e.g., MGC-803 cells in vitro, and gastric cancer tissues in mice in vivo) thus highlight distinct tissue structures. In addition to CT imaging, fluorescent dye-labeled gold nanoparticles are used as a dual imaging modality (Zhou et al. 2013; Zhang et al. 2015). Besides gold, the nanoparticles composed of high atomic number elements such as bismuth, bromine, tantalum, platinum, ytterbium, yttrium, gadolinium, and tungsten have also shown suitability for CT imaging. However, the potential toxicity of these materials has limited clinical translation. Several strategies have been suggested to address this drawback, and the formulation of core-shell structured nanoparticles is one of the widely adopted way. The CT contrast agent "core" is encapsulated in the biocompatible "shell" such as liposome, silica, or polymeric nanoparticles to reduce adverse side effects (Jakhmola et al. 2012). Moreover, hybrid nanostructures based on the combination of different domains exhibit various functionalities and exploit CT imaging and other diagnostic or therapeutic techniques.

In parallel to the development of new CT contrast agents, there have also been progresses in instrumentation for efficient and safe preparation and detection of the contrast agents. For example, novel X-ray sources such as synchrotrons provide tremendously more flux and a higher degree of coherence, leading to new approaches such as phase-contrast CT, diffraction enhanced imaging, and holotomography (Momose et al. 1996; Chapman et al. 1997; Cloetens et al. 1999). Exploiting the intrinsic nature of the X-ray wave contributes to better imaging and increased sensitivity. In case of the phase contrast CT, it provides enhanced contrast ratio over 200 times higher than conventional CT imaging based on absorption (Mohammadi et al. 2014; Beltran et al. 2010). Therefore, high spatial resolution target-specific 3D organ imaging can be obtained (Dullin et al. 2015). It has been achieved up to 9 μm-scale in resolution, which enables to visualize the macrophage function within the local 3D environment (Cloetens et al. 1999). The spatial resolution could be further improved to 400 nm by using holotomography (Krenkel et al. 2015).

The CT imaging-based clinical diagnosis is also further changed in trend and potentially improved by utilizing the nanoparticles as contrast agents. As described above, a large dosage of iodine or barium is currently composed in CT contrast agents to improve the signaling sensitivity and imaging resolution, although those elemental compounds are not tolerated for patients. In addition, these traditional contrast agents are hard to specifically target the diseased sites and/or bind to cellular receptors for efficient accumulation in the pathological tissues such as tumors owing to lack of homing motifs. However, by applying the nanoparticles-based contrast agents which is functionalized with targeting moieties such as antibody or peptide, the energy-resolved detectors have allowed to achieve more clear discrimination and multitude imaging of various target in a single CT scan (Cormode et al. 2010). For this purpose, the direction for challenging the current limitations and drawbacks is toward design and engineering the biocompatible nanoparticles with high atomic number elements. Conjugation of targeting moieties that specifically homes cellular receptors, which is strongly related to diseased or pathological target tissues is another promising strategy for next-generation CT imaging and corresponding early diagnostics in clinic. Moreover, since novel CT techniques are not only limited to focusing on X-ray attenuation but also utilizing the phase shift or scattering as sources of contrast, various structures of nanoparticles such as hollow spheres or core/shell formation will become more important (Tang et al. 2011). Despite the unexpected side effects of the applied X-ray dosage, the CT is still the most highly adopted imaging technique for clinical diagnosis of deep tissues. Such a high demand leads to the development of improved instrumentation with the implementation of novel algorithms and detectors, especially newly developed reconstruction techniques, resulting in reduced X-ray dosage used in patients (Silva et al. 2010).

Therapeutic Nanomedicine

Applications of nanoparticles for therapeutic treatment is also a rapidly growing area, although only a few nanoparticle-based formulations reach the market for safe and effective clinical translation. Magnetic nanoparticles, for example, are one of the leading formulations for therapeutic applications because they are already widely employed for MRI as discussed (see Section "Photothermal Therapy"). Magnetic nanoparticles are further adopted as a magnetic hyperthermia agent for cancer treatment, which has been highly investigated for clinical trials (Johannsen et al. 2007; van Landeghem et al. 2009). By applying an external magnetic field, intravenously injected magnetic nanoparticles can cause local temperature increment up to the therapeutically effective temperature (e.g., 43–45 °C). However, a critical issue is the minimum concentration of the nanoparticles to obtain the significant increase of temperature because the intravenous administration of the nanoparticles hardly guarantees the appropriate amount of accumulation which essentially determines whether enough temperature can be achieved or not. Although the most effective approach to achieve the significant concentration of nanoparticles is the direct injection to the pathological site such as cancer or infected tissue, this limits clinical uses due to potential intracellular hyperthermia.

Photothermal and photodynamic treatments are also commonly used techniques for therapeutic nanomedicine. Photothermal therapy (PTT) is based on the several metallic nanoparticles, which is readily prepared through solution-based synthesis. Compared to magnetic hyperthermia, relatively facile equipment is required to trigger the PTT stimulus: an infrared laser is simply irradiated to induce the local heating at the site of nanoparticle accumulation. Furthermore, co-existence of two or more domains in a single nanoparticulated system (e.g., core-shell structure), which can perform different functions at the same time, enabling both diagnosis and treatment of a diseased tissue. The combination of diagnosis and therapy is referred to as "theranostics," indicating the tools that can be employed for a simultaneous role in the body: diagnosis (e.g., MRI and CT) and direct therapy (e.g., PTT) or delivery of therapeutic drugs at a time. In this section, therapeutic applications and corresponding principles of nanomedicine are thoroughly discussed.

Magnetic Hyperthermia

Hyperthermia treatment aims to raise the local temperature of cells or tissues. For example, to treat the cancerous tissue, heating up to 40–43 °C has been known to cause cytotoxic effect, thus eventually induces cancer cell death (Fig. 9) (Wust et al. 2002). Clinical and preclinical studies have extensively shown that hyperthermia has a great potential to improve the anticancer performance while combining with chemo- or radiotherapy while noticeable side effects to healthy region of the body are suppressed (De Haas-Kock et al. 2009; Lutgens et al. 2010). However, hyperthermia is rarely incorporated into anticancer treatment in clinic owing to limited effectiveness as a single modality treatment despite the promising potential. In addition, it should be noted that not only malignant cells, but also nonmalignant cells are quite sensitive to heating. This is a major drawback of conventional hyperthermia whereas chemo- or radiotherapy has more cytotoxic toward malignant cells (Dewey et al. 1977; Roizintowle and Pirro 1991). Therefore, targeted hyperthermia using magnetic nanoparticles as a local heating probe have attracted significant interest. Thanks to the recent progress in rational design and synthesis of iron oxide

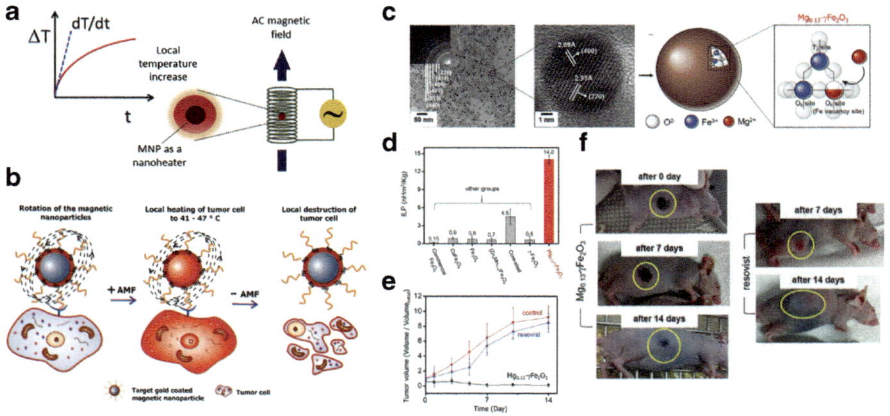

Fig. 9 (**a**) Schematic illustration of a magnetic nanoparticle (MNP)-based hyperthermia setup (Rivas et al. 2016). Application of an alternating current (AC) magnetic field on the nanoparticle induces a local heating that can be measured in temperature increase as a function of the time. (**b**) Principle of magnetic hyperthermia-mediated cell death. Targeted magnetic nanoparticles delivered to tumor cells are exposed to an AC magnetic field (AMF), then AMF energy is converted into heat by the magnetic nanoparticles, leading to local temperature increase over the tolerant range of tumor cells, typically between 41 and 47 °C (Belyanina et al. 2017). (**c**) TEM images of Mg-doped γFe_2O_3 nanoparticles (Inset: selected area electron diffraction patterns) and a schematic diagram of spinel structure (Jang et al. 2018). (**d**) Comparison of the intrinsic loss power (ILP) value of as-developed Mg-doped Fe_2O_3 nanoparticles to previously reported SPIONs. Two types of commercial Fe_3O_4 (Feridex and Combidex) are used for baseline (Jang et al. 2018). (**e**) Tumor volume changes upon magnetic hyperthermia treatment with the Mg-doped Fe_2O_3 nanoparticles and commercially available Resovist® (Jang et al. 2018). (**f**) Photographs of xenografted nude mice after magnetic hyperthermia treatment with Mg-doped Fe_2O_3 nanoparticles and commercially available Resovist® for comparison (Jang et al. 2018)

nanoparticles, systemically administrated magnetic nanoparticles can be applied to heat the tumorous tissues under alternating magnetic field.

Ferrite nanoparticles including magnetite (Fe_3O_4) or maghemite (γ-Fe_2O_3) are commonly used for magnetic hyperthermia. They are typically in the size range between 10 and 100 nm on the basis of a single domain, and the magnetic properties arise from the presence of ions with different valencies in the crystalline structure (e.g., Fe_3O_4 consists of two trivalent iron (III) ions and one divalent iron (II) ion). Unpaired ions result in oppositely aligned parallel magnetic moments, subjecting to strong and spontaneous magnetization. When exposed to an external alternating magnetic field, the iron oxide nanoparticles generate heat mainly via two mechanisms, in general, hysteresis loss and relaxational loss. Firstly, the hysteresis losses occur in large-sized nanoparticles possessing multiple magnetic domains. The orientation of magnetic moments in such large nanoparticles will align continuously with the direction of the magnetic field, and by applying alternating magnetic field, this results in an energy difference followed by the heat generation (Kirschning et al. 2012). Therefore, the size of a magnetic nanoparticle is critical to generate heat. Smaller size that contains a reduced number of magnetic domains will also decrease heat generation until a single magnetic domain remains at a threshold size of approximately 30 nm (Houlding and Rebrov 2012). In addition, below this size, the magnetic iron oxide nanoparticles have superparamagnetic properties and heat is mainly produced by Néel relaxation and Brownian relaxation under the alternating magnetic field. Néel relaxation refers to rapid changes in the magnetic moment of nanoparticles when exposed to alternating magnetic fields, in which rapid realignment is opposed due to the crystalline structures, finally resulting in heat generation. Brownian relaxation refers to the frictional heat induced by the physical rotation of nanoparticles within a surrounding medium in the course of realignment under alternating magnetic field (Suto et al. 2009; Ng and Kumar 2017; Ruta et al. 2015).

Since the magnetic iron oxide nanoparticles have the advantageous features (long-term chemical stability, biocompatibility, and facile surface functionalization), compared to other types of magnetic susceptible materials such as certain metals (e.g., iron, nickel, or cobalt) or metal alloys (e.g., FePt and FeCo), (Dunn et al. 2014), they have promising potential as a tool for clinical anticancer treatment. Moreover, magnetic iron oxide nanoparticles can be utilized as a contrast agent for computed tomography (CT) and magnetic resonance imaging (MRI) as discussed in the Section "Therapeutic Nanomedicine". A few of these iron oxide nanoparticles have been highly investigated and approved by the FDA for the imaging purpose (Anselmo and Mitragotri 2015). This is particularly very useful since the concentration of the magnetic nanoparticle accumulated in the tumorous tissue can be estimated via CT scan, aiding the estimation of hyperthermia dosimetry (Johannsen et al. 2007). Magnetic nanoparticles have also been shown the enhanced therapeutic index of anticancer radio treatment without applying alternating magnetic field, because of incremental generation of reactive oxygen species through the Fenton reaction (Huang et al. 2010; Klein et al. 2012; Khoei et al. 2014; Bouras et al. 2015). Furthermore, localized heating induced by intracellular magnetic nanoparticles has known to drive lysosomal damage and cell death in the absence of bulk heating

(Creixell et al. 2011; Domenech et al. 2013). As an essential component of the human body, iron can be systemically administrated to large extent and safely demonstrated in clinical settings with minimal risk of toxicity (Hetzel et al. 2014).

Despite the promising advancement of magnetic hyperthermia using iron oxide nanoparticles, preclinical studies often apply field strengths, frequencies, or quantities of MIONs that are beyond the clinically feasible ranges. This indicates that further research should be warranted in wide areas of design, synthesis, functionalization, delivery, and the heating of nanoparticles to achieve clinical translation as a form of nanomedicine in the future.

Photothermal Therapy

Photothermal therapy (PTT) is another strategy utilizing local heating to treat cancer or infection, which gained a great deal of attention as an alternative for magnetic hyperthermia (Fig. 10). As most photothermal-inducing materials are nanometer scaled, photothermal therapy constitutes a category of nanomedicine as well. The PTT agents absorb light energy upon exposure to near-infrared radiation, which causes electron excitation and subsequent nonradiative relaxation (Melancon et al. 2011; Kennedy et al. 2011; Yang et al. 2013). This process leads to gain kinetic energy, resulting in heat generation on the surrounding medium of PTT agents (Camerin et al. 2005a, b; He and Bischof 2003). The PTT is a noninvasive thermal therapy in which the induced thermal energy causes cell membrane disruption or protein denaturation of the surrounding cells (van der Zee 2002).

Coupled with the PTT for anticancer treatment, a huge breakthrough in nanotechnology has opened the significant contribution of nanomaterials as a potential PTT agent. Due to the unique optical properties, noble metal nanoparticles are promising for PTT applications (El-Sayed 2001; Niemeyer 2001; West and Halas 2003; Katz and Willner 2004). Gold and silver nanoparticles exhibit the localized surface plasmon resonance (LSPR), which is a particular photo-physical phenomenon enhancing the absorption of photons (Dreaden et al. 2012, 2011; Huang et al. 2009; Lee et al. 2014). Consequently, they are considered effective PTT agents because hyperthermia can be correspondingly stimulated upon exposure to light, especially under near-infrared irradiation (Jain et al. 2008; Huang et al. 2007, 2008). Since the noble metal nanoparticles are eligible to efficiently absorb the light energy in 600–900 nm of wavelengths owing to the LSPR phenomenon, the absorbed photons are then converted to heat for cellular damage and necrosis of tissues in the particular site of nanoparticles targeted (Day et al. 2009). The surface plasmon resonance effect is that the free electrons in the conduction band oscillate around the nanoparticle's surface upon exposure to the electromagnetic field of light. The oscillation of the electron then leads to charge separation and dipole oscillation formation. Finally, the oscillation reaches the maximum, which is known as surface plasmon resonance at a certain frequency. This supports the increased light absorption of noble metal nanoparticles, followed by enhanced heat generation at the surface (Jain et al. 2006; Xie et al. 2017).

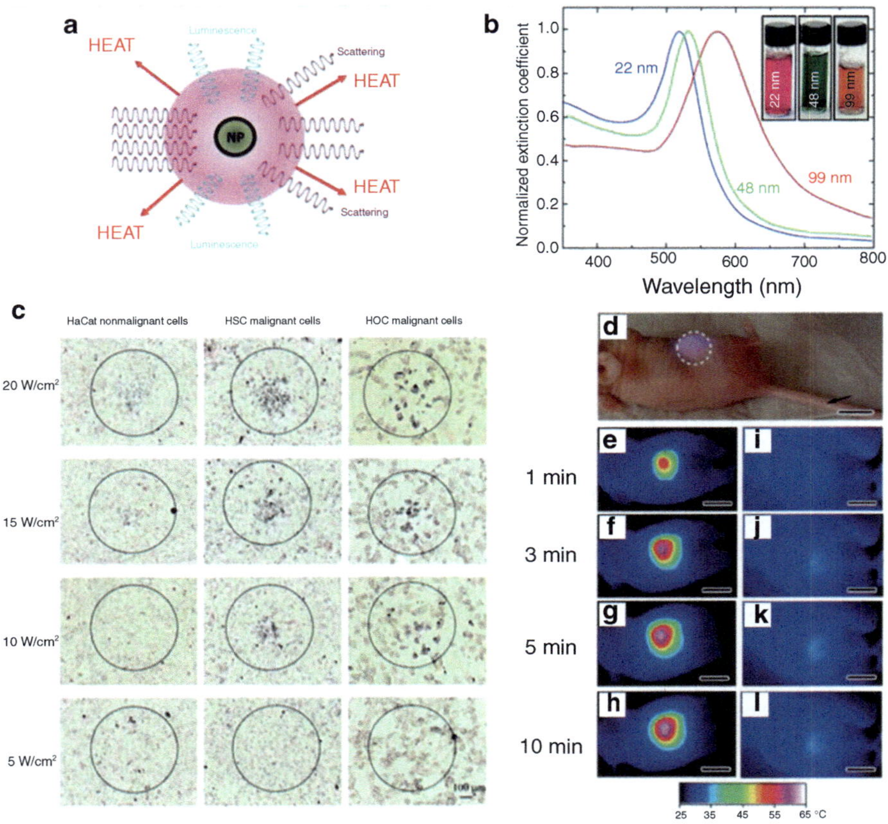

Fig. 10 (**a**) Schematic representation of the different processes activated when a light beam interacts with a nanoparticle (Jaque et al. 2014). Co-existence of scattering, luminescence, and heat generation should be indicated. Heat and luminescence occur as a result of light absorption. (**b**) Normalized extinction coefficients of three aqueous solutions of spherical gold nanoparticles with diameters of 22, 48, and 99 nm, respectively (Link and El-Sayed 2000). (**c**) Selective PTT for cancer cells in the NIR region by using anti-EGFR conjugated gold nanorods. After incubation with anti-EGFR conjugated gold nanorods, HaCat normal cells are destroyed at a laser power threshold of 20 W/cm^2, while HSC and HOC cancerous cells are destroyed at a much lower threshold of 10 W/cm^2. The difference reflects the much larger density of gold nanorods on the surface of the cancer cells compared to that on the normal cells. A dashed circle indicates the spatial extension of the laser spot. Cell damage is evidenced by the appearance of blue fluorescence emission generated by a damaged cell marker (Huang et al. 2006). (**d**) Photograph of a tumor-bearing mouse undergoing photothermal treatment after intravenous injection with a colloidal solution containing gold nanoparticles (Chen et al. 2010). (**e-l**) Thermographic images of (**e-h**) nanoparticle-injected and (**i-l**) saline-injected tumor-bearing mice at different irradiation times, indicating local temperature increase at the tumor (Chen et al. 2010)

Conversion of the absorbed light energy to thermal energy involves sequential physicochemical processes: the formation of a hot metallic lattice followed by cooling off for heat dissipation from the nanoparticle to the surrounding environment. Accordingly, the dissipated energy will heat up the cancerous cells around the noble metal nanoparticles (e.g., gold nanoparticles) causing cell destruction. However, such energy conversion and dissipation only occur when the frequency of the incident light is overlapped with the SPR absorption band (Xie et al. 2017; Link and Ei-Sayed 2003; Webb and Bardhan 2014). The incident laser light is generally in the visible to near-infrared region, and especially for in vivo treatment, the near-infrared irradiation has the advantageous feature to penetrate the skin without causing serious damage to normal tissues. Therefore, this noninvasive heating process has a great potential to kill the desired cells, upon low energy light illumination with minimal concerns about the safety (Kim et al. 2006). The near-infrared light is also very weakly absorbed by the normal tissue components such as water, hemoglobin, and melanin. With these features, many studies have opened the door for the PTT application of several nanoparticles, which strongly absorb the NIR and, as a result of SPR effect, generate sufficient heat to cause surrounding cell death.

This PTT has been considered as a promising anticancer treatment since it can be controlled spatiotemporally, thus avoiding damage to non-targeted regions (Fig. 10) (Melancon et al. 2011; Choi et al. 2011; Gobin et al. 2007; Lal et al. 2008; Yang et al. 2010). For effective PTT treatment, several fundamental prerequisites are required in depth. First, the template nanomaterials (e.g., gold or nanoparticles) should possess a photothermal effect that can efficiently convert light energy into heat energy (i.e., high photothermal conversion efficiency). Second, the photothermal effects should occur in response to near-infrared light to ensure deep tissue penetration as discussed above. Finally, the surface of the nanoparticles needs to be easily functionalized to targeting moieties while maintaining the efficient photothermal conversion efficient. As a result, the noble metal nanoparticles decorated with targeting ligands (e.g., antibody and peptide) can aid in specific homing and intracellular uptake, followed by near-infrared light irradiation to induce the photothermal effects within the target cell. A few nanoparticles that fulfill the aforementioned criteria have been employed for PTT. Not limited to noble metal nanoparticles (Chen et al. 2010; Link and El-Sayed 2000; Rycenga et al. 2009; Boca et al. 2011), sp^2 domain rich carbon nanoparticles as a form of nanotubes or graphene are also well known for photothermal converting agents (Yang et al. 2013; Marches et al. 2011; Liu and Liang 2012; Robinson et al. 2011). In addition, single layered transition metal dichalcogenides (TMDCs) (Cheng et al. 2014b; Liu et al. 2014a, b; Qian et al. 2015) and melanin-based compounds also have recently shown great potential for application as photothermal therapeutic agents (Liu et al. 2015, 2013; Song et al. 2015; Kim et al. 2016).

Photothermal therapy has a promising potential as translational medicine, thus tremendously applied with other therapeutic modalities. Since the primary purpose of photothermal therapy is eradication of the target cells only at the light-subjected region, various treatment strategies such as chemo- or radiotherapy have been combined with photothermal ablation for the synergistic therapeutic index (Lee et al.

2010). Combinations of PTT and chemotherapy have shown particular interest as it shows great advancement, thus many studies are focused on the optimization of the synergistic effect. For example, hydrophobic interactions between drug molecules and photothermal gold nanoparticles are commonly employed to load small molecular anticancer drugs such as doxorubicin or paclitaxel, then this hybrid nanomedicine formulation allows the chemotherapy and hyperthermia therapy simultaneously under light irradiation in vivo (Tang et al. 2012; Wang et al. 2013b; Zhang et al. 2011; Zheng et al. 2013). However, clinical translation is still a bit far to proceed because the potential risks of ablation are remained in question with ultimate fate of those noble metal nanoparticles inside the body. Recently, mild and non-ablative photothermal effect has been discovered and applied to overcome the limitations of conventional drug delivery systems, resulting in superior therapeutic efficacy (Weigel and Oka 1981; Wolkers et al. 2003). In contrast to the combination of chemotherapy and PTT, a synergistic therapeutics along with gene therapy and photothermal ablation remains at preliminary stages, and this will open up great potential of nanomedicine as a translational medical platform as it has impactful application area including personalized medicine.

Drug Delivery Systems Using Nanoparticles

Developments in the nanomedicine have provided enormous ways to improve delivery efficiency of therapeutic drugs to the target tissues. For instance, drugs with very low solubility possess various biopharmaceutical delivery concerns including limited cellular uptake upon administration, less diffusion capacity throughout the body, and unwanted side effects preceding traditional drug formulation process. These limitations could be overcome by the application of nanomedicine strategies in the drug delivery mechanism using nanoparticles. Nanoparticles are widely studied and applied in the drug delivery systems (DDS), and systemically administered therapeutic drugs show enhanced targetability and clinically feasible biodistribution. There are many examples of nanomedicine formulations for the DDS, such as liposomes, polymers, micelles, inorganic nanoparticles, and their hybrid structures. By delivering pharmacologically active agents more effectively and specifically while maintaining the intact activity to the pathological target site and by guiding them away from potentially endangered healthy tissues (e.g., liver or kidney), the nanomedicine as a drug delivering carrier aims to improve the balance between the efficacy and the toxicity of systemic (chemo) therapeutic interventions (Peer et al. 2007; Sanhai et al. 2008; Davis et al. 2008). Besides the targeted drug delivery to diseased sites for therapeutic purposes, nanomedicine formulations have been also used for diagnostic imaging and therapeutic monitoring systems. The theranostic (therapeutic + diagnostic) platform is the strategy where disease diagnosis and therapeutic treatments are combined (Caruthers et al. 2007; Xie et al. 2010; Sun 2010; Lammers et al. 2011). To this end, the drug delivering nanoparticles are co-loaded with both drugs and contrast agents. On the other hand, nanoparticles that have intrinsic capability to be used for imaging purposes as discussed in the previous section (see Section "Therapeutic Nanomedicine") are increasingly implemented for theranostics. Theranostic nanoparticles can help to report the location, identifying the stage of the disease, and providing information about the treatment responses. In addition, such nanoparticles can carry therapeutic agents and release desired amount of the therapeutic agent via molecular and/or external stimuli. In this section, several exemplary efforts will be presented, and suitable nanomedicine systems for monitoring drug delivery, release, and efficacy are also highlighted. Despite tremendous investigation and successful employment of various drug delivery systems in recent studies, there are still certain challenges that need to be addressed and a few requests for developing advanced platforms to be highly translated for clinically relevant drug delivery strategy to its target sites. Hence, the nanomedicine for targeted drug delivery systems is also currently in struggling field of study that facilitates the advanced system of therapeutics.

Design strategy of therapeutic drug has been extensively studied, and it is also one of the most attractive research fields on the applications of nanoparticles due to the potential advantages such as the possibility to modify physicochemical properties including the solubility, releasing profile, diffusivity, bioavailability, and immunogenicity (Fig. 11). Nanoparticles can consequently lead to support convenient administration, reduced toxicity, fewer side effects, improved biodistribution, and extended lifetime of the drug. Engineered drug delivery systems based on nanoparticles are intended for (1) the targeting to a specific location, (2) the controlled release of therapeutic agents at the target site, or (3) both targeted delivery and

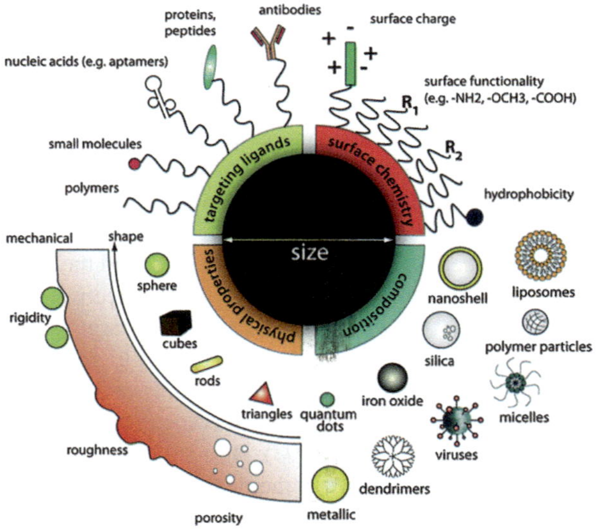

Fig. 11 Designing nanoparticles for intracellular applications: Nanoparticles can be modularly synthesized from different materials composition with various physical and chemical properties and functionalized with a myriad of ligands for biological targeting. Such flexibility in design freedom enables to prepare tailor-made nanoparticle for specific intracellular applications as contrast agents, drug delivery carriers, as well as theranostic platforms (Chou et al. 2011)

controlled release. Targetability is the most significant aspect of nanoparticles, which can deliver the therapeutic drugs to the specific site of the body. Typically, the delivery strategy using nanoparticles are classified as (1) passive accumulation and (2) active targeting. For active targeting, homing moieties, such as antibodies and peptides are coupled with nanoparticles to anchor them to the specific receptors (or derivative structures) that are highly expressed at the target site. The main targets are commonly the receptors on cell membranes, lipid components of the cell membrane, and antigens or proteins on the cell surfaces. In passive targeting, the drug-containing nanoparticles circulate through the bloodstream and are driven to accumulate in the diseased tissue by certain affinity or microenvironmental conditions such as local pH, temperature, molecular site, and shape. Cancer or inflammation causes the formation of complicated vascular structures, which induce enhanced permeation and retention (EPR) effect for the passive accumulation of the nanoparticles. This is certainly an effective strategy for improved drug delivery efficiency of nanoparticle-based formulation over intact drug compound itself.

Since the nanoparticles are devised as drug carriers, various types of nanoparticles are applied for enhanced drug delivery through improved biodistribution and reduced toxicological effects of therapeutics (Mirza and Siddiqui 2014; Thrall 2004). This has expanded the availability of many therapeutic drugs that pose serious challenges of delivery concerns. For example, doxorubicin is hardly cross the blood–brain barrier, but shows significantly incremental penetration to the brain by integrating with polybutylcyanoacrylate nanoparticles (Gulyaev et al. 1999). Moreover, nanoparticles have widened the scope of pharmacokinetics for insoluble drugs. For example, the trans-retinoic acid forms a nanoparticle structure by encapsulating with $CaCO_3$ and shows enhanced water-dispersity. This enables improved delivery and circulate half-life in blood stream, resulting in long-term stimulation of insulin secretion from islets. Owing to the tunability of size, shape, and function, the nanoparticles also play a pivotal role in the creation of DNA delivery vectors (Han et al. 2007). They can penetrate deep into tissues and are absorbed by the cells efficiently, followed by enhanced gene transduction (Kabanov et al. 2002). Moreover, modification and functionalization of nanoparticles to deliver therapeutic agents have led to an excellent breakthrough in drug delivery systems.

Generally, nanoparticles for drug delivery are composed of polymeric or inorganic materials. The nanoparticle-based drug delivery systems have been gaining great importance and attractive potential, because they can promote the controlled release of active ingredients in the body (Fig. 12). Interestingly, each type of nanoparticles has its own chemical, physical, and morphological characteristics, and have a particular affinity for specific drugs or related motifs through chemical (e.g., covalent bonds and hydrogen bonds) or physical (e.g., electrostatic and van der Waals interactions) interactions. For instance, the release profile of chemically grafted (or loaded) drug molecules to the nanoparticles is typically slower than that of physically adsorbed drug molecules. Hence, chemical and physical factors can determine the interactions of nanoparticles with biological systems, as well as the pharmacokinetics and pharmacodynamics of the active ingredient in the body (Wong and Choi 2015). Apart from this, other parameters, such as the composition

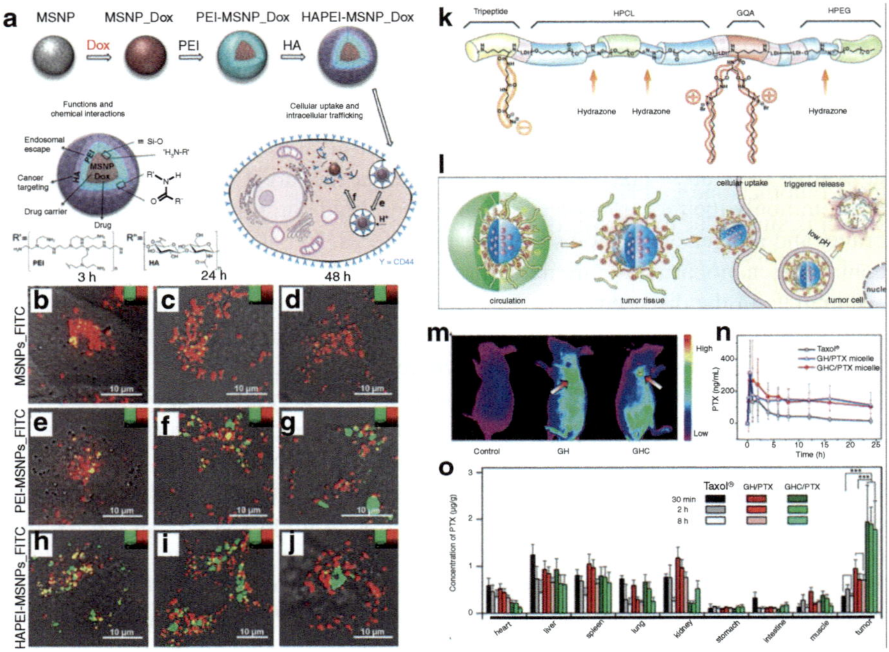

Fig. 12 (**a**) Multifunctional drug delivery system based on mesoporous silica nanoparticles (MSNPs): As-synthesized MSNPs are loaded with anticancer drugs (Doxorubicin, Dox), coated polyethyleneimine (PEI) layer, and further functionalized with hyaluronic acid (HA) (Fortuni et al. 2019). (**b–j**) Cellular uptake and intracellular trafficking. The nanoparticles show specific interaction and binding to CD44 onto the plasma membrane, followed by cellular uptake via receptor-mediated endocytosis and wrapping in endosomes. Endosomal escape is then induced upon proton sponge effect-derived rupture; thus drug payloads are released into the cytoplasm. (**b–j**) Influence of surface modification on the intracellular trafficking of MSNPs at different time points. Fluorescence images of A549 cells incubated with MSNPs_FITC (**b–d**), PEI-MSNPs_FITC (**e–g**), and HAPEI-MSNPs_FITC (**h–j**) at 3, 24, and 48 h post-incubation. The lysosomes were stained using LysoTracker Red® (Fortuni et al. 2019). (**k–l**) Schematics molecular structure of multifunctional multiblock polyurethane (**k**), and mechanism of targeted intracellular delivery of anticancer drugs into tumor cells using the multiblock polyurethanes nanocarriers (**l**). (**m-o**) In vivo fluorescence images of control mice and tumor-bearing nude mice receiving intravenous injection of fluorescent-labeled polyurethanes nanocarriers after 60 min. Arrows show A431 tumor in mice. (**n**) Plasma concentration–time profiles of Paclitaxel (PTX) after intravenous injection of Taxol®, GH/PTX, and GHC/PTX formulations in rats. (**o**) Tissue distribution profiles of PTX after intravenous administration of Taxol®, and micellar formulations in nude mice (Ding et al. 2013b)

of the nanoparticles (e.g., organic, inorganic, or hybrid materials) and the form in which drugs are associated with them (either core-shell or matrix-embedded system) are also very important factors for understanding the drug loading and release mechanism/kinetics (Mattos et al. 2017; Siepmann et al. 2008). Taken together, vigorous studies have been conducted to investigate the release mechanisms of drugs loaded in the nanoparticles. Typically, the release of drug payloads from the nanoparticles comes with diffusion, chemical reaction, stimuli-controlled release, or a combination of those mechanisms (Ding and Li 2017).

Along with understanding the different mechanisms of action for drug loading and release in nanoparticulate-based drug delivery systems, improving the specificity of the nanoparticles to target regions of the organism is also highly investigated with diverse strategies (Sawant and Torchilin 2012). In addition, surface coating or functionalization of nanoparticles with additional substances such as polymers, natural products, antibodies, cell membrane, tunable ligands, and peptides is another important direction to reduce the immunogenicity of nanomedicine for successful delivery inside the body (Pelaz et al. 2015; Almalik et al. 2017; Martens et al. 2015; Kolhar et al. 2013; Muller et al. 2017). The surface-modified nanoparticles allow a programmed drug delivery in a particular environment by displaying a specific binding affinity to target cells or capability to cross certain biological barriers such as blood–brain barrier (BBB). For example, hyaluronic acid (a polysaccharide naturally found in the extracellular matrix) is widely used as a ligand to target several tumors including melanoma stem-like cells, breast cancer, and pulmonary adenocarcinoma (Shen et al. 2015; Gao et al. 2017; Kim et al. 2017). Moreover, hyaluronic acid has been known to facilitate intravitreal drug delivery for retinal gene therapy and to reduce the immunogenicity of the nanoparticles by suppressing protein corona formation. However, the formation of ligand-appended nanomedicine for specific drug delivery is relatively labor-intensive, and several targeting strategies must be designed previously, taking into account the physiological variables such as blood flow, the stage of disease, and tissue architecture (Muro 2012). There are many efforts to investigate the interactions of ligand-appended nanoparticles with cell membrane or tissues, and eventually to understand the complete mechanism of cellular uptake, which is still unclear. Furthermore, it is also hardly understanding the mechanisms of intracellular uptake of nanoparticles, which occurs via either phagocytic or non-phagocytic pathways (e.g., clathrin- and caveolae-mediated endocytosis). Since the physicochemical characteristics of each nanoparticle vary in wide range, the mechanism of action and the interactions between the nanoparticles and cells are also hardly standardized. This is the reason why the study on nanoparticle–cell interaction is critical to understand the fundamental phenomena and to translate the nanomedicine to the clinic (Fig. 13).

Regarding the interactions occurred between the nanoparticles and cells including binding, uptake, intracellular trafficking, degradation and exocytosis, there are increasing number of studies that reveals the mechanism and kinetics on the cellular behavior of nanoparticles. For example, it has been discovered that the endocytosis is the mainly responsible mechanism for cellular uptake of polysaccharide nanoparticles whether it contains drug payloads or not (Salatin and Khosroushahi 2017). On the other hand, the nanoparticles that have external stimuli-responsive characteristics show the ability of controlled release of drug payloads. In this "triggered release" system, the drug-loaded nanoparticles show modulated drug release kinetics using external factors such as ultrasound, heat, light, magnetic field, pH, and ionic strength, thus eventually improve the targeted delivery and superior dosage control in the pathological site (Anirudhan and Nair 2018; Al-Ahmady and

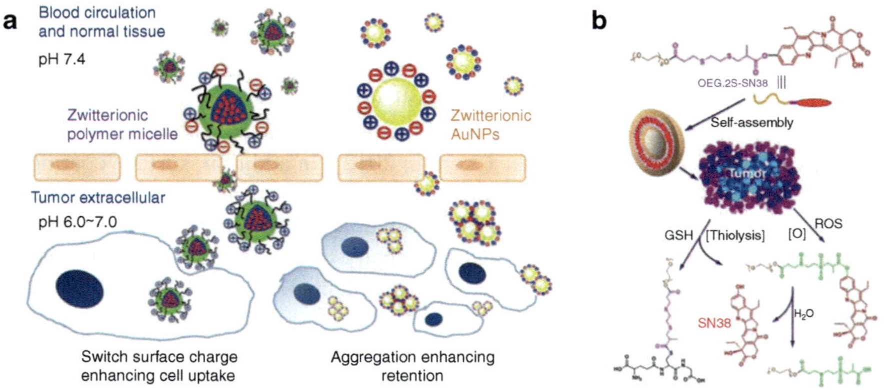

Fig. 13 (**a**) Interchange between "stealthy and sticky" for zwitterionic polymer micelles (left) and the mix-charged AuNPs (right) (Yuan et al. 2012). Many proteins behave stealthily to most cells and they can have long circulation times in vivo. However, they can alternate between "stealthy and sticky" under even a slight stimulus such as a minor change in pH. Although the detailed mechanism of this change has not been elucidated and might indeed be complex, the zwitterionic properties of amino acids and the mixed charge structures of proteins might contribute to this feature. By exploiting pH changes in different structures (e.g., in tumor tissues, 6.5–6.8; in endosomes, 5.5–6.8; and in lysosomes, 4.0–5.0), zwitterionic polymer-based NPs, capable of altering their charge densities in different tissues, can be prepared (Yuan et al. 2012). (**b**) SN38 prodrug formed a responsive nanoparticle upon exposure to tumorous GSH/ROS heterogeneity, resulting in release of the parent drug SN38 via thiolysis in the presence of GSH or via enhanced hydrolysis due to ROS oxidation of the linker. This process thereby gives rise to high anticancer therapeutic activity (Wang et al. 2013c)

Kostarelos 2016; Guo et al. 2018; Hervault and Thanh 2014; Xu et al. 2018; Ma et al. 2017). For instance, superparamagnetic iron oxide nanoparticles are associated with polymeric shell or lipid, and stimulated a controlled release of therapeutic drug payloads upon applying external magnetic field (Grillo et al. 2016; Alonso et al. 2016). In this case, the origin of drug loading, particularly whether the payloads are covalently conjugated or noncovalently attached, also critically affects release kinetics (Ulbrich et al. 2016). As mentioned in the photothermal therapy (see Section "Biodistribution and Toxicity"), the gold nanoparticles are responsible for near-infrared heating, thus can be utilized for use in light-induced drug release. Furthermore, combinational therapeutic treatment with chemo-drug release and photothermal hyperthermia is another strategy for gold-based hybrid nanomedicine (Chen et al. 2017). Hybrid nanoparticles for medical applications are currently among the most promising field of study in the nanomedicine as they present a synergistic mixture of featured properties in a single system. In particular, theranostic systems that perform both diagnostic and therapeutic applications in a single platform have attracted great attention as the accuracy and sensitivity enhanced along with advancement in nanotechnology. Despite this, there are still far beyond the real mechanisms of action and toxicity of nanoparticle-based drug delivery systems, which open another opportunity for new studies in the field of nanomedicine. Along with this trend, a number of studies are focusing on the synthesis of nanoparticles through environmentally safe and chemically nontoxic processes by implementing natural products such as plant extracts and microorganisms.

Perspectives: Challenges and Future of Nanomedicine

Traditional clinical methods have been extensively studied to increase the specificity and accuracy of diagnostics and therapeutics. In addition to their own way, numerous new materials, platforms, and procedures have been also proposed with the tremendous progression of nanomedicine. For instance, new routes of drug administration are being explored, and newly developed methods are focusing to ensure the targeted action in specific pathological sites not only to reduce the toxicity but also to increase their bioavailability in the cell, tissues, and whole body as well (Mignani et al. 2013). In this context, designer nanomedicine has been a promising feature characterizing the discovery of novel diagnostic and/or therapeutic systems based on deep understanding of the biological targets with nanomaterials. Experimental discovery on the structural functions of innate biological components such as proteins and peptides are essential for the progression and development of the nanomedicine, along with the theoretical proof helped from the advancements in computer sciences (Mavromoustakos 2011). Upon incorporation of various functional groups such as ligands and biomolecules, the nanoparticles have shown a wide range of potential applications as a platform of nanomedicine. Tremendous advancements in design and synthetic strategy have been studied to obtain the promising nanomedicine that has specific targetability and controlled drug delivery capability toward practical uses in clinic. For this purpose, multi aspects of nanoparticles such as dimensional scales (size, shape, etc.) have been extensively investigated, related with the targeting efficiency in physiological environment to study the surface chemistry and ligands.

As discussed in the previous sections, nanoparticles are promising candidates as contrast agents for anatomical and functional imaging through noninvasive platforms. In general, those nanoparticles, particularly called as nanomedicine, enable the visualization of inner structures of the human body by measuring unique signals such as fluorescence or magnetic dipolar momentum. The noninvasive imaging is particularly a very powerful technique that helps clinicians determining pathological and/or healthy tissues and functions in a patient-friendly manner. It is very important to choose the proper element and composition of nanoparticles to design and engineer the contrast agent for each imaging modality. In this chapter, several common imaging modalities (e.g., luminescence imaging, MRI, and CT) are presented as widely applicable diagnostic visualization methods, where physical characteristics such as quantum yield or magnetic momentum determine the sensitivity and efficiency of visualization. In addition to those techniques, an increasing number of nanomedicines has been extensively studied and adopted for other imaging modalities which is also widely used in clinic such as photoacoustic imaging, and radioactive imaging including positron emission tomography (PET) and single-photon emission computed tomography (SPECT). By utilizing the unique features of each nanoparticle, biological components are readily visualized, thus it opens up a promising future in the field of clinical imaging methods and instrumentations. Since various types of nanoparticles are extensively studied as a drug carrier, particularly in combination with a myriad of imaging capability, drug payloads, and targeting moieties on the surface, and it leads to the perspective nano particles for clinically translational medicine: targeted therapeutics concomitant with diagnosis. Overall, the nanoparticles can provide an optimistic future in the field of diagnostic imaging and therapeutics while the size domain lies in the range of biological components such as proteins, peptides, and oligonucleotides. In this section, important features of nanomaterials facing the scientific challenges and clinical translation are thoroughly presented.

Preparation: Synthesis and Characterization

Consistent and reproducible synthesis of nanoparticles is one of the major requirements for successful clinical translation. Manufacturing them in (1) uniform size and shape, (2) controlled surface charge ligand conjugation, and (3) consistent functionality is the critical components, since those characteristics strongly determine the performance of the nanomedicine, as well as causing unexpected side effects inside the body (Wolfram et al. 2015). In particular, the reproducibility has become into one of the critical issues because several cases of nanomedicine has been revoked by the reliability of reproducibe synthesis and clinical performance as it is certainly sensitive to microenvironment or handling skills approval of nanomedicine. Moreover, the addition of every new function or moiety to the approved nanoparticles for clinical uses should be thoroughly investigated as the complexity increases accordingly.

For reproducible preparation of nanoparticles for medical uses, a variety of synthetic (or fabrication) methods has been presented throughout the literatures over the decades. Advancement in analytical instrumentations allows investigating the physicochemical characteristics such as microstructures, crystallinity, and optical properties to support the reproducible preparation of the nanoparticles for their appropriate purposes (Cho et al. 2013). For example, the size and size distribution can be readily measured by dynamic light scattering that utilizes the Brownian motion of colloidal nanoparticles in liquid media, while zeta-potential that reports surface charge and stability of the nanoparticles is simultaneously obtained upon exposure to different environment varying ionic strength or pH. Scanning electron microscopy (SEM) and transmission electron microscopy (TEM) enable the morphological analysis, and further provide elemental compositions of the nanoparticles with additional instrumentation such as energy dispersive X-ray spectroscopy (EDX) or electron energy loss spectroscopy (EELS). X-ray diffraction analysis provides crystalline domain size and phase of the nanoparticles, which is closely related to physical characteristics such as quantum yield, bandgap energy, and magnetic momentum. In the early generation of nanomedicine, the lack of physicochemical characterization standards that report the synthetic parameters of nanoparticles for medical uses has limited reproducibility and difficulty for desired uses and FDA approval. However, to overcome the challenges, the Nanotechnology Characterization Laboratory, a part of the National Cancer Institute (NCI), provides standard protocols of synthesis and characterization of nanoparticle preparation for medical purposes to support the translation of nanomedicine (Ehlerding et al. 2018; Crist et al. 2013; Adiseshaiah et al. 2010). Although the standardized characterization methods and reports are generally agreed in the field of nanomedicine, further regulations over the NCI's protocols will be required to mature the nanotechnology-based platform for translational medicine worldwide. In addition, large-scale, reproducible methods for nanoparticle preparation are also required to apply the desired nanomedicine for clinical trials and approval (Beyer et al. 2015). It should be also noted that biological synthesis of nanoparticles gains increasing interest owing to its potential for low-cost and environment-friendly processes (Czapar and Steinmetz 2017), while most of the synthetic strategies, particularly for large-scale production, are based on chemical and physical productions.

Biodistribution and Toxicity

Besides the promising potential of nanomedicine, possible toxicity and adverse effects in the body are a major concern on the FDA approval and clinical translation. Indeed, nanoscale materials have certainly different functionality and behavior compared to bulk materials, thus their characteristic features (e.g., biodistribution, solubility, and chemical reactivity) are hardly generalized. In this regard, the biocompatibility and tolerance are critical to determining whether the nanomedicine can be applied to medical purposes or not. Complete evaluation of the in vivo nanoparticle behaviors may be certainly limited, however, several forms of in vitro testing for preclinical study can provide sufficiently important information prior to animal model demonstration or human trials. For in vitro assessment, a set of simple biological assays such as MTT and Ames provides basic data on biocompatibility and cytotoxicity of a nanomedicine candidate (Almutary and Sanderson 2016; Mortelmans and Zeiger 2000). Organoid, a type of advanced 3D cell culture models, that mimics the in vivo microenvironmental conditions on a culture dish or microfluidic chip is also widely employed to demonstrate biological assessment of newly developed nanomedicine (Edmondson et al. 2014; Lancaster and Knoblich 2014; Chambers et al. 2013; Cruz et al. 2017). To study the biodistribution of nanoparticles administrated via intravenous, intraperitoneal, oral, or subcutaneous injection in a clinically relevant environment, noninvasive imaging helps to visualize accumulated nanoparticles and corresponding biodistribution. This allows the investigation on dosing, accumulation, excretion, and metabolic profiles both in acute tracking and in long-term monitoring. Particularly, as the patients ideally live many more years post-administration of nanomedicine, preclinical or clinical evaluation, or estimation need to take place on an appropriate time scale to fully inform any unknown adverse effects. This includes a comprehensive study and complete understanding of the biodistribution, biodegradation, and excretion of the nanoparticles in the human body (Ehlerding et al. 2016).

In Vivo Validation and Preclinical Models

The validation of nanomedicine in animal models and humans should be performed as a course toward clinical trials and approval. To successfully investigate the efficacy and possible side effect of nanomedicine, it is required to choose proper preclinical models to validate the potential capability of both diagnostic and therapeutic nanoparticles. This includes the use of positive targets and negative controls in cell lines, clinically relevant pathological models such as tumor, and representative treatment strategies (dosing condition, imaging protocol, etc.). On the other hand, inadequate choice of preclinical models results in clinically non-relevant and redundant false hope. For example, it should be noted that the xenograft tumor models in preclinical studies showed surprisingly high level of EPR effect of the nanoparticle accumulation to tumorous tissue, which cannot represent the realistic level of tumor targeting in human clinical trials. Therefore, different types of tumors (with different growth rates, vasculature properties, etc.) are essential to be utilized for meaningful (pre)clinical validation. Some early reports on tumor targeting of nanomedicine have dealt with great tumor homing efficiency in a simple tumor model of mice, but this will certainly cause bias in the results. Advanced tumor model such as orthotopic xenograft provides more relevant validation reports, indicating clinical insight into potential translation from bench to bedside (Day et al. 2015). Biodistribution of the nanomedicine during the course of circulation is strongly determined by the preclinical models, thus the choice of proper model strictly matters. In addition, it is noteworthy that the metabolic pathways and clearance of nanoparticles varies across different animal species. For example, human shows primary uptake of larger nanoparticles in the liver, which is very consistent in mice, rat, monkey, and rabbit, while pig, goat, and cat have a higher uptake of those large nanoparticles in pulmonary vasculatures of the lung (Zamboni et al. 2012).

To systematically validate the performance of nanomedicine through imaging modalities, the quantitative analysis is crucial. The Radiological Society of North America (RSNA) have established the Quantitative Imaging Biomarkers Alliance (QIBA) to give a guide for the quantitative imaging analysis and interpretation in clinical trials and practice. However, those guidelines for clinical applications of nanomedicine are still not clearly declared. Although much effort has been made to correlate the featured signals with the concentration of nanoparticles in preclinical studies through establishing calibration and standard curves, those calibrations may not necessarily represent the realistic observations found in humans. In addition, several molecular markers are widely used to detect the target cells both in vitro and in vivo, through imaging based visualization of nanoparticles that targets the molecular markers; however, those have not been sufficiently validated for clinical decisions. Therefore, it is also required to validate the nanoparticle-based imaging biomarkers and to establish consensus in instrumentation along with analysis for human applications. Overall, proper preclinical models and corresponding analytical systems should thus be carefully adopted.

References

Adiseshaiah PP, Hall JB, McNeil SE (2010) Nanomaterial standards for efficacy and toxicity assessment. Wires Nanomed Nanobi 2:99–112

Al-Ahmady Z, Kostarelos K (2016) Chemical components for the design of temperature-responsive vesicles as cancer therapeutics. Chem Rev 116:3883–3918

Aldakov D, Reiss P (2019) Safer-by-design fluorescent nanocrystals: metal halide perovskites vs semiconductor quantum dots. J Phys Chem C 123:12527–12541

Ali Z et al (2011) Multifunctional nanoparticles for dual imaging. Anal Chem 83:2877–2882

Almalik A et al (2017) Hyaluronic acid coated chitosan nanoparticles reduced the immunogenicity of the formed protein Corona. Sci Rep-UK 7:1–9

Almutary A, Sanderson BJS (2016) The MTT and crystal violet assays: potential confounders in nanoparticle toxicity testing. Int J Toxicol 35:454–462

Alonso J et al (2016) Superparamagnetic nanoparticles encapsulated in lipid vesicles for advanced magnetic hyperthermia and biodetection. J Appl Phys 119:083904

Anchordoquy TJ et al (2017) Mechanisms and barriers in cancer nanomedicine: adressing challenges, looking for solutions. ACS Nano 11:12–18

Anirudhan TS, Nair AS (2018) Temperature and ultrasound sensitive gatekeepers for the controlled release of chemotherapeutic drugs from mesoporous silica nanoparticles. J Mater Chem B 6:428–439

Anselmo AC, Mitragotri S (2015) A review of clinical translation of inorganic nanoparticles. AAPS J 17:1041–1054

Baig S, Xu J, Wu P, Chen B, Wang M (2008) Formation of visible single-mode light sources using quantum dots. Proceedings of the SPIE 7058:7

Baker SN, Baker GA (2010) Luminescent carbon nanodots: emergent nanolights. Angew Chem Int Ed 49:6726–6744

Bao CC et al (2014) A promising road with challenges: where are gold nanoparticles in translational research? Nanomedicine-UK 9:2353–2370

Bao Y, Sherwood JA, Sun Z (2018) Magnetic iron oxide nanoparticles as T-1 contrast agents for magnetic resonance imaging. J Mater Chem C 6:1280–1290

Baughman RH, Zakhidov AA, de Heer WA (2002) Carbon nanotubes - the route toward applications. Science 297:787–792

Beltran MA, Paganin DM, Uesugi K, Kitchen MJ (2010) 2D and 3D X-ray phase retrieval of multi-material objects using a single defocus distance. Opt Express 18:6423–6436

Belyanina I et al (2017) Targeted magnetic nanotheranostics of Cancer. Molecules 22:975

Betzer O et al (2014) Nanoparticle-based CT imaging technique for longitudinal and quantitative stem cell tracking within the brain: application in neuropsychiatric disorders. ACS Nano 8:9274–9285

Beyer S et al (2015) Bridging laboratory and large scale production: preparation and in vitro evaluation of photosensitizer-loaded Nanocarrier devices for targeted drug delivery. Pharm Res-Dordr 32:1714–1726

Boca SC et al (2011) Chitosan-coated triangular silver nanoparticles as a novel class of biocompatible, highly effective photothermal transducers for in vitro cancer cell therapy. Cancer Lett 311:131–140

Bouras A, Kaluzova M, Hadjipanayis CG (2015) Radiosensitivity enhancement of radioresistant glioblastoma by epidermal growth factor receptor antibody-conjugated iron-oxide nanoparticles. J Neurooncol 124:13–22
Bruchez M, Moronne M, Gin P, Weiss S, Alivisatos AP (1998) Semiconductor nanocrystals as fluorescent biological labels. Science 281:2013–2016
Camerin M et al (2005a) Photothermal sensitisation as a novel therapeutic approach for tumours: studies at the cellular and animal level. Eur J Cancer 41:1203–1212
Camerin M, Rodgers MAJ, Kenney ME, Jori G (2005b) Photothermal sensitisation: evidence for the lack of oxygen effect on the photosensitising activity. Photochem Photobiol Sci 4:251–253
Caruthers SD, Wickline SA, Lanza GM (2007) Nanotechnological applications in medicine. Curr Opin Biotechnol 18:26–30
Chambers SM, Tchieu J, Studer L (2013) Build-a-brain. Cell Stem Cell 13:377–378
Chan WCW, Nie SM (1998) Quantum dot bioconjugates for ultrasensitive nonisotopic detection. Science 281:2016–2018
Chang YP, Pinaud F, Antelman J, Weiss S (2008) Tracking bio-molecules in live cells using quantum dots. J Biophotonics 1:287–298
Chapman D et al (1997) Diffraction enhanced x-ray imaging. Phys Med Biol 42:2015–2025
Chen JY et al (2010) Gold nanocages as photothermal transducers for cancer treatment. Small 6:811–817
Chen CW et al (2017) Encapsulation of au/Fe3O4 nanoparticles into a polymer nanoarchitecture with combined near infrared-triggered chemo-photothermal therapy based on intracellular secondary protein understanding. J Mater Chem B 5:5774–5782
Cheng K et al (2014a) Hybrid Nanotrimers for dual T-1 and T-2-weighted magnetic resonance imaging. ACS Nano 8:9884–9896
Cheng L et al (2014b) PEGylated WS2 Nanosheets as a multifunctional Theranostic agent for in vivo dual-modal CT/Photoacoustic imaging guided Photothermal therapy. Adv Mater 26:1886–1893
Chhour P et al (2016) Labeling monocytes with gold nanoparticles to track their recruitment in atherosclerosis with computed tomography. Biomaterials 87:93–103
Cho EJ et al (2013) Nanoparticle characterization: state of the art, challenges, and emerging technologies. Mol Pharm 10:2093–2110
Choi JS et al (2010) Self-confirming "AND" logic nanoparticles for fault-free MRI. J Am Chem Soc 132:11015–11017
Choi WI et al (2011) Tumor regression in vivo by photothermal therapy based on gold-nanorod-loaded, functional nanocarriers. ACS Nano 5:1995–2003
Chou LYT, Ming K, Chan WCW (2011) Strategies for the intracellular delivery of nanoparticles. Chem Soc Rev 40:233–245
Chow EKH, Ho D (2013) Cancer Nanomedicine: from drug delivery to imaging. Sci Transl Med 5:216
Clarke S et al (2010) Covalent Monofunctionalization of peptide-coated quantum dots for single-molecule assays. Nano Lett 10:2147–2154
Cloetens P et al (1999) Holotomography: quantitative phase tomography with micrometer resolution using hard synchrotron radiation x rays. Appl Phys Lett 75:2912–2914
Colombo M et al (2012) Biological applications of magnetic nanoparticles. Chem Soc Rev 41:4306–4334
Cormode DP et al (2010) Atherosclerotic plaque composition: analysis with multicolor CT and targeted gold nanoparticles. Radiology 256:774–782
Cormode DP, Naha PC, Fayad ZA (2014) Nanoparticle contrast agents for computed tomography: a focus on micelles. Contrast Media Mol I 9:37–52
Creixell M, Bohorquez AC, Torres-Lugo M, Rinaldi C (2011) EGFR-targeted magnetic nanoparticle heaters kill Cancer cells without a perceptible temperature rise. ACS Nano 5:7124–7129
Crist RM et al (2013) Common pitfalls in nanotechnology: lessons learned from NCI's nanotechnology characterization laboratory. Integr Biol-UK 5:66–73

Cruz NM et al (2017) Organoid cystogenesis reveals a critical role of microenvironment in human polycystic kidney disease. Nat Mater 16:1112
Czapar AE, Steinmetz NF (2017) Plant viruses and bacteriophages for drug delivery in medicine and biotechnology. Curr Opin Chem Biol 38:108–116
Davis ME, Chen Z, Shin DM (2008) Nanoparticle therapeutics: an emerging treatment modality for cancer. Nat Rev Drug Discov 7:771–782
Day ES, Morton JG, West JL (2009) Nanoparticles for thermal cancer therapy. J Biomech Eng-T Asme 131:7
Day CP, Merlino G, Van Dyke T (2015) Preclinical mouse cancer models: a maze of opportunities and challenges. Cell 163:39–53
De Haas-Kock DFM et al (2009) Concomitant hyperthermia and radiation therapy for treating locally advanced rectal cancer. Cochrane Db Syst Rev 3:CD006269. https://doi.org/10.1002/14651858.CD006269.pub2
Delcea M, Mohwald H, Skirtach AG (2011) Stimuli-responsive LbL capsules and nanoshells for drug delivery. Adv Drug Deliv Rev 63:730–747
Derfus AM, Chan WCW, Bhatia SN (2004) Probing the cytotoxicity of semiconductor quantum dots. Nano Lett 4:11–18
Dewey WC, Hopwood LE, Sapareto SA, Gerweck LE (1977) Cellular responses to combinations of hyperthermia and radiation. Radiology 123:463–474
Ding CZ, Li ZB (2017) A review of drug release mechanisms from nanocarrier systems. Mat Sci Eng C-Mater 76:1440–1453
Ding D, Li K, Liu B, Tang BZ (2013a) Bioprobes based on AIE Fluorogens. Acc Chem Res 46:2441–2453
Ding MM et al (2013b) Toward the next-generation nanomedicines: design of multifunctional multiblock polyurethanes for effective cancer treatment. ACS Nano 7:1918–1928
Ding D et al (2014) Precise and Long-term tracking of adipose-derived stem cells and their regenerative capacity via superb bright and stable organic Nanodots. ACS Nano 8:12620–12631
Domenech M, Marrero-Berrios I, Torres-Lugo M, Rinaldi C (2013) Lysosomal membrane Permeabilization by targeted magnetic nanoparticles in alternating magnetic fields. ACS Nano 7:5091–5101
Dreaden EC, Mackey MA, Huang XH, Kang B, El-Sayed MA (2011) Beating cancer in multiple ways using nanogold. Chem Soc Rev 40:3391–3404
Dreaden EC, Alkilany AM, Huang XH, Murphy CJ, El-Sayed MA (2012) The golden age: gold nanoparticles for biomedicine. Chem Soc Rev 41:2740–2779
Du JJ, Xu N, Fan JL, Sun W, Peng XJ (2019) Carbon dots for in vivo bioimaging and theranostics. Small 15:1805087
Dullin C et al (2015) Functionalized synchrotron in-line phase-contrast computed tomography: a novel approach for simultaneous quantification of structural alterations and localization of barium-labelled alveolar macrophages within mouse lung samples. J Synchrotron Radiat 22:143–155
Edmondson R, Broglie JJ, Adcock AF, Yang LJ (2014) Three-dimensional cell culture systems and their applications in drug discovery and cell-based biosensors. Assay Drug Dev Technol 12:207–218
Ehlerding EB, Chen F, Cai WB (2016) Biodegradable and renal clearable inorganic nanoparticles. Adv Sci 3:1500223
Ehlerding EB, Grodzinski P, Cai WB, Liu CH (2018) Big potential from small agents: nanoparticles for imaging-based companion diagnostics. ACS Nano 12:2106–2121
Eliseeva SV, Bunzli JCG (2010) Lanthanide luminescence for functional materials and biosciences. Chem Soc Rev 39:189–227
El-Sayed MA (2001) Some interesting properties of metals confined in time and nanometer space of different shapes. Acc Chem Res 34:257–264
Escudero A et al (2016) Synthesis and functionalization of monodisperse near-ultraviolet and visible excitable multifunctional Eu3+, Bi3+: REVO4 nanophosphors for bioimaging and biosensing applications. Nanoscale 8:12221–12236

Etheridge ML et al (2013) The big picture on nanomedicine: the state of investigational and approved nanomedicine products. Nanomed-Nanotechnol 9:1–14
Fan QL et al (2014) Transferring biomarker into molecular probe: melanin nanoparticle as a naturally active platform for multimodality imaging. J Am Chem Soc 136:15185–15194
Field LD, Delehanty JB, Chen YC, Medintz IL (2015) Peptides for specifically targeting nanoparticles to cellular organelles: quo Vadis? Acc Chem Res 48:1380–1390
Formica D, Silvestri S (2004) Biological effects of exposure to magnetic resonance imaging: an overview. Biomed Eng Online 3:11
Fortuni B et al (2019) Polymeric engineering of nanoparticles for highly efficient multifunctional drug delivery systems. Sci Rep-Uk 9:1–13
Frangioni JV (2003) In vivo near-infrared fluorescence imaging. Curr Opin Chem Biol 7:626–634
Gao XH, Cui YY, Levenson RM, Chung LWK, Nie SM (2004) In vivo cancer targeting and imaging with semiconductor quantum dots. Nat Biotechnol 22:969–976
Gao XH et al (2005) In vivo molecular and cellular imaging with quantum dots. Curr Opin Biotechnol 16:63–72
Gao X et al (2017) Hyaluronic acid-coated cationic nanostructured lipid carriers for oral vincristine sulfate delivery. Drug Dev Ind Pharm 43:661–667
Gobin AM et al (2007) Near-infrared resonant nanoshells for combined optical imaging and photothermal cancer therapy. Nano Lett 7:1929–1934
Grillo R et al (2016) Sub-micrometer magnetic Nanocomposites: insights into the effect of magnetic nanoparticles interactions on the optimization of SAR and MRI performance. ACS Appl Mater Interfaces 8:25777–25787
Gulyaev AE et al (1999) Significant transport of doxorubicin into the brain with polysorbate 80-coated nanoparticles. Pharm Res-Dordr 16:1564–1569
Guo PX (2010) The emerging field of RNA nanotechnology. Nat Nanotechnol 5:833–842
Guo YX et al (2018) Light/magnetic hyperthermia triggered drug released from multi-functional thermo-sensitive magnetoliposomes for precise cancer synergetic theranostics. J Control Release 272:145–158
Han G, Ghosh P, Rotello VM (2007) Functionalized gold nanoparticles for drug delivery. Nanomedicine-UK 2:113–123
Hartmann R, Weidenbach M, Neubauer M, Fery A, Parak WJ (2015) Stiffness-dependent in vitro uptake and Lysosomal acidification of colloidal particles. Angew Chem Int Ed 54:1365–1368
Hartwig V et al (2009) Biological effects and safety in magnetic resonance imaging: a review. Int J Env Res Pub He 6:1778–1798
He X, Bischof JC (2003) Quantification of temperature and injury response in thermal therapy and cryosurgery. Biomed Eng 31:67
Hervault A, Thanh NTK (2014) Magnetic nanoparticle-based therapeutic agents for thermo-chemotherapy treatment of cancer. Nanoscale 6:11553–11573
Hetzel D et al (2014) A phase III, randomized, open-label trial of ferumoxytol compared with iron sucrose for the treatment of iron deficiency anemia in patients with a history of unsatisfactory oral iron therapy. Am J Hematol 89:646–650
Ho YP, Leong KW (2010) Quantum dot-based theranostics. Nanoscale 2:60–68
Houlding TK, Rebrov EV (2012) Application of alternative energy forms in catalytic reactor engineering. Green Process Synth 1:19–31
Hu FQ et al (2006) Preparation of biocompatible magnetite nanocrystals for in vivo magnetic resonance detection of cancer. Adv Mater 18:2553
Huang XH, El-Sayed IH, Qian W, El-Sayed MA (2006) Cancer cell imaging and photothermal therapy in the near-infrared region by using gold nanorods. J Am Chem Soc 128:2115–2120
Huang XH, Jain PK, El-Sayed IH, El-Sayed MA (2007) Gold nanoparticles: interesting optical properties and recent applications in cancer diagnostic and therapy. Nanomedicine-UK 2:681–693
Huang XH, Jain PK, El-Sayed IH, El-Sayed MA (2008) Plasmonic photothermal therapy (PPTT) using gold nanoparticles. Lasers Med Sci 23:217–228

Huang XH, Neretina S, El-Sayed MA (2009) Gold Nanorods: from synthesis and properties to biological and biomedical applications. Adv Mater 21:4880–4910
Huang FK et al (2010) Enhancement of irradiation effects on cancer cells by cross-linked dextran-coated iron oxide (CLIO) nanoparticles. Phys Med Biol 55:469–482
Huang XN et al (2013) Multi-chromatic pH-Activatable F-19-MRI Nanoprobes with binary ON/OFF pH transitions and chemical-shift barcodes. Angew Chem Int Ed 52:8074–8078
Jain PK, Lee KS, El-Sayed IH, El-Sayed MA (2006) Calculated absorption and scattering properties of gold nanoparticles of different size, shape, and composition: applications in biological imaging and biomedicine. J Phys Chem B 110:7238–7248
Jain PK, Huang XH, El-Sayed IH, El-Sayed MA (2008) Noble metals on the nanoscale: optical and photothermal properties and some applications in imaging, sensing, biology, and medicine. Acc Chem Res 41:1578–1586
Jakhmola A, Anton N, Vandamme TF (2012) Inorganic nanoparticles based contrast agents for X-ray computed tomography. Adv Healthc Mater 1:413–431
James ML, Gambhir SS (2012) A molecular imaging primer: modalities, imaging agents, and applications. Physiol Rev 92:897–965
Jang JT et al (2018) Giant magnetic heat induction of magnesium-doped gamma-Fe2O3 superparamagnetic nanoparticles for completely killing tumors. Adv Mater 30:1704362
Jaque D et al (2014) Nanoparticles for photothermal therapies. Nanoscale 6:9494–9530
Johannsen M et al (2007) Thermotherapy of prostate cancer using magnetic nanoparticles: feasibility, imaging, and three-dimensional temperature distribution. Eur Urol 52:1653–1662
Johnson NJJ, Oakden W, Stanisz GJ, Prosser RS, van Veggel FCJM (2011) Size-tunable, ultrasmall NaGdF4 nanoparticles: insights into their T-1 MRI contrast enhancement. Chem Mater 23:3714–3722
Kabanov AV, Lemieux P, Vinogradov S, Alakhov V (2002) Pluronic((R)) block copolymers: novel functional molecules for gene therapy. Adv Drug Deliv Rev 54:223–233
Kantner K et al (2015) Particle-based optical sensing of intracellular ions at the example of calcium - what are the experimental pitfalls? Small 11:896–904
Katz E, Willner I (2004) Integrated nanoparticle-biomolecule hybrid systems: synthesis, properties, and applications. Angew Chem Int Ed 43:6042–6108
Kennedy LC et al (2011) A new era for Cancer treatment: gold-nanoparticle-mediated thermal therapies. Small 7:169–183
Khawaja AZ et al (2015) Revisiting the risks of MRI with gadolinium based contrast agents—review of literature and guidelines. Insights Imaging 6:553–558
Khoei S, Mahdavi SR, Fakhimikabir H, Shakeri-Zadeh A, Hashemian A (2014) The role of iron oxide nanoparticles in the radiosensitization of human prostate carcinoma cell line DU145 at megavoltage radiation energies. Int J Radiat Biol 90:351–356
Kim S et al (2004) Near-infrared fluorescent type II quantum dots for sentinel lymph node mapping. Nat Biotechnol 22:93–97
Kim J et al (2006) Designed fabrication of multifunctional magnetic gold nanoshells and their application to magnetic resonance imaging and photothermal therapy. Angew Chem Int Ed 45:7754–7758
Kim BYS, Rutka JT, Chan WCW (2010) Current concepts: Nanomedicine. New Engl J Med 363:2434–2443
Kim J, Kim J, Jeong C, Kim WJ (2016) Synergistic nanomedicine by combined gene and photothermal therapy. Adv Drug Deliv Rev 98:99–112
Kim D et al (2017) Two-photon in vivo imaging with porous silicon nanoparticles. Adv Mater 29:1703309
Kirchner C et al (2005) Cytotoxicity of colloidal CdSe and CdSe/ZnS nanoparticles. Nano Lett 5:331–338
Kirschning A, Kupracz L, Hartwig J (2012) New synthetic opportunities in miniaturized flow reactors with inductive heating. Chem Lett 41:562–570
Kleijn A et al (2011) Distinguishing inflammation from tumor and peritumoral edema by myeloperoxidase magnetic resonance imaging. Clin Cancer Res 17:4484–4493

Klein S, Sommer A, Distel LVR, Neuhuber W, Kryschi C (2012) Superparamagnetic iron oxide nanoparticles as radiosensitizer via enhanced reactive oxygen species formation. Biochem Bioph Res Co 425:393–397
Kobayashi H, Ogawa M, Alford R, Choyke PL, Urano Y (2010) New strategies for fluorescent probe design in medical diagnostic imaging. Chem Rev 110:2620–2640
Kolhar P et al (2013) Using shape effects to target antibody-coated nanoparticles to lung and brain endothelium. P Natl Acad Sci USA 110:10753–10758
Kotov NA (2010) Inorganic nanoparticles as protein mimics. Science 330:188–189
Kovalenko MV et al (2015) Prospects of nanoscience with nanocrystals. ACS Nano 9:1012–1057
Krenkel M et al (2015) Phase-contrast zoom tomography reveals precise locations of macrophages in mouse lungs. Sci Rep-UK 5:9973
Kwok RTK, Leung CWT, Lam JWY, Tang BZ (2015) Biosensing by luminogens with aggregation-induced emission characteristics. Chem Soc Rev 44:4228–4238
Lal S, Clare SE, Halas NJ (2008) Nanoshell-enabled Photothermal Cancer therapy: impending clinical impact. Acc Chem Res 41:1842–1851
Lammers T, Aime S, Hennink WE, Storm G, Kiessling F (2011) Theranostic nanomedicine. Acc Chem Res 44:1029–1038
Lancaster MA, Knoblich JA (2014) Organogenesis in a dish: modeling development and disease using organoid technologies. Science 345:6194
Langereis S et al (2009) A temperature-sensitive liposomal H-1 CEST and F-19 contrast agent for MR image-guided drug delivery. J Am Chem Soc 131:1380
Laurent S et al (2008) Magnetic iron oxide nanoparticles: synthesis, stabilization, vectorization, physicochemical characterizations, and biological applications. Chem Rev 108:2064–2110
Leawoods JC, Yablonskiy DA, Saam B, Gierada DS, Conradi MS (2001) Hyperpolarized He-3 gas production and MR imaging of the lung. Concept Magnetic Res 13:277–293
Lee Y et al (2007) A protein nanocarrier from charge-conversion polymer in response to endosomal pH. J Am Chem Soc 129:5362
Lee SM, Park H, Yoo KH (2010) Synergistic cancer therapeutic effects of locally delivered drug and heat using multifunctional nanoparticles. Adv Mater 22:4049–4053
Lee WW et al (2012) PET/MRI of inflammation in myocardial infarction. J Am Coll Cardiol 59:153–163
Lee J, Chatterjee DK, Lee MH, Krishnan S (2014) Gold nanoparticles in breast cancer treatment: promise and potential pitfalls. Cancer Lett 347:46–53
Li HT et al (2010) Water-soluble fluorescent carbon quantum dots and photocatalyst design. Angew Chem Int Ed 49:4430–4434
Li HX et al (2020) Recent advances in carbon dots for bioimaging applications. Nanoscale Horiz 5:218–234
Liang J, Tang B, Liu B (2015) Specific light-up bioprobes based on AIEgen conjugates. Chem Soc Rev 44:2798–2811
Lidke DS et al (2004) Quantum dot ligands provide new insights into erbB/HER receptor-mediated signal transduction. Nat Biotechnol 22:198–203
Lin CAJ et al (2009) Synthesis, characterization, and bioconjugation of fluorescent gold Nanoclusters toward biological labeling applications. ACS Nano 3:395–401
Link S, Ei-Sayed MA (2003) Optical properties and ultrafast dynamics of metallic nanocrystals. Annu Rev Phys Chem 54:331–366
Link S, El-Sayed MA (2000) Shape and size dependence of radiative, non-radiative and photothermal properties of gold nanocrystals. Int Rev Phys Chem 19:409–453
Liu Z, Liang XJ (2012) Nano-carbons as theranostics. Theranostics 2:235–237
Liu YL et al (2013) Dopamine-melanin colloidal nanospheres: an efficient near-infrared photothermal therapeutic agent for in vivo cancer therapy. Adv Mater 25:1353–1359
Liu T et al (2014a) Combined photothermal and photodynamic therapy delivered by PEGylated MoS2 nanosheets. Nanoscale 6:11219–11225
Liu T et al (2014b) Drug delivery with PEGylated MoS2 Nano-sheets for combined photothermal and chemotherapy of cancer. Adv Mater 26:3433–3440

Liu MY et al (2015) Self-polymerization of dopamine and polyethyleneimine: novel fluorescent organic nanoprobes for biological imaging applications. J Mater Chem B 3:3476–3482
Liu Q et al (2018) Single upconversion nanoparticle imaging at sub-10 W cm(−2) irradiance. Nat Photonics 12:548–553
Lutgens L et al (2010) Combined use of hyperthermia and radiation therapy for treating locally advanced cervix carcinoma. Cochrane Db Syst Rev 1:CD006377. https://doi.org/10.1002/14651858.CD006377.pub2
Ma N, Marshall AF, Gambhir SS, Rao JH (2010) Facile synthesis, Silanization, and biodistribution of biocompatible quantum dots. Small 6:1520–1528
Ma XF et al (2016) Fluorescence aggregation-caused quenching versus aggregation-induced emission: a visual teaching Technology for Undergraduate Chemistry Students. J Chem Educ 93:345–350
Ma GL et al (2017) Development of ionic strength/pH/enzyme triple-responsive zwitterionic hydrogel of the mixed L-glutamic acid and L-lysine polypeptide for site-specific drug delivery. J Mater Chem B 5:935–943
Mahmoudi M et al (2011) Magnetic resonance imaging tracking of stem cells in vivo using Iron oxide nanoparticles as a tool for the advancement of clinical regenerative medicine. Chem Rev 111:253–280
Marches R et al (2011) The importance of cellular internalization of antibody-targeted carbon nanotubes in the photothermal ablation of breast cancer cells. Nanotechnology 22:095101
Martens TF et al (2015) Coating nanocarriers with hyaluronic acid facilitates intravitreal drug delivery for retinal gene therapy. J Control Release 202:83–92
Martynenko IV et al (2017) Application of semiconductor quantum dots in bioimaging and biosensing. J Mater Chem B 5:6701–6727
Mattos BD, Tardy BL, Magalhaes WLE, Rojas OJ (2017) Controlled release for crop and wood protection: recent progress toward sustainable and safe nanostructured biocidal systems. J Control Release 262:139–150
Mavromoustakos T (2011) Hot topic: methodologies and applied strategies in the rational drug design. Curr Med Chem 18:2516–2516
McHugh KJ et al (2018) Biocompatible semiconductor quantum dots as cancer imaging agents. Adv Mater 30:1706356
Medintz IL, Uyeda HT, Goldman ER, Mattoussi H (2005) Quantum dot bioconjugates for imaging, labelling and sensing. Nat Mater 4:435–446
Mei J et al (2014) Aggregation-induced emission: the whole is more brilliant than the parts. Adv Mater 26:5429–5479
Melamed JR, Riley RS, Valcourt DM, Day ES (2016) Using gold nanoparticles to disrupt the tumor microenvironment: an emerging therapeutic strategy. ACS Nano 10:10631–10635
Melancon MP, Zhou M, Li C (2011) Cancer Theranostics with near-infrared light-Activatable multimodal nanoparticles. Acc Chem Res 44:947–956
Mignani S, El Kazzouli S, Bousmina M, Majoral JP (2013) Expand classical drug administration ways by emerging routes using dendrimer drug delivery systems: a concise overview. Adv Drug Deliv Rev 65:1316–1330
Min YZ, Caster JM, Eblan MJ, Wang AZ (2015) Clinical translation of nanomedicine. Chem Rev 115:11147–11190
Mirza AZ, Siddiqui FA (2014) Nanomedicine and drug delivery: a mini review. International Nano Letters 4:94
Mohammadi S et al (2014) Quantitative evaluation of a single-distance phase-retrieval method applied on in-line phase-contrast images of a mouse lung. J Synchrotron Radiat 21:784–789
Momose A, Takeda T, Itai Y, Hirano K (1996) Phase-contrast X-ray computed tomography for observing biological soft tissues. Nat Med 2:473–475
Mortelmans K, Zeiger E (2000) The Ames salmonella/microsome mutagenicity assay. Mutat Res-Fund Mol M 455:29–60

Motomura K et al (2011) SPIO-enhanced magnetic resonance imaging for the detection of metastases in sentinel nodes localized by computed tomography Lymphography in patients with breast cancer. Ann Surg Oncol 18:3422–3429
Muller J et al (2017) Coating nanoparticles with tunable surfactants facilitates control over the protein corona. Biomaterials 115:1–8
Mulvaney P, Parak WJ, Caruso F, Weiss PS (2016) Standardizing nanomaterials. ACS Nano 10:9763–9764
Muro S (2012) Challenges in design and characterization of ligand-targeted drug delivery systems. J Control Release 164:125–137
Na HB, Song IC, Hyeon T (2009) Inorganic nanoparticles for MRI contrast agents. Adv Mater 21:2133–2148
Ng EYK, Kumar SD (2017) Physical mechanism and modeling of heat generation and transfer in magnetic fluid hyperthermia through Neelian and Brownian relaxation: a review. Biomed Eng Online 16:1–22
Niemeyer CM (2001) Nanoparticles, proteins, and nucleic acids: biotechnology meets materials science. Angew Chem Int Ed 40:4128–4158
Nurunnabi M, Khatun Z, Reeck GR, Lee DY, Lee YK (2014) Photoluminescent graphene nanoparticles for cancer phototherapy and imaging. ACS Appl Mater Interfaces 6:12413–12421
Parak WJ, Pellegrino T, Plank C (2005) Labelling of cells with quantum dots. Nanotechnology 16:R9–R25
Park JH et al (2009) Biodegradable luminescent porous silicon nanoparticles for in vivo applications. Nat Mater 8:331–336
Park YI, Lee KT, Suh YD, Hyeon T (2015) Upconverting nanoparticles: a versatile platform for wide-field two-photon microscopy and multi-modal in vivo imaging. Chem Soc Rev 44:1302–1317
Park Y, Yoo J, Kang MH, Kwon W, Joo J (2019) Photoluminescent and biodegradable porous silicon nanoparticles for biomedical imaging. J Mater Chem B 7:6271–6292
Pasternak JJ, Williamson EE (2012) Clinical pharmacology, uses, and adverse reactions of iodinated contrast agents: a primer for the non-radiologist. Mayo Clin Proc 87:390–402
Peer D et al (2007) Nanocarriers as an emerging platform for cancer therapy. Nat Nanotechnol 2:751–760
Pelaz B et al (2015) Surface functionalization of nanoparticles with polyethylene glycol: effects on protein adsorption and cellular uptake. ACS Nano 9:6996–7008
Pelaz B et al (2017) Diverse applications of nanomedicine. ACS Nano 11:2313–2381
Pinaud F, Clarke S, Sittner A, Dahan M (2010) Probing cellular events, one quantum dot at a time. Nat Methods 7:275–285
Qian XX, Shen SD, Liu T, Cheng L, Liu Z (2015) Two-dimensional TiS2 nanosheets for in vivo photoacoustic imaging and photothermal cancer therapy. Nanoscale 7:6380–6387
Qin CX et al (2013) Tyrosinase as a multifunctional reporter gene for photoacoustic/MRI/PET triple modality molecular imaging. Sci Rep-UK 3:1490
Qiu PY et al (2013) Recent advances in lanthanide-doped upconversion nanomaterials: synthesis, nanostructures and surface modification. Nanoscale 5:11512–11525
Rabin O, Perez JM, Grimm J, Wojtkiewicz G, Weissleder R (2006) An X-ray computed tomography imaging agent based on long-circulating bismuth sulphide nanoparticles. Nat Mater 5:118–122
Riedinger A et al (2010) Ratiometric optical sensing of chloride ions with organic Fluorophore-gold nanoparticle hybrids: a systematic study of design parameters and surface charge effects. Small 6:2590–2597
Rivas J, Kolen'ko YV, Bañobre-López M (2016) Magnetic nanocolloids. Nanomedicine 2016:75–129
Rivlin M, Horev J, Tsarfaty I, Navon G (2013) Molecular imaging of tumors and metastases using chemical exchange saturation transfer (CEST) MRI. Sci Rep-UK 3:1–7
Robinson JT et al (2011) Ultrasmall reduced graphene oxide with high near-infrared absorbance for photothermal therapy. J Am Chem Soc 133:6825–6831

Roizintowle L, Pirro JP (1991) The response of human and rodent cells to hyperthermia. Int J Radiat Oncol 20:751–756

Ros PR et al (1995) Hepatic Mr-imaging with Ferumoxides - a multicenter clinical-trial of the safety and efficacy in the detection of focal hepatic-lesions. Radiology 196:481–488

Ruta S, Chantrell R, Hovorka O (2015) Unified model of hyperthermia via hysteresis heating in systems of interacting magnetic nanoparticles. Sci Rep-UK 5:9090

Rycenga M et al (2009) Probing the Photothermal effect of gold-based nanocages with surface-enhanced raman scattering (SERS). Angew Chem Int Ed 48:9924–9927

Saha S et al (2016) Gold nanoparticle reprograms pancreatic tumor microenvironment and inhibits tumor growth. ACS Nano 10:10636–10651

Sahu S, Behera B, Maiti TK, Mohapatra S (2012) Simple one-step synthesis of highly luminescent carbon dots from orange juice: application as excellent bio-imaging agents. Chem Commun 48:8835–8837

Salatin S, Khosroushahi AY (2017) Overviews on the cellular uptake mechanism of polysaccharide colloidal nanoparticles. J Cell Mol Med 21:1668–1686

Sanhai WR, Sakamoto JH, Canady R, Ferrari M (2008) Seven challenges for nanomedicine. Nat Nanotechnol 3:242–244

Sapsford KE, Tyner KM, Dair BJ, Deschamps JR, Medintz IL (2011) Analyzing nanomaterial bioconjugates: a review of current and emerging purification and characterization techniques. Anal Chem 83:4453–4488

Sawant RR, Torchilin VP (2012) Multifunctional nanocarriers and intracellular drug delivery. Curr Opin Solid St M 16:269–275

Schaak RE, Williams ME (2012) Full disclosure: the practical side of nanoscale total synthesis. ACS Nano 6:8492–8497

Semmling M et al (2008) A novel flow-Cytometry-based assay for cellular uptake studies of polyelectrolyte microcapsules. Small 4:1763–1768

Serkova NJ (2017) Nanoparticle-based magnetic resonance imaging on tumor-associated macrophages and inflammation. Front Immunol 8:590

Shen HX, Shi SJ, Zhang ZR, Gong T, Sun X (2015) Coating solid lipid nanoparticles with hyaluronic acid enhances antitumor activity against melanoma stem-like cells. Theranostics 5:755–771

Shilo M, Reuveni T, Motiei M, Popovtzer R (2012) Nanoparticles as computed tomography contrast agents: current status and future perspectives. Nanomedicine-UK 7:257–269

Shin TH et al (2014) T-1 and T-2 dual-mode MRI contrast agent for enhancing accuracy by engineered Nanomaterials. ACS Nano 8:3393–3401

Siepmann F, Herrmann S, Winter G, Siepmann J (2008) A novel mathematical model quantifying drug release from lipid implants. J Control Release 128:233–240

Silva AC, Lawder HJ, Hara A, Kujak J, Pavlicek W (2010) Innovations in CT dose reduction strategy: application of the adaptive statistical iterative reconstruction algorithm. Am J Roentgenol 194:191–199

Silva ACA et al (2014) Controlling the cytotoxicity of CdSe magic-sized quantum dots as a function of surface defect density. Nano Lett 14:5452–5457

Singh N, Jenkins GJS, Asadi R, Doak SH (2010) Potential toxicity of superparamagnetic iron oxide nanoparticles (SPION). Nano Reviews 1:5358

Song XJ, Chen Q, Liu Z (2015) Recent advances in the development of organic photothermal nano-agents. Nano Res 8:340–354

Sperling RA, Rivera Gil P, Zhang F, Zanella M, Parak WJ (2008) Biological applications of gold nanoparticles. Chem Soc Rev 37:1896–1908

Sun DX (2010) Nanotheranostics: integration of imaging and targeted drug delivery. Mol Pharm 7:1879–1879

Sun TM et al (2014) Engineered nanoparticles for drug delivery in cancer therapy. Angew Chem Int Ed 53:12320–12364

Suto M et al (2009) Heat dissipation mechanism of magnetite nanoparticles in magnetic fluid hyperthermia. J Magn Mater 321:1493–1496

Tang RB et al (2011) Microbubble-based synchrotron radiation phase contrast imaging: basic study and angiography applications. Phys Med Biol 56:3503–3512
Tang HY et al (2012) Gold nanorods@mSiO(2) with a smart polymer shell responsive to heat/near-infrared light for chemo-photothermal therapy. J Mater Chem 22:16095–16103
Tassa C, Shaw SY, Weissleder R (2011) Dextran-coated Iron oxide nanoparticles: a versatile platform for targeted molecular imaging, molecular diagnostics, and therapy. Acc Chem Res 44:842–852
Theis T et al (2006) Nan'o.Tech.nol'o.Gy n. Nat Nanotechnol 1:8–10
Thrall JH (2004) Nanotechnology and medicine. Radiology 230:315–318
Tian BN et al (2018) Low irradiance multiphoton imaging with alloyed lanthanide nanocrystals. Nat Commun 9:3082
Torchilin VP (2007) Micellar nanocarriers: pharmaceutical perspectives. Pharm Res-Dordr 24:1–16
Ulbrich K et al (2016) Targeted drug delivery with polymers and magnetic nanoparticles: covalent and noncovalent approaches, release control, and clinical studies. Chem Rev 116:5338–5431
Vallet-Regi M, Colilla M, Izquierdo-Barba I, Manzano M (2018) Mesoporous silica nanoparticles for drug delivery: current insights. Molecules 23:47
van der Zee J (2002) Heating the patient: a promising approach? Ann Oncol 13:1173–1184
van Landeghem FKH et al (2009) Post-mortem studies in glioblastoma patients treated with thermotherapy using magnetic nanoparticles. Biomaterials 30:52–57
Wang YXJ, Idee JM (2017) A comprehensive literatures update of clinical researches of superparamagnetic resonance iron oxide nanoparticles for magnetic resonance imaging. Quant Imaging Med Surg 7:88–122
Wang GF, Peng Q, Li YD (2011) Lanthanide-doped Nanocrystals: synthesis, optical-magnetic properties, and applications. Acc Chem Res 44:322–332
Wang YC, Hu R, Lin GM, Roy I, Yong KT (2013a) Functionalized quantum dots for biosensing and bioimaging and concerns on toxicity. ACS Appl Mater Interfaces 5:2786–2799
Wang Y et al (2013b) Multifunctional Mesoporous silica-coated Graphene Nanosheet used for chemo-Photothermal synergistic targeted therapy of Glioma. J Am Chem Soc 135:4799–4804
Wang JR et al (2013c) Tumor redox heterogeneity-responsive Prodrug Nanocapsules for Cancer chemotherapy. Adv Mater 25:3670–3676
Wang YG et al (2014) A nanoparticle-based strategy for the imaging of a broad range of tumours by nonlinear amplification of microenvironment signals. Nat Mater 13:204–212
Webb JA, Bardhan R (2014) Emerging advances in nanomedicine with engineered gold nanostructures. Nanoscale 6:2502–2530
Wei H et al (2017) Exceedingly small iron oxide nanoparticles as positive MRI contrast agents. P Natl Acad Sci USA 114:2325–2330
Weigel PH, Oka JA (1981) Temperature-dependence of endocytosis mediated by the Asialoglycoprotein receptor in isolated rat hepatocytes - evidence for 2 potentially rate-limiting steps. J Biol Chem 256:2615–2617
Wen SH et al (2018) Advances in highly doped upconversion nanoparticles. Nat Commun 9:2415
West JL, Halas NJ (2003) Engineered nanomaterials for biophotonics applications: improving sensing, imaging, and therapeutics. Annu Rev Biomed Eng 5:285–292
Wilhelm S et al (2015) Water dispersible upconverting nanoparticles: effects of surface modification on their luminescence and colloidal stability. Nanoscale 7:1403–1410
Wisser MD et al (2015) Strain-induced modification of optical selection rules in lanthanide-based Upconverting nanoparticles. Nano Lett 15:1891–1897
Wolfbeis OS (2015) An overview of nanoparticles commonly used in fluorescent bioimaging. Chem Soc Rev 44:4743–4768
Wolfram J et al (2015) Safety of nanoparticles in medicine. Curr Drug Targets 16:1671–1681
Wolkers WF et al (2003) Temperature dependence of fluid phase endocytosis coincides with membrane properties of pig platelets. BBA-Biomembranes 1612:154–163
Wong PT, Choi SK (2015) Mechanisms of drug release in Nanotherapeutic delivery systems. Chem Rev 115:3388–3432

Wu X et al (2015) Upconversion nanoparticles: a versatile solution to multiscale biological imaging. Bioconjug Chem 26:166–175
Wust P et al (2002) Hyperthermia in combined treatment of cancer. Lancet Oncol 3:487–497
Xie J, Lee S, Chen XY (2010) Nanoparticle-based theranostic agents. Adv Drug Deliv Rev 62:1064–1079
Xie T, Jing C, Long YT (2017) Single plasmonic nanoparticles as ultrasensitive sensors. Analyst 142:409–420
Xu L et al (2018) Biodegradable pH-responsive hydrogels for controlled dual-drug release. J Mater Chem B 6:510–517
Xue XD et al (2014) Spatiotemporal drug release visualized through a drug delivery system with tunable aggregation-induced emission. Adv Mater 26:712–717
Yang K et al (2010) Graphene in mice: ultrahigh in vivo tumor uptake and efficient Photothermal therapy. Nano Lett 10:3318–3323
Yang K, Feng LZ, Shi XZ, Liu Z (2013) Nano-graphene in biomedicine: theranostic applications. Chem Soc Rev 42:530–547
Yu SJ, Kang MW, Chang HC, Chen KM, Yu YC (2005) Bright fluorescent nanodiamonds: no photobleaching and low cytotoxicity. J Am Chem Soc 127:17604–17605
Yu YX et al (2017) Size-dependent photoluminescence efficiency of silicon Nanocrystal quantum dots. J Phys Chem C 121:23240–23248
Yuan YY et al (2012) Surface charge switchable nanoparticles based on Zwitterionic polymer for enhanced drug delivery to tumor. Adv Mater 24:5476–5480
Zamboni WC et al (2012) Best practices in Cancer nanotechnology: Perspective from NCI nanotechnology Alliance. Clin Cancer Res 18:3229–3241
Zhang W et al (2011) Synergistic effect of chemo-photothermal therapy using PEGylated graphene oxide. Biomaterials 32:8555–8561
Zhang R, Yang JY, Sima M, Zhou Y, Kopecek J (2014a) Sequential combination therapy of ovarian cancer with degradable N-(2-hydroxypropyl)methacrylamide copolymer paclitaxel and gemcitabine conjugates. P Natl Acad Sci USA 111:12181–12186
Zhang CQ et al (2014b) Imaging intracellular anticancer drug delivery by self-assembly micelles with aggregation-induced emission (AIE micelles). ACS Appl Mater Interfaces 6:5212–5220
Zhang CL et al (2015) Gold Nanoclusters-based Nanoprobes for simultaneous fluorescence imaging and targeted photodynamic therapy with superior penetration and retention behavior in tumors. Adv Funct Mater 25:1314–1325
Zhao ZH et al (2013) Octapod iron oxide nanoparticles as high-performance T-2 contrast agents for magnetic resonance imaging. Nat Commun 4:1–7
Zheng MB et al (2013) Single-step assembly of DOX/ICG loaded lipid-polymer nanoparticles for highly effective chemo-photothermal combination therapy. ACS Nano 7:2056–2067
Zhou ZX, Lu ZR (2013) Gadolinium-based contrast agents for magnetic resonance cancer imaging. Wires Nanomed Nanobi 5:1–18
Zhou KJ et al (2012) Multicolored pH-tunable and Activatable fluorescence Nanoplatform responsive to physiologic pH stimuli. J Am Chem Soc 134:7803–7811
Zhou ZJ et al (2013) Folic acid-conjugated silica capped gold nanoclusters for targeted fluorescence/X-ray computed tomography imaging. J Nanobiotechnol 11:17
Zhou J, Yang Y, Zhang CY (2015) Toward biocompatible semiconductor quantum dots: from biosynthesis and bioconjugation to biomedical application. Chem Rev 115:11669–11717
Zhu XL et al (2015) Body temperature sensitive micelles for MRI enhancement. Chem Commun 51:9085–9088
Zhu XH, Zhang J, Liu JL, Zhang Y (2019) Recent Progress of rare-earth doped Upconversion nanoparticles: synthesis, optimization, and applications. Adv Sci 6:1901358
Zrazhevskiy P, Sena M, Gao XH (2010) Designing multifunctional quantum dots for bioimaging, detection, and drug delivery. Chem Soc Rev 39:4326–4354

Semiconductor Nanocrystals for Biological Imaging and Fluorescence Spectroscopy

Fumihiko Fujii

Introduction

Semiconductor nanocrystals (SNCs), referred to as quantum dots (QDs), are nano-sized crystals with diameters ranging from 2 to 10 nm (Alivisatos 1996). As such, they are larger than conventional organic dyes in size, and comparable to the size of fluorescent proteins (Giepmans et al. 2006), other large biomolecules, and the exciton Bohr radius that is the distance of the excited electron–hole pair (Fig. 1). Since the motion of the electron–hole pair is confined in the nano-sized space, SNCs possess unique optical and electrical properties that differ from bulk materials (Norris and Bawendi 1996). Thus, SNCs are expected to be applied to the next generation of solar cells, lasers, light-emitting diodes, computers, and bioimaging probes (Bera et al. 2010).

There are two types of SNCs, epitaxial and colloidal SNCs (Bayer 2019). Epitaxial SNCs grow on substrate by molecular beam epitaxy from the vapor phase, in which the reaction vessel is filled with expensive high-purity materials and maintained at ultrahigh vacuum conditions (Goldstein et al. 1985). On the other hand, colloidal SNCs are synthesized by simple chemical processes at relatively moderate conditions. Unlike substrate-bounded epitaxial SNCs, colloidal SNCs are freestanding and thus can be moved to post-chemical processes, such as water solubilization (Bruchez et al. 1998; Chan and Nie 1998). This is notable, as water solubilization is an important step for the biological application of SNCs (Michalet et al. 2005).

SNCs possess unique optical properties for biological application. Due to high resistance to photobleaching, they are suitable for high-sensitive and long-term imaging (Koneswaran and Narayanaswamy 2015). All SNCs with different emission spectra can be excited by a single light source, and the different emission

F. Fujii (✉)
Faculty of Pharmaceutical Sciences, Kobe Gakuin University, Kobe, Japan
e-mail: fumihiko.fujii@pharm.kobegakuin.ac.jp

J. K. Kim et al. (eds.), *Advanced Imaging and Bio Techniques for Convergence Science*, Advances in Experimental Medicine and Biology,
https://doi.org/10.1007/978-981-33-6064-8_16

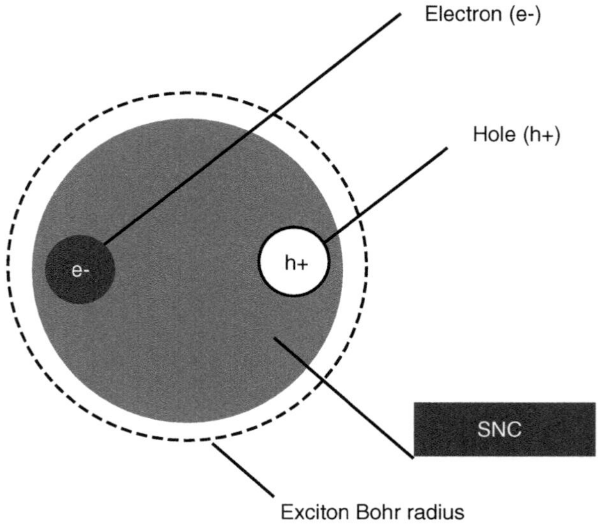

Fig. 1 SNC and the exciton Bohr radius

spectra that result minimally overlap each other. Further, SNCs, by changing their sizes and compositions, possess wide emission colors that range from ultraviolet (UV) to near infrared (NIR) (Michalet et al. 2005). Thus, SNCs are suitable for multicolored imaging (Kobayashi et al. 2007). This review mainly addresses the unique properties and synthetic procedures of colloidal SNCs, and the biological imaging and fluorescence spectroscopies that are enabled using these water-soluble colloidal SNCs.

Synthesis of SNCs and Preparation of Water-Soluble SNCs

In the synthesis of colloidal SNCs, a series of chemical reactions proceed in the flask. A typical procedure for high quality SNCs with a narrow-sized distribution and a high quantum yield is based on heating specific organic solvents and rapidly injecting of semiconductor precursors (Murray et al. 1993). The rapid injection at high temperatures initiates decomposition and reduction of precursors, which then frees metal ions in the organic solvents. The accumulation of metal ions allows the first small nanocrystals to form, followed by further growth of the crystals at high temperatures. At the high temperature at which growth occurs, capping ligands, such as tri-*n*-octylphosphine oxide (TOPO), rapidly adsorb to and desorb from the surface of the nanocrystals, and the free metal ions adsorb to the surface without nanocrystal aggregation. By selecting suitable semiconductor precursors, we can obtain desirable SNCs (Table 1). This section addresses the structures, typical synthetic procedures, shape control of SNCs, and preparation methods of water-soluble SNCs.

Table 1 Classification of SNCs

Groups in the periodic table (IUPAC numbering)		Examples of SNCs
II–VI	(12 and 16 groups)	CdTe, CdSe, CdS, ZnTe, ZnSe, ZnS
III–V	(13 and 15 groups)	InP, InAs
I–III–VI_2	(11, 13, and 16 groups)	$CuInS_2$, $AgInS_2$
IV–VI	(14 and 16 groups)	PbSe, PbS, PbTe
IV	(14 group)	Si, Ge

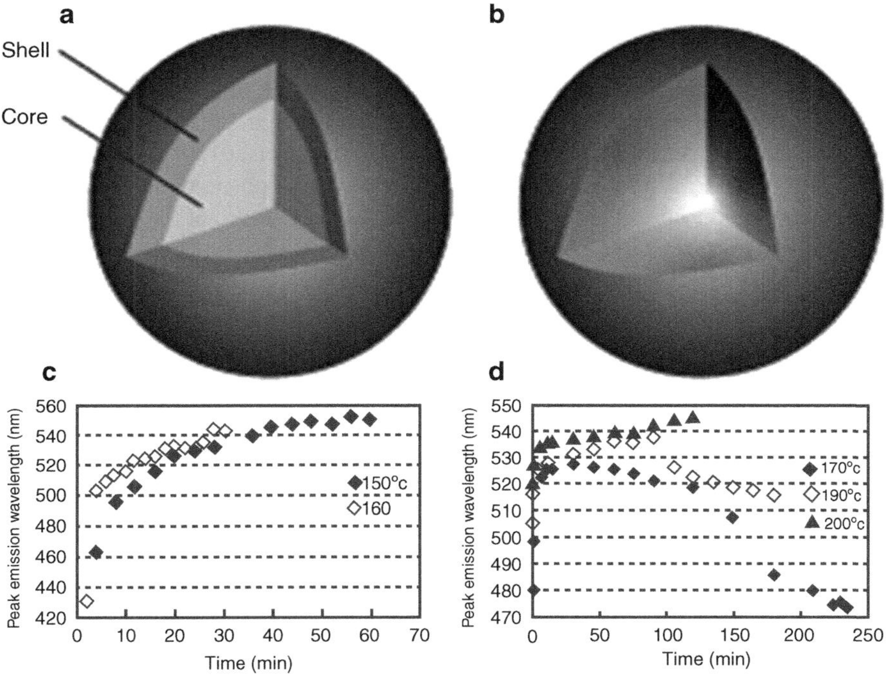

Fig. 2 Core/Shell and gradient-alloyed SNCs. (**a**) Structure of core/shell SNC. (**b**) Structure of gradient-alloyed SNC. (**c**) Time course of peak emission wavelength of core/shell SNCs. (**d**) Time course of peak emission wavelength of gradient-alloyed SNCs

Structure and Synthesis of SNCs

Typical colloidal SNCs are composed of core and shell structures (Fig. 2a). Bare SNCs, namely core-only SNCs, are sensitive to environmental changes and photo-oxidation, due to surface defects that occur frequently because of an overly large surface-to-volume ratio of the core. By covering the core with the shell structure of a higher bandgap semiconductor material, the core/shell SNCs possess high quantum yield and high stability against both photobleaching and blinking (Baranov et al. 2003; Mahler et al. 2008). However, these parameters are also affected by a lattice mismatch between core and shell materials (Smith et al. 2009). To overcome the lattice mismatch problem, gradient-alloyed SNCs have been developed (Fig. 2b) (Zhang et al. 2016; Cho et al. 2017). The chemical composition of the gradient-alloyed SNCs gradually changes from the inner part to the outer one without interfacial separation.

The gradient-alloyed SNCs, which emit blue light, are greatly challenging, because in the synthetic procedure of the core/shell SNCs, the growth kinetics of the core component is rapid, and the red shift of the emission spectrum occurs during the epitaxial growth of the shell component onto the core (Dabbousi et al. 1997). Gradient-alloyed SNCs that emit below 480 nm have also been synthesized by our research team. In our experimental conditions at a temperature of 170 °C, the peak emission wavelength of the gradient-alloyed SNCs ($CdSe_XS_{1-X}$) increased for 30 minutes, but gradually decreased and reached 475 nm after a reaction time of 225 min (Fig. 2d). In contrast, the peak emission wavelength of the core/shell SNCs (CdSe) increased and did not reduce to below 480 nm, even at 150 °C (Fig. 2c). The blue shift of the gradient-alloyed SNCs represents an opposite trend compared to the patterns of the core/shell SNCs. This suggests the existence of the intradiffusion process of cations or anions from the outer component into the inner one (Groeneveld et al. 2013).

Shape Control of SNCs

There are various shapes that exist for SNCs, such as rod-shaped (Peng et al. 2000) and tetrapod-shaped SNCs (Jun et al. 2001). The spherical-, rod-, and tetrapod-shaped SNCs have also been synthesized by our research team. The TEM images of the heterostructured SNCs are shown in Fig. 3.

Shape control of the SNCs was achieved by manipulating the growth kinetics and changing the crystal structure type of the core. Comparing the spherical- and rod-shaped SNCs (CdSe/CdS), the crystal structure of the core (CdSe) and the core and shell components are the same (Fig. 5a, b). However, the shapes and structures

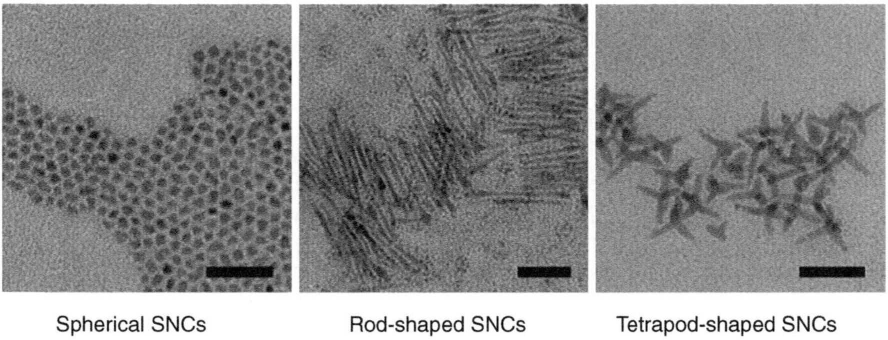

Fig. 3 Heterostructured SNCs. Scale bars: 40 nm (TEM images by Minkyo Jung and Chan-Gi Pack at Asan Medical Center, Korea)

of the shells differ, due to the growth rate of the shell and the anisotropy of the crystal structure of the core. Further, the core (CdSe) of the SNC is a wurtzite-type crystal and has a *c*-axis, whereby the growth rate differs for each axis (Peng et al. 2000). At a slow growth rate, the spherical SNC preferentially generates due to surface area minimization of the shell (Fig. 5a). In contrast, when the growth rate is increased, the shell grows along the *c*-axis of the wurtzite-type crystal in a preferential manner (Fig. 5b). To control the growth rate, it is necessary to change the capping ligands. The utilization of longer alkyl chains, such as octadecylphosphonic acid (ODPA), significantly reduces the growth rate. The typical synthesis procedure of the rod-shaped SNCs (CdSe/CdS) is described below (Fig. 4), and the details of this procedure are reported in the studies by Deka et al. (2009) and Watanabe et al. (2013).

1. Cadmium oxide, tri-*n*-octylphosphine oxide (TOPO), and octadecylphosphonic acid (ODPA) are placed in a three-necked round bottom flask, degassed under an atmosphere of N_2 and stirring.
2. To synthesize core CdSe, the mixture is heated and then tri-n-octylphosphine (TOP) is injected. After heating, selenium solution is quickly added to the mixture using a syringe, and then annealed.
3. After cooling to room temperature, the CdSe are precipitated by the addition of methanol, and separated and re-dispersed in TOP.
4. CdO, TOPO, ODPA, and hexylphosphonic acid (HPA) are placed in the flask and degassed under N_2 atmosphere with stirring. The mixture is heated and TOP is added to the mixture.
5. To synthesize core/shell CdSe/CdS, after heating, the mixture of sulfur in TOP and the CdSe/TOP solution is quickly added to the mixture. Once the fluores-

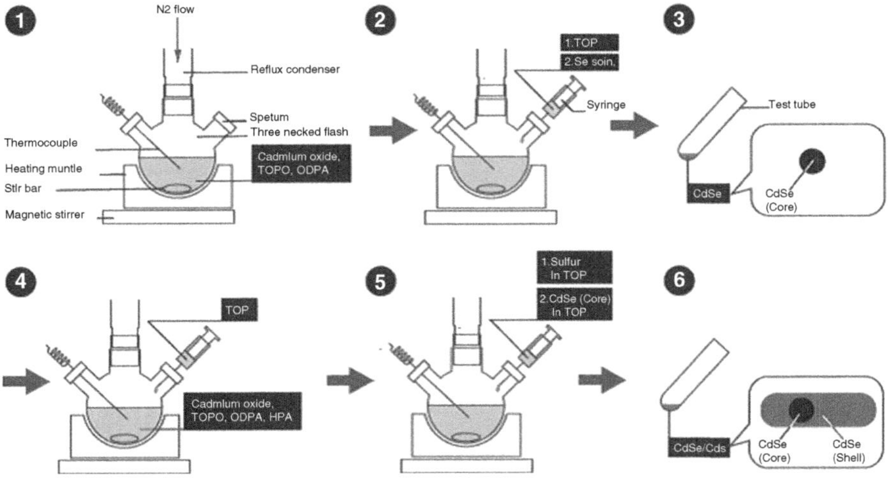

Fig. 4 Synthetic procedure of rod-shaped SNCs

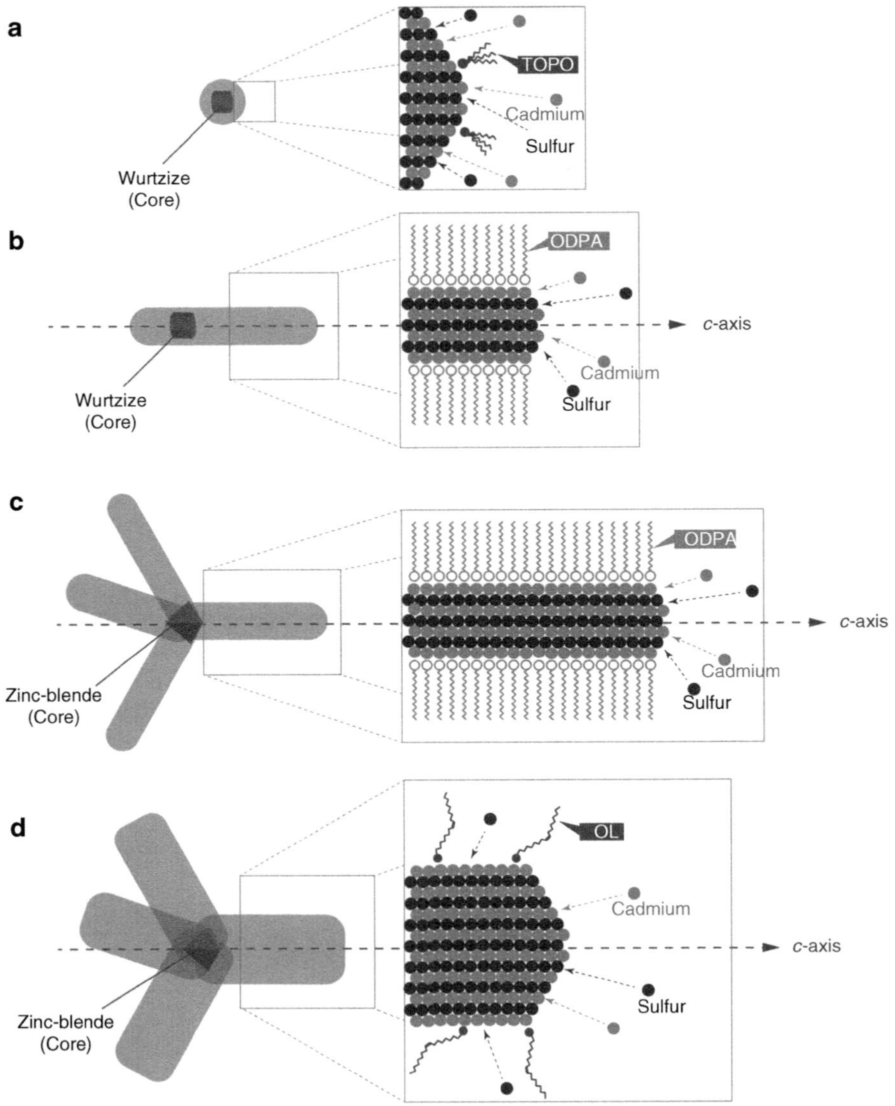

Fig. 5 Schematic growth diagram of heterostructured SNCs. (**a**) Spherical SNC. (**b**) Rod-shaped SNC. (**c**) Tetra pod-shaped SNC with long arm. (**d**) Tetrapod-shaped SNC with thick arm

cence spectrum of the solution reaches the purpose one, the solution is lowered to 100 °C and stirred.

6. After the temperature is lowered to 60 °C, CdSe/CdS are precipitated by the addition of excess methanol and separated by centrifugation. The precipitated CdSe/CdS are dissolved in cyclohexane and stored in a dark place.

To compare rod- and tetrapod-shaped SNCs (CdSe/CdS), the core and shell components are the same, while the shapes of the shells are different (namely containing two and four, respectively) (Fig. 5b, c). The difference in the number of arms is derived from the crystal structure of the core. The rod- and tetrapod-shaped SNCs, respectively, have wurtzite- and zinc blende-type crystals as cores, (Talapin et al. 2007). Since the zinc blende-type crystals are characterized by tetrahedral coordination, the four arms grow on the crystallographic surface of the core. Further, on the core surface, the nucleation and growth of the shell structure occur based on the role of surface-selective capping ligands (Mishra et al. 2016). In the presence of strongly binding capping ligands, such as ODPA, the shell grows solely along the *c*-axis and develops long arms (Fig. 5c). This is because the ligands preferentially bind to the face, preventing the approach of ions to the sides of the face. In contrast, weakly bound capping ligands, such as oleic acid (OL), allow shell growth at both the side and end faces, which results in thick arms (Fig. 5d).

The rod- and tetrapod-shaped SNCs feature unique photophysical phenomena. Since the rod-shaped SNCs exhibit polarized emission due to a large aspect ratio (Hu et al. 2001), these SNCs have been utilized for the orientation measurement of the biomolecules (see Sections "Fluorescence Correlation Spectroscopy Using SNCs and Single Particle Tracking and Orientation Measurement Using SNCs"). In contrast, the tetrapod-shaped SNCs exhibit dual-color emission (Lutich et al. 2010; Wong et al. 2014). The dual-color emission—red and green—originates from the core and shell (arms), respectively. This is because two excitons generate and recombine independently in the spatially distinct regions of the core and shell. Thus, the intensity of the green emission depends on the length and thickness of the arms; SNCs with thicker arms exhibit stronger green emission (Mishra et al. 2017). These phenomena of rod- and tetrapod-shaped SNCs allow sophisticated fluorescence imaging to investigate the dynamics of biomolecules *in vitro* and in cells.

Preparation of Water-Soluble SNCs and Conjugation of Biomolecules with SNCs

The SNCs synthesized in nonpolar organic solvents are capped with hydrophobic ligands, such as TOPO and TOP (Fig. 6a). Thus, the hydrophobic SNCs must become hydrophilic by surface modification to allow biological application in aqueous buffers or bodily fluids. The strategies for preparing water-soluble SNCs are classified into two broad categories: ligand exchange and encapsulation methods.

In the ligand exchange method, the hydrophobic ligands, such as TOPO and TOP, on the SNCs surfaces are replaced with amphiphilic ligands (Fig. 6b). The amphiphilic ligands contain anchoring groups, such as thiol, at one end, which allows binding to the SNCs surfaces, and hydrophilic functions at the other end, which promotes affinity to aqueous solutions. Monothiolated amphiphilic ligands, such as mercaptoacetic acid (Chan and Nie 1998), mercaptoundecanoic acid (Jiang et al. 2006), cysteine (Liu et al. 2007), and glutathione (Jin et al. 2008a, b), have been utilized to prepare water-soluble SNCs. The drawbacks of the ligand-exchanged

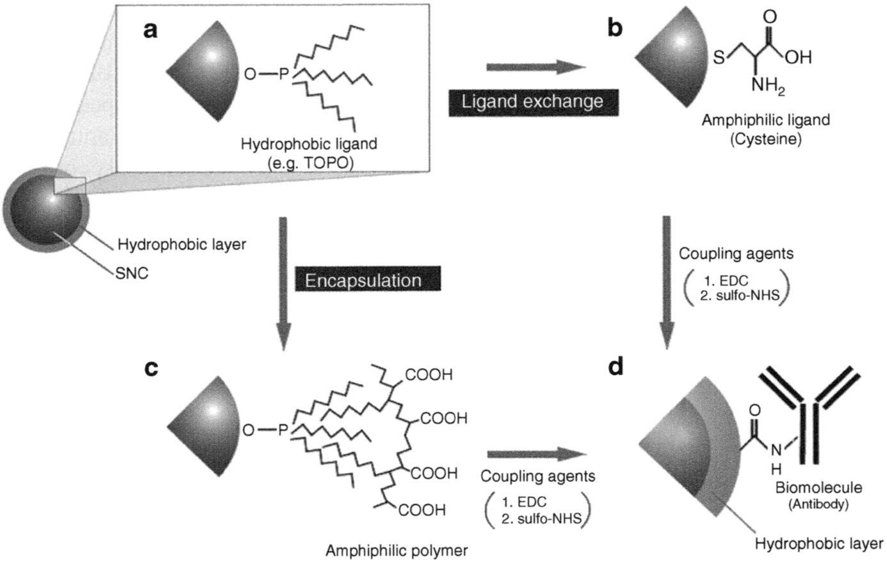

Fig. 6 Water solubilization and bioconjugation of SCNs. (**a**) Hydrophobic SCN. (**b**) Hydrophilic SCN with amphiphilic ligand. (**c**) Hydrophilic SCN with amphiphilic polymer. (**d**) Hydrophilic SCN with antibody

SNCs include relatively poor stability in the biological buffer due to desorption of the amphiphilic ligands from the SNCs surfaces. Thus, dithiol-terminated amphiphilic ligands, such as dihydrolipoic acid and their derivatives, have been utilized to improve the long-term stability of the anchoring groups on the SNCs surfaces (Susumu et al. 2009).

In the encapsulation method, an amphiphilic polymer, which interacts with a hydrophobic alkyl chains, is introduced, while hydrophobic ligands, such as TOPO and TOP, are retained on the SNC surface (Fig. 6c). Thus, the SNCs are protected from the outside environment by the hydrocarbon bilayer. Several polymers, such as octylamine-modified polyacrylic acid (Wu et al. 2003), polyethylene glycol derivatized phospholipids (Dubertret et al. 2002), and amphiphilic polyanhydrides (kirchner et al. 2005), have been utilized to prepare water-soluble SNCs. Although the diameters of the encapsulated SNCs are much larger than those of the ligand-exchanged SNCs, the encapsulated SNCs are stable over a broader range of salt concentrations (Gao et al. 2004). The typical procedure for the encapsulated SNCs is described below (Fig. 7). The details of the procedure are reported in the studies by Yu et al. (2007) and Watanabe et al. (2013).

1. To prepare amphiphilic polymers (PMAO–PEG) as coating agents, poly(maleic anhydride-alt-1-octadecene (PMAO) is reacted with primary amino PEG in chloroform overnight at room temperature.
2. The PMAO–PEG is mixed with the SNCs in the chloroform solution and stirred for several days at room temperature.

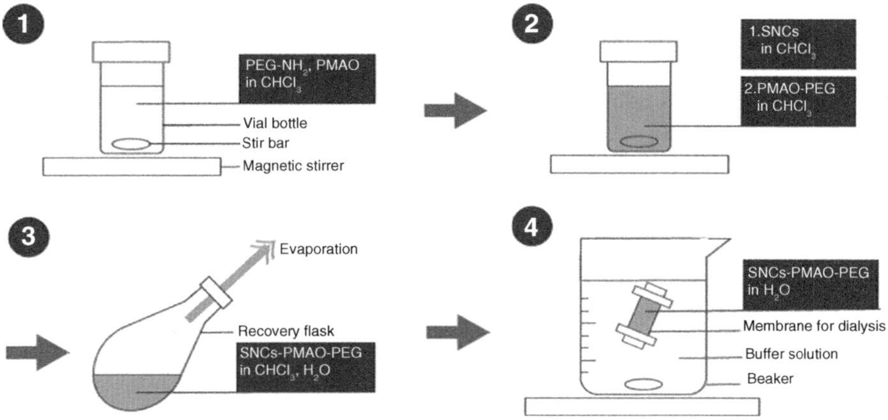

Fig. 7 Preparation method of encapsated SNCs

3. Deionized distilled water is added to the solution and the chloroform is gradually evaporated at room temperature.
4. Excess PMAO–PEG is removed by dialysis in buffer.

To perform the specific binding of SNCs to target molecules, the water-soluble SNCs must be conjugated to biomolecules, such as peptides, oligonucleotides, and proteins, including antibodies (Fig. 6d). Since most water-soluble SNCs have carboxyl, amine, and inherent hydrophobic groups on their surfaces, the link to biomolecules is enabled through the following conjugation methods: electrostatic binding, noncovalent binding, and covalent crosslinking. In the electrostatic binding method, an adaptor protein is utilized as a bifunctional adaptor with its two domains, namely a leucine zipper domain to electrostatically bind to SNCs and a protein G domain to bind to the antibodies (Goldman et al. 2002). In noncovalent binding methods, a pair of streptavidin–biotin is utilized, because its dissociation constant is the strongest among the noncovalent interactions (Xiao and Barker 2004). The covalent conjugation that utilizes both 1-ethyl-3-[3-dimethylaminopropyl] carbodiimide (EDC) and N-hydroxysulfosuccinimide (Sulfo-NHS) is the most common method, and is based on cross-linking reactions between amines of proteins and carboxylic acid groups of SNCs, as most proteins have primary amines that do not require chemical modifications before covalent conjugation with both EDC and Sulfo-NHS (Guan et al. 2012).

Fluorescence Spectroscopy and Biological Imaging Using SNCs

The unique optical properties of SNCs allow various advantages for biological applications over traditional fluorophores, such as organic dyes and fluorescent proteins. Since the molar extinction coefficients of SNCs are approximately 10–50 times larger than those of organic dyes (Leatherdale et al. 2002), the fluorescence brightness of SNCs is much higher than that of the organic dyes. While the traditional fluorophores have narrow absorption and broad emission spectra with asymmetric shapes, SNCs allow extremely broad absorption and comparatively narrow emission spectra with symmetric shapes (Resch-Genger et al. 2008). Thus, SNCs are suitable for multicolored imaging with minimal overlap of emission spectra (Kobayashi et al. 2007). In addition, as SNCs are highly stable in response to photobleaching (Wu et al. 2003), they are also suitable for continuous tracking studies across a long period of time (Koneswaran and Narayanaswamy 2015). Therefore, SNCs have become one of the ideal probes for sensing, labeling, and *in vivo* fluorescence imaging. This section addresses fluorescence spectroscopy and biological imaging using SNCs.

Fluorescence Correlation Spectroscopy Using SNCs

Fluorescence correlation spectroscopy (FCS) is a fluorescence spectroscopy based on the confocal system. The concept of FCS was introduced in the early 1970s (Magde et al. 1972, 1974; Elson and Magde 1974), and its feasibility in biological studies was confirmed in the 1990s (Rigler et al. 1993; Kinjo and Rigler 1995). FCS detects fluorescence intensity fluctuations in the confocal area, and then the fluctuations are analyzed using the autocorrelation function. If the size of the confocal area is defined using a standard fluorophore, the hydrodynamic radius and the concentration and interaction of fluorescent-labeled molecules are determined at very low concentrations (~ 1 pM) in a tiny area (~ 1 fL) using FCS.

Using FCS, it is possible to determine the important properties of SNCs, such as hydrodynamic radius and quantum yield. The size and quantum yield of SNCs are typically measured by transmission electronic microscopy in dry conditions and with an integrating sphere in submicromolar concentrations, respectively. For biological applications, such as in cellular studies, the hydrodynamic radius, and quantum yield of water-soluble SNCs must be determined in aqueous solution at low concentrations. In FCS measurement, both the apparent hydrodynamic radius and the quantum yield of the SNCs are underestimated due to their blinking phenomena (Yao et al. 2005). Thus, the real hydrodynamic radius (de Thomaz et al. 2015) and quantum yield (Abbandonato et al. 2018) in the absence of the blinking must be estimated by a calibration procedure both multiple measurements at various laser powers.

FCS allows sensitive detection *in vitro* of the interactions between proteins and nanomaterials. When nanomaterials are introduced into a biological milieu, the surfaces of nanomaterials absorb several proteins, which results in the formation of a protein layer, namely the protein corona (Monopoli et al. 2012). The formation of the protein corona depends on the nanomaterial characteristics, such as size, morphology, and surface charge (Lundqvist et al. 2008). Thus, understanding the process of the corona formation is one of the most important aims in nanomedicine, in which nanomaterials are utilized for visualization, diagnosis, and therapy. FCS can evaluate the corona formation *in vitro* using SNCs as model nanomaterials (Wang et al. 2016). This FCS study revealed that the thickness of the protein layer and its binding affinity strongly depends on the type of SNC surface ligands.

In FCS studies, translational diffusion of fluorescent-labeled molecules has been primarily utilized to determine the hydrodynamic radius and the concentration and interaction of biomolecules. Although translational diffusion is proportional to the cube root of the molecular weight, rotational diffusion is directly proportional to the molecular weight. The term "rotational diffusion" may be separated from that of "translational diffusion" in single autocorrelation functions by assuming spherical-shape diffusors with polarization and fluorescence lifetimes that are shorter than the rotational correlation times (Kask et al. 1989; Widengren et al. 1999). Tsay et al. (2006) concluded that analysis of rotational diffusion using both polarized FCS and rod-shaped SNCs is more sensitive and informative to the interactions of biomolecules, compared to translational diffusion. In the polarized-FCS system, the cross-correlation analysis was performed using dual-channel detectors to extinguish the after-pulse noise in the sub-microsecond range of the autocorrelation functions (Tsay et al. 2006; Lee et al. 2014; Oura et al. 2016), while the single after-pulse free detector was also utilized to simplify the polarized-FCS system (Yamamoto et al. 2015).

Fluorescence cross-correlation spectroscopy (FCCS) is based on the same principles as FCS. Further, FCCS was proposed in the early 1990s (Eigen and Rigler 1994) and was first realized in the late 1990s (Schwille et al. 1997). In FCCS, the cross-correlation of the two independent signals is determined instead of the autocorrelation of one signal in FCS. In addition, FCCS is utilized to detect the interaction between two molecular species that are labeled with two color fluorophores *in vitro* (Kettling et al. 1998; Winkler et al. 1999) and in cells (Saito et al. 2004; Kogure et al. 2006). In addition, FCCS was applied to the highly sensitive detection of pathogenic antigens (Fujii et al. 2007). The pathogenic antigens were bound to two fluorescent monoclonal antibodies, and the doubly labeled antigens are detected by FCCS with two laser lines (TL-FCCS) (Fig. 8a–c). However, the sensitivity of FCCS is often reduced due to the difficulty that is associated with precise and stable overlap of two laser lines (Fig. 8b).

The utilization of SNCs overcomes this issue in the FCCS measurement. Due to the broad absorption spectra of SNCs, FCCS may be performed with a single laser line (SL-FCCS) (Fujii and Kinjo 2007) (Fig. 8d–f). In SL-FCCS analysis (excitation wavelength 473 nm), the antigens were bound to two types of labeling of monoclonal antibodies with Alexa Fluor-488 and SNCs (Quantum dots-655), whereby the two types of fluorescent antibodies were excited with a single laser line. In contrast, in TL-FCCS analysis (excitation wavelengths 473 and 635 nm), the antigens were bound to two types of labeling of monoclonal antibodies labeling with Alexa Fluor-488 and Alexa Fluor-647 (Fig. 8a–c). Consequently, the method using SL-FCCS and SNCs was approximately ten times more sensitive than that with TL-FCCS and two organic dyes.

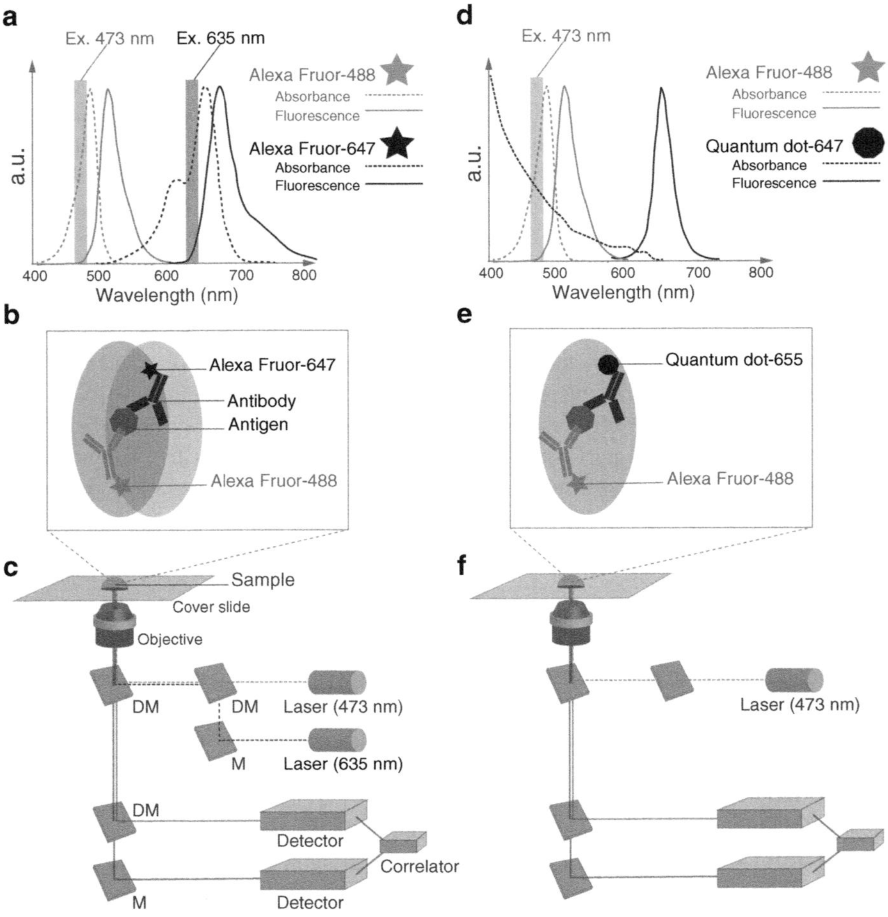

Fig. 8 Two lasers- and single laser- FCCS. (**a**) Absorbance and fluorescence spectra of Alexa Fruor-488 and -647. (**b**) Absorbance and fluorescence spectra of Alexa Fruor-488 and Quantum dot-655. (**c**) Antigen labeled fluorescent antibodies in two confocal area. (**d**) Antigen labeled fluorescent antibodies in single confocal area. (**e**) Schematic diagrum of two lasers FCCS. (**f**) Schematic diagrum of single laser FCCS

Single Particle Tracking and Orientation Measurement Using SNCs

Single particle tracking (SPT) is an observation method used to track individual fluorophores with nanometer precision. The spatial precision of SPT depends on the number of photons from the fluorophores (Kubitscheck et al. 2000; Ichimura et al. 2014). SNCs exhibit strong brightness and resistance to photobleaching (Koneswaran and Narayanaswamy 2015), while traditional fluorophores, such as organic dyes and fluorescent proteins, exhibit poor brightness and photobleaching. In fact, GFP-tagged peroxisomes in cultured cells were tracked with 1.5 nm accuracy and 1 ms time resolution (Kural et al. 2005), while endocytic vesicles that contain SNCs can be tracked with a brief period of less than 0.3 ms (Nan et al. 2005). Thus, SNCs became one of the ideal fluorescent probes for SPT and further provided precise information that concerns protein localization and dynamics with high spatial and temporal resolution.

In addition to two- and three-dimensional (2 and 3D) SPT, the orientation measurement of biomolecules is required to precisely reveal dynamics, since the biomolecules shift the X, Y, and Z positions and rotate and interact with partners at a particular orientation (Adachi et al. 2000; Brunnbauer et al. 2012; Can et al. 2014). Spherical SNCs contain isotropic photophysical properties, since their excitons are confined in three dimensions. In contrast, rod-shaped SNCs contain anisotropic photophysical properties, namely polarized emission, as the excitons are confined in two dimensions due to elongated structure along the *c*-axis (Hu et al. 2001). Thus, rod-shaped SNCs can be utilized as an orientation sensor to detect the orientation of biomolecules using polarized fluorescence microscopy.

A study by Ohmachi et al. (2012) used rod-shaped SNCs to record the simultaneous SPT at 2D positions (X and Y) and two types of orientations (θ and ϕ) of the myosin-V stepping mechanism on the actin filaments *in vitro* (Fig. 9). Since the SNC was bound to the myosin V at the C terminus via two HaloTags, the motion of the myosin V was completely synchronized with a single SNC (Fig. 9a). To determine the two orientations (θ and ϕ), the fluorescent signals of the SNCs are first split into two beams using a beam splitter (Fig. 9b). Thereafter, each beam is further divided into two perpendicular polarization components (0° and 90°, 45°, and 135°, respectively) by a pair of Wollaston prisms. Based on these fluorescent intensities, angular values of the in-plane and tilt angles (θ and ϕ, respectively) can be determined (Fig. 9b). Using the polarized fluorescence microscopy, Ohmachi et al. recorded the two types of orientations of myosin V with an accuracy of 10° at a time resolution of 33 ms (Fig. 9c).

Further, a study by Watanabe et al. (2013) recorded the simultaneous SPT in 3D positions (X, Y, Z) and orientation (θ) of a scavenger receptor, known as CD36 (Fig. 10). The receptors on the plasma membrane of the isolated macrophages were labeled with anti-CD36 antibody-conjugated rod-shaped SNCs (Fig. 10a). For 3D measurement, concave (CnC) and convex cylindrical (CvC) lenses were established behind polarization optics (Fig. 10b). Two cylindrical lenses were found to produce different optical path lengths along the *X*- and *Y*-axes, which results in a measurable relationship between the Z-position of the fluorophore and the ellipticity of the image. For orientation measurement, the polarizing beam splitter (BS) was established in front of the cylindrical lenses to split the fluorescent signals into S- and P-polar channels (Fig. 10b). Using polarized fluorescence microscopy, the orientation of the scavenger receptors can be determined by the ratio of the intensities of the S- and P-polarized signals (Fig. 10c). The most precise calculations for the X-, Y-, Z-positions and the angle were 5, 5, 12 nm, and < 6°, respectively, at 1500 photons per 100 ms.

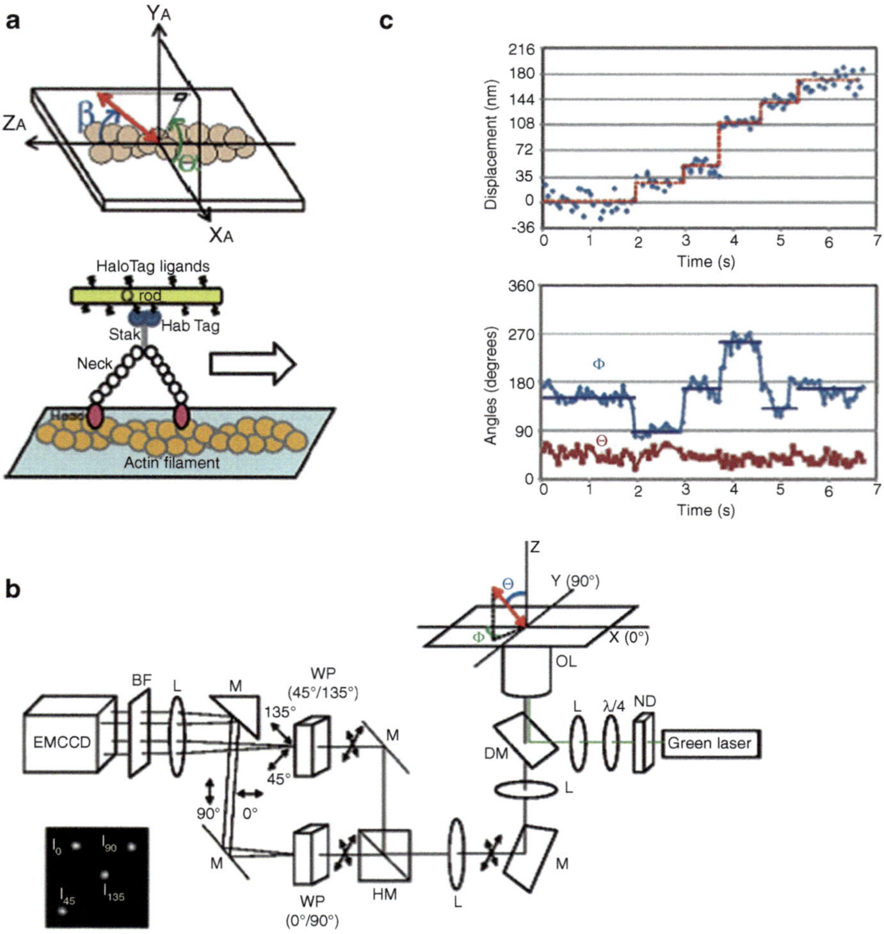

Fig. 9 Orientation measurements of myosin V using rod-shaped SNCs.(**a**) Reference frame of actin filament (upper), myosin V labeled with rodshaped SNC (lower). (**b**) Position (upper) and orientation (lower) of myosin V. (**c**) Shcematic diagram of the optics for simultaneous 2D and two types of orientations measurement [From Ohmachi M. et al. (2012) Proc. Nat. Acad. Sd. U S A., 106(14):5294–5298. With permission]

Using rod-shaped SNCs, a study by Sugawa et al. (2018) recorded *in vitro* the stepwise rotation of F_1-ATPase, namely the water-soluble portion of F_0F_1-ATPase (Fig. 11). F_0F_1-ATPase is a nanosized rotational motor in the mitochondrial inner membrane and creates ATP using an electrochemical gradient of protons. To monitor the rotational motion of F_1-ATPase, a single SNC was attached to the axle of the F_1-ATPase, referred to as the γ subunit, via streptavidin (Fig. 11a). A pair of two quarter-wave plates (QWP1&2), which bend the excitation light, and a pair of two wedge prisms (WP1&2) were established and rotated to achieve circular orientation

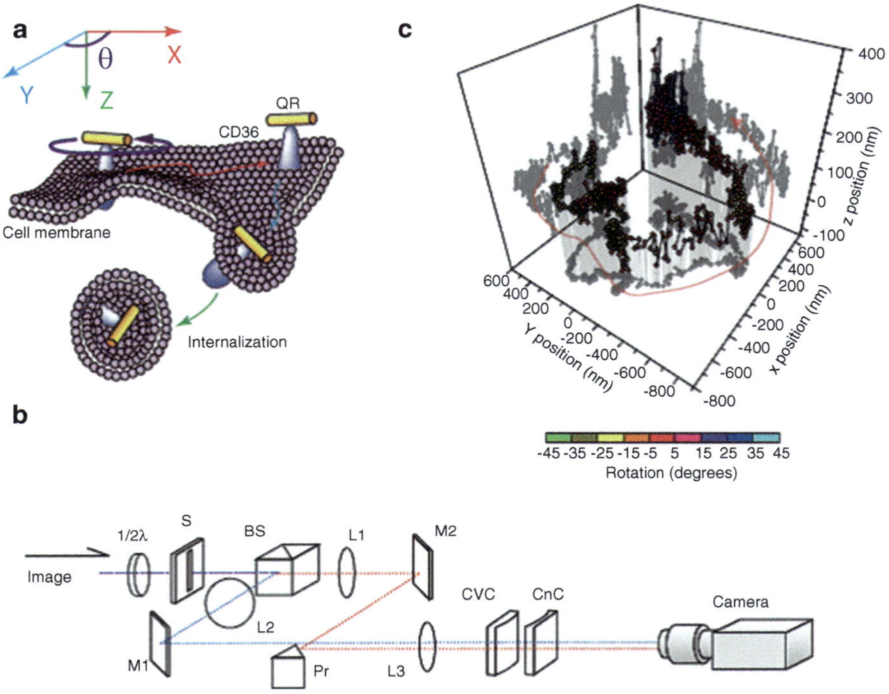

Fig. 10 Orientation measurements of CD36 using rod-shaped SNCs.(**a**) Reference frame and CD36 labeled with rodshaped SNC. (**b**) Position and orientation of CD36. (**c**) Shcematic diagram of the optics for simultaneous 3D and orientation measurement. [From Watanabe TM. et al. (2013) Biophys J. 105(3):565-E64. With permission]

fluorescence emitter imaging (COFEI) as a measurement system (Fig. 11b). The COFEI images exhibited orientations of the dipole moment of the rod-shaped SNCs that were attached to the γ subunit of F_1-ATPase. Using this fluorescence polarization measurement system, it was confirmed that the subunit, in the presence of ATP, rotates 80° + 40°for the substeps (Fig. 11c).

In vivo Fluorescence Imaging Using SNCs

A study by Jöbsis (1977) suggested that near-infrared (NIR) light penetrates deep living tissues and can be utilized to noninvasively monitor oxygenated and deoxygenated hemoglobin concentration changes. Since the ratio of the oxygenated and deoxygenated hemoglobin is related to oxygenation of the tissues, NIR spectroscopy has been applied to clinical studies (Sen et al. 2016; Brogan et al. 2017), allowing diffuse optical tomography to develop (Hoshi and Yamada 2016). NIR spectroscopy can also monitor changes in the oxygenation of myoglobin in skeletal muscles (Hamaoka and McCully 2019) and the redox state of cytochrome c oxidase

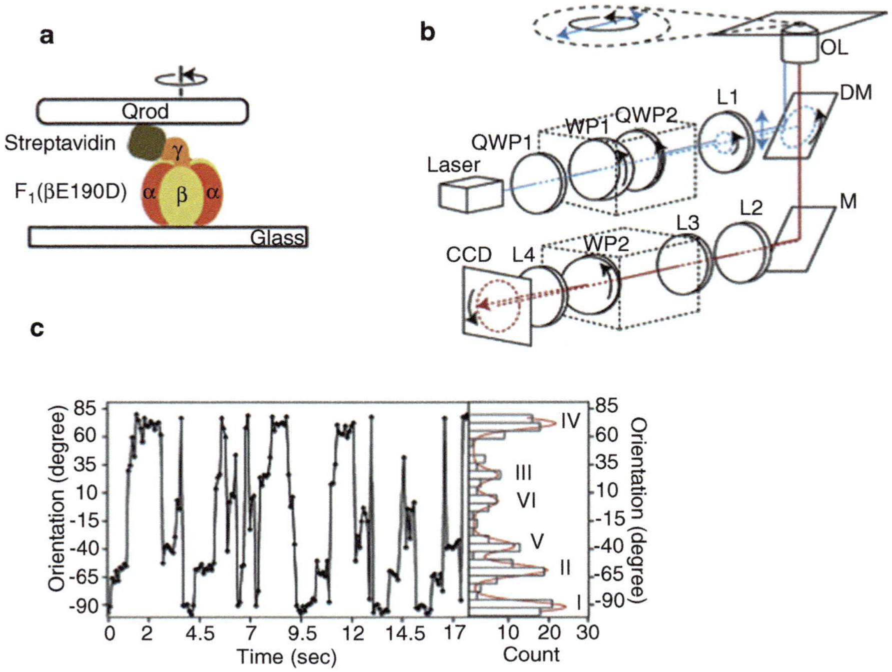

Fig. 11 Orientation measurements of F1-ATPase using rod-shaped SNCs. (**a**) F1-ATPase labeled with rodshaped SNC. (**b**) Shcematic diagram of the optics for COFEI. (**c**) Time course and histgrams of orientation of F1-ATPase. [From Sugawa M. et al. (2018) Biochem Biophys Res Commun. 504(4):709–714. With permission]

in mitochondria (Bale et al. 2016), as both the heme group in myoglobin and the Cu–Cu dimer in cytochrome c oxidase show strong absorption in the NIR region. As explained above, NIR spectroscopy utilizes light absorption to measure oxygenation and redox states of these endogenous chromophores.

In contrast, NIR fluorescence imaging utilizes fluorescence from exogenous contrast agents including SNCs *in vivo* to visualize target molecules and cells. The signal-to-noise ratio of NIR fluorescence imaging strongly depends on autofluorescence signals that arise from biological tissues. Nearly no autofluorescence signals are found above the NIR region, which is above 700 nm, though the fluorescence of the skin and viscera are detectable in the visible light region (400–700 nm) (Frangioni 2003; Diao et al. 2015). In addition, utilizing NIR light has the following two advantages over visible light: low light absorption and reduced light scattering. The sum of absorption with endogenous chromophores is decreased in the NIR region; however, the light absorption with both water and lipids is gradually increased from the visible region the infrared region (Nachabé et al. 2011; Cao et al. 2013; Sabino et al. 2016). The light scattering coefficient in tissues is dependent on several physical parameters, including cell shape, organelle size, and their refractive index, and is gradually decreased toward the infrared region (Beauvoit and Chance

1998; Fujii et al. 2004; Jacques 2013; Golovynskyi et al. 2018). Thus, NIR light penetrates deep tissue and is suitable for *in vivo* fluorescence imaging (Fig. 12 and Table 2).

Further, two-photon microscopy is a fluorescence imaging method that utilizes NIR light to allow visualizing fluorescent-labeled cells several hundred microns deep in living animals. While conventional microscopy, which includes confocal microscopy, excites fluorophores using one photon of visible light, two-photon microscopy excites fluorophores using two photons of NIR light and then detects photons of visible light from deep tissues (Denk et al. 1990) (Fig. 12). The efficiency of two-photon absorption of the fluorophores depends on the following two factors: quality of the excitation source and the two-photon absorption cross-section of fluorophores. The femtosecond-pulsed laser with a high peak intensity has been adapted as an excitation source in two-photon microscopy, as two-photon absorption only occurs when two photons arrive at the fluorophore in approximately 0.5 femtoseconds (Helmchen and Denk 2005). The two-photon absorption cross-section

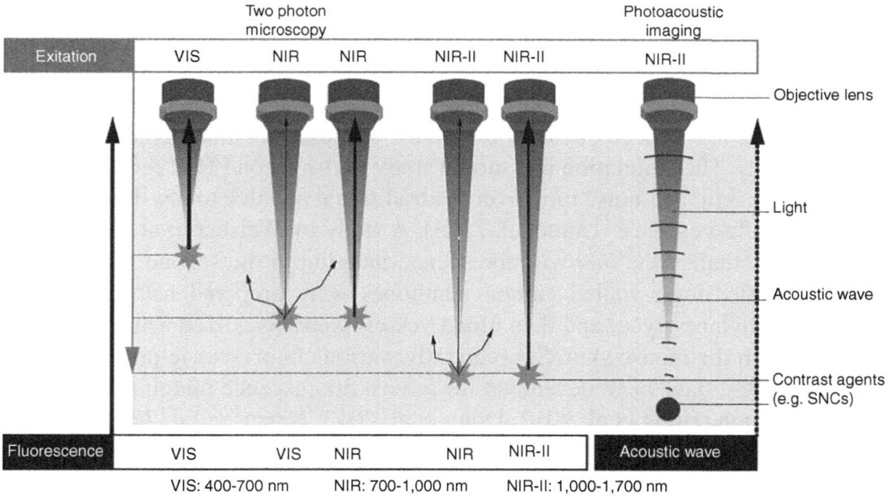

Fig. 12 Fluorescence and photoacoustic imaging using from visible to NIR-II light

Table 2 Optical properties from visible to NIR-II region

Region	Autofluorescence (a.u.)	Absorption coefficient (cm^{-1})	Scattering coefficient (cm^{-1})
Visible (600 nm)	>1000	1–2	ca. 17
NIR(900 nm)	ca. 300	0.7–1.2	ca. 10
NIR-II (1300 nm)	ca. 17	1–1.7	ca. 7
Reference (Specimen)	Diao et al. (2015) (Mice liver, ex.808 nm)	Sabino et al. (2016) (Mice skin)	Golovynskyi et al. (2018) (Rat brain)

of SNCs (CdSe) was orders of magnitude larger than those of organic dyes and fluorescent proteins (Larson et al. 2003). Thus, SNCs that emit in visible region is useful for both conventional fluorescence microscopy and two-photon microscopy.

In addition to the visible-emitting SNCs, NIR-emitting SNCs have been developed for *in vivo* fluorescence imaging (Fig. 12). Although indocyanine green, which is a NIR organic dye that is approved by the Food and Drug Administration, is utilized in clinical studies, it is not sufficient to visualize target molecules and cells in deep tissues due to its low quantum yield in aqueous solutions (< 1%) (Benson and Kues 1978). Moreover, it is difficult to precisely control the excitation and emission wavelengths of the organic dyes, as the wavelengths are dependent on their chemical structures. In contrast, NIR SNCs feature a relatively high quantum yield (10–80%), compared to organic dyes (<40%) and fluorescent proteins (5–20%) in the NIR region (Martinic et al. 2017). Further, their emission wavelengths can be shifted to the NIR region by selecting the appropriate compositions (Zhao et al. 2018), and their absorption wavelengths are broadband and reached the NIR region. These properties allow SNCs to be an ideal fluorescence probe for *in vivo* fluorescence imaging. Studies by Kim et al. (2004) and Gao et al. (2004) demonstrate that NIR SNCs may be utilized as a contrast agent for cancer mapping and surgery. Since then, various NIR SNCs that emit light at 700–1000 nm have been applied for *in vivo* fluorescence imaging (Jin et al. 2008a; Jin et al. 2008b).

In addition to the traditional NIR region (700–1000 nm), a second NIR region (1000–1700 nm) has been explored for *in vivo* fluorescence imaging in deeper tissues (Fig. 12). The simulation and model study in the second NIR region suggest an improved signal-to-noise ratio over the traditional one due to low-light scattering and autofluorescence (Lim et al. 2003). A study by Welsher et al. (2009) was the first to actualize the *in vivo* fluorescence imaging in the second NIR region. Biocompatible single-walled carbon nanotubes were prepared using phospholipid–polyethylene glycol and then blood vessels were visualized using the nanotubes through the mouse skin. Subsequently, various fluorescence probes, one of which is SNCs, have been developed for *in vivo* fluorescence imaging in the second NIR region (Hong et al. 2012; Dong et al. 2013; Nakane et al. 2013; Li et al. 2014). Recently, bright InAs-based SNCs were synthesized and their quantum yield (30% in best sample) was found to be significantly higher than that of commonly utilized carbon nanotubes (Bruns et al. 2017). In addition to the fluorescence probes, optical imaging systems have been developed for *in vivo* fluorescence imaging in the second NIR region. Light sheet fluorescence microscopy is based on orthogonally arranged planar illumination and wide-field detection, and enables high-speed 3D sectioning. A study by Wang et al. (2019) extended the light sheet fluorescence microscopy to include the second NIR region, and excitation and emission wavelength were approximately 1320 and 1700 nm, respectively. Using both sophisticated microscopy and SNCs (PbS), the localization of programmed cell death protein 1 in tumors was visualized in living animals at a cellular resolution level.

SNCs are also considered a candidate of contrast agents for photoacoustic imaging. In fact, photoacoustic imaging is based on the combination of optical excitation and ultrasonic detection (Bell 1880). Since the optical imaging systems, such as confocal and two-photon microscopy, depend on ballistic photons for imaging, their imaging depths are limited to the below-millimeter range due to strong light scattering in biological tissues. On the other hand, photoacoustic imaging enables deep tissue imaging, as ultrasonic waves feature a weaker scattering of 2 to 3 orders of magnitude compared to the optical waves of tissues (Wang and Yao 2016; Yao and Wang 2018) (Fig. 12). Although SNCs are an ideal fluorescent probe, a portion of the light energy absorbed by SNCs is converted to heat through nonradiative relaxation pathways. Thus, the nanomaterials that include SNCs have been developed as exogenous contrast agents for photoacoustic imaging in parallel with exploring endogenous probes (Shashkov et al. 2008; Guo et al. 2017; Xie et al. 2020).

Summary

The colloidal SNCs have been developed as an ideal fluorescent probe. The procedure for high-quality SNCs is based on heating of specific organic solvents and rapidly adding semiconductor precursors. Both the selection of appropriate precursors and the controlling of the SNC size enable the synthesis of the desirable SNCs that emit wavelengths from the visible to NIR regions. Although typical SNCs consist of core and shell structures, the gradient-alloyed SNCs were also developed to overcome the lattice mismatch problem. The heterostructured SNCs, such as rod- and tetrapod-shaped SNCs, can be obtained by both manipulating the growth kinetics and changing the crystal structure type of the core. The SNCs that are capped with hydrophobic ligands can become hydrophilic via surface modifications, and are then conjugated with biomolecules for biological application.

Due to the quantum confinement effect, the SNCs demonstrate the following unique optical properties: extremely broad absorption and relatively narrow emission spectra, strong brightness and high stability to photobleaching, and polarized emission of the rod-shaped SNCs. The extremely broad absorption enables single-laser FCCS, which overcomes the difficulties that are associated with precise and stable overlap of the two laser lines in conventional FCCS. The strong brightness and high stability to photobleaching enable SPT with both high-sensitive and long-term imaging. The polarized emission in the rod-shaped SNCs allows the orientation measurement of biomolecules, which thus more precisely estimates the motion of the biomolecules. In addition, the emission wavelengths of the SNCs can be shifted to the NIR region, in which the quantum yield of NIR SNCs is relatively higher compared to organic dyes and proteins in the conventional NIR region. Further, the novel SNCs that emit wavelengths at the second NIR region have also been developed for deeper *in vivo* fluorescence imaging. The unique optical properties that are addressed in this review enable colloidal SNCs to be applied toward more sophisticated optical methods in the future.

References

Abbandonato G, Hoffmann K, Resch-Genger U (2018) Determination of quantum yields of semiconductor nanocrystals at the single emitter level via fluorescence correlation spectroscopy. Nanoscale 10(15):7147–7154

Adachi K, Yasuda R, Noji H, Itoh H, Harada Y, Yoshida M, Kinosita K Jr (2000) Stepping rotation of F1-ATPase visualized through angle-resolved single-fluorophore imaging. Proc Natl Acad Sci U S A 97(13):7243–7247

Alivisatos AP (1996) Semiconductor clusters, nanocrystals, and quantum dots. Science 271(5251):933–937

Bale G, Elwell CE, Tachtsidis I (2016) From Jöbsis to the present day: a review of clinical near-infrared spectroscopy measurements of cerebral cytochrome-c-oxidase. J Biomed Opt 21(9):091307

Baranov AV, Rakovich YP, Donegan JF, Perova TS, Moore RA, Talapin DV, Rogach AL, Masumoto Y, Nabiev I (2003) Effect of ZnS shell thickness on the phonon spectra in CdSe quantum dots. Phys Rev B 68(16):165306–165306-7

Bayer M (2019) Bridging two worlds: colloidal versus epitaxial quantum dots. Ann Phys (Berlin) 531:1900039

Beauvoit B, Chance B (1998) Time-resolved spectroscopy of mitochondria, cells and tissues under normal and pathological conditions. Mol Cell Biochem 184(1-2):445–455

Bell AG (1880) Upon the production and reproduction of sound by light. Am J Sci 20:305–324

Benson RC, Kues HA (1978) Fluorescence properties of indocyanine green as related to angiography. Phys Med Biol 23(1):159–163

Bera D, Qian L, Tseng TK, Holloway PH (2010) Quantum dots and their multimodal applications: a review. Materials 3(4):2260–2345

Brogan RJ, Kontojannis V, Garara B, Marcus HJ, Wilson MH (2017) Near-infrared spectroscopy (NIRS) to detect traumatic intracranial haematoma: a systematic review and meta-analysis. Brain Inj 31(5):581–588

Bruchez MJR, Moronne M, Gin P, Weiss S, Alivisatos AP (1998) Semiconductor nanocrystals as fluorescent biological labels. Science 281(5385):2013–2016

Brunnbauer M, Dombi R, Ho TH, Schliwa M, Rief M, Ökten Z (2012) Torque generation of kinesin motors is governed by the stability of the neck domain. Mol Cell 46(2):147–158

Bruns OT, Bischof TS, Harris DK, Franke D, Shi Y, Riedemann L, Bartelt A, Jaworski FB, Carr JA, Rowlands CJ, Wilson MWB, Chen O, Wei H, Hwang GW, Montana DM, Coropceanu I, Achorn OB, Kloepper J, Heeren J, So PTC, Fukumura D, Jensen KF, Jain RK, Bawendi MG (2017) Next-generation in vivo optical imaging with short-wave infrared quantum dots. Nat Biomed Eng 1:0056

Can S, Dewitt MA, Yildiz A (2014) Bidirectional helical motility of cytoplasmic dynein around microtubules. elife 3:e03205

Cao Q, Zhegalova NG, Wang ST, Akers WJ, Berezin MY (2013) Multispectral imaging in the extended near-infrared window based on endogenous chromophores. J Biomed Opt 18(10):101318

Chan WC, Nie S (1998) Quantum dot bioconjugates for ultrasensitive nonisotopic detection. Science 281(5385):2016–2018

Cho J, Jung YK, Lee JK, Jung HS (2017) Highly efficient blue-emitting CdSe-derived core/shell gradient alloy quantum dots with improved photoluminescent quantum yield and enhanced photostability. Langmuir 33(15):3711–3719

Dabbousi BO, Rodriguezviejo J, Mikulec FV, Heine JR, Mattoussi H, Ober R, Jensen KF, Bawendi MG (1997) (CdSe)ZnS core-shell quantum dots: synthesis and characterization of a size series of highly luminescent nanocrystallites. J Phys Chem B 101:9463–9475

de Thomaz AA, Almeida DB, Pelegati VB, Carvalho HF, Cesar CL (2015) Measurement of the hydrodynamic radius of quantum dots by fluorescence correlationspectroscopy excluding blinking. J Phys Chem B 119(11):4294–4299

Deka S, Quarta A, Lupo MG, Falqui A, Boninelli S, Giannini C, Morello G, De Giorgi M, Lanzani G, Spinella C, Cingolani R, Pellegrino T, Manna L (2009) CdSe/CdS/ZnS double shell nanorods with high photoluminescence efficiency and their exploitation as biolabeling probes. J Am Chem Soc 131(8):2948–2958

Denk W, Strickler JH, Webb WW (1990) Two-photon laser scanning fluorescence microscopy. Science 248:73–76

Diao S, Hong G, Antaris AL, Blackburn JL, Cheng K, Cheng Z, Dai H (2015) Biological imaging without autofluorescence in the second near-infrared region. Nano Res 8:3027–3034

Dong B, Li C, Chen G, Zhang Y, Zhang Y, Deng M, Wang Q (2013) Facile synthesis of highly Photoluminescent Ag2Se quantum dots as a new fluorescent probe in the second near-infrared window for in vivo imaging. Chem Mater 25:2503–2509

Dubertret B, Skourides P, Norris DJ, Noireaux V, Brivanlou AH, Libchaber A (2002) In vivo imaging of quantum dots encapsulated in phospholipid micelles. Science 298(5599):1759–1762

Eigen M, Rigler R (1994) Sorting single molecules: application to diagnostics and evolutionary biotechnology. Proc Natl Acad Sci U S A 91:5740–5747

Elson EL, Magde D (1974) Fluorescence correlation spectroscopy: I. conceptual basis and theory. Biopolymers 13:1–27

Frangioni JV (2003) In vivo near-infrared fluorescence imaging. Curr Opin Chem Biol 7(5):626–634

Fujii F, Nodasaka Y, Nishimura G, Tamura M (2004) Anoxia induces matrix shrinkage accompanied by an increase in light scattering in isolated brain mitochondria. Brain Res 999(1):29–39

Fujii F, Horiuchi M, Ueno M, Sakata H, Nagao I, Tamura M, Kinjo M (2007) Detection of prion protein immune complex for bovine spongiform encephalopathy diagnosis using fluorescence correlation spectroscopy and fluorescence cross-correlation spectroscopy. Anal Biochem 370(2):131–141

Fujii F, Kinjo M (2007) Detection of antigen protein by using fluorescence cross-correlation spectroscopy and quantum-dot-labeled antibodies. Chembiochem 8(18):2199–2203

Gao X, Cui Y, Levenson RM, Chung LW, Nie S (2004) In vivo cancer targeting and imaging with semiconductor quantum dots. Nat Biotechnol 22(8):969–976

Giepmans BN, Adams SR, Ellisman MH, Tsien RY (2006) The fluorescent toolbox for assessing protein location and function. Science 312(5771):217–224

Goldman ER, Anderson GP, Tran PT, Mattoussi H, Charles PT, Mauro JM (2002) Conjugation of luminescent quantum dots with antibodies using an engineered adaptor protein to provide new reagents for fluoroimmunoassays. Anal Chem 74(4):841–847

Goldstein L, Glas F, Marzin JY, Charasse MN, Le Roux G (1985) Growth by molecular beam epitaxy and characterization of InAs/GaAs strained-layer superlattices. Appl Phys Lett 47(10):1099–1101

Golovynskyi S, Golovynska I, Stepanova LI, Datsenko OI, Liu L, Qu J, Ohulchanskyy TY (2018) Optical windows for head tissues in near-infrared and short-wave infrared regions: approaching transcranial light applications. J Biophotonics 11(12):e201800141

Groeneveld E, Witteman L, Lefferts M, Ke X, Bals S, Van Tendeloo G, Donega Cde M (2013) Tailoring ZnSe-CdSe colloidal quantum dots via cation exchange: from core/shell to alloy nanocrystals. ACS Nano 7(9):7913–7930

Guan LY, Li YQ, Lin S, Zhang MZ, Chen J, Ma ZY, Zhao YD (2012) Characterization of CdTe/CdSe quantum dots-transferrin fluorescent probes for cellular labeling. Anal Chim Acta 741:86–92

Guo W, Qiu Z, Guo C, Ding D, Li T, Wang F, Sun J, Zheng N, Liu S (2017) Multifunctional theranostic agent of $Cu_2(OH)PO_4$ quantum dots for photoacoustic image-guided photothermal/photodynamic combination cancer therapy. ACS Appl Mater Interfaces 9(11):9348–9358

Hamaoka T, McCully KK (2019) Review of early development of near-infrared spectroscopy and recent advancement of studies on muscle oxygenation and oxidative metabolism. J Physiol Sci 69(6):799–811

Helmchen F, Denk W (2005) Deep tissue two-photon microscopy. Nat Methods 2(12):932–940

Hong G, Robinson JT, Zhang Y, Diao S, Antaris AL, Wang Q, Dai H (2012) In vivo fluorescence imaging with Ag2S quantum dots in the second near-infrared region. Angew Chem Int Ed Eng 51(39):9818–9821

Hoshi Y, Yamada Y (2016) Overview of diffuse optical tomography and its clinical applications. J Biomed Opt 21(9):091312

Hu J, Li LS, Yang W, Manna L, Wang L, Alivisatos AP (2001) Linearly polarized emission from colloidal semiconductor quantum rods. Science 292(5524):2060–2063

Ichimura T, Jin T, Fujita H, Higuchi H, Watanabe TM (2014) Nano-scale measurement of biomolecules by optical microscopy and semiconductor nanoparticles. Front Physiol 5:273

Jacques SL (2013) Optical properties of biological tissues: a review. Phys Med Biol 58(14):5007–5008

Jiang W, Mardyani S, Fischer H, Chan WCW (2006) Design and characterization of lysine cross-linked mercapto-acid biocompatible quantum dots. Chem Mater 18:872–878

Jin T, Fujii F, Komai Y, Seki J, Seiyama A, Yoshioka Y (2008a) Preparation and characterization of highly fluorescent, glutathione-coated near infrared quantum dots for in vivo fluorescence imaging. Int J Mol Sci 9(10):2044–2061

Jin T, Yoshioka Y, Fujii F, Komai Y, Seki J, Seiyama A (2008b) Gd3+−functionalized near-infrared quantum dots for in vivo dual modal (fluorescence/magnetic resonance) imaging. Chem Commun (Camb) 44:5764–5766

Jöbsis F (1977) Noninvasive, infrared monitoring of cerebral and myocardial oxygen sufficiency and circulatory parameters. Science 198(4323):1264–1267

Jun YW, Lee SM, Kang NJ, Cheon J (2001) Controlled synthesis of multi-armed CdS nanorod architectures using monosurfactant system. J Am Chem Soc 123(21):5150–5151

Kask P, Piksarv P, Pooga M, Mets U, Lippmaa E (1989) Separation of the rotational contribution in fluorescence correlation experiments. Biophys J 55(2):213–220

Kettling U, Koltermann A, Schwille P, Eigen M (1998) Real-time enzyme kinetics monitored by dual-color fluorescence cross-correlation spectroscopy. Proc Natl Acad Sci U S A 95(4):1416–1420

Kim S, Lim YT, Soltesz EG, De Grand AM, Lee J, Nakayama A, Parker JA, Mihaljevic T, Laurence RG, Dor DM, Cohn LH, Bawendi MG, Frangioni JV (2004) Near-infrared fluorescent type II quantum dots for sentinel lymph node mapping. Nat Biotechnol 22(1):93–97

Kinjo M, Rigler R (1995) Ultrasensitive hybridization analysis using fluorescence correlation spectroscopy. Nucleic Acids Res 23:1795–1799

Kirchner C, Liedl T, Kudera S, Pellegrino T, Muñoz Javier A, Gaub HE, Stölzle S, Fertig N, Parak WJ (2005) Cytotoxicity of colloidal CdSe and CdSe/ZnS nanoparticles. Nano Lett 5(2):331–338

Kobayashi H, Hama Y, Koyama Y, Barrett T, Regino CAS, Urano Y, Choyke PL (2007) Simultaneous multicolor imaging of five different lymphatic basins using quantum dots. Nano Lett 7(6):1711–1716

Kogure T, Karasawa S, Araki T, Saito K, Kinjo M, Miyawaki A (2006) A fluorescent variant of a protein from the stony coral Montipora facilitates dual-color single-laser fluorescence cross-correlation spectroscopy. Nat Biotechnol 24(5):577–581

Koneswaran M, Narayanaswamy R (2015) Ultrasensitive detection of vitamin B6 using functionalised CdS/ZnS Core−shell quantum dots. Sensors Actuators B Chem 210:811–816

Kubitscheck U, Kückmann O, Kues T, Peters R (2000) Imaging and tracking of single GFP molecules in solution. Biophys J 78(4):2170–2179

Kural C, Kim H, Syed S, Goshima G, Gelfand VI, Selvin PR (2005) Kinesin and dynein move a peroxisome in vivo: a tug-of-war or coordinated movement? Science 308(5727):1469–1472

Larson DR, Zipfel WR, Williams RM, Clark SW, Bruchez MP, Wise FW, Webb WW (2003) Water-soluble quantum dots for multiphoton fluorescence imaging in vivo. Science 300(5624):1434–1436

Leatherdale CA, Woo WK, Mikulec FV, Bawendi MG (2002) On the absorption cross section of CdSe nanocrystal quantum dots. J Phys Chem B 106:7619–7622

Lee J, Fujii F, Kim SY, Pack CG, Kim SW (2014) Analysis of quantum rod diffusion by polarized fluorescence correlation spectroscopy. J Fluoresc 24(5):1371–1378

Li C, Zhang Y, Wang M, Zhang Y, Chen G, Li L, Wu D, Wang Q (2014) In vivo real-time visualization of tissue blood flow and angiogenesis using Ag2S quantum dots in the NIR-II window. Biomaterials 35(1):393–400

Lim YT, Kim S, Nakayama A, Stott NE, Bawendi MG, Frangioni JV (2003) Selection of quantum dot wavelengths for biomedical assays and imaging. Mol Imaging 2(1):50–64

Liu W, Choi HS, Zimmer JP, Tanaka E, Frangioni JV, Bawendi M (2007) Compact cysteine-coated CdSe(ZnCdS) quantum dots for in vivo applications. J Am Chem Soc 129(47):14530–14531

Lundqvist M, Stigler J, Elia G, Lynch I, Cedervall T, Dawson KA (2008) Nanoparticle size and surface properties determine the protein corona with possible implications for biological impacts. Proc Natl Acad Sci U S A 105(38):14265–14270

Lutich AA, Mauser C, Da Como E, Huang J, Vaneski A, Talapin DV, Rogach AL, Feldmann J (2010) Multiexcitonic dual emission in CdSe/CdS tetrapods and nanorods. Nano Lett 10(11):4646–4650

Magde D, Elson EL, Webb WW (1972) Thermodynamic fluctuations in a reacting system—measurement by fluorescence correlation spectroscopy. Phys Rev Lett 29:705–708

Magde D, Elson E, Webb WW (1974) Fluorescence correlation spectroscopy. II. An experimental realization. Biopolymers 4:29–61

Mahler B, Spinicelli P, Buil S, Quelin X, Hermier JP, Dubertret B (2008) Towards non-blinking colloidal quantum dots. Nat Mater 7:659–664

Martinic I, Eliseeva SV, Petoud S (2017) Near-infrared emitting probes for biological imaging: organic fluorophores, quantum dots, fluorescent proteins, lanthanide(III) complexes and nanomaterials. J Lumin 189:19–43

Michalet X, Pinaud FF, Bentolila LA, Tsay JM, Doose S, Li JJ, Sundaresan G, Wu AM, Gambhir SS, Weiss S (2005) Quantum dots for live cells, in vivo imaging and diagnostics. Science 307(5709):538–544

Mishra N, Wu WY, Srinivasan BM, Hariharaputran R, Zhang YW, Chan Y (2016) Continuous shape tuning of nanotetrapods: toward shape-mediated self-assembly. Chem Mater 28:1187–1195

Mishra N, Orfield NJ, Wang F, Hu Z, Krishnamurthy S, Malko AV, Casson JL, Htoon H, Sykora M, Hollingsworth JA (2017) Using shape to turn off blinking for two-colour multiexciton emission in CdSe/CdS tetrapods. Nat Commun 8:15083

Monopoli MP, Aberg C, Salvati A, Dawson KA (2012) Biomolecular coronas provide the biological identity of nanosized materials. Nat Nanotechnol 7(12):779–786

Murray CB, Norris DJ, Bawendi MG (1993) Synthesis and characterization of nearly monodisperse CdE (E = sulfur, selenium, tellurium) semiconductor nanocrystallites. J Am Chem Soc 115:8706–8715

Nachabé R, Evers DJ, Hendriks BH, Lucassen GW, van der Voort M, Rutgers EJ, Peeters MJ, Van der Hage JA, Oldenburg HS, Wesseling J, Ruers TJ (2011) Diagnosis of breast cancer using diffuse optical spectroscopy from 500 to 1600 nm: comparison of classification methods. J Biomed Opt 16(8):087010

Nakane Y, Tsukasaki Y, Sakata T, Yasuda H, Jin T (2013) Aqueous synthesis of glutathione-coated PbS quantum dots with tunable emission for non-invasive fluorescence imaging in the second near-infrared biological window (1000-1400 nm). Chem Commun (Camb) 49(69):7584–7586

Nan X, Sims PA, Chen P, Xie XS (2005) Observation of individual microtubule motor steps in living cells with endocytosed quantum dots. J Phys Chem B 109(51):24220–24224

Norris DJ, Bawendi MG (1996) Measurement and assignment of the size-dependent optical spectrum in CdSe quantum dots. Phys Rev B 53(24):16338–16346

Ohmachi M, Komori Y, Iwane AH, Fujii F, Jin T, Yanagida T (2012) Fluorescence microscopy for simultaneous observation of 3D orientation and movement and its application to quantum rod-tagged myosin V. Proc Natl Acad Sci U S A 109(14):5294–5298

Oura M, Yamamoto J, Ishikawa H, Mikuni S, Fukushima R, Kinjo M (2016) Polarization-dependent fluorescence correlation spectroscopy for studying structural properties of proteins in living cell. Sci Rep 6:31091

Peng X, Manna L, Yang W, Wickham J, Scher E, Kadavanich A, Alivisatos AP (2000) Shape control of CdSe nanocrystals. Nature 404(6773):59–61
Resch-Genger U, Grabolle M, Cavaliere-Jaricot S, Nitschke R, Nann T (2008) Quantum dots versus organic dyes as fluorescent labels. Nat Methods 5(9):763–775
Rigler R, Mets U, Widengren J, Kask P (1993) Fluorescence correlation spectroscopy with high count rates and low background analysis of translational diffusion. Eur Biophys J 22:169–175
Sabino CP, Deana AM, Yoshimura TM, da Silva DF, França CM, Hamblin MR, Ribeiro MS (2016) The optical properties of mouse skin in the visible and near infrared spectral regions. J Photochem Photobiol B 160:72–78
Saito K, Wada I, Tamura M, Kinjo M (2004) Direct detection of caspase-3 activation in single live cells by cross-correlation analysis. Biochem Biophys Res Commun 324:849–854
Schwille P, Meyer-Almes FJ, Rigler R (1997) Dual-color fluorescence cross-correlation spectroscopy for multicomponent diffusional analysis in solution. Biophys J 72(4):1878–1886
Sen AN, Gopinath SP, Robertson CS (2016) Clinical application of near-infrared spectroscopy in patients with traumatic brain injury: a review of the progress of the field. Neurophotonics 3(3):031409
Shashkov EV, Everts M, Galanzha EI, Zharov VP (2008) Quantum dots as multimodal photoacoustic and photothermal contrast agents. Nano Lett 8(11):3953–3958
Smith AM, Mohs AM, Nie S (2009) Tuning the optical and electronic properties of colloidal nanocrystals by lattice strain. Nat Nanotechnol 4(1):56–63
Sugawa M, Masaike T, Mikami N, Yamaguchi S, Shibata K, Saito K, Fujii F, Toyoshima YY, Nishizaka T, Yajima J (2018) Circular orientation fluorescence emitter imaging (COFEI) of rotational motion of motor proteins. Biochem Biophys Res Commun 504(4):709–714
Susumu K, Mei BC, Mattoussi H (2009) Multifunctional ligands based on dihydrolipoic acid and polyethylene glycol to promote biocompatibility of quantum dots. Nat Protoc 4(3):424–436
Talapin DV, Nelson JH, Shevchenko EV, Aloni S, Sadtler B, Alivisatos AP (2007) Seeded growth of highly luminescent CdSe/CdS nanoheterostructures with rod and tetrapod morphologies. Nano Lett 7(10):2951–2959
Tsay JM, Doose S, Weiss S (2006) Rotational and translational diffusion of peptide-coated CdSe/CdS/ZnS nanorods studied by fluorescence correlation spectroscopy. J Am Chem Soc 128(5):1639–1647
Wang F, Wan H, Ma Z, Zhong Y, Sun Q, Tian Y, Qu L, Du H, Zhang M, Li L, Ma H, Luo J, Liang Y, Li WJ, Hong G, Liu L, Dai H (2019) Light-sheet microscopy in the near-infrared II window. Nat Methods 16(6):545–552
Wang H, Shang L, Maffre P, Hohmann S, Kirschhöfer F, Brenner-Weiß G, Nienhaus GU (2016) The nature of a hard protein Corona forming on quantum dots exposed to human blood serum. Small 12(42):5836–5844
Wang LV, Yao J (2016) A practical guide to photoacoustic tomography in the life sciences. Nat Methods 13(8):627–638
Watanabe TM, Fujii F, Jin T, Umemoto E, Miyasaka M, Fujita H, Yanagida T (2013) Four-dimensional spatial nanometry of single particles in living cells using polarized quantum rods. Biophys J 105(3):555–564
Welsher K, Liu Z, Sherlock SP, Robinson JT, Chen Z, Daranciang D, Dai H (2009) A route to brightly fluorescent carbon nanotubes for near-infrared imaging in mice. Nat Nanotechnol 4(11):773–780
Widengren J, Mets U, Rigler R (1999) Photodynamic properties of green fluorescent proteins investigated by fluorescence correlation spectroscopy. Chem Phys 250:171–186
Winkler T, Kettling U, Koltermann A, Eigen M (1999) Confocal fluorescence coincidence analysis: an approach to ultra high-throughput screening. Proc Natl Acad Sci U S A 96(4):1375–1378
Wong JI, Mishra N, Xing G, Li M, Chakrabortty S, Sum TC, Shi Y, Chan Y, Yang HY (2014) Dual wavelength electroluminescence from CdSe/CdS tetrapods. ACS Nano 8(3):2873–2879
Wu X, Liu H, Liu J, Haley KN, Treadway JA, Larson JP, Ge N, Peale F, Bruchez MP (2003) Immunofluorescent labeling of cancer marker Her2 and other cellular targets with semiconductor quantum dots. Nat Biotechnol 21(1):41–46

Xiao Y, Barker PE (2004) Semiconductor nanocrystal probes for human metaphase chromosomes. Nucleic Acids Res 32(3):e28

Xie H, Liu M, You B, Luo G, Chen Y, Liu B, Jiang Z, Chu PK, Shao J, Yu XF (2020) Biodegradable Bi_2O_2Se quantum dots for photoacoustic imaging-guided cancer photothermal therapy. Small 16(1):e1905208

Yamamoto J, Oura M, Yamashita T, Miki S, Jin T, Haraguchi T, Hiraoka Y, Terai H, Kinjo M (2015) Rotational diffusion measurements using polarization-dependent fluorescence correlation spectroscopy based on superconducting nanowire single-photon detector. Opt Express 23(25):32633–32642

Yao J, Larson DR, Vishwasrao HD, Zipfel WR, Webb WW (2005) Blinking and nonradiant dark fraction of water-soluble quantum dots in aqueous solution. Proc Natl Acad Sci U S A 102(40):14284–14289

Yao J, Wang LV (2018) Recent progress in photoacoustic molecular imaging. Curr Opin Chem Biol 45:104–112

Yu WW, Chang E, Falkner JC, Zhang J, Al-Somali AM, Sayes CM, Johns J, Drezek R, Colvin VL (2007) Forming biocompatible and nonaggregated nanocrystals in water using amphiphilic polymers. J Am Chem Soc 129(10):2871–2879

Zhang J, Yang Q, Cao H, Ratcliffe CI, Kingston D, Chen QY, Ouyang JY, Wu XH, Leek DM, Riehle FS, Yu K (2016) Bright gradient-alloyed $CdSe_xS_{1-x}$ quantum dots exhibiting cyan-blue emission. Chem Mater 28:618–625

Zhao P, Xu Q, Tao J, Jin Z, Pan Y, Yu C, Yu Z (2018) Near infrared quantum dots in biomedical applications: current status and future perspective. Wiley Interdiscip Rev Nanomed Nanobiotechnol 10(3):e1483

DNA/RNA Fluorescence Imaging by Synthetic Nucleic Acids

Akimitsu Okamoto

Introduction

The mechanism of gene expression in cell development, differentiation, and evolution is the most important research target in molecular biology and cell biology researches. Imaging of events within individual cells (Dirks and Tanke 2006; Silverman and Kool 2005a), and even within organisms (Weigert et al. 2013), is less invasive and more suitable for observation of events than the traditional biochemical methods, and thus, it is frequently used in molecular biology and cell biology researches. Nucleic acid is a key biomolecular target in cell imaging because all important events involved in gene information control, such as storage, transfer, exchange, and expression, are realized via nucleic acids. Observing the behavior of nucleic acids in cells is the best way to monitor and understand ongoing cellular function expression (Fig. 1). Because hybridization of synthetic nucleic acid fragments complementary to the target nucleic acid is very sequence specific, flexible design of nucleic acid fragments give us a direct method for the analysis and visualization of single-stranded nucleic acids in cells and facilitates various nucleic acid analyses when sequence information is available. For the effective design of fluorescent nucleic acid probes for nucleic acid imaging in living cells, the review written by Venyaminova and colleagues lists the requirements that must be taken into account, such as (Boutorine et al. 2013):

- High affinity and specificity for the target nucleic acid sequence
- Good cellular penetration and ability to reach its nucleic acid target in cells

A. Okamoto (✉)
Research Center for Advanced Science and Technology, The University of Tokyo, Tokyo, Japan
e-mail: okamoto@chembio.t.u-tokyo.ac.jp

J. K. Kim et al. (eds.), *Advanced Imaging and Bio Techniques for Convergence Science*, Advances in Experimental Medicine and Biology,
https://doi.org/10.1007/978-981-33-6064-8_17

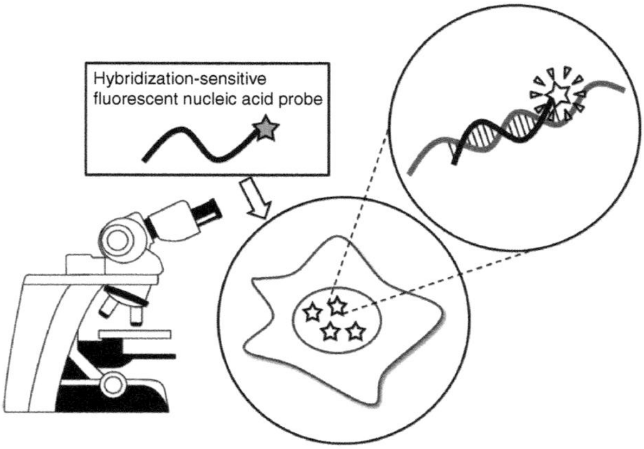

Fig. 1 What are the requirements for effective nucleic acid fluorescence imaging?

- Stability of probes under physiological conditions
- Minimal cytotoxicity
- Minimal distortion of cellular functions
- Simple detection in living cells after excitation with non-harmful visible light
- Modulation of fluorescence spectra upon interaction with a target
- High signal-to-background ratio

This review addresses fluorescent nucleic acid probes for direct observation of nucleic acids in living cells using a fluorescent microscope. It also covers the principles of probe design for this purpose.

Choice of Probe Structures

Synthetic DNA, which can bind very specifically to target nucleic acid sequences, is easily available by the use of a DNA autosynthesizer; thus, a DNA structure is frequently used as the backbone structure of fluorescent probes which sequence-specifically bind to nucleic acids. However, a general disadvantage of synthetic DNA probes is their poor stability in biological media due to containing DNA degrading enzymes. In order to avoid this problem, artificial nucleic acids having modified backbone structures have been developed, such as peptide nucleic acid (PNA) (Molenaar et al. 2003a; Flierl et al. 2003; Röthlingshöfer et al. 2011a, b; Sadhu and Winssinger 2013; Grossmann and Seitz 2006; Grossmann et al. 2008; Chen et al. 2012), locked nucleic acid (LNA) (Silahtaroglu et al. 2004, 2003; Thomsen et al. 2005; Kim et al. 2007; Yang et al. 2007; Wang et al. 2005; Martinez et al. 2009; Astakhova et al. 2008), and 2'-O-methyl RNA (Krasheninina et al. 2011; Novopashina et al. 2008; Kholodar et al. 2009; Dirks et al. 2003; Rhee and Bao 2010; Molenaar et al. 2001; Chen et al. 2009; Okabe et al. 2011;

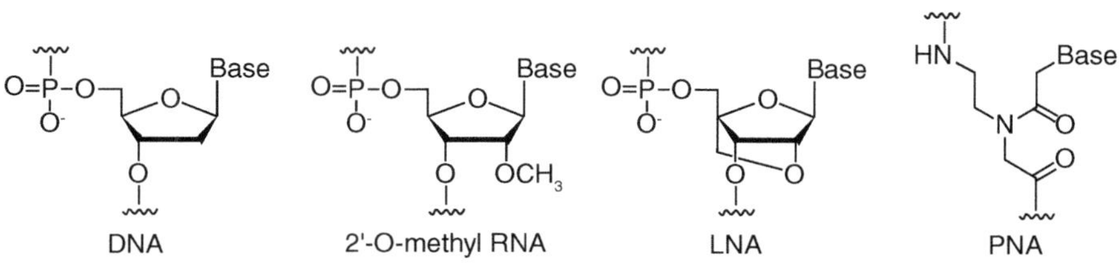

Fig. 2 Probe backbones

Martí et al. 2006; Sando and Kool 2002; Abe and Kool 2006; Silverman et al. 2006; Silverman and Kool 2005b) (Fig. 2).

PNA probes, which have a peptide backbone instead of a sugar phosphate backbone, form very stable hybrids with complementary nucleic acids in vitro and are considered good candidates for alternative nucleic acid probes (Egholm et al. 1993). Fluorescent PNA has been used for detecting and tracking the spatial localization and dynamics of telomere sequences in human osteosarcoma cells (Molenaar et al. 2003b). Telomeres were observed as discrete dots distributed throughout the cell nucleus, and most telomeres moved to very short distances but some were also observed to move over quite long distances.

Similar to PNA, 2'-O-methyl RNA probes have not been degraded by cellular nucleases and are reported to form very stable hybrids with their target sequences (Majlessi et al. 1998). 2'-O-methyl RNA probes were used to study the kinetic properties of poly(A) RNA localization in the cell nucleus in detail (Molenaar et al. 2004).

LNA is also useful for RNA detection. LNA nucleotides are nucleotide analogs that have an ethylene bond between the 2′ oxygen and 4′ carbons of the ribose ring and form a stable hybrid with RNA sequences. The melting temperature of the hybrid is raised to 10 °C per modification (Bondensgaard et al. 2000). Therefore, LNA has the ability to hybridize to highly structured RNA and is resistant to nuclease digestion (Darfeuille et al. 2004).

Simple Fluorescence-Labeled Probes for In Situ Hybridization

The use of fluorescent nucleic acid probes is very effective for sequence-specifically visualizing single-stranded nucleic acids, such as mRNA. Flexible design of fluorescent nucleic acid probes facilitates the detection of RNA in cells as long as the sequence information is available. A number of groups have developed various hybridization probes with unique designs to visualize RNA dynamics in living cells. Sequence-specific DNA imaging has greatly advanced by the development of fluorescence in situ hybridization (FISH) methods (Nath and Johnson 2000; Levsky and Singer 2003; Smolina et al. 2007; Jaco et al. 2008). This fluorescence imaging method is based on chemical fixation of cells and denaturation of DNA, followed by hybridization of the denatured single-stranded DNA in the cells with a fluorescence-labeled nucleic acid probe. The FISH method has greatly contributed to improving our understanding of gene expression in a cell, although it is not compatible with observations in live cells. Labeled PCR fragments and synthetic oligonucleotides, and nucleic acid analogs, such as PNA (Paulasova and Pellestor 2004; Pellestor and Paulasova 2004a, b) or LNA (Silahtaroglu et al. 2004, 2003; Thomsen et al. 2005; Astakhova et al. 2011), have been used as hybridization probes. Although a single hybridization probe with only a single fluorophore is not sufficient to visualize a low copy number of intracellular RNA, it is possible to image abundant RNA sequences, such as polyA RNA, ribosomal RNA, and β-actin mRNA. Tyagi and colleagues have synthesized many different fluorescent nucleic acid probes that fluoresce brightly on target mRNA in fixed cells of several eukaryotic cell lines (Raj et al. 2008). In order to increase the sensitivity of imaging, they inserted the repetitive sequences complementary to the hybridization probe into the 3′-untranslated region of the target mRNA. COMBO-FISH also increased the sensitivity of imaging, using a combinatorial mixture of labeled oligonucleotides targeting one gene (Bridger et al. 2010; Hausmann et al. 2003; Schwarz-Finsterle et al. 2007).

Conventional Methods for Nucleic Acid Detection in Living Cells

Genetic fusion of fluorescent and target proteins facilitates the visualization of target proteins and understanding of the protein's dynamic behavior. However, it is not easy to covalently link a fluorescent protein and the target RNA in living cells. Therefore, if the use of fluorescent proteins is planned, they must be delivered to target RNA using other strategies. Singer and colleagues applied the RNA–protein interaction found in bacteriophage MS2 to fluorescent labeling of RNA. In this tagging strategy, an artificial mRNA containing MS2 RNA motif in the 3′-untranslated region and a fusion protein of MS2 binding protein and green fluorescent protein (GFP) were expressed in cells (Beach et al. 1999; Bertrand et al. 1998; LeCuyer et al. 1996). The interaction between this protein and RNA is bioorthogonal. Multiple copies of the MS2 stem-loop motif were introduced into the gene encoding the target mRNA to enhance imaging sensitivity. Introduction of 24 tandem repeats of the MS2 RNA motif has enabled single molecule detection of the target RNA in living cells (Fusco et al. 2003). Although the unbound MS2-GFP protein maintains fluorescence and produces a bright background, the noise signal in the cytoplasm was reduced by adding a nuclear localization signal to the MS2-GFP protein.

There is another GFP-based live cell RNA imaging method called "split GFP." When a GFP is split into two fragments at appropriate positions, the formation of fluorophores in the GFP is inhibited. However, when a pair of tag molecules that associate with each other are conjugated to GFP fragments, they acquire fluorescence by forming a properly folded GFP structure. This biomolecular fluorescence complementation (BiFC) was applied to RNA imaging (Ozawa et al. 2007; Rackham and Brown 2004; Valencia-Burton et al. 2007). GFP or related proteins are split into two nonfluorescent fragments and genetically fused with RNA-binding proteins.

Placing them close to each other in two different RNA regions causes folding of the fluorescent protein. An advantage of the BiFC approach is that reconstitution of fragmented fluorescent protein does not occur in the absence of target RNA sequences in cells, resulting in a lower background signal than methods using unsplit GFP.

Broude and colleagues developed an RNA imaging method that utilizes aptamer–protein interactions to visualize lacZ mRNA and 5S rRNA in *Escherichia coli*. In this technology, target RNA recognition is achieved using an RNA aptamer for the eukaryotic initiation factor 4A (eIF4A), a dumbbell-shaped protein containing two globular domains. Although each globular domain of eIF4A contributes to high affinity by binding to one side of the aptamer (Oguro et al. 2003), the two split eIF4A fragments do not interact with the aptamer by themselves. Each fragment of eIF4A was fused to either of the two split GFP fragments. When these two protein fragments and aptamer-tagged mRNA were coexpressed in cells, GFP complementation occurred.

Ozawa and colleagues have developed a split GFP method using proprietary RNA-binding proteins to visualize intracellular RNA molecules (Ozawa et al. 2007). The human RNA-binding protein, PUMILIO1, which can alter sequence specificity in a predictable manner, was used to monitor the migration and degradation of mRNA-encoding NADH dehydrogenase subunit 6 in the mitochondria of HeLa cells.

Alternative strategies for RNA imaging that are not dependent on fluorescent proteins have also been developed. In these cases, fluorescent dyes, being nonfluorescent in the free state, fluoresce upon binding. An aptamer that binds to malachite green (MG) increased the fluorescence of MG by more than 2000-fold (Babendure et al. 2003). Jaffrey and colleagues used a unique dye–aptamer pair, which was an RNA mimic of GFP (Paige et al. 2011). RNA aptamers bind to synthetic fluorescent molecules that resemble the chromophores of GFP, thereby providing various dye–aptamer combinations with different emission wavelengths. One of the RNA–fluorophore complexes exhibits higher quantum yield (0.72) than the enhanced GFP. Another complex called “Spinach” is more resistant to photobleaching. Several engineered versions of Spinach, which have additional small molecule-binding domains at a location adjacent to the dye-binding domain, were developed for live-cell imaging and in vitro detection of cellular metabolites (Nakayama et al. 2012; Paige et al. 2012). The advantage of this dye–aptamer method compared to GFP tagging is that no cumbersome GFP expression is required and that fluorescence emission occurs earlier than mature GFP, allowing on-time imaging of RNA dynamics.

FRET-Based Probes

Fluorescent probes that emit fluorescence or change emission wavelength during hybridization to a complementary sequence to reduce the background signal have been developed with the support of quenchers or Förster resonance energy transfer

(FRET). Fluorescence quenching and FRET are similar in that a donor fluorophore excited by light irradiation transfers the energy via photochemical energy or electron transfer mechanism to a nearby acceptor molecule. In FRET, the acceptor dye fluoresces from the energy it receives, while in fluorescence quenching the acceptor molecule consumes the energy it receives as heat. One of the earliest methods of using a quenching mechanism for nucleic acid detection is called "competitive hybridization"; double-stranded oligonucleotide probes have a fluorophore at the 5′ end of an antisense strand (probe sequence) and a quencher at the 3′ end of a sense strand (Morrison et al. 1989) (Fig. 3). In the duplex state, the fluorophore attached to the antisense strand remains quenched due to the proximity between the fluorophore and the quencher. In the presence of the RNA sequence complementary to the antisense strand, strand substitution occurs slowly after the short double-stranded oligonucleotide equilibrates between the associated and dissociated states, which in turn produce a labeled complex of the target RNA. Intracellular detection of synthetic oligonucleotides was first achieved by Chin and colleagues in 1994, using a competitive hybridization method (Sixou et al. 1994).

Asanuma and colleagues introduced one or more perylene nucleobase analogues as acceptors and one or more pyrene analogues as donors into D-threoninol as a backbone. All artificial bases were separated with one or more natural bases to avoid excimer fluorescence and to observe only FRET (Kashida et al. 2010). These probes show very high Stokes shifts, and their fluorescence spectra depend on the number of donors and acceptors introduced, the distance between them, and the hybridization state of the probes (Kato et al. 2012).

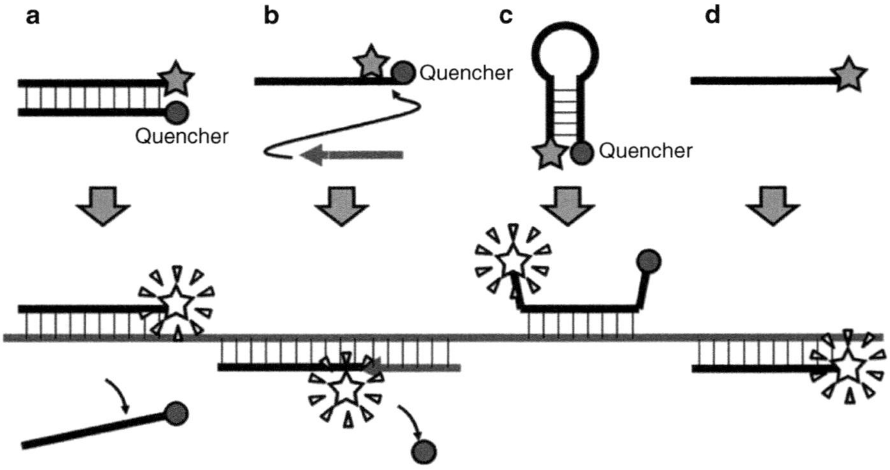

Fig. 3 Molecular designs for hybridization-sensitive fluorescent switching. (**a**) Competitive hybridization; (**b**) Parallel probe (QUAL); (**c**) Conformational change (molecular beacon); (**d**) Emission intensity change depending on the microenvironment

A combination of two probes that bind side by side on the target RNA is also effective for fluorescence imaging. In 1988, Cardullo and colleagues developed a strategy using FRET between two adjacent hybridizing probes (Cardullo et al. 1988). Parallel FRET probes have been applied to live cell imaging for c-fos mRNA in Cos7 cells (Tsuji et al. 2000). Quench autoligation (QUAL) method also use parallel probe system, in which one probe has a FRET donor and quencher coupled to a leaving group and the other probe has a nucleophile thiol group (Fig. 3). When two probes that bind side by side on the target RNA sequence are covalently linked, the quencher molecule is released and the fluorescence is emitted (Sando and Kool 2002). Further examples of nucleic acid templated chemical reactions using parallel probe system have been reported for the fluorescent labels of target RNA (Furukawa et al. 2009a, b; Pianowski et al. 2009).

Molecular Beacons

Molecular beacons are also used for cell imaging as turn-on fluorescent nucleic acid probes (Tyagi and Kramer 1996; Tan et al. 2004; Huang and Martí 2012), which have a stem-loop hairpin structure that contains a fluorophore at one end and a quencher at the other end (Fig. 3). In this conformation, the fluorophore and the quencher are in close proximity to one another and the fluorescence is quenched. The hairpin loop is complementary to the target RNA region (Tsourkas et al. 2002a, b, 2003). The stem-loop hairpin structure opens by hybridization with the target RNA and the distance between the fluorophore and the quencher increases to become too large for energy transfer. Therefore, the fluorescence intensity increases up to 200-fold (Tyagi and Kramer 1996) and the ratio of signal to background is much higher. Several examples of molecular beacon systems in live cell imaging have been provided (Rhee and Bao 2010, 2009; Molenaar et al. 2001; Chen et al. 2010). Molenaar and colleagues applied the molecular beacons with a 2'-O-methyl RNA backbone to the detection of different RNA types such as rRNA, snRNA, and mRNAs (Molenaar et al. 2001).

Molecular beacons labeled with two pyrene moieties at different ends (Fujimoto et al. 2004; Huang et al. 2011), or at the same end (Yamana et al. 2008), are also reported. When the two conjugated pyrenes are close to one another, they emit excimer fluorescence in the form of a hairpin. On the other hand, the fluorescence of the monomer is predominant during the hybridization (Yamana et al. 2008).

In the case of pH-sensitive molecular beacons, pH-sensitive fluorophore 7-hydroxycoumarin was inserted into the stem region as a nucleobase analog (Kashida et al. 2012). The pK_a of 7-hydroxycoumarin depends on the microenvironment: the pK_a is 8.8 in the single-strand status, but upon insertion, it changes its pK_a to 10. Thus, whereas the fluorescence is quenched in the hairpin stem, the 7-hydroxycoumarin deprotonates and emits a fluorescence upon excitation when the stem dissociates in complex with the complementary strand.

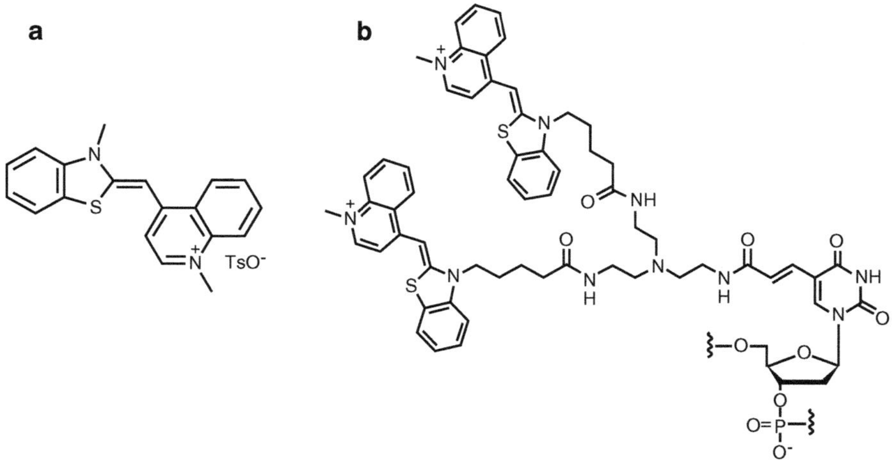

Fig. 4 ECHO probe. (**a**) Thiazol orange (TO); (**b**) Doubly TO-labeled nucleoside in ECHO probe

Thiazole Orange-Labeled Probes

Thiazole orange (TO, 1-methyl-4-[(3-methyl-2(3H)-benzothiazolylidene)methyl]quinolinium p-tosylate), which contains a benzothiazole ring covalently linked to a quinoline ring through a monomethine bridge (Nygren et al. 1998), is used for the labelling of nucleic acid probes (Fig. 4). TO has low fluorescence in aqueous solution because of intramolecular twisting around the methine bond in the excited state (Svanvik et al. 2000; Kummer et al. 2012; Karunakaran et al. 2006; Silva et al. 2007; Carlsson et al. 1994; Ghasemi et al. 2008; Netzel et al. 1995). TO is cationic and has high affinity for DNA through intercalation. Therefore, upon intercalation with a DNA duplex and consequent restriction of rotation around the methine bond, the nonradiative decay channel of TO is closed and the fluorescence intensity significantly increases (Rye et al. 1992; Ikeda and Okamoto 2007). In a significant amount of research, TO dyes were linked to artificial DNA strands or DNA analogues and applied to nucleic acid analysis (Svanvik et al. 2000; Wang and Krull 2005; Privat and Asseline 2001; Köhler et al. 2005; Berndl and Wagenknecht 2009; Hara et al. 2010).

For example, Krull and colleagues linked TO dyes to the 5′-end of DNA strands (Wang and Krull 2005, 2002; Hanafi-Bagby et al. 2000; Algar et al. 2006). The TO-tethered DNA strands exhibited enhancement of fluorescence intensity upon hybridization with the complementary DNA strands at the surfaces of optical fibers. Asseline and colleagues developed αT20–TO conjugates as hybridization-sensitive fluorescent probes for mRNA in living cells (Privat and Asseline 2001; Asseline et al. 2006; Lartia and Asseline 2006; Privat et al. 2001). The TO-labeled α-DNA was applied to the in situ time-resolved detection of mRNA in adherent fibroblasts, resulting in the high intensity of fluorescent signals. Kubista and colleagues prepared TO-labeled PNA strands (Svanvik et al. 2000, 2001). The TO-PNA probe contained both the excellent hybridization properties of PNA and the large fluorescence enhancement of TO upon binding to DNA. Whereas free probes showed low fluorescence, the fluorescence of probes increased almost 50-fold upon hybridization to the complementary DNA. Seitz and colleagues developed forced intercalation (FIT) probes, which have TO as a fluorescent base in a PNA strand (Kummer et al. 2012; Karunakaran et al. 2006; Köhler et al. 2005, 2004; Jarikote et al. 2007, 2005; Socher et al. 2008; Köhler and Seitz 2003). TO in FIT probes has the characteristics of a universal base while maintaining duplex stability. The emission of FIT probes is attenuated when they are forced to intercalate next to a mismatched base pair; thus, FIT probes were used to distinguish the target DNA strand from its single nucleotide polymorphism mutant under nonstringent hybridization conditions. Wagenknecht and colleagues developed an excimer-type TO probe (Berndl and Wagenknecht 2009; Menacher et al. 2008). A single TO dye shows typical green emission when used as an artificial DNA base, whereas the interstrand TO dimer in a DNA duplex exhibits an orange excimer-type emission. Their TO dimer showed a large Stokes shift of nearly 100 nm with a brightness that is comparable to that of a single TO label in DNA. Asanuma and colleagues synthesized a highly sensitive in-stem molecular beacon (ISMB) by combining TO with a second dye (Hara et al. 2010; KamiyaY et al. 2013). For example, TO and Cy3 on D-threoninols are incorporated as a pseudo base pair in ISMB probes. Minimization of the difference between the λ_{max} of the fluorophore and quencher in ISMB probes is critical for maximizing the efficiency of quenching before hybridization with the target nucleic acid.

ECHO Probes

TO-labeled nucleic acids show strong hybridization-dependant fluorescence emission. However, fluorescence suppression is insufficient in the single-stranded state, and a more specific design of TO-labeled DNA derivatives is required for effective nucleic acid sensing with negligible background fluorescence.

TO dyes are well-known for showing little emission by the exciton coupling effect (Khairutdinov and Serpone 1997; Simon et al. 1998; Cosa et al. 2001; Sagawa et al. 2004; Fürstenberg et al. 2006) when they are arranged parallely (H-aggregation) (West and Pearce 1965; Czikkely et al. 1970; Harrison et al. 1996; Zeena and Thomas 2001; Hannah and Armitage 2004; Rösch et al. 2006). Interaction between TO dyes is explained in terms of exciton coupling theory, in which the excited state of the TO dimer splits into two energy levels (Kasha 1963; Kasha et al. 1965; Levinson et al. 1957; McRae and Kasha 1958). The transition to the upper excitonic state is allowed for the H-aggregates, which then rapidly deactivate to a lower state. Emission from the lower state is theoretically forbidden, and thus, the singlet excited state of the aggregate becomes trapped in a nonemissive state. According to the principle described above, we designed a DNA strand containing a nucleotide in which two TO subunits are linked covalently (Fig. 4). The doubly TO-labeled DNA strands showed characteristic absorption, excitation, and emission spectra before and after hybridization with the complementary nucleic acids (Ikeda and Okamoto 2008). The blueshift of the absorption band of TO in the unhybridized state indicates excitation of H-aggregate of TO. When the doubly TO-labeled DNA strand hybridized with the complementary DNA strand, strong emission was observed, whereas the emission intensity was strongly suppressed before hybridization. The doubly TO-labeled DNA strand, in which the fluorescence emission is controlled by such interdye excitonic interaction, was named the ECHO (exciton-controlled hybridization-sensitive fluorescent oligonucleotide) probe (Wang and Okamoto 2012; Okamoto 2011). The duplex stability of ECHO/DNA duplexes increased compared with DNA/DNA duplexes of the same sequence. The thermodynamics of ECHO/DNA duplexes was reported by Hayashizaki and colleagues (Kimura et al. 2012). The thermodynamic parameters were obtained from the melting curves of 64 ECHO/DNA duplexes measured by fluorescence.

FISH is a powerful tool used in karyotyping, cytogenotyping, cancer diagnosis, species specification, and gene expression analysis, but, as described above, conventional FISH is a cumbersome and time-consuming method containing repetitive washing processes. Because of the higher signal-to-noise ratio of ECHO probes compared with conventional fluorescence-labeled nucleic acid probes, a high-resolution FISH has become possible by adapting ECHO probes (ECHO–FISH) (Wang et al. 2012). The ECHO–FISH method includes no stringent washing steps, is accomplished in a 25-min procedure, from fixation to mounting, and can make clear fluorescence detection of the intracellular RNA possible. ECHO–FISH is highly reproducible, stringent, and compatible with other fluorescent cellular labeling techniques. The resolution allows detection of intranuclear speckles of poly(A) RNA in HeLa cells and dissociated hippocampal primary cultures, and mRNAs in the distal dendrites of hippocampal neurons.

To monitor intracellular RNA effectively and to avoid background fluorescence and washing processes, high hybridization sensitivity of the fluorescent emission is essential. The character of fluorescence switching by ECHO probes is suitable for this purpose. In addition, the fluorescence of ECHO probes increases immediately after mixing with the target RNA (Kubota et al. 2009a). ECHO probes were applied to living HeLa cells by using manipulator-assisted microinjection, lipofection, or electropolation to visualize intracellular mRNA localization. For example, immediately after the injection of 5′-$T_6D_{514}T_6$-3′ (D_{514} is a thymine derivative doubly labeled with two TO dyes) into the nucleus of a living HeLa cell, fluorescence was observed from the nucleus.

In efforts to understand spatiotemporal correlations between gene expression and the interaction of nucleic acids in living cells, a series of new fluorescent nucleotides were designed in which the TO moiety was substituted by its derivatives. A series of derivatives were developed with at least 15 emission wavelengths in ECHO probes, from blue to the near-infrared region, which were based on excitonic interaction chemistry and showed on/off performance levels as high as those of TO-labeled ECHO probes (Ikeda et al. 2009, 2011a). In a model experiment with simultaneous live cell miRNA imaging using the mixture of ECHO probes with different emission wavelengths, each ECHO probe recognized the corresponding target miRNA in the cell and emitted different fluorescence color with a miRNA-specific emission wavelength.

Although the use of a diverse range of fluorescent wavelengths has expanded the ECHO probe series, the simultaneous incorporation of suitable dyes into one probe molecule is complex because post-synthetic incorporation of dyes is used for probe synthesis. The key to solving this problem is to design a dye-tethering nucleoside unit that is soluble in several organic solvents used during the synthesis from a nucleoside to a DNA fragment, thus, providing an alternative method of incorporating the dye into the probe. Therefore, desmethyl TO dye unit has been developed (Okamoto et al. 2013). The use of uncharged desmethyl TO made the preparation of the nucleoside unit and the reactions in the cycles of DNA synthesis more efficient because both the dye and the dye-tethering nucleoside are soluble in the organic solvents used in the synthetic processes. The DNA probe doubly modified with desmethyl TO dye units showed effective on/off switching of fluorescence at pH 5–7, which is similar to the behavior of conventional ECHO probes. Because acidic pH is important for the protonation and emission of desmethyl TO dye units, the fluorescence from the probe at higher pH (pH 8–10) was weak even in the presence of the complementary RNA strand. The greatest advantage of the incorporation into the ECHO probes of a "ready-made" fluorescent nucleotide with the function of hybridization-sensitive fluorescence switching is to make it possible to incorporate a further, different fluorescence color in combination with the conventional ECHO probe procedure. For example, a D'_{505}–D600-double-colored ECHO probe exhibited on/off switching of fluorescence at the emission wavelengths of either D'_{505} or D600 depending on the addition of the complementary RNA strand (D'_{505} is a thymine derivative doubly labeled with two desmethyl TO dyes). The partially hybridized D'_{505}–D600 probe emitted fluorescence corresponding to the hybridized region.

ECHO probes with a DNA backbone are digested by intracellular nucleases. In order to add nuclease resistance to ECHO probes for the long-term intracellular RNA observation, ECHO probes replaced with a 2′-O-methyl RNA backbone were synthesized (Kubota et al. 2009b). The high nuclease resistance of 2′-O-methyl RNA probes makes long-term intracellular RNA imaging possible. Intracellular mRNA behavior during cell life events, such as migration and division, was observed through long-term monitoring of the cells into which a 2′-O-methyl RNA probe $5'\text{-}{}^{OMe}U_6D_{514}{}^{OMe}U_6\text{–}3'$ was transfected. The fluorescence of the probe in the mother cells was evenly distributed to their two daughter cells after cell division (Fig. 5).

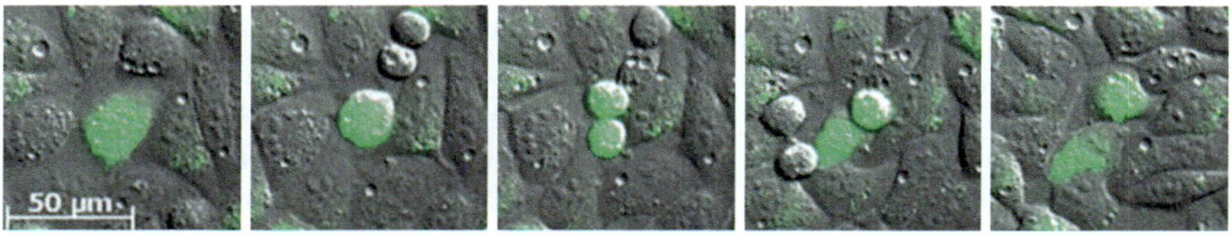

Fig. 5 Monitoring of division of HeLa cells using an ECHO probe (2′-O-methyl RNA backbone). From left: at the initial state, 1, 1.5, 2.5, and 4 h later. Bar: 50 μm (Kubota et al. 2009b)

Incorporation of LNA trinucleotides into ECHO probes enhanced the thermostability of duplexes formed with complementary sequences, whereas the duplex stability with mismatched sequences was much lower (Sugizaki and Okamoto 2010). Another 16-mer ECHO–LNA conjugate designed for the TAR RNA stem-loop region provided high binding affinity despite the higher-order structure of TAR stem-loop RNA. The combination of an ECHO probe and an SNP-recognizing LNA trinucleotide was also effective for mix-and-read fluorescent SNP typing (You et al. 2006). ECHO–LNA conjugate probes enabled the fluorescent SNP detection of placenta-specific 4 (PLAC4) rs130833, which is an SNP located in the transcribed regions of placental-expressed mRNA in maternal plasma (Lo et al. 2007; Go et al. 2007).

Caged ECHO probes were also synthesized (Ikeda et al. 2011b). Caging technology is a useful method for the activation of substances in a spatiotemporal manner. The ECHO probe, protected by a photolabile nitrobenzyl unit, suppressed the function of the probes by decreasing the hybridization ability of probes. After photolysis of the nitrobenzyl unit, the probes recovered the property of hybridization-sensitive fluorescence emission. Area-specific uncaging of the caged ECHO probe in a living cell using a blue laser was applicable to visualization of subnuclear mRNA diffusion (Fig. 6). The caged ECHO probe offers an alternative to the most popular area-specific photocontrolled fluorescence recovery after photobleaching assay or fluorescence loss in photobleaching assay.

Finally, we developed a high-resolution fluorescence RNA imaging method, ECHO-live FISH, to label endogenous nuclear RNA in living mice and chicks (Oomoto et al. 2015). Upon in vivo electroporation, ECHO probes revealed focally concentrated endogenous 28S rRNA and U3 snoRNA at nucleoli and poly(A) RNA at nuclear speckles. Time-lapse imaging reveals steady state stability of these RNA

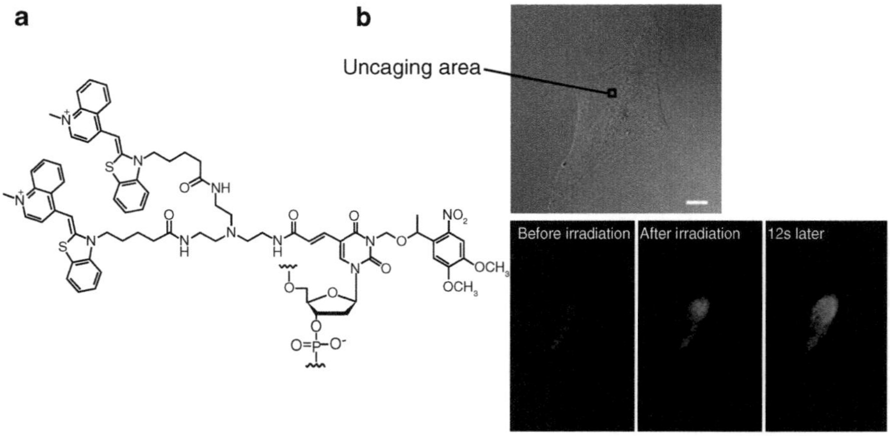

Fig. 6 Point uncaging. (**a**) Caged ECHO probe; (**b**) Diffusion of the fluorescence after uncaging at an irradiation area (square) in the nucleus of the cell into which a caged ECHO probe was injected. Bar: 10 μm (Ikeda et al. 2011b)

foci and dynamic dissipation of 28S rRNA concentrations upon polymerase I inhibition in native brain tissue.

Conclusion

Fluorescent imaging of intracellular nucleic acids made big progress in the last few decades and increasing attention is now being paid to direct dynamic imaging in living cells. Despite the development of various methods with fused fluorescent and RNA-binding proteins, only nucleic acid probes can provide sequence-specific and sensitive RNA detection and visualization. Molecular beacons and different binary probes, such as FRET pairs, template-directed fluorophore activation or synthesis, excimers, and exciplexes, fluorophore-binding aptamers, and other systems have largely contributed toward solutions in the imaging of intracellular nucleic acids. The ECHO probe technology, one of the TO probes, made it possible to detect the target nucleic acid and to analyze both the spatiotemporal dimensions of the diverse RNA dynamics in a living cell. Further expansion of the diversity of TO-tethered nucleic acids should lead to further improvements in the methodology of nucleic acid imaging.

References

Abe H, Kool ET (2006) Flow cytometric detection of specific RNAs in native human cells with quenched autoligating FRET probes. Proc Natl Acad Sci U S A 103:263–268

Algar WR, Massey M, Krull UJ (2006) Fluorescence resonance energy transfer and complex formation between thiazole orange and various dye-DNA conjugates: implications in signaling nucleic acid hybridization. J Fluoresc 16:555–567

Asseline U, Chassignol M, Aubert Y, Roig V (2006) Detection of terminal mismatches on DNA duplexes with fluorescent oligonucleotides. Org Biomol Chem 4:1949–1957

Astakhova IV, Korshun VA, Jahn K, Kjems J, Wengel J (2008) Perylene attached to 2′-amino-LNA: synthesis, incorporation into oligonucleotides, and remarkable fluorescence properties in vitro and in cell culture. Bioconjug Chem 19:1995–2007

Astakhova IV, Ustinov AV, Korshun VA, Wengel J (2011) LNA for optimization of fluorescent oligonucleotide probes: improved spectral properties and target binding. Bioconjug Chem 22:533–539

Babendure JR, Adams SR, Tsien RY (2003) Aptamers switch on fluorescence of triphenylmethane dyes. J Am Chem Soc 125:14716–14717

Beach DL, Salmon ED, Bloom K (1999) Localization and anchoring of mRNA in budding yeast. Curr Biol 9:569–578

Berndl S, Wagenknecht HA (2009) Fluorescent color readout of DNA hybridization with thiazole orange as an artificial DNA base. Angew Chem Int Ed 48:2418–2421

Bertrand E, Chartrand P, Schaefer M, Shenoy SM, Singer RH, Long RM (1998) Localization of ASH1 mRNA particles in living yeast. Mol Cell 2:437–445

Bondensgaard K, Peterson M, Singh SK, Rajwanshi VK, Kumar R, Wengel J, Jacobsen JP (2000) Structural studies of LNA:RNA duplexes by NMR: conformations and implications for RNase H activity. Chemistry 6:2687–2695

Boutorine AS, Novopashina DS, Krasheninina OA, Nozeret K, Venyaminova AG (2013) Molecules 18:15357–15397

Bridger JM, Volpi EV, Schmitt E, Schwarz-Finsterle J, Stein S, Boxler C, Müller P, Mokhir A, Krämer R, Cremer C et al (2010) Methods in molecular biology: fluorescence in situ hybridization (FISH), Combinatorial Oligo FISH: directed labeling of specific genome domains in differentially fixed cell material and live cells, vol 659. Humana Press, New York, pp 185–202
Cardullo RA, Agrawal S, Flores C, Zamecnik PC, Wolf DE (1988) Detection of nucleic acid hybridization by nonradiative fluorescence resonance energy transfer. Proc Natl Acad Sci U S A 85:8790–8794
Carlsson C, Larsson A, Jonsson M, Albinsson B, Norden B (1994) Optical and photophysical properties of the oxazole yellow DNA probes YO and YOYO. J Phys Chem 98:10313–10321
Chen AK, Behlke MA, Tsourkas A (2009) Sub-cellular trafficking and functionality of 2'-O-methyl and 2'-O-methyl-phosphorothioate molecular beacons. Nucleic Acids Res 37:e149
Chen AK, Davydenko O, Behlke MA, Tsourkas A (2010) Ratiometric bimolecular beacons for the sensitive detection of RNA in single living cells. Nucleic Acids Res 38:e148
Chen XH, Roloff A, Seitz O (2012) Consecutive signal amplification for DNA detection based on de novo fluorophore synthesis and host-guest chemistry. Angew Chem Int Ed 51:4479–4483
Cosa G, Focsaneanu KS, McLean JRN, McNamee JP, Scaiano JC (2001) Photophysical properties of fluorescent DNA-dyes bound to single- and double-stranded DNA in aqueous buffered solution. Photochem Photobiol 73:585–599
Czikkely V, Forsterling HD, Kuhn H (1970) Light absorption and structure of aggregates of dye molecules. Chem Phys Lett 6:207–210
Darfeuille K, Hansen JB, Orum H, Di Primo C, Toulme JJ (2004) LNA/DNA chimeric oligomers mimic RNA aptamers targeted to the TAR RNA element of HIV-1. Nucleic Acids Res 32:3101–3107
Dirks RW, Tanke HJ (2006) Advances in fluorescent tracking of nucleic acids in living cells. BioTechniques 40:489–496
Dirks RW, Molenaar C, Tanke HJ (2003) Visualizing RNA molecules inside the nucleus of living cells. Methods 29:51–57
Egholm M, Buchardt O, Christensen L, Behrens C, Freier SM, Driver DA, Berg RH, Kim SK et al (1993) PNA hybridizes to complementary oligonucleotides obeying the Watson-Crick hydrogen-bonding rules. Nature 365:566–568
Flierl A, Jackson C, Cottrell B, Murdock D, Seibel P, Wallace DC (2003) Targeted delivery of DNA to the mitochondrial compartment via import sequence-conjugated peptide nucleic acid. Mol Ther 7:550–557
Fujimoto K, Shimizu H, Inouye M (2004) Unambiguous detection of target DNAs by excimer-monomer switching molecular beacons. J Organomet Chem 69:3271–3275
Fürstenberg A, Julliard MD, Deligeorgiev TG, Gadjev NI, Vasilev AA, Vauthey E (2006) Ultrafast excited-state dynamics of DNA fluorescent intercalators: new insight into the fluorescence enhancement mechanism. J Am Chem Soc 128:7661–7669
Furukawa K, Abe H, Hibino K, Sako Y, Tsuneda S, Ito Y (2009a) Reduction-triggered fluorescent amplification probe for the detection of endogenous RNAs in living human cells. Bioconjug Chem 20:1026–1036
Furukawa K, Abe H, Wang J, Uda M, Koshino H, Tsuneda S, Ito Y (2009b) Reduction-triggered red fluorescent probes for dual-color detection of oligonucleotide sequences. Org Biomol Chem 7:671–677
Fusco D, Accornero N, Lavoie B, Shenoy SM, Blanchard JM, Singer RH, Bertrand E (2003) Single mRNA molecules demonstrate probabilistic movement in living mammalian cells. Curr Biol 13:161–167
Ghasemi J, Ahmadi S, Ahmad AI, Ghobadi S (2008) Spectroscopic characterization of thiazole orange-3 DNA interaction. Appl Biochem Biotechnol 149:9–22
Go ATJI, Visser A, Mulders MAM, Blankenstein MA, van Vugt JMG, Oudejans CBM (2007) 44 single-nucleotide polymorphisms expressed by placental RNA: assessment for use in noninvasive prenatal diagnosis of trisomy 21. Clin Chem 53:2223–2224
Grossmann TN, Seitz O (2006) DNA-catalyzed transfer of a reporter group. J Am Chem Soc 128:15596–15597

Grossmann TN, Roglin L, Seitz O (2008) Target-catalyzed transfer reactions for the amplified detection of RNA. Angew Chem Int Ed 47:7119–7122

Hanafi-Bagby D, Piunno PAE, Wust CC, Krull UJ (2000) Concentration dependence of a thiazole orange derivative that is used to determine nucleic acid hybridization by an optical biosensor. Anal Chim Acta 411:19–30

Hannah KC, Armitage BA (2004) DNA-templated assembly of helical cyanine dye aggregates: a supramolecular chain polymerization. Acc Chem Res 37:845–853

Hara Y, Fujii T, Kashida H, Sekiguchi K, Liang X, Niwa K, Takase T, Yoshida Y, Asanuma H (2010) Coherent quenching of a fluorophore for the design of a highly sensitive in-stem molecular beacon. Angew Chem Int Ed 49:5502–5506

Harrison WJ, Mateer DL, Tiddy GJT (1996) Liquid-crystalline J-aggregates formed by aqueous ionic cyanine dyes. J Phys Chem 100:2310–2321

Hausmann M, Winkler R, Hildenbrand G, Finsterle J, Weisel A, Rapp A, Schmitt E, Janz S, Cremer C (2003) COMBO-FISH: specific labeling of nondenatured chromatin targets by computer-selected DNA oligonucleotide probe combinations. BioTechniques 35:564–577

Huang K, Martí A (2012) Recent trends in molecular beacon design and applications. Anal Bioanal Chem 402:3091–3102

Huang J, Wu Y, Chen Y, Zhu Z, Yang X, Yang CJ, Wang K, Tan W (2011) Pyrene-excimer probes based on the hybridization chain reaction for the detection of nucleic acids in complex biological fluids. Angew Chem Int Ed 50:401–404

Ikeda S, Okamoto A (2007) pH-dependent fluorescence of uncharged benzothiazole-based dyes binding to DNA. Photochem Photobiol Sci 6:1197–1201

Ikeda S, Okamoto A (2008) Hybridization-sensitive on-off DNA probe: application of the exciton coupling effect to effective fluorescence quenching. Chem Asian J 3:958–968

Ikeda S, Kubota T, Yuki M, Okamoto A (2009) Exciton-controlled hybridization-sensitive fluorescent probes: multicolor detection of nucleic acids. Angew Chem Int Ed 48:6480–6484

Ikeda S, Yanagisawa H, Nakamura A, Wang DO, Yuki M, Okamoto A (2011a) Hybridization-sensitive fluorescence control in the near-infrared wavelength range. Org Biomol Chem 9:4199–4204

Ikeda S, Kubota T, Wang DO, Yanagisawa H, Umemoto T, Okamoto A (2011b) Design and synthesis of caged fluorescent nucleotides and application to live cell RNA imaging. Chembiochem 12:2871–2880

Jaco I, Canela A, Vera E, Blasco MA (2008) Centromere mitotic recombination in mammalian cells. J Cell Biol 181:885–892

Jarikote DV, Köhler O, Socher E, Seitz O (2005) Divergent and linear solid-phase synthesis of PNA containing thiazole orange as artificial base. Eur J Org Chem 2005:3187–3195

Jarikote DV, Krebs N, Tannert S, Röder B, Seitz O (2007) Exploring base-pair-specific optical properties of the DNA stain thiazole orange. Chem Eur J 13:300–310

KamiyaY IA, Ito H, Urushihara M, Takai J, Fujii T, Liang X, Kashida H, Asanuma H (2013) Selective labeling of mature RISC using a siRNA carrying fluorophore-quencher pair. Chem Sci 4:4016–4021

Karunakaran V, Lustres JLP, Zhao L, Ernsting NP, Seitz O (2006) Large dynamic stokes shift of DNA intercalation dye thiazole orange has contribution from a high-frequency mode. J Am Chem Soc 128:2954–2962

Kasha M (1963) Energy transfer mechanisms and the molecular exciton model for molecular aggregates. Radiat Res 20:55–70

Kasha M, Rawls HR, El-Bayoumi MA (1965) The exciton model in molecular spectroscopy. Pure Appl Chem 11:371–392

Kashida H, Takatsu T, Sekiguchi K, Asanuma H (2010) An efficient fluorescence resonance energy transfer (FRET) between pyrene and perylene assembled in a DNA duplex and its potential for discriminating single-base changes. Chem Eur J 16:2479–2486

Kashida H, Yamaguchi K, Hara Y, Asanuma H (2012) Quencher-free molecular beacon tethering 7-hydroxycoumarin detects targets through protonation/deprotonation. Bioorg Med Chem 20:4310–4315

Kato T, Kashida H, Kishida H, Yada H, Okamoto H, Asanuma H (2012) Development of a robust model system of FRET using base surrogates tethering fluorophores for strict control of their position and orientation within DNA duplex. J Am Chem Soc 135:741–750

Khairutdinov RF, Serpone N (1997) Photophysics of cyanine dyes: subnanosecond relaxation dynamics in monomers, dimers, and H- and J-aggregates in solution. J Phys Chem B 101:2602–2610

Kholodar SA, Novopashina DS, Meschaninova MI, Lomzov AA, Venyaminova AG (2009) Multipyrene tandem probes for detection of C677T polymorphism in MTHFR gene. Nucleic Acids Symp Ser 53:143–144

Kim Y, Yang CJ, Tan W (2007) Superior structure stability and selectivity of hairpin nucleic acid probes with an L-DNA stem. Nucleic Acids Res 35:7279–7287

Kimura Y, Hanami T, Tanaka Y, de Hoon MJL, Soma T, Harbers M, Lezhava A, Hayashizaki Y, Usui K (2012) Effect of thiazole orange doubly labeled thymidine on DNA duplex formation. Biochemistry 51:6056–6067

Köhler O, Seitz O (2003) Thiazole orange as fluorescent universal base in peptide nucleic acids. Chem Commun 23:2938–2939

Köhler O, Jarikote DV, Seitz O (2004) Ensemble hybridization-a new method for exploring sequence dependent fluorescence of dye-nucleic acid conjugates. Chem Commun 23:2674–2675

Köhler O, Jarikote DV, Seitz O (2005) Forced intercalation probes (FIT probes): Thiazole orange as a fluorescent base in peptide nucleic acids for homogeneous single-nucleotide-polymorphism detection. Chembiochem 6:69–77

Krasheninina OA, Novopashina DS, Venyaminova AG (2011) Oligo(2'-O-methylribonucleotides) containing insertions of 2′-bispyrenylmethylphosphoro-diamidate nucleoside derivatives as prospective fluorescent probes for RNA detection. Russ. J. Bioorg Chem 37:244–248

Kubota T, Ikeda S, Okamoto A (2009a) Doubly thiazole orange-labeled DNA for live cell RNA imaging. Bull Chem Soc Jpn 82:110–117

Kubota T, Ikeda S, Yanagisawa H, Yuki M, Okamoto A (2009b) Hybridization-sensitive fluorescent probe for long-term monitoring of intracellular RNA. Bioconjug Chem 20:1256–1261

Kummer S, Knoll A, Socher E, Bethge L, Herrmann A, Seitz O (2012) PNA FIT-probes for the dual color imaging of two viral mRNA targets in influenza H1N1 infected live cells. Bioconjug Chem 23:2051–2060

Lartia R, Asseline U (2006) New cyanine–oligonucleotide conjugates: relationships between chemical structures and properties. Chem Eur J 12:2270–2281

LeCuyer KA, Behlen LS, Uhlenbeck OC (1996) Mutagenesis of a stacking contact in the MS2 coat protein–RNA complex. EMBO J 15:6847–6853

Levinson GS, Simpson WT, Curtis W (1957) Electronic spectra of pyridocyanine dyes with assignments of transitions. J Am Chem Soc 79:4314–4320

Levsky JM, Singer RH (2003) Fluorescence in situ hybridization: past, present and future. J Cell Sci 116:2833–2838

Lo YMD, Tsui NBY, Chiu RWK, Lau TK, Leung TN, Heung MMS, Gerovassili A, Jin Y, Nicolaides KH, Cantor CR, Ding C (2007) Plasma placental RNA allelic ratio permits noninvasive prenatal chromosomal aneuploidy detection. Nat Med 13:218–223

Majlessi M, Nelson NC, Becker MM (1998) Advantages of 2'-O-methyl oli-goribonucleotide probes for detecting RNA targets. Nucleic Acids Res 26:2224–2229

Martí AA, Li X, Jockusch S, Li Z, Raveendra B, Kalachikov S, Russo JJ, Morozova I, Puthanveettil SV, Ju J et al (2006) Pyrene binary probes for unambiguous detection of mRNA using time-resolved fluorescence spectroscopy. Nucleic Acids Res 34:3161–3168

Martinez K, Estevez MC, Wu Y, Phillips JA, Medley CD, Tan W (2009) Locked nucleic acid based beacons for surface interaction studies and biosensor development. Anal Chem 81:3448–3454

McRae EG, Kasha M (1958) Enhancement of phosphorescence ability upon aggregation of dye molecules. J Chem Phys 28:721–722

Menacher F, Rubner M, Berndl S, Wagenknecht HA (2008) Thiazole orange and Cy3: improvement of fluorescent DNA probes with use of short range electron transfer. J Organomet Chem 73:4263–4266

GPSR Compliance

The European Union's (EU) General Product Safety Regulation (GPSR) is a set of rules that requires consumer products to be safe and our obligations to ensure this.

If you have any concerns about our products, you can contact us on ProductSafety@springernature.com

In case Publisher is established outside the EU, the EU authorized representative is:

Springer Nature Customer Service Center GmbH
Europaplatz 3
69115 Heidelberg, Germany

Batch number: 10732076

Printed by Printforce, the Netherlands

Molenaar C, Marras SA, Slats JCM, Truffert JC, Lemaître M, Raap AK, Dirks RW, Tanke HJ (2001) Linear 2'-O-methyl RNA probes for the visualization of RNA in living cells. Nucleic Acids Res 29:e89

Molenaar C, Wiesmeijer K, Verwoerd NP, Khazen S, Eils R, Tanke HJ, Dirks RW (2003a) Visualizing telomere dynamics in living mammalian cells using PNA probes. EMBO J 22:6631–6641

Molenaar C, Wiesmeijer K, Verwoerd NP, Khazen ESR, Tanke HJ, Dirks RW (2003b) Visualizing telomere dynamics in living mammalian cells using PNA probes. EMBO J 22:6631–6641

Molenaar C, Abdulle A, Gena A, Tanke HJ, Dirks RW (2004) Poly(A)$^+$ RNAs roam the cell nucleus and pass through speckle domains in transcriptionally active and inactive cells. J Cell Biol 165:191–202

Morrison LE, Halder TC, Stols LM (1989) Solution-phase detection of polynucleotides using interacting fluorescent labels and competitive hybridization. Anal Biochem 183:231–244

Nakayama S, Luo Y, Zhou J, Dayie TK, Sintim HO (2012) Nanomolar fluorescent detection of c-di-GMP using a modular aptamer strategy. Chem Commun 48:9059–9061

Nath J, Johnson KL (2000) A review of fluorescence in situ hybridization (FISH): current status and future prospects. Biotech Histochem 75:54–78

Netzel TL, Nafisi K, Ziao M, Lenhard JR, Johnson I (1995) Base-content dependence of emission enhancements, quantum yields, and lifetimes for cyanine dyes bound to double-strand DNA: Photophysical properties of monomeric and bichromophoric DNA stains. J Phys Chem 99:17936–17947

Novopashina DS, Meschaninova MI, Kholodar SA, Lomzov AA, Venyaminova AG (2008) New eximer-based tandem systems for SNP detection. Nucleic Acids Symp Ser 52:229–230

Nygren J, Svanvik N, Kubista M (1998) The interactions between the fluorescent dye thiazole orange and DNA. Biopolymers 46:39–51

Oguro A, Ohtsu T, Svitkin YV, Sonenberg N, Nakamura Y (2003) RNA aptamers to initiation factor 4A helicase hinder cap-dependent translation by blocking ATP hydrolysis. RNA 9:394–407

Okabe K, Harada Y, Zhang J, Tadakuma H, Tani T, Funatsu T (2011) Real time monitoring of endogenous cytoplasmic mRNA using linear antisense 2'-O-methyl RNA probes in living cells. Nucleic Acids Res 39:e20

Okamoto A (2011) ECHO probes: a concept of fluorescence control for practical nucleic acid sensing. Chem Soc Rev 40:5815–5828

Okamoto A, Sugizaki K, Yuki M, Yanagisawa H, Ikeda S, Sueoka T, Hayashi G, Wang DO (2013) A nucleic acid probe labeled with desmethyl thiazole orange: a new type of hybridization-sensitive fluorescent oligonucleotide for live-cell RNA imaging. Org Biomol Chem 11:362–371

Oomoto I, Suzuki-Hirano A, Umeshima H, Han YW, Yanagisawa H, Carlton P, Harada Y, Kengaku M, Okamoto A, Shimogori T, Wang DO (2015) ECHO-liveFISH: in vivo RNA labeling reveals dynamic regulation of nuclear RNA foci in living tissues. Nucleic Acids Res 43:e126

Ozawa T, Natori Y, Sato M, Umezawa Y (2007) Imaging dynamics of endogenous mitochondrial RNA in single living cells. Nat Methods 4:413–419

Paige JS, Wu KY, Jaffrey SR (2011) RNA mimics of green fluorescent protein. Science 333:642–646

Paige JS, Nguyen-Duc T, Song WJ, Jaffrey SR (2012) Fluorescence imaging of cellular metabolites with RNA. Science 335:1194

Paulasova P, Pellestor F (2004) The peptide nucleic acids (PNAs): a new generation of probes for genetic and cytogenetic analyses. Ann Genet Paris 47:349–358

Pellestor F, Paulasova P (2004a) The peptide nucleic acids, efficient tools for molecular diagnosis (review). Int J Mol Med 13:521–525

Pellestor F, Paulasova P (2004b) The peptide nucleic acids (PNAs), powerful tools for molecular genetics and cytogenetics. Eur J Hum Genet 12:694–700

Pianowski Z, Gorska K, Oswald L, Merten CA, Winssinger N (2009) Imaging of mRNA in live cells using nucleic acid-templated reduction of azidorhodamine probes. J Am Chem Soc 131:6492–6497

Privat E, Asseline U (2001) Synthesis and binding properties of Oligo-2′-deoxyribonucleotides covalently linked to a thiazole orange derivative. Bioconjug Chem 12:757–769

Privat E, Melvin T, Asseline U, Vigny P (2001) Oligonucleotide-conjugated thiazole orange probes as "light-up" probes for messenger ribonucleic acid molecules in living cells. Photochem Photobiol 74:532–541
Rackham O, Brown CM (2004) Visualization of RNA–protein interactions in living cells: FMRP and IMP1 interact on mRNAs. EMBO J 23:3346–3355
Raj A, van den Bogaard P, Rifkin SA, van Oudenaarden A, Tyagi S (2008) Imaging individual mRNA molecules using multiple singly labeled probes. Nat Methods 5:877–879
Rhee WJ, Bao G (2009) Simultaneous detection of mRNA and protein stem cell markers in live cells. BMC Biotechnol 9:30
Rhee WJ, Bao G (2010) Slow non-specific accumulation of 2′-deoxy and 2'-O-methyl oligonucleotide probes at mitochondria in live cells. Nucleic Acids Res 38:e109
Rösch U, Yao S, Wortmann R, Würthner F (2006) Fluorescent H-aggregates of merocyanine dyes. Angew Chem Int Ed 45:7026–7030
Röthlingshöfer M, Gorska K, Winssinger N (2011a) Nucleic acid-templated energy transfer leading to a photorelease reaction and its application to a system displaying a nonlinear response. J Am Chem Soc 133:18110–18113
Röthlingshöfer M, Gorska K, Winssinger N (2011b) Nucleic acid templated uncaging of fluorophores using Ru-catalyzed photoreduction with visible light. Org Lett 14:482–485
Rye HS, Yue S, Wemmer DE, Quesada MA, Haugland RP, Mathies RA, Glazer AN (1992) Stable fluorescent complexes of double-stranded DNA with bis-intercalating asymmetric cyanine dyes: properties and applications. Nucleic Acids Res 20:2803–2812
Sadhu KK, Winssinger N (2013) Detection of miRNA in live cells by using templated Ru(II)-catalyzed unmasking of a fluorophore. Chem Eur J 19:8182–8189
Sagawa T, Tobata H, Ihara H (2004) Exciton interactions in cyanine dye-hyaluronic acid (HA) complex: reversible and biphasic molecular switching of chromophores induced by random coil-to-double-helix phase transition of HA. Chem Commun 18:2–4
Sando S, Kool ET (2002) Imaging of RNA in bacteria with self-ligating quenched probes. J Am Chem Soc 124:9686–9687
Schwarz-Finsterle J, Stein S, Grossmann C, Schmitt E, Trakhtenbrot L, Rechavi G, Amariglio N, Cremer C, Hausmann M (2007) Comparison of triple helical COMBO-FISH and standard FISH by means of quantitative microscopic image analysis of ABL/BCR positions in cell nuclei. J Biochem Biophys Methods 70:397–406
Silahtaroglu AN, Tommerup N, Vissing H (2003) FISHing with locked nucleic acids (LNA): evaluation of different LNA/DNA mixmers. Mol. Cell Probes 17:165–169
Silahtaroglu A, Pfundheller H, Koshkin A, Tommerup N, Kauppinen S (2004) LNA-modified oligonucleotides are highly efficient as FISH probes. Cytogenet Genome Res 107:32–37
Silva GL, Ediz V, Yaron D, Armitage BA (2007) Experimental and computational investigation of unsymmetrical cyanine dyes: understanding torsionally responsive fluorogenic dyes. J Am Chem Soc 129:5710–5718
Silverman AP, Kool ET (2005a) Quenched probes for highly specific detection of cellular RNAs. Trends Biotechnol 23:225–230
Silverman AP, Kool ET (2005b) Quenched autoligation probes allow discrimination of live bacterial species by single nucleotide differences in rRNA. Nucleic Acids Res 33:4978–4986
Silverman AP, Baron EJ, Kool ET (2006) RNA-templated chemistry in cells: discrimination of Escherichia, Shigella and Salmonella bacterial strains with a new two-color FRET strategy. Chembiochem 7:1890–1894
Simon LD, Abramo KH, Sell JK, McGown LB (1998) Oxazole yellow dye interactions with short DNA oligomers of homogeneous base composition and their hybrids. Biospectroscopy 4:17–25
Sixou S, Szoka FC, Green GA, Giusti B, Zon G, Chin DJ (1994) Intracellular oligonucleotide hybridization detected by fluorescence resonance energy transfer (FRET). Nucleic Acids Res 22:662–668
Smolina I, Lee C, Frank-Kamenetskii M (2007) Detection of low-copy-number genomic DNA sequences in individual bacterial cells by using peptide nucleic acid-assisted rolling-circle amplification and fluorescence in situ hybridization. Appl Environ Microbiol 73:2324–2328

Socher E, Jarikote DV, Knoll A, Röglin L, Burmeister J, Seitz O (2008) FIT probes: peptide nucleic acid probes with a fluorescent base surrogate enable real-time DNA quantification and single nucleotide polymorphism discovery. Anal Biochem 375:318–330

Sugizaki K, Okamoto A (2010) ECHO-LNA conjugates: hybridization-sensitive fluorescence and its application to fluorescent detection of various RNA strands. Bioconjug Chem 21:2276–2281

Svanvik N, Westman G, Wang D, Kubista M (2000) Light-up probes: Thiazole orange-conjugated peptide nucleic acid for detection of target nucleic acid in homogeneous solution. Anal Biochem 281:26–35

Svanvik N, Nygren J, Westman G, Kubista M (2001) Free-probe fluorescence of light-up probes. J Am Chem Soc 123:803–809

Tan W, Wang K, Drake TJ (2004) Molecular beacons. Curr Opin Chem Biol 8:547–553

Thomsen R, Nielsen PS, Jensen TH (2005) Dramatically improved RNA in situ hybridization signals using LNA-modified probes. RNA 11:1745–1748

Tsourkas A, Behlke MA, Bao G (2002a) Hybridization of 2'-O-methyl and 2'-deoxy molecular beacons to RNA and DNA targets. Nucleic Acids Res 30:5168–5174

Tsourkas A, Behlke MA, Bao G (2002b) Structure–function relationships of shared-stem and conventional molecular beacons. Nucleic Acids Res 30:4208–4215

Tsourkas A, Behlke MA, Rose SD, Bao G (2003) Hybridization kinetics and thermodynamics of molecular beacons. Nucleic Acids Res 31:1319–1330

Tsuji A, Koshimoto H, Sato Y, Hirano M, Sei-Iida Y, Kondo S, Ishibashi K (2000) Direct observation of specific messenger RNA in a single living cell under a fluorescence microscope. Biophys J 78:3260–3274

Tyagi S, Kramer FR (1996) Molecular beacons: probes that fluoresce upon hybridization. Nat Biotechnol 14:303–308

Valencia-Burton M, McCullough RM, Cantor CR, Broude NE (2007) RNA visualization in live bacterial cells using fluorescent protein complementation. Nat Methods 4:421–427

Wang X, Krull UJ (2002) Tethered thiazole orange intercalating dye for development of fibre-optic nucleic acid biosensors. Anal Chim Acta 470:57–70

Wang X, Krull UJ (2005) Synthesis and fluorescence studies of thiazole orange tethered onto oligonucleotide: development of a self-contained DNA biosensor on a fiber optic surface. Bioorg Med Chem Lett 15:1725–1729

Wang DO, Okamoto A (2012) ECHO probes: fluorescence emission control for nucleic acid imaging. J Photochem Photobiol C 13:112–123

Wang L, Yang CJ, Medley CD, Benner SA, Tan W (2005) Locked nucleic acid molecular beacons. J Am Chem Soc 127:15664–15665

Wang DO, Matsuno H, Ikeda S, Nakamura A, Yanagisawa H, Hayashi Y, Okamoto A (2012) A quick and simple FISH protocol with hybridization-sensitive fluorescent linear oligodeoxynucleotide probes. RNA 18:166–175

Weigert R, Porat-Shliom N, Amornphimoltham P (2013) Imaging cell biology in live animals: ready for prime time. J Cell Biol 201:969–979

West W, Pearce S (1965) The dimeric state of cyanine dyes. J Phys Chem 69:1894–1903

Yamana K, Ohshita Y, Fukunaga Y, Nakamura M, Maruyama A (2008) Bis-pyrene-labeled molecular beacon: a monomer–excimer switching probe for the detection of DNA base alteration. Bioorg Med Chem 16:78–83

Yang CJ, Wang L, Wu Y, Kim Y, Medley CD, Lin H, Tan W (2007) Synthesis and investigation of deoxyribonucleic acid/locked nucleic acid chimeric molecular beacons. Nucleic Acids Res 35:4030–4041

You Y, Moreira BG, Behlke MA, Owczarzy R (2006) Design of LNA probes that improve mismatch discrimination. Nucleic Acids Res 34:e60

Zeena S, Thomas KG (2001) Conformational switching and exciton interactions in hemicyanine-based bichromophores. J Am Chem Soc 123:7859–7865

Mesenchymal Stem Cell-derived Extracellular Vesicles for Skin Wound Healing

Soo Kim, Joonghoon Park, and Tae Min Kim

Introduction

Skin wound healing is a highly coordinated event involving various cells such as keratinocytes, fibroblasts, platelets, inflammatory, and endothelial cells. This fine-tuned process can be broadly divided into three phases; namely, inflammatory, re-epithelialization/granulation, and remodeling phases (Landen et al. 2016; Reinke and Sorg 2012). Re-epithelialization/granulation process is mediated by keratinocytes and fibroblasts interaction, which shifts the inflammatory wound to a highly proliferative phase, in order to re-establish the lost connective tissue (Shaw and Martin 2016). However, such well-orchestrated process may become dysfunctional in serious chronic pathological conditions of severe burns, trauma, and diabetes (Frykberg and Banks 2015). Therefore, there is deep interest in developing novel methods to enhance wound healing, such as the use of mesenchymal stem cells (MSCs) (Chen et al. 2016).

The MSCs are multipotent stem cells that can be isolated from various tissues including bone marrow, adipose tissue, and umbilical cord. They can also be isolated from solid organs such as the liver, kidney, lung, or amnion, synovial fluid, muscle, and periodontal ligaments, with distinct biological traits from each other (Elahi et al. 2016; Marquez-Curtis et al. 2015). The innate role of MSCs in tissue homeostasis is believed to be due to their potential to differentiate into mesodermal

S. Kim
Brexogen Inc., Seoul, South Korea
e-mail: sue.kim@brexogen.com

J. Park · T. M. Kim (✉)
Graduate School of International Agricultural Technology and Institute of Green-Bio Science and Technology, Pyeongchang Daero 1447, Seoul National University, Pyeongchang, Gangwon-do, South Korea
e-mail: joonghoon@snu.ac.kr; taemin21@snu.ac.kr

J. K. Kim et al. (eds.), *Advanced Imaging and Bio Techniques for Convergence Science*, Advances in Experimental Medicine and Biology,
https://doi.org/10.1007/978-981-33-6064-8_18

tissues including bone, adipose tissue, cartilage, and muscles (Pittenger et al. 1999). However, ample evidence shows that they also produce various bioactive molecules that can exert paracrine effects on surrounding cells. Additionally, it has been recently accepted that the trophic properties of MSCs are largely attributed to these secretome (Vizoso et al. 2017). The therapeutic potential of MSCs in tissue repair has been demonstrated by the angiogenic, anti-fibrotic, anti-inflammatory, and anti-apoptotic functions of secretome (Fitzsimmons et al. 2018). Moreover, MSCs are also considered immunologically privileged due to the lack of expression of MHC class II molecules, endothelial markers—CD31, and hematopoietic markers—CD34 or CD45, as compared with pluripotent stem cells (ESCs/iPSCs) (Jacobs et al. 2013). Thus, since its discovery in the 1970s, there has been increasing interest in MSCs for use in various target diseases such as autoimmune disease and, ischemic tissue injuries including skin wound healing (Langrzyk et al. 2018; Rad et al. 2019).

The MSCs also play an immunomodulatory role. Several mechanisms have been reported, among which the most attributed are the direct cell contact and secretion of various factors. MSCs interact with various immune cells and inhibit pathways involving both, innate and acquired immunity. For instance, MSCs establishes direct contact with dendritic cells to inhibit their maturation and, their ability to stimulate lymphocytes is reduced by upregulation of IL10 and downregulation of interferon-γ (IFN-γ) and TNF-α expression, thus providing a therapeutic advantage in immune disorders (Gao et al. 2017).

The co-culture of macrophages and MSCs induced the production of M2-macrophages and IL-10, alongside the inhibition of inflammatory cytokines (i.e., IL-12, IL-1ß, and TNF-α) in a study, suggesting that MSCs indirectly plays a trophic role in tissue regeneration via preferentially stimulating M2-macrophages. Additionally, MSCs can inhibit the activity of lymphocytes. The MSCs express B7-H4 or HLA-G on their surface, which are responsible for T-cell suppressive role, that may differ depending on the types of inflammatory stimuli and T-cell subsets (Wang et al. 2018). It has also been reported that MSCs can suppress the B-cell proliferation and immunoglobulin production via regulating several pathways such as STAT3/PAX5, Blimp1, and AKT signaling (Fan et al. 2016; Feng et al. 2014). Owing to these aforementioned characteristics in tissue regeneration and immune regulation, it has been suggested that the classically used term "MSCs" attributing to the mesenchymal properties, be refined to "medicinal signaling cells (MSCs)" (Caplan 2017).

MSCs for Wound Healing

The therapeutic potentials of MSCs have been demonstrated in various preclinical models of cardiovascular and inflammatory diseases, and tissue injuries including cutaneous wounds (Kucharzewski et al. 2019; Samsonraj et al. 2017). The culture supernatant of MSCs was found to be enriched with trophic mediators like TGF-ß1,

IL6, IL-8, MCP-1, RANTES, collagen type 1, and fibronectin (Samsonraj et al. 2017). The therapeutic function of these molecules has been elucidated using various wound models, such as severe burn (Liu et al. 2014), excisional splint (Kim et al. 2012; Nie et al. 2011), or a combination of various pathological condition including skin excision (Maharlooei et al. 2011) or foot ulceration (Kato et al. 2014) in diabetic animals.

However, there are several hurdles in the clinical use of MSCs including its efficacy, possibility of tumor formation and rejection, which remains unsolved (Phan et al. 2018; Wu et al. 2018). Also, the stable maintenance of MSCs using standard protocols may not be easy due to donor-to-donor variability, and the decline in functional characteristics due to subculturing over several generations (Denu and Hematti 2016). Also, MSCs are generally accumulated in the lungs upon being injected into the circulation, and becomes non-detectable after a few days (Leibacher and Henschler 2016; Makela et al. 2015). Thus, an alternative means that can overcome these problems is needed. One possible option to eliminate these risks is to use the conditioned medium (CM) from MSCs culture. A study demonstrated that the concentrations of growth factors (bFGF, VEGF-A, IL-6, IL-8) were higher in the CM harvested from human BM-MSC cultured under hypoxic condition. Functionally, the proliferation of fibroblast, endothelial cells, and keratinocyte, all of which are involved in wound healing, was enhanced (Chen et al. 2014). Most importantly, an increased wound contraction was seen together with the neovascularization and decreased inflammatory macrophages and collagen deposition. Another study reported that hypoxia enhanced the function of CM collected from amniotic stem cells. Authors also found that the increased amount of VEGF and TGF-ß1 are responsible for the increased wound closure ability (Jun et al. 2014).

Studies have been recently conducted to determine the mechanism of reparative function of stem cells. Increasing data have shown that, indirect communication by paracrine players such as growth factors, cytokines, or extracellular vesicles (EVs) are essential for the therapeutic function of replacing the damaged cells, than via direct cell-to-cell manner. It is now becoming evident that, EVs play a crucial role among the secretome in regulating the function of stem cells. Thus, EVs derived from MSCs (MSC-EVs) play a diverse role in immune-modulation and regeneration of various tissues including wounds (Bjorge et al. 2017; Ren 2018) (Fig. 1).

Extracellular Vesicles

Extracellular vesicles have long been recognized as a means of removing cellular waste; however, it is now evident that they have a capacity to exchange specific cargo materials between the cells, suggesting their role in cell-to-cell communication (Rani et al. 2015; Valadi et al. 2007). Classically, EVs can be classified into exosomes, microvesicles, and apoptotic bodies based on their size and biogenesis. Exosomes are first formed as intraluminal vesicles (ILV) within the multivesicular

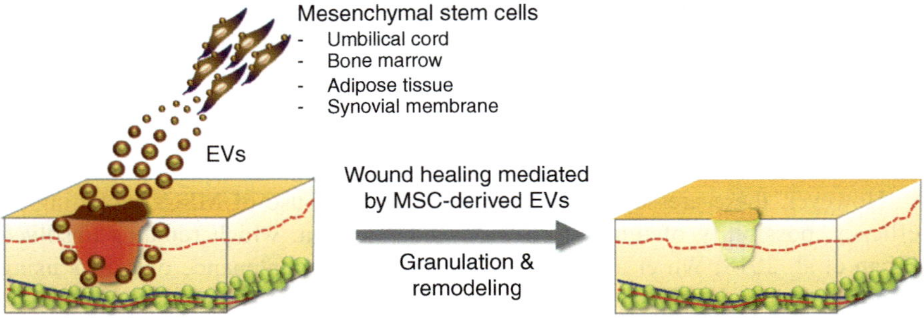

Fig. 1 The trophic role of MSC-EVs in wound healing. Studies have shown that EVs secreted by MSCs derived from various tissue including umbilical cord, bone marrow, adipose tissue, and synovial membrane are capable of enhancing wound repair in various preclinical models

body (MVB), and then released upon fusion with plasma membrane; while, microvesicles originate from outward budding of plasma membrane. Apoptotic bodies have a relatively large size (>1 μm) and are produced by fragmentation of the plasma membrane during apoptosis. Typically, exosomes have a diameter of 30–150 nm, and exhibit surface markers such as tetraspanins (CD63, CD9, CD81), TSG101, flotillin, and Alix. Microvesicles are relatively larger (100–1000 nm) and express selectins, integrin, and CD40 molecules (Tang et al. 2019). The biogenesis of EVs occur in a complex way depending on its types; for example, various classes of ESCRT (endosomal sorting complex required for transport), which are responsible for the budding and release of ILV into the lumen of endosome, are involved in the generation of both exosomes and microvesicles. However, the role of each class differs with their function (Harada et al. 2019; Juan and Furthauer 2018; van Niel et al. 2018). Owing to its complexity and unknown mechanisms of the biogenesis, identification of membrane molecules, specific for each EV class is not possible. Thus, the terminology "EV" has been generally accepted for its use (Akers et al. 2013). As EVs contain various biomolecule cargos such as proteins, RNA, lipids, and carbohydrates (Lai et al. 2016; Maas et al. 2017), the contents of EVs reflect the physiology of parental cells. Thus, in addition to being promising candidate of cell-free therapy, the EVs found in serum, urine, or saliva is now a recognized biomarker for analysis of various diseases including cancer and metabolic diseases (He et al. 2018). From a practical point of view, EVs have several advantages over MSCs, such as they can be produced robustly, filtered, and stored without cryoprotectant. Further, EVs are less immunogenic and tumorigenic compared to replicating viable MSCs, and have less chance of thrombus formation within the vasculature in vivo (Keshtkar et al. 2018; Toh et al. 2018). Besides, EVs have tremendous potential as a natural drug carrier, and various techniques are now being developed for loading biomolecular cargos within the EVs. Currently, various methods such as the addition of molecules during culture, transfection, electroporation, genetic engineering of parental cells are used for loading proteins, small molecules, nucleic acids (Tang et al. 2019), thus further expanding its application.

MSC-Derived EVs for Wound Healing

The therapeutic function of MSC-EVs was initially demonstrated in a mice model of hypoxic pulmonary hypertension. Specifically, systemic injection of MSC-EVs significantly reduced the pulmonary influx of macrophages and the level of inflammatory mediators, by inhibiting STAT3 signaling (Lee et al. 2012). After this work, the therapeutic role of MSC-EVs has been intensively reported in various preclinical disease models including tissue ischemia, immune regulation, fibrosis, inflammation, wound, and angiogenesis (Chan et al. 2019; Keshtkar et al. 2018; Rad et al. 2019).

So far, the therapeutic function of MSC-EVs has been demonstrated in various wound models using MSCs of different origins (Table 1). First, it was shown that EVs from human umbilical cord mesenchymal stem cells (hucMSC-exo) were

Table 1 Extracellular vesicles for wound healing

Cell source	EV types	Animal model	Dose, route, times	Mechanism	Refs.
Human umbilical cord MSC	Exosome	Rat model of severe burn wounds	200 μg, subcutaneous, three sites around the wound	Wnt/ß-catenin-mediated wound re-epithelialization	Zhang et al. (2015)
Human umbilical cord MSC	Exosome	Full thickness wound	Hydrogel mixture of exosomes at 48 hours after wounding	Inhibition of myofibroblast differentiation by suppression of TGF-ß2/SMAD2 by microRNAs (miR-21, −23a, −125b, −145)	Fang et al. (2016)
Human umbilical cord MSC	Exosome	30% TBSA full thickness burn	800 μg (RNA amount), i.v.	Reduction of inflammation, acceleration of wound recovery, downregulation of TLR4 in wound tissue mediated by miR-181c delivered by exosomes	Li et al. (2016)
Human adipose mesenchymal stem cells	Exosome	Full thickness wound	200 μg, subcutaneous or i.v.	Increased collagen I/III at early healing stages (detection of injected exosomes at D7 and D21 post-wounding)	Hu et al. (2016)
Human umbilical cord MSC	Exosome	Rat, dorsal skin burns	200 μg, subcutaneous, three sites around the wound	Regulation of Wnt/ß-catenin signaling by inhibiting hippo-YAP pathway via exosomal 14-3-3ζ	Zhang et al. (2016)

(continued)

Table 1 (continued)

Cell source	EV types	Animal model	Dose, route, times	Mechanism	Refs.
Human adipose mesenchymal stem cells	UC-concentrated EVs (100,000xg, 70 min)	Rat, dorsal wound splint model	2.0×10^8 EVs in hydroxyethyl cellulose gel, daily on wounds	Accelerated wound closure In vitro results showed activated AKT in fibroblasts and keratinocytes.	Ferreira et al. (2017)
Human adipose mesenchymal stem cells	Exosome	Mice, full thickness	At 48 hours, 200 μg, i.v.	Enhanced wound healing, increase of collagen III to I ratio In vitro studies showed that increased MMP3 in fibroblast by activation of ERK/ MAPK pathway, regulation of ECM remodeling (collagen type III: Type I, TGF-β3:TGF-β1 and MMP3:TIMP1)	Wang et al. (2017)
Human adipose mesenchymal stem cells overexpressing Nrf2	Exosomes	Rat, diabetic foot ulcer	Co-injection of EPCs and exosomes	Increased granulation tissue formation and angiogenesis in wound, reduction in inflammation and oxidative stress markers	Li et al. (2018)
Synovium mesenchymal stem cells overexpressing miRNA-126	Exosomes	Diabetic rat, full thickness dorsal skin wound	Applying the dressing of chitosan-based hydrogel mixed with exosomes	Accelerated wound closure by increasing re-epithelialization and angiogenesis, collagen maturity	Tao et al. (2017)
Adipose-derived MSCs	Exosomes	Mice, diabetic, full thickness wound	Application of FHE hydrogel—Exosome mixture	Enhanced wound closure, angiogenesis, re-epithelialization, collagen deposition	Wang et al. (2019)
DFO-treated bone marrow MSCs	Exosomes	Diabetic rat, skin wound on the back	100 μg, subcutaneous, four sites around the wound	Enhanced wound closure and angiogenesis, miR-126-mediated PTEN downregulation and subsequent activation of PI3K/ AKT by DFO-exo in HUVEC study	Zhu et al. (2019)

DFO deferoxamine, *EPCs* endothelial progenitor cells, *FHE* Pluronic *F127* (F127), *OHA* oxidative hyaluronic acid, *EPL* Poly-ε-L-lysine, *TBSA* total body surface area, *UC* ultracentrifugation capable of promoting wound re-epithelialization and cell proliferation in a rat model with severe burns. Further, the expression of cytokeratin 19, PCNA, and collagen I was seen to be increased (Zhang et al. 2015). Mechanistically, the upstream signaling from Wnt4 in the hucMSC-exo promoted the nuclear translocation of ß-catenin and subsequent skin cell proliferation in vitro. Importantly, shRNA-mediated blockade of Wnt4 abrogated such role, indicating the Wnt4 existing in hucMSC-exo plays a critical role in wound healing. The role of hucMSC-exo has also been shown in mice model of full thickness wound model. Authors reported that hucMSC-exo mixed with hydrogel mixture was able to reduce the formation of myofibroblasts, responsible for excessive scar formation. Interestingly, this role was found to be mediated by inhibition of the TGF-ß2/SMAD2 pathway by a subset of microRNAs (miR-21, −23a, −125b, and − 145) enriched in hucMSC-exo. Antagomirs against these microRNAs abolished the therapeutic function of hucMSC-exo, and these results indicate that MSCs can exert their anti-fibrotic activity by posttranscriptional regulation of recipient cells' mRNA using microRNAs within the exosomes (Fang et al. 2016).

The role of hucMSC-exo in regulating skin cell growth was also reported. Zhang et al. found that when hucMSC-exo were treated with high-density keratinocytes culture, the expression of PCNA and Cyclin-D3, α-SMA, and collagen I and III expressions decreased. Further examination showed that 14-3-3ζ protein in hucMSC-exo had ability to bind phosphorylation-large tumor suppressor (p-LATS) kinases with YAP, thereby enhancing the YAP phosphorylation. The Wnt/β-catenin pathway was thus controlled, which overall led to a decreased scar formation during wound healing (Zhang et al. 2016).

The recovery of wound using EVs from adipose tissue was also reported (Ferreira et al. 2017; Hu et al. 2016; Wang et al. 2017). Hu et al. reported that exosomes from adipose-derived MSCs (ASC-exo) could be mobilized to the wound area, as shown by in vivo imaging, to promote the tissue recovery. Interestingly, ASC-exos injected intravenously showed better outcomes of wound closure as compared with subcutaneous local injection, although the reason was unclear (Hu et al. 2016). Similarly, Ferreira et al. found that ASC-exo enhanced the migration of keratinocytes and skin fibroblasts, and also showed that phosphorylated AKT was increased by ASC-exo treatment. Phosphorylation of Histone H3, a mitosis marker, was also increased indicating that ASC-exo was effective in enhancing the proliferation of skin cells. Authors demonstrated that ASC-exos co-treated in the mixture with hydroxyethyl cellulose aqueous gel was more effective in wound closure using in vivo wound model, providing a better concept on translational application (Ferreira et al. 2017). Investigation was also conducted with the focus on the effect of ASC-exo in regulating ECM molecules that are associated with skin tissue remodeling. It was demonstrated that ASC-exo enhanced ECM remodeling by increasing the ratios of several ECM molecules (collagen type III/type I, TGF-β3/β1, and MMP3/TIMP1), and was beneficial in the regulation of differentiation of fibroblast to myofibroblast type, leading to the reduction in scar formation during cutaneous wound repair (Wang et al. 2017).

As the composition of biomolecules of EVS can be changed depending on the parental cells, we can modify the contents of MSCs to enhance their potential therapeutic uses (Phinney and Pittenger 2017). Thus, an advanced approach was conducted using exosomes harvested from engineered MSCs. The ASCs, that were engineered to overexpress Nrf2, which plays an important role in enhancing antioxidant enzymes and glutathione biosynthesis, were used for producing exosomes in evaluating wound healing activity using foot ulcer model of diabetic rats. Administration of EPCs (endothelial progenitor cells) together with EVs from Nrf2-overexpressing ADSCs significantly enhanced the wound recovery as compared with those treated only with EPCs. In vitro, ASC-exo was able to reduce the senescence induced by high glucose culture, as shown by the reduction of apoptosis, ROS generation, capillary tube formation, and the expression of vegf, nox1, nox4, IL-6, IL-1b, and TNF-a (Li et al. 2018). The contents of secretome and EVs from MSCs can also be changed by treatment with small molecules and chemicals; for example, 3,3′-diindolylmethane (DIM), a natural small compound that is present in cruciferous fruits, was utilized to treat hucMSCs and then their therapeutic effects were analyzed. The DIM treatment stimulated the expression of genes related to pluripotency concomitantly, with the activation of Wnt/β-catenin signaling. The wound healing outcome was enhanced by DIM-treated hucMSCs compared with those from untreated cells using a burn model, while the therapeutic effect in vivo was reduced by the blockade of Wnt11. Also, the DIM treatment led to higher expression of Wnt11 in the exosomes from hucMSCs, and these Wnt11-enriched exosomes were able to promote the proliferation and migration of keratinocytes (Shi et al. 2017).

Sustained release of nanoparticles would be an ideal method for increasing the therapeutic effect of EVs. Using MSCs lentivirally transfected with miR-126, which was identified as an angiogenic microRNA, Toe et al. reported that the mixture made of exosomes from miR-126-overexpressing synovium-derived MSCs and chitosan (CS) was able to enhance the re-epithelialization, microvessel formation, and collagen formation, leading to an accelerated the wound closure than those treated with CS only in a diabetic rat model of skin wound (Tao et al. 2017). Another study demonstrated that, diabetic ulcer also decreased upon application of ASC-exo mixed with antibacterial peptide-based hydrogel. Interestingly, this application enhanced capillary angiogenesis, epithelialization, and collagen deposition in the full thickness chronic wound. Also, such treatment enabled sustained release of exosomes, depending on the pH, overall led to a better healing ability (Wang et al. 2019).

MicroRNAs are non-coding RNAs that play a pivotal role in regulating gene expression, and various studies have shown that inflammation can be modulated by several miRNAs (Raisch et al. 2013). Li et al. found that miR-181c, which has been known to alleviate inflammation via inhibiting TLR4 (He et al. 2014), was enriched in exosomes from hucMSCs. In a skin burn wound model, animals treated with hucMSC-exo alleviated the inflammation compared with those treated with exosomes from dermal fibroblasts as shown by total WBC counts, serum level of inflammatory cytokines, and histopathology. Importantly, exosomes from hucMSCs that had been transiently transfected with miR-181c mimic, further enhanced the anti-inflammatory role and tissue recovery. It was also found that hucMSC-exo led to a decreased TLR expression in the wound tissue, while TLF blockade became more significant on treatment with exosomes from hucMSCs, that had been transiently transfected with miR-181c mimic (Li et al. 2016). More recently, an enhanced function of bone marrow (BM)-derived exosomes on wound healing was reported. Based on the fact that the secretome of MSCs can be altered via preconditioning, Ding et al. reported that BM-MSCs cultured with deferoxamine, a hypoxic mimetic, enhanced the ability of exosomes in wound recovery. Mechanistically, in vitro examination revealed that, DFO treatment led to increased delivery of exosomal miR-126 into HUVEC (human umbilical cord vein endothelial cells), activating the PI3K/AKT signaling via miR-126-mediated decrease of PTEN (Zhu et al. 2019).

Future Perspectives and Conclusions

Over past decades, preclinical studies have shown that the EVs from MSCs have capacity to restore dysfunctional tissues or regulate the inflammation, leading to tissue recovery. The biomolecular contents within EVs from various sources of MSCs are now being identified, and it is becoming evident that, the cargoes from parental MSCs can be delivered to damaged tissue itself, or to the resident stem/progenitor cells to rebuild the injured tissue. In addition to its own biological characteristics, such functions of EVs are now being expanded in combination with various technologies. An ideal option would be using engineered sheets consisting of EV-producing cells as biologic scaffolds or mucosal sheets (Sjoqvist et al. 2019; Watt and Pleat 2018). Also, an innovative dressing with biocompatible matrix, e.g., chitosan hydrogels, would be another option that can combine the current concept of EV-based therapy (Zarrintaj et al. 2017). Other approaches should be considered to modify the surface of EVs, so that the trafficking and uptake of EVs could be maximized in vivo (Liu and Su 2019). The most relevant approach in this is to engineer EV-producing cells, so that EVs either conjugated with specific surface proteins or loaded with drug can be produced. Direct modification of MSC-EVs via nanoparticle technologies such as hydrophobic intake or physical conjugation can also be considered (Murphy et al. 2019). Ideally, the epithelial autografts with defined materials, or EVs designed for specific purposes will be required to improve the prognosis of severe wounds (Watt and Pleat 2018).

Lastly, it should be noted that several issues are critical in making EVs therapeutically feasible. Primarily, the preclinical testing of EVs should be conducted in larger animals to better translate the results of clinical purpose. Additionally, the studies that can clearly determine the optimal dosage and administration route should be conducted, and a proper model for examining the absorption and distribution of EVs under normal physiology or pathological conditions should be developed. Through these efforts, MSC-EVs may become an efficient alternative for cutaneous wound repair.

References

Akers JC, Gonda D, Kim R, Carter BS, Chen CC (2013) Biogenesis of extracellular vesicles (EV): exosomes, microvesicles, retrovirus-like vesicles, and apoptotic bodies. J Neuro-Oncol 113:1–11. https://doi.org/10.1007/s11060-013-1084-8

Bjorge IM, Kim SY, Mano JF, Kalionis B, Chrzanowski W (2017) Extracellular vesicles, exosomes and shedding vesicles in regenerative medicine - a new paradigm for tissue repair. Biomater Sci 6:60–78. https://doi.org/10.1039/c7bm00479f

Caplan AI (2017) Mesenchymal stem cells: time to change the name! Stem Cells Transl Med 6:1445–1451. https://doi.org/10.1002/sctm.17-0051

Chan BD, Wong WY, Lee MM, Cho WC, Yee BK, Kwan YW, Tai WC (2019) Exosomes in inflammation and inflammatory disease. Proteomics 19(8):e1800149. https://doi.org/10.1002/pmic.201800149

Chen L et al (2014) Conditioned medium from hypoxic bone marrow-derived mesenchymal stem cells enhances wound healing in mice. PLoS One 9:e96161. https://doi.org/10.1371/journal.pone.0096161

Chen D, Hao H, Fu X, Han W (2016) Insight into Reepithelialization: how do mesenchymal stem cells perform? Stem Cells Int 2016:6120173. https://doi.org/10.1155/2016/6120173

Denu RA, Hematti P (2016) Effects of oxidative stress on mesenchymal stem cell biology. Oxidative Med Cell Longev 2016:2989076. https://doi.org/10.1155/2016/2989076

Elahi KC, Klein G, Avci-Adali M, Sievert KD, MacNeil S, Aicher WK (2016) Human mesenchymal stromal cells from different sources diverge in their expression of cell surface proteins and display distinct differentiation patterns. Stem Cells Int 2016:5646384. https://doi.org/10.1155/2016/5646384

Fan L, Hu C, Chen J, Cen P, Wang J, Li L (2016) Interaction between mesenchymal stem cells and B-cells. Int J Mol Sci 17:650. https://doi.org/10.3390/ijms17050650

Fang S et al (2016) Umbilical cord-derived Mesenchymal stem cell-derived Exosomal MicroRNAs suppress Myofibroblast differentiation by inhibiting the transforming growth factor-beta/SMAD2 pathway during wound healing. Stem Cells Transl Med 5:1425–1439. https://doi.org/10.5966/sctm.2015-0367

Feng X et al (2014) Restored immunosuppressive effect of mesenchymal stem cells on B cells after olfactory 1/early B cell factor-associated zinc-finger protein down-regulation in patients with systemic lupus erythematosus. Arthritis Rheum 66:3413–3423. https://doi.org/10.1002/art.38879

Ferreira ADF et al (2017) Extracellular vesicles from adipose-derived Mesenchymal stem/stromal cells accelerate migration and activate AKT pathway in human keratinocytes and fibroblasts independently of miR-205 activity. Stem Cells Int 2017:9841035. https://doi.org/10.1155/2017/9841035

Fitzsimmons REB, Mazurek MS, Soos A, Simmons CA (2018) Mesenchymal stromal/stem cells in regenerative medicine and tissue engineering. Stem Cells Int 2018:8031718. https://doi.org/10.1155/2018/8031718

Frykberg RG, Banks J (2015) Challenges in the treatment of chronic wounds. Adv Wound Care (New Rochelle) 4:560–582. https://doi.org/10.1089/wound.2015.0635

Gao WX et al (2017) Effects of mesenchymal stem cells from human induced pluripotent stem cells on differentiation, maturation, and function of dendritic cells. Stem Cell Res Ther 8(48):1–16. https://doi.org/10.1186/s13287-017-0499-0

Harada Y et al (2019) Generation of the heterogeneity of extracellular vesicles by membrane organization and sorting machineries. Biochim Biophys Acta Gen Subj 1863:681–691. https://doi.org/10.1016/j.bbagen.2019.01.015

He X, Jing Z, Cheng G (2014) MicroRNAs: new regulators of toll-like receptor signalling pathways. Biomed Res Int 2014:945169. https://doi.org/10.1155/2014/945169

He C, Zheng S, Luo Y, Wang B (2018) Exosome Theranostics: biology and translational medicine. Theranostics 8:237–255. https://doi.org/10.7150/thno.21945

Hu L et al (2016) Exosomes derived from human adipose mensenchymal stem cells accelerates cutaneous wound healing via optimizing the characteristics of fibroblasts. Sci Rep 6:32993. https://doi.org/10.1038/srep32993

Jacobs SA, Roobrouck VD, Verfaillie CM, Van Gool SW (2013) Immunological characteristics of human mesenchymal stem cells and multipotent adult progenitor cells. Immunol Cell Biol 91:32–39. https://doi.org/10.1038/icb.2012.64

Juan T, Furthauer M (2018) Biogenesis and function of ESCRT-dependent extracellular vesicles. Semin Cell Dev Biol 74:66–77. https://doi.org/10.1016/j.semcdb.2017.08.022

Jun EK et al (2014) Hypoxic conditioned medium from human amniotic fluid-derived mesenchymal stem cells accelerates skin wound healing through TGF-beta/SMAD2 and PI3K/Akt pathways. Int J Mol Sci 15:605–628. https://doi.org/10.3390/ijms15010605

Kato J et al (2014) Mesenchymal stem cells ameliorate impaired wound healing through enhancing keratinocyte functions in diabetic foot ulcerations on the plantar skin of rats. J Diabetes Complicat 28:588–595. https://doi.org/10.1016/j.jdiacomp.2014.05.003

Keshtkar S, Azarpira N, Ghahremani MH (2018) Mesenchymal stem cell-derived extracellular vesicles: novel frontiers in regenerative medicine. Stem Cell Res Ther 9(63):777–780. https://doi.org/10.1186/s13287-018-0791-7

Kim SW, Zhang HZ, Guo L, Kim JM, Kim MH (2012) Amniotic mesenchymal stem cells enhance wound healing in diabetic NOD/SCID mice through high angiogenic and engraftment capabilities. PLoS One 7:e41105. https://doi.org/10.1371/journal.pone.0041105

Kucharzewski M, Rojczyk E, Wilemska-Kucharzewska K, Wilk R, Hudecki J, Los MJ (2019) Novel trends in application of stem cells in skin wound healing. Eur J Pharmacol 843:307–315. https://doi.org/10.1016/j.ejphar.2018.12.012

Lai RC et al (2016) MSC secretes at least 3 EV types each with a unique permutation of membrane lipid, protein and RNA. J Extracell Vesicles 5:29828. https://doi.org/10.3402/jev.v5.29828

Landen NX, Li D, Stahle M (2016) Transition from inflammation to proliferation: a critical step during wound healing. Cell Mol Life Sci 73:3861–3885. https://doi.org/10.1007/s00018-016-2268-0

Langrzyk A, Nowak WN, Stepniewski J, Jazwa A, Florczyk-Soluch U, Jozkowicz A, Dulak J (2018) Critical view on Mesenchymal stromal cells in regenerative medicine. Antioxid Redox Signal 29:169–190. https://doi.org/10.1089/ars.2017.7159

Lee C et al (2012) Exosomes mediate the cytoprotective action of mesenchymal stromal cells on hypoxia-induced pulmonary hypertension. Circulation 126:2601–2611. https://doi.org/10.1161/CIRCULATIONAHA.112.114173

Leibacher J, Henschler R (2016) Biodistribution, migration and homing of systemically applied mesenchymal stem/stromal cells. Stem Cell Res Ther 7(7):1–12. https://doi.org/10.1186/s13287-015-0271-2

Li X et al (2016) Exosome derived from human umbilical cord Mesenchymal stem cell mediates MiR-181c attenuating burn-induced excessive inflammation. EBioMedicine 8:72–82. https://doi.org/10.1016/j.ebiom.2016.04.030

Li X et al (2018) Exosomes from adipose-derived stem cells overexpressing Nrf2 accelerate cutaneous wound healing by promoting vascularization in a diabetic foot ulcer rat model. Exp Mol Med 50(29):1–14. https://doi.org/10.1038/s12276-018-0058-5

Liu C, Su C (2019) Design strategies and application progress of therapeutic exosomes. Theranostics 9:1015–1028. https://doi.org/10.7150/thno.30853

Liu L et al (2014) Human umbilical cord mesenchymal stem cells transplantation promotes cutaneous wound healing of severe burned rats. PLoS One 9:e88348. https://doi.org/10.1371/journal.pone.0088348

Maas SLN, Breakefield XO, Weaver AM (2017) Extracellular vesicles: unique intercellular delivery vehicles. Trends Cell Biol 27:172–188. https://doi.org/10.1016/j.tcb.2016.11.003

Maharlooei MK et al (2011) Adipose tissue derived mesenchymal stem cell (AD-MSC) promotes skin wound healing in diabetic rats. Diabetes Res Clin Pract 93:228–234. https://doi.org/10.1016/j.diabres.2011.04.018

Makela T et al (2015) Safety and biodistribution study of bone marrow-derived mesenchymal stromal cells and mononuclear cells and the impact of the administration route in an intact porcine model. Cytotherapy 17:392–402. https://doi.org/10.1016/j.jcyt.2014.12.004

Marquez-Curtis LA, Janowska-Wieczorek A, McGann LE, Elliott JA (2015) Mesenchymal stromal cells derived from various tissues: biological, clinical and cryopreservation aspects. Cryobiology 71:181–197. https://doi.org/10.1016/j.cryobiol.2015.07.003

Murphy DE, de Jong OG, Brouwer M, Wood MJ, Lavieu G, Schiffelers RM, Vader P (2019) Extracellular vesicle-based therapeutics: natural versus engineered targeting and trafficking. Exp Mol Med 51(32):1–12. https://doi.org/10.1038/s12276-019-0223-5

Nie C, Yang D, Xu J, Si Z, Jin X, Zhang J (2011) Locally administered adipose-derived stem cells accelerate wound healing through differentiation and vasculogenesis. Cell Transplant 20:205–216. https://doi.org/10.3727/096368910X520065

Phan J, Kumar P, Hao D, Gao K, Farmer D, Wang A (2018) Engineering mesenchymal stem cells to improve their exosome efficacy and yield for cell-free therapy. J Extracell Vesicles 7:1522236. https://doi.org/10.1080/20013078.2018.1522236

Phinney DG, Pittenger MF (2017) Concise review: MSC-derived Exosomes for cell-free therapy. Stem Cells 35:851–858. https://doi.org/10.1002/stem.2575

Pittenger MF et al (1999) Multilineage potential of adult human mesenchymal stem cells. Science 284:143–147

Rad F, Ghorbani M, Mohammadi Roushandeh A, Habibi Roudkenar M (2019) Mesenchymal stem cell-based therapy for autoimmune diseases: emerging roles of extracellular vesicles. Mol Biol Rep 46:1533–1549. https://doi.org/10.1007/s11033-019-04588-y

Raisch J, Darfeuille-Michaud A, Nguyen HT (2013) Role of microRNAs in the immune system, inflammation and cancer. World J Gastroenterol 19:2985–2996. https://doi.org/10.3748/wjg.v19.i20.2985

Rani S, Ryan AE, Griffin MD, Ritter T (2015) Mesenchymal stem cell-derived extracellular vesicles: toward cell-free therapeutic applications. Mol Ther 23:812–823. https://doi.org/10.1038/mt.2015.44

Reinke JM, Sorg H (2012) Wound repair and regeneration. Eur Surg Res 49:35–43. https://doi.org/10.1159/000339613

Ren K (2018) Exosomes in perspective: a potential surrogate for stem cell therapy. Odontology 107(3):271–284. https://doi.org/10.1007/s10266-018-0395-9

Samsonraj RM, Raghunath M, Nurcombe V, Hui JH, van Wijnen AJ, Cool SM (2017) Concise review: multifaceted characterization of human Mesenchymal stem cells for use in regenerative medicine. Stem Cells Transl Med 6:2173–2185. https://doi.org/10.1002/sctm.17-0129

Shaw TJ, Martin P (2016) Wound repair: a showcase for cell plasticity and migration. Curr Opin Cell Biol 42:29–37. https://doi.org/10.1016/j.ceb.2016.04.001

Shi H et al (2017) 3,3'-Diindolylmethane stimulates exosomal Wnt11 autocrine signaling in human umbilical cord mesenchymal stem cells to enhance wound healing. Theranostics 7:1674–1688. https://doi.org/10.7150/thno.18082

Sjoqvist S et al (2019) Exosomes derived from clinical-grade oral mucosal epithelial cell sheets promote wound healing. J Extracell Vesicles 8:1565264. https://doi.org/10.1080/20013078.2019.1565264

Tang TT, Lv LL, Lan HY, Liu BC (2019) Extracellular vesicles: opportunities and challenges for the treatment of renal diseases. Front Physiol 10:226. https://doi.org/10.3389/fphys.2019.00226

Tao SC, Guo SC, Li M, Ke QF, Guo YP, Zhang CQ (2017) Chitosan wound dressings incorporating Exosomes derived from MicroRNA-126-overexpressing Synovium Mesenchymal stem cells provide sustained release of Exosomes and heal full-thickness skin defects in a diabetic rat model. Stem Cells Transl Med 6:736–747. https://doi.org/10.5966/sctm.2016-0275

Toh WS, Zhang B, Lai RC, Lim SK (2018) Immune regulatory targets of mesenchymal stromal cell exosomes/small extracellular vesicles in tissue regeneration. Cytotherapy 20:1419–1426. https://doi.org/10.1016/j.jcyt.2018.09.008

Valadi H, Ekstrom K, Bossios A, Sjostrand M, Lee JJ, Lotvall JO (2007) Exosome-mediated transfer of mRNAs and microRNAs is a novel mechanism of genetic exchange between cells. Nat Cell Biol 9:654–659. https://doi.org/10.1038/ncb1596

van Niel G, D'Angelo G, Raposo G (2018) Shedding light on the cell biology of extracellular vesicles. Nat Rev Mol Cell Biol 19:213–228. https://doi.org/10.1038/nrm.2017.125

Vizoso FJ, Eiro N, Cid S, Schneider J, Perez-Fernandez R (2017) Mesenchymal stem cell Secretome: toward cell-free therapeutic strategies in regenerative medicine. Int J Mol Sci 18:1852. https://doi.org/10.3390/ijms18091852

Wang L et al (2017) Exosomes secreted by human adipose mesenchymal stem cells promote scarless cutaneous repair by regulating extracellular matrix remodelling. Sci Rep 7:13321. https://doi.org/10.1038/s41598-017-12919-x

Wang M, Yuan Q, Xie L (2018) Mesenchymal stem cell-based immunomodulation: properties and clinical application. Stem Cells Int 2018:3057624. https://doi.org/10.1155/2018/3057624

Wang C et al (2019) Engineering bioactive self-healing antibacterial Exosomes hydrogel for promoting chronic diabetic wound healing and complete skin regeneration. Theranostics 9:65–76. https://doi.org/10.7150/thno.29766

Watt SM, Pleat JM (2018) Stem cells, niches and scaffolds: applications to burns and wound care. Adv Drug Deliv Rev 123:82–106. https://doi.org/10.1016/j.addr.2017.10.012

Wu P, Zhang B, Shi H, Qian H, Xu W (2018) MSC-exosome: a novel cell-free therapy for cutaneous regeneration. Cytotherapy 20:291–301. https://doi.org/10.1016/j.jcyt.2017.11.002

Zarrintaj P et al (2017) Can regenerative medicine and nanotechnology combine to heal wounds? The search for the ideal wound dressing. Nanomedicine (London) 12:2403–2422. https://doi.org/10.2217/nnm-2017-0173

Zhang B et al (2015) HucMSC-exosome mediated-Wnt4 signaling is required for cutaneous wound healing. Stem Cells 33:2158–2168. https://doi.org/10.1002/stem.1771

Zhang B et al (2016) HucMSC exosome-delivered 14-3-3zeta orchestrates self-control of the Wnt response via modulation of YAP during cutaneous regeneration. Stem Cells 34:2485–2500. https://doi.org/10.1002/stem.2432

Zhu Z et al (2019) Exosomes derived from human umbilical cord mesenchymal stem cells accelerate growth of VK2 vaginal epithelial cells through MicroRNAs in vitro. Hum Reprod 34:248–260. https://doi.org/10.1093/humrep/dey344

Mass Spectrometry-based Metabolomics in Translational Research

Su Jung Kim, Ha Eun Song, Hyo Yeong Lee, and Hyun Ju Yoo

Introduction

The metabolome represents the collection of all metabolites in a cell and includes all biomolecules, except for the genome, transcriptome, and proteome, and metals. Conventionally, low molecular weight biomolecules (<1500 Da) involved in endogenous metabolism are called metabolites. They play important roles as energy sources, signaling molecules, and metabolic intermediates in complex biological systems. Metabolite levels provide collective information on biomedical states and an instantaneous snapshot of biological responses caused by genetic and environmental perturbations. Metabolite signals result from the interplay of biochemical reactions across the genome, transcriptome, and proteome, and serve as biological modulators across multilayer omics to maintain cellular homeostasis (Yugi and Kuroda 2018). Thus, the metabolome is crucial to understand biological responses to diseases, and genetic and environmental changes.

Metabolites and their related features have been used as diagnostic markers since ancient times. Ancient people monitored the characteristic odor of an individual's breath in eastern traditional medicine for thousands of years. Recently, the sweet odor of a diabetic patient's breath was shown to be caused by the presence of acetone, which is related to high blood glucose level, the gold standard currently used in the diagnosis of diabetes (Wang and Wang 2013). In the Middle Ages, a urine chart was used to link the color, smell, and taste of urine to various medical conditions (Nicholson and Lindon 2008). Such features are based on the chemical patterns in response to biological conditions, and metabolites can represent these biochemical patterns with quantifiable data. Metabolites can define the molecular

S. J. Kim · H. E. Song · H. Y. Lee · H. J. Yoo (✉)
Department of Convergence Medicine, Asan Institute for Life Sciences, Asan Medical Center, University of Ulsan College of Medicine, Seoul, South Korea
e-mail: hes1290@snu.ac.kr; yoohyunju@amc.seoul.kr

J. K. Kim et al. (eds.), *Advanced Imaging and Bio Techniques for Convergence Science*, Advances in Experimental Medicine and Biology,
https://doi.org/10.1007/978-981-33-6064-8_19

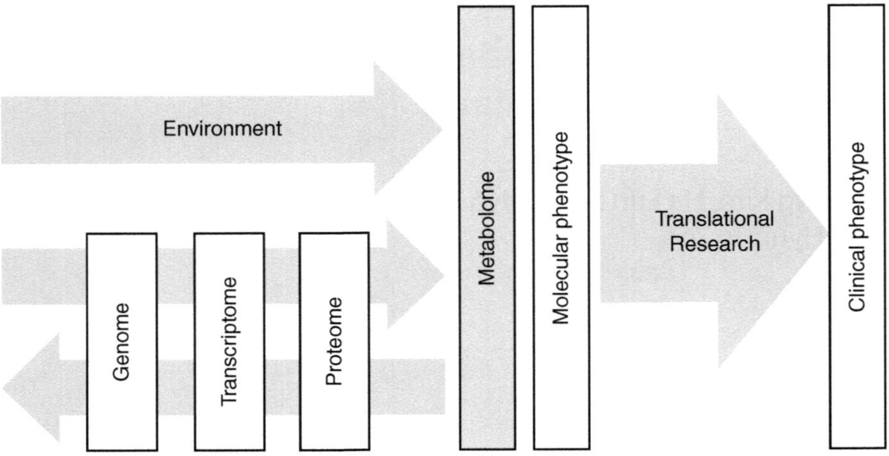

Fig. 1 Metabolomics provides the integrated information caused by complex biochemical reactions and environment, thus plays an important role in translational research

phenotypes of living organisms and be directly associated with the biological outcomes of diseases. In addition, the concentrations of metabolites are often highly correlated with biochemically related genetic variations (Gieger et al. 2008; Fendt et al. 2010; Bondia-Pons et al. 2011; Suhre et al. 2011a, b; Vander Heiden 2011; Jain et al. 2012). These characteristics allow metabolome alterations to be easily translated into disease states and can help us understand the pathophysiological mechanisms that contribute to various biomedical symptoms. Thus, translational research has included the metabolome and metabolomics (Fig. 1).

Metabolomics involves the systematic identification and quantification of metabolites. The physicochemical properties of metabolites vary widely, including in their polarity, acidity, and volatility. In addition, most metabolites consist of four to five atoms, including carbon, nitrogen, oxygen, hydrogen, and phosphorous, and many metabolites have the same elemental composition with slight variations in chemical bonds. Thus, the identification and quantification of metabolites are both difficult and challenging. Metabolomics has been implemented with various analytical platforms to identify and quantify diverse intra- and extracellular metabolites.

Methods for metabolite analysis include nuclear magnetic resonance (NMR) spectroscopy, mass spectrometry (MS), Fourier transform-infrared (FT-IR) spectroscopy, high performance liquid chromatography (HPLC), gas chromatography (GC), Raman spectroscopy, and other analytical platforms (Fig. 2) (Bogdanov and Smith 2005; Want et al. 2005; Villas-Boas et al. 2005; Boskey and Mendelsohn 2005; Defernez and Colquhoun 2003; Deleris and Petibois 2003). FT-IR and Raman spectroscopies are nondestructive and rapid techniques that can analyze various types of biological samples. Absorption spectra at specific wavelengths provide important clues crucial for determining the structure of unknown metabolic features, whereas the peak area under the curve of absorption spectra is used for quantitation. The sensitivity and selectivity of FT-IR and Raman spectroscopy, however,

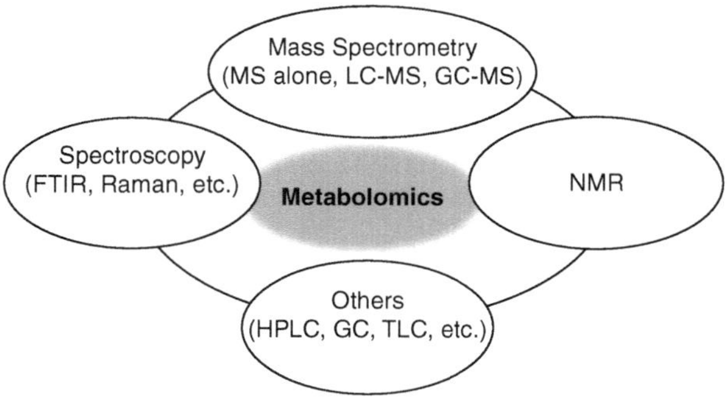

Fig. 2 Analytical instrumentations for metabolomics should be able to detect various kinds of metabolites present in biological systems

are not as high as those of other methods (Boskey and Mendelsohn 2005; Deleris and Petibois 2003). The detection methods used in HPLC and GC, such as UV absorption, electrochemical detection, and flame ionization, are non-selective, thus complete separation of metabolites is crucial in accurate quantitation of metabolites, especially from complex biological samples (Gathungu et al. 2014; Keyfi and Varasteh 2016; Richins et al. 2018). NMR spectroscopy, another rapid and nondestructive analytical method, requires negligible sample preparation. Chemical shifts dependent on the nucleus' chemical environment can be used to identify metabolites. NMR spectroscopy, however, is relatively insensitive, and millimolar to high micromolar concentrations are often required. Thus, only the most abundant metabolites are detected (Emwas 2015). Because of its high sensitivity and broad dynamic range, MS has become the most suitable analytical tool in metabolomics research (Johnson et al. 2016). MS can also detect various types of metabolites because mass is a universal property. Selectivity among isobaric (identical masses) metabolites can be ensured by various MS/MS techniques and by combination with separation modules such as GC or LC (Metz et al. 2007; Lei et al. 2011; Xiao et al. 2012; Fiehn 2016). MS-based metabolomics platforms have become more popular than NMR-based platforms in recent research (Fig. 3). Thus, this chapter will emphasize MS-based metabolomics.

Analytical Platforms for Mass Spectrometry-based Metabolomics

MS is an analytical method that measures charged molecules based on their mass-to-charge ratios. The signal intensities of charged metabolites reflect the amounts of them present in biological samples. The mass analyzer, a primary component of the MS system, separates molecules based on their mass to charge ratios. Several types of mass analyzers are currently available, including quadrupole, time of flight (TOF), ion-trap, orbitrap, and Fourier transform-ion cyclotron resonance (FT-ICR). High resolution mass spectrometry (HRMS) methods, such as TOF, orbitrap, and

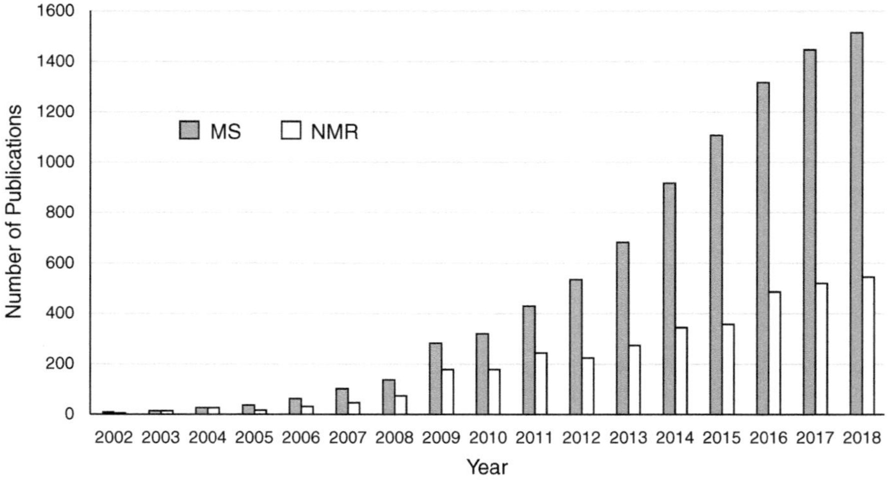

Fig. 3 Publications related to MS- or NMR-based metabolomics. PubMed search was performed with keywords of (MS and metabolomics) or (NMR and metabolomics)

FT-ICR, measure the exact masses of charged molecules. The exact mass is the theoretical mass of specific isotopic composition of a charged molecule (Murray et al. 2013). By contrast, low resolution mass spectrometry methods, such as quadrupole and ion-trap, measure the nominal mass, defined by the integer mass of the most abundant stable isotope of a molecular ion (Murray et al. 2013).

Various tandem mass spectrometric techniques (MS/MS) have been used to generate unique fragmentation patterns from different metabolites, which enables the specific characterization of unknown metabolites or differentiation among isobaric metabolites (Ceglarek et al. 2009). Upon collision with a neutral gas or interaction with activated electrons, any specific metabolite isolated from a mass analyzer produces characteristic fragment ions which are measured by a second mass analyzer. Sequential fragmentation (MS^n) can be performed as needed. Several MS/MS methods are available for fragmenting molecular ions for tandem mass spectrometry (Fig. 4). Collision induced dissociation (CID) is the most common MS/MS technique, and fragment ions are generated from collision with neutral gases (Xiao et al. 2012; Wang et al. 2008). By contrast, electron capture dissociation (ECD), electron-induced dissociation (EID), and electron transfer dissociation (ETD) are electron-based MS/MS methods, and their fragmentation patterns are normally different from those generated by CID (Johnson et al. 2016; Ongay et al. 2013; Yoo et al. 2007, 2011). The structural information collected from different MS/MS methods is often complementary and crucial in revealing the identities of unknown metabolites (Yoo et al. 2007; Liang et al. 2007; Liu et al. 2008).

Chromatography-based separation modules are often combined with MS systems in MS-based metabolomics. Liquid chromatography-mass spectrometry (LC-MS) and gas chromatography-mass spectrometry (GC-MS) have several advantages over MS alone. The matrix effect and ion suppression, caused by salts, ion pairing

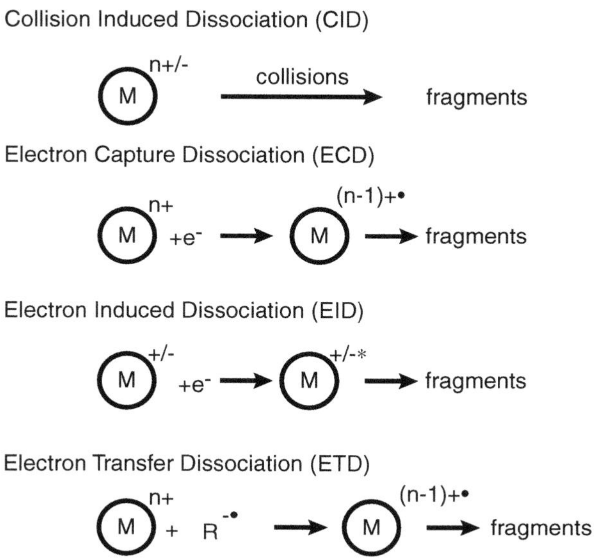

Fig. 4 Various tandem mass spectrometric techniques used for the structural characterization of metabolites

Table 1 Comparison of GC-MS and LC-MS based metabolomics

	GC-MS based metabolomics	LC-MS based metabolomics
Metabolites	Relatively nonpolar, small metabolites	Wide range of metabolites
Chemical derivatization	Generally necessary	Generally not necessary
Total run time	Longer run time (generally >1 h)	Shorter run time (generally 10–40 min)
Retention time drift	Rarely observed	Often observed
Ion source	EI (Electron impact)	ESI (Electrospray ionization)
Metabolite identification	Mass fingerprint generated from one metabolite is used	Molecular ion and/or its specific fragment ions during MS/MS (CID) are used
Dynamic range for quantitation	Shorter dynamic range ($\tilde{\ }10^2$)	Wide dynamic range ($\tilde{\ }10^{3-4}$)

agents, and any other interfering endogenous compounds in a complex biological matrix, can be alleviated by appropriate chromatographic separation among analytes and interfering compounds. In addition, metabolites in biological samples can be quantitated by measuring their chromatographic peak areas. GC-MS has been utilized to profile relatively more volatile and lower molecular weight metabolites, and chemical derivatization of metabolites is commonly required (Fiehn 2016; Papadimitropoulos et al. 2018; Garcia and Barbas 2011). By contrast, metabolites being analyzed by LC-MS are generally not subjected to chemical derivatization, and LC-MS has been applied to various types of metabolites with broad ranges of physicochemical properties and molecular weights (Table 1) (Zhou et al. 2012;

Lu et al. 2008; Theodoridis and Wilson 2008). Capillary electrophoresis (CE) can be used with MS, especially for highly polar metabolites (Ramautar et al. 2009, 2017; Zhang et al. 2017).

An ion source is a device that produces molecular ions in a mass spectrometer. Commonly used ion sources in metabolomics are electrospray ionization (ESI), electron impact ionization (EI), and desorption electrospray ionization (DESI). ESI, which links LC to MS, uses high temperature to evaporate solvents and high voltage to generate charged molecular ions from metabolites. ESI is a soft ionization technique and produces molecular ions with very little fragmentation, which is advantageous for database searches to identify unknown metabolites (Ho et al. 2003; Fenn et al. 1989; Yamashita and John 1984). EI is a harsh ionization technique in which high energy electrons (typically 70 eV) interact with molecules in the gas phase, and extensive fragmentation specific to each metabolite is generated. EI is used as an ion source in GC-MS, and is generally useful for small organic molecules (<600 Da) (Mark and Dunn 2013). DESI is a recently developed ionization technique, in which a fast-moving charged solvent stream extracts metabolites from the surfaces of samples (Takats et al. 2004). Like matrix-assisted laser desorption ionization (MALDI) and secondary ion mass spectrometry (SIMS), the DESI source is useful in revealing the location of selected metabolites on tissue samples (Petras et al. 2017). These techniques have allowed spatial localization and visualization of molecular distribution of target molecules (Fessenden 2016; Claude et al. 2017; Sun et al. 2019).

Metabolomics can be performed using two different approaches (Table 2). Nontargeted metabolomics can be performed to identify as many metabolites as possible in a biological sample, with the results generating novel hypotheses and/or metabolite targets. By contrast, targeted metabolomics can provide quantitative information for target metabolites involved in specific metabolic pathways. Targeted metabolomics is used to answer specific biochemical questions and hypotheses. These two approaches will be discussed in greater detail below.

Table 2 Nontargeted metabolomics and targeted metabolomics

	Nontargeted metabolomics	Targeted metabolomics
Purpose	To generate hypothesis	To prove hypothesis
	To find metabolic features related to disease pathophysiology	To confirm metabolite levels in any disease states
Targets	Not necessary	Necessary
Analytical platform	Universal methods are preferable to observe as many metabolic features as possible	Specific quantification methods are needed to measure target metabolites
Outcomes	Statistically meaningful changes in any metabolic features or metabolic pathways related to disease states or perturbations	Quantitative result of target metabolites for the specific disease states or perturbations
Validation	Necessary	Not necessary, but related biological experiments may be useful to interpret the result

Nontargeted Metabolomics

The purpose of nontargeted metabolomics is to measure the entire metabolome in a biological sample, and to generate a hypothesis. However, it is not possible to measure all constituents of an entire metabolome from a biological sample, because their amounts and physicochemical properties are quite diverse. Thus, the number of metabolites identified in a sample is dependent on sample preparation and analytical instrumentation. Because accurate measurement of mass is critical in identifying unknown metabolites, high resolution MS, such as TOF, orbitrap, or FT-ICR, should be utilized in nontargeted metabolomics. Advanced instrumentation and bioinformatics tools are important in nontargeted metabolomics. The general workflow of global metabolome profiling involves sample preparation, instrumental analysis, data analysis using various bioinformatics tools, and verification including biological interpretation (Fig. 5) (Want et al. 2013; Dunn et al. 2011; Osborn et al. 2013).

Sample preparation is designed to reduce the metabolome pool or sample matrices, based on the physicochemical properties of metabolites. Metabolites from a complex biological sample can be separated into several fractions with similar physicochemical properties. Initially, a cold organic solvent is added to a biological fluid to precipitate large biomolecules, such as the genome and proteome. This is followed by liquid–liquid extraction, which separates a sample solution into two aliquots, one containing hydrophobic and the other containing hydrophilic metabolites. When necessary, solid particulates from biological samples can be eliminated using a membrane filter.

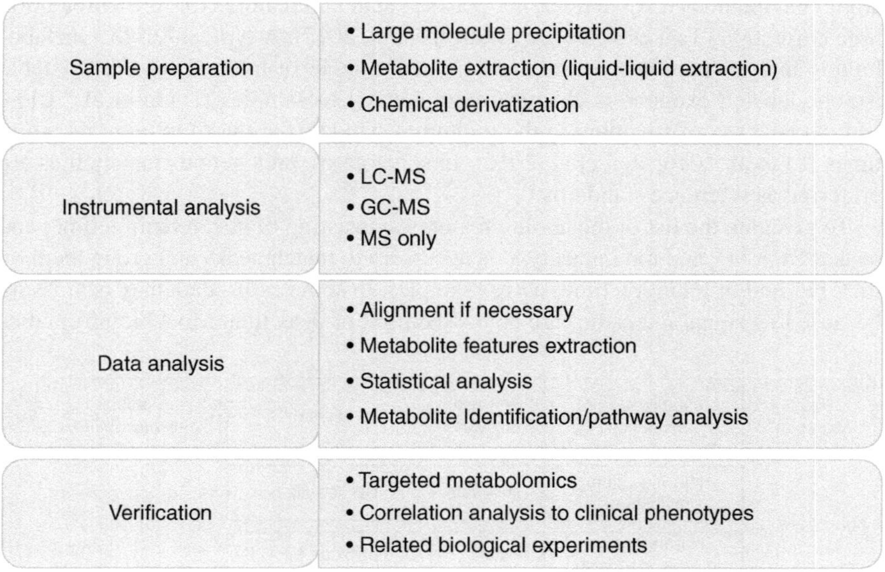

Fig. 5 Workflow of nontargeted metabolomics

Although nontargeted metabolomics using multiple metabolomics platforms may maximize the number of metabolites detected, access to multiple analytical platforms is often difficult. The choice of analytical platforms for nontargeted metabolomics depends on their availability and applications. If LC-MS or CE-MS is available for global metabolome profiling, metabolome profiles can be obtained from both positive and negative ion modes to maximize the number of metabolites observed. Basic and acidic metabolites can be detected better using positive and negative ion mode, respectively.

The data quality of global metabolome profiling is dependent on the types of sample, sample preparation, chromatographic separation, and types of mass analyzer. The data acquired from nontargeted metabolomics include a list of the mass-to-charge ratio (*m/z*) with/without its retention time and signal intensity of each metabolic feature. If GC-MS is used for global metabolome profiling, extensive fragment ion peaks are obtained, rather than a single molecular ion peak of an intact metabolite. A list of metabolite features, consisting of *m/z*, retention time, and peak area or *m/z* and ion intensity, should be generated from the intensive processing of all detected peaks in global metabolome profiling.

The workflow of data processing for global metabolome profiling begins with peak alignment (Fig. 6). Inevitable retention time shifts in chromatograms over runs are often observed and need to be corrected. Two types of peak alignment can be performed, profile-based and feature-based peak alignment (Vandenbogaert et al. 2008; Smith et al. 2015; Katajamaa and Oresic 2007). Profile-based methods align peaks using raw total ion chromatograms (Vahamaa et al. 2011; Tsai et al. 2013a), whereas feature-based methods align peaks after peak detection. Most feature-based alignment methods involve reference peaks as standard features for further retention time correction (Tsai et al. 2013b; Watrous et al. 2017). A typical LC-MS metabolomics analysis can involve spiking a limited number of internal standards (stable isotope-labeled exogenous chemical compounds) to samples (Dunn et al. 2011), and the peaks corresponding to these internal standards are used to align retention times (Li et al. 2016; Ren et al. 2016). In some cases, endogenous metabolites are preferred as reference standards.

To generate the list of metabolite features, consisting of *m/z*, retention time, and peak area or *m/z* and ion intensity, it is necessary to match peaks and group them by their *m/z* and/or retention times across samples. In some cases, data may be missing because low signal intensities are below the limit of detection. Most bioinformatics

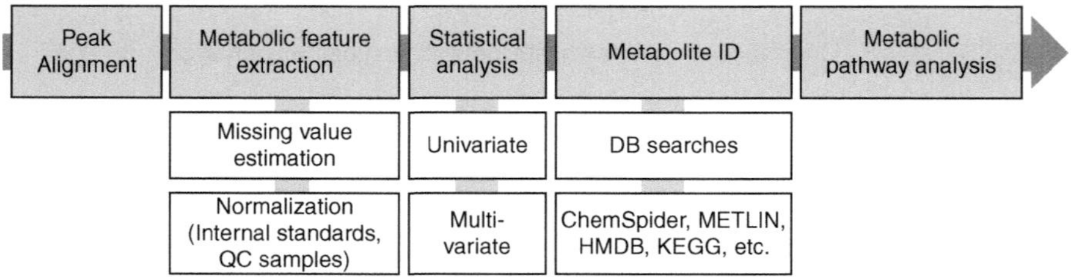

Fig. 6 Workflow of data processing in nontargeted metabolomics

tools used for global metabolomics provide several options to deal with missing values (Katajamaa et al. 2006; Smith et al. 2006; Xia and Wishart 2011; Chong et al. 2018; Forsberg et al. 2018). For example, missing values can be replaced by a very small value, generally half of the minimum value found in the data set, or by estimated values, such as the mean or median. If too many values are missing, the missing values can be automatically excluded. Data filtering can be performed to remove background noises and non-informative features in a data set, improving the statistical power of subsequent data analysis. Non-informative features, consisting of features with constant intensities throughout the experiments, can be identified by comparing statistical analyses, such as standard deviation or the coefficient of variance.

Data are subsequently normalized to reduce any systematic error or bias over analytical experiments, and to allow meaningful biological comparisons. Data normalization can be combined with data transformation and/or data scaling. Data can be normalized using several characteristics, including sum, median, reference sample, or a reference feature (Chong et al. 2018; Forsberg et al. 2018). If all samples are spiked with internal standards, then the internal standards can be used for data normalization as reference features. Data transformation, e.g., log transformation or cube-root transformation, is a method to alter features so that they exhibit a normal or Gaussian distribution. Data scaling is useful when features have very different orders of magnitude. Various scaling methods can be applied, such as auto-scaling, Pareto scaling, and range scaling. There is no optimal strategy for all types of metabolomics data. Users should perform trial-and-error testing by combining various options for data normalization, transformation, and scaling to obtain a Gaussian-shaped data distribution. Overall data quality should be assessed and apparent outliers should be checked prior to statistical analysis of metabolomics data.

Statistical analysis of metabolomics data can be univariate or multivariate, depending on the number of metabolic features. Univariate analysis evaluates differences among groups on a parameter-by-parameter basis, with statistical comparisons including t-tests and analysis of variance (ANOVA). By contrast, multivariate analysis assesses not only differences in single metabolite among groups of metabolites, but any relevant structures among individual metabolites. Commonly used multivariate analysis techniques are principal component analysis (PCA), cluster analysis, and partial least squares (PLS) regression. PCA and cluster analysis are unsupervised techniques, whereas PLS regression is a supervised technique. Multivariate analysis is able to visualize samples according to their groups or intrinsic similarities. The PCA and PLS techniques use a data matrix to evaluate the response vector using a linear regression model. The variables and metabolic features that maximize the discrimination among sample groups can be determined, and these variables should be confirmed further as significant metabolic features to discriminate among sample groups using targeted metabolomics or other biochemical approaches (Worley 2013; Smkmrmstt 2012).

Metabolite identification is challenging in global metabolomics (Matsuda 2016). Successful metabolite identification requires accurate measurement of the masses of the metabolic features observed, and database searches to find accurate matches.

If available, MS/MS spectral libraries can be used for metabolite identification (Bowen and Northen 2010; Vaniya and Fiehn 2015). However, this approach does not result in a definitive identification of compounds, because a single chemical formula may have many valid chemical structures. In addition, the metabolite databases (e.g., KEGG, HMDB, METLIN, and ChemSpider) are not complete, and many of the observed metabolic features remain unidentified (Sas et al. 2015; Sreekumar et al. 2009). Thus, a list of potential candidates would be obtained, and the identification should be confirmed using authentic standards or additional analysis. The process for the identification of all observed metabolic features requires considerable time and effort, suggesting that metabolite identification should be performed on the metabolic features that show statistical significance (Dunn et al. 2011; Evans et al. 2009; Want et al. 2010; Theodoridis et al. 2012).

Another challenge in global metabolome profiling is the changes in signal intensities or retention times over time. Various approaches have been applied to ensure that the results of metabolic profiling studies are valid. These include spiking of internal standards into all samples, use of QC samples, and randomized order of analysis. QC samples can be either pooled samples containing equal aliquots of all samples, or test mixtures containing a limited number of commercially available standards in solutions of a matrix similar to a real sample. QC samples should be used to assess experimental performance such as mass accuracy, signal response, retention time stability, and peak shape. Translational research would include at least tens of clinical samples for global metabolome profiling and may require long-term experimental periods. In that case, QC samples should be analyzed at regular intervals, e.g., every five to ten samples, throughout the analysis, and the reproducibility of the data should be evaluated to ensure that metabolome profiling is reliable (Want et al. 2010; Broadhurst et al. 2018).

Targeted Metabolomics

Targeted metabolomics can be used to obtain quantitative information on target metabolites involved in specific metabolic pathways. Targeted metabolomics is used to answer any specific biochemical questions and hypotheses (Yuan et al. 2012; Quehenberger et al. 2010; Quehenberger and Dennis 2011; Roberts et al. 2012). Sample preparation, chromatographic conditions, and specific mass fragmentation methods can be established or optimized to identify any specific metabolites in an analytical platform of targeted metabolomics. Low-level metabolites or a specific metabolite among several isobaric metabolites can be quantified selectively in complex biological samples. Greater effort is required to develop targeted than nontargeted metabolomics platforms, but, once targeted metabolomics platforms have been established, metabolites in specific metabolic pathways can be quantitated relatively easily. As a result, targeted metabolomics can provide a better understanding and interpretation of a specific hypothesis.

The metabolome from human serum contains about 20 biochemical classes, including amino acids, carnitines, carbohydrates, fatty acids, and bile acids

(Psychogios et al. 2011; Boudah et al. 2014). Many metabolomics platforms should be needed to quantify various metabolites, and novel targeted metabolomics platforms have been developed (Yoo et al. 2007; Yoo and Hakansson 2011; Strassburg et al. 2012; Han et al. 2013; Kim et al. 2017, 2012). Specific MS/MS and chromatographic separation can be useful to discriminate metabolites from isobaric metabolites or compounds having very similar chemical structures (Kim et al. 2014).

To outline the practical procedures encountered in targeted metabolomics, targeted metabolomics platforms focusing on fatty acids and their oxidized derivatives are highlighted. Fatty acids and their oxidized derivatives include short-chain fatty acids, medium and long-chain fatty acids, and eicosanoids. Short-chain fatty acids have different physicochemical and biological characteristics comparing to other fatty acids (Zeng and Cao 2018; Gao et al. 2009a; Cooper et al. 1995; Fushimi et al. 2006). Because short-chain fatty acids are more volatile and more water-soluble than medium and long-chain fatty acids, short-chain fatty acids are usually prepared from biological samples by aqueous extraction (Huda-Faujan et al. 2010). Organic phase extraction can be used to prepare longer chain fatty acids, with >5 carbons, from biological samples. Methyl esterification or trimethylsilylation is a commonly used chemical derivatization method for GC-MS analysis of medium- and long-chain fatty acids. These chemical derivatives of medium- and long-chain fatty acids are generally quantified by GC-MS. Intact molecular ions are rarely observed due to the harsh ionization conditions used by the electron impact ion source of GC-MS. Thus, the specific fragmentation pattern of each metabolite is used for identification and quantification (Antolin et al. 2008; Jayasinghe and Dias 2013). By contrast, short-chain fatty acids can be chemically derivatized using alanine, O-benzylhydroxylamine, methyl-/ethyl-/propyl-chloroformate, or a trimethylsilylation agent, prior to GC-MS or LC-MS analysis (Zeng and Cao 2018; Tao et al. 2008; Gao et al. 2009b; Zheng et al. 2013; Kvitvang et al. 2011; Qiu et al. 2007; Perez et al. 2016; Han et al. 2015; Chan et al. 2017).

Eicosanoids are oxidized derivatives of polyunsaturated fatty acids (PUFAs), and play a key role in human diseases related to inflammation and immune responses. The amounts of eicosanoids and their imbalances affect various biomedical processes and the pathophysiology of human diseases (Wenzel et al. 2007; Sanak et al. 2011; Pavord et al. 1999; Higashi et al. 2002; Dennis and Norris 2015; Huang and Peters-Golden 2008). In general, fatty acids are present at micromolar concentrations in human blood, whereas many eicosanoids are present with less than nanomolar concentrations (Psychogios et al. 2011). Thus, these lipid mediators from biological samples should be specifically extracted and concentrated (Yang et al. 2009, 2011). Eicosanoids share very similar chemical structures and some of them have the same molecular weight (isobaric). Thus, LC separation is crucial to discriminate against these isobaric eicosanoids (Yang et al. 2009). For example, prostaglandin D_2 (PGD_2) and prostaglandin E_2 (PGE_2) have the same elemental compositions, differing only in their stereochemistry. Elaborate optimization of chromatographic separation conditions allows discrimination between PGD_2 and PGE_2. LC-MS/MS has been applied to specifically profile about 40 eicosanoids (Yang et al. 2009).

Targeted metabolomics without authentic standard metabolites can also be useful in exploring human diseases and their pathophysiological mechanisms, despite obtaining relative quantitative results. Quantitative profiling of phospholipids or sphingomyelins is a good example. In CID, fragment ions corresponding to the headgroups of phospholipids or sphingomyelins are commonly obtained in positive ion mode, whereas fragment ions corresponding to fatty acyl chains are obtained in negative ion mode. Thus, LC-MS/MS in both positive and negative ion modes can provide valuable structural information about phospholipids and sphingomyelins (Li et al. 2016; Knittelfelder et al. 2014; Anand et al. 2016). In addition, MS/MS strategies such as neutral loss or precursor ion scan can be useful in profiling different types of phospholipids (Kim et al. 2017, 2012, 2014).

Metabolomics Applications in Translational Research

Metabolites are intermediates involved in various biochemical processes of biological systems. Metabolite levels change rapidly, depending on genomic or environmental perturbations, and can provide instant snapshots of changes throughout the human body. Metabolomics can be used to investigate metabolite–biological interactions and elucidate their roles in biomedical and clinical environments. The metabolome plays a crucial role in clinical applications, with about 95% of clinical assays based on metabolites and their related features (Wu 2006). One representative application is the blood glucose level in diabetic patients and sweetness in their urine samples (Karamanou et al. 2016). Human diseases are associated with adverse interactions of the genome or proteome with the metabolome. For example, many inherited metabolic disorders were found to be related to a specific organic acid or amino acid deficiency (Pitt et al. 2002). Every molecule in a human body may respond to a perturbation, and a system-wide approach may be needed to explain the biological complexity. Recent tangible advances in analytical platforms have resulted in metabolomics becoming the optimal strategy for translational research. Like other omics, metabolomics has a great impact on biomedical and translational research, especially in the areas of biomarker discovery and drug development. Metabolomics analyses can enhance understanding of the mechanisms underlying diseases or the adverse influence of exposure to toxic substances, and provide crucial clues to identify novel targets in drug development or treatment strategies. Thus, metabolomics should be a valuable tool in translational studies of novel drugs and treatments, of safety assessments, and in identifying clinical biomarkers for monitoring or diagnosing disease.

Diabetes

One of the earliest metabolomics studies in diabetes was the comparison of phospholipids in the plasma of patients with diabetes mellitus type 2 and normal controls (Wang et al. 2005). The study reported that metabolites could distinguish between patients and normal controls using multivariate statistical analysis such as PLS and PCA. Moreover, the study identified several metabolites as potential biomarkers for group discrimination. A metabolomics analysis using LC-MS/MS and H-NMR to investigate metabolic pathways changed by insulin deficiency found that metabolic pathways related to amino acid oxidation, mitochondrial bioenergetics, and gluconeogenesis were altered due to insulin deficiency (Lanza et al. 2010). In another study, metabolome profiling revealed that metabolic signatures related to branched chain amino acids were changed due to obesity and contributed to insulin resistance (Newgard et al. 2009). Furthermore, monitoring of metabolome profiles in control and prediabetic groups for 12 years identified five branched chain and aromatic amino acids as possible prognostic biomarkers for an earlier risk of developing diabetes (Wang et al. 2011). The significance of these metabolites was confirmed using independent groups of patients. An evaluation of the predictive performance of these amino acids as prognostic biomarkers for the risk of diabetes involved the quantification of amino acids, nucleotide metabolites, and urea cycle metabolites using targeted metabolomics platforms. Use of nontargeted metabolomics to explore metabolome changes in the urine and plasma of prediabetic subjects showed that tryptophan, uric acid, fatty acids, bile acids, and lysophospholipids were major metabolites related to the prediabetes-associated alterations (Zhao et al. 2010). Analysis of volatile organic compounds (VOCs) from human exhaled breath has been also investigated in the diagnosis of diabetes. Acetone is produced in mammals by lipolysis or amino acid degradation, and elevated levels of acetone have been reported in the exhaled breath of diabetes mellitus patients (Das et al. 2016; Lebovitz 1995).

Brain

The high-throughput quantitative nature of metabolomics studies can lead to a comprehensive mapping of mammalian brain function. These results are expected to lead to the development of sensitive and accurate diagnostic tools and the design of personalized therapeutic treatments. Metabolomics has been applied to the study of central nervous system (CNS) physiology and pathophysiology to better understand the metabolic complexity of the CNS and the onset, progression, and treatment of multifactorial neurodegenerative diseases (Vasilopoulou et al. 2016). The blood–brain barrier (BBB) is a contraindication to the analysis of blood in brain biology. Rather, cerebrospinal fluid (CSF) may better reflect brain physiology. Metabolomics studies and genetic validation using CSF and serum samples from hundreds of patients and healthy controls found that cerebral metabolism of tryptophan was closely related to outcomes in patients with tuberculous meningitis (van Laarhoven et al. 2018). The finding suggested that the tryptophan metabolic pathway may be a novel drug target in patients with tuberculous meningitis, and interventions targeting tryptophan metabolism may benefit these patients. Metabolome changes in multiple sclerosis have also been explored. Global metabolome profiling identified that metabolomics signatures, including hormones, lipids, and amino acids, were associated with the severity of multiple sclerosis (Villoslada et al. 2017). Vitamin D supplementation had different effects on metabolomes induced by oxidative stress and xenobiotic metabolism in multiple sclerosis patients and healthy controls (Bhargava et al. 2017). Metabolomics also revealed that chronic inflammation activated the kynurenine pathway, exacerbating the progression of multiple sclerosis. These results were validated in an independent cohort, suggesting that kynurenine metabolites in patient serum were biomarkers of multiple sclerosis (Lim et al. 2017). Glycerosphingolipids, including cerebrosides, which are important myelin lipids in the brain, were assessed to determine whether myelin membrane lipids interacted with infection and immune responses in demyelinating neurological diseases (Yang et al. 2011; Bergholt et al. 2018).

Fibrosis

Fibrosis results in the excess formation of fibrous connective tissues and interferes with the normal architecture and/or function of affected organs and tissues (Birbrair et al. 2014; Neary et al. 2015). Fibrosis develops differently in different individuals and is often related to wound healing response or disease severity (Vilar-Gomez et al. 2018; Wynn 2008; Leask and Abraham 2004). However, reliable molecular biomarkers to predict or diagnose fibrosis susceptibility are still lacking. Disturbance in mitochondrial homeostasis was reported in radiation-induced fibrosis, suggesting that mitochondrial dysfunction was closely related to fibrosis through metabolic perturbations such as lipid accumulation due to reduced fatty acid oxidation (Maeda 1982). Fatty acids and bile acids in fibrotic plasma were changed in cystic fibrosis, suggesting abnormal lipid metabolism in patients with cystic fibrosis (Guilbault et al. 2009; Teichgraber et al. 2008). In addition, the lipid mediator sphingosine-1-phosphate was found to be elevated in the human fibrotic liver, due to upregulation of sphingosine kinase (Li et al. 2011). Disturbed lipid metabolism in the liver was found to result in inflammation and fibrosis (Moustafa et al. 2012). Moreover, a high-fat diet resulted in cardiac fibrosis even before the development of obesity and hyperlipidemia (Aubin et al. 2008). In addition, eicosanoids were reported to be involved in cardiac fibrosis (Levick et al. 2007). Changes in phosphatidic acid and lysophosphatidic acid were observed in the bronchoalveolar lavage fluids of patients with idiopathic pulmonary fibrosis (Crow and Wakeland 2012). Branched chain amino acids such as leucine and valine were reduced in liver fibrosis via TGF_{β} inhibition (Cha et al. 2013). Exhaled ethane and pentane of systemic origin are produced by lipid peroxidation, a chain reaction induced by reactive oxygen species (ROS) (Miekisch et al. 2004; Larstad et al. 2007; Sarbach et al. 2013). Destructive oxidative stress was found to damage cells (Ross et al. 2011). Elevated levels of exhaled ethane in human breath have been reported in patients with asthma, chronic obstructive pulmonary diseases (COPDs), and cystic fibrosis (Paredi et al. 2000a, b, c). In addition, other VOCs, such as saturated hydrocarbons, were found in exhaled breath condensates of patients with COPD and idiopathic pulmonary fibrosis (Psathakis et al. 2006; Cazzola et al. 2015). These studies indicate that metabolomics is likely a promising tool in identifying biomarkers of various lung diseases and in elucidating disease mechanisms.

Cancer

Most cancer cells utilize highly activated glycolysis to produce energy or substrates for rapid cell proliferation. This pathway is characterized by lactic acid accumulation, also called the Warburg effect (Liberti and Locasale 2016), which is responsible for adaptation to low oxygen and damage to mitochondria in a cancer environment (Warburg 1956). Metabolic modifications have been observed in tumors (Kim et al. 2009; Griffin and Shockcor 2004), and metabolic changes in a cancer environment may involve energy metabolism, lipid metabolism, and nucleotide metabolism (Kim et al. 2015). In addition, investigation of eicosanoids would be very useful in understanding tumor development, progression, and metastasis (Wang and Dubois 2010). However, the roles of lipid mediators in tumors remain incompletely understood. Lipidome profiling showed that fatty acid synthesis was activated in breast cancer, possibly for the synthesis of membrane phospholipids (Hilvo et al. 2011). Exhaled breath was explored as a noninvasive tool for early detection of lung cancer and characterization of suspicious lung nodules (Nardi-Agmon and Peled 2017; Phillips et al. 2015; Capuano et al. 2015; Peled et al. 2012; Broza et al. 2013; Shehada et al. 2016). In one study, metabolome profiles were assessed in 968 breath samples from 484 patients with gastric cancer to identify diagnostic markers for reduced cancer incidence and mortality. Eight significant VOCs in breath were found to correlate with cancer risk, suggesting the use of breath tests for follow-up surveillance of high-risk patients (Amal et al. 2015).

Conclusion

Metabolomics is a relatively new omics strategy, and is increasingly explored in translational research. Genomics can predict overall disease risk and potential drug responses, but has limitations in assessing alterations due to changes in diet and environmental factors. By contrast, metabolomics can quickly detect biochemical changes associated with specific disease states or detrimental environments, and provide valuable clues for novel drug targets or therapeutic interventions.

Significant time and effort are needed to identify metabolite biomarkers for diagnosis and surveillance that can replace traditional methods. These metabolite biomarkers may constitute a powerful tool to aid or enhance the early detection of primary or recurrent disease as well as for the reliable characterization of suspected disease signatures. In addition, an in-depth understanding of metabolism related to disease progression may help identify novel targets for treatment. Thus, metabolomics is expected to flourish as a valuable research platform in translational research.

Acknowledgments This study was supported by a grant (2018IL0540, 2019IL0540) from the Asan Institute of Life Sciences, Asan Medical Center, Seoul and by the National Research Foundation of Korea (NRF) grant funded by the Korea government (MSIT) (2019R1F1A1056286).

Conflicts of Interest The authors declare no competing interests relevant to this chapter.

References

Amal AL, Funka K, Skapars R, Sivins A, Ancans G, Liepniece-Karele I, Kikuste I, Lasina I, Haick H (2015) Detection of precancerous gastric lesions and gastric cancer through exhaled breath. Gut 65:1–8

Anand S, Young S, Esplin MS, Peaden B, Tolley HD, Porter TF et al (2016) Detection and confirmation of serum lipid biomarkers for preeclampsia using direct infusion mass spectrometry. J Lipid Res 57(4):687–696

Antolin EM, Delange DM, Canavaciolo VG (2008) Evaluation of five methods for derivatization and GC determination of a mixture of very long chain fatty acids (C24:0-C36:0). J Pharm Biomed Anal 46(1):194–199

Aubin MC, Lajoie C, Clement R, Gosselin H, Calderone A, Perrault LP (2008) Female rats fed a high-fat diet were associated with vascular dysfunction and cardiac fibrosis in the absence of overt obesity and hyperlipidemia: therapeutic potential of resveratrol. J Pharmacol Exp Ther 325(3):961–968

Bergholt MS, Serio A, McKenzie JS, Boyd A, Soares RF, Tillner J et al (2018) Correlated Heterospectral Lipidomics for biomolecular profiling of Remyelination in multiple sclerosis. ACS Cent Sci 4(1):39–51

Bhargava P, Fitzgerald KC, Calabresi PA, Mowry EM (2017) Metabolic alterations in multiple sclerosis and the impact of vitamin D supplementation. JCI Insight 2(19):e95302

Birbrair A, Zhang T, Files DC, Mannava S, Smith T, Wang ZM et al (2014) Type-1 pericytes accumulate after tissue injury and produce collagen in an organ-dependent manner. Stem Cell Res Ther 5(6):122

Bogdanov B, Smith RD (2005) Proteomics by FTICR mass spectrometry: top down and bottom up. Mass Spectrom Rev 24(2):168–200

Bondia-Pons I, Nordlund E, Mattila I, Katina K, Aura AM, Kolehmainen M et al (2011) Postprandial differences in the plasma metabolome of healthy Finnish subjects after intake of a sourdough fermented endosperm rye bread versus white wheat bread. Nutr J 10:116

Boskey AL, Mendelsohn R (2005) Infrared spectroscopic characterization of mineralized tissues. Vib Spectrosc 38(1-2):107–114

Boudah S, Olivier MF, Aros-Calt S, Oliveira L, Fenaille F, Tabet JC et al (2014) Annotation of the human serum metabolome by coupling three liquid chromatography methods to high-resolution mass spectrometry. J Chromatogr 966:34–47

Bowen BP, Northen TR (2010) Dealing with the unknown: metabolomics and metabolite atlases. J Am Soc Mass Spectrom 21(9):1471–1476

Broadhurst D, Goodacre R, Reinke SN, Kuligowski J, Wilson ID, Lewis MR et al (2018) Guidelines and considerations for the use of system suitability and quality control samples in mass spectrometry assays applied in untargeted clinical metabolomic studies. Metabolomics 14(6):72

Broza YY, Kremer R, Tisch U, Gevorkyan A, Shiban A, Best LA et al (2013) A nanomaterial-based breath test for short-term follow-up after lung tumor resection. Nanomedicine 9(1):15–21

Capuano R, Santonico M, Pennazza G, Ghezzi S, Martinelli E, Roscioni C et al (2015) The lung cancer breath signature: a comparative analysis of exhaled breath and air sampled from inside the lungs. Sci Rep 5:16491

Cazzola MS, Capuano R, Bergamini A, Martinelli E, Calzetta L, Rogliani P, Ciaprini C, Ora J, Paolesse R, Di Natale C, D'Amico A (2015) Analysis of exhaled breath fingerprints and volatile organic compounds in COPD. COPD Research and Practice 10:1
Ceglarek U, Leichtle A, Brugel M, Kortz L, Brauer R, Bresler K et al (2009) Challenges and developments in tandem mass spectrometry based clinical metabolomics. Mol Cell Endocrinol 301(1–2):266–271
Cha JH, Bae SH, Kim HL, Park NR, Choi ES, Jung ES et al (2013) Branched-chain amino acids ameliorate fibrosis and suppress tumor growth in a rat model of hepatocellular carcinoma with liver cirrhosis. PLoS One 8(11):e77899
Chan JC, Kioh DY, Yap GC, Lee BW, Chan EC (2017) A novel LCMSMS method for quantitative measurement of short-chain fatty acids in human stool derivatized with (12)C- and (13) C-labelled aniline. J Pharm Biomed Anal 138:43–53
Chong J, Soufan O, Li C, Caraus I, Li S, Bourque G et al (2018) MetaboAnalyst 4.0: towards more transparent and integrative metabolomics analysis. Nucleic Acids Res 46(W1):W486–WW94
Claude E, Jones EA, Pringle SD (2017) DESI mass spectrometry imaging (MSI). Methods Mol Biol 1618:65–75
Cooper SD, Kyriazakis I, Nolan JV (1995) Diet selection in sheep: the role of the rumen environment in the selection of a diet from two feeds that differ in their energy density. Br J Nutr 74(1):39–54
Crow YJ, Wakeland EK (2012) Autoimmunity. Curr Opin Immunol 24(6):649–650
Das S, Pal S, Mitra M (2016) Significance of exhaled breath test in clinical diagnosis: a special focus on the detection of diabetes mellitus. J Med Biol Eng 36(5):605–624
Defernez M, Colquhoun IJ (2003) Factors affecting the robustness of metabolite fingerprinting using 1H NMR spectra. Phytochemistry 62(6):1009–1017
Deleris G, Petibois C (2003) Applications of FT-IR spectrometry to plasma contents. Vib Spectrosc 32:129–136
Dennis EA, Norris PC (2015) Eicosanoid storm in infection and inflammation. Nat Rev Immunol 15(8):511–523
Dunn WB, Broadhurst D, Begley P, Zelena E, Francis-McIntyre S, Anderson N et al (2011) Procedures for large-scale metabolic profiling of serum and plasma using gas chromatography and liquid chromatography coupled to mass spectrometry. Nat Protoc 6(7):1060–1083
Emwas A-HM (2015) The strengths and weaknesses of NMR spectroscopy and mass spectrometry with particular focus on metabolomics research. Methods Mol Biol 1227:161–193
Evans AM, DeHaven CD, Barrett T, Mitchell M, Milgram E (2009) Integrated, nontargeted ultrahigh performance liquid chromatography/electrospray ionization tandem mass spectrometry platform for the identification and relative quantification of the small-molecule complement of biological systems. Anal Chem 81(16):6656–6667
Fendt SM, Buescher JM, Rudroff F, Picotti P, Zamboni N, Sauer U (2010) Tradeoff between enzyme and metabolite efficiency maintains metabolic homeostasis upon perturbations in enzyme capacity. Mol Syst Biol 6:356
Fenn JB, Mann M, Meng CK, Wong SF, Whitehouse CM (1989) Electrospray ionization for mass spectrometry of large biomolecules. Science 246(4926):64–71
Fessenden M (2016) Metabolomics: small molecules, single cells. Nature 540(7631):153–155
Fiehn O (2016) Metabolomics by gas chromatography-mass spectrometry: combined targeted and untargeted profiling. Curr Protoc Mol Biol 114:30–42
Forsberg EM, Huan T, Rinehart D, Benton HP, Warth B, Hilmers B et al (2018) Data processing, multi-omic pathway mapping, and metabolite activity analysis using XCMS online. Nat Protoc 13(4):633–651
Fushimi T, Suruga K, Oshima Y, Fukiharu M, Tsukamoto Y, Goda T (2006) Dietary acetic acid reduces serum cholesterol and triacylglycerols in rats fed a cholesterol-rich diet. Br J Nutr 95(5):916–924
Gao Z, Yin J, Zhang J, Ward RE, Martin RJ, Lefevre M et al (2009a) Butyrate improves insulin sensitivity and increases energy expenditure in mice. Diabetes 58(7):1509–1517

Gao X, Pujos-Guillot E, Martin JF, Galan P, Juste C, Jia W et al (2009b) Metabolite analysis of human fecal water by gas chromatography/mass spectrometry with ethyl chloroformate derivatization. Anal Biochem 393(2):163–175
Garcia A, Barbas C (2011) Gas chromatography-mass spectrometry (GC-MS)-based metabolomics. Methods Mol Biol 708:191–204
Gathungu RM, Bird SS, Sheldon DP, Kautz R, Vouros P, Matson WR et al (2014) Identification of metabolites from liquid chromatography-coulometric array detection profiling: gas chromatography-mass spectrometry and refractionation provide essential information orthogonal to LC-MS/microNMR. Anal Biochem 454:23–32
Gieger C, Geistlinger L, Altmaier E, Hrabe de Angelis M, Kronenberg F, Meitinger T et al (2008) Genetics meets metabolomics: a genome-wide association study of metabolite profiles in human serum. PLoS Genet 4(11):e1000282
Griffin JL, Shockcor JP (2004) Metabolic profiles of cancer cells. Nat Rev Cancer 4(7):551–561
Guilbault C, Wojewodka G, Saeed Z, Hajduch M, Matouk E, De Sanctis JB et al (2009) Cystic fibrosis fatty acid imbalance is linked to ceramide deficiency and corrected by fenretinide. Am J Respir Cell Mol Biol 41(1):100–106
Han J, Tschernutter V, Yang J, Eckle T, Borchers CH (2013) Analysis of selected sugars and sugar phosphates in mouse heart tissue by reductive amination and liquid chromatography-electrospray ionization mass spectrometry. Anal Chem 85(12):5965–5973
Han J, Lin K, Sequeira C, Borchers CH (2015) An isotope-labeled chemical derivatization method for the quantitation of short-chain fatty acids in human feces by liquid chromatography-tandem mass spectrometry. Anal Chim Acta 854:86–94
Higashi N, Taniguchi M, Mita H, Osame M, Akiyama K (2002) A comparative study of eicosanoid concentrations in sputum and urine in patients with aspirin-intolerant asthma. Clin Exp Allergy 32(10):1484–1490
Hilvo M, Denkert C, Lehtinen L, Muller B, Brockmoller S, Seppanen-Laakso T et al (2011) Novel theranostic opportunities offered by characterization of altered membrane lipid metabolism in breast cancer progression. Cancer Res 71(9):3236–3245
Ho CS, Lam CW, Chan MH, Cheung RC, Law LK, Lit LC et al (2003) Electrospray ionisation mass spectrometry: principles and clinical applications. Clin Biochem Rev 24(1):3–12
Huang SK, Peters-Golden M (2008) Eicosanoid lipid mediators in fibrotic lung diseases: ready for prime time? Chest 133(6):1442–1450
Huda-Faujan N, Abdulamir AS, Fatimah AB, Anas OM, Shuhaimi M, Yazid AM et al (2010) The impact of the level of the intestinal short chain fatty acids in inflammatory bowel disease patients versus healthy subjects. Open Biochem J 4:53–58
Jain M, Nilsson R, Sharma S, Madhusudhan N, Kitami T, Souza AL et al (2012) Metabolite profiling identifies a key role for glycine in rapid cancer cell proliferation. Science 336(6084):1040–1044
Jayasinghe NS, Dias DA (2013) A robust GC-MS method for the quantitation of fatty acids in biological systems. Methods Mol Biol 1055:39–56
Johnson CH, Ivanisevic J, Siuzdak G (2016) Metabolomics: beyond biomarkers and towards mechanisms. Nature Reviews 17(7):451–459
Karamanou M, Protogerou A, Tsoucalas G, Androutsos G, Poulakou-Rebelakou E (2016) Milestones in the history of diabetes mellitus: the main contributors. World J Diabetes 7(1):1–7
Katajamaa M, Oresic M (2007) Data processing for mass spectrometry-based metabolomics. J Chromatogr A 1158(1-2):318–328
Katajamaa M, Miettinen J, Oresic M (2006) MZmine: toolbox for processing and visualization of mass spectrometry based molecular profile data. Bioinformatics 22(5):634–636
Keyfi F, Varasteh A (2016) Development and validation of a GC-FID method for diagnosis of Methylmalonic Acidemia. Rep Biochem Mol Biol 4(2):104–109
Kim K, Aronov P, Zakharkin SO, Anderson D, Perroud B, Thompson IM et al (2009) Urine metabolomics analysis for kidney cancer detection and biomarker discovery. Mol Cell Proteomics 8(3):558–570
Kim SJ, Kim N, Koh EH, Yoo HJ (2012) Identification of ethanolamine plasmalogens from complex lipid mixtures by MS/MS and Ag adduction. Anal Sci 28(12):1207–1212

Kim SJ, Back SH, Koh JM, Yoo HJ (2014) Quantitative determination of major platelet activating factors from human plasma. Anal Bioanal Chem 406(13):3111–3118

Kim S, Rhee JK, Yoo HJ, Lee HJ, Lee EJ, Lee JW et al (2015) Bioinformatic and metabolomic analysis reveals miR-155 regulates thiamine level in breast cancer. Cancer Lett 357(2):488–497

Kim SH, Song HE, Kim SJ, Woo DC, Chang S, Choi WG et al (2017) Quantitative structural characterization of phosphatidylinositol phosphates from biological samples. J Lipid Res 58(2):469–478

Knittelfelder OL, Weberhofer BP, Eichmann TO, Kohlwein SD, Rechberger GN (2014) A versatile ultra-high performance LC-MS method for lipid profiling. J Chromatogr 951–952:119–128

Kvitvang HF, Andreassen T, Adam T, Villas-Boas SG, Bruheim P (2011) Highly sensitive GC/MS/MS method for quantitation of amino and nonamino organic acids. Anal Chem 83(7):2705–2711

Lanza IR, Zhang S, Ward LE, Karakelides H, Raftery D, Nair KS (2010) Quantitative metabolomics by H-NMR and LC-MS/MS confirms altered metabolic pathways in diabetes. PLoS One 5(5):e10538

Larstad MA, Toren K, Bake B, Olin AC (2007) Determination of ethane, pentane and isoprene in exhaled air--effects of breath-holding, flow rate and purified air. Acta Physiol (Oxf) 189(1):87–98

Leask A, Abraham DJ (2004) TGF-beta signaling and the fibrotic response. FASEB J 18(7):816–827

Lebovitz HE (1995) Diabetic ketoacidosis. Lancet 345(8952):767–772

Lei Z, Huhman DV, Sumner LW (2011) Mass spectrometry strategies in metabolomics. J Biol Chem 286(29):25435–25442

Levick SP, Loch DC, Taylor SM, Janicki JS (2007) Arachidonic acid metabolism as a potential mediator of cardiac fibrosis associated with inflammation. J Immunol 178(2):641–646

Li C, Zheng S, You H, Liu X, Lin M, Yang L et al (2011) Sphingosine 1-phosphate (S1P)/S1P receptors are involved in human liver fibrosis by action on hepatic myofibroblasts motility. J Hepatol 54(6):1205–1213

Li L, Zhao J, Zhao Y, Lu X, Zhou Z, Zhao C et al (2016) Comprehensive investigation of tobacco leaves during natural early senescence via multi-platform metabolomics analyses. Sci Rep 6:37976

Liang X, Liu J, LeBlanc Y, Covey T, Ptak AC, Brenna JT et al (2007) Electron transfer dissociation of doubly sodiated glycerophosphocholine lipids. J Am Soc Mass Spectrom 18(10):1783–1788

Liberti MV, Locasale JW (2016) The Warburg effect: how does it benefit cancer cells? Trends Biochem Sci 41(3):211–218

Lim CK, Bilgin A, Lovejoy DB, Tan V, Bustamante S, Taylor BV et al (2017) Kynurenine pathway metabolomics predicts and provides mechanistic insight into multiple sclerosis progression. Sci Rep 7:41473

Liu H, Yoo HJ, Hakansson K (2008) Characterization of phosphate-containing metabolites by calcium adduction and electron capture dissociation. J Am Soc Mass Spectrom 19(6):799–808

Lu W, Bennett BD, Rabinowitz JD (2008) Analytical strategies for LC-MS-based targeted metabolomics. J Chromatogr 871(2):236–242

Maeda S (1982) Pathology of experimental radiation pancarditis. II. Correlation between ultrastructural changes of the myocardial mitochondria and succinic dehydrogenase activity in rabbit heart receiving a single dose of X-ray irradiation. Acta Pathol Jpn 32(2):199–218

Mark TDD, Dunn GH (2013) Electron impact ionization. Springer, Cham

Matsuda F (2016) Technical challenges in mass spectrometry-based metabolomics. Mass Spectrom (Tokyo) 5(2):S0052

Metz TO, Zhang Q, Page JS, Shen Y, Callister SJ, Jacobs JM et al (2007) The future of liquid chromatography-mass spectrometry (LC-MS) in metabolic profiling and metabolomic studies for biomarker discovery. Biomark Med 1(1):159–185

Miekisch W, Schubert JK, Noeldge-Schomburg GF (2004) Diagnostic potential of breath analysis--focus on volatile organic compounds. Clin Chim Acta 347(1-2):25–39

Moustafa T, Fickert P, Magnes C, Guelly C, Thueringer A, Frank S et al (2012) Alterations in lipid metabolism mediate inflammation, fibrosis, and proliferation in a mouse model of chronic cholestatic liver injury. Gastroenterology 142(1):140-51–140e12

Murray KK, Boyd RK, Eberlin MN, Langley GJ, Li L, Naito Y (2013) Definitions of terms relating to mass spectrometry (IUPACRecommendations 2013). Pure Appl Chem 85(5):1515–1609

Nardi-Agmon I, Peled N (2017) Exhaled breath analysis for the early detection of lung cancer: recent developments and future prospects. Lung Cancer (Auckl) 8:31–38

Neary R, Watson CJ, Baugh JA (2015) Epigenetics and the overhealing wound: the role of DNA methylation in fibrosis. Fibrogenesis Tissue Repair 8:18

Newgard CB, An J, Bain JR, Muehlbauer MJ, Stevens RD, Lien LF et al (2009) A branched-chain amino acid-related metabolic signature that differentiates obese and lean humans and contributes to insulin resistance. Cell Metab 9(4):311–326

Nicholson JK, Lindon JC (2008) Systems biology: Metabonomics. Nature 455(7216):1054–1056

Ongay S, Hermans J, Bruins AP, Nieuwendijk AM, Overkleeft H, Bischoff R (2013) Electron transfer and collision induced dissociation of non-derivatized and derivatized desmosine and isodesmosine. J Am Soc Mass Spectrom 24(1):83–91

Osborn MP, Park Y, Parks MB, Burgess LG, Uppal K, Lee K et al (2013) Metabolome-wide association study of neovascular age-related macular degeneration. PLoS One 8(8):e72737

Papadimitropoulos MP, Vasilopoulou CG, Maga-Nteve C, Klapa MI (2018) Untargeted GC-MS metabolomics. Methods Mol Biol 1738:133–147

Paredi P, Kharitonov SA, Barnes PJ (2000a) Elevation of exhaled ethane concentration in asthma. Am J Respir Crit Care Med 162(4 Pt 1):1450–1454

Paredi P, Kharitonov SA, Leak D, Shah PL, Cramer D, Hodson ME et al (2000b) Exhaled ethane is elevated in cystic fibrosis and correlates with carbon monoxide levels and airway obstruction. Am J Respir Crit Care Med 161(4 Pt 1):1247–1251

Paredi P, Kharitonov SA, Leak D, Ward S, Cramer D, Barnes PJ (2000c) Exhaled ethane, a marker of lipid peroxidation, is elevated in chronic obstructive pulmonary disease. Am J Respir Crit Care Med 162(2 Pt 1):369–373

Pavord ID, Ward R, Woltmann G, Wardlaw AJ, Sheller JR, Dworski R (1999) Induced sputum eicosanoid concentrations in asthma. Am J Respir Crit Care Med 160(6):1905–1909

Peled N, Hakim M, Bunn PA Jr, Miller YE, Kennedy TC, Mattei J et al (2012) Non-invasive breath analysis of pulmonary nodules. J Thorac Oncol 7(10):1528–1533

Perez ER, Knapp JA, Horn CK, Stillman SL, Evans JE, Arfsten DP (2016) Comparison of LC-MS-MS and GC-MS analysis of benzodiazepine compounds included in the drug demand reduction urinalysis program. J Anal Toxicol 40(3):201–207

Petras D, Jarmusch AK, Dorrestein PC (2017) From single cells to our planet-recent advances in using mass spectrometry for spatially resolved metabolomics. Curr Opin Chem Biol 36:24–31

Phillips M, Bauer TL, Cataneo RN, Lebauer C, Mundada M, Pass HI et al (2015) Blinded validation of breath biomarkers of lung Cancer, a potential ancillary to chest CT screening. PLoS One 10(12):e0142484

Pitt JJ, Eggington M, Kahler SG (2002) Comprehensive screening of urine samples for inborn errors of metabolism by electrospray tandem mass spectrometry. Clin Chem 48(11):1970–1980

Psathakis K, Mermigkis D, Papatheodorou G, Loukides S, Panagou P, Polychronopoulos V et al (2006) Exhaled markers of oxidative stress in idiopathic pulmonary fibrosis. Eur J Clin Investig 36(5):362–367

Psychogios N, Hau DD, Peng J, Guo AC, Mandal R, Bouatra S et al (2011) The human serum metabolome. PLoS One 6(2):e16957

Qiu Y, Su M, Liu Y, Chen M, Gu J, Zhang J et al (2007) Application of ethyl chloroformate derivatization for gas chromatography-mass spectrometry based metabonomic profiling. Anal Chim Acta 583(2):277–283

Quehenberger O, Dennis EA (2011) The human plasma lipidome. N Engl J Med 365(19):1812–1823

Quehenberger O, Armando AM, Brown AH, Milne SB, Myers DS, Merrill AH et al (2010) Lipidomics reveals a remarkable diversity of lipids in human plasma. J Lipid Res 51(11):3299–3305

Ramautar R, Somsen GW, de Jong GJ (2009) CE-MS in metabolomics. Electrophoresis 30(1):276–291

Ramautar R, Somsen GW, de Jong GJ (2017) CE-MS for metabolomics: developments and applications in the period 2014-2016. Electrophoresis 38(1):190–202

Ren S, Shao Y, Zhao X, Hong CS, Wang F, Lu X et al (2016) Integration of metabolomics and Transcriptomics reveals major metabolic pathways and potential biomarker involved in prostate cancer. Mol Cell Proteomics 15(1):154–163

Richins RD, Rodriguez-Uribe L, Lowe K, Ferral R, O'Connell MA (2018) Accumulation of bioactive metabolites in cultivated medical cannabis. PLoS One 13(7):e0201119

Roberts LDS, L A, Gerszten RE, Clish CB (2012) Targeted metabolomics. Curr Protoc Mol Biol 98(1):30–32

Ross BM, Maxwell R, Glen I (2011) Increased breath ethane levels in medicated patients with schizophrenia and bipolar disorder are unrelated to erythrocyte omega-3 fatty acid abundance. Prog Neuro-Psychopharmacol Biol Psychiatry 35(2):446–453

Sanak M, Gielicz A, Bochenek G, Kaszuba M, Nizankowska-Mogilnicka E, Szczeklik A (2011) Targeted eicosanoid lipidomics of exhaled breath condensate provide a distinct pattern in the aspirin-intolerant asthma phenotype. J Allergy Clin Immunol 127(5):1141–7 e2

Sarbach C, Stevens P, Whiting J, Puget P, Humbert M, Cohen-Kaminsky S et al (2013) Evidence of endogenous volatile organic compounds as biomarkers of diseases in alveolar breath. Ann Pharm Fr 71(4):203–215

Sas KM, Karnovsky A, Michailidis G, Pennathur S (2015) Metabolomics and diabetes: analytical and computational approaches. Diabetes 64(3):718–732

Shehada N, Cancilla JC, Torrecilla JS, Pariente ES, Bronstrup G, Christiansen S et al (2016) Silicon nanowire sensors enable diagnosis of patients via exhaled breath. ACS Nano 10(7):7047–7057

Smith CA, Want EJ, O'Maille G, Abagyan R, Siuzdak G (2006) XCMS: processing mass spectrometry data for metabolite profiling using nonlinear peak alignment, matching, and identification. Anal Chem 78(3):779–787

Smith R, Ventura D, Prince JT (2015) LC-MS alignment in theory and practice: a comprehensive algorithmic review. Brief Bioinform 16(1):104–117

Smkmrmstt M (2012) Bioinformatics tools for mass spectroscopy-based Metabolomic data processing and analysis. Curr Bioinforma 7(1):96–108

Sreekumar A, Poisson LM, Rajendiran TM, Khan AP, Cao Q, Yu J et al (2009) Metabolomic profiles delineate potential role for sarcosine in prostate cancer progression. Nature 457(7231):910–914

Strassburg K, Huijbrechts AM, Kortekaas KA, Lindeman JH, Pedersen TL, Dane A et al (2012) Quantitative profiling of oxylipins through comprehensive LC-MS/MS analysis: application in cardiac surgery. Anal Bioanal Chem 404(5):1413–1426

Suhre K, Shin SY, Petersen AK, Mohney RP, Meredith D, Wagele B et al (2011a) Human metabolic individuality in biomedical and pharmaceutical research. Nature 477(7362):54–60

Suhre K, Wallaschofski H, Raffler J, Friedrich N, Haring R, Michael K et al (2011b) A genome-wide association study of metabolic traits in human urine. Nat Genet 43(6):565–569

Sun C, Li T, Song X, Huang L, Zang Q, Xu J et al (2019) Spatially resolved metabolomics to discover tumor-associated metabolic alterations. Proc Natl Acad Sci U S A 116(1):52–57

Takats Z, Wiseman JM, Gologan B, Cooks RG (2004) Mass spectrometry sampling under ambient conditions with desorption electrospray ionization. Science 306(5695):471–473

Tao X, Liu Y, Wang Y, Qiu Y, Lin J, Zhao A et al (2008) GC-MS with ethyl chloroformate derivatization for comprehensive analysis of metabolites in serum and its application to human uremia. Anal Bioanal Chem 391(8):2881–2889

Teichgraber V, Ulrich M, Endlich N, Riethmuller J, Wilker B, De Oliveira-Munding CC et al (2008) Ceramide accumulation mediates inflammation, cell death and infection susceptibility in cystic fibrosis. Nat Med 14(4):382–391

Theodoridis H, Wilson I (2008) LC-MS-based methodology for global metabolite profiling in metabonomics/metabolomics. TrAC Trends Anal Chem 27:251–260

Theodoridis GA, Gika HG, Want EJ, Wilson ID (2012) Liquid chromatography-mass spectrometry based global metabolite profiling: a review. Anal Chim Acta 711:7–16

Tsai TH, Tadesse MG, Wang Y, Ressom HW (2013a) Profile-based LC-MS data alignment--a Bayesian approach. IEEE/ACM Trans Comput Biol Bioinform 10(2):494–503
Tsai TH, Tadesse MG, Di Poto C, Pannell LK, Mechref Y, Wang Y et al (2013b) Multi-profile Bayesian alignment model for LC-MS data analysis with integration of internal standards. Bioinformatics 29(21):2774–2780
Vahamaa H, Koskinen VR, Hosia W, Moulder R, Nevalainen OS, Lahesmaa R et al (2011) PolyAlign: a versatile LC-MS data alignment tool for landmark-selected and -automated use. Int J Proteomics 2011:450290
van Laarhoven A, Dian S, Aguirre-Gamboa R, Avila-Pacheco J, Ricano-Ponce I, Ruesen C et al (2018) Cerebral tryptophan metabolism and outcome of tuberculous meningitis: an observational cohort study. Lancet Infect Dis 18(5):526–535
Vandenbogaert M, Li-Thiao-Te S, Kaltenbach HM, Zhang R, Aittokallio T, Schwikowski B (2008) Alignment of LC-MS images, with applications to biomarker discovery and protein identification. Proteomics 8(4):650–672
Vander Heiden MG (2011) Targeting cancer metabolism: a therapeutic window opens. Nat Rev Drug Discov 10(9):671–684
Vaniya A, Fiehn O (2015) Using fragmentation trees and mass spectral trees for identifying unknown compounds in metabolomics. Trends Analyt Chem 69:52–61
Vasilopoulou CG, Margarity M, Klapa MI (2016) Metabolomic analysis in brain research: opportunities and challenges. Front Physiol 7:183
Vilar-Gomez E, Calzadilla-Bertot L, Wai-Sun Wong V, Castellanos M, Aller-de la Fuente R, Metwally M et al (2018) Fibrosis severity as a determinant of cause-specific mortality in patients with advanced nonalcoholic fatty liver disease: a multi-National Cohort Study. Gastroenterology 155(2):443–457
Villas-Boas SG, Mas S, Akesson M, Smedsgaard J, Nielsen J (2005) Mass spectrometry in metabolome analysis. Mass Spectrom Rev 24(5):613–646
Villoslada P, Alonso C, Agirrezabal I, Kotelnikova E, Zubizarreta I, Pulido-Valdeolivas I et al (2017) Metabolomic signatures associated with disease severity in multiple sclerosis. Neurol Neuroimmunol Neuroinflamm 4(2):e321
Wang D, Dubois RN (2010) Eicosanoids and cancer. Nat Rev Cancer 10(3):181–193
Wang Z, Wang C (2013) Is breath acetone a biomarker of diabetes? A historical review on breath acetone measurements. J Breath Res 7(3):037109
Wang C, Kong H, Guan Y, Yang J, Gu J, Yang S et al (2005) Plasma phospholipid metabolic profiling and biomarkers of type 2 diabetes mellitus based on high-performance liquid chromatography/electrospray mass spectrometry and multivariate statistical analysis. Anal Chem 77(13):4108–4116
Wang Y, Chen X, Li Q, Zhong D (2008) Metabolite identification of arbidol in human urine by the study of CID fragmentation pathways using HPLC coupled with ion trap mass spectrometry. J Mass Spectrom 43(8):1099–1109
Wang TJ, Larson MG, Vasan RS, Cheng S, Rhee EP, McCabe E et al (2011) Metabolite profiles and the risk of developing diabetes. Nat Med 17(4):448–453
Want EJ, Cravatt BF, Siuzdak G (2005) The expanding role of mass spectrometry in metabolite profiling and characterization. Chembiochem 6(11):1941–1951
Want EJ, Wilson ID, Gika H, Theodoridis G, Plumb RS, Shockcor J et al (2010) Global metabolic profiling procedures for urine using UPLC-MS. Nat Protoc 5(6):1005–1018
Want EJ, Masson P, Michopoulos F, Wilson ID, Theodoridis G, Plumb RS et al (2013) Global metabolic profiling of animal and human tissues via UPLC-MS. Nat Protoc 8(1):17–32
Warburg O (1956) On the origin of cancer cells. Science 123(3191):309–314
Watrous JD, Henglin M, Claggett B, Lehmann KA, Larson MG, Cheng S et al (2017) Visualization, quantification, and alignment of spectral drift in population scale untargeted metabolomics data. Anal Chem 89(3):1399–1404
Wenzel SE, Busse WW, National Heart L (2007) Blood Institute's severe asthma research P. severe asthma: lessons from the severe asthma research program. J Allergy Clin Immunol 119(1):14–21

Worley BP (2013) R. Multivariate analysis in metabolomics. Curr Metabolomics 1(1):92–107
Wu AHB (2006) Clinical guide to laboratory tests, 4th edn. Saunders/Elsevier, St. Louis, MO
Wynn TA (2008) Cellular and molecular mechanisms of fibrosis. J Pathol 214(2):199–210
Xia J, Wishart DS (2011) Web-based inference of biological patterns, functions and pathways from metabolomic data using MetaboAnalyst. Nat Protoc 6(6):743–760
Xiao JF, Zhou B, Ressom HW (2012) Metabolite identification and quantitation in LC-MS/MS-based metabolomics. Trends Analyt Chem. 32:1–14
Yamashita MF, John B (1984) Electrospray ion source. Another variation on the free-jet theme. J Phys Chem 88(20):4451–4459
Yang J, Schmelzer K, Georgi K, Hammock BD (2009) Quantitative profiling method for oxylipin metabolome by liquid chromatography electrospray ionization tandem mass spectrometry. Anal Chem 81(19):8085–8093
Yang J, Dong H, Hammock BD (2011) Profiling the regulatory lipids: another systemic way to unveil the biological mystery. Curr Opin Lipidol 22(3):197–203
Yoo HJ, Hakansson K (2011) Determination of phospholipid regiochemistry by Ag(I) adduction and tandem mass spectrometry. Anal Chem 83(4):1275–1283
Yoo HJ, Liu H, Hakansson K (2007) Infrared multiphoton dissociation and Electron-induced dissociation as alternative MS/MS strategies for metabolite identification. Anal Chem 20:7858–7866
Yoo HJ, Wang N, Zhuang S, Song H, Hakansson K (2011) Negative-ion electron capture dissociation: radical-driven fragmentation of charge-increased gaseous peptide anions. J Am Chem Soc 133(42):16790–16793
Yuan M, Breitkopf SB, Yang X, Asara JM (2012) A positive/negative ion-switching, targeted mass spectrometry-based metabolomics platform for bodily fluids, cells, and fresh and fixed tissue. Nat Protoc 7(5):872–881
Yugi K, Kuroda S (2018) Metabolism as a signal generator across trans-omic networks at distinct time scales. Curr Opin Syst Biol 8:59–66
Zeng M, Cao H (2018) Fast quantification of short chain fatty acids and ketone bodies by liquid chromatography-tandem mass spectrometry after facile derivatization coupled with liquid-liquid extraction. J Chromatogr 1083:137–145
Zhang W, Hankemeier T, Ramautar R (2017) Next-generation capillary electrophoresis-mass spectrometry approaches in metabolomics. Curr Opin Biotechnol 43:1–7
Zhao X, Fritsche J, Wang J, Chen J, Rittig K, Schmitt-Kopplin P et al (2010) Metabonomic fingerprints of fasting plasma and spot urine reveal human pre-diabetic metabolic traits. Metabolomics 6(3):362–374
Zheng X, Qiu Y, Zhong W, Baxter S, Su M, Li Q et al (2013) A targeted metabolomic protocol for short-chain fatty acids and branched-chain amino acids. Metabolomics 9(4):818–827
Zhou B, Xiao JF, Tuli L, Ressom HW (2012) LC-MS-based metabolomics. Mol BioSyst 8(2):470–481

The Application of Mass Spectrometry in Drug Metabolism and Pharmacokinetics

Ji-Yoon Lee, Sang Kyum Kim, Kiho Lee, and Soo Jin Oh

Introduction

Drug development is time-consuming, labor-intensive, and expensive process. And, since approximately 10% of new chemical entities (NCEs) successfully make it to the market, it is an inefficient process. In the late 1900s, costly late-stage drug development failures were related to pharmacokinetics and toxicity (Kola and Landis 2004). Therefore, it has become common practice to consider these parameters as early as possible in the drug discovery process. The term "drug-like" describes compounds that have sufficiently acceptable absorption, distribution, metabolism, and excretion (ADME) properties and toxicities low enough to successfully complete human Phase I clinical trials (Lipinski 2000). A balance between pharmacological activity, safety, and drug metabolism and pharmacokinetics (DMPK) determines the success of a new chemical entity (NCE) during the drug discovery process. Early

J.-Y. Lee
Convergence Medicine Research Center, Asan Institute for Life Sciences, Asan Medical Center, Seoul, South Korea

S. K. Kim
College of Pharmacy, Chungnam National University, Daejeon, South Korea
e-mail: sangkim@cnu.ac.kr

K. Lee
College of Pharmacy, Korea University, Sejong, South Korea
e-mail: kiholee@korea.ac.kr

S. J. Oh (✉)
Convergence Medicine Research Center, Asan Institute for Life Sciences, Asan Medical Center, Seoul, South Korea

College of Medicine, University of Ulsan, Seoul, South Korea
e-mail: diatree@amc.seoul.kr

J. K. Kim et al. (eds.), *Advanced Imaging and Bio Techniques for Convergence Science*, Advances in Experimental Medicine and Biology,
https://doi.org/10.1007/978-981-33-6064-8_20

DMPK characterization is essential for identifying potential drug candidates. DMPK covers all aspects of a drug's (ADME). The fundamental goals of early DMPK characterization are: (1) to understand how structural features affect drug ADME, (2) to screen multiple compounds, and (3) to perform, additional metabolism studies to optimize leads that may proceed to drug development (Korfmacher 2009; Riley et al. 2002). Thus, high-throughput analysis is essential during DMPK studies to determine whether an end product of a screening assay or the parent compound (and/or metabolites) is a viable drug candidate.

Mass spectrometry has been an important analytical tool for scientists working in the DMPK field for several decades (Ackermann et al. 2008; Hsieh 2008). Liquid chromatography tandem mass spectrometry (LC-MS/MS) allows for high-throughput and high sensitivity analyses for large numbers of compounds. Since the ADME of an increasing number of new chemicals is being evaluated, there is a need to increase throughput. Because of its speed, sensitivity, and selectivity, MS is now the primary method used for DMPK characterization during drug discovery. This review focuses on the use of MS as a bioanalytical tool to support DMPK studies in drug discovery.

Bioanalysis to Support ADME Studies

A drug's action (desired or undesired) on human body is dictated by its interactions with the specific target sites, and such action can be modulated by transfer processes and the pharmacokinetic variables of the drug's ADME. Pharmacokinetics (PK) represents how a drug and/or its metabolites enter the body, reach their site of action, and leave the body. PK has been broadly divided into two categories of study: absorption and disposition. Disposition is further subdivided into the study of distribution and elimination. The term elimination includes metabolism and excretion. ADME properties have a tremendous impact on the success of drug candidates. They represent the fundamental physicochemical and biochemical properties of drug molecules that can be affected in vivo.

A successful drug must demonstrate good potency and possess drug-like properties. Pharmacokinetic problems such as poor absorption, very short or very long half-life, enzyme inhibition/induction, and a high first pass effect can be improved through structural modification by medicinal chemists (Caldwell 1996). Therefore, preclinical DMPK studies play a critical role in the identification and optimization of drug candidates during the early discovery. Particularly, it is important to predict and evaluate drug metabolism at this stage because many pharmacokinetic defects arise from drug metabolism (Laine 2008; Rodrigues 2008). It is important to characterize the metabolic stability, metabolic routes, metabolizing enzymes, and kinetics of pharmaceuticals, so that the "metabolic soft spots" of compounds, which exhibit low metabolic stability, can be identified. Altering the structure of a compound according to its metabolic soft spot can reduce its metabolic rate in the early discovery stage. Also, promising NCEs should be identified before toxicological studies in animals are performed. During the discovery stage, drug metabolism studies provide a basis for choosing chemical structures and lead compounds with desirable DMPK and safety profiles.

As mentioned above, rapid and reliable prediction evaluation methods are important during drug discovery. Recent advancements in such methods, particularly in analytical techniques, make it possible to test NCEs during the early stages of drug discovery and development (Korfmacher 2005). Mass spectrometry was introduced to drug metabolism laboratories in the 1990s. Since then, it has been widely used in many aspects of pharmaceutical research and development including structural characterization of NCEs, metabolite identification, and pharmacokinetic studies (Hoke et al. 2001; Hsieh and Korfmacher 2006). LC-MS/MS is a well-established quantitative and qualitative technique in pharmaceutical research (Kostiainen et al. 2003; Prakash et al. 2007). Due to its sensitivity, speed, and specificity, LC-MS/MS has been the workhorse analytical tool for in vitro and in vivo screening. Typical quantitative LC-MS/MS applications include in vitro ADME assays, for example, drug permeability and metabolic stability (Chu and Nomeir 2006). Profiling metabolites from different biological matrices is a prime example of a qualitative LC-MS/MS analysis conducted during drug discovery.

In this review, bioanalysis is defined as the chemical analysis of pharmaceutical substances in biological samples such as blood, plasma, urine, tissues, and cells. A pharmaceutical substance of interest is often termed an "analyte." The identification and quantification of analytes and their metabolites are very important in drug discovery and development. Thus, the objectives of the chemical analysis are to both qualify (identification) and quantify (quantification) an analyte during drug discovery. A general bioanalysis workflow is depicted in Fig. 1.

Most samples obtained from DMPK studies are analyzed by LC-MS/MS. Biological fluids like plasma and whole blood are extremely complex samples and are usually not injected directly into an analytical instrument when analyzing a drug substance. In some cases, a sample may contain matrix components with similar mass and retention time as the target analyte. Under this undesirable condition, the matrix components can directly interfere with the detection and measurement of the target analyte. This may cause false identification of the analyte in the sample or inaccurate quantitative results. Although the matrix components have molecular masses that are different from the target analyte and will not be detected by the mass spectrometer, their co-elution with the target analyte can affect the ionization of the latter (Ikonomou et al. 1991; Kebarle and Tang 1993). In this case, the signal is partially suppressed and does not provide an accurate quantitative determination. This is called ion suppression. Therefore, sample preparation is required before analysis.

In general, sampling and sample preparation constitute over 80% of the total analysis time, and these steps are highly important for the overall success of the bioanalytical measurement. The most important sample preparation techniques are protein precipitation (PP), liquid–liquid extraction (LLE), and solid-phase extraction (SPE) (Bylda et al. 2014). PP is the most common sample preparation

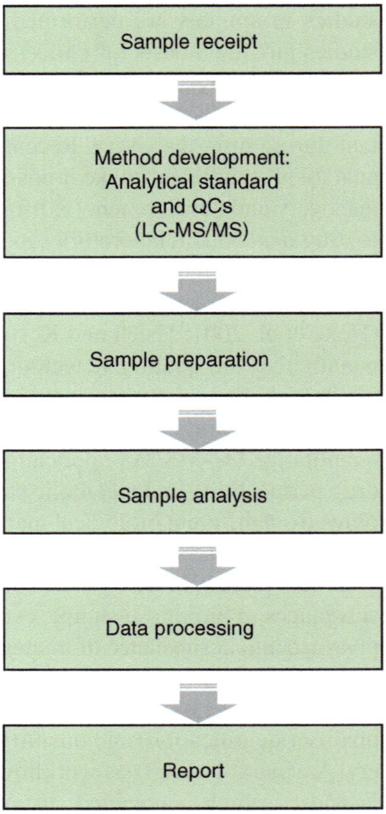

Fig. 1 The common workflow of bioanalysis in DMPK studies

procedure used in ADME bioanalyses. It uses a generic organic solvent such as methanol and acetonitrile to reduce sample processing time. PP is not an extraction technique, however, extraction involves the transfer of the target analyte from the sample medium into another medium. This is not the case in PP since the target analyte remains in the aqueous medium of the plasma sample.

Several steps, such as target identification, screening, lead generation, and optimization are involved in the drug discovery process. DMPK studies, supported by bioanalysis, play a critical role in screening, evaluation, and optimization of new compounds during drug discovery. Drug candidates that successfully pass the discovery phase are subjected to preclinical and clinical studies before final registration as a drug. The goal of the drug discovery stage is to generate a novel lead candidate with drug-like properties for preclinical evaluation. Various types of studies that are conducted during the different stages of drug discovery are depicted in Table 1 (Baranczewski et al. 2006).

Table 1 DMPK study in drug discovery

Hit-to-lead	Lead optimization	Preclinical candidate selection
• In vivo rodent PK • In vitro hepatic clearance across species • Caco-2 permeability/PAMPA • In vitro–in vivo correlation • Plasma/target tissue exposures in pharmacological models • Support to enhance exposures for proof of concept (PoC) in model systems	• In vivo PK optimization: Cassette/discrete PK screen in appropriate species • Metabolic profiles • Use of in vitro screen to overcome PK liabilities • In vitro cytochrome P450 (CYP) involvement and inhibition • Transporters' involvement/inhibition • Plasma protein binding • PK/PD relationship in pharmacology models	• In vivo PK in higher species • Modeling human PK and dose projections • Toxicokinetics • ID metabolites in Tox species • Validation of bioanalytical methods • Initiation of bioanalytical method development for metabolites of pharmacological or toxicological interest

Metabolite Profiling and Identification in Drug Discovery

The main purpose of metabolite screening during early drug discovery is to identify metabolic soft spots that cause poor metabolic stability. This information can guide synthetic chemistry efforts for refining molecules with desirable pharmacokinetic properties. Until recently, most metabolite profiling studies focused on qualitative information (Yao et al. 2009; Zhu et al. 2011). This approach often identifies species differences in metabolism during the early stages of drug discovery. However, quantitative interspecies comparisons of metabolites can provide information that is important for predicting the pharmacological and toxicological effects of metabolites in humans.

Recently, the US Food and Drug Administration (Food and Drug Administration 2008) and the International Conference on Harmonization of Technical Requirements for Registration of Pharmaceuticals for Human Use (ICH Harmonised Tripartite Guideline 2009) issued a formal regulatory guidelines on metabolites in safety testing (MIST). These guidelines emphasize the quantitative monitoring of metabolites in preclinical and clinical studies and established a strategy for the toxicological evaluation of circulating metabolites. The objective of in vitro metabolite profiling is to tentatively predict not only the various metabolites formed via biotransformation reactions, but also the amounts of the formed metabolites; this helps to identify metabolic soft spots and predict in vivo metabolism. The level of the metabolite that triggers an assessment for nonclinical safety testing is defined as 10% of the parent drug's exposure at the steady state (FDA) or 10% of total drug-related exposure (ICH). Therefore, the early identification of qualitative and quantitative differences in drug metabolism between animals used in nonclinical safety assessment and humans is critical to avoid delays in the drug development process.

Liquid chromatography (LC) coupled with electrospray-ionization mass spectrometry (LC-ESI-MS) is an important method for detecting metabolites in complex biological samples. Chromatographic separation improves the sensitivity of mass spectrometry and its signal reproducibility by decreasing matrix interference during ionization and reducing sample complexity and background noise. Recent advancements in LC such as capillary monolithic chromatography and ultra-performance LC (UPLC) have improved peak resolution and have expedited analysis (Plumb et al. 2004).

Several different ionization techniques are used in mass spectrometry. The type of ionization is selected based on the molecular weight, structure, and polarity of the analyte being tested. Atmospheric pressure ionization (API) is the most widely used ionization method for LC-MS coupling because it involves soft ionization, which induces little or no fragmentation. There are three major API techniques: electrospray ionization (ESI) (Fenn et al. 1989; Ho et al. 2003; Kebarle 2000), atmospheric pressure chemical ionization (APCI) (Byrdwell 2001; Rosenberg 2003), and atmospheric pressure photoionization (APPI) (Guo et al. 2004; Leinonen et al. 2002; Raffaelli and Saba 2003). For API, both positive and negative ion modes can be used for the ionization of various kinds of metabolites. ESI is the first choice for most analyses because it can analyze nonvolatile and polar molecules. It is also preferred when profiling unknown metabolites since this "soft" ionization approach forms intact molecular ions and aids initial identification. However, ESI is more susceptible to matrix effects than APCI.

Mass spectrometers can be differentiated by their mass resolution capabilities. A low-resolution mass spectrometer such as the quadrupole MS and ion-trap (IT) MS systems can distinguish compounds that differ by at least 1 Da. Thus, it can distinguish a compound detected at m/z 455 from compounds that are at least 1 Da higher (m/z 456) or 1 Da lower (m/z 454). A high-resolution mass spectrometer including time-of-flight (TOF) and orbital trap (Orbitrap, Thermo Fisher Scientific) systems can measure the mass of a compound down to four decimal places (for example, 455.2832). Therefore, high-resolution MS (HRMS) systems can therefore differentiate compounds that differ in mass by as little as 0.01 Da. MS-MS systems combine various types of mass spectrometers into a single system (Hopfgartner et al. 2004; King and Fernandez-Metzler 2006). For metabolite identification, tandem or hybrid mass analyzers are used to obtain highly resolved and accurate MS/MS spectra. The triple-quadrupole MS-MS system is commonly used. It is a reference tool for the absolute quantification of small molecules due to its sensitivity and specificity using selected reaction monitoring (SRM). Hybrid MS-MS systems combine two types of MS systems into one instrument. A QTOF MS-MS system combines a quadrupole system and a TOF system, and a QIT MS-MS system combines a quadrupole system and a linear IT system (QTRAP, AB Sciex). The advantages and disadvantages of each MS system are summarized in Table 2.

LC-MS/MS using triple quadrupole MS in the multiple reaction monitoring (MRM) scan mode has become the most widely used analytical technique for quantitative drug metabolism studies due to its superior sensitivity and selectivity. The product ion scanning and other full scan data acquisition modes such as neutral loss (NL) and precursor ion (PI) scanning are used to screen drug metabolites (Clarke et al. 2001; Hakala et al. 2006; Liu and Hop 2005). These structure-specific acquisitions result in true positives only if metabolites display similar MS/MS

Table 2 The comparisons of each mass spectrometer

Mass spectrometer	Advantages	Disadvantages
Triple quadrupole MS	• Easy to operate • Very sensitive for quantitative bioanalysis using SRM or MRM mode • Several useful modes (e.g., neutral loss and precursor ion scans)	• Unit mass resolution • Limited full–scan sensitivity • Poor mass accuracy
QTOF	• Improved duty cycle • Increase the sensitivity in full-scan MS and MS/MS mode • Accurate mass determination (within 5-10 ppm)	• Harder to operate • More expensive • Less suitable for quantitation (MRM is not feasible) • Constant neutral loss and precursor ion mode scanning not feasible
QTRAP	• Triple quadrupole MS advantages • MS^3 capabilities • Enhanced scan mode	• Unit mass resolution • Limited full–scan sensitivity • Poor mass accuracy

fragmentation behavior as the parent drug: thus, many unexpected metabolites may be missed along with metabolites with more than one biotransformation site at opposite ends of the molecule. LC-MS/MS is most commonly used to scan for the neutral loss of phase II conjugated metabolites (i.e., the NL of 176 Da to detect glucuronide and the NL of 129 Da for detection GSH adducts) (Liu and Hop 2005; Holcapek et al. 2008). Moreover, the use of these precursor ion related fragment ion patterns for screening metabolites requires laborious pre-adjustment of the instrument and typically several LC/MS/MS runs to cover all of the desired NL and PI acquisition types.

The use of QTOF in drug metabolite screening is preferred in many cases because of its high full scan detection sensitivity and mass accuracy (usually within 5–10 ppm) (Peterman et al. 2006). The improved duty cycle and accurate mass detection are unique advantages of QTOF. The enhanced duty cycle increases the sensitivity in full-scan MS and MS/MS spectra. Although product ion MS/MS data can be acquired, these QTOF hybrid mass spectrometers do not allow constant NL and PI scan modes. However, it does allow for easy simultaneous detection of both expected and unexpected metabolites in a single run while acquiring fragment ion data. This approach does not require metabolite prediction, pre-adjusted detection reactions, or changes in the settings between different test compounds. The ability to accurately determine the mass of ions allows the user to significantly reduce the number of possible molecular formulas of metabolites. Obtaining the best mass accuracy requires internal calibration. Accurate mass data can also be obtained for fragment ions in MS/MS mass spectra, and this greatly enhances the ability to interpret MS/MS mass spectra and assign more specific structures to metabolites. Mass defect filter (MDF) refers to a post-acquisition data filtering technique based on the mass defect of the parent drug and its metabolites. This type of filtering can be performed in real-time or post-analysis. The high resolution and accurate mass capability of this technique also enable mass defect filtering for metabolite identification (Zhang et al. 2003, 2008, 2009). The application of MDF in drug metabolite profiling studies is centered on the observation that the core structure of a drug does not change significantly during biotransformation. Phase 1 and 2 metabolites generally result in mass defect shifts of <50 mDa from the parent drug and its metabolites (Zhang et al. 2003). Thus, the full-scan high-resolution MS data can be filtered for a mass window close to that of the fractional mass. In the case of Phase 2 conjugates, a different mass defect filter is required because of the addition of multiple C, H, N, O, and S atoms (Zhu et al. 2007). Therefore, different mass defect filters are usually set up to detect different types of metabolites. This technique is particularly useful in detecting unanticipated metabolites that can easily escape detection if non-radiolabeled material is used.

Recently, high-resolution MS techniques such as QTOF have been used for qualitative and quantitative analysis (Chen et al. 2015). To some extent, the enhanced mass resolution compensates for the significant drop in selectivity in the MS mode. Nevertheless, the dynamic range and accuracy and precision do not meet the high standards set by triple quadrupole mass spectrometers for quantitation.

A QTRAP is typically not suitable for full scan screening because the detection sensitivity is rather poor when compared to other types of MS instruments when full scan acquisition is used. However, this instrument has multiple reaction monitoring-enhanced product ion (MRM-EPI) and multiple ion monitoring-enhanced product ion (MIM-EPI) scanning functions that are not available on other types of mass spectrometers (Rousu et al. 2010; Zhang et al. 2005).

MRM scans have been used as survey scans because of their higher selectivity and sensitivity than enhanced MS scan (EMS), NL, MIM, and PI scans, and they can trigger an EPI scan to profile predicted metabolites in samples (Yao et al. 2009; Rousu et al. 2010; Gao et al. 2010). The main limitation of this method is its inability to detect unexpected metabolites not included in the expected metabolite list used to create the MRM transitions. The detection of metabolites by MRM is based on predicted molecular weights and fragmentation patterns. Therefore, unknown metabolites cannot be detected if neither the molecular weights nor the fragmentation patterns are predictable. However, MRM is suitable for multi-target screening. Also, MRM and EPI scans can be combined to identify proposed metabolites, although MRM transitions are not sufficiently specific to determine metabolites in unknown samples without synthetic standards. For effective metabolite screening, predictive MRM should be designed by selecting the correct theoretical metabolite ions and their product ions. Some studies reported that multiple reaction monitoring-information dependent aquisition-enhanced product ion (MRM-IDA-EPI) quantitative analysis was not affected by the addition of EPI scans used to obtain qualitative information during the same chromatographic run, compared to MRM-only method (Huang et al. 2010; Shou et al. 2005). In drug discovery, chemical synthesis of metabolite standards is often difficult and costly. Therefore, the MRM-IDA-EPI method is a useful approach for obtaining qualitative and semi-quantitative information about expected metabolites in a single LC-MS/MS analysis without authentic metabolite standards. However, structural changes between a parent and metabolite molecule may cause dramatic differences in ionization efficiency.

Although MRM transitions are not sufficiently specific to determine metabolites in unknown samples without synthetic standards, combined MRM and EPI scans can identify the proposed metabolites.

In most metabolite profiling studies, samples are generally collected at two time points, including time zero and one time point after 60 or 120 min of incubation. Many drugs and xenobiotics have a chemical structure suitable for biotransformation mediated by one or more drug metabolizing enzymes. Therefore, some metabolites that are eliminated rapidly or are produced slowly are missed if the sampling time is incorrect. In our previous study, to evaluate the relative abundance of each metabolite formed from amitriptyline, the area under the peak area ratio–time curve (AUC) of each metabolite from LC–MS/MS data was calculated (Lee et al. 2015). An important principle of metabolite kinetics is that the in vivo disposition of a metabolite is dependent on its formation and elimination. The relative importance of a metabolite when compared to the parent compound and other metabolites in vivo can be predicted using the metabolite kinetic profile in vitro. The measurement of each metabolite's AUC in a time series reflects both its formation and elimination. Although most studies examining metabolite profiling focus on qualitative analysis, the kinetic analysis of metabolites can provide useful information about species-specific differences in metabolism and lead to useful predictions regarding in vivo metabolism. This methodology provides a more sensitive, comprehensive, and meaningful display of drug metabolism than measuring the disappearance of the parent drug or identifying the metabolites at a single time point. A semi-quantitative interspecies comparison of metabolites in liver microsomes can be accomplished even in the absence of authentic metabolite standards.

Reactive Metabolite Detection in Drug Discovery

Drug-induced liver injury (DILI) is one of the most common reasons for drug withdrawal from the market, and it accounts for up to 50% of acute liver failure cases.

Some examples of drugs withdrawn from the market because of idiosyncratic hepatotoxicity include sitaxentan, lumiracoxib, alatrofloxacin, ximelagatran, pemoline, nefazodone, troglitazone, and bromfenac sodium, while drugs that have received black box warnings for hepatotoxicity include bosentan, trovafloxacin, tolcapone, and felbamate. According to FDA Black Box warnings, 71% of drugs are known to produce reactive metabolites. The formation of reactive metabolites is the main cause of DILI. In general, the reactive metabolites are electrophiles. Their structures can contain polarized double bonds, epoxides, carbocation ions, carbenes, nitrenium ions, or nitrenes. The metabolism of drugs into reactive intermediates followed by their covalent binding to cellular components such as protein and DNA is the basis for drug-induced acute toxicities and idiosyncratic adverse drug reactions (Coles 1984–1985). Therefore, it is very important to evaluate whether the drug candidate is converted into a reactive metabolite that may cause the drug to be withdrawn from the market due to drug-induced toxicity. Three methodologies are discussed: trapping method with glutathione (GSH), mechanism-based CYP inhibition, and methods for evaluating acyl glucuronide (AG) formation (Table 3).

Table 3 Methods for the evaluation of reactive metabolites

Approach	Analysis	Scan mode or calculated value	Feature
Trapping method (GSH adduct)	LC-MS/MS	Neutral loss of 129 Da	Qualitative test provide structure information
Mechanism-based CYP inhibition	LC-MS/MS	IC_{50} shift (≥1.5-fold)	Quantitative test
Acyl glucuronide	LC-MS/MS	Neutral loss of 176 Da, $t_{1/2}$ of acyl glucuronide	Qualitative test provide relative reactivity information

GSH is an endogenous tripeptide (γ-glutamyl-cysteinyl-glycine) with a molecular weight of 307 Da and is present at high concentrations in the liver. It is most commonly used as a trapping agent to assess reactive metabolites. GSH conjugates in a complex matrix are selectively detected by positive ion mode ESI mass spectrometry and neutral loss scanning for components that liberate the γ-glutamyl moiety (129 Da) of GSH. The main limitations of this technique are its low sensitivity and its rate of false-positive detections due to endogenous components in biological matrices, which may also result in a neutral loss of 129 Da. To solve this problem, Yan and Caldwell (Yan and Caldwell 2004) developed a novel method that greatly improved the selectivity and reliability of high-throughput screens searching for reactive metabolites by using a stable isotope-labeled trapping agent. An equimolar ratio of GSH and $^{13}C_2/^{15}N$-labeled GSH (stable-isotope label in glycine residue) was used to trap reactive metabolites produced from microsomal incubation with NADPH. An unambiguous GSH conjugate was identified by the presence of a unique double isotopic peak with a mass spectra difference of 3 Da (Fig. 2). Triple-labeling is an optimal choice considering the rare likelihood of finding a mass difference of 3 Da between a test compound and its metabolite or derivatives. Trapping methods are amenable to high-throughput screening and can be performed earlier in the drug discovery stage. Also, trapping methods combined with LC-MS/MS provide structural information on reactive metabolites and mechanistic insight into metabolic activation pathways. However, they have some limitations: they provide only qualitative information and a relationship between GSH conjugate formation and toxicity has not been discovered. Therefore, trapping methods with GSH are typically used for structural optimization of drug candidates to minimize this problem.

Members of the cytochrome P450 (CYP) family are the most important drug-metabolizing enzymes involved in drug clearance. Clinical drug–drug interactions (DDI) involving inhibition or time-dependent inactivation of these enzymes can result in serious clinical consequences caused by reduced clearance and increased exposure of the victim drug (Wienkers and Heath 2005). Some fatal adverse reactions of drugs are mediated by metabolism-dependent DDIs, which have caused the withdrawal of some drugs from the market (Onakpoya et al. 2016). Because of this, CYP inhibition is now investigated earlier in the drug discovery process. This inhibitory effect can be either a reversible or an irreversible (time-dependent) interaction.

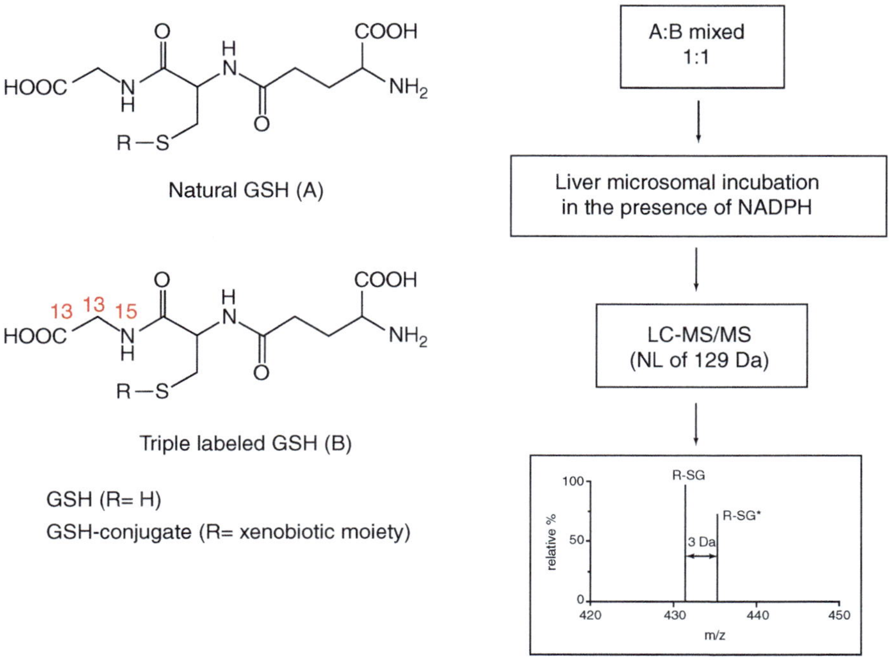

Fig. 2 Trapping methods using stable-isotope labeled GSH

Time-dependent inhibition (TDI) of CYP is of particular concern because de novo synthesis of the enzyme is typically required in order to restore activity. TDI can lead to drug development termination, withdrawal of a drug from the market, or serious restrictions of use. TDI of CYP refers to a change in potency during an in vitro incubation or an in vivo dosing period. Potential mechanisms include the formation of inhibitory metabolites and mechanism-based inhibition (MBI). Mechanism-based CYP inhibitors display NADPH-, concentration-, and time-dependent (quasi-) irreversible inactivation.

Three pathways lead to the inactivation of CYPs by the reactive intermediate: (1) reaction with nucleophilic amino acids at the active site, (2) reaction with the nitrogen atoms in heme, and (3) coordination of the iron in heme to form a metabolite–intermediate complex (Polasek and Miners 2007). In metabolite–intermediate complexes, the hemoprotein is not destroyed, and the inactivation of CYP can be reversed through oxidation with potassium ferricyanide. MBI has also been associated with idiosyncratic toxicity (Tucker et al. 2001). Traditionally, screening for MBI of CYP isoforms requires the detection of enzymatic activity alterations with preincubation of the test article in the presence or absence of NADPH in liver microsomes or recombinant CYPs. This is followed by the addition of probe substrates. The rate of disappearance of the probe substrate is then measured. Thus, MBI experiments require two steps (a preincubation step with NADPH, but not a probe substrate, and an incubation step with NADPH and a probe substrate) and

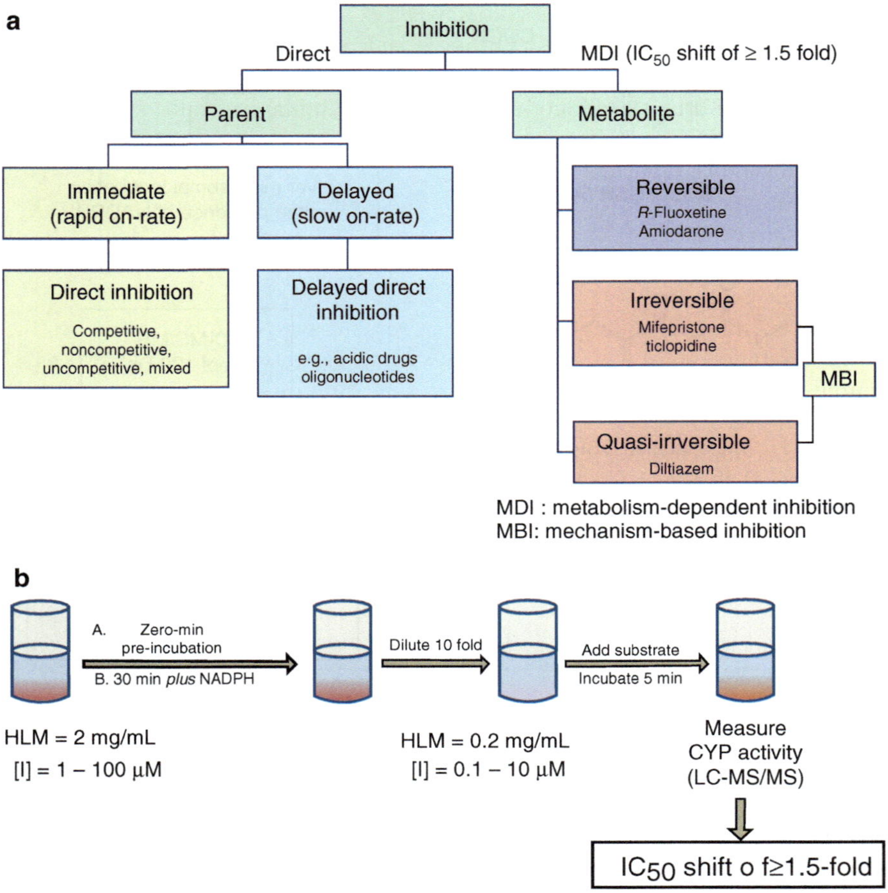

Fig. 3 CYP inhibition study (**a**) CYP inhibition mode and (**b**) method of metabolism-dependent CYP inhibition

several preincubation times to generate inactivation rate constants. Berry and Zhao (Berry and Zhao 2008) suggested that metabolism dependent inhibitors show at least a 1.5-fold shift in IC_{50} after a 30-min preincubation (Fig. 3).

Metabolism-dependent CYP inhibition can be categorized into three types (reversible, quasi-irreversible, and irreversible inhibition) based on the reversibility of CYP inhibition. Quasi-irreversible/irreversible inhibition is considered to be MBI. To distinguish among reversible, quasi-irreversible, and irreversible inhibition of CYP enzymes, the reversibility CYP isoform inactivation was tested after ultracentrifugation or, in the case of quasi-irreversible inhibition, oxidation with potassium ferricyanide ($K_3Fe\ (CN)_6$) (Lee et al. 2011, 2012).

Although metabolic activations are usually mediated by CYPs in the liver, phase II conjugation pathways, including glucuronidation, are also known to transform latent functional groups such as carboxylic acids into reactive intermediates

(Spahn-Langguth and Benet 1992; Stachulski et al. 2006). The formation of acyl glucuronides (AGs) has been implicated in hepatic injury via covalent binding to proteins, resulting in the withdrawal of drugs from the market. Many carboxylic acid-containing drugs such as diclofenac, a nonsteroidal anti-inflammatory drug, and clofibrate, a lipid-lowering agent, are metabolized into AGs by UDP-glucuronosyltransferases. Reactive AGs are chemically unstable at physiological pHs and are prone to hydrolysis, intramolecular rearrangements, and covalent interactions with cellular target molecules. AGs can bind covalently to plasma and cellular proteins in vitro and in vivo via two mechanisms: transacylation and glycation (Spahn-Langguth et al. 1996).

In direct transacylation, the glucuronic acid moiety is displaced from the AG by protein nucleophilic groups, such as hydroxyl groups, sulfhydryl groups, or amine groups (Ding et al. 1995; Williams and Dickinson 1994; Williams et al. 1995). In glycation, rearranged AG isomers exist transiently in an open chain form and contain a reactive aldehyde group that can yield a Schiff base (Spahn-Langguth et al. 1996; Williams and Dickinson 1994; Williams et al. 1995; Wang and Dickinson 1998). The covalent binding of cellular proteins by AGs has been associated with idiosyncratic hepatotoxicity in susceptible patients (Spahn-Langguth et al. 1996; Boelsterli 2002; Boelsterli et al. 1995). AGs are a matter of concern in metabolism studies because they have the potential to covalently bind with other biological molecules, including proteins. These reactions have been implicated in severe toxicity, resulting in the withdrawal of drugs from the market. The production of AGs from carboxylic acid-containing drugs is a cause of DILI, and the chemical stability of AGs is related to their reactivity. Benet et al. showed a correlation between the extent of covalent binding to human serum albumin (HSA) and the apparent first-order degradation rate of AGs of six drugs, suggesting that the half-life of AG corresponds with its chemical reactivity (Benet et al. 1993). Several groups have also evaluated the reactivity of AGs by determining their half-lives in buffer, HSA solution, or fresh human plasma (Boelsterli 2002; Bolze et al. 2002; Sallustio et al. 1997). Recently, the Daiichi group determined the relationship between idiosyncratic drug toxicity risk and the half-life of AG in potassium phosphate buffer, HSA solution, and fresh human plasma (Sawamura et al. 2010). The authors suggested that the chemical stability of AG is important for the assessment of DILI during the drug discovery stage. In in vivo experiments that assessed the stability of AGs, AGs were added (final conc. 10 μM) to 0.1 M potassium phosphate buffer (pH 7.4) and incubated at 37 °C (Fig. 4). Sample aliquots were taken at four or more time points from the start of the incubation followed by the immediate addition of a two-fold volume of 5% formic acid in acetonitrile including an internal standard on ice. The samples were then stored at −20 °C until analyzed. The incubated sample of AG was analyzed to obtain the MRM transition using an LC-MS/MS. The concentrations of AG at four or more time points were used to determine the degradation rate constants of the AGs. The degradation rate constant was determined from the AG concentration versus time curve by linear regression of the semi-logarithmic plot.

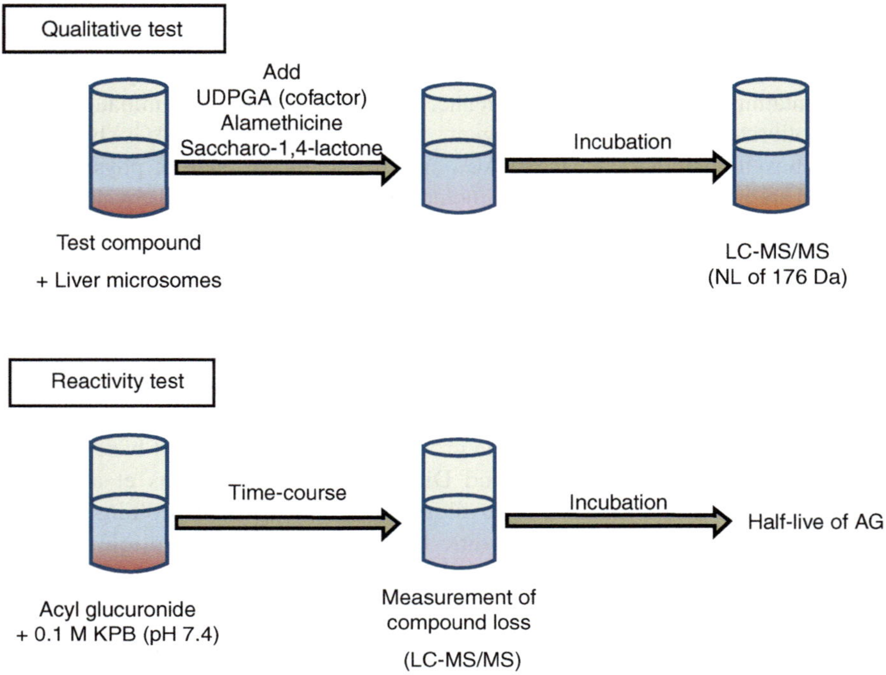

Fig. 4 Qualitative test for acyl glucuronide (AG) formation and reactivity test of AG

Conclusion

LC-MS technologies play an important role in studying the DMPK of NCEs during drug discovery and development. Recently, many advancements in LC-MS have enabled high-throughput and high-quality bioanalyses to be conducted in the drug discovery area. Triple quadrupole MS and triple quadrupole-linear ion mass spectrometers (QTRAP) are and will continue to be the main LC-MS platforms for in vitro and in vivo ADME screening. High-resolution MS, QTOF-MS does not require the development of chemical-specific methods for quantitative analysis. This is a very attractive feature for efficient ADME screening. QTOF continuously has been developed to provide high resolution, high accuracy, high sensitivity, and high linearity. Recent advancements in QTOF will be able to bring about a paradigm shift in this field. DMPK studies are heavily utilized by a variety of mass spectrometers, including triple quadrupole MS, QTRAP, ion trap, and QTOF. QTOF will become an invaluable LC-MS platform and likely the platform of choice for DMPK studies in drug discovery.

References

Ackermann BL, Berna MJ, Eckstein JA, Ott LW, Chaudhary AK (2008) Current applications of liquid chromatography/mass spectrometry in pharmaceutical discovery after a decade of innovation. Annu Rev Anal Chem 1:357–396

Baranczewski P, Stańczak A, Sundberg K, Svensson R, Wallin A, Jansson J, Garberg P, Postlind H (2006) Introduction to *in vitro* estimation of metabolic stability and drug interactions of new chemical entities in drug discovery and development. Pharmacol Rep 58(4):453–472

Benet LZ, Spahn-Langguth H, Iwakawa S, Volland C, Mizuma T, Mayer S, Mutschler E, Lin ET (1993) Predictability of the covalent binding of acidic drugs in man. Life Sci 53:PL141–PL146

Berry LM, Zhao Z (2008) An examination of IC50 and IC50-shift experiments in assessing time-dependent inhibition of CYP3A4, CYP2D6 and CYP2C9 in human liver microsomes. Drug Metab Lett 2(1):51–59

Boelsterli UA (2002) Xenobiotic acyl glucuronides and acyl CoA thioesters as protein-reactive metabolites with the potential to cause idiosyncratic drug reactions. Curr Drug Metab 3(4):439–450

Boelsterli UA, Zimmerman HJ, Kretz-Rommel A (1995) Idiosyncratic liver toxicity of nonsteroidal anti-inflammatory drugs: molecular mechanisms and pathology. Crit Rev Toxicol 25(3):207–235

Bolze S, Bromet N, Gay-Feutry C, Massiere F, Boulieu R, Hulot T (2002) Development of an in vitro screening model for the biosynthesis of acyl glucuronide metabolites and the assessment of their reactivity toward human serum albumin. Drug Metab Dispos 30(4):404–413

Bylda C, Thiele R, Kobold U, Volmer DA (2014) Recent advances in sample preparation techniques to overcome difficulties encountered during quantitative analysis of small molecules from biofluids using LC-MS/MS. Analyst 139(10):2265–2276

Byrdwell WC (2001) Atmospheric pressure chemical ionization mass spectrometry for analysis of lipids. Lipids 36(4):327–346

Caldwell J (1996) The role of drug metabolism studies for efficient drug discovery and development: opportunities to enhance time- and cost-efficiency. Pharm Sci 2:117–119

Chen D, Lin S, Xu W, Huang M, Chu J, Xiao F, Lin J, Peng (2015) Qualitative and quantitative analysis of the major constituents in Shexiang Tongxin dropping pill by HPLC-Q-TOF-MS/MS and UPLC-QqQ-MS/MS. Molecules 20(10):18597–18619

Chu I, Nomeir AA (2006) Utility of mass spectrometry for in vitro ADME assays. Curr Drug Metab 7(5):467–477

Clarke NJ, Rindge D, Korfmacher WA, Cox KA (2001) Systematic LC/MS metabolite identification in drug discovery. Anal Chem 73(15):430A–439A

Coles B (1984-1985) Effects of modifying structure on electrophilic reactions with biological nucleophiles. Drug Metab Rev 15(7):1307-1334

Ding A, Zia-Amirhosseini P, McDonagh AF, Burlingame AL, Benet LZ (1995) Reactivity of tolmetin glucuronide with human serum albumin. Identification of binding sites and mechanisms of reaction by tandem mass spectrometry. Drug Metab Dispos 23(3):369–376

Fenn JB, Mann M, Meng CK, Wong SF, Whitehouse CM (1989) Electrospray ionization for mass spectrometry of large biomolecules. Science 246(4926):64–71

FoodandDrugAdministration(2008)Guidanceforindustry:safetytestingofdrugmetabolitesAvailable at: http://www.fda.gov/downloads/Drugs/GuidanceComplianceRegulatoryInformation/Guidances/ucm079266.pdf. Last accessed 13th January 2015

Gao H, Deng S, Obach RS (2010) A simple liquid chromatography-tandem mass spectrometry method to determine relative plasma exposures of drug metabolites across species for metabolite safety assessments. Drug Metab Dispos 38(12):2147–2156

Guo T, Chan M, Soldin SJ (2004) Steroid profiles using liquid chromatography-tandem mass spectrometry with atmospheric pressure photoionization source. Arch Pathol Lab Med 128:469–475

Hakala KS, Kostiainen R, Ketola RA (2006) Feasibility of different mass spectrometric techniques and programs for automated metabolite profiling of tramadol in human urine. Rapid Commun Mass Spectrom 20(14):2081–2890

Ho CS, Lam CWK, Chan MHM, Cheung RCK, Law LK, Lit LCW, Ng KF, Suen MWM, Tai HL (2003) Electrospray ionisation mass spectrometry: principles and clinical applications. Clin Biochem Rev 24(1):3–12

Hoke SH, Morand KL, Greis KD, Baker TR, Harbol KL, Dobson RLM (2001) Transformations in pharmaceutical research and development, driven by innovations in multidimensional mass spectrometry-based technologies. Int J Mass Spectrom 212(1–3):135–196

Holcapek M, Kolárová L, Nobilis M (2008) High-performance liquid chromatography-tandem mass spectrometry in the identification and determination of phase I and phase II drug metabolites. Anal Bioanal Chem 391(1):59–78

Hopfgartner G, Varesio E, Tschäppät V, Grivet C, Bourgogne E, Leuthold LA (2004) Triple quadrupole linear ion trap mass spectrometer for the analysis of small molecules and macromolecules. J Mass Spectrom 39(8):845–855

Hsieh Y (2008) HPLC-MS/MS in drug metabolism and pharmacokinetics screening. Expert Opin Drug Metab Toxicol 4(1):93–101

Hsieh Y, Korfmacher W (2006) Increasing speed and throughput when using HPLC-MS/MS systems for drug metabolism and pharmacokinetic screening. Curr Drug Metab 7(5):479–489

Huang J, Bathena SP, Alnouti Y (2010) Metabolite profiling of praziquantel and its analogs during the analysis of *in vitro* metabolic stability using information-dependent acquisition on a hybrid triple quadrupole linear ion trap mass spectrometer. Drug Metab Pharmacokinet 25(5):487–499

ICH Harmonised Tripartite Guideline (2009) Guidance on nonclinical safety studies for the conduct of human clinical trials and marketing authorization for pharmaceuticals M3 (R2). Available at: http://www.ich.org/fileadmin/Public_Web_Site/ICH_Products/Guidelines/Multidisciplinary/M3_R2/Step4/M3_R2_Guideline.pdf. Last Accessed 13 January 2015

Ikonomou MG, Blades AT, Kebarle P (1991) Electrospray-ion spray: a comparison of mechanisms and performance. Anal Chem 63(18):1989–1998

Kebarle P (2000) A brief overview of the present status of the mechanisms involved in electrospray mass spectrometry. J Mass Spectrom 35(7):804–817

Kebarle P, Tang L (1993) From ions in solution to ions in the gas phase-the mechanism of electrospray mass spectrometry. Anal Chem 65(22):972A–986A

King R, Fernandez-Metzler C (2006) The use of Qtrap technology in drug metabolism. Curr Drug Metab 7(5):541–545

Kola I, Landis J (2004) Can the pharmaceutical industry reduce attrition rates? Nat Rev Drug Discov 3:711–715

Korfmacher WA (2005) Using mass spectrometry for drug metabolism studies. CRC Press, Boca Raton, FL, pp 83–93

Korfmacher WA (2009) Advances in the integration of drug metabolism into the lead optimization paradigm. Mini-Rev Med Chem 9(6):703–716

Kostiainen R, Kotiaho T, Kuuranne T, Auriola S (2003) Liquid chromatography/atmospheric pressure ionization-mass spectrometry in drug metabolism studies. J Mass Spectrom 38(4):357–372

Laine R (2008) Metabolic stability: main enzymes involved and best tools to assess it. Curr Drug Metab 9(9):921–927

Lee KS, Oh SJ, Kim HM, Lee KH, Kim SK (2011) Assessment of reactive metabolites in drug–induced liver injury. Arch Pharm Res 34(11):1879–1886

Lee JY, Lee SY, Oh SJ, Lee KH, Jung YS, Kim SK (2012) Assessment of drug-drug interactions caused by metabolism-dependent cytochrome P450 inhibition. Chem Biol Interact 198(1-3):49–56

Lee JY, Lee SY, Lee K, Oh SJ, Kim SK (2015) Determination of species-difference in microsomal metabolism of amitriptyline using a predictive MRM-IDA-EPI method. Chem Biol Interact 5(229):109–118

Leinonen A, Kuuranne T, Kostiainen R (2002) Liquid chromatography/mass spectrometry in anabolic steroid analysis--optimization and comparison of three ionization techniques:

electrospray ionization, atmospheric pressure chemical ionization and atmospheric pressure photoionization. J Mass Spectrom 37(7):693–698
Lipinski CA (2000) Drug-like properties and the causes of poor solubility and poor permeability. J Pharmacol Toxicol Methods 44(1):235–249
Liu DQ, Hop CE (2005) Strategies for characterization of drug metabolites using liquid chromatography-tandem mass spectrometry in conjunction with chemical derivatization and on-line H/D exchange approaches. J Pharm Biomed Anal 37(1):1–18
Onakpoya IJ, Heneghan CJ, Aronson JK (2016) Post-marketing withdrawal of 462 medicinal products because of adverse drug reactions: a systematic review of the world literature. BMC Med 14:10
Peterman SM, Duczak N Jr, Kalgutkar AS, Lame ME, Soglia JR (2006) Application of a linear ion trap/orbitrap mass spectrometer in metabolite characterization studies: examination of the human liver microsomal metabolism of the non-tricyclic anti-depressant nefazodone using data-dependent accurate mass measurements. J Am Soc Mass Spectrom 7(3):363–375
Plumb R, Castro-Perez J, Granger J, Beattie I, Joncour K, Wright A (2004) Ultra-performance liquid chromatography coupled to quadrupole-orthogonal time-of-flight mass spectrometry. Rapid Commun Mass Spectrom 18(19):2331–2337
Polasek TM, Miners JO (2007) In vitro approaches to investigate, mechanism-based inactivation of CYP enzymes. Expert Opin Drug Metab Toxicol 3(3):321–329
Prakash C, Shaffer CL, Nedderman A (2007) Analytical strategies for identifying drug metabolites. Mass Spectrom Rev 26(3):340–369
Raffaelli A, Saba A (2003) Atmospheric pressure photoionization mass spectrometry. Mass Spectrom Rev 22:318–331
Riley RJ, Martin IJ, Cooper AE (2002) The influence of DMPK as an integrated partner in modern drug discovery. Curr Drug Metab 3(5):527–550
Rodrigues AD (2008) Drug–drug interactions. In: Rodrigues AD (ed) Drugs and the pharmaceutical sciences, 2nd edn. Informa Healthcare, New York, pp 1–719
Rosenberg E (2003) The potential of organic (electrospray- and atmospheric pressure chemical ionisation) mass spectrometric techniques coupled to liquid-phase separation for speciation analysis. J Chromatogr A 1000(1–2):841–889
Rousu T, Herttuainen J, Tolonen A (2010) Comparison of triple quadrupole, hybrid linear ion trap triple quadrupole, time-of-flight and LTQ-Orbitrap mass spectrometers in drug discovery phase metabolite screening and identification in vitro-amityptiline and verapamil as model compounds. Rapid Commun Mass Spectrom 24(7):939–957
Sallustio BC, Fairchild BA, Pannall PR (1997) Interaction of human serum albumin with the electrophilic metabolite 1-O-gemfibrozil-β-D-glucuronide. Drug Metab Dispos 25(1):55–60
Sawamura R, Okudaira N, Watanabe K, Murai T, Kobayashi Y, Tachibana M, Ohnuki T, Masuda K, Honma H, Kurihara A, Okazaki O (2010) Predictability of idiosyncratic drug toxicity risk for carboxylic acid-containing drugs based on the chemical stability of acyl glucuronide. Drug Metab Dispos 38(10):1857–1864
Shou WZ, Magis L, Li AC, Naidong W, Bryant MS (2005) A novel approach to perform metabolite screening during the quantitative LC-MS/MS analyses of in vitro metabolic stability samples using a hybrid triple-quadrupole linear ion trap mass spectrometer. J Mass Spectrom 40(10):1347–1356
Spahn-Langguth H, Benet LZ (1992) Acyl glucuronides revisited: is the glucuronidation process a toxification as well as a detoxification mechanism? Drug Metab Rev 24(1):5–47
Spahn-Langguth H, Dahms M, Hermening A (1996) Acylglucuronides: covalent binding and its potential relevance. Adv Exp Med Biol 387:313–328
Stachulski AV, Harding JR, Lindon JC, Maggs JL, Park BK, Wilson ID (2006) Acyl glucuronides: biological activity, chemical reactivity, and chemical synthesis. J Med Chem 49(24):6931–6945
Tucker GT, Houston JB, Huang SM (2001) Optimizing drug development: strategies to assess drug metabolism/transporter interaction potential–toward a consensus. Br J Clin Pharmacol 52(1):107–117

Wang M, Dickinson RG (1998) Disposition and covalent binding of diflunisal and diflunisal acyl glucuronide in the isolated perfused rat liver. Drug Metab Dispos 26(2):98–104

Wienkers LC, Heath TG (2005) Predicting in vivo drug interactions form in vitro drug discovery data. Nat Rev Drug Discov 4(10):825–833

Williams AM, Dickinson RG (1994) Studies on the reactivity of acyl glucuronides-VI. Modulation of reversible and covalent interaction of diflunisal acyl glucuronide and its isomers with human plasma protein in vitro. Biochem Pharmacol 47(3):457–467

Williams AM, Worrall S, De Jersey J, Dickinson RG (1995) Studies on the reactivity of acyl glucuronides-VIII. Generation of an antiserum for the detection of diflunisal-modified proteins in diflunisal-dosed rats. Biochem Pharmacol 49(2):209–217

Yan Z, Caldwell GW (2004) Stable-isotope trapping and high-throughput screenings of reactive metabolites using the isotope MS signature. Anal Chem 76(23):6835–6847

Yao M, Ma L, Duchoslav E, Zhu M (2009) Rapid screening and characterization of drug metabolites using multiple ion monitoring dependent product ion scan and postacquisition data mining on a hybrid triple quadrupole-linear ion trap mass spectrometer. Rapid commun Mass Spectorm 23(11):1683–1693

Zhang H, Zhang D, Ray K (2003) A software filter to remove interference ions from drug metabolites in accurate mass liquid chromatography/mass spectrometric analyses. J Mass Spectrom 38(10):1110–1112

Zhang MY, Pace N, Kerns EH, Kleintop T, Kagan N, Sakuma T (2005) Hybrid triple quadrupole linear ion trap mass spectrometry in fragmentation mechanism studies: application to structure elucidation of buspirone and one of its metabolites. J Mass Spectrom 40(8):1017–1029

Zhang H, Zhu M, Ray KL, Ma L, Zhang D (2008) Mass defect profiles of biological matrices and the general applicability of mass defect filtering for metabolite detection. Rapid Commun Mass Spectrom 22(13):2082–2088

Zhang H, Zhang D, Ray K, Zhu M (2009) Mass defect filter technique and its applications to drug metabolite identification by high-resolution mass spectrometry. J Mass Spectrom 44(7):999–1016

Zhu M, Ma L, Zhang H, Humphreys WG (2007) Detection and structural characterization of glutathione-trapped reactive metabolites using liquid chromatography-high-resolution mass spectrometry and mass defect filtering. Anal Chem 79(21):8333–8341

Zhu M, Zhang H, Humphreys WG (2011) Drug metabolite profiling and identification by high resolution mass spectrometry. J Biol Chem 286(29):25419–25425

The Genetic Basis of Reporter Mouse Strains

Gyeong-Nam Kim and Young Hoon Sung

Introduction

Our understanding of in vivo gene functions has been accelerated by analyzing the phenotypes in genetically engineered mouse (GEM) models. Among various GEM model types, reporter mouse strains permit the scrutiny of information linked to the endogenous gene expression patterns in mouse tissues. Reporter mice harbor a reporter gene that is directly or indirectly activated by a specific promoter and are useful for measuring gene expression levels in specific cells and tissues under normal and specific experimental conditions. With gene trapping strategies intercepting normal gene expression, reporter mice were initially generated through genetic screens identifying developmental genes using mouse embryonic stem (ES) cells (Friedrich and Soriano 1991). If a reporter mouse strain is verified to express a distinct pattern of reporter gene activity, the phenomenon can be used to distinguish

G.-N. Kim
Department of Medical Science and Asan Medical Institute of Convergence Science and Technology, Asan Medical Center, University of Ulsan College of Medicine, Seoul, Republic of Korea

ConveRgence mEDIcine research cenTer (CREDIT), Asan Institute for Life Sciences, Asan Medical Center, Seoul, Republic of Korea

Y. H. Sung (✉)
Department of Medical Science and Asan Medical Institute of Convergence Science and Technology, Asan Medical Center, University of Ulsan College of Medicine, Seoul, Republic of Korea

ConveRgence mEDIcine research cenTer (CREDIT), Asan Institute for Life Sciences, Asan Medical Center, Seoul, Republic of Korea

Department of Convergence Medicine, Bio-Medical Institute of Technology, University of Ulsan College of Medicine, Seoul, Republic of Korea
e-mail: yhsung@amc.seoul.kr

J. K. Kim et al. (eds.), *Advanced Imaging and Bio Techniques for Convergence Science*, Advances in Experimental Medicine and Biology 1310,
https://doi.org/10.1007/978-981-33-6064-8_21

individual cells and specific lineages or tissues in mice under specific experimental conditions (e.g., stressful or diseased state). If a reporter gene (e.g., the green fluorescent protein, GFP) is inserted in-frame with the coding sequence at the target locus, the reporter signal will allow us to trace the molecular dynamics of an endogenously expressed protein at the subcellular level.

Until recently, genetic manipulation using ES cells had been the only way to produce GEM models with a targeted genomic modification, the principles of which were discovered by Mario R. Capecchi, Martin J. Evans, and Oliver Smithies; they were jointly awarded Nobel Prize in Physiology or Medicine in 2007 (The Nobel Prize in Physiology or Medicine 2007). However, gene targeting using ES cells is an inefficient process. Occasional double-strand breaks (DSBs) that are in part repaired by homologous recombination (HR) are essential for gene targeting using ES cells and these DSBs occur at an extremely low frequency (Deng and Capecchi 1992). Further, even after correctly targeted ES cell clones are obtained, chimeric mice or chimera should be generated by microinjecting the targeted ES cells into mouse blastocysts and then by transferring them into the uterus of pseudo-pregnant foster mothers. The germline transmission of the targeted allele is one of the major obstacles in this process.

It is noteworthy that HR frequency is significantly increased by a DSB at the target locus in mammalian cells (Rouet et al. 1994). With regard to this, the application of engineered endonucleases, such as zinc-finger nucleases (ZFNs), Transcription Activator-Like Effector (TALE) Nucleases (TALENs), and Clustered Regularly Interspaced Short Palindromic Repeats (CRISPR)/CRISPR-associated protein 9 (Cas9) system, is useful to facilitate HR by modifying the target genomic locus. Now, it is well recognized that the efficiency of the process is enough to generate GEM models by directly microinjecting an engineered endonuclease and the targeting construct into the pronuclear-stage mouse embryos, thus eliminating the annoying germline transmission step for obtaining a chimera.

Now, we can establish novel reporter mouse strains relatively easily by grafting the currently available engineered endonucleases such as Cas9 over genetic strategies developed in ES cells. In this chapter, we will sequentially explain the genetic basis of reporter genes and the application of engineered endonucleases in generating reporter mouse strains.

Generation of Reporter Mouse Strains Using ES Cells

Initially, reporter mouse strains were generated to identify developmentally regulated genes. For example, Gossler et al. (Gossler et al. 1989) devised a gene trap vector harboring promoterless *lacZ* reporter gene possessing a 5′ *En-2* splice acceptor site and a 3′ polyadenylation signal, and then introduced the construct into ES cells (Fig. 1). When the construct is integrated into introns of genes in the correct orientation and the reading frame is maintained, a functional fusion protein is detected from a spliced *lacZ* fusion transcript in mouse embryos (Skarnes et al. 1992). The reporter gene expressions reflect expressions of genes with which the

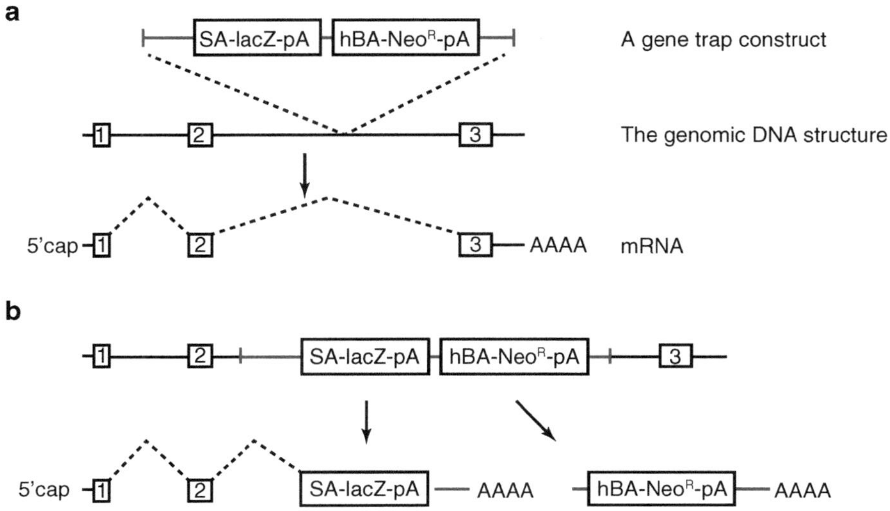

Fig. 1 *Basic principles for the generation of lacZ reporter mouse strains by gene trapping.* (**a**) A gene trap construct can randomly integrate into the mouse genome. This figure depicts a functional mRNA generated from a model gene. SA, splice acceptor site; pA, poly A signal; hBA, human *β-actin* promoter; *Neo^R*, neomycin resistance gene. (**b**) When the gene trap construct is integrated into an intron of the gene, pA from the SA-lacZ-pA cassette stops the transcription from the endogenous promoter. The existence of the SA contributes to the generation of a hybrid mRNA enabling the expression of *lacZ* reporter. The *Neo^R* expression is independently driven by hBA, which is used to select gene trapped ES cell clones

reporter gene was integrated, demonstrating the capacity of this gene trap vector to accurately report endogenous gene expression (Skarnes et al. 1992). As a polyadenylation signal terminates the transcription of a gene (Proudfoot 2011), this type of gene trap vector is highly mutagenic and thus is useful to obtain gene-trapped ES cell clones for the generation of gene knockout mice with *lacZ* reporter gene activity (Friedrich and Soriano 1991).

As the gene trap strategy is suitable for high-throughput generation of mutagenized ES cells, with the construction of a considerable number of ES cell resources, there has been an international effort to centralize access to these valuable resources. It has eventually established the International Gene trap Consortium (IGTC), a large repository of publicly available gene-trap ES cell lines covering almost 40% of known mouse genes (Nord et al. 2006). These days, a more versatile mutant mouse resource is also available. The International Mouse Phenotyping Consortium (IMPC; formerly, the International Knockout Mouse Consortium, IKMC) is aiming to mutate all protein-encoding genes using a combined genetic strategy of gene trapping and gene targeting using ES cells derived from C57BL/6 (Collins et al. 2007). As shown in Fig. 2, the IMPC created null-first conditional ready alleles, and thus, upon generation of mutant mice using the targeted ES cells, endogenous gene expression patterns can be examined through an SV40 splice acceptor site-*lacZ* reporter gene fused to the neomycin-resistance gene (βgeo)

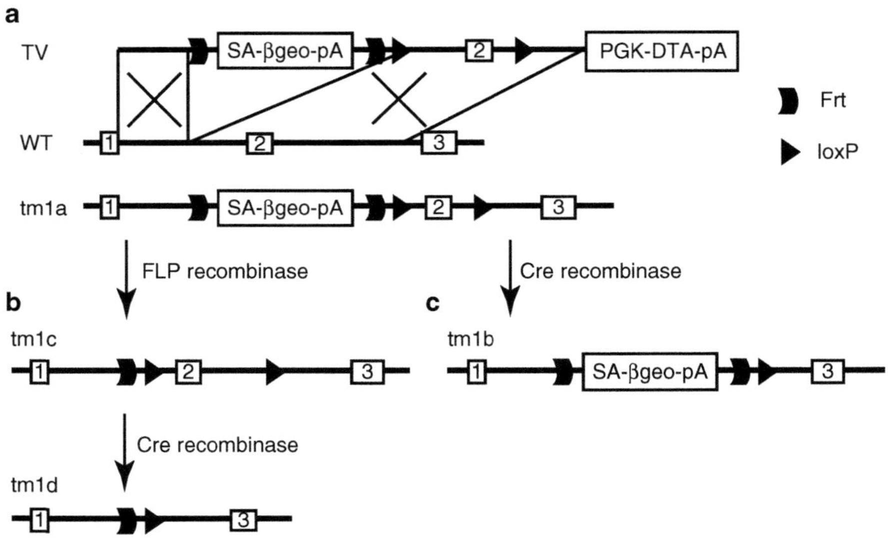

Fig. 2 *The IMPC conditional allele.* (**a**) Generation of tm1a allele. TA, targeting vector; SA, splice acceptor site; *βgeo*, a fused gene of *lacZ* and *Neo*^{R}; pA, poly A signal *Phosphoglycerate kinase (PGK)* promoter-driven *diphtheria toxin A (DTA)* cassette is a negative selection marker. (**b**) The tm1c allele is a floxed allele produced by removing SA-*βgeo*-pA cassette using FLP recombinase specific for Frt sequences. When an appropriate Cre recombinase is expressed in mice, a null allele (tm1d) will be generated by deleting the critical exon 2. (**c**) If tm1a allele is exposed to Cre recombinase before FLP recombinase, the reporter mouse allele (tm1b) will be generated

flanked with Frt sequences. In addition to generating mutant mouse strains, IMPC analyzes the basic phenotypes such as the *lacZ* reporter gene expressions and makes the data publicly available through their website (www.mousephenotype.org). These resources are helping us know where and when a gene of interest is expressed in vivo.

Marking Specific Cell Lineages Using ROSA26 Locus-Based Reporter Mice

Some reporter mouse strains exhibit discrete *lacZ* expression patterns and are useful to mark and trace specific tissues and cell lineages in vivo. Furthermore, these defined *lacZ* expression patterns also provide enough rationale for the development of Cre recombinase-expressing transgenic mouse strains essential for generating Cre-loxP-mediated conditional gene-knockout mice. After Cre transgenic mice generation, their tissue or cell lineage-specific Cre recombinase activities should be confirmed using a conditional reporter mouse strain. In addition to conditional knockout, if the Cre recombinase activity is reproduced in tissues or cells in mice, the combined use of Cre transgenic mice and a conditional reporter allele will be critical for delineating cell fates in vivo by marking the cell lineage. In this regard, diverse conditional reporter strains generated by integrating versatile conditional reporter genes into the *ROSA26* locus have significantly contributed to our ability to discriminate specific cells in mouse tissues.

The *ROSA26* locus is one of the most useful genomic loci in mice. Initially, *ROSA26* locus was found by a genetic screen to identify and mutate developmentally important genes in mice (Friedrich and Soriano 1991). In this study, they introduced a retroviral gene trap vector, ROSAβgeo, into ES cells and in turn, established mouse models harboring the ROSAβgeo gene trap vector (Fig. 3a) (Friedrich and Soriano 1991). The *lacZ* reporter gene activity from βgeo (the in-frame fusion of *lacZ* and *Neo*^{R} genes) was examined, and one of these mouse strains, ROSAβgeo 26 (*ROSA26* locus harboring a βgeo expression cassette), showed a ubiquitous tissue expression in both embryonic and adult stages (Friedrich and Soriano 1991; Zambrowicz et al. 1997). *ROSA26* locus encodes three transcripts, of which two transcripts sharing a common origin are lost in homozygous *ROSA26* animals (Zambrowicz et al. 1997). No significant open reading frame (ORF) exists in these transcripts, and no overt phenotype is observed in homozygous mutant mice (Zambrowicz et al. 1997). Along these lines, a conditional reporter allele, *R26R*, was created by targeting *ROSA26* locus in ES cells (Fig. 3b) (Soriano 1999).

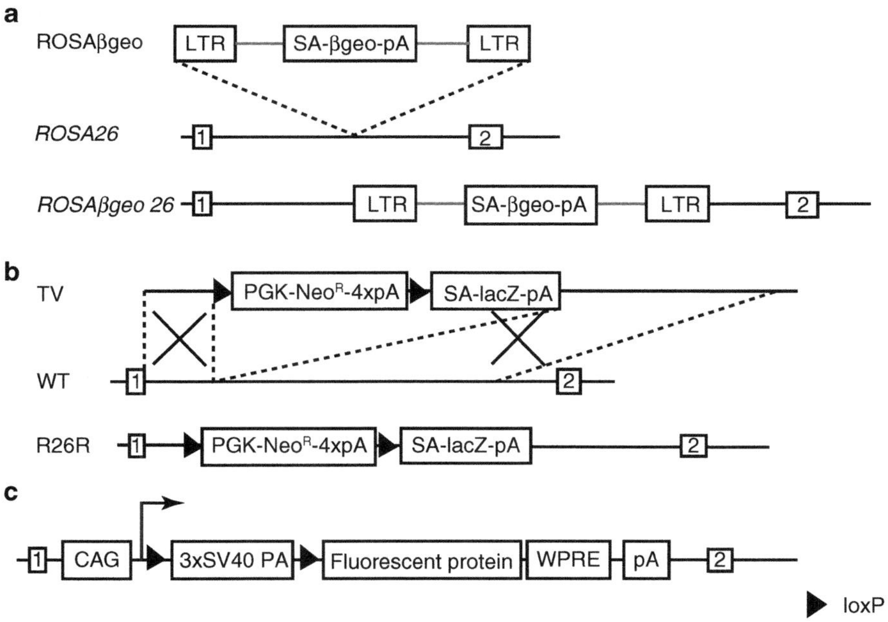

Fig. 3 *ROSA26 locus-based reporter alleles.* (**a**) The structure depicting ROSAβgeo gene trap retrovirus, *ROSA26* locus, and gene-trapped *ROSOβgeo 26* allele. LTR, long terminal repeat. (**b**) The genetic strategy used to generate the *R26R* reporter mouse strain through homologous recombination in ES cells. PGK, human *phosphoglycerate kinase* promoter. The floxed PGK-*Neo*^{R}-4 × pA cassette blocks the transcription of the SA-*lacZ*-pA cassette. (**c**) The CAG promoter is strong and ubiquitous and thus dramatically improves the induction of fluorescent reporter genes from the *ROSA26* locus. WPRE, woodchuck hepatitis virus posttranscriptional regulatory element

Although the overall reporter gene structure is similar to the original gene trap vector (Friedrich and Soriano 1991), a floxed (flanked by two loxP) sequence containing a *Neo*^{R} cDNA and a quadruple polyadenylation signal (4 × pA) was placed at the 5′ end of a *lacZ* gene with an upstream splice acceptor (SA) sequence (Fig. 3b) (Soriano 1999). Notably, 4 × pA sequence existing at the 3′ end of the *Neo*^{R} cDNA blocks the *ROSA26* transcription ((Soriano 1999); Fig. 3b), and thus the floxed sequence is called a loxP-STOP-loxP sequence. Therefore, when the mice were only bred with *R26Cre* mice, which is a transgenic mouse ubiquitously expressing Cre recombinase by inserting Cre recombinase expression cassette into the *ROSA26* locus, the floxed *Neo*^{R}-4 × pA sequence is removed and then generalized lacZ activities are observed throughout mouse embryonic tissues (Soriano 1999). A similar approach also reported and verified the advantages of the *ROSA26*-based conditional *lacZ* reporter strain in both embryonic and adult mouse tissues (Mao et al. 1999). These results demonstrated that *R26R* allele is only activated by Cre recombination; further, the subsequently induced lacZ signal can mark specific tissues that express Cre recombinase in mice.

Replacing the *lacZ* reporter gene with diverse fluorescent proteins makes Cre-loxP systems particularly useful for visualizing living cells in embryos or organs. In fact, Srinivas et al. (2001) used a similar gene-targeting strategy in the *ROSA26* locus and replaced the *lacZ* reporter gene with *enhanced yellow fluorescent protein (EYFP)* and *enhanced cyan fluorescent protein (ECFP)* genes to generate *R26R-EYF*P and *R26R-ECFP* strains, respectively. When these mouse strains were mated to a Cre transgenic mouse strain carrying a *β-actin-Cre* transgene, ubiquitous EYFP or ECFP signals were displayed in E8.5 embryos (Srinivas et al. 2001). When tissue-specific Cre transgenic mouse strains were used, the signals were detected only in the restricted embryonic tissues expressing Cre recombinase. For example, *Isl1-Cre* mice were generated by targeting Cre recombinase into the *Isl1* locus encoding a LIM homeodomain transcription factor, which is expressed in motor neurons and dorsal root and cranial sensory ganglia (Srinivas et al. 2001). As expected, the mice carrying both *R26R-YFP* and *Isl1-Cre* alleles faithfully labelled motor neurons and dorsal root ganglia with the EYFP signal at embryonic day 14.5 (E14.5) (Srinivas et al. 2001). The utility of *R26R-EYFP* mice was also proved by using *Engrailed-1 (En-1)*-Cre transgenic mice expressing the Cre recombinase specifically at the mid-hindbrain junction in E8.5 embryos (Srinivas et al. 2001).

In all aforementioned *R26R* strains, the reporter gene units were directly driven by the endogenous promoter activity of the *ROSA26* locus and a splice acceptor site located at the 5′ end of the reporter gene. To achieve robust induction of fluorescent reporter genes, the strong and ubiquitously expressed CAG promoter was inserted at the 5′ end of the reporter gene and the effect of the woodchuck hepatitis virus posttranscriptional regulatory element (WPRE), which is known to stabilize mRNAs at the 3′ end, was examined (Fig. 3c) (Madisen et al. 2010). In this case, there is no need to add an upstream artificial splice acceptor site. Compared with the EYFP signal with the *R26R-EYFP* reporter (Srinivas et al. 2001), that after the addition of CAG promoter was enhanced dramatically (Madisen et al. 2010). When the WPRE sequence was added, an even more significant increase in the EYFP signal was observed compared with that with the reporter gene only or with CAG promoter added to the reporter gene (Madisen et al. 2010). Madisen et al. (2010) also produced *R26R* reporter strains expressing brightest fluorescent proteins including ZsGreen and tdTomato. In their study, they successfully generated consistent and versatile reporter mouse strains strongly expressing fluorescent proteins, and these mouse strains are now prevalently used to study the labeled cells in vivo.

Advantages of Cre-ER^{T2} Mouse Strains

For achieving temporal regulation of Cre recombinase in vivo, a ligand-dependent chimeric Cre recombinase, Cre-ER, was generated (Metzger et al. 1995). This is a fusion protein consisting of Cre recombinase linked to the ligand-binding domain (LBD) of the human estrogen receptor (ER) that is activated upon ligand binding (Metzger et al. 1995). Cre recombinase activity of Cre-ER was evoked by the natural ligand 17β-estradiol (E_2) and the synthetic ER antagonist 4-hydroxytamoxifen (4-OHT), but not by ICI 182,780 (ICI) (Metzger et al. 1995). To prevent Cre-ER from responding to the natural ligand E_2, Cre-ER^{T2}, a derivative containing the G400V/M543A/L540A mutations in the LBD of human ER, was developed (Feil et al. 1997). These mutations block Cre recombinase activation by the natural ligand E_2 but allow activation of Cre-ER^{T2} by 4-OHT and ICI (Feil et al. 1997).

As Cre-mediated excision of a floxed sequence is irreversible, the use of Cre transgenic mice in an adult stage might be limited by Cre recombinase expression during the fetal development. For example, the glial fibrillary acidic protein (GFAP) is an intermediate filament protein expressed almost exclusively in mature astrocytes and is commonly used as a specific marker for astrocytes in the brain (reviewed in (Wang and Bordey 2008)). Accordingly, a human *GFAP (hGFAP)*-Cre transgenic mouse strain was established, and the Cre recombinase activity was examined using reporter mouse strains including the previously described *R26R* reporter mice (Soriano 1999; Zhuo et al. 2001). Cre expression appears to be limited to astrocytes in the adult brain but begins to be detected in the forebrain by E13.5 (Zhuo et al. 2001). Although lacZ expression was primarily restricted to the central nervous system (CNS), its widespread expression was also detected in neurons and ependyma (Zhuo et al. 2001), indicating that *hGFAP*-Cre expression is developmentally regulated in a multipotent neural stem cell.

In addition to the astroglial cells, the *GFAP* gene is also considerably expressed in neurogenic radial glial cells during fetal development and at early postnatal stages, and *hGFAP-Cre-ER*^{T2} transgenic mice were used to decide the developmental fates of GFAP-expressing cells in the CNS by temporarily treating mice with tamoxifen. When adult *hGFAP-Cre-ER*^{T2} transgenic mice were treated with tamoxifen, which is converted to 4-OHT in the liver, astrocyte-selective Cre activation was detected (Hirrlinger et al. 2006; Koeppen et al. 2018). In contrast, when GFAP-expressing cells are pulse-labelled by treating *hGFAP*-Cre-ER^{T2}; *R26R* double transgenic mice with 4-OHT at postnatal day 5 (P5), lacZ-positive cells are quiescent astroglial cells with stem cell properties (Ganat et al. 2016). If these cells are chased for a longer period, they will eventually produce diverse cell lineages including neural precursors and their mature neuronal and oligodendrocytic progenies in the brain (Ganat et al. 2016).

Taken together, these reports indicate that, in contrast to constitutively active Cre transgenic mice, Cre-ER^{T2} transgenic mice allow timely control of selective labeling of a specific cell lineage in adult mice, the expression of which is developmentally dynamic.

Multicolor Labeling of Specific Cell Lineages and Individual Cells in Live Mice

Until now, we described reporter mouse strains by simply turning on the expression of *lacZ* or a fluorescent protein using Cre-loxP system. A reporter mouse strain, Z/EG was established to express either lacZ or enhanced green fluorescent protein (EGFP) mutually exclusively in a cell from one transgene construct (Fig. 4a). The reporter gene possesses a floxed *lacZ* expression cassette driven by CAG promoter and a subsequent *EGFP* cDNA-pA and thus ubiquitously expresses *lacZ* gene without Cre recombinase (Fig. 4a) (Novak et al. 2000). When Cre recombinase is

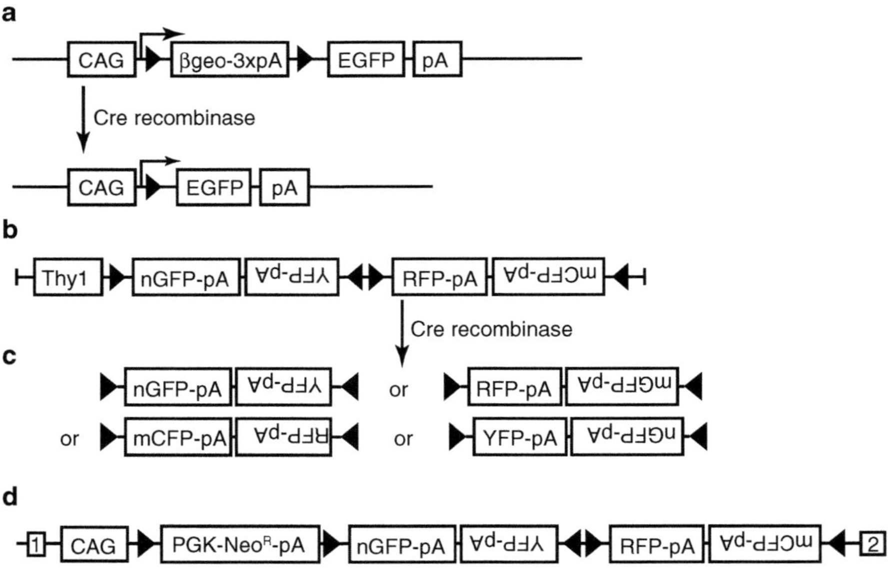

Fig. 4 *Reporter mouse strains conditionally expressing multiple reporter genes.* (**a**) The Z/EG reporter mouse strain. Z/EG mice basically express *βgeo* gene, and EGFP expression is turned on by Cre-mediated deletion of the floxed *βgeo*-3 × pA cassette. (**b,c**) *Thy1-Brainbow-2.1* reporter mice. The *Thy-1* promoter initially drives nGFP expression (**b**), and the Cre-mediated inversion and excision events can induce RFP, mCFP, or YEP expression (**c**). (**d**) *R26R-Confetti* reporter mouse strain. CAG promoter-driven Brainbow-2.1 reporter was inserted into the *ROSA26* locus

expressed, EGFP expression is induced by Cre-mediated excision of the floxed *lacZ* expression cassette (Fig. 4a) (Novak et al. 2000).

By ingeniously applying the Cre-loxP system, a novel strategy was introduced with the Brainbow reporter vectors for simultaneously marking multiple neurons by many different colored hues generated from the combinatorial expression of several fluorescent proteins in the mouse brain (Fig. 4b, c) (Livet et al. 2007). The Brainbow-2.1 system, one of the Brainbow reporters, sequentially contains coding sequences and polyA signals for a nuclear GFP (nGFP), YFP, red fluorescent protein (RFP), and a plasma membrane-targeting CFP (M-CFP), but YFP and M-CFP are inverted and positioned in a tail-to-tail orientation to nGFP and RFP, respectively (Fig. 4b). This configuration makes the Brainbow-2.1 basically express nGFP, and as each tail-to-tail unit is flanked with inversely oriented loxP sites, Cre recombinase can induce four different combinations of excision and inversion events switching the expression of nGFP to YFP, RFP, or M-CFP (Fig. 4c). Neuron-specific *Thy1-Brainbow* reporter mouse strains were generated by microinjecting the constructs into the pronucleus of mouse zygotes and were crossed with *CAGGS-Cre-ER*^{T2} animals (Livet et al. 2007). As pronuclear injections typically result in a tandem integration of multiple transgene copies that are independently and stochastically recombined within each transgene copy upon tamoxifen treatment, the combinatorial expression of different fluorescent proteins can generate many different hues in individual neurons (Livet et al. 2007). Thus, polychromatic neurons are generally observed in diverse *Thy1-Brainbow* reporter mouse strains (Livet et al. 2007).

Owing to the ability of uniquely labeling many individual cells within a complex population, the Brainbow strategy is also useful for studies in other tissues. However, *Thy1-Brainbow* reporter mouse strains are only expressed in neuronal cell lineages. Further, as the transgenes are randomly integrated into the mouse genome, the level and extent of reporter gene expressions also considerably vary among reporter mouse strains established using the same transgene construct (Livet et al. 2007). To make the strategy more versatile, *R26R-Confetti* reporter mouse strain was generated (Fig. 4d) (Snippert et al. 2010). In *R26R-Confetti* reporter mice, the Brainbow-2.1 reporter is driven by a CAG promoter, inducing strong and ubiquitous expression, but the expression is initially blocked by a floxed *Neo*^{R}-cassette serving as transcriptional roadblock (Fig. 4d). Therefore, unlike *Thy1-Brainbow* reporter mice basically manifesting nGFP signal, *R26R-Confetti* reporter mice do not express any fluorescent proteins before the removal of floxed *Neo*^{R}-cassette by Cre-mediated excision. Furthermore, *R26R-Confetti* reporter mice have only one copy of the transgene at the *ROSA26* locus and thus display reporter signals in a mutually exclusive manner rather than a combinatorial manner observed in *Thy1-Brainbow* reporter mice harboring multiple transgene copies (Snippert et al. 2010).

These strategies have been proved to be useful and therefore are prevalently adopted in diverse studies analyzing distinct cell lineages and individual cells in vivo. This success will trigger the generation of many more improved reporter mouse strains as spectral variants of fluorescent proteins and the mechanical systems for their detection are now developing rapidly.

Reporter Mouse Strains Generated Using Engineered Endonucleases

Based on the availability of validated materials and methods, *ROSA26*-based reporter mouse strains can be generated relatively efficiently using ES cells. However, the rapid development of engineered endonuclease tools makes it feasible to directly target *ROSA26* locus in mouse zygotes. We summarize the initial attempts of generating diverse reporter mouse strains using engineered endonucleases such as ZFN, TALEN, and CRISPR-Cas9 in both ES cells and mouse zygotes.

When a ZFN pair targeting the *ROSA26* locus (ZFN^{Rosa}) is co-transfected into ES cells with a targeting vector harboring a hygromycin/puromycin resistance cassette for the *ROSA26* locus, approximately 32% of the analyzed clones are correctly recombined, with no recombined clone being detected without ZFN^{Rosa} (Meyer et al. 2010). Although the HR efficiency (approximately 1.7%–4.5%) is much lower than that in ES cells, Meyer et al. (2010) successfully generated two *ROSA26* reporter mouse strains expressing *lacZ* and *Venus*, a yellow fluorescent protein, by directly applying ZFN-assisted HR to mouse zygotes. Compared to ZFNs, TALENs are highly active with an improved specificity and are less toxic in mouse zygotes (Sung et al. 2013). Compared to ZFN-assisted HR, TALEN-assisted targeted transgenesis into the *ROSA26* locus is more efficient in mouse zygotes (8.8%, (Kasparek et al. 2014)). Chu et al. (2016) reported much more improved CRISPR-Cas9-assisted targeted transgenesis in mouse *ROSA26* locus. These results indicate that engineered endonuclease-stimulated HR, especially that induced using CRISPR-Cas9 in mouse zygotes is an effective shortcut alternative for the generation of novel *ROSA26*-based reporter mouse strains by conventional gene targeting in ES cells.

Based on the ease of application, engineered endonucleases can be used to insert a reporter element into a certain gene other than *ROSA26* locus. For example, by co-injecting a targeting construct and a pair of ZFN into pronuclear-stage fertilized eggs, Cui et al. (2010) successfully integrated a GFP expression cassette into the *Mdr1a* locus in both Sprague Dawley (SD) rats and FVB mice. The GFP expression, however, did not reflect the endogenous expression of *Mdr1a* gene because a 1.5-kb human *phosphoglycerate kinase (PGK)* promoter drives the GFP cassette (Cui et al. 2010). Sommer et al. (2014) modified the mouse *Satb1* gene using TALENs and a targeting vector. The start codon of *Satb1* gene resides in exon 2 and the targeting vector consists of the normal exon 2 and an inversely oriented artificial exon 2 containing the endogenous 5′ untranslated region (UTR), EGFP ORF, and the endogenous 3′ UTR. Two pairs of loxP and lox2272, a derivative of loxP, encompass the whole element in the opposite orientation (Fig. 5a). As homotypic recombinations (e.g., recombinations of loxP-loxP and lox2272-lox2272, respectively), but not heterotypic recombinations (e.g., loxP-lox2272), are almost exclusively induced, a Cre-mediated recombination induces the exchange of the endogenous exon 2 with the artificial exon 2, and then the *Satb1* promoter drives the EGFP expression (Fig. 5b).

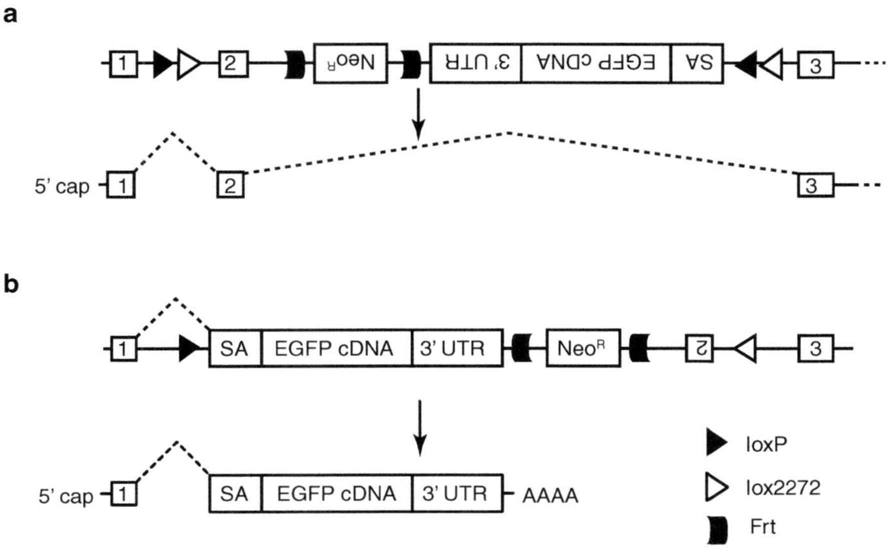

Fig. 5 *A conditional reporter gene integrated into the Satb1 locus using TALENs.* (**a**) A SA-*EGFP* cDNA-3′UTR reporter gene cassette is floxed with two pairs of loxP and lox2272 in an opposite orientation and is inversely inserted into the *Satb1* locus. *Satb1* mRNA is expected to be normally produced from the reporter allele without the Cre recombinase. (**b**) Cre-mediated inversion and excision produces a functional reporter allele expressing EGFP protein

Owing to the robust activity in the cells and the ease of design and production, CRISPR-Cas9 is offering a bright future for producing valuable reporter mouse strains. Using sgRNA targeting the sequence in the vicinity of the stop codon and a targeting vector aiming an in-frame fusion of the *Nanog* gene to *mCherry* cDNA, the *Nanog* locus was engineered to express Nanog protein with C-terminal mCherry tagging (Yang et al. 2013). In addition to monitoring the expression pattern of a gene in vivo, the tagging of endogenous proteins with fluorescent proteins enables us to study the molecular dynamics of a specific protein in vivo. They also generated *Oct4-an internal ribosome entry site (IRES)-EGFP* allele by inserting an *IRES-EGFP* unit with an sgRNA recognizing the target sequence at the 3′ UTR of the *Oct4* gene (Yang et al. 2013). These beneficial aspects of engineered endonucleases, especially CRISPR-Cas9, should prevalently accelerate the establishment of various reporter mouse strains.

Conclusions and Perspectives

Beginning with strategies for identifying developmental genes that manifest distinct gene expression patterns, reporter mouse strains have been broadly used to denote the expression of a gene and to measure its expression levels in specific cells and tissues in mice. The verified reporter gene expressions in turn allow us to label individual cells and lineages in mouse tissues. Furthermore, the molecular dynamics of an endogenously expressed protein can be analyzed at a subcellular level by inserting an epitope tagging or an ORF encoding a fluorescent reporter gene in-frame (e.g., a fluorescent protein).

As engineered endonuclease-mediated reporter gene knock-in requires appropriate targeting vectors with sequences homologous to a target gene, the targeting vectors generated by IMPC can be conjugated to provide materials needed for producing novel reporter mouse strains. For example, *βgeo* cDNA can be replaced with *EGFP* cDNA through relatively simple molecular works. Furthermore, IMPC targeting vectors typically contain 10 kb of homology to the endogenous locus (Skarnes et al. 2011), which is long enough to find efficient sgRNAs of CRISPR-Cas9 for gene targeting. As these vectors are generated in a high-throughput system, if an appropriate series of molecular works is designed, IMPC targeting vectors can be systematically converted to others containing a new reporter gene other than *βgeo*. Furthermore, rapidly developing algorithms or software predicting specificity and on-target efficiency of sgRNAs will be helpful to select sgRNAs for actual gene targeting in mouse zygotes. Undoubtedly, the development of both novel reporter genes and newly devised apparatuses that sensitively detect the signals emanating from expressed reporter genes in vivo are essential for supporting the design and potential applications for reporter mouse strains.

Acknowledgments This study was supported by the Ministry of Science, ICT, and Future Planning (MSIP) through the National Research Foundation of Korea (NRF) (2018R1A2B6002192).

References

Chu VT, Weber T, Graf R et al (2016) Efficient generation of Rosa26 knock-in mice using CRISPR/Cas9 in C57BL/6 zygotes. BMC Biotechnol 16:4. https://doi.org/10.1186/s12896-016-0234-4

Collins FS, Rossant J, Wurst W et al (2007) A mouse for all reasons. Cell 128(1):9–13. https://doi.org/10.1016/j.cell.2006.12.018

Cui X, Ji D, Fisher DA et al (2010) Targeted integration in rat and mouse embryos with zinc-finger nucleases. Nat Biotechnol 29(1):64–67. https://doi.org/10.1038/nbt.1731

Deng C, Capecchi MR (1992) Reexamination of gene targeting frequency as a function of the extent of homology between the targeting vector and the target locus. Mol Cell Biol 12(8):3365–3371. https://doi.org/10.1128/mcb.12.8.3365

Feil R, Wagner J, Metzger D (1997) Regulation of Cre recombinase activity by mutated estrogen receptor ligand-binding domains. Biochem Biophys Res Commun 237(3):752–757. https://doi.org/10.1006/bbrc.1997.7124

Friedrich G, Soriano P (1991) Promoter traps in embryonic stem cells: a genetic screen to identify, and mutate developmental genes in mice. Genes Dev 5:1513–1523. https://doi.org/10.1101/gad.5.9.1513

Ganat YM, Silbereis J, Cave C (2016) Early postnatal astroglial cells produce multilineage precursors and neural stem cells in vivo. J Neurosci 26(33):8609–8621. https://doi.org/10.1523/JNEUROSCI.2532-06.2006

Gossler A, Joyner AL, Rossant J, Skarnes WC (1989) Mouse embryonic stem cells and reporter constructs to detect developmentally regulated genes. Science 244(4903):463–465. https://doi.org/10.1126/science.2497519

Hirrlinger PG, Scheller A, Braun C (2006) Temporal control of gene recombination in astrocytes by transgenic expression of the tamoxifen-inducible DNA recombinase variant CreERT2. Glia 54(1):11–20. https://doi.org/10.1002/glia.20342

Kasparek P, Krausova M, Haneckova R et al (2014) Efficient gene targeting of the Rosa26 locus in mouse zygotes using TALE nucleases. FEBS Lett 588(21):3982–3988. https://doi.org/10.1016/j.febslet.2014.09.014

Koeppen J, Nguyen AQ, Nikolakopoulou AM (2018) Functional consequences of synapse remodeling following astrocyte-specific regulation of Ephrin-B1 in the adult hippocampus. J Neurosci 38(25):5710–5726. https://doi.org/10.1523/JNEUROSCI.3618-17.2018

Livet J, Weissman TA, Kang H et al (2007) Transgenic strategies for combinatorial expression of fluorescent proteins in the nervous system. Nature 450(7166):56–62. https://doi.org/10.1038/nature06293

Madisen L, Zwingman TA, Sunkin SM (2010) A robust and high-throughput Cre reporting and characterization system for the whole mouse brain. Nat Neurosci 13(1):133–140. https://doi.org/10.1038/nn.2467

Mao X, Fujiwara Y, Orkin SH (1999) Improved reporter strain for monitoring Cre recombinase-mediated DNA excisions in mice. PNAS 96(9):5037–5042. https://doi.org/10.1073/pnas.96.9.5037

Metzger D, Clifford J, Chiba H et al (1995) Conditional site-specific recombination in mammalian cells using a ligand-dependent chimeric Cre recombinase. PNAS 92(15):6991–6995. https://doi.org/10.1073/pnas.92.15.6991

Meyer M, de Angelis MH, Wurst W et al (2010) Gene targeting by homologous recombination in mouse zygotes mediated by zinc-finger nucleases. PNAS 107(34):15022–15026. https://doi.org/10.1073/pnas.1009424107

Nord AS, Chang PJ, Conklin BR et al (2006) The international gene trap consortium website: a portal to all publicly available gene trap cell lines in mouse. Nucleic Acids Res 1(34):D642–D648. https://doi.org/10.1093/nar/gkj097

Novak A, Guo C, Yang W et al (2000) Z/EG, a double reporter mouse line that expresses enhanced green fluorescent protein upon Cre-mediated excision. Genesis 28(3–4):147–155. https://doi.org/10.1002/1526-968X(200011/12)28:3/4<147::AID-GENE90>3.0.CO;2-G

Proudfoot NJ (2011) Ending the message: poly(A) signals then and now. Genes Dev 25(17):1770–1782. https://doi.org/10.1101/gad.17268411

Rouet P, Smih F, Jasin M (1994) Expression of a site-specific endonuclease stimulates homologous recombination in mammalian cells. PNAS 91(13):6064–6068. https://doi.org/10.1073/pnas.91.13.6064

Skarnes WC, Anna Auerbach B, Joyner AL (1992) A gene trap approach in mouse embryonic stem cells: the lacZ reported is activated by splicing, reflects endogenous gene expression, and is mutagenic in mice. Genes Dev 6(6):903–918. https://doi.org/10.1101/gad.6.6.903

Skarnes WC, Rosen B, West AP et al (2011) A conditional knockout resource for the genome-wide study of mouse gene function. Nature 474(7351):337–342. https://doi.org/10.1038/nature10163

Snippert HJ, van der Flier LG, Sato T et al (2010) Intestinal crypt homeostasis results from neutral competition between symmetrically dividing Lgr5 stem cells. Cell 143(1):134–144. https://doi.org/10.1016/j.cell.2010.09.016

Sommer D, Peters A, Wirtz T et al (2014) Efficient genome engineering by targeted homologous recombination in mouse embryos using transcription activator-like effector nucleases. Nat Commun 5:3045. https://doi.org/10.1038/ncomms4045

Soriano P (1999) Generalized lacZ expression with the ROSA26 Cre reporter strain. Nat Genet 21(1):70–71. https://doi.org/10.1038/5007

Srinivas S, Watanabe T, Lin CS et al (2001) Cre reporter strains produced by targeted insertion of EYFP and ECFP into the ROSA26 locus. BMC Dev Biol 2001:1–4. https://doi.org/10.1186/1471-213X-1-4

Sung YH, Baek IJ, Kim DH et al (2013) Knockout mice created by TALEN-mediated gene targeting. Nat Biotechnol 31(1):23–24. https://doi.org/10.1038/nbt.2477

The Nobel Prize in Physiology or Medicine (2007) NobelPrize.org. Nobel Media AB 2019. Sat. 22 Jun 2019. https://www.nobelprize.org/prizes/medicine/2007/summary/

Wang DD, Bordey A (2008) The astrocyte odyssey. Prog Neurobiol 86(4):342–367. https://doi.org/10.1016/j.pneurobio.2008.09.015

Yang H, Wang H, Shivalila CS et al (2013) One-step generation of mice carrying reporter and conditional alleles by CRISPR/Cas-mediated genome engineering. Cell 154(6):1370–1379. https://doi.org/10.1016/j.cell.2013.08.022

Zambrowicz BP, Imamoto A, Fiering S et al (1997) Disruption of overlapping transcripts in the ROSA beta geo 26 gene trap strain leads to widespread expression of beta-galactosidase in mouse embryos and hematopoietic cells. PNAS 94(8):3789–3794. https://doi.org/10.1073/pnas.94.8.3789

Zhuo L, Theis M, Alvarez-Maya (2001) IhGFAP-cre transgenic mice for manipulation of glial and neuronal function in vivo. Genesis 31(2):85–94. https://doi.org/10.1002/gene.10008